Mathematical
Surveys
and
Monographs

Volume 145

Simple Groups of Finite Morley Rank

Tuna Altınel
Alexandre V. Borovik
Gregory Cherlin

American Mathematical Society
Providence, Rhode Island

2000 *Mathematics Subject Classification.* Primary 03C60, 03C98, 03C45, 20A15, 20D05, 20E42, 20G05, 20G10, 20G20, 20G40, 20B07, 20B15, 20B20, 20B22, 20B25, 20B27, 20P05, 51E12, 51E24; Secondary 17B20, 17B25, 17B45, 19C20.

The graphics used to create Table 1 of Chapter II were taken from the website:
`http://www.math.ubc.ca/~cass/frivs/frivolities.html`
©by Bill Casselman.

For additional information and updates on this book, visit
www.ams.org/bookpages/series-surv-145

Library of Congress Cataloging-in-Publication Data

Altinel, Tuna, 1966–
Simple groups of finite Morley rank / Tuna Altinel, Alexandre V. Borovik, Gregory Cherlin.
p. cm. — (Mathematical surveys and monographs, ISSN 0076-5376 ; v. 145)
Includes bibliographical references and index.
ISBN 978-0-8218-4305-5 (alk. paper)
1. Model theory. 2. Finite groups. I. Borovik, Alexandre. II. Cherlin, Gregory L., 1948– III. Title.

QA9.7.A48 2007
511.3′4–dc22   2007060551

∞ The paper used in this book is acid-free and falls within the guidelines
established to ensure permanence and durability.
Visit the AMS home page at `http://www.ams.org/`

10 9 8 7 6 5 4 3 2 1   13 12 11 10 09 08

Sami, Zuhal, Kuban ve Evren Altınel'e
— Tuna Altınel

To Anna
— Alexandre Borovik

Contents

Introduction

Quand est-ce que vous allez démontrer un vrai théorème?
— Bruno Poizat, printemps 2000

We will be concerned here with the following conjecture.

ALGEBRAICITY CONJECTURE. *An infinite simple group of finite* Morley rank *is algebraic, over an algebraically closed field.*

This conjecture arises in Model Theory, where *Morley rank* is an abstract notion of dimension which generalizes the notion of the dimension of an algebraic variety in some of its usual formulations. The conjecture asserts that any infinite simple group which can be equipped with such a dimension function must be isomorphic, as an abstract group, to a Chevalley group: the group of F-rational points of a simple algebraic group, over some algebraically closed field F. It remains open.

The main result to be proved here can be stated as follows.

MAIN THEOREM. *Let G be a simple group of finite Morley rank. Then G satisfies one of the following two conditions.*

(1) *G is algebraic over an algebraically closed field of characteristic* 2.
(2) *G has finite* 2*-rank.*

The 2*-rank* of G, denoted $m_2(G)$, is the dimension of the largest elementary abelian 2-subgroup of G.

The condition that the 2-rank is finite can be reformulated in more useful but somewhat more technical ways, notably as follows: the Sylow 2-subgroups contain divisible abelian subgroups of finite index and finite 2-rank. Such groups are said to be of "odd type", when the divisible abelian subgroup is nontrivial, and of "degenerate type" when it is trivial. We therefore prefer the following formulation.

MAIN THEOREM. *Let G be a simple group of finite Morley rank, and nonalgebraic. Then G is of odd or degenerate type.*

This is somewhat more than we had set out to do here. We had expected to confine our results to the analysis of *minimal counterexamples* to the Algebraicity Conjecture. The turning point came in [**2**], when it became clear that methods for achieving "absolute" results like the foregoing could be envisioned using many of the techniques already developed for the treatment of minimal cases. This is startling, as it is the analog in our subject of a classification of finite simple groups of characteristic 2 type *without the Feit-Thompson (Odd Order) theorem.* Indeed, groups without involutions fall in the degenerate class—and conversely, simple groups of degenerate type contain no involutions, as the Algebraicity Conjecture predicts (Theorem IV 4.1). About such groups we say nothing, and of course they may occur, in principle, as subgroups or sections of the groups we do study. We work around them.

Much of our approach will be modeled closely on the methods of finite group theory. The Algebraicity Conjecture is analogous to the classification of the finite simple groups as Chevalley groups, possibly twisted, together with the alternating groups and 26 "sporadic" finite simple groups. The methods we use are largely those which were involved in the two proofs of that classification discussed above, combined with certain additional ingredients, namely: (1) the *amalgam method*, which is part of a proposed *third generation* approach to the classification of the finite simple groups, and is very effective in our context; (2) more elementary ideas modeled on the theory of algebraic groups and lacking a finite analog; (3) specific properties of algebraic groups. To this list, a fourth category must be appended, relating to the body of techniques which enables us to work around the presence of degenerate sections. This is a very geometrical theory, based ultimately on dimension computations, and which is developed in Chapter IV. Everything we do there could be done in the category of algebraic groups, but is not—primarily, it seems, because stronger results based on properties of complete varieties are available. In our category, there is no coherent notion of complete variety, and we see no obvious parallel with the methods of Chapter IV, but we observe that the results go in the same general direction. Both model theorists and algebraists may find this chapter of particular interest (though it really has to be seen in action, as in Chapter VI, to be appreciated)—model theorists because the material is model theoretic in character, and group theorists because the line of argument varies considerably from the accustomed lines of group theory, both finite and algebraic, while at the same time having a clear meaning within the algebraic category. The closest model for this kind of analysis is found in the so-called black box group theory (randomized finite group theory), where properties of "most" elements play an important role.

The Main Theorem and some additional results which will be detailed in the final chapter, relating to groups of odd type, impose sharp limitations on the structure of a possible counterexample to the Algebraicity Conjecture, and suggest that such a group is unlikely to contain any involutions at all. When we first began this project, it seemed entirely possible that exceptions to the conjecture do occur in nature, or not far removed from nature; in the finite case one has both the "twisted" Chevalley groups and the sporadic ones to deal with, and possible analogs of both could be envisioned in our case. This possibility now appears to be rapidly receding. On the other hand, one can imagine various model theoretic constructions which would most naturally produce torsion free examples, and our results say nothing about that possibility, except to suggest that the groups so constructed would look more like free groups than like conventional matrix groups.

The final chapter is where we expose a detailed summary of concrete applications of the main theorem of this book. The analysis of permutation groups of finite Morley rank outlined in that chapter illustrates, in a way reminiscent of the applications of the classification of the finite simple groups, how the main result of this book can be put into action to obtain results not directly related to classification issues. It is worth noting that the proofs of some results (e.g. generic equations) in Chapter IV, in their first incarnations, used the classification of simple groups of even type. It later turned out that the full classification was not necessary for these results.

When we set out on this project, we looked forward to the possibility of extracting from it, as a byproduct, a "skeletal" version of the classification of the finite simple groups, showing roughly what the core of that proof would look like in the absence of such complications as sporadic groups, very small base fields, and wreath products. In other words, we aimed to give a *reading* of the very long classification proof of the finite simple groups that imparts some particular structure to it, while providing a rigorous proof in a different context. What we do here, supplemented by the other material to be described below, could be taken as such a reading, but that is not how we see it after the fact. Rather, what emerges from this analysis is that the methods used to prove the classification of the finite simple groups are *more than adequate* to the task, and there is an embarrassment of riches. At various points, and indeed at the level of global strategy, one is confronted with several approaches, all apparently adequate, though differing in their efficiency. The theory in the finite case, and the fragment given here, sufficient for our purposes, can be read as involving a number of large and not very intimately connected theories, which have been developed simultaneously, and in some cases, it seems, only as far as a particular approach to the classification requires. We have made a selection from among these theories, which works particularly efficiently in the case of groups of finite Morley rank, but which might not represent a particularly efficient, or even viable, way of handling the finite case. Most strikingly, the theme of "standard components", which plays a large role in the finite case, almost disappears from view in our work, simply because at a key point more efficient methods appear on the scene. We welcomed this—we had no desire to pursue standard components, and a lingering suspicion that a lifetime (or three) might not be sufficient, though it is in fact likely the theory would collapse to reasonable proportions, adapted to the finite Morley rank context. Had we taken the conventional route, what we do here would look very much like the two proofs known in the finite case—whether it would be more difficult than the one we give remains unclear, but it would certainly be longer! We will suggest at the end that we may be following a different line of proof which makes sense in the finite case, not necessarily as a classification of all finite simple groups, but as an independent approach to a narrower subclass, including the Chevalley groups in characteristic 2. There is an analog with work of Timmesfeld in the finite case; while what we do here is not strictly parallel to that, the relationship seems real.

The difference between our problem and the finite problem seems to have less to do with sporadic groups than with small fields. Indeed, we make considerable use of tori, which over the field $\mathbb{F}_2$ reduce to the identity. We do have some trouble laying our hands on nontrivial tori sometimes, but in the end they can be produced when needed.

One point which does work out largely as we anticipated is the following: the theories that we do develop are applied here in much the same way that they are used in finite group theory, but with considerably less "background noise", and as a result the connection between methods adapted from the finite case, and the situation in algebraic groups, becomes more transparent. However, even here there is a nuance. The starting point for our main analysis (in Part C) is the classical theory of groups with strongly embedded subgroups, and its neo-classical revival, groups with weakly embedded subgroups. If one consults the original papers [**1, 124**] which deal with the K^*-case, one finds lines of argument which are certainly different from those used at the corresponding point in the theory of finite simple groups, but which nonetheless have very much the flavor of finite group theory, and in particular rely heavily on the theory of solvable groups, which runs in important respects closely parallel to the theory of finite solvable groups. All of the latter goes away when one drops the inductive hypothesis (K^*) and it is here that our Chapter IV comes into its own. As a result, this particular piece of the theory blows up considerably, and the chapter is a long one.

At the opposite extreme, our Chapter IX is a direct adaptation of work of Stellmacher to the finite Morley rank context. The subject would be rather dull if this chapter were typical. But the bulk of the developments have a different character: the main results achieved are closely parallel to results in the theory of finite simple groups, and the methods used owe much to the theory of finite group theory—but not to the proofs of the corresponding results! The dominant theme in these more typical parts of the theory is the adaptation to the context of connected groups of the fundamental notions of finite group theory, which in many cases brings them much closer to the notions of algebraic group theory which inspired them.

In any case, the pursuit of this classification problem has led those involved to develop a set of theories for groups of finite Morley rank which provide useful extensions of the theories developed in the finite case, and the specific requirements and challenges of the classification project have suggested some lines of development which were not immediately obvious; we mention particularly work by Corredor, Frécon, Poizat, and Wagner in this connection. About half of the present volume is devoted to the treatment of general topics of this kind, continuing in the vein of [**52**], and the other half to its applications to classification theorems in the simple case.

Our Main Theorem contains roughly half, or perhaps somewhat more, of what is currently known about the Algebraicity Conjecture (at least, as far as 2-local structure is concerned). To explain the present state of affairs more fully requires a little more background.

There is a Sylow theory for groups of finite Morley rank, for the prime 2. In addition to the conjugacy of the Sylow 2-subgroups, there is a very particular structure theory, considerably more reminiscent of the situation in algebraic groups than the situation in finite groups, which is summarized by the following cryptic formula.

$$(*) \qquad S^\circ = U * T$$

Using the language of algebraic groups, this formula may be read as follows: *"The connected component of a Sylow 2-subgroup is a central product of a unipotent 2-group and the 2-torsion from a split torus."* For a precise interpretation of the statement in our more general context, see §I 6. The point to bear in mind is that if we actually were dealing with an algebraic group, this result would hold in a considerably sharper form, depending on the characteristic of the base field:

$$S^\circ = U \text{ in characteristic 2; } S^\circ = T \text{ in all other characteristics.}$$

In particular the Algebraicity Conjecture predicts that this strong form should hold for simple groups of finite Morley rank, and the Main Theorem can be reformulated more lucidly as stating that this is, in large measure, the case. According to formula $(*)$, there are four possible structures for S°, depending on which of the factors U and T are present, and they correspond in some sense to hypotheses on the characteristic of the, as yet, unidentified base field: If $U \neq 1$ and $T = 1$, we say the group has *even type*; if $U = 1$ and $T \neq 1$, we speak of *odd type*, thereby inadvertently taking 0 to be odd; when U and T are both nontrivial we speak of *mixed type*, and finally when both are absent—which means the full Sylow 2-subgroup is finite, and possibly trivial—we speak of *degenerate type*. It will be seen that this terminology is consistent with the abbreviated account with which we began.

The Algebraicity Conjecture therefore breaks up naturally into four cases; in mixed and degenerate types we seek a contradiction, and in odd and even types we seek an identification of the group as an algebraic group (or, to put the matter both more concretely and more accurately, as a Chevalley group) over a field of appropriate characteristic. The Main Theorem can then be put in a third and very natural form as follows.

MAIN THEOREM, VERSION II.

(1) *There are no simple groups of finite Morley rank of mixed type.*

(2) *A simple group of finite Morley rank of even type is isomorphic to a Chevalley group over a field of characteristic* 2.

In view of the formula $(*)$, this is equivalent to the previous versions, and it is in this form that we will prove it. One can see now the sense in which we deal with "half" of the problem; but actually the deepest problem lies in the degenerate case. Since we know that there are no involutions in this case, 2-local analysis ends there, but the problem remains. In odd type there is now a substantial theory, which we omit.

The state of knowledge in odd type was covered until recently by the thesis of Jeff Burdges [**60**]. In odd type one has the following, which is limited to the inductive framework of K^*-groups, where a K^*-group is a group of finite Morley rank all of whose proper definable infinite simple sections are Chevalley groups, or in practical terms, as we suggested earlier, a group which is a putative minimal counterexample to the Algebraicity Conjecture.

ODD TYPE. *A simple K^*-group G of finite Morley rank and odd type satisfies one of the following conditions, where S° is the connected component of a Sylow* 2*-subgroup.*

(1) *G is algebraic.*
(2) $m_2(S^\circ) \leq 2$.

Can this approach actually prove the Algebraicity Conjecture in full? This seems very unlikely, for reasons well known to model theorists. The critical case is that in which there are no involutions, the most degenerate case in our taxonomy. Here the methods of the present text are not helpful, though the methods used in odd type have a certain force even in the absence of involutions, and we hope that further exploration of the degenerate case will lead to the further development of such methods. The focus of attention in the degenerate case is on Borel subgroups (maximal connected solvable subgroups) and the pattern of their intersections; they may, however, intersect trivially, at which point group theoretic analysis appears to come to a final halt. In any case, we are not yet this far.

The conjecture antipodal to the Algebraicity Conjecture runs as follows.

ANTI-ALGEBRAICITY CONJECTURE. *There is a simple torsion free group of finite Morley rank.*

The conventional wisdom at present is Manichaean: one of the two extremes ought to be correct. Beyond that, there seem to be few strong opinions as to how the matter should stand, though model theory has certainly clarified the issues involved over time. In particular, our work here relies crucially on some clarification of the model theoretic issues by Frank Wagner, as will be seen in Chapter IV at a preparatory level, and in Chapter VI in the context of a concrete application.

One striking difference between our subject and the theory of finite groups is our ability to prove a general result on groups of even type without first disposing of the case of degenerate groups. This would be analogous to disposing of characteristic 2 type finite simple groups without first proving the Feit-Thompson theorem. Evidently, the two situations differ substantially.

The proof of the Main Theorem evolved gradually, as we have mentioned. At first, we dealt with K^*-groups, that is, with minimal potential counterexamples, under the additional assumption (called *tameness*) of the noninvolvement of "bad fields" (cf. §I 4), though with the intention of reexamining the latter hypothesis at a later stage. After Jaligot's thesis [**125**] (cf. also [**123, 124**]), it became clear that the time had already come to proceed in the mixed and even type cases without reliance on this simplifying hypothesis (and to a large extent Burdges' thesis [**60**] has performed a comparable service for odd type). At this stage the K^*-hypothesis remained an integral part of the project. The program aiming at the full classification by adjusting the inductive framework was initiated in [**2**]. In this connection, methods derived from Wagner's work on the model theory of fields of finite rank have been essential.

Finally, one should not lose sight of two trivial but important points:

- The class of algebraic groups over algebraically closed fields of characteristic 2 is already a rich class, in the sense that the classification of Dynkin diagrams is an interesting, though relatively direct, classification, with its own "sporadic" (non-classical) members.
- At a deeper level, there are many nonalgebraic simple groups of finite Morley rank, because there are many fields of finite Morley rank with pathological structure, furnished by the Hrushovski construction—and all of this structure is visible in the associated groups. This is an important point, and more than once we have been confronted with the fact that we do not actually know the properties of "algebraic groups" when they are endowed with a finite Morley rank different from the usual dimension theory. Strictly speaking, algebraic groups (in this broad sense) are not even known to be K-groups!; though this does hold in positive characteristic, via work of Poizat.

Our main theorem says that in the presence of a fairly strong dimension concept—and nothing further—the underlying *group structure* is governed by the same finite combinatorics as in the algebraic case (Coxeter groups), at least in the case which corresponds to characteristic 2 in the algebraic setting; furthermore, this holds regardless of what pathology is allowed a priori in definable sections. We do not actually show that our groups are algebraic; we show that, like simple algebraic groups, they are Chevalley groups, which from our point of view means that they are amalgams of copies of SL_2 governed by the "recipe" encoded in a Dynkin diagram.

From this point of view, the reader should not be surprised to see considerable space devoted to the "tiny" group SL_2: all of Chapters VI and

VII, and much of Chapters V and VIII. On the other hand, the finite group theorist may be surprised to see that so little space is taken up with the remaining groups. By the standards of finite group theory, our inductive analysis is instantaneous.

The proof of the classification of the finite simple groups has given rise to a polemic between some who feel that the complexity of the proof *must* be due more to a poor choice of methods than to the nature of the problem, and those who feel that this is not at all the case—including, of course, most of those who have worked on the proof. This is not a polemic into which we feel a need to enter. We find the methods used extremely attractive. We also feel largely fortunate that we are not obliged to follow them too closely, and at the same time a bit unfortunate that we have no access to character theory or transfer methods—either one would be enormously helpful. Possibly our present work can make a modest contribution to the discussion underlying the polemic, by giving a demonstration of the flavor of a substantial portion of the finitistic methods in a context which lies somewhere in between the conceptual theory of algebraic groups and the more combinatorial theory of finite simple groups, and whose complexity in the primitive measure theoretic sense of length (or volume) is in the vicinity of the geometric mean of the two.

Acknowledgements

Our personal debts are at this point too numerous to record in full. We owe a particular debt to our closest collaborators on this project, namely Ayşe Berkman, Luis-Jaime Corredor, Olivier Frécon, Eric Jaligot, and Ali Nesin, and also to Poizat and Wagner, who have provided some fundamental insights relating to the relevant model theory. In a similar vein, we have benefited enormously from the insights of various finite group theorists. Stellmacher and Stroth in particular pointed out the viability of the "third generation" approach in our context (a line of thought vigorously seconded by Aschbacher), and Lyons and Solomon have clarified a number of points over the years. At a considerably earlier stage, one of us benefited from the general insight of Chat Ho into the finite theory, as well as his recipe for noodles.

Among the programs which have supported this work and advanced it in significant ways, we would like to mention the conference on Finite and Locally Finite Groups held in Istanbul in 1994, where the first concrete classification results were presented, the MSRI model theory program in 1997-1998 and specifically the support offered in January 1998, the Euroconference on Groups of finite Morley rank at the University of Crete's charming and tranquil Anogia Academic Village (June 1998), where finite group theorists and model theorists mixed with particular fluency, the conference Groupes, Géométrie et Logique at CIRM, Luminy (September, 2004), and the Newton Institute's program in model theory and its applications, for the month devoted to groups of finite Morley rank (March, 2005).

This work has benefited from continuing support from the NSF (current grant DMS-0600940) as well as from the Agence Nationale de la Recherche (grant :JC05_47037:jaligot:eric:). The Institut Camille Jordan at the Université Claude Bernard Lyon-1 has repeatedly offered its hospitality to those of the authors not already employed there.

Substantial portions of the work on this book were carried out at the Villa Amélie, Fons-par-Aubenas, under the auspices of Christiane and Fernand Lemay, whose hospitality is warmly acknowledged.

And we thank Jeff Burdges for some ingenious and artful Latex coding (perhaps masked by some of our own less skillful coding).

The first author acknowledges the everlasting presence of his Rutgers friends which makes negligible the growing distance in time and space: Isabel, Rita ve Oluş; Kerem, Esra ve Alper; Şelale, Eda, Özlem ve Yüksel, hepinize teşekkürler.

Le premier auteur remercie Bruno Poizat et Christine Charretton pour leur amitié sincère et enrichissante.

Les charmants ardéchois, Christiane et Fernand Lemay, ont une place particulière dans le coeur du premier auteur. Chez eux, il se sent chez lui. Le premier auteur a une multitude de raisons pour remercier Chantal Cherlin. Heureusement, il existe un mot miraculeux qui les résume toutes: Limoncello.

Haso Arkadaşım, bir parça sohbet edebiliriz artık. Sait Faik'in Simitle Çay öyküsünü kendi elyazımla yazıp gönderdiğim mektubun üzerinden on beş yıl geçti. Sanırım aynı mektupta da yazmıştım, amacım başkalarından hırsızlamadığım bir öyküyü gönderebilmek oldu hep. Bu kitap, bu amacıma en yakın olduğum nokta. Yazılmasına katkım olmadı. İş bölümü(!) böyle gerektirdi. Ama, içinde anlatılan öyküye bir katkım oldu. Ne ilginçtir ki, katkımın matematiğe olan herhangi bir yetenekle ilgisi yok. Yalnızca inat... Bolca da şans... Pilavdan dönmedim, kaşığı kırdırmadım.

In a historical vein

The history of the subject is bound up not only with the history of the theory of finite simple groups, but much of the history of pure model theory, which underwent a revolution beginning in the late sixties, and even (or perhaps, particularly) for those who lived through much of the latter, is not easy to reconstruct in a balanced way. We offer just a few scattered remarks, first from the second author:

> Vladimir Nikanorovich Remeslennikov in 1982 drew my attention to Gregory Cherlin's paper [**68**] on groups of finite Morley rank and conjectured that some ideas from my work [on periodic linear groups] could be used in this then new area of algebra. A year later Simon Thomas sent to me the manuscripts of his work on locally finite groups of finite Morley rank. Besides many interesting results and observations his manuscripts contained also

> an exposition of Boris Zilber's fundamental results on $\aleph_1$-categorical structures which were made known to many western model theorists in Wilfrid Hodges' translation of Zilber's paper [**196**] but which, because of the regrettably restricted form of publication of the Russian original, remained unknown to me.

The third author came to the subject by a rather different route, the common point of origin being the work of Zilber, which seems to have become more rapidly known in the West than in his country of origin, and in the original Russian. This was the subject of considerable interest (notably in Paris and Jerusalem) in the summer and fall of 1980, where as a result of their relationships with a notorious open problem in pure model theory, the broader conjectures of Zilber began to reach a wide audience. At the outset, work on the algebraic content of stability theory was stimulated in the West by Macintyre's work on $\aleph_1$-categorical fields [**137**], and in the East by a suggestion of Taitslin. For the third author, coming into model theory via the Robinson school, the question of the algebraic content of stability theory was both natural and inevitable. The notions of $\aleph_1$-categoricity and model completeness, characteristic of the two main schools of model theory at that time, had both arisen from considerations of the double-edged question: What is so special about the theory of algebraically closed fields, and is anything in fact special about this theory? The text [**128**] also arrived at a timely moment; in particular, this text made use of a notion of connectivity close to the one adopted in the present text, and for similar reasons.

The first author came to the subject from Mecidiyeköy, İstanbul.

The complex and provocative Poizat has played a complex and provocative role in the development of this theory. In particular, his early intervention brought the more "algebraic" formulation of the rank notion into its proper form, and generally he has been very attentive to foundational issues, some of great practical importance.

The complex and vigorous Nesin has played a complex and vigorous role in the development of this theory, entering at an early phase and, with his collaborators, treating a number of key configurations. In a historical vein, we remark that a period of forced confinement gave him the necessary leisure to familiarize himself with the contents of [**183**]; whether he wishes to convey his thanks for this we do not know, but it may be doubted. ("Si j'avais quelque chose à adresser aux grands de ce monde, je jure, ça ne serait pas des remerciements!" - Galois)

Model theorists will be aware that the subject has also grown in other directions—geometrical stability theory, applications to number theory in the hands of Hrushovski and several others, and that in these developments the structure of abelian groups turns out to be central, and not at all as trivial as might appear at first glance. We have also found close attention to the structure of abelian subgroups and their definable subgroups valuable, notably in connection with the theory of "good tori".

Part A

Methods

CHAPTER I

Tools

Oh me dolente! come mi riscossi
quando mi prese dicendomi: "Forse
tu non pensavi ch'io loico fossi!".
— *l'Inferno,* Canto XXVII

Introduction

This chapter contains material relating to the general theory of groups of finite Morley rank. Much of this material may be found in [**52**], and some of it in considerably more general form in [**186**]. For the fundamental principles, particularly those involving a certain amount of model theory, the source [**156**] is excellent. We avoid anything involving particular classes of groups, such as Chevalley groups. Everything we need in that line of a general character will be given in the following chapter, with additional specialized topics in the third.

Our treatment is compact but reasonably full. We begin with a few points from abstract group theory, notably the basic commutator laws and the classical Schur-Zassenhaus splitting which we transfer later in the chapter to the context of finite Morley rank. We then lay out the rank axioms on which everything we do will depend, and derive the theory of connected groups.

In §4 we take up the theory of fields of finite Morley rank. Recent advances in this direction play a major role in our work. We give Macintyre's theorem, that infinite fields of finite rank are algebraically closed, Zilber's results on the interpretation of fields in groups, results of Wagner and Newelski limiting the nongeneric definable subsets of fields of finite rank, and a more recent result of Wagner on fields of finite Morley rank in positive characteristic: torsion is dense in any definable subgroup of a torus. In the last part of this book we will be in a context in which all fields should have characteristic 2, and this will be a critical property; so we encapsulate it in the term "good torus" which is introduced in this section.

The next section deals with the theory of nilpotent groups of finite Morley rank: their structure, the existence in general of the Fitting subgroup, two notions of Frattini subgroup (both useful, but quite distinct), and other staples of general group theory like the normalizer condition. In the presence of connectedness, a number of these points take on noticeably strengthened forms.

A subject which is closely related to the theory of nilpotent groups is the Sylow theory, which also exists in two forms, connected and general, but only for the prime 2. It is most convenient to work with the connected theory as the main variant, but it occasionally misses something relevant, in which case one passes to the general theory, but for the most part after first profiting from the connected version. We write "Sylow°" and "Sylow" for the two theories respectively, so the reader should expect to encounter "Sylow°" rather than "Sylow" more or less throughout (and heuristically one does little damage by ignoring the distinction).

Up to a point, the reader knows what to expect from Sylow theory. However, one remarkable, and relatively elementary point, is that the structure of a Sylow° 2-subgroup of a group of finite Morley rank is remarkably limited. This can be explained by invoking the theory of algebraic groups, a topic we leave for the next chapter. The structures of Sylow° 2-subgroups in algebraic groups are remarkably limited, and depend mainly on whether the characteristic is 2 or not (in this context, 0 is odd, or in any case not equal to 2). The structure of a Sylow° 2-subgroup of a group of finite Morley rank in general is close to a direct product of the two types of Sylow° 2-subgroups occurring in algebraic groups. Since a product of algebraic groups over various fields is an example, possibly typical, of a group of finite Morley rank, this is not a completely unexpected result.

This structural result plays a fundamental role in our approach to the subject. It gives us a way of distinguishing groups of "characteristic 2" type from the others, at the outset. Something similar is done in finite group theory, but in a more complicated way, using the structure of centralizers of involutions instead, which complicates the treatment of small groups (groups of low Lie rank). On the other hand, this is inevitable in the finite case, as there are various isomorphisms between quasisimple groups over small fields of different characteristics, and the characteristic of a very small group may depend on the group in which it is located. In our case, we have a clearcut distinction at the outset, unless the 2-Sylow° subgroup is trivial; this would mean that the ordinary 2-Sylow subgroup is finite, and should not occur in a connected group of finite Morley rank. This last point would be the analog of the Feit-Thompson theorem in the finite Morley rank context.

In §7 we take up Bender's generalized Fitting subgroup $F^*(G)$. This theory plays a central role in the classification of the finite simple groups. It will be less visible in our treatment, but only because we stay on the "characteristic 2" side of the theory. On the other side, this notion plays an absolutely central role, and indeed the same role as in the finite case. The theory in the finite Morley rank context is exactly parallel to the finite theory, once one takes into account the slight variations associated with taking connected components of everything, which we tend to do on every possible occasion.

In the following section we give some of the theory of solvable groups of finite Morley rank, notably the Hall and Carter theories. The Carter theory is more important here than in the finite case; indeed, it is reminiscent of the theory of maximal tori in the algebraic case. We also discuss the solvable radical, along with the socle and the p-unipotent radicals $U_p(H)$, which naturally accompany the solvable radical. This section also includes some results on "lifting" centralizers which are extremely useful in practice, and are well known in the finite case. There is also an early form of Schur-Zassenhaus, in a minimal case.

In the following section we come back to the Schur-Zassenhaus theory in general, one of the leading themes of [**52**].

The next section collects some useful information about automorphisms, in a general setting. There are four topics: (1) the actions of automorphisms of finite order, notably order 2 or order p where the group being acted on contains no p-elements; (2) action of a group of even type on a degenerate type group; (3) automorphisms of p-tori; (4) "continuously characteristic" subgroups. With the exception of (2) these provide useful general principles of an elementary nature. However, (2) is the motor for much of the present work. It turns out that the action of a connected 2-group on a degenerate type group must be trivial, or in other words: if a connected 2-group normalizes a degenerate type group, then the two groups commute. In practice this is what allows us to prove a classification theorem for groups of even type without first proving a Feit-Thompson theorem; this result uncouples any degenerate type sections of the groups in question from the more interesting parts of the group—most easily in proper subgroups, where induction applies.

In §11 we take up some matters connected with modules, that is definable actions of groups on abelian groups. For the irreducible case, the main points were dealt with under the interpretability of fields, and for the most part we record some generalities here concerning Clifford theory and composition series. We also point out one situation in which a group action must involve a good torus. The subject of modules, or representation theory, is certainly an important one in our subject, and we will return to it in various more specialized contexts, but in a general setting it is largely exhausted by Zilber's results on interpretability of fields and subsequent elaborations by Wagner and Poizat already given in §4.

Our last two sections are more technical. We first discuss the Thompson $A \times B$-lemma, which goes over into our context very naturally, and can be used to "kill cores" (and a bit more than cores, actually) in 2-local subgroups in an inductive even type setting. As it happens, our approach uses fewer 2-local subgroups than the standard approaches of finite group theory, and any cores we need killed are more or less dead on arrival anyway, but in any case we give this theory, and show later how it may be applied to simplify the situation.

The other point concerns the theory of complex reflection groups, which is our main route toward the identification of Coxeter groups, just prior to final identification of a generic simple group of even type. We will use various representations of our "Weyl group" on the torsion subgroup of a maximal torus to build a representation in characteristic zero (an ultraproduct). This can be viewed as a complex representation, and retain enough of the character of the natural reflection representations (in finite characteristic) of the Weyl group to be viewed as irreducible complex reflection groups. Fortunately, these have been classified explicitly, and this material is given here, with specific information needed later on to eliminate the non-Coxeter "interlopers".

Overview

We review the main points of the chapter here for the general reader unfamiliar with the foundations of the subject (as given in [**52**]) and anxious to move along rapidly.

The section on general group theory (§1) consists largely of points we will call on occasionally in the sequel, and which may not be familiar in the precise form we require. Our development actually begins in §2 with a discussion of *rank* as a "dimension function" on definable sets. This notion of rank coincides with Morley rank on the class of groups, but not in general. We do not give a separate definition of Morley rank; our rank notion is adequate not only in groups, but in any structures which can be interpreted into groups. One should be a little cautious though in looking beyond groups. We will deal later with the theory of buildings, for example, and it takes some work to show that our rank notion can again be called "Morley rank" in that context—none of which affects any of our applications, but does raise some doubts about our terminology, for those who take the model theory seriously.

Rank behaves for practical purposes like an estimate on the logarithm of the cardinality, a point that can be rendered rigorous in some contexts (e.g. for families of groups defined, uniformly, over finite fields), and also behaves like the Zariski dimension (which by the Lang-Weil theorem agrees with the former in the large finite case).

Thus the rank of a Cartesian product is the sum of the ranks. Less obvious, by far, is the following property: in any uniformly definable family of definable sets, the sizes of the finite sets are bounded. This is connected with the notion of rank as the sets of rank zero are the finite ones, and it is related to *definability of the rank.*

The four axioms for rank are called *Monotonicity, Additivity, Definability, and Uniform bounds*, and we have touched on the three nontrivial axioms already, so at this point the reader should have a fairly precise sense of the notion. As a consequence of these axioms, one gets a *Fubini principle* governing the rank of a subset of a product, or more generally of a subset of a disjoint union of sets of constant rank (a fibering). In particular, a subset S of such a disjoint union is "generic" (i.e., of full rank), if its fibers are *generically generic*; that is, the generic fiber of the ambient set meets S in a generic subset.

With rank providing a notion of dimension, there is also a notion of *multiplicity*, known as *Morley degree* in this context. The multiplicity is the number of irreducible components *of maximal dimension*. Unlike the algebraic case, these components are well defined only modulo sets of lower rank, and thus the multiplicity can only be defined in the top dimension. The fine structure of the rank notion is hard to exploit; things are always clearest in the top dimension.

The rank (and degree) provides a *descending chain condition* for definable subgroups. There is a more subtle *uniform chain condition* for uniformly definable families of subgroups, due to Baldwin and Saxl; see Lemma 2.8 for details.

From the descending chain condition on definable subgroups we derive a kind of Zariski closure for subgroups, namely the smallest definable subgroup containing a given group. A point which seems modest, but eventually plays a large role, is found in Lemma 2.16: the definable hull of a cyclic group is the direct sum of a divisible group and a finite cyclic group. A defect in the theory is the fact that the function $d(a)$ taking an element a to the definable hull of the cyclic group $\langle a\rangle$ is typically not a definable function; this defect is largely remedied in §IV 4 by introducing a definable approximation $\hat{d}$ to d with very similar properties. Another apparently innocuous result is *lifting of torsion*: the preimage of p-torsion under a definable homomorphism contains p-torsion. In view of the natural map $\mathbb{Z} \to \mathbb{Z}/n\mathbb{Z}$ this has to be considered a particularly agreeable circumstance, and one that keeps our theory reasonably close to the finite theory; it is of course true in the algebraic context as well, when one restricts attention to Zariski closed groups and algebraic homomorphisms.

In a very similar vein, we have the *Basic Fusion Lemma* 2.20: Two involutions are either conjugate or commute with a third. This is familiar in the finite case, much less familiar but still true in the algebraic case, and true in our case as well, and essential if nontrivial parts of the theory of finite groups are to be applied in our context to classification problems.

Everything so far is essential for subsequent developments. We may then pass on (§3) to the critical notion of *connected* group and the *connected component* (of the identity) in a definable group. Obviously we are no longer following the motivation of the finite case, but the finite notions will mix well with this notion. In the algebraic category, while definable sets and Zariski closed sets are decidedly not the same thing (more's the pity), definable subgroups and Zariski closed subgroups do in fact coincide, and thus in our more general context we may expect definable groups to behave well. We call a definable group *disconnected* if it contains a proper definable subgroup of finite index, and connected otherwise. The existence of the connected component G° is then purely formal, in view of the descending chain condition on definable subgroups, but there is a useful property which lies deeper: a connected group must be *irreducible*, that is, of Morley degree one—or in other words, to put the thing more usefully: a connected group cannot contain two disjoint generic subsets. This is the foundation for a geometrical line of argument in our subject, as will be seen in Chapter IV. For the proof, see Lemma 3.6. The idea is to consider the action of G on its irreducible components of top dimension (not, admittedly, a well-defined notion, but manageable nonetheless). In carrying through this idea one makes use of the Fubini principle.

At this point we have a substantial corpus of basic ingredients to work with, and we can derive some concrete consequences, notably: Any definable action of a connected group of finite Morley rank on a finite set is trivial; any definable endomorphism with finite kernel of a group of finite Morley rank is surjective; the additive and multiplicative groups of an infinite field (or division ring) of finite Morley rank are connected; an infinite group of finite Morley rank contains an infinite definable abelian subgroup. The reader new to the subject might want to try his hand at these.

Another tool of wide applicability is the following definability result, due to Zilber: If G is a group of finite Morley rank, and H a subgroup generated by connected definable subgroups, then H is also connected. This is one of several reasons that it is useful to adapt most of the notions of group theory to *connected versions*; thus we work very often with the *connected normalizer* $N^\circ(X)$, the *connected centralizer* $C^\circ(X)$, and so on.

Before proceeding, we extend the notion of connected component to subgroups which are not necessarily definable: $H^\circ = H \cap [d(H)^\circ]$ when H is not necessarily definable. The most interesting case of this arises in conjunction with the Sylow theory, as Sylow subgroups are typically not definable. We note also that definable hulls of abelian, nilpotent, or solvable groups are again abelian, nilpotent, or solvable, respectively.

For definable groups H it follows from what we have said already that the Morley degree of H is the index $[H : H^\circ]$. A considerably more delicate point is the following: if $\mathcal{F}$ is a uniformly definable family of definable groups, then these indices $[H : H^\circ]$ are uniformly bounded—this would be obvious if the family

$$\{H^\circ : H \in \mathcal{F}\}$$

were itself known to be uniformly definable. For the proof, it seems to be necessary to go back into the foundations of the subject.

Next we take up the theory of *fields* of finite Morley rank (§4), which contains both the earliest results in this area, and some of the most recent, all of them very useful in applications. This consists of the following ingredients.

- Macintyre's theorem: An infinite field of finite Morley rank is algebraically closed.
- Zilber's Linearization Theorem: Let G be a connected group of finite Morley rank acting definably, faithfully, and irreducibly on an abelian group V, and let $T \triangleleft G$ be infinite abelian. Then the

subring of End(A) generated by T is a an algebraically closed field, with respect to which V becomes a finite dimensional vector space on which G acts linearly.

- The Newelski-Wagner Genericity Lemma: A definable subset of a field of finite Morley rank which contains an infinite field is generic.
- The Good Torus Principle: Any definable subgroup of the multiplicative group of a field of finite Morley rank, in *positive characteristic* is the definable hull of its torsion subgroup.

Most of this can be extracted reasonably directly based on the foundations established in the previous two sections, with the exception of the final point, which is a reformulation of results of Wagner which require some further foundational work, but in the specific context of fields. As will be seen in Chapter VI (and to some extent already in Chapter IV) this last result provides the basis for our geometrical lines of argument which fill the gap which would otherwise result from our inability to handle nonsolvable groups without involutions—not by casting any light on the class of groups, but by allowing us to work around it entirely.

For Macintyre's theorem, we have already laid sufficient foundations. We know that the additive and multiplicative groups of our field are connected, and in Galois theoretic terms it follows that the field has no Kummer or (in characteristic p) Artin-Schreier extensions, and is perfect. While this by itself will not yield algebraic closure, these properties are inherited by finite algebraic extensions, which "live inside" the universe of definable sets. And indeed, a Galois theoretic argument then shows that a field with these properties holding *hereditarily* is algebraically closed.

Zilber's theorem (found in this form in [**52**]) is a distant cousin of Schur's lemma, coupled with the following point (Lemma 4.5) any definable group of automorphisms of a field of finite Morley rank is trivial. *"But what of the Frobenius automorphism?"* The alert reader will object—observe the distinction between a definable group of automorphisms and a group of definable automorphisms. The triviality of such definable automorphism groups is largely a consequence of Macintyre's theorem (the fixed field must be either finite or algebraically closed, and we need only concern ourselves with the former case, which indeed requires a little attention).

The Newelski-Wagner Genericity Lemma would be obvious if the "infinite field" referred to were itself definable, as the structure consisting of a pair of algebraically closed fields, nested, has *infinite rank*. So this becomes largely a matter of trading a not necessarily definable field in for a definable one.

Finally, the Good Torus Principle is both essential and subtle, and not in the direct line of thought we have followed to this point. The underlying model theoretic result is the following: if F is a field of finite Morley rank, then the subfield F_{alg} of *model theoretically algebraic* elements is an elementary substructure. In positive characteristic one may show (via the Frobenius automorphism, which respects both the field structure and whatever multiplicative subgroups may be definable) that these algebraic elements are algebraic in the conventional sense, and thus our field has a *locally finite* elementary substructure. This then decodes into the Good Torus Principle. The point of this principle will be seen considerably later, starting with §IV 1 (Proposition IV 1.15).

At this point we have the foundations well in hand, and we can develop various standard group theoretic topics on that basis (§§5–9): the structure of nilpotent groups, the Fitting subgroup (and later, the generalized Fitting subgroup), and the theory of solvable groups, which includes the theory of Hall and Carter subgroups, the solvable radical, and the important Schur-Zassenhaus lemma. The latter requires a considerable development. The solvable theory has some special features which are reminiscent of the algebraic theory. It is much simpler than the theory in the finite case, because of our attention to connected groups. The most striking parallel to algebraic group theory is a version of the Lie-Kolchin theorem: if H is a connected solvable group of finite Morley rank, then the quotient $H/F^\circ(H)$ modulo the connected Fitting subgroup is a divisible abelian group (Lemma 8.3). The idea of the proof is to use the Zilber Linearization Theorem on a composition series for H to get H acting as a subgroup of a product of fields, where the kernel of the action has a nilpotent action on H and hence lies in the Fitting subgroup (after which one may pass to the connected Fitting subgroup with a little more argument).

One can also define a reasonable notion of p-unipotent subgroup, and show in the solvable case that these groups necessarily lie in the Fitting subgroup. We would have more trouble introducing a notion of 0-unipotence; this has been done, for use in groups of odd type, but for us the critical case is 2-unipotence, and we may be spared these interesting refinements.

In the Introduction we took special note of the Sylow theory, for good reason. On the one hand it presents several subtleties: the groups in question are typically not definable, which makes their management a delicate point; furthermore, while we have a good Hall theory, in general, inside solvable groups, we have no real Sylow theory at all for any prime other than 2. And indeed the treatment of Sylow theory in the finite case is resolutely arithmetical, whichever of the variety of approaches one adopts. We must adopt an entirely different approach, working inductively, and relying ultimately on the Basic Fusion Lemma when all else fails.

One may deal with either Sylow subgroups or their connected analog, Sylow° subgroups; the latter is the workhorse of the subject, but the former intervenes on occasion. Sylow 2-subgroups are maximal 2-subgroups, and 2-subgroups are those whose elements have order a power of 2. Existence is therefore not an issue; conjugacy is, but can be proved by an inductive argument. Structure is also an issue. An essential point is the *local finiteness* of the Sylow 2-subgroups. If one wishes to extend the theory to p-subgroups, one should probably build local finiteness into the definition; but even so, problems will remain.

Considerably more can be said about the structure of a Sylow° 2-subgroup S of a group of finite Morley rank, and indeed the situation runs closely parallel to what one sees in the algebraic case. The structure is as follows:

$$S = U * T$$

where:

(1) The $*$ represents a central product, with a finite intersection.
(2) U is 2-*unipotent*: definable, connected, solvable of bounded exponent (and hence nilpotent, by Lemma 5.5.
(3) T is a 2-*torus*: divisible abelian but presumably not definable, as the definable hull should contain elements of infinite order.

This structural result gives us a way of distinguishing groups of "characteristic 2" type from the others, at the outset. Either or both of U and T may be trivial; in a simple group one expects exactly one of the two to be absent. There are all told four possibilities, and hence four possible "types": *mixed type*, with both factors present; *degenerate type*, with $S = 1$; and *even* or *odd* type, with, respectively, $S = U$ or $S = T$. Each of these types is approached differently, though as we shall see the two types with $U > 1$ are approached similarly, and in the end the case of mixed type reduces to the case of even type.

The chapter contains four further sections, dealing with automorphism, modules, the Thompson $A \times B$ theorem, and complex reflection groups (§§10–13). This is a motley collection, ending up with topics that lie somewhere between general tools and the sort of more specialized developments treated in Chapter III.

To begin with, a number of general principles involving automorphisms should be considered part of the basic tools of the trade, and have been collected together. More specialized topics relating to automorphisms of Chevalley groups will be found under that more specialized heading. But we have buried one topic of the first importance in this section, namely Proposition 10.13: *If a* 2*-unipotent group acts definably on a definable subgroup without* 2*-unipotent subgroups (for example, on a group of degenerate type), then the action is trivial.* This result furnishes one of the key mechanisms for *neutralizing* degenerate sections of a group. And in the course of the analysis we make our first acquaintance with the important notion of strong embedding, to be investigated in detail in Chapter VI.

The topics considered under "modules" do not form a particularly coherent whole. Logically, this section could equally well include the Zilber Linearization Lemma, some at least of the Schur-Zassenhaus analysis, and other topics. But we collect here a few points for which the language of modules is particularly convenient, and which do not belong to any more notable category.

The last two topics ($A \times B$ theorem, complex reflection groups) are of a very specific character. They are close in spirit to the specialized topics which are treated on their own in Chapter III, but seem to us to have more of the character of general group theory.

The Thompson $A \times B$ Lemma goes over smoothly from finite group theory to our context, and it plays much the same role for us that it does classically; and it could easily be hidden in the section on nilpotent groups, or modules, or automorphisms, which are themselves three faces of a single thing. But it plays a distinguished role in the subject, and we let it stand alone. The theory of complex reflection groups, on the other hand, is a topic which clearly belongs, as far as the content is concerned, in our first section, which was devoted to topics in general group theory which we take over and use—we use this one rarely, but to great effect.

In other words, we deal at the end with four afterthoughts (excepting the fundamental Proposition 10.13) perhaps out of their proper places, but all playing an important role in the sequel.

If the reader is still with us, he has been here long enough and should either browse the chapter or move on to points of greater interest (probably Chapters III or IV, as taste may dictate).

1. General group theory

In this section we record some general group theoretic facts which are occasionally useful, and establish our group theoretic notations, particularly in areas in which conventions are not entirely standardized. We note at the outset that in model theoretic contexts, the notation X^n stands for the Cartesian power $X \times \cdots \times X$ of the set X, but the notation G^n, for G a group, will also refer to a term in the upper central series; this usage should not result in any substantial ambiguity. But we will avoid the use of the same notation for the set of n-th powers. In abelian contexts, that set may be denoted nG, and in nonabelian contexts we have no special notation for this set.

When we work with a group H having a normal subgroup K, we will sometimes avoid passing to the quotient $\bar{H} = H/K$ by a standard notational device: We write for example $Z(H \bmod K)$ for the pullback to H of $Z(\bar{H})$, $C_H(X \bmod K)$ for the pullback to H of $C_{\bar{H}}(\bar{X})$, and $N_H(L \bmod K)$ for the pullback to H of $N_{\bar{H}}(\bar{L})$.

1.1. Notations.

NOTATION 1.1. *Let G be a group, and π a set of primes.*

(1) *For $a, b \in G$, we set $a^b = b^{-1}ab$ and $[a, b] = a^{-1}a^b$.*
(2) *For $X, Y \subseteq G$, $[X, Y]$ denotes the* subgroup *generated by commutators $[x, y]$ with $x \in X$, $y \in Y$; but for $x \in X$, $[x, Y]$ denotes the* set *of commutators $\{[x, y] : y \in Y\}$.*

(3) *G^i and $G^{(i)}$ are defined inductively by:*

$$\begin{aligned} G^0 = G^{(0)} &= G, \\ G^{i+1} &= [G, G^i], \\ G^{(i+1)} &= [G^{(i)}, G^{(i)}]. \end{aligned}$$

The series G^i is called the descending central series*; the series $G^{(i)}$ is called the* commutator series.

(4) *The* ascending central series *$Z_i(G)$ is defined inductively by*

$$\begin{aligned} Z_0(G) &= 1, \\ Z_{i+1}(G)/Z_i(G) &= Z(G/Z_i(G)). \end{aligned}$$

(5) *A π-number is a positive integer all of whose factors belong to π.*
(6) *For $a \in G$, $H \le G$, a^H denotes the set $\{a^h : h \in H\}$.*
(7) *A π-element of G is an element whose order is a π-number.*
(8) *A π-group is a group all of whose elements are π-elements.*
(9) *A $\pi^\perp$-group is a group none of whose elements other than 1 is a π-element.*
(10) *π' is the complement of π in the set of prime numbers.*
(11) *We write $G^\times$ for $G \setminus \{1\}$.*
(12) *G is* π-radicable *if for every $g \in G$ and every π-number n, g has an n-th root in G, and if G is abelian, the term* π-divisible *means the same thing (but may be expressed additively).*
(13) *If P is a $\{p\}$-group, then $\Omega_i(P)$ is the subgroup generated by elements of order at most p^i.*

Note that a π'-group is a $\pi^\perp$ group, but that π'-groups are necessarily *periodic* (all elements are of finite order) whereas $\pi^\perp$-groups may contain elements of infinite order.

When $\pi = \{p\}$ for a single prime p, we lighten the notation accordingly: p-group, p'-group, $p^\perp$-group, and so on.

1.2. Commutator laws.

LEMMA 1.2. [**108**, ...] *Let G be a group, and $a, b, c, g \in G$.*

(L) $[ab, g] = [a, g]^b [b, g]$,

(R) $[g, ab] = [g, b][g, a]^b$,

(J) $[[a, b^{-1}], c]^b [[b, c^{-1}], a]^c [[c, a^{-1}], b]^a = 1$.

We will refer to (J) as the Jacobi identity.

COROLLARY 1.3. *Let $i \in I(G)$ (that is, an involution), $x \in G$, $\gamma = [i, x]$. Then $\gamma^i = \gamma^{-1}$.*

COROLLARY 1.4. *Let G be a group, A an abelian subgroup, and $g \in N_G(A)$. Then the commutator map $\gamma_g : A \to A$ defined by $\gamma_g(a) = [g, a]$ is an endomorphism of A.*

LEMMA 1.5. *Let G be a group, and X, Y subgroups. Then X and Y normalize $[X, Y]$.*

PROOF. Let $x_1, x \in X$, $y_1 \in Y$, $\gamma = [x_1, y_1]$. Then $[x_1 x, y_1] = [x_1, y_1]^x [x, y_1]$, hence $\gamma^x \in [X, Y]$. □

LEMMA 1.6 (Three subgroups lemma). *Let G be a group, H, K, L three subgroups, and $N \triangleleft G$. If two of the three subgroups*

$$[[H, K], L], [[K, H], L], [[L, H], K]$$

are contained in N, then so is the third.

PROOF. Suppose the first two are contained in N. By the Jacobi identity, for $l \in L$, $h \in H$, $k \in K$, we find

$$[[l, h], k]^{h^{-1}} \in [[H, K], L]^K [[K, L], H]^L \subseteq N$$

and thus $[[l, h], k] \in N$. □

DEFINITION 1.7. *Let G be a group.*

(1) *G is* quasisimple *if $G' = G$ and $G/Z(G)$ is simple.*
(2) *G is* quasisemisimple *if G is a central product of quasisimple groups.*
(3) *For any group G, $E(G)$ is the subgroup of G generated by its subnormal quasisimple subgroups.*

LEMMA 1.8. *Let G be a quasisemisimple group. If $H \triangleleft G$ then H' is quasisemisimple, $H = H'Z(H)$, $Z(H) = H \cap Z(G)$, and the quasisimple normal subgroups of H are normal in G.*

PROOF. Let $\bar{G} = G/Z(G)$. Then $\bar{G}$ is a direct product of simple groups and $\bar{H}$ is normal in $\bar{G}$, so the same applies to $\bar{H}$. Thus $H = H'Z(H)$ and $Z(H) = H \cap Z(G)$, and it follows that $H' = H''$. It remains to check the last claim.

If H_1 is a quasisimple normal subgroup of H, then $\bar{H}_1 \triangleleft \bar{G}$ and hence $H_1 Z(G) \triangleleft G$. Then $H_1 = H_1' = (H_1 Z(G))'$ is normal in G. □

LEMMA 1.9. *Let G be a group.*

(1) *If H, K are subnormal in G and quasisimple, then either $[H, K] = 1$ or $H = K$.*
(2) *$E(G)$ is the central product of the subnormal quasisimple subgroups of G.*

PROOF. As the second claim follows from the first, we concern ourselves with the first.

Let $H = H_0 \triangleleft H_1 \triangleleft \dots H_m = G$ and $K = K_0 \triangleleft K_1 \triangleleft \dots K_n = G$, and proceed by induction on $\max(m, n)$. Then the conjugates of H in G lie inside H_{m-1}, so by induction distinct conjugates of H commute with one another, and generate a subgroup $\hat{H}$ which is their central product; similarly the conjugates of $\hat{K}$ generate a subgroup $\hat{K}$ which is their central product.

Suppose first that $L = \hat{H} \cap \hat{K}$ is abelian. Then $L \leq Z(\hat{K})$ so $[[H,K],K]=1$ and by the three subgroups lemma, $[H,K] = [H,[K,K]] = 1$.

Now suppose that $L = \hat{H} \cap \hat{K}$ is nonabelian, and let L_1 be a quasisimple normal factor of L. Then L_1 is also quasisimple normal in $\hat{H}$ and in $\hat{K}$. Accordingly there are conjugates H_1 of H and K_1 of K such that $L_1 Z(\hat{H}) = H_1 Z(\hat{H})$ and similarly for K_1. Then $L_1 = L_1' = (L_1 Z(\hat{H}))' = (H_1 Z(\hat{H}))' = H_1$ and similarly $L_1 = K_1$, so H and K are conjugate, and the claim follows as already remarked. □

LEMMA 1.10. *Let G be a group, and H a solvable normal subgroup. Then* $[E(G), H] = 1$.

PROOF. The commutator $[E(G), H]$ is a solvable normal subgroup of $E(G)$, hence is contained in $Z(E(G))$, or in other words $[E(G),[E(G),H]]=1$ and hence $[E(G), H] = 1$ since $E(G)' = E(G)$. □

LEMMA 1.11. *Let G be a group, and H a normal subgroup. Then $E(H) = (E(G) \cap H)'$.*

PROOF. By the definitions, $E(H) \leq E(G)$ and hence $E(H) = E(H)' \leq (E(G) \cap H)'$. Conversely $E(G) \cap E(H) \lhd E(G)$, hence $(E(G) \cap H)'$ is quasisemisimple. □

1.3. Commutator subgroup.

LEMMA 1.12 ([**52**, Ex. 21, p. 7]). *Let G be a group with $G/Z(G)$ finite. Then G' is finite.*

PROOF. Let $X = \{[a,b] : a, b \in G\}$. Then $|X|$ is finite. Let $X = \{x_1, \ldots, x_N\}$ in some definite order. We claim

(∗) Any element of G' may be written in the form $x_1^* \cdot \cdots \cdot x_N^*$ with x_i^* a positive power of x_i

For this, let $g \in G'$ have the representation $g = x_{i_1} \cdot \cdots \cdot x_{i_l}$. We proceed by induction on l.

If $i_1 \leq \cdots \leq i_l$, we have our claim. Assuming the contrary, let j be minimal so that $i_j > i_k$ for some $k > j$, and with l fixed, choose the representation of g to maximize j. Choose $k > j$ to minimize i_k. Let $g_1 = x_{i_1} \cdot \cdots \cdot x_{i_{j-1}}$, $g_2 = x_{i_j} \cdot \cdots \cdot x_{i_{k-1}}$, and $g_3 = x_{i_{k+1}} \cdot \cdots \cdot x_{i_l}$. Note that $g_2^{x_{i_k}}$ can be expressed as a product of length $k-j$ since X is invariant under conjugation. Consider the representation of g as $g_1 x_{i_k} g_2^{x_{i_k}} g_3$. This again has length l, but has i_k in place of i_j in the j-th position. Thus relative to this second representation, the value of j has increased, a contradiction. This proves (∗).

Note also that the transformation of an expression for g into the standard form of (∗) does not increase length. Let $n = |G/Z(G)|$. We claim now that the representation can always be taken with length at most $|X| \cdot n$. To see this, it suffices to show that any expression of the form x^{n+1} (considered as length $n+1$), with $x = [a,b] \in G'$, can be shortened by rewriting. Since $x^n \in Z(G)$, this may be done as follows:

$$x^{n+1} = a^{-1}x^n b^a = [(x^{n-1})^a x^a a^{-1}]b^a = (x^{n-1})^a [a,b]^a [a,b] = (x^{n-1})^a [a^2, b]$$

where the last expression can be construed as length n, since $[a^2, b] \in X$. □

LEMMA 1.13 ([**52**, Ex. 22, p. 8]). *Let G be a group, and H_1, H_2 subgroups of G which normalize each other. If the set of commutators $X = \{[h_1, h_2] : h_1 \in H_1, h_2 \in H_2\}$ is finite, then the commutator subgroup $H = [H_1, H_2]$ is finite.*

PROOF. We may take $G = H_1H_2$ and thus $C_G(X) = C_G(H)$ is a normal subgroup of finite index. Hence $Z(H)$ has finite index in H, and it follows that H' is finite. Accordingly we may factor out H' without loss of generality, and assume that H is abelian.

Therefore the commutator maps $\gamma_{h_1} : H_1 \to H$ defined by $\gamma_{h_1}(h_2) = [h_1, h_2]$ are homomorphisms, so X is closed under taking powers. Thus the elements of X are of finite order, and H is generated by a finite set of elements of finite order; H is finite. □

1.4. Abelian and nilpotent groups.

NOTATION 1.14. *Let A be an abelian group:*

(1) *For $n \geq 1$, $A[n] = \{a \in A : na = 0\}$.*
(2) *A subgroup B of A is* pure *in A if $nA \cap B = nB$ for all $n \geq 1$.*

LEMMA 1.15 ([**96**, 28.2, Kulikov]). *If A is an abelian group, and B a pure subgroup with A/B of bounded exponent, then A splits as $B \oplus C$ for some complement C.*

PROOF. By Zorn's Lemma one can find a maximal pure subgroup $\hat{B}$ of A containing B, such that $\hat{B}/B$ splits over B. We claim $\hat{B} = A$. If not, take $a \in A$ of maximal prime power order $q = p^n$ modulo $\hat{B}$, and use the purity of $\hat{B}$ to write $\langle \hat{B}, a\rangle$ as $\langle \hat{B}\rangle \oplus C$ with C cyclic. To reach a contradiction it suffices to check that $\hat{B} \oplus C$ is again pure in A. This reduces easily to checking that $pA \cap (B \oplus C) = pB \oplus pC$ and follows from the maximization of q. □

LEMMA 1.16 ([**96**, Theorem 17.2]). *Let A be an abelian group of bounded exponent. Then A is a direct sum of cyclic groups.*

PROOF. Take a maximal direct sum of cyclic subgroups which is pure in A, and apply the preceding lemma. This reduces the problem to one of finding a single cyclic direct factor of A, and for this one takes a cyclic subgroup of maximal prime power order, which is again pure in A and hence is a direct factor. □

LEMMA 1.17. *If H is a nilpotent group and π is a set of primes, then the set H_π of π-elements of H is a subgroup, and is the direct sum of the* *subgroups H_p (i.e., $H_{\{p\}}$) for $p \in \pi$. In particular the set H_{tor} of elements of finite order in H is a subgroup, and is the direct sum of all the subgroups H_p.*

LEMMA 1.18 ([**71, 72**]). *Let H be a π-radicable nilpotent group. Then:*

(1) *$H_\pi \leq Z(H)$.*
(2) *$Z_i(H)/Z_{i-1}(H)$ is π-torsion free and π-divisible for $i > 1$.*

PROOF.
Ad (1) We show for each prime $p \in \pi$ that

$$(*) \qquad \text{If } h^p \in Z(H) \text{ with } h \in H, \text{ then } h \in Z(H).$$

We may suppose that the corresponding statement holds in $H/Z(H)$, and accordingly if $h^p \in Z(H)$, then $h \in Z_2(H)$.

Fix such an h. Then for $x \in H$ we have $1 = [h^p, x] = [h, x]^p = [h, x^p]$, and as H is p-radicable we find $h \in Z(H)$ as required.

Ad (2) In $\bar{H} = H/Z_{i-1}(H)$, our claim is that $Z(\bar{H})$ is π-torsion free and π-divisible. It is π-torsion free by part (1) applied to $H/Z_{i-2}(H)$, and it is π-divisible by $(*)$ applied to $\bar{H}$. □

LEMMA 1.19. *Let G be a group and H a normal nilpotent subgroup such that G/H' is nilpotent. Then G is nilpotent.*

PROOF. Let $H_0 = H$, $H_{i+1} = [G, H_i]$, and take n minimal so that $H_n \leq H'$. We proceed by induction on n. It suffices therefore to show that G/H_1' is nilpotent.

Let $K_i = [H_i, H]$. As G/H' is nilpotent, it suffices to show that $[G, K_i] \leq K_{i+1}H_1'$ for all i. We apply the three subgroups lemma: $[[G, H_i], H] = K_{i+1}$ and $[[G, H], H_i] = [H_1, H_i]$, which is contained in H_1' for $i \geq 1$ and is equal to K_{i+1} for $i = 0$. Thus $[G, K_i] \leq K_{i+1}H_1'$ for all i. □

LEMMA 1.20. *If H is a nilpotent group, π a set of primes, and H has a central series H_i such that every section H_i/H_{i+1} is π-radicable, then H is π-radicable.*

LEMMA 1.21. *If H, K are normal nilpotent subgroups of the group G, then HK is nilpotent.*

LEMMA 1.22. *Let G be a nilpotent by finite p-group for some prime p.*

(1) *$Z(G) \neq 1$.*
(2) *For $H < G$, we have $N_G(H) > H$.*

PROOF.
Ad (1) Let $G_0 \lhd G$ be nilpotent and normal, with $\bar{G} = G/G_0$ finite. Let $A = \Omega_1(Z(G_0))$. Then the group $A \rtimes \bar{G}$ is an elementary abelian by finite p-group. Take $a \in A^\times$ and consider the subgroup $A_0 = \langle a^{\bar{G}} \rangle$ of A; this is a finite $\bar{G}$-invariant group.

The finite p-group $A_0 \rtimes \bar{G}$ is nilpotent, so its center is nontrivial and meets the normal subgroup A_0 nontrivially. Thus $C_A(\bar{G}) \neq 1$ and hence $Z(G) \neq 1$.

Ad (2) Let H_0 be the largest normal subgroup of G contained in H, and factor it out. We may assume then that H contains no normal subgroup of G. In particular, $H \cap Z(G) = 1$, and as $Z(G) \leq N(H)$ the claim follows. □

LEMMA 1.23. *Let G be an infinite abelian by finite p-group of bounded exponent for some prime p. Then $Z(G)$ is infinite.*

PROOF. We may think of G as giving an action of a finite p-group P on an infinite abelian p-group A of bounded exponent, and we may restrict attention to the elementary abelian p-group $\Omega_1(A)$. So take A to be elementary abelian. Then taking $x \in Z(P)$ of order p, and proceeding by induction on $|P|$, it suffices to check that the extension of A by $\langle x \rangle$ is nilpotent.

In End(A), which is a ring of characteristic p, we may perform the calculation: $(1-x)^p = 1 - x^p = 0$, which proves our claim. □

LEMMA 1.24 (Maschke). *Let $H = A \times B$ be a group, π a set of primes, and G a finite group acting on H. Suppose that A is an abelian G-invariant π-group, B is a $\pi^\perp$-group, and G is a π'-group. Then H splits as $A \times \tilde{B}$ for some G-invariant complement $\tilde{B}$.*

PROOF. This is a variant of Maschke's Theorem. Let $n = |G|$, and let $\pi : H \to A$ be the projection map with respect to the decomposition $A \times B$. The group G acts naturally on End(H), and we let $\tilde{\pi} = \frac{1}{n}\sum_{g \in G} \pi^g$ (since π^g takes values in A, the summation notation makes sense).

Then $\tilde{\pi} : H \to A$, $\tilde{\pi} \restriction A = \text{id}$, and $\tilde{\pi}$ is G-invariant, so $\ker \tilde{\pi}$ is a G-invariant complement to A. □

1.5. Schur-Zassenhaus. The next two results are standard, with substantial proofs which we will not reproduce here.

LEMMA 1.25 ([**101**, 2.1, page 221]). *Let G be a finite group, H a normal solvable subgroup, with $|H|$ and $|G/H|$ relatively prime. Then H has a complement K in G: $G = H \rtimes K$; and any two such complements are conjugate under H.*

A generalization of part of the above, not limited to the finite case, is given by Suzuki [**177**],

LEMMA 1.26. *Let G be a group with a normal abelian subgroup A, and let L be a subgroup of G such that $A \leq L \leq G$ and $[G : L] = m < \infty$. Assume the following:*

1. *A is m-divisible and m-torsion free.*
2. *A has a complement in L.*

Then A has a complement in G, and any two such are conjugate. Furthermore, any complement of A in L is contained in a complement of A in G.

Similarly, in the nilpotent case:

LEMMA 1.27. *Let G be a group with a normal nilpotent subgroup A of finite index m, and assume that A is m-radicable and m-torsion free. Then A has a complement in G, and any two such are conjugate.*

PROOF. We can argue inductively since $A/Z(A)$ is m-torsion free by Lemma 1.18. So assume $G = AT$ with $T \cap A = Z(A)$. As $Z(A)$ is m-divisible we can split T by the preceding lemma. Thus G splits as $A \rtimes T_0$ for some T_0.

For conjugacy one argues similarly that the complements may be taken in $Z(A)T_0$, and the previous lemma applies. □

We also need a more technical version of the Schur-Zassenhaus Theorem proved in [**109**, p. 172] and [**110**, p. 267].

PROPOSITION 1.28. *Let G be a group, π a set of primes, and suppose $A \lhd G$ with A an abelian, π-divisible, $\pi^{\perp}$-group. Suppose that G/A is a locally finite π-group and A satisfies the descending chain condition for centralizers of subsets of G. Then the following hold:*

(1) *G splits over A, that is there is a complement to A in G.*
(2) *Any two such complements are conjugate.*
(3) *Any group $H \leq G$ with $H \cap A = 1$ is contained in a complement to A.*

PROOF. We will show that (2) and (3) follow directly from (1), and we prove (1) by induction on the cardinality of G/A.

Proof of (2) and (3) assuming G splits. (Cf. [**109**, p. 172, proof of Lemma 4.2].)

Let $G = A \rtimes H$, and let H_1 be a subgroup with $H_1 \cap A = (1)$. We will show that H_1 is conjugate to a subgroup of H.

By the descending chain condition on centralizers in G, there is a finite subset K_0 of H_1 such that $C_A(K_0) = C_A(H_1)$. Let K_1 be a finite subgroup of H_1 containing K_0, and let K_2 be a finite subgroup of H such that $K_1 \subseteq A \rtimes K_2$. By Lemma 1.26, K_1 is conjugate to a subgroup of K_2. In particular, K_1 is conjugate to a subgroup of H, and since $G = AH$, there is an element $a \in A$ such that $K_1^a \leq H$.

This element a is *unique* modulo $C_A(H_1)$: if $K_1^b \leq H$ and $b \in A$, then $K_0^a, K_0^b \leq H$ and $[a^{-1}b, K_0^a] \leq A \cap H = (1)$, forcing $a^{-1}b \in C_A(K_0^a) = C_A(K_0) = C_A(H_1)$. In other words, if $K_0^a \leq H$ then also $K_1^a \leq H$ for all finite groups K_1 with $K_0 \leq K_1 \leq H_1$, and thus $H_1^a \leq H$, as desired.

Proof of (1). (Cf. [**110**, p. 267].)

We proceed by induction on the cardinality κ of G/A. If κ is finite, then Lemma 1.26 applies. If κ is infinite, then G can be written as an increasing continuous union of groups G_i $(i < \kappa)$ with $A \leq G_i$ and $|G_i/A| < \kappa$. Hence by induction, G_i splits as $A \rtimes H_i$. We can now replace the groups H_i by groups H_i^* which are again complements to A in G_i, and are increasing:

$i < j$ implies $H_i^* \le H_j^*$, proceeding by induction on i. At limit ordinals we take unions, and at successor stages we take H_{i+1}^* to be a suitable conjugate of H_{i+1}, using (2) and (3). □

1.6. Central extensions. We include some extremely general background on central extensions here. The connections with cohomology are not used here; what will matter is the very substantial theory of central extensions of algebraic groups, which will be quoted later.

DEFINITION 1.29. *Let G be a group.*

(1) *A* central extension *of G is a group $\hat{G}$ together with a surjective homomorphism $h : \hat{G} \to G$ whose kernel is in the center of $\hat{G}$.*
(2) *A* universal central extension *of G is a central extension $h : \hat{G} \to G$ which is universal in the sense that it factors uniquely through any other: if $h_0 : G_0 \to G$ is a central extension, there is a unique homomorphism $\hat{G} \to G_0$ making the diagram commute.*

As is always true in such cases, universal central extensions, if they exist, are unique up to canonical isomorphisms. The question of existence will be discussed here. It is elementary and standard. The connection with cohomology is also standard, and illuminating, but is not needed here.

The substantial theory of central extensions as it applies to Chevalley groups will be simply quoted at a later point. There one has to distinguish central extensions in the algebraic sense, which are well understood, from central extensions in the abstract sense, which are more subtle. The following turn out to be true: for any Chevalley group over an algebraically closed field, there is a universal central extension in any of the following categories: algebraic groups; abstract groups; groups of finite Morley rank. There is no reason a priori to expect the third point to be valid, but it is proved by showing that the universal algebraic central extension is also universal in the finite Morley rank category, which is natural enough. The following result is given just to establish the framework in which one works, at the level of abstract groups.

LEMMA 1.30 (Universal Extensions). *Let G be an abstract group.*

(1) *G has a universal central extension iff G is perfect, i.e. $G = G'$.*
(2) *If G is perfect, then a central extension $\hat{G} \to G$ of G is universal iff:*
   (a) *$\hat{G}$ is perfect.*
   (b) *For any central extension $e : E \to G$ of G, the extension $\hat{G} \to G$ factors through E.*
(3) *If G is perfect, and*

$$1 \to R \to F \to G \to 1$$

is a presentation of G (i.e., a short exact sequence with F is free, so that the kernel R is the set of "relations" holding in G), then the *group $F'/[F,R]$, with the natural map onto $G' = G$, is a universal central extension of G.*

PROOF.

Ad (1) Suppose first that $\hat{G}$ has a universal central extension. We will apply the uniqueness condition to show that G is perfect.

If $A = G/G'$ is nontrivial, then $\hat{G} = G \times A$ is a central extension of G with respect to projection on the first factor. Any central extension $e : E \to G$ then factors through $\hat{G}$ in at least two ways, via the map $(e,1)$ and $(e,\bar{e})$ where $\bar{e} : E \to A$ is induced by e.

The converse will follow from part (3).

Ad (2) We claim that the condition: *$\hat{G}$ is perfect,* is equivalent to the uniqueness of the desired map.

If $\hat{G}$ is not perfect, then we consider the central extension $E = \hat{G} \times A$ with $A = \hat{G}/\hat{G}'$, as in the first part, and we clearly do not have uniqueness here. So the conditions are necessary.

Conversely, if $\hat{G}$ is perfect and we have two maps $\pi_1, \pi_2 : \hat{G} \to E$ compatible with $\hat{G} \to G$, with E a central extension, then π_1 and π_2 differ only by central elements of E and hence coincide on $\hat{G} = \hat{G}'$.

Ad (3) If G is perfect, and $1 \to R \to F \to G \to 1$ is a presentation of G, set $\hat{G} = F'/[F,R]$ (depending somewhat on the fixed presentation), with the natural map $h : \hat{G} \to G$. Since the kernel of this map is $R/[F,R]$, even $F/[F,R] \to G$ would be a central extension, and of course as G is perfect the restriction to $F'/[F,R]$ remains surjective. So $\hat{G} \to G$ is a central extension, and we claim it is universal.

We now apply part (2). Let us check that $\hat{G}$ is perfect (condition (a)). Since the natural map $F' \to G$ is surjective, we have $F = F'R$. Hence $F' = [F'R, F'R] \leq F''R'$ and as $R' \leq [F,R]$ we find $\hat{G}' = \hat{G}$.

Now let us check condition (b). For any central extension $E \to G$ of G, we can lift $F \to G$ to a compatible map $\pi : F \to E$. This then restricts to a map $\pi_0 : F' \to E$ with $[F,R]$ in the kernel, since R goes into the center of E. So we have at least one compatible map $\bar{\pi} : \hat{G} \to E$. □

Some other points of the general theory worth noting here, that the reader may wish to check, are the following:

(1) If we have $\hat{G} \to G$ a universal central extension, and any subgroup Z of $Z(\hat{G})$, then $\hat{G}$ is the universal central extension of $\hat{G}/Z$.
(2) Just as the "uniqueness" condition was translated by the condition of perfection, the "existence" condition may be translated into the following: any central extension of $\hat{G}$ splits; or equivalently, $\hat{G}$ has no nontrivial perfect central extension.

This last condition is very natural: in the category of perfect central extensions of G, we require maximality; and the intermediate extensions have a natural lattice structure.

What we have done here does not, of course, prove that simple algebraic groups have universal central extensions in the category of algebraic groups. That requires more concrete considerations. But as a point of terminology, it may be mentioned that the universal central extension in the algebraic category is called *simply connected*, and the whole theory is analogous to, bound up with, and illuminated by, the consideration of universal covers of topological groups.

1.7. The class $\mathfrak{U}$.

DEFINITION 1.31. *Let G be a group. We say that G is* locally solvable *if every finitely generated subgroup is solvable.*

LEMMA 1.32. *A locally solvable periodic group is locally finite.*

LEMMA 1.33 ([**135**]). *Let G be a locally solvable group, and suppose that H is a minimal normal subgroup of G. Then H is abelian.*

PROOF. If H is nonabelian, take $h \in H^\times$ arbitrary. Then $\langle h^G \rangle = H$ and $H' = H$. Take G_0 finitely generated so that $h \in \langle h^{G_0} \rangle'$. Let $H_0 = \langle h^{G_0} \rangle$. Then as G_0 normalizes H_0', we find $h^{G_0} \subseteq H_0'$ and hence $H_0' = H_0$. But $H_0 \leq \langle h, G_0 \rangle$, which is solvable, a contradiction. □

DEFINITION 1.34. *$\mathfrak{U}$ is the class of locally finite groups G such that every subgroup H of G satisfies the following:*

(1) *There is a finite series*

$$1 = H_0 \lhd H_1 \lhd \ldots \lhd H_n = H$$

with locally nilpotent quotients.

(2) *For each set π of primes, the maximal π-subgroups of H are conjugate.*

LEMMA 1.35 ([**59**]). *Let G be a locally finite solvable group satisfying the minimum condition on centralizers. Then $G \in \mathfrak{U}$.*

DEFINITION 1.36. *Let G be a group. A* Sylow basis *for G is a collection of Sylow subgroups S_p, one for each prime, such that for any set π of primes, the group*

$$S_\pi = \langle S_p : p \in \pi \rangle$$

is a π-group.

PROPOSITION 1.37 ([**97**, Corollary 2.6, Theorem 2.10]). *Let $G \in \mathfrak{U}$. Then*

(1) *G contains a Sylow basis.*
(2) *Any two Sylow bases of G are conjugate.*
(3) *If $(S_p : p$ prime$)$ is a Sylow basis for G, then $S_pS_q = S_qS_p$ for all p, q and hence $S_\pi = \prod_{p\in\pi} S_p$.*

1.8. Groups of bounded exponent.

LEMMA 1.38. *Let* $1 \to H \to K \to L \to 1$ *be a short exact sequence. Then* K *has bounded exponent if and only if* H *and* L *do also.*

LEMMA 1.39. *A group of exponent* 2 *is abelian.*

LEMMA 1.40. *A group* G *of exponent three is locally nilpotent.*

PROOF. We may suppose that G is finitely generated.
For $x, y \in G$ we have $(xy)^3 = 1$ and hence

$$xyx = (yxy)^{-1} = (yxy)^2 = yxy^{-1}xy$$

and hence $x^y x = x x^y$. Thus $x^G \leq C(x)$ for each x. Letting $H_x = \langle x^G \rangle$, if G is finitely generated, it follows that the intersection of H_{x_i} over a generating set is central, and may be factored out. Then G embeds into the product of the factor groups G/H_{x_i}, which may be supposed nilpotent by induction on the number of generators. □

1.9. Strongly real elements.

LEMMA 1.41. *If* i, j *are involutions in a group* G *and* $a = ij$, *then* $\langle i, j \rangle = \langle a \rangle \rtimes \langle i \rangle$, *a dihedral group (possibly infinite).*

DEFINITION 1.42. *An element of a group* G *is* strongly real *if it is the product of two involutions.*

REMARK 1.43. *An element is strongly real if and only if it is inverted by some involution.*

2. Rank

We present the rank axioms, and discuss the resulting descending chain conditions, and the notion of definable hull of a group. Some key foundational properties have delicate model theoretic proofs, which are treated at a more satisfactory level of generality in [**156**], and from other points of view in [**186**].

2.1. Axioms. We consider a group G equipped with additional structure, such as an algebraic group which may be considered also as a variety. We will suppose that G carries a *rank function* in the sense of Borovik and Poizat, namely a function "rk" which assigns to each set S definable over G, its "dimension" $\text{rk}(S)$, and which satisfies the following four axioms.

The Rank Axioms

Monotonicity: For any n, $\mathrm{rk}(S) > n$ if and only if S contains an infinite family of disjoint definable subsets S_i of rank n.

Additivity: If $f : A \to B$ is definable and surjective, and if the fibers $f^{-1}(b)$ have constant rank r for $b \in B$, then $\mathrm{rk}(A) = r + \mathrm{rk}(B)$.

Definability: For any uniformly definable family $\{S_b : b \in B\}$ of definable sets, and for any $n \in \mathbb{N}$, the set

$$\{b \in B : \mathrm{rk}(S_b) = n\}$$

is also definable.

Finite Bounds: For any uniformly definable family $\mathcal{F}$ of finite subsets, the sizes of the sets in $\mathcal{F}$ are bounded.

The definable subsets over G are, more precisely, quotients by definable equivalence relations of definable subsets of G^n for some n. A family $\{S_b : b \in B\}$ is *uniformly definable* if, first, B is itself definable, and second, the relation "$x \in S_y$" is definable. For example, the set $\{C_G(a, b) : a, b \in G\}$ would be such a family.

Sooner or later one encounters some technical difficulties in working with these definitions that need to be addressed in a systematic way, so we will point them out at the start. The underlying group G here is a fixed abstract group. In model theory, however, just as in the algebraic theory, the "groups" that interest us are not abstract groups but are really variable (functors, interpretations, models). Just as one may vary the base field in an algebraic group, in model theory one may replace the group G by any elementarily equivalent group G^*, that is, by any model of the same complete theory.

Now when one replaces the abstract group G by an elementarily equivalent group G^*, we claim that the rank function on G gives rise to a rank function on G^*, satisfying the same axioms. This essential and nontrivial fact is not a formal consequence of the axioms, and would not hold if we were to consider algebraic structures of a more general type; it depends on the assumption that G is, specifically, a group. This is proved in detail in [**156**], where it is also proved that these groups are exactly the groups of *finite Morley rank* in the sense of model theory.

We do not want to develop the underlying model theory at great length, so we will take it as known, and we will take note of the few places where this kind of principle is applied in an essential way. We remark that in algebraic geometry it is also well known that group varieties are better behaved than arbitrary varieties; for example, it is immediate that they are smooth. For model theorists they are also smooth .... The uniformity given in a very rudimentary form in the fourth rank axiom is a symptom of this.

As a direct consequence of the axioms we find:

LEMMA 2.1. *Let G be a group of finite Morley rank. Then:*

(1) Algebraic sets: *The sets of rank 0 are the finite sets.*

(2) Invariance: *If* $f : A \leftrightarrow B$ *is a definable bijection between definable sets, then* $\text{rk}(A) = \text{rk}(B)$.

(3) Finite Unions: *The rank of a finite union* $S = \bigcup_i S_i$ *is the maximum of* $\text{rk}(S_i)$.

(4) Extended Additivity: *If* $f : A \to B$, *with* f, A, *and* B *definable, then setting*

$$r_b^* = \text{rk}(f^{-1}(b)); B_k = \{b \in B : r_b^* = k\}$$

we have

$$\text{rk}(A) = \max_k(\text{rk}(B_k) + k) \ .$$

(5) Fubini: *Let* $X \subseteq \dot{\bigcup}_{b \in B} A_b$ *(disjoint union) with* X, B, *and* $(A_b : b \in B)$ *all definable, with* $\text{rk}(A_b)$ *independent of* b. *Then the following are equivalent:*
  (a) $\text{rk}(X) = \text{rk}(A) + \text{rk}(B)$;
  (b) $\text{rk}(\{b \in B : \text{rk}(X \cap A_b) = \text{rk}(A)\}) = \text{rk}(B)$.
  In particular, if $X \subseteq A \times B$, *then writing* $A_b = A \times \{b\}$ *and* $B_a = \{a\} \times A$, *the following three conditions are all equivalent:*
  (a) $\text{rk}(X) = \text{rk}(A) + \text{rk}(B)$;
  (b) $\text{rk}(\{b \in B : \text{rk}(X \cap A_b) = \text{rk}(A)\}) = \text{rk}(B)$.
  (c) $\text{rk}(\{a \in A : \text{rk}(X \cap B_a) = \text{rk}(B)\}) = \text{rk}(A)$;
  The equivalence of the last two points is sometimes called Symmetry.

PROOF. The first point follows directly from the Monotonicity property.

The second follows from additivity, viewing the bijection f as a covering with fibers of rank 0.

The third can be proved by induction on rank, using Monotonicity.

For the fourth point, observe first that to make it fully meaningful we should introduce the convention:

$$\text{rk}(\emptyset) = -\infty$$

which has the effect of eliminating the values of k for which B_k is empty (in particular, limiting k to at most $\text{rk}(A)$), and is generally an appropriate convention.

Now divide A up into the sets $A_k = \{a \in A : r_{f(a)}^* = k\}$; in more geometrical language, this is the union of the fibers of dimension k. We have $\text{rk}(A) = \max_k \text{rk}(A_k)$, by part (2), and by additivity $\text{rk}(A_k) = \text{rk}(B_k) + k$.

The final point is an application of the fourth. □

An essential point which is not particularly evident is the following:

FACT 2.2. *Let* G *be a group of finite Morley rank, and* H *a group definable over* G. *Then* H *has finite Morley rank.*

PROOF. This is fairly clear with the customary model theoretic notion of finite Morley rank, but not with the definitions (or axioms) we have provided—particularly, with respect to reduction of the language. So we refer to [**156**] for a thorough discussion of the model theoretic basis for the equivalence of the various definitions. □

We may also introduce a notion of *multiplicity* for definable sets, more precisely "multiplicity in the top dimension", denoted degree(S):

DEFINITION 2.3. *Let S be a definable set over a group G of finite Morley rank, with* $\mathrm{rk}(S) = r$.

(1) *A subset X of S is* generic *if it contains a definable subset of rank* $\mathrm{rk}(S)$.
(2) *S is* irreducible *if S contains no definable subset A such that both A and its complement have rank r; in other words, S does not contain two disjoint generic subsets.*
(3) *If S is the union of d irreducible definable sets S_i of rank r, we set* $\mathrm{degree}(S) = d$.

To justify this terminology, we require a few more facts:

LEMMA 2.4. *Let S be a definable set over a group G of finite Morley rank, with* $\mathrm{rk}(S) = r$.

(1) Finiteness of Degree: *S is a union of finitely many definable irreducible subsets of rank r.*
(2) Uniqueness*: If S is the union of d definable irreducible subsets S_i of rank r, and of d' definable irreducible subsets S'_j of rank r, then $d = d'$, and the relation $i \sim j$ defined by*

$$\mathrm{rk}(S_i \cap S'_j) = r$$

defines a bijection between the indices i and j.
(3) Additivity of Degree: *If $S = \bigcup S_i$ is a finite union with* $\mathrm{rk}(S_i) =$ *r constant, and* $\mathrm{rk}(S_i \cap S_j) < r$ *for $i \neq j$, then* $\mathrm{degree}(S) = \sum \mathrm{degree}(S_i)$.
(4) Multiplicativity: *For A and B definable, we have*

$$\mathrm{degree}(A \times B) = \mathrm{degree}(A) \times \mathrm{degree}(B) .$$

(5) Invariance: *Degree is invariant under definable bijections.*

PROOF.

Ad (1) Suppose S is not a finite union of definable irreducible sets of rank r. Then there is $S_1 \subseteq S$ such that S_1 and $S \setminus S_1$ both have rank r; furthermore, one of these sets, which we may suppose to be $S \setminus S_1$, is again not a finite union of definable irreducible sets of rank r. Hence we may iterate this procedure, defining $S_1, S_2, \ldots$ disjoint definable subsets of S of rank r, and contradicting the monotonicity.

Ad (2) With i fixed and j varying, we have

$$S_i = \bigcup_j (S_i \cap S'_j),$$

a disjoint union, and as S_i is irreducible, exactly one of the intersections has rank r. The same applies with the roles of i and j reversed.

Ad (3) Replacing S_i by $S_i \setminus \bigcup_{j<i} S_j$, we take the S_i to be a finite partition of S, and the claim is obvious.

We note that it is not possible to speak of the *irreducible components* of S in the above situation, except as *equivalence classes* of irreducible subsets, where irreducible subsets of rank r are equivalent if their intersection has rank r.

Ad (4) We have $\mathrm{rk}(A \times B) = \mathrm{rk}(A) + \mathrm{rk}(B)$ and the same applies to a product of irreducible definable subsets $A_0 \subseteq A$, $B_0 \subseteq B$, so it suffices to check that if A and B are irreducible, then so is their product.

Let A and B be irreducible, and $S \subseteq A \times B$ with $\mathrm{rk}(S) = \mathrm{rk}(A) + \mathrm{rk}(B)$. By Fubini, the set $\{b \in B : \mathrm{rk}(S \cap (A \times \{b\})) = \mathrm{rk}(A)\}$ has rank equal to $\mathrm{rk}(B)$.

If the complement $S' = (A \times B) \setminus S$ also has full rank in $A \times B$, then the set $\{b \in B : \mathrm{rk}((A \times \{b\}) \setminus S) = \mathrm{rk}(A)\}$ also has full rank in B. Thus these two sets meet, by irreducibility of B, and with $b \in B$ chosen to lie in both, we get two disjoint definable subsets of $A \times \{b\}$ of full rank, and a contradiction to the irreducibility of A.

(5) is immediate. □

Ranks and degrees behave well with respect to definable subgroups:

LEMMA 2.5. *Let G be a group of finite Morley rank, and H a definable subgroup.*

(1) $\mathrm{rk}(G) = \mathrm{rk}(G/H) + \mathrm{rk}(H)$.
(2) *If* $[G : H] = n$ *is finite, then* $\mathrm{rk}(H) = \mathrm{rk}(G)$ *and* $\mathrm{degree}(G) = n\,\mathrm{degree}(H)$.

PROOF. It suffices to note that the cosets of H in G are of equal rank and degree, as there are definable bijections between them. Then the general properties of rank mentioned above suffice: for (1) consider the surjection $G \to G/H$, and for (2) apply the additivity of degree. □

With these definitions, irreducible sets are those of *Morley degree one* (in geometrical terms, they are irreducible in the top dimension).

2.2. Chain conditions.

LEMMA 2.6. *Let G be a group of finite Morley rank. Then G satisfies the* descending chain condition *for definable subgroups: any properly descending chain*

$$G > G_1 > G_2 > \dots$$

with all G_i definable, is finite.

PROOF. Apply Lemma 2.5 to see that at each stage, either the rank or the degree of G_i decreases. □

There is another chain condition, rather more specialized but occasionally very powerful, due to Baldwin and Saxl, in considerably greater generality. It depends on a combinatorial consequence of the rank axioms:

LEMMA 2.7. *Let G be a group of finite Morley rank, S a definable set over G, and let $R(x,y)$ be a definable binary relation on S. Then there is a finite bound $n = n_R$ on the lengths of finite subsets of S which are linearly ordered by R.*

PROOF. If this fails, then we have arbitrarily long finite sequences linearly ordered by the relation R, and thus in some group G^* elementarily equivalent to G, we can find an infinite subset $X \subseteq S$, indeed one of order type $\mathbb{Q}$, on which R defines a linear ordering. In more detail, this passage from the unbounded finite to the infinite runs as follows. Extend the language by constants denoting elements of such a set X, extend the theory of G by the properties we wish to have holding on X, and check the consistency of the resulting theory with the complete theory of G—by hypothesis, every finite subset of the extended theory is actually satisfied by G, under a suitable interpretation of the additional constants. Then by the Completeness Theorem, the theory has a model, which is the desired group G^*.

Now take a definable subset S of minimal rank and degree such that $S \cap X$ contains a subset of order type $\mathbb{Q}$. Fixing an element $0 \in X \cap S$, we may split S into two subsets $S^< = \{x \in S : R(x,0)\}$ and $S^> = S \setminus S^<$, both of which contain subsets of X of order type $\mathbb{Q}$; but at least one of these two sets has either lesser rank, or lesser degree, than S, contradicting the choice of S. □

LEMMA 2.8. *Let G be a group of finite Morley rank, and $\mathcal{H} = \{H_b : b \in B\}$ a uniformly definable family of subgroups of G. Then:*

(1) *There is an absolute bound $n = n_H$ on the length of any chain of subgroups in $\mathcal{H}$.*
(2) *The set $\bigcap \mathcal{H}$ of arbitrary intersections of subgroups from $\mathcal{H}$ is again uniformly definable, and any such group can be represented as an intersection of bounded length.*

PROOF. Define $R(x,y)$ as follows: "$H_x \supseteq H_y$". Then the previous lemma proves (1).

We turn to (2). By the descending chain condition for definable subgroups, any intersection of definable groups is an intersection of finitely many of them; but in the case of intersections of groups in the uniformly definable family $\mathcal{H}$, we claim we can bound the length of a chain of such intersections.

Supposing the contrary, we have for every n an intersection $H_{b_1} \cap \ldots \cap H_{b_n}$ which cannot be shortened. Now we can again replace the group G by an elementarily equivalent one in which we have an *infinite series* of groups H_n ($n \in \mathbb{N}$), all belonging to the family $\mathcal{H}$. This is achieved by introducing constants denoting the *defining parameters* of the groups H_n

(using the hypothesis that $\mathcal{H}$ is a uniformly definable family). In checking the consistency, we use the following: if an intersection $H_1 \cap \cdots \cap H_n$ cannot be shortened, then the sequence $(H_1 \cap \cdots \cap H_k : 1 \leq k \leq n)$ is strictly decreasing.

So in a group G^* elementarily equivalent to G we find groups $H_n \in \mathcal{H}$ such that the entire sequence $(H_1 \cap \cdots \cap H_k : 1 \leq k < \infty)$ is strictly decreasing, violating the descending chain condition for definable subgroups. □

COROLLARY 2.9. *Let G be a group of finite Morley rank, and X an arbitrary subset. Then $C(X)$ is definable. In fact there is a bound c such that for any $X \subseteq G$ there is $X_0 \subseteq X$ with $|X_0| \leq c$ and $C_G(X_0) = C_G(X)$.*

PROOF. $C(X) = \bigcap_{a \in X} C(a)$, which is an intersection taken from a family of uniformly definable subgroups. □

LEMMA 2.10. *Let G be a group of finite Morley rank, and $\{H_i : i \in I\}$ a family of definable pairwise commuting nonabelian subgroups of G. Then I is finite.*

PROOF. Let $X = C(\langle H_i : i \in I\rangle$. Then By Lemma 2.6 we have $X = C(\langle H_i : i \in I_0\rangle)$ for some finite set I_0. Then $I_0 = I$, as for $i \in I \setminus I_0$ we would have H_i centralizing $\langle H_j : j \in I_0\rangle$, but not H_i itself. □

LEMMA 2.11. *Let A be a p-divisible abelian group of finite Morley rank for some prime p. Then $A[p]$ is finite.*

PROOF. Supposing the contrary, let A be a counterexample of minimal rank, and $\bar{A} = A/A[p]$. As $A[p]$ is infinite, $\mathrm{rk}(A) > \mathrm{rk}(\bar{A})$. Thus by induction $\bar{A}[p]$ is finite, i.e. $A[p^2]/A[p]$ is finite. But $pA[p^2] = A[p]$, a contradiction. □

COROLLARY 2.12. *Let A be a divisible group of finite Morley rank. Then $A[n]$ is finite for all n.*

PROOF. The case of prime powers reduces to that of primes, and the general case then follows. □

LEMMA 2.13. *Let A be an abelian group of finite Morley rank, π a set of primes. Then*

(1) *$A = U \oplus T$ with U a π-group of bounded exponent, and T π-divisible and definable.*
(2) *$A = V + T$ with V a π-group of bounded exponent and T π-divisible, with both V and T definable, and with $V \cap T$ finite.*

PROOF.

Ad (1) Apply the DCC for definable subgroups: take T minimal definable in A such that A/T is a π-group of bounded exponent. Then by construction T is π-divisible. In particular, T is a pure subgroup of A.

It follows that A splits as $A = U \oplus T$ by Lemma 1.15, where U is not necessarily definable.

Ad (2) If U has exponent n, replace U by $V = A[n]$. □

2.3. Definable hull. Another consequence of the descending chain condition is the existence of the *definable hull* of an arbitrary set (and, in particular, of an arbitrary subgroup).

DEFINITION 2.14. *Let G be a group of finite Morley rank, $X \subseteq G$ a subset, not necessarily definable. Then $d(X)$ denotes the smallest definable subgroup of G containing X.*

Since the intersection of all definable subgroups containing X is again definable, by the DCC, the existence of $d(X)$ is assured. This notion is used mainly in two special cases: (1) if X consists of a single element $\{a\}$, then $d(X)$ is sometimes denoted $d(a)$, and is in a sense "cyclic", or at least 1-generated; (2) if X is itself a group, but is not definable, then $d(X)$ is a group which is analogous to the "Zariski closure" in an algebraic group.

Note also that $d(X) = d(\langle X \rangle)$ where $\langle X \rangle$ denotes the group generated by X.

LEMMA 2.15. *Let G be a group of finite Morley rank, and X, Y subgroups, not assumed definable, and $H \lhd G$ definable.*

(1) $C(X) = C(d(X))$.
(2) $d(N(X)) \leq N(d(X))$.
(3) $d(XH) = d(X)H$; $d(X)H/H = d(XH/H)$.
(4) $[d(X), d(Y)] = d([X, Y])$.
(5) $d(X^i) = d(X)^i$, $d(X^{(i)}) = d(X)^{(i)}$.
(6) $Z_i(X) \leq Z_i(d(X))$.
(7) *If X is nilpotent of class c, then $d(X)$ is nilpotent of class c.*
(8) *If X is solvable of class c, then $d(X)$ is solvable of class c.*

PROOF.

Ad (1) $X \leq C(C(X))$, and the latter is definable by Corollary 2.9. So $d(X) \leq C(C(X))$, as claimed.

Ad (2) If $g \in N(X)$, then $X \leq d(X)^g$, so $d(X) \leq d(X)^g$, and the same applies to g^{-1}. Thus $N(X) \leq N(d(X))$ and hence $d(N(X)) \leq N(d(X))$.

Ad (3) Both claims follow directly from the definitions. The second claim can be written more clearly as $d(\bar{X}) = \overline{d(X)}$, working in the group $\bar{N}_G(H) = N_G(H)/H$.

Ad (4) Part of the claim is that $[d(X), d(Y)]$ is definable, which will be proved in §3. Assuming this, we prove the rest here. We use the elementary group theoretic fact that X and Y normalize $[X, Y]$ (see 1.5).

Let $Z = [X, Y]$. As $X, Y \leq N(d(Z))$, we have $d(X), d(Y) \leq N(d(Z))$. Let $N = N_G(d(Z))$, and $\bar{N} = N/d(Z)$. Then $[\bar{X}, \bar{Y}] = 1$, so $[\bar{d}(X), \bar{d}(Y)] = [d(\bar{X}), d(\bar{Y})] = 1$ by (1). In other words, $[d(X), d(Y)] \leq d([X, Y])$. Given that $[d(X), d(Y)]$ is definable, the reverse inclusion is clear.

Ad (5) The commutator series G^i and $G^{(i)}$ are defined by iterated commutation, so this is clear.

Ad (6) Proceed by induction, starting with $i = 0$, $Z_0(X) = 1$. Assume inductively that $Z_i(X) \le Z_i(d(X))$. Then $[Z_{i+1}(X), X] \le Z_i(d(X))$, so working in $N = N_G(Z_i(d(X)))$, with $\bar{N} = N/Z_i(d(X))$, we have $[\overline{Z_{i+1}(X)}, X] = 1$ and hence $[\overline{Z_{i+1}(X)}, d(X)] = 1$, that is $[Z_{i+1}(X), d(X)] \le Z_i(d(X))$, and $Z_{i+1}(X) \le Z_{i+1}(d(X))$.

Now (7) follows from (5) or (6), and (8) follows from (5). □

LEMMA 2.16 ([**52**, Ex. 10, p. 93]). *The definable hull of a cyclic subgroup of a group G of finite Morley rank is the direct sum of a divisible group and a finite cyclic group.*

PROOF. Let $C = \langle x \rangle$ be a cyclic subgroup of G. Then $d(C)$ is an abelian subgroup by Lemma 2.15. By Lemma 2.13, $d(C) = U \oplus T$, with U of bounded exponent, and T divisible and definable. Let u be the projection of x to U; then $x \in \langle u \rangle \oplus T$ and since u is of finite order, the latter group is definable. Hence $d(C) = d(x) \le \langle u \rangle \oplus T$, that is, $d(C) = \langle u \rangle \oplus T$. □

LEMMA 2.17. *Let G be a $p^\perp$-group of finite Morley rank. Then G is* uniquely p-radicable*: that is, every element of G has a unique p-th root.*

PROOF. Let $a \in G$, and let $A = d(a)$. As A contains no elements of order p, by Lemma 2.16 it is p-divisible. So $a = b^p$ with $b \in A$, and in particular $b \in Z(C(a))$.

If in addition $c^p = a$, then $c \in C(a)$ and c commutes with b, hence $(bc^{-1})^p = 1$ and $b = c$. □

LEMMA 2.18. *Let G be a group of finite Morley rank, π a set of primes, and $h : H \to K$ a definable homomorphism with H, K definable. If $a \in H$ and $h(a)$ is a π-element, then there is a π-element $a' \in H$ so that $h(a) = h(a')$.*

PROOF. Let $H_0 = \ker h$. Then $a^n \in H_0$ for some π-number n. By Lemma 2.16 the group $d(a^n) \le H_0$ factors as $C \times D$ with C finite cyclic of order a π-number, and D π-divisible. Note that a centralizes $d(a^n)$ and hence after replacing a by ad for a suitable $d \in D$, we have $a^n \in C$ and hence a is a π-element. □

LEMMA 2.19. *Let G be a group of finite Morley rank, π a set of primes, and H, K two definable $\pi^\perp$ subgroups of G with K normalizing H. Then HK is a $\pi^\perp$-subgroup of G.*

PROOF. Let $\overline{HK} = HK/H$. By Lemma 2.18, this is $\pi^\perp$-group. □

LEMMA 2.20 (Basic Fusion Lemma). *Let G be a group of finite Morley rank, and i, j two involutions of G. Then either i and j are conjugate in $d(\langle i, j \rangle)$, or there is an involution $k \in d(\langle (ij) \rangle)$ commuting with both i and j.*

PROOF. Let $a = ij$. Then $\langle i, j \rangle = \langle a \rangle \rtimes \langle i \rangle$ (Lemma 1.41) and hence $d(\langle i, j \rangle) = d(\langle a \rangle) \rtimes \langle i \rangle$. Here $d(\langle a \rangle)$ is the direct product of a divisible group and a finite cyclic group C. It follows that either a is a square in $d(\langle a\rangle)$, or C contains an involution.

Now i, j invert $\langle a\rangle$ and the subgroup of $d(\langle a\rangle)$ consisting of all elements inverted by i, j is definable, so i, j invert $d(\langle a\rangle)$.

If $a = b^2$ with $b \in d(\langle a\rangle)$, then $i^b = j$, which gives the first alternative.

If C contains an involution k, then k commutes with i and j, giving the second alternative. $\square$

It would be reasonable to investigate the relationship between X and $d(X)$ more systematically. Some specific aspects have been illuminated thoroughly, and in great generality, by Wagner in [**186**], but there is a great deal more that remains unexplored in this direction. The following further example can be obtained by simply combining results proved above.

LEMMA 2.21. *Suppose $X = X_1 * X_2$ is a central product. Then $d(X) = d(X_1) * d(X_2)$.*

3. Connected groups

3.1. Connectivity and irreducibility. We will develop an extensive theory of connected definable groups (and a less extensive theory of connected groups in general, not necessarily definable, with divisible abelian torsion groups providing some of the main examples of the latter).

There are two ways to define connectivity, a coarse algebraic way which we take as our definition, and an alternative similar to irreducibility in the algebraic case; fortunately, the two coincide (Lemma 3.6). One of the important applications of connectivity is the following: any subgroup of a group of finite Morley rank which is generated by connected definable subgroups is itself definable. This is generalized in Proposition 3.19 below.

The theme of connectivity will accompany us throughout the book.

DEFINITION 3.1. *Let G be a group of finite Morley rank.*

(1) *G is* connected *if G has no proper definable subgroup of finite index.*
(2) *A subgroup H of G is* definably characteristic *in G if it is invariant under all definable automorphisms of G.*

Observe that a definably characteristic subgroup of a normal subgroup H of G will be normal in G, since the inner automorphisms induced by G on H are definable in H (in the language inherited from G).

LEMMA 3.2. *Let G be a group of finite Morley rank. Then:*

(1) *G contains a unique connected definable subgroup of finite index, denoted G°.*
(2) *G° is definably characteristic in G.*
(3) *G° is definable without parameters.*

PROOF.

Ad (1) By the descending chain condition there is a minimal definable subgroup of G of finite index; and this is then connected. If there are two such, then their intersection is of finite index in each, and hence they are equal. Thus G° exists, and is unique.

Ad (2) Clear.

Ad (3) If G° is defined by the formula $\phi(x, \bar{a})$ with the parameter $\bar{a}$, it is the unique group of this form with the correct index. So one can quantify out the parameter $\bar{a}$. □

We mention some useful consequences of connectivity.

LEMMA 3.3. *Let G be a connected group of finite Morley rank.*

(1) *Any definable action of G on a finite set is trivial.*
(2) *If $h : G \to G$ is a definable endomorphism with finite kernel, then h is surjective.*

PROOF.

Ad (1) The point stabilizers are definable subgroups of finite index.

Ad (2) As the kernel is finite, $\mathrm{rk}(\mathrm{im}\, h) = \mathrm{rk}(G)$ by additivity. Hence $[G : \mathrm{im}\, h] < \infty$ by monotonicity, and $G = \mathrm{im}\, h$ by connectivity. □

There is also a useful standard convention relating to "connected analogues" of notions of finite group theory. We write $N_G{}^\circ(H)$ for $(N_G(H))^\circ$, and similarly:

NOTATION 3.4. *Let G be a group of finite Morley rank, H a definable subgroup, and Ξ any group theoretic operation (such as N_G, C_G, F). Then Ξ° is the group theoretic operation defined by $\Xi^\circ(H) = (\Xi(H))^\circ$.*

This will bring out certain analogies more forcefully, as well as lightening the notation in some cases.

At one extreme, if G is finite, then $G^\circ = 1$. At the other:

EXAMPLE 3.5. *Let F be an infinite division ring of finite Morley rank, not necessarily associative. Then the additive group of F is connected.*

PROOF. F° is not only an additive subgroup of F, but an *ideal*, since nonzero multiplication maps are definable automorphisms of the additive group. As F is infinite, we will not have $F^\circ = (0)$, so we have $F^\circ = F$. □

The case of the multiplicative group is more subtle, and requires a more sophisticated result, given in a more general form in [**156**]:

LEMMA 3.6. *Let G be a group of finite Morley rank. Then:*

(1) $\mathrm{degree}(G) = [G : G^\circ]$*; in particular:*
(2) *G is connected if and only if G is irreducible*

PROOF. First, the general case (1) follows at once from the special case (2), and additivity of degree. So one may concentrate on (2), and one direction is trivial: an irreducible group must be connected, as the cosets of G° in G have constant rank.

Assuming now that G is connected, of rank r, we claim that $\mathrm{degree}(G) = 1$. Supposing the contrary, let $d = \mathrm{degree}(G) > 1$ and let $A_1, \ldots, A_d$ be irreducible definable subsets of G; let $\bar{A}_i$ be the equivalence class of all irreducible subsets of G meeting A_i in a set of rank r (and hence differing from A_i by a set of lower rank). Then G acts naturally on the set $\{\bar{A}_i : 1 \leq i \leq d\}$, in two ways, by right translation and by left translation (by g^{-1} in the latter case).

This action is clearly definable, since rank is definable. As the orbits under the action are finite, and G is connected, G acts trivially under both left and right translation.

Now if $d > 1$, consider the multiplication map $\mu : A_1 \times A_2 \to G$. There is an index j such that $\mu^{-1}(A_j)$ has full rank in $A_1 \times A_2$, that is; $2r$. By the Fubini principle, Lemma 2.1, this means

$$\text{rk}\{a \in A_2 : \bar{A}_1 a = \bar{A}_j\} = r \,, \tag{1}$$

$$\text{rk}\{a \in A_1 : a\bar{A}_2 = \bar{A}_j\} = r \,. \tag{2}$$

However, we have already seen that $\bar{A}_1 g = \bar{A}_1$ and $g\bar{A}_2 = \bar{A}_2$ for all $g \in G$; so $1 = j = 2$, a contradiction. □

COROLLARY 3.7. *Let F be an infinite field of finite Morley rank. Then the multiplicative group of F is connected.*

PROOF. We saw above that the additive group of F is connected, and hence, by the foregoing, also irreducible. As the multiplicative group differs from the additive group, as a set, by removal of a single point, it is also irreducible, and hence connected. □

It is worth noticing at this point that the real field $\mathbb{R}$ provides a canonical "counterexample" to the foregoing; its additive group is connected, and its multiplicative group is not. This means of course that $\mathbb{R}$ does not admit a rank function in our sense, though the natural notion of real dimension has many of the desired properties. Monotonicity fails: $\mathbb{R}$ contains infinitely many disjoint intervals, all of dimension 1.

LEMMA 3.8. *Let G be a connected group of finite Morley rank, and K a finite normal subgroup such that G/K is abelian. Then G is abelian, and if G/K is divisible, G is also divisible.*

PROOF. As G is connected, $K \leq Z(G)$. As $[G, G] \leq K \leq Z(G)$, the commutator map $\gamma_g : G \to K$ induced by commutation with an element of G is a homomorphism, whose image is both finite and connected, hence trivial. This proves that G is abelian.

Now $G = B + D$ with B of bounded exponent, D divisible, and both factors connected. With $\bar{G} = G/K$, we have $\bar{G} = \bar{B} + \bar{D}$ and evidently if $B > 1$, then $\bar{B} > 1$. □

LEMMA 3.9. *Let G be a group of finite Morley rank, and $H \leq G$ a connected subgroup, not necessarily definable. If $Z(H)$ is finite, then $H/Z(H)$ is centerless.*

PROOF. We claim $Z_2(H) = Z(H)$.

Take $h \in Z_2(H)$. Then commutation with h gives a homomorphism

$$H \to Z(H)$$

and as the image is finite, we have

$$[H : C_H(h)] < \infty .$$

By connectivity $h \in Z(H)$. □

LEMMA 3.10. *Let G be an infinite group of finite Morley rank. Then G contains an infinite definable abelian subgroup.*

PROOF. Suppose G is a counterexample. Using the DCC for definable subgroups, we may suppose that G has no proper infinite definable subgroup. Then G is connected, and $Z(G)$ is finite, so $\bar{G} = G/Z(G)$ is centerless. Replace G by $G/Z(G)$; then our assumption is that G is centerless, and contains no proper infinite definable subgroup.

For $a \in G^\times$ we have $C_G(a) < G$ and hence $C_G(a)$ is finite. This implies that the order of a is finite. Furthermore, as there is a definable bijection between the conjugacy class a^G and the space of cosets $G/C_G(a)$, we have $\mathrm{rk}(a^G) = G$. By irreducibility, then, $G^\times$ consists of a single conjugacy class. In particular, taking a to be of prime order p, all elements of $G^\times$ are of order p. But a is conjugate to a^{-1} in G, and this implies that $p = 2$. So, after all, G is abelian (Lemma 1.39). □

The following is in a similar vein.

LEMMA 3.11. *Let G be a connected group of finite Morley rank containing no involution. Then every element of G has an infinite centralizer.*

PROOF. Suppose $a \in G$ has a finite centralizer. Then arguing as in the previous lemma, we find $\mathrm{rk}(a^G) = \mathrm{rk}(G)$ and also $\mathrm{rk}((a^{-1})^G) = \mathrm{rk}(G)$, and as G is connected these two conjugacy classes meet, and hence coincide: $a^g = a^{-1}$ for some $g \in G$. Furthermore, $g^2 \in C(a)$, which by assumption is finite, so g is of finite order. As G contains no involution, g is of odd order; but then evidently $a = a^{-1}$ and we have a contradiction. □

3.2. Nondefinable groups and connectivity. The notion of connectivity applies to undefinable subgroups as well as definable ones.

DEFINITION 3.12. *Let G be a group of finite Morley rank, and X an arbitrary subgroup.*

(1) *A subgroup Y of X is* relatively definable *in G, if $Y = X \cap H$ for some definable subgroup H of G.*
(2) *X is* connected *if X has no proper relatively definable subgroup of finite index.*
(3) *X° is the smallest relatively definable subgroup of X of finite index.*

The basic properties of connected components go over to this setting. To begin with, one essential point is the following:

LEMMA 3.13. *Let G be a group of finite Morley rank, and X an arbitrary subgroup. Then X satisfies* (1) *the DCC on relatively definable subgroups, as well as* (2) *the existence of uniform bounds on chains of uniformly relatively definable subgroups.*

PROOF.

Ad (1) Let $H_i \cap X$ be a strictly decreasing sequence of relatively definable subgroups of X. Then we may replace H_i by $H_i^* = \bigcap_{j \le i} H_j$. Then the sequence (H_i) is also strictly decreasing, and the usual DCC applies in G.

Ad (2) Similarly, noting that if $\mathcal{H}$ is a uniformly definable family of subgroups of G, then the set $\mathcal{H}^*$ of arbitrary intersections of subgroups from $\mathcal{H}$ is also uniformly definable, so the construction used above is again available. □

While the foregoing is extremely useful, it can often be avoided by recourse to the following indirect approach, and more generally by systematic use of $d(X)$ in place of X:

LEMMA 3.14. *Let G be a group of finite Morley rank, and X an arbitrary subgroup. Then $X^\circ = d^\circ(X) \cap X$.*

PROOF. Certainly $d^\circ(X) \cap X$ is relatively definable and of finite index, so $X^\circ \le d^\circ(X) \cap X$.

Conversely, suppose $X^\circ = X \cap H$. Let $H_1 = \bigcap_{x \in X} H^x$. Then $X^\circ = X \cap H_1$ and H_1 is definable. Furthermore, X normalizes H_1 and XH_1 is a finite extension of H_1, hence definable. So $d(X) \le H_1 X$, and $d^\circ(X) \le H_1$. Thus $d^\circ(X) \cap X \le X \cap H_1 = X^\circ$. □

LEMMA 3.15 ([**52**, Ex. 10, p. 78]). *Let G be a group of finite Morley rank, X an arbitrary subgroup. Then X° contains all connected relatively definable subgroups of X.*

PROOF. Let $H \cap X$ be connected and relatively definable. Replacing H by $H^\circ \cap d(X)$, we may suppose that H is connected and contained in $d(X)$. Then $H \cap d^\circ(X)$ has finite index in H, so $H \le d^\circ(X)$. □

LEMMA 3.16 ([**52**, Lemma 5.41]). *Let G be a group of finite Morley rank. If X is an arbitrary nilpotent by finite subgroup of G, then X° is nilpotent. Similarly, if X is solvable by finite, then X° is solvable.*

PROOF. Let $Y \le X$ be a subgroup of finite index, either nilpotent or solvable, according to the case under consideration. Then $d(Y)$ is, correspondingly, nilpotent or solvable, and $X^\circ \le X \cap d(Y)$. □

The following fact is more model theoretic in nature, and is not an immediate consequence of the rank axioms.

FACT 3.17 ([**155**]). *Let $\mathcal{F}$ be a uniformly definable family of definable subgroups of a group of finite Morley rank. Then the indices $[H : H^\circ]$ are bounded for $H \in \mathcal{F}$.*

This should be taken as information coming directly from model theory. However, as it is not given in [**156**], we will indicate the line of proof without developing the underlying machinery.

PROOF. Poizat shows in [**156**] (reworking an analysis of Lascar to use the rank axioms as given here) that there are finitely many strongly minimal sets D_i $(i = 1, \ldots, n)$ such that every complete type is nonorthogonal to one of them. Fixing such a finite family, we may add defining parameters to the language and suppose they are definable without parameters.

Pillay then shows that for every uniformly definable collection of definable *sets*, the Morley degree is bounded. (This is analogous to definability of rank, which was built into the axioms after being derived similarly from finiteness of Morley rank and the Lascar analysis.)

One shows first that U-rank equals Morley rank (again, this is done by Poizat), and then one shows a similar uniform bound on degree for families realizing a complete type. Namely, if $p(x, b)$ is a complete type of Morley rank n, then we claim that for some formula $\varphi \in p$, every nonempty set defined by a formula of the form $\varphi(x, b')$ has rank n and bounded degree.

Having treated types, one passes to formulas by a compactness argument.

This is the general line of argument; the details follow.

One proceeds by induction on n. Using the nonorthogonality one reduces the rank: there is, by assumption, a strongly minimal set D among the D_i, and a parameter c, such that for some a satisfying p and d satisfying D, independent over b, c, we have $d \in \operatorname{acl}(a, bc) \setminus \operatorname{acl}(bc)$. Let p' be the type of a over bcd. Then the Morley rank of p' (or the U-rank) is $n-1$. The induction hypothesis furnishes a formula $\psi(x, bcd)$ associated with p' in our sense (and lying in p'). As $d \in \operatorname{acl}(abc)$ there is also a formula $\mu(w, abc)$ controlling the multiplicity of d over abc. But $d \notin \operatorname{acl}(bc)$, so there is a further formula $\mu'(bc)$ true of bc, and implying that the subset of D defined by the following is infinite:

$$\exists y, \psi(y, bcw) \& \mu(w, ybc) \ .$$

(The free variable here is w.)

We can now define a suitable formula $\varphi'(x, yz)$ for the type p', namely

$$\exists w \in D\, \psi(x, y, z, w) \& \mu(w, x, y, z) \& \mu'(y, z) \ .$$

This says that w belongs to a certain set whose rank is $n-1$, which is fibered over D (or most of D, according to the final clause). With $yz = b'c'$ fixed, and assuming the set defined by $\varphi(x, b'c')$ is nonempty, the set of pairs (x, w) satisfying

$$\psi(x, b', c', w,) \& \mu(w, x, b', c')$$

is a disjoint union of sets of fixed rank $n-1$ and bounded degree, n over a set of rank 1, hence has rank n and bounded degree. Finally, the set defined by $\varphi'(x, b'c')$ is the projection of this set by a boundedly finite-to-one map. So it has the same rank, and bounded degree.

We finally need to descend to the type p. Let q be the type of c over b. We consider the formula $\varphi_0(x,b)$ which expresses the following:

$$\text{For generic } y \text{ satisfying } q,\ \varphi'(x,b,y) \text{ holds} .$$

By stability theory this is definable. Furthermore, the set defined by $\varphi'(x,b,c)$ is generic in the set defined by $\varphi(x,b)$ in the sense that the complement is of lower rank. This last point is a property of b:

$$\exists z \text{ The sets defined by } \varphi_0(x,b) \text{ and } \varphi'(x,b,z) \text{ differ by a set of lower rank} .$$

So this is expressible by a formula $\alpha(b)$. Consider $\varphi(x,y) = \varphi_0(x,y)\&\alpha(y)$. This asserts that the set defined by $\varphi(x,b')$ for any b', if nonempty, has the same rank and degree as a set of the form $\varphi(x,b',c')$. So we descend to p.

This completes the bound in the case of types. □

3.3. Generation by indecomposable subsets. Zilber extends the notion of connectivity to arbitrary definable sets as follows.

DEFINITION 3.18. *Let G be a group of finite Morley rank, and X a subset of G. Then X is* indecomposable *if there is no definable subgroup H of G for which X/H is finite, with $|X/H| > 1$.*

Evidently a connected group is an indecomposable subset.

The following is an analog of a basic lemma in the theory of algebraic groups. Its main effect for us will be to lead directly to the definability of a number of groups of interest to us.

PROPOSITION 3.19 ([**52**, Theorem 5.26]). *Let G be a group of finite Morley rank, and $\mathcal{X}$ a collection of indecomposable definable subsets of G, each of which contains the identity of G. Then the group $H_{\mathcal{X}} = \langle X : X \in \mathcal{X}\rangle$ is definable and connected, and is the setwise product of finitely many of the groups $X \in \mathcal{X}$ and their inverses, possibly with repetitions.*

PROOF. Let Y be a product of finitely many groups $X \in \mathcal{X}$, allowing repetitions, chosen so as to maximize $r = \text{rk}(Y)$. The main point is to show that $H_{\mathcal{X}} = YY^{-1}$. This yields the stated representation of $H_{\mathcal{X}}$, which includes the definability. Once one has the definability, connectivity follows. Namely, for any definable group K of finite index in $H_{\mathcal{X}}$, one has that X/K is finite, hence by hypothesis $|X/K| = 1$; as $1 \in X$, this means $X \subseteq K$, and thus $H_{\mathcal{X}} \leq K$.

So it suffices to show that $H_{\mathcal{X}} = YY^{-1}$, which can be rewritten as follows: for $h \in H_{\mathcal{X}}$, hY meets Y. In fact we make a stronger claim: for $h \in H_{\mathcal{X}}$, Y and hY differ by a set of rank less than r. The advantage of this stronger claim is that it suffices to prove it for $h \in X$, where $X \in \mathcal{X}$, as it passes easily to inverses and finite products.

For $X \in \mathcal{X}$, we have $\text{rk}(XY) = r$. Let $A \subseteq Y$ be irreducible. Let $G_A = \{g \in G : gA \sim A\}$, where for A, B irreducible of rank r, $A \sim B$ means that $\text{rk}(A \cap B){=}r$, equivalently A and B differ by a set of rank less than r. Then G_A is a definable subgroup of G, and as $\text{rk}(XY) = r$, there are, up to equivalence, only finitely many irreducible definable subsets of XY of rank r. Hence X/G_A is finite, and as $1 \in X$ we conclude that $X \subseteq G_A$. As this holds for each irreducible definable subset of Y, we find that gY and Y differ by a set of rank less than r for all $g \in X$, as claimed. □

The following corollary of Proposition 3.19 provides a quick way of proving the existence of *connected* analogs of notions arising in finite group theory, whose strict analogs may be problematic. In some cases one can dispense entirely with the strict analogs; in others the simpler connected analogs can be used effectively as a first step toward the strict notions.

COROLLARY 3.20. *Let G be a group of finite Morley rank, and let $\mathcal{X}$ be any collection of connected subgroups of G. Then $\langle X : X \in \mathcal{X} \rangle$ is definable, and connected.*

DEFINITION 3.21. *Let G be a group of finite Morley rank. Then:*

(1) *$O(G)$ is the largest definable connected normal solvable $2^{\perp}$-subgroup of G (invoking Lemma 2.19).*
(2) *$\hat{O}(G)$ is the largest definable connected normal subgroup of G of degenerate type.*

LEMMA 3.22. *Let G be a group of finite Morley rank and P a definable 2-subgroup of G. Then $O(N_G(P)) = O(C_G(P))$.*

PROOF. We have $[O(N_G(P)), P] \leq O(N_G(P)) \cap P = 1$, so $O(N_G(P)) \leq C_G(P)$ and hence $O(N_G(P)) \leq O(C_G(P))$. The reverse inclusion holds since $C_G(P)$ is normal in $N_G(P)$ and $O(C_G(P))$ is definably characteristic in $C_G(P)$. □

LEMMA 3.23. *Let G be a group of finite Morley rank and A and B definable subgroups with A connected. Then $\langle A, B \rangle$ is a definable subgroup of G.*

PROOF. $\langle A, B \rangle = \langle A^b : b \in B \rangle B$ and $\langle A^b : b \in B \rangle$ is definable by Corollary 3.20. □

This result is rendered more useful by criteria for the existence of indecomposable subsets.

LEMMA 3.24. *Let G be a group of finite Morley rank, and X a definable subset.*

(1) *If $N = \{g \in G : X^g = X\}$, and we do not have $1 < |X/H| < \infty$ for any N-invariant definable normal subgroup H of G, then X is indecomposable.*
(2) *If X is irreducible, then there is a finite subset F of X such that $X \setminus F$ is indecomposable.*
(3) *If X is indecomposable and $g \in G$, then gX and Xg are indecomposable.*

PROOF.

Ad (1) Suppose toward a contradiction that $1 < |X/H| < \infty$ for some definable subgroup H of G. Let $H_0 = \bigcap_{n \in N} H^n$. For each $n \in N$ we have $|X/H^n| = |X/H|$ and as H_0 is a *finite* intersection of such conjugates (by the DCC), it follows that $1 < |X/H_0| < \infty$ as well, contradicting our hypothesis.

Ad (2) Take H minimal definable such that $|X/H| < \infty$. Then for any other definable subgroup K with $|X/K| < \infty$, we have $|X/H \cap K| < \infty$ and hence $H \leq K$, that is, $|X/H|$ is maximal.

As X is irreducible, there is exactly one coset C in G/H meeting X in an infinite set; let F be $X \setminus C$.

Ad (3) This is purely formal. There are bijections $gX/H \leftrightarrow X/H$, and $Xg/H \leftrightarrow gX/H^{g^{-1}}$. □

PROPOSITION 3.25. *Suppose that G is a group of finite Morley rank with no nontrivial proper normal definable subgroups. Then G is simple.*

PROOF. Let C be a nontrivial conjugacy class in G. Then C is G-invariant, so by the criterion given in the last lemma, as G has no nontrivial G-invariant proper definable subgroups, C is indecomposable.

For $g \in C$, the set $g^{-1}C = [g, G]$ is also indecomposable, and contains 1. So by Proposition 3.19 $[G, C] = \langle [g, G] : g \in C \rangle$ is definable; since this group is normal in G, it coincides with G: G is generated by each of its conjugacy classes. □

LEMMA 3.26. *Let G be a group of finite Morley rank, and H a minimal normal definable subgroup. If H is nonabelian, then H is a finite direct sum of definable simple subgroups.*

PROOF. Let S be minimal normal definable in H. Then $H = d(\langle S^g : g \in G \rangle)$. The conjugates of S are all minimal normal in H and hence commute pairwise. If S is abelian, it follows that H is abelian, a contradiction. So S is nonabelian and hence has finitely many conjugates by Lemma 2.10. Thus $\langle S^g : g \in G \rangle$ is definable and $H = \langle S^g : g \in G \rangle$ is a finite direct sum.

Now any nontrivial normal definable subgroup of S will be normal in H and hence equal to S. By Proposition 3.25, S is simple. □

LEMMA 3.27. *Let G be a locally solvable group of finite Morley rank. Then G is solvable.*

PROOF. We proceed by induction on the rank and degree of G. We may suppose that G is connected and contains no infinite abelian normal subgroup. So $Z(G)$ is finite and by Lemma 3.9 if we replace G by $G/Z(G)$, we may suppose $Z(G) = 1$. As G is connected, it has no finite nontrivial normal subgroup.

Now take $H \lhd G$ minimal normal definable. Then H is a finite direct sum of simple subgroups, which are again locally solvable, contradicting Lemma 1.33. □

LEMMA 3.28. *Let G be a group of finite Morley rank, H a connected definable subgroup, and $g \in G$. Then the sets $[g, H]$ and $g^H = \{g^h : h \in H\}$ are indecomposable.*

PROOF. Since $[g, H]$ is a translate by g^{-1} of the set g^H, we deal only with the latter. Now g^H is H-invariant, so it suffices to show that if K is an H-invariant definable subgroup of G, then we do not have $1 < |g^H/K| < \infty$.

Suppose g^H/K is finite; then as H acts definably on this set by conjugation, and H is connected, the action is trivial. Hence $g^H/K = (g/K)^H = (g/K)$ and $|g^H/K| = 1$. □

COROLLARY 3.29. *Let G be a group of finite Morley rank, H a connected definable subgroup, and $X \subseteq G$ an arbitrary subset. Then $[X, H]$ is definable and connected.*

PROOF. $[X, G]$ is generated by the family $\{[x, G] : x \in X\}$ of indecomposable definable subsets of G, each of which contains 1. So Proposition 3.19 applies. □

This is a case in which we will not remain satisfied with the "connected" version of the result. We prove:

LEMMA 3.30 ([**52**, Corollary 5.31]). *Let G be a group of finite Morley rank and H, K definable subgroups of G that normalize one another. Then $[H, K]$ is definable.*

PROOF. Working in $N_G(H) \cap N_G(K)$, we may suppose that H and K are normal in G.

The subgroups $[H^\circ, K]$ and $[H, K^\circ]$ are then definable normal subgroups of G, and are contained in $[H, K]$, so we may pass to $G/[H^\circ, K][H, K^\circ]$ or, in other words, assume that

(*). $$[H, K^\circ] = [K, H^\circ] = 1$$

After these reductions, we will show that $[H, K]$ is finite, hence definable.

Let C be the *set* of commutators $[h, k]$ with $h \in H$, $k \in K$. By (*) the commutation map $\gamma : H \times K \to C$ factors through $H/H^\circ \times K/K^\circ$, so C is finite. By Lemma 1.13, $[H, K]$ is finite. □

LEMMA 3.31. *Let G be a group of finite Morley rank, and H a normal subgroup of G not commuting with G°. Then the following are equivalent:*

(1) *H is a minimal normal definable subgroup of G.*
(2) *H is a minimal normal subgroup of G*

PROOF. If H is minimal normal and $[H, G^\circ] \neq 1$, then $H = [H, G^\circ]$ is definable.

Conversely, suppose H is minimal normal definable and $K \leq H$ is normal in G. Then $[K, G^\circ] \leq K$ is normal and definable. Thus either $K = H$ or $[K, G^\circ] = 1$, in which case H meets $C(G^\circ)$ and hence is contained in $C(G^\circ)$. □

The exceptional case is that in which $H \leq Z(G^\circ)$, with H minimal normal definable. Such subgroups exist by the descending chain condition for definable subgroups, but may not be minimal normal. On the other hand, minimal normal subgroups of $Z(G^\circ)$ may not exist, but will be definable if they do exist, as shown in the following.

LEMMA 3.32. *Let G be a group of finite Morley rank, and H a minimal normal subgroup of G. Then H is definable, and if $[H, G^\circ] = 1$, then H is finite.*

PROOF. If $[H, G^\circ] > 1$, then the previous lemma applies, so it suffices to deal with the last point. So suppose $[H, G^\circ] = 1$. Of course, if $H \cap G^\circ = 1$ then H is finite, so we may suppose $H \leq Z(G^\circ)$. In particular, H is abelian and may be considered as a G/G°-module.

Now H is either an elementary abelian p-group for some p, or a divisible abelian group. Furthermore, being irreducible for the action of a finite group, H is finitely generated as a group. Thus it is not divisible, and must be a finite p-group for some p. □

We can now cast some further light on $E(G)$ as defined in Definition 1.7.

LEMMA 3.33. *Let G be a group of finite Morley rank and L a subnormal quasisimple subgroup. Then L is definable.*

PROOF. We may suppose that L is infinite. We may suppose, after replacing G by $C^\circ(Z(L))/Z(L)$, that L is simple and G is connected. We proceed by induction on the rank of G.

If $LZ(G)$ is definable then $L = (LZ(G))'$ is definable. In particular, if $Z(G)$ is infinite, then induction applies in $G/Z(G)$ and we conclude.

If $Z(G)$ is finite, then $G/Z(G)$ is centerless and it suffices to treat the image of L in $G/Z(G)$. So we may suppose now that $Z(G) = 1$.

Let $H \lhd G$ be a minimal definable normal subgroup. As G is centerless, H is infinite and hence connected. By hypothesis there is a series

$$L = K_0 \lhd K_1 \lhd \cdots \leq K_n = G \ .$$

Take i minimal with $H \cap K_i > 1$. If $i = 0$, then $L \leq H$; if $H < G$, we conclude by induction, while if $H = G$, then G is simple (Proposition 3.25), forcing $L = G$. If $i > 0$, then $[K_{i-1}, K_i \cap H] \leq K_{i-1} \cap H = 1$, and so $L \leq C(K_i \cap H) < G$. As $C(K_i \cap H)$ is definable, we may conclude by induction. □

LEMMA 3.34. *Let G be a group of finite Morley rank. Then for every n, G^n and $G^{(n)}$ are definable, and G^∞ and $G^{(\infty)}$ are definable.*

PROOF. The case of G^n and $G^{(n)}$ follows from Lemma 3.30 by induction. By the DCC these sequences both stabilize at some point. □

LEMMA 3.35. *Let A be an abelian group of finite Morley rank and π a set of primes. Then:*

(1) *There is a decomposition*

$$A = A_\pi \oplus A_{\pi^\perp}$$

with A_π *all the* π*-torsion, and* $A_{\pi^\perp}$ *some complement to* A_π *in* A.

(2) *If* G *is a group acting on* A*, and* A *satisfies the DCC with respect to* G*-centralizers, then so does* A/A_π.

PROOF.

Ad (1) First write $A = B \oplus T$ with B having bounded exponent and T divisible (Lemma 2.13). Then each of B and T splits as claimed, and the result follows.

Ad (2) Let $\bar{A} = A/A_\pi$. We claim that for any subset X of G we have

$$(*) \qquad C_{\bar{A}}(X) = C_A(X)A_\pi\,.$$

Then the DCC will be inherited from A to $\bar{A}$.

We prove $(*)$. Fix $X \subseteq G$ and $\bar{a} \in C_{\bar{A}}(X)$, represented by $a \in A$. By the DCC for G-centralizers in A, there is $X_0 \subseteq X$ finite so that $C_A(X) = C_A(X_0)$. For $x \in X_0$ we have $[a, x] \in A_\pi$ and hence for some π-number n we have $1 = [a, x]^n = [a^n, x]$, hence $a^n \in C_A(X_0) = C_A(X)$. Now by Lemma 2.18, there is a π-element a' representing the coset $aC_A(X)$. So $a \in a'C_A(X) \le A_\pi C_A(X)$. □

LEMMA 3.36. *Let* G *be a solvable group of finite Morley rank and* π *a set of primes. Then there is a series* $G = G_0 > G_1 > \cdots > G_n = 1$ *with all* G_i *characteristic in* G*, having abelian quotients* $A_i = G_i/G_{i+1}$*, so that for each* i *either*

(1) A_i *is a* π*-group; or*

(2) A_i *is a* π*-divisible* $\pi^\perp$*-group satisfying the DCC for* G*-centralizers.*

PROOF. We refine the series $G^{(i)}$, which has definable terms. Let B_i be the quotient $G^{(i)}/G^{(i+1)}$, a definable abelian group. As such, B_i itself has the DCC for G-centralizers, and splits as $B_{i,\pi} \oplus B_{i,\pi^\perp}$ according to Lemma 3.35. Let $G_{i,\pi}$ be the preimage of $B_{i,\pi}$ in $G^{(i)}$ and insert the terms $G_{i,\pi}$ into the series; these are typically not definable.

Now the quotients are, alternately, π-groups $G_{i,\pi}/G_{i+1}$, and π-divisible $\pi^\perp$-groups $G_i/G_{i,\pi}$. By the preceding lemma, the DCC for G-centralizers is inherited by $G_i/G_{i,\pi}$. □

DEFINITION 3.37. *A group* G *is* semisimple *if* G *is a direct sum of simple groups.*

LEMMA 3.38. *Let* G *be a group of finite Morley rank and suppose that* $G' = G$*. Then the following are equivalent:*

(1) $G/Z(G)$ *is semisimple.*

(2) G *is quasisemisimple.*

PROOF. It suffices to show that $1 \implies 2$. So suppose that $\bar{G} = G/Z(G)$ is semisimple, with simple factors $\bar{L}_i$ $(i \in I)$ covered by $L_i \le G$. Then G is the central product of the L_i.

Let $K_i = L_i^{(\infty)}$, $G_0 = \langle K_i : i \in I \rangle$. Then G_0 is the central product of the K_i, and as K_i covers $\bar{L}_i$ we have $G = G_0 Z(G)$. Thus $G = G' = G_0' = G_0$. □

LEMMA 3.39. *Let G be a connected solvable group of finite Morley rank, and $H \le G$ a minimal infinite definable normal subgroup. Then H is abelian.*

PROOF. H is connected. Hence H' is connected and definable. By the minimality of H, either $H' = H$, which contradicts solvability, or $H' = 1$, as claimed. □

4. Fields

4.1. Fields of finite Morley rank. The theory of groups of finite Morley rank is intimately bound up with the theory of fields of finite Morley rank, and not just because of the conjectured relation with algebraic groups. Under general conditions, an action of one group of finite Morley rank on another tends to give rise to a field (Proposition 4.11), which when infinite is algebraically closed (Proposition 4.2). Difficulties arise when the a proper connected subgroup of the multiplicative subgroup appears in such an action; these difficulties are partly neutralized by results of Wagner, which take on a group theoretic cast in the theory of "good tori" (§4.4).

LEMMA 4.1. *Let R be a commutative ring of finite Morley rank without nilpotent elements. Then R is a finite product of fields.*

PROOF. We do not assume at the outset that R contains an identity.

The principal ideals aR of R form a uniformly definable family, hence satisfy a uniform chain condition by Lemma 2.8. Let I be a minimal nontrivial principal ideal of R. Then I is an integral domain, since if $ab = 0$ with $b \ne 0$ then $aI = (0)$ by minimality, so $a^2 = 0$ and finally $a = 0$. Again by minimality $aI = I$ for all nonzero $a \in I$, and I is a field. Let $I^\perp = \mathrm{Ann}_R(I)$.

We show $R = I \oplus I^\perp$. Fix $e \in I$ as its multiplicative identity, and consider $r \in R$. Then $re = xe$ for some $x \in I$, and hence $(r - x)e = 0$. Accordingly, $(r - x)I = 0$ and $r - x \in I^\perp$, as required.

Now let R_0 be minimal among definable ideals contained in R such that $R = R_0 \oplus R_0^\perp$ and $R_0^\perp$ is a finite sum of fields. Note that any ideal of R_0 is an ideal of R, so if $R_0 \ne (0)$ we can apply our initial construction to enlarge $R_0^\perp$ and shrink R_0, a contradiction. Thus R is a finite sum of fields. □

PROPOSITION 4.2 ([**137**]). *Let F be an infinite field of finite Morley rank. Then F is algebraically closed.*

PROOF. The polynomial maps of the form $p(x) = x^n$ and $\tau(x) = x^p - x$, the latter in characteristic $p > 0$, are endomorphisms of the additive and multiplicative groups of F, respectively, with finite kernel. Both of these groups are connected by Example 3.5 and Corollary 3.7. So Lemma 3.3 implies that all of these maps are surjective; so the multiplicative group of F is divisible, and in particular F is perfect, and all polynomials of the form $x^p - x - a = 0$ split over F in characteristic $p > 0$.

More importantly, all of these conclusions apply to any *finite extension* K of F, as such an extension can be identified, definably, with F^n for some n, and hence K inherits a rank function from F.

From this point on, Galois theory suffices. Assuming F is not algebraically closed, take a finite extension K of F and a proper Galois extension $\hat{K}$ of K, chosen so as to minimize the dimension $[\hat{K} : K]$. Then there are no intermediate fields between $\hat{K}$ and K, so the extension is of some prime order l. Furthermore, the l-th roots of unity are in K, as otherwise one can lower the dimension by replacing $\hat{K}$ by $K[1^{1/l}]$ and K by an intermediate field.

Now since K has an adequate supply of roots of unity, the theory of cyclic extensions tells us that if $l \neq \mathrm{char}(K)$, then $\hat{K}$ is a *Kummer extension* $K[a^{1/l}]$ for some $a \in K$, while if $l = \mathrm{char}(K)$, then $\hat{K}$ is an *Artin-Schreier extension* $K[\alpha]$ with $\tau(\alpha) \in K$. However, Kummer extensions are excluded by the divisibility of $K^\times$, and Artin-Schreier extensions by the surjectivity of τ on K. □

LEMMA 4.3. *Let F be a field of finite Morley rank, and K a proper definable subfield. Then K is finite.*

PROOF. Otherwise, both K and F are algebraically closed. Let $V \leq F$ be a K-subspace of F of dimension n. Fixing a basis of V, we have a bijection $V \leftrightarrow K^n$; hence $\mathrm{rk}(F) \geq \mathrm{rk}(V) = n\,\mathrm{rk}(K) \geq n$. This is a contradiction for large n. □

This has the following nice consequence.

LEMMA 4.4. *Let F be a field of finite Morley rank and characteristic zero. Then the additive group of F is minimal: it contains no nontrivial proper definable subgroup.*

PROOF. Let $A \leq F$ be a definable subgroup and let $R = \{r \in F : rA = A\}$. Then $\mathbb{Z} \leq R$ and thus R is an infinite definable subring, hence also a subfield, of F. So $R = F$ and $A = 0$ or F. □

LEMMA 4.5. *Suppose that F is an infinite field of finite Morley rank, and G is a definable group of automorphisms of F. Then $G = 1$.*

PROOF. For $\sigma \in G$, let F_σ be the fixed field. By the Lemma 4.3, F_σ is finite. For each n, F_{σ^n} is an extension of degree n, and by compactness there is $\tau \neq 1 \in G$ (at least, in an elementary extension) such that $[F_\tau : F_\sigma] = \infty$ and, in particular, F_τ is infinite. This contradicts Lemma 4.3. □

4.2. Linearization.

Definition 4.6. *Let G be a group of finite Morley rank acting definably on an abelian group V. Then V is* definably G-irreducible *if it has no nontrivial proper definable G-invariant subgroup.*

Lemma 4.7 (Schur's Lemmma). *Let G be a group acting definably on a definably G-irreducible abelian group V. Then the ring of definable endomorphisms of V is a division ring.*

Proof. Kernels and images of definable endomorphisms are definable, hence 0 or V. □

Lemma 4.8. *Let A be an abelian group of finite Morley rank acting definably on a definably A-irreducible abelian group V. Then the subring K of* End(V) *generated by A is definable.*

Proof. By Schur's Lemma, K is a field.

Let $v \in V$ be a nontrivial element. In the semidirect product $V \rtimes A$, we have $[v, A] = \{(a-1).v : a \in A\}$. By Lemma 3.28 these sets are indecomposable, and hence by Proposition 3.19, V is a finite sum of such sets, or equivalently $v^{A+\cdots+A} = v^K$ for some finite length sum $A+\cdots+A$. As K is a field, this relation forces $A+\cdots+A = K$, and proves the definability of K. □

Definition 4.9. *Let G be a group of finite Morley rank acting definably on an abelian group V. Then V is* G-minimal *if V is infinite and contains no G-invariant proper infinite definable subgroup.*

Lemma 4.10. *Let A be an infinite abelian group of finite Morley rank and V an A-minimal abelian group on which A has a nontrivial action. Then V is A-irreducible.*

Proof. By minimality, $C_V(A)$ is finite. Let $\bar{V} = V/C_V(A)$. There is a map from the commutative ring of definable endomorphisms of V generated by A to the corresponding subring of End$(\bar{V})$. The kernel of this map is trivial, since no nonzero definable endomorphism can take the connected group V into $C_V(A)$. Thus A generates an infinite field F, in End(V). Taking $v \in V \setminus C_V(A)$, we see that $V = F.v$ and that $C_V(A) = (0)$. □

Proposition 4.11 (Linearization). *Let G be a connected group of finite Morley rank acting definably, faithfully, and irreducibly on an abelian group V, and let $T \lhd G$ be infinite abelian. Then the subring K of* End(A) *generated by T is a field which is definable over G, and under the action of K V is a finite dimensional vector space on which G acts K-linearly.*

Proof. Let $A = C_V(T^\circ)$. We claim $A = 1$; otherwise, by irreducibility, we have $A = V$ and $T^\circ = 1$, T is finite, a contradiction.

Let U be a minimal infinite definable T-invariant subgroup of V. Then U is T-irreducible, as otherwise there is a finite T-invariant subgroup U_0 of U and then $U_0 \leq C_V(H^\circ)$, a contradiction.

By Clifford theory (see §11) V is a finite direct sum of conjugates $U_i = U^{g_i}$ of U. Let K be the subring of $\mathrm{End}(V)$ generated by T. As the U_i are T-submodules, we have a natural map with trivial kernel $K \to \sum \mathrm{End}(U_i)$, and the image in each factor generates a field K_i by Schur's Lemma. That is, we may view K as a subring of $\sum K_i$ where the field K_i is generated by T in $\mathrm{End}(U_i)$. Such a ring is itself isomorphic to a direct sum of fields. By Lemma 4.8, the fields involved are definable in G, and hence K is also definable.

On the other hand, G acts by conjugation as a group of definable automorphisms of the ring K, permuting the factors, and as G is connected it stabilizes the factors (Lemma 3.3). Thus G acts as a definable group of automorphisms on each field, hence trivially by Lemma 4.5. In other words, G commutes with K.

On the other hand, by construction, G permutes the fields K_i transitively, thus there is only one such, and K is a field. Note that the irreducible module U can now be construed as a 1-dimensional subspace. □

PROPOSITION 4.12. *Let G be a connected solvable group of finite Morley rank, and suppose that G is not nilpotent. Then there are definable sections K, T of G, with an action of T on K induced by conjugation in G, such that the pair (K, T), together with the action, is definably isomorphic to a pair (F_+, T^*) where F is a field, T^* is a multiplicative subgroup of F which generates F_+ additively, and the action of T on K corresponds to the natural action of T^* on F_+ by multiplication.*

PROOF. Fix i so that $\mathrm{rk}(Z_i(G))$ is maximal, and replace G by $G/Z_i(G)$. Then $Z(G)$ is finite, so $G/Z(G)$ is centerless; factoring out again, assume that G is centerless.

Let A be a minimal infinite definable normal subgroup of G. As G is centerless, the action of G on A is faithful. We claim that A is G-irreducible. Suppose $B \leq A$ is a nontrivial G-submodule. Then $[B, G]$ is a definable connected G-submodule by Corollary 3.29. Since the action of G is faithful, $[B, G]$ is nontrivial. As this module is connected, it is infinite, and hence by minimality $[B, G] = A$, and $B = A$. So A is irreducible.

Let T be a minimal infinite definable normal subgroup of $G/C_G(A)$. By Lemma 3.39, T is abelian. By the preceding proposition, T generates a field $F \subseteq \mathrm{End}(A)$ which gives A the structure of a finite dimensional vector space. We can represent F_+ by any 1-dimensional subspace of A. The generation of F by T is additive since T is closed under multiplication. □

4.3. Bad fields. One difficulty arises in applying the Linearization Lemma: we may have $T < K$, something which does not occur in algebraic examples.

DEFINITION 4.13. *A* bad field *is a structure $(K; T)$ of finite Morley rank in which K is a field and T is a proper infinite subgroup of its multiplicative group.*

While no examples of such fields have actually been constructed, methods of Hrushovski suffice for the construction, in characteristic 0, of analogous structures of infinite rank; cf. [**158**]. On the other hand, model theoretic arguments due to Wagner provide some measure of control in positive characteristic (enough, in fact, to show that their existence would require number theoretic restrictions on primes which are heuristically unlikely).

The main model theoretic result in positive characteristic is as follows. (This result does not immediately give number theoretic restrictions, but it is all we require.)

DEFINITION 4.14. *Let M be an arbitrary structure. An element a of M is* algebraic *if it belongs to a finite set which is definable without parameters; equivalently, assuming M is reasonably homogeneous, the condition is*

$$a^{\operatorname{Aut} M} \text{ is finite} .$$

The set of algebraic elements is denoted M_{alg}.

PROPOSITION 4.15 ([**188**]). *Let F be a field of finite Morley rank. Then F_{alg} is an elementary substructure of F (in its full language).*

For the proof, we first prepare some machinery.

LEMMA 4.16 ([**187, 148**]). *Let K be a field of finite Morley rank, and X a definable subset of K which contains an infinite subfield F of K, not assumed definable. Then:*

(1) *X is* generic *in K in the sense that* $\operatorname{rk}(K \setminus X) < \operatorname{rk}(K)$.
(2) *If in addition X is a multiplicative subgroup of K, then $X = K^\times$.*

PROOF. The second point follows at once from the first: a generic subgroup of a group, in the sense given, must be the whole group. So we concentrate on the first point.

We may suppose that X is a counterexample of minimal rank r and degree d. Then for any $a \in F^\times$, as $a + X$ and aX are also counterexamples, and contain the same field F, we have $a + X \sim X$ and $aX \sim X$ in the sense that in each case the symmetric difference has rank less than r.

Let $K_1 = \{a \in K : a + X \sim X\}$. Then $F \subseteq K_1$ and K_1 is a definable subgroup of K_+. Let $K_2 = \{a \in K_1 : aX \sim X \text{ or } a = 0\}$. Then $F \subseteq K_2$ and K_2 is a definable subring of K. In particular, K_2 is an integral domain and hence by Lemma 4.1, K_2 is a subfield of K; but as K_2 is infinite, by Lemma 4.3 to Proposition 4.2, $K_2 = K$.

Now let $R \subseteq X \times K$ be defined as

$$\{(x, a) \in X \times K : xa \in X\} .$$

For all $a \in K$, $R \cap (X \times \{a\})$ is generic in X. Hence by the Fubini principle,

$$\{x \in X : R \cap (\{x\} \times K) \text{ is generic in } K\}$$

is generic in X; pick one such element x. Then

$$\operatorname{rk}(X) \geq \operatorname{rk}(xK \cap X) = \operatorname{rk}(R \cap \{x\} \times K) = \operatorname{rk}(K)$$

and hence X is generic in K. □

PROOF OF PROPOSITION 4.15. We aim to show that F_{alg} is an elementary substructure of F. For this, we need to show that every F_{alg}-definable nonempty set X has a point in F_{alg}. Evidently, an elementary substructure will satisfy this condition, and the converse also holds: this is known as the *Tarski-Vaught test* for elementary substructures.

Toward a contradiction take a counterexample X with $\text{rk}(X)$ and then degree(X) minimized. Then X is infinite, as otherwise $X \subseteq F_{\text{alg}}$. Furthermore, by minimization, X has no nonempty proper F_{alg}-definable subset.

Let $f = \text{rk}(F)$. In an elementary extension, we may take an element $a \in F$ such that a belongs to no F_{alg}-definable subset of rank less than f. We claim that there is no proper nonempty $F_{\text{alg}}(a)$-definable subset of X. Suppose on the contrary that S_a is definable from a and parameters in F_{alg}, and both $X \cap S_a$ and $X \setminus S_a$ are nonempty. Then we may consider the set

$$F_0 = \{y \in F : X \cap S_y \neq \emptyset \text{ and } X \setminus S_y \neq \emptyset\}$$

of rank f since it contains a. By Lemma 4.16, $F \setminus F_0$ cannot contain F_{alg}. Hence F_{alg} meets F_0 and there is a parameter $a^* \in F_{\text{alg}}$ such that $X \cap S_{a^*}$ and $X \setminus S_{a^*}$ are both nonempty, contradicting the choice of X. So there is no proper nonempty a-definable subset of X.

Now there is a finite sequence $a_1, a_2, \ldots$ of elements of F_{alg} such that $F = \sum a_i X$. In particular,

$$\text{rk}\left(\sum a_i X\right) = f \ . \tag{1}$$

Choose a sequence $a_1, \ldots, a_n$ in F_{alg}, of minimal length, satisfying condition (1). Then $a \in \sum a_i X$:

$$a = a_1 x_1 + \cdots + a_n x_n \text{ with } x_i \in X \ .$$

Consider $X_a = \{x \in X : a \in a_1 x + \sum_{i>1} a_i X\}$. Here X is definable from a and the parameters $a_1, \ldots, a_n$. This set is nonempty and definable over a and F_{alg}, hence contains all of X.

Consider the set $R \subseteq F \times X$ of pairs (a, x) for which $a \in a_1 x + \sum_{i>1} a_i X$. Then a belongs to the F_{alg}-definable set $S = \{u : \text{rk}(R \cap \{u\} \times X) = \text{rk}(X)\}$, and hence $\text{rk}(S) = f$; so $\text{rk}(R) = \text{rk}(F) + \text{rk}(X)$. Consider the map $R \to F$ given by $(a, x) \mapsto a - a_1 x$. The fibers have rank at most $\text{rk}(X)$ since a given $x \in X$ determines at most one corresponding a. Hence the image has rank at least $\text{rk}(R) - \text{rk}(X) = \text{rk}(F)$; in other words, $\text{rk}(\{a - a_1 x : (a, x) \in R\}) = f$, but this implies that

$$\text{rk}(\sum_{i>1} a_i X) = f$$

and this contradicts the minimization of the length n.

This contradiction completes the argument. □

LEMMA 4.17. *Let F be a field of finite Morley rank and X an infinite definable subset. Then for some finite set of elements a_i in F_{alg},*

$$(*) \qquad F = a_1X + \cdots + a_kX \ .$$

PROOF. We can select a subset of X which is indecomposable relative to the additive group F_+, for example by Lemma 3.24. So after translation, we may suppose that X is indecomposable and contains 0. Rescaling, we may also suppose that $1 \in X$. Then by Proposition 3.19, $Y = \langle aX : a \in F_{\mathrm{alg}} \rangle$ is both definable, and a finite sum as in $(*)$. On the other hand, the group $T_Y = \{a \in F^\times : aY = Y\}$ contains F_{alg} by construction, hence coincides with $F^\times$ by Lemma 4.16. Hence $Y = F$ and our claim follows. □

PROPOSITION 4.18. *Let F be an infinite field of finite Morley rank and positive characteristic p, whose language consists of the usual language of fields augmented by certain subgroups of $(F_+)^n$ and $(F^\times)^n$. Let F_0 be the subfield of elements algebraic over $\mathbb{F}_p$, viewed as a substructure for the extended language (note that F_0 is the algebraic closure $\tilde{\mathbb{F}}_p$ of $\mathbb{F}_p$, as a field). Then*

$$F_0 \prec F \ .$$

PROOF. It suffices to check that $F_0 = F_{\mathrm{alg}}$ in this case. For this we use the Frobenius automorphism Frob; this is an automorphism of F, by our assumption on the language, and for $a \in F_{\mathrm{alg}}$ the orbit of a under powers of Frob must be finite, so $a \in F_0$. □

4.4. Good tori.

DEFINITION 4.19. *Let G be a group of finite Morley rank.*

(1) *A* torus *in G is a definable connected divisible abelian group.*
(2) *A* good subgroup *in G is a definable subgroup H of G group such that every definable subgroup of H is the definable hull of its torsion.*
(3) *A* good torus *is a torus in G which is a good subgroup of G.*
(4) *An* absolutely good torus *in G is a torus in G which is good in every elementary extension of G.*

By a combinatorial argument, every good torus is absolutely good [**69**]. However, the good tori which arise in groups of finite Morley rank are all absolutely good a priori, so we do not require this result, which however simplifies the terminology.

PROPOSITION 4.20. *Let F be an infinite field of finite Morley rank and of positive characteristic. Then $F^\times$ is an absolutely good torus.*

PROOF. The hypotheses are unaffected by passage to elementary extensions, so we work in the given model. Of course, $F^\times$ is a torus. Let $T \le F^\times$ be definable. We claim that T is the definable hull of its torsion.

Let T_{tor} be the torsion subgroup, $T_0 = d(T_{\mathrm{tor}})$, and consider the structure $\tilde{F}$ consisting of F equipped with the field operations, as well as predicates denoting T and T_0. By Fact 2.2 $\tilde{F}_+$ is again a group of finite Morley rank with respect to this language. Furthermore, $\tilde{F}$ is a structure of the sort to which Wagner's result (Proposition 4.15) applies. Hence the structure $\tilde{F}_{\text{alg}}$ derived by restricting everything to F_{alg} is an elementary substructure. However, in this structure, $T = T_{\text{tor}}$ (since F_{alg} is locally finite) and *a fortiori* $T = T_0$; however, the property "$T = T_0$" is expressible in the given language, hence passes to K. $\square$

LEMMA 4.21. *Let G be a group of finite Morley rank.*

(1) *If H is a definable subgroup of G such that every connected subgroup of H is the definable hull of its torsion, then H is a good subgroup of G.*
(2) *If $1 \to H \to K \to L \to 1$ is a short exact sequence of definable groups, then K is good if and only if H and L are good.*

PROOF.
Ad (1) Consider a definable subgroup K of H.

For $x \in K$, let $A_x = Z(C_K(x))$. This is an abelian group containing x, and it will be sufficient to show that A_x is the definable hull of its torsion. So we may suppose that $H = A_x$, and in particular H is abelian.

Then by Lemma 2.13, $H = H_1 + H_2$ with H_1 of bounded exponent and H_2 divisible, hence connected. Thus $H_1 \le H_{\text{tor}}$ and $H_2 \le d(H_{\text{tor}})$, and our claim follows.

Ad (2) The property of goodness clearly passes to definable subgroups. We consider homomorphic images. Let $h : K \to L$ be a surjection, and L^* a subgroup of L, K^* its preimage in K. Then $K^* = d(K^*_{\text{tor}})$ and by Lemma 2.15 we find that $L^* = h[d(K^*_{\text{tor}})] = d(h[K^*_{\text{tor}}]) \le h[L^*_{\text{tor}}]$, as required.

Now suppose that H and L are good. Any definable subgroup of K fits into a similar short exact sequence, so it suffices to show that $K = d(K_{\text{tor}})$. Now $d(K_{\text{tor}})$ contains $d(H_{\text{tor}}) = H$. We need to show that $d(K_{\text{tor}})$ also covers L; since its image is definable, it suffices to show that it covers L_{tor}.

Let $\bar{x} \in L_{\text{tor}}$, and let x be a lifting of $\bar{x}$ to K. If $\bar{x}^n = 1$, then $x^n \in H$. Let $H^+ = \langle H, x \rangle$. Then $[H^+ : H]$ is finite, so H^+ is definable. As H is good, every connected definable subgroup of H^+ is the definable hull of its torsion; so H^+ is good. Thus $x \in d(H^+_{\text{tor}}) \le d(K_{\text{tor}})$, and $\bar{x} \in h[d(K_{\text{tor}})] = d(h[K_{\text{tor}}])$. $\square$

COROLLARY 4.22.

(1) *Let G be a group of finite Morley rank, and T a connected definable subgroup of a finite product of good tori. Then T is a good torus.*
(2) *Let G be a connected group of finite Morley rank. Then there is a unique minimal definable normal subgroup K of G such that G/K is good torus.*

PROOF.
Ad (1) By the formal properties of goodness, T is a good subgroup of G. It is also an abelian group of finite Morley rank and hence can be written

as a sum $T_1 + T_2$ with T_1 of bounded exponent, and T_2 divisible, and both factors definable. By Lemma 2.11, T_1 is finite. Hence $[T : T_2] < \infty$. By connectivity, $T = T_2$ is divisible.

Ad (2) If G/K_1 and G/K_2 are good tori, then $G/(K_1 \cap k_2)$ embeds as a connected subgroup of the product, hence is a good torus. By the minimal condition on definable subgroups, the claim follows. □

LEMMA 4.23 (Rigidity Lemma). *Let G be a group of finite Morley rank and T a definable abelian subgroup of G with T° an absolutely good torus. Then the following hold.*

(1) $N^\circ{}_G(T) = C^\circ{}_G(T)$.
(2) *If $\mathcal{H}$ is a uniformly definable family of subgroups of T, then $\mathcal{H}$ is finite.*
(3) *If H is a definable subgroup of G, and $\mathcal{F}$ is a uniformly definable subgroup of homomorphisms from H to T, then $\mathcal{F}$ is finite.*

The first of these requires only goodness.

PROOF.

Ad (1) Let $N = N^\circ{}_G(T)$. Then N acts on T_{tor} and as $T[n]$ is finite for all n, and N is connected, N centralizes T_{tor} (Lemma 3.3). As $C(T_{\text{tor}}) = C(d(T_{\text{tor}})) = C(T)$, our first claim follows.

Ad (2) Supposing the contrary, in some elementary extension we can have arbitrarily many uniformly definable subgroups of T (specifically, more than $2^{\aleph_0}$). However, each such group is determined by its intersection with T_{tor}, and as $T[n]$ is finite for each n, T_{tor} is a countable set.

Ad (3) We may suppose H is connected, as once a homomorphism is determined on H° it has finitely many extensions to H. If $K \lhd H$ is minimal definable normal with H/K a good torus, then for every homomorphism in $\mathcal{F}$ factors through H/K, so we may suppose that H is itself a good torus. Then each homomorphism is determined by its action on the torsion of H, so the cardinality of $\mathcal{F}$ is bounded, and hence finite. □

LEMMA 4.24. *Let G be a group of finite Morley rank, K a definable subgroup, and $T \leq Z(K)$ a good torus. Then $N_G{}^\circ(K) \leq C_G{}^\circ(T)$.*

In particular, if $K = C_G{}^\circ(T)$, then K is almost self-normalizing.

PROOF. There is a unique maximal good torus $\hat{T}$ contained in $Z(K)$. Hence $N^\circ(K) \leq N^\circ(\hat{T}) = C^\circ(\hat{T}) \leq C^\circ(T)$, proving the first point, and the second is then immediate. □

4.5. Division rings and related structures. The following results have geometrical consequences; see §III 6.

PROPOSITION 4.25. *A division ring of finite Morley rank is commutative, hence a finite or algebraically closed field.*

PROOF. The finite case is classical ([**190**]). Suppose therefore that D is an infinite division ring of finite Morley rank, and is noncommutative.

We may suppose also that every proper definable division subring of D is commutative. Let $K = Z(D)$. If D is finite dimensional over K, then K is algebraically closed and it follows that $D = K$. So we suppose that D is infinite dimensional over K. It follows that K is finite, as otherwise by considering finite dimensional K-subspaces of D we find $\mathrm{rk}(D) \geq d\,\mathrm{rk}(K)$ for all d.

By Lemma 3.10, $D^\times$ contains an infinite definable abelian subgroup A. As the center of D is finite, there is $a \in A$ noncentral, and then $L = C_D(a)$ is a proper infinite definable division subring of D, hence commutative. But a division ring which is finite dimensional over a commutative subfield is also finite dimensional over its center (cf. [**131**, 15.8]), and we have a contradiction. □

Our next result concerns the more specialized subject of *alternative division rings.* These are not necessarily associative rings with left and right multiplicative inverses, satisfying a weak associative law:

$$(A) \qquad (ax)x^{-1} = a = x^{-1}(xa)$$

We will follow [**185**], where in addition the *Moufang identities* are taken as part of the definition:

$$(ab \cdot a)c = a(b \cdot ac), b(a \cdot ca) = (ba \cdot c)a; ab \cdot ca = a(bc \cdot a) = (a \cdot bc)a\ .$$

For a derivation of these conditions from the axiom (A) see [**140**, App. B]. Alternative division rings coordinatize *Moufang projective planes*; see Chapter III, §III 7.

In this context, the *centralizer* $C(X)$ of a subset X of an alternative division ring D is defined as the set of elements a which both centralize and associate with elements of X: $ax = xa$, $(ax)y = a(xy)$, and in particular the *center* $Z(D) = C_D(D)$ is a field.

The structure of *non-associative* alternative division rings is completely determined by the Bruck-Kleinfeld Theorem ([**58, 129**], cf. [**140, 185**]), and from this one can see that alternative division rings of finite Morley rank are fields. One can also derive this from more general principles, in a way that is analogous to the treatment of the finite case. For this one begins with the following.

LEMMA 4.26. *Let D be a nonassociative ring of finite Morley rank, and $R \subseteq D$ an associative subring. Then R is contained in a definable associative subring of D.*

PROOF. Viewing R and D as additive groups, let $\hat{R} = d(R)$. Easily $R \cdot \hat{R} = \hat{R}$ and hence $\hat{R} \cdot \hat{R} = \hat{R}$, so it suffices to check that $\hat{R}$ is associative.

Let

$$A = \{d \in D : r_1r_2 \cdot d = r_1 \cdot r_2d \text{ for } r_1, r_2 \in R\}\ .$$

For fixed r_1, r_2 the corresponding set is definable and contains R, hence $\hat{R} \leq A$.

Let

$$S = \{a \in \hat{R} : \text{For } r \in R \text{ and } b \in A \text{ we have } r_1 b \cdot a = r_1 \cdot ba\} \ .$$

Arguing as above, $\hat{R} \subseteq S$ and hence $\hat{R} = S$.

This means that elements in R satisfy $r \cdot s_1 s_2 = r s_1 \cdot s_2$ for $s_1, s_2 \in \hat{R}$, and hence the same applies to $\hat{R}$; $\hat{R}$ is associative. □

One can show similarly that if we drop the assumption on Morley rank, then any *definable* subset that generates an associative subring is contained in a definable associative subring, which is equally useful for the following.

PROPOSITION 4.27. *An alternative division ring D of finite Morley rank is associative and commutative.*

PROOF. As noted, this can be seen from the explicit classification given by the Bruck-Kleinfeld Theorem ([**185**, 20.1, 20.2]), according to which such a division ring is eight-dimensional over its center K, and has a "norm" function which is an anisotropic quadratic form on K in eight variables, which is incompatible with the two possible forms of K, finite or algebraically closed.

One may also argue in another way using a theorem of Artin which applies to the broader class of alternative rings (not necessarily division rings): every subring generated by two elements is associative [**58**, p. 888], as well as the relatively elementary result that commutative alternative rings are associative.

By Artin's result and our Lemma 4.26, any two elements $a, b \in D$ are contained in a definable associative subring R of D. This subring is clearly a division ring, and hence by Proposition 4.25 it is commutative. So D is commutative. □

In the classification of *Moufang generalized quadrangles* (cf. §III 6), we also encounter more exotic structures of the type known as an *involutory set* that may be involved. These are triples of the form (K, K_0, σ) in which K is a division ring, σ is an involutory anti-automorphism of K, and K_0 is an additive subgroup of K with the following properties:

(1) $1 \in K_0$.
(2) $K_\sigma \subseteq K_0 \subseteq \mathrm{Fix}_K(\sigma)$ where K_σ denotes

$$\{a + a^\sigma : a \in K\}$$

(3) $a^\sigma K_0 a \subseteq K_0$ for $a \in K$.

LEMMA 4.28. *Let K be a an infinite division ring with an involutory antiautomorphism σ. Then either K_σ is infinite, or σ is trivial; in the latter case K is a field of characteristic 2.*

PROOF. We may assume that K_σ is finite. Let $h(a) = a + a^\sigma$ and let $K_1 = \ker(h)$. Then K_1 has finite index in K.

Suppose first that

$$\mathrm{Fix}_K(\sigma) \text{ is infinite.}$$

Then $\mathrm{Fix}_K(\sigma)$ meets K_1 nontrivially, since the intersection has finite index in $\mathrm{Fix}_K(\sigma)$. But on $\mathrm{Fix}_K(\sigma) \cap K_1$ we have $2a = a + a^\sigma = 0$ and thus the characteristic of K is 2. In this case,

$$K_1 = \mathrm{Fix}_K(\sigma) \ .$$

Now for $a \in K_1$, it follows that $aK_1 \cap K_1$ also has finite index in K. But if $ab \in aK_1 \cap K_1$, we have

$$ba = b^\sigma a^\sigma = (ab)^\sigma = ab \ .$$

Thus $C(a)$ contains $K_1 \cap a^{-1}K_1$ and so $C(a)$ is of finite index in K. But then $C(a)$ is an infinite division ring, and $K/C(a)$ is a $C(a)$-vector space, which being finite can only be trivial. So this forces $K = C(a)$ for $a \in K_1$, that is, $K_1 \leq Z(K)$. Then similarly $K/Z(K)$ is trivial, $K = Z(K) = \mathrm{Fix}_K(\sigma)$, and our claim follows.

So now suppose

$$\mathrm{Fix}_K(\sigma) \text{ is finite.}$$

Consider the action of σ on $Z(K)$. This gives an involutory automorphism of the field $Z(K)$ with finite fixed field, so the field $Z(K)$ is also finite.

Let $x \in \mathrm{Fix}_K(\sigma)$, $x \neq 0$. Then $K_1x \cap K_1$ has finite index in K. Take $ax \in K_1x \cap K_1$. Then

$$-ax = (ax)^\sigma = xa^\sigma = -xa$$

and thus $a \in C(x)$. Hence $C(x)$ has finite index in K, and as $\mathrm{Fix}_K(\sigma)$ is finite, $C_K(\mathrm{Fix}_K(\sigma))$ has finite index in K. So $\mathrm{Fix}_K(\sigma) \leq Z(K)$.

Now K_1 is a vector space over $\mathrm{Fix}_K(\sigma)$. Consider the map

$$q : K_1 \to \mathrm{Fix}_K(\sigma)$$

given by $q(a) = a^2$. Then $q(xa) = x^2 q(a)$ for $x \in \mathrm{Fix}_K(\sigma)$ and $q(a+b) = q(a) + q(b) + (ab + ba)$. Since the function $ab + ba$ is bilinear on K_1 over $\mathrm{Fix}_K(\sigma)$, q is a quadratic form on K_1. As K_1 is infinite, its dimension over $\mathrm{Fix}_K(\sigma)$ is greater than two and therefore the quadratic form has a nontrivial zero, that is $a^2 = 0$ for some $a \in K_1$, $a \neq 0$, a contradiction. □

LEMMA 4.29. *Let (K, K_0, σ) be an involutory set, and let $\mathcal{K}$ be the structure consisting of K as an additive group together with K_0 as a distinguished subset, and the multiplication map restricted to $K_0 \times K$. If $\mathcal{K}$ has finite Morley rank and K is infinite, then K is an algebraically closed field and σ is trivial.*

PROOF. Let $K_1 = K_0{}^\circ$, which is nontrivial by the preceding lemma. Let R be the subring of K generated by K_1. As $R = RK_1$, if we consider R as an additive group, it is generated by the connected subgroups rK_1 for $r \in R$, and is therefore definable, and is the sum of finitely many of these subgroups: $R = \sum_i r_i K_1$. On the other hand, each multiplication map

$\mu_{r_i} : K \to K$ is definable in $\mathcal{K}$, and hence the action of R on K is definable. In particular, the ring structure on R itself is definable.

So R is a ring without zero divisors, and has finite Morley rank. It follows easily that the nontrivial multiplication maps are surjective, and thus R is a division ring, hence an algebraically closed field. Since the vector space structure on K is definable and K has finite Morley rank, the dimension of K over R is finite. Hence the dimension of K over its center Z is finite [**131**, 15.8], and the dimension of K over $Z \cap R$ is finite. It follows that the multiplication on K is definable, and hence K is also an algebraically closed field. As σ has order two, its fixed field cannot be finite, and hence σ is trivial (Lemma 4.5). □

4.6. Pure fields. We refer to a field F which carries no additional structure beyond the field operations as a *pure field.* In a pure algebraically closed field, the definable sets are the constructible sets in the sense of algebraic geometry.

FACT 4.30 ([**156**, Théorème 4.15]). *An infinite field definable in a pure algebraically closed field F is definably isomorphic to F.*

This is a delicate point. One uses the theory of algebraic groups, and the result that a group definable (or interpretable) in an algebraically closed field is itself algebraic, which is a relative of the Weil group chunk theorem, later much generalized by Hrushovski and Zilber. We give an indication of the second phase of the analysis.

Let the field K be interpreted in the algebraically closed field F, and suppose that the group $G = K_+ \rtimes K^\times$ (with the natural action of $K^\times$ on K_+) is known to be algebraic over F. By Corollary 3.7 this group is connected, and it is solvable. From the general theory of algebraic groups it follows that it is linear (being centerless) and may be identified with a Zariski closed subgroup of the upper triangular matrices, with K_+ unipotent.

One may then distinguish cases according to the characteristic of F.

If the characteristic is zero, we have Lemma 4.4. As K_+ is unipotent, this easily proves that K_+ is 1-dimensional and definably isomorphic with F_+, with which we identify it. The same Lemma 4.4 implies that K acts F-linearly on F_+ and hence $K^\times$ becomes identified with $F^\times$.

If the characteristic is positive, then evidently K_+, being unipotent, is also an elementary p-group. Furthermore, by inspection of the torsion in $K^\times$, the latter group is a 1-dimensional torus. Hence K_+ is also 1-dimensional. It then follows that K_+ can be identified (definably) with F_+, and it remains to understand $K^\times$ and its action on F_+, which requires some brief consideration of the multiplication maps induced by $K^\times$ as definable functions on F.

5. Nilpotent groups

One of the most useful results of the present section is the structural analysis of connected nilpotent groups of finite Morley rank, Proposition 5.8, which provides a satisfying generalization of the abelian case. Also of great importance is the existence and definability of the *Fitting subgroup*, Proposition 5.13. The other topics covered in the present section are more specialized.

5.1. The normalizer condition.

LEMMA 5.1. *Let H be a nilpotent group of finite Morley rank and P an infinite normal subgroup. Then $P \cap Z(H)$ is infinite.*

PROOF. Take i minimal so that $P \cap Z_i(H)$ is infinite. Then we may suppose that $P \leq Z_i(H)$.

We may also suppose that P is connected. For $x \in H$, the commutator map $\gamma_x : P \to Z_{i-1}(H) \cap P$ has finite image and has fibers of constant rank $C_P(x)$. Hence $\mathrm{rk}(C_P(x)) = \mathrm{rk}(P)$ and as P is irreducible by Lemma 3.6, we have $[x, P] = 1$. Thus $P \leq Z(H)$, and the claim follows. □

COROLLARY 5.2. *Let H be an infinite nilpotent group of finite Morley rank. Then $Z(H)$ is infinite.*

PROPOSITION 5.3. *Let H be a nilpotent group of finite Morley rank, and suppose P is a definable subgroup of infinite index. Then $[N_H(P) : P]$ is infinite.*

PROOF. We know that $Z^\circ(H) \neq 1$. If $Z^\circ(H)$ is not contained in P, then $[Z^\circ(H) : P \cap Z^\circ(H)]$ is infinite and hence $[Z^\circ(H)P : P] = \infty$, so the claim follows in this case.

If $Z^\circ(H) \leq P$, then we may pass to $H/Z^\circ(H)$, and conclude by induction on the rank. □

5.2. Structure.

DEFINITION 5.4.

(1) *Let G be a group of finite Morley rank. Then G is* unipotent *if G is connected, has bounded exponent, and is solvable. It is* π-unipotent *if it is a unipotent π-group.*

(2) *$U_2(G)$ is the subgroup of G generated by the unipotent 2-subgroups of G.*

LEMMA 5.5. *Let G be a unipotent group of finite Morley rank. Then G is nilpotent.*

PROOF. Otherwise, by Proposition 4.12, there is an infinite section T of G isomorphic to a subgroup of the multiplicative group of a field. As G is unipotent, T° has bounded exponent, and is therefore trivial, a contradiction. □

LEMMA 5.6. *Let H be a connected nilpotent group of finite Morley rank. Then the series $Z_i^\circ(H)$ is a central series.*

PROOF. We know that $Z^\circ(H)$ is nontrivial, and by iterating this it follows that the series terminates at H after at most $\mathrm{rk}(H)$ steps.

Furthermore, $[G, Z_i^\circ(G)] \leq Z_{i-1}(G)$ is connected by Corollary 3.29. Thus $[G, Z_i^\circ(G)] \leq Z_{i-1}{}^\circ(G)$. □

LEMMA 5.7. *Let H be a connected nilpotent group of finite Morley rank. Suppose that H contains no nontrivial normal unipotent subgroup. Then H is radicable.*

PROOF. It suffices to show that each section $A_i =: Z_i^\circ(H)/Z_{i-1}{}^\circ(H)$ is divisible (Lemma 1.20). By Lemma 2.13 it suffices to show that A_i contains no nontrivial unipotent subgroup B.

Suppose toward a contradiction that i is minimal such that A_i contains a nontrivial unipotent subgroup B. Let $\hat{B}$ be the preimage of B in $Z_i^\circ(H)$. For $g \in G$, commutation induces a homomorphism $\gamma_g : \hat{B} \to A_{i-1}$ which factors through $\hat{B} \to B$. Thus the image, a homomorphic image of B, is unipotent, Thus the image of the subgroup $[G, \hat{B}]$ in the section A_{i-1} is unipotent, hence trivial. This means that $[G, \hat{B}] \leq Z_{i-2}{}^\circ(H)$ and thus $\hat{B} \leq Z_{i-1}{}^\circ(H)$, a contradiction. □

PROPOSITION 5.8. *Let H be a connected nilpotent group of finite Morley rank. Then there are definable normal subgroups U, T of H such that:*

(1) *U is unipotent and T is radicable;*
(2) *$H = U * T$, a central product;*
(3) *$U \cap T$ is finite.*

PROOF. Note that the second and third points are actually consequences of the first. The third point follows by Lemma 1.18, while for the second point we note that $[U, T] \leq U \cap T$ is connected and finite, hence trivial.

Now choose U maximal definable unipotent and normal in H. By Lemma 5.7, H/U is radicable.

If $U = 1$, we are done. Otherwise, we proceed by induction on $\mathrm{rk}(H)$. Let $A \leq U$ be a minimal infinite definable normal subgroup of H; then A is connected. In $\bar{H} = H/A$ we have $\bar{H} = \bar{U} * \bar{D}$ with D/A radicable. We may suppose $H = D$ and $D > A$.

Now $\bar{D}' < \bar{D}$ so $D'A \neq D$. As $D'A$ is also connected, we have inductively $D' = A * D_1$ with D_1 radicable.

If $D_1 > 1$, then in $\tilde{D} = D/D_1$ we have $\tilde{D} = \tilde{A} * \tilde{T}$ with T/D_1 radicable, and hence T radicable. Hence $D = AT$ with A unipotent, T radicable, both normal, connected, and definable, and the same follows for H.

Now suppose $D_1 = 1$, that is, $D' \leq A$. Then for $x \in D$, the map $\gamma_x : D \to A$ induced by commutation is a homomorphism, whose image is radicable and of bounded exponent, hence trivial. This means that D is abelian and our claim follows from Lemma 2.13. □

REMARK 5.9. *In the preceding proposition, the subgroups U and T are characteristic.*

PROOF. If n is the exponent of U, then $T = \{h^n : n \in H\}$, and U is the connected component of the subgroup generated by $\{h \in H : h^n = 1\}$. □

We proceed further with the structure of radicable nilpotent groups.

LEMMA 5.10. *Let H be a π-radicable nilpotent group of finite Morley rank. Then H' is π-torsion-free.*

PROOF. Suppose this fails, and let i be minimal such that $[Z_i(H), H]$ contains torsion. Set $H_i = [Z_i(H), H]$. Then H_{i-1} is π-torsion-free. Let

$$h = [x_1, h_1] \cdot \dots \cdot [x_n, h_n]$$

be a π-torsion element with $x_1, \dots, x_n \in Z_i(H)$.

The maps $\gamma_x : Z_i(H)/Z_{i-1}(H) \to H_i/H_{i-1}$ induced by commutation are homomorphisms, and so is their sum

$$\gamma = \sum_i \gamma_{x_i} .$$

As the domain of γ is π-torsion-free (Lemma 1.18), so is the image, by Lemma 2.18. But h is in the image, a contradiction. □

PROPOSITION 5.11. *Let H be a radicable nilpotent group of finite Morley rank. Then H factors as a direct product*

$$H_{\text{tor}} \times H_0$$

with H_0 radicable and torsion-free (and not usually definable).

PROOF. Let $\bar{H} = H/H'$. This factors as $\bar{H}_{\text{tor}} \times \bar{H}_0$ where the group H_0 contains H'. As H' is torsion free and $\bar{H}_0$ is torsion free, we find that H_0 is torsion free. Furthermore, $H = H_{\text{tor}}H_0$ with both factors normal, and with trivial intersection, so we have the desired representation. □

LEMMA 5.12. *Let G be a group of finite Morley rank, and H a definable connected nilpotent subgroup. Then the p-torsion subgroup H_p of H is connected.*

PROOF. $H = U * T$ with T radicable, U is connected of bounded exponent, and both are definable. The p-torsion of U is connected by Lemma 1.17, and the p-torsion of T is divisible. □

5.3. Fitting subgroup.

PROPOSITION 5.13. *Let G be a group of finite Morley rank. Then there is a unique maximal normal nilpotent subgroup of G, and it is definable.*

PROOF. If H, K are definable connected normal nilpotent subgroups of G, then HK is again definable connected normal, and nilpotent (Lemma 1.21). So granted the existence, uniqueness is clear. Again, granted the existence, the definability is immediate by Lemma 2.15.

So the problem is one of existence. We proceed by induction on the rank of G.

The problem reduces at once to the case of G connected, since if H is the maximal normal nilpotent subgroup of G°, then H is normal in G, and for all normal nilpotent subgroups $\hat{H}$ of G containing H, the index $[\hat{H} : H]$ is bounded by $[G : G^\circ]$.

Again, if $Z(G)$ is infinite, then by induction $G/Z(G)$ has a maximal nilpotent normal subgroup, which then lifts to the desired subgroup of G.

If $Z(G)$ is finite and nontrivial, we may first treat $G/Z(G)$, which is centerless by Lemma 3.9, then return to G. So we will suppose

$$Z(G) = 1 \ .$$

Now let A be the centralizer in G of all nilpotent normal subgroups of G. Then A is normal in G, and definable by the definable DCC, Lemma 2.6. Now by induction $C(A)$ has a unique maximal normal nilpotent subgroup H, and H contains all nilpotent normal subgroups of G. On the other hand, H is nilpotent and normal in G, so H is the desired subgroup. □

DEFINITION 5.14. *Let G be a group of finite Morley rank. The maximal nilpotent normal subgroup of G is called the* Fitting subgroup, *and is denoted* $F(G)$.

We note that $F^\circ(G)$ is the largest definable connected nilpotent normal subgroup of G, or, in other words, is the *connected analogue* of the Fitting subgroup from the finite case. As we saw in the argument above, the construction of the connected analogue can be less problematic than the construction of the strict analogue; in some cases only the connected analogue has been given a sense.

The Fitting subgroup is of particular use in the study of solvable groups (§8), and also via its extension to the *generalized Fitting subgroup* F^* (§7).

LEMMA 5.15. *Let* G *be a group of finite Morley rank, and* H *a normal subgroup. Then* $F(H) = F(G) \cap H$.

PROOF. As $F(H)$ is characteristic in H, it is normal in G. □

5.4. Stabilizers of chains.

LEMMA 5.16. *Let* G *be a group of finite Morley rank,* P *a nilpotent* p*-group contained in* G *but not necessarily definable, and* H *a definable group acting faithfully on* P*; this action is assumed to extend definably to* $d(P)$*. Suppose there is a series* $P = P_0 > P_1 > \cdots > P_n = 1$ *of* H*-invariant subgroups with* $P_{i+1} \lhd P_i$*, so that* H *acts trivially on each quotient* P_i/P_{i+1}*. Then* H *is a* p*-group.*

PROOF. Let $h \in H$ and choose q a power of p to maximize $C_P(h^q)$, using Lemma 3.13. Replace h by h^q, then $C_P(h) = C_P(h^{p^n})$ for all n. We claim $h = 1$. Supposing the contrary, take i maximal so that h acts nontrivially on P_i, and take $x \in P_i$ so that $[h, x] \neq 1$. As $[h, x] \in P_{i+1}$, h commutes with $[h,x]$. So $[h^{p^n},x]=[h,x]^{p^n}=1$ for n large, and hence $[h,x]=1$, a contradiction. □

5.5. The Frattini subgroups.

DEFINITION 5.17. *Let G be a group of finite Morley rank. The* connected Frattini subgroup *of G, denoted $\Phi(G)$, is the intersection of all maximal connected proper definable subgroups of G°.*

We do not claim that the "connected Frattini" subgroup is itself connected. Still it is a straightforward analog of the ordinary Frattini subgroup relativized to connected subgroups. One could also replace this group by its connected component, and perhaps this is the better notion. But it behaves quite well with the definition we are using.

LEMMA 5.18. *Let G be a group of finite Morley rank, and $H \triangleleft G$ a subgroup of $\Phi(G)$. Then $\Phi(G/H)=\Phi(H)/H$.*

PROOF. It is easy to see that the maximal connected subgroups of G and of G/H correspond. □

LEMMA 5.19. *Let G be a connected group of finite Morley rank, and H a definable subgroup such that $H\Phi(G)=G$. Then $H=G$.*

PROOF. Note that $[H\Phi(G):H^\circ\Phi(G)]$ is finite, and hence $H^\circ\Phi(G)=G$. If $H<G$, then let $M\le G$ be a maximal definable connected subgroup of G containing H°. Then by definition $\Phi(G)\le M$, and $G=H^\circ\Phi(G)\le M$, a contradiction. □

LEMMA 5.20. *Let G be a unipotent p-group of finite Morley rank, and G_p the set of p-th powers of elements of G. Then $\langle G'\cup G_p\rangle\le\Phi(G)$.*

PROOF. Let M be a maximal definable connected subgroup of G. By nilpotence, $M\triangleleft G$, and by maximality G/M has no proper infinite definable subgroups, hence is abelian. $G'\le M$. As G/M is a minimal infinite abelian p-group of finite exponent, it has exponent p, that is, $G_p\subseteq M$. So $\langle G'\cup G_p\rangle\le\Phi(G)$. □

One expects $\Phi(G)=\langle G'\cup G_p\rangle$ but this is not necessarily correct. In any case one may pass to the quotient $\bar G=G/\langle G'\cup G_p\rangle$, which is an elementary abelian p-group. As it is not clear what the pattern of definable subgroups will be, we cannot say anything more. In some cases one might prefer to substitute the group $\langle G'\cup G_p\rangle$ for the Frattini subgroup: it is a definable subgroup which gives the largest possible elementary abelian quotient.

We take note of the "ordinary" Frattini subgroup, which also has its uses.

DEFINITION 5.21. *Let P be a p-group. Then the* ordinary Frattini subgroup *of P, denoted $\phi(P)$, is the subgroup generated by P' and $\{x^p : x\in P\}$.*

LEMMA 5.22. *Let P be a nilpotent p-group of finite Morley rank. Then $\phi(P)$ is definable, and is the intersection of the maximal subgroups of P.*

PROOF. Let $\bar{P} = P/P'$. Then $\phi(P)$ is the preimage in P of $p\bar{P}$, so is definable.

Let $\phi_0(P)$ be the intersection of its maximal subgroups (which would be taken to be P if there are none). Now $\hat{P} = P/\phi(P)$ is an elementary abelian p-group, and the intersection of the maximal abelian subgroups of $\hat{P}$ is trivial, thus $\phi_0(P) \leq \phi(P)$.

Let Q be a maximal subgroup of P. By the normalizer condition in nilpotent groups, Q is normal in P, and P/Q is a simple nilpotent group, hence cyclic of prime order. Thus $\phi(P) \leq Q$, and $\phi(P) \leq \phi_0(P)$. □

5.6. p-Tori.

DEFINITION 5.23. *A p-torus is a divisible abelian p-group.*

LEMMA 5.24. *Let G be a group of finite Morley rank. Then there is a constant w such that for any prime p and any p-torus T we have $[N_G(T) : C_G(T)] \leq w$.*

PROOF. Let us write $\mathrm{Inn}\, T$ for $N_G(T)/C_G(T)$, thought of as the induced group of G-inner automorphisms.

First, since $T[p^n]$ is finite for all n, $N_G{}^\circ(T)$ centralizes T. Hence $\mathrm{Inn}(T)$ is finite.

Now we need to relate these finite sets to a family of uniformly definable finite sets, to get the uniform bound.

For T a p-torus, let $\hat{T} = Z(C_G(T))$. Let S be the maximal p-torus in $\hat{T}$. Then $N_G(\hat{T})$ acts on S, and as $C_G(S) \leq C_G(T) \leq C_G(\hat{T}) \leq C_G(S)$, the quotient $\mathrm{Inn}(\hat{T}) = N_G(\hat{T})/C_G(\hat{T})$ embeds in $\mathrm{Inn}(S) = N_G(S)/C_G(S)$, which is finite as seen at the outset.

On the other hand, as centralizers are uniformly definable (Corollary 2.9) the groups $\hat{T}$ and hence also $\mathrm{Inn}(\hat{T})$ are uniformly definable, and hence of bounded order.

Finally, $\mathrm{Inn}(T)$ embeds into $\mathrm{Inn}(\hat{T})$ naturally, and this yields a uniform bound on the size of $\mathrm{Inn}(T)$. □

COROLLARY 5.25. *Let G be a group of finite Morley rank, T a p-torus contained in G, and H a connected subgroup of G which normalizes T. Then H centralizes T.*

5.7. Locally finite p-groups.

LEMMA 5.26. *Let H be a group of finite Morley rank, and $P \leq H$ an infinite locally finite connected p-subgroup for some prime p. Then $Z(P)$ is infinite.*

PROOF. We have $Z(P) = C_P(P_0)$ for some finite subgroup of P by Lemma 2.8. In particular, $Z(P_0) \leq Z(P)$, so the center is at least nontrivial.

Suppose it is finite; then working modulo the center we have $Z_2(P) > Z(P)$, contradicting Lemma 3.9. □

LEMMA 5.27. *Let G be a group of finite Morley rank, and P a locally finite p-subgroup of G. Then P is nilpotent by finite.*

PROOF. We may suppose that P is connected, in which case we need to prove that P is nilpotent. By the previous lemma, if P is nontrivial, then $Z(P)$ is infinite, and we may work inductively, replacing G by $C(Z(P))/Z(P)$. □

PROPOSITION 5.28. *Let G be a group of finite Morley rank, and P a p-subgroup. Then the following are equivalent.*

(1) *P is locally finite.*
(2) *P is nilpotent by finite.*
(3) *P° is nilpotent.*

5.8. Generic equations.

DEFINITION 5.29. *Let G be a group. A* commutator term *over G is a term composed by iterating the commutator operation on variables x_i, their inverses x_i^{-1}, and constants $c \in G$. A commutator term is* proper *if it involves at least one variable.*

Note that the inverse operation is applied exclusively to variables. This is convenient, but not limiting, in view of the identity $[a,b]^{-1} = [b,a]$.

LEMMA 5.30. *Let G be a group, and γ a commutator term over G involving the variables $x_1, \ldots, x_n$ explicitly. Then there is an identity of the form*

$$\gamma(xy, x_2, \ldots, x_n) = \gamma(x, x_2, \ldots, x_n)\gamma(y, x_2, \ldots, x_n)\Gamma$$

where Γ is a product of commutator terms over G involving all $n+1$ variables $x, y, x_2, \ldots, x_n$ explicitly.

PROOF. We note at the outset that it will suffice to expand the term

$$\gamma(xy, x_2, \ldots, x_n)$$

as a product in which the terms $\gamma(x, x_2, \ldots, x_n)$ and $\gamma(y, x_2, \ldots, x_n)$ are present, and all other terms are commutators involving all the variables $x, y, x_2, \ldots, x_n$, as then the desired rearrangement simply involves adding some additional commutators of the same general type.

If γ is a constant or a variable, then Γ is the empty product. If γ has the form x^{-1}, then Γ is a commutator. For the rest, we proceed inductively, taking $\gamma = [\gamma_1, \gamma_2]$. Here the variable x_1 may appear in one or both of the terms γ_i, and we will consider the case in which it occurs in both, leaving consideration of the other similar and simpler case to the reader. Writing $\bar{x} = (x, x_2, \ldots, x_n)$, $\bar{y} = (y, x_2, \ldots, x_n)$, and $\bar{z} = (xy, x_2, \ldots, x_n)$, we suppose inductively $\gamma_i(\bar{z}) = \gamma_i(\bar{x})\gamma_i(\bar{y})\Gamma_i(\bar{z})$ with Γ_i a product of commutator terms, each of which involves all variables $x, y, x_2, \ldots, x_n$.

Now

$$\gamma(\bar{z}) = [\gamma_1(\bar{x})\gamma_1(\bar{y})\Gamma_1, \gamma_2(\bar{x})\gamma_2(\bar{y})\Gamma_2] \ .$$

We may expand the right side using a variant of Lemma 1.2, namely

$$[ab, g] = [a, g] \cdot [[a, g], b] \cdot [b, g] \ .$$

A little thought shows that the result involves the two desired terms

$$[\gamma_1(\bar{x}), \gamma_2(\bar{x})] = \gamma(\bar{x}), \quad [\gamma_1(\bar{y}), \gamma_2(\bar{y})] = \gamma(\bar{y}),$$

a number of higher order commutators involving terms from Γ_1 or Γ_2, automatically of the desired form, and two "cross terms" $[\gamma_1(\bar{x}), \gamma_2(\bar{y})]$ and $[\gamma_1(\bar{y}), \gamma_2(\bar{x})]$, which we notice also involve all the variables. □

LEMMA 5.31 ([**120**]). *Let G be a group of finite Morley rank, H a definable normal nilpotent subgroup of G, and w a word which can be expressed as a product of proper commutator terms, each defined over G. If w is generically constant on H, then $w = 1$ generically on H.*

PROOF. The assumption on w is that there is at least one value c such that as the variables $x_1, \ldots, x_n$ run over H, the set of values $\bar{x}$ for which $w(\bar{x}) = c$ has rank $n\,\mathrm{rk}(H)$, in other words, forms a set of full rank in H^n (Cartesian power). It then follows that there are n cosets of H° in H such that w is generically constant on the Cartesian product of these cosets. We do not insist that this take place on H° itself.

Let k be the minimum number of variables occurring in a factor γ of w, and let the variables of w be $x_1, \ldots, x_n$, where x_1 is present in γ. Applying the previous lemma we arrive at an identity of the form

$$w(xy, x_2, \ldots, x_n) = w(x, x_2, \ldots, x_n)w(y, x_2, \ldots, x_n)W(x, y, x_2, \ldots, x_n)$$

where W is a product of commutator terms, each involving at least $k+1$ variables.

Now take elements $g, h, g_2, \ldots, g_n$ in H lying in the appropriate cosets of H° which are generic and independent over the base set B consisting of the parameters of w together with the element $c \in G$ which gives a generic value of w on H. In other words, each of the elements chosen should be generic in its coset over all the others, together with the base set B. We then have $w(gh, g_2, \ldots, g_n) = w(g, g_2, \ldots, g_n) = w(h, g_2, \ldots, g_n) = c$, and thus

$$W(g, h, g_2, \ldots, g_n) = c^{-1} \ .$$

This produces a new generic identity for H in which the minimum number of variables occurring in a factor is at least $k+1$. Continuing inductively, we arrive eventually at a generic equation for H of the form

$$\hat{w}(\bar{x}) = c^{\pm 1}$$

in which $\hat{w}$ is a product of commutator terms each of which involve at least $n+1$ variables, with n the nilpotency class of H. But since H is normal in G and nilpotent, the factors of $\hat{w}$ vanish identically on H, and thus $c = 1$. □

PROPOSITION 5.32. *Let $w = 1$ be an equation holding generically on a connected nilpotent group G of finite Morley rank. Then $w = 1$ identically on G.*

PROOF. We wish to write w as far as possible as a product of proper commutator terms. For this, we move the constant parameters at the cost of some commutator terms; thus $xcy = 1$ may be written in the form $cx[x, c]y = 1$ or $x\gamma y = c^{-1}$, and the general form is

$$w = c$$

where w is now a product of proper commutator terms and c is a constant. By Lemma 5.31, the parameter c equals 1.

Now take $a_1, \ldots, a_n \in G$ arbitrary, and take $x_1, \ldots, x_n$ in G generic and independent over the a_i. We have the equation

$$w(a_1x_1, \ldots, a_nx_n) = w(\bar{a})w(\bar{x})W(\bar{x})$$

where W depends on both $\bar{a}$ and $\bar{x}$, but may be thought of as a product of commutator terms in which the a_i occur as parameters. This equation reduces to

$$1 = w(\bar{a})W(\bar{x})$$

and thus W is itself generically constant, with value $w(\bar{a})^{-1}$. So again Lemma 5.31, $w(\bar{a}) = 1$. □

6. Sylow theory

We do not have a good Sylow theory for arbitrary primes, and certainly not in "characteristic 0" (what would be wanted in that case is a theory of unipotence). However, things will work out well for the prime 2, and this leads to a distinction between "characteristic two" and all other cases, which will be subdivided into a classification into four "types" which is fundamental for the present work.

6.1. 2-Subgroups. The theory of Sylow subgroups develops along lines which are a mixture of the finite and algebraic cases. The theory only works well for the prime 2, in general—in the case of solvable groups it works well for all primes, and is subsumed under the Hall theory; see §8. It is not quite true that the Sylow 2-subgroups are nilpotent, but nearly so; their connected components are nilpotent, and while they are not necessarily definable, they satisfy much the same structure theory as definable nilpotent groups—by inheritance from the definable closure.

The proof of conjugacy of Sylow 2-subgroups is arduous. Once it is in hand, it has the same sort of immediate applications as are familiar in the finite case.

The theory of Sylow subgroups also allows us to recognize the "characteristic" of our group, or more properly to distinguish the case of characteristic two from all other characteristics, directly from the structure of the connected component of a Sylow subgroup. This is a very convenient organizing principle, and indeed the bulk of the present work will be aimed at proving the algebraicity of simple groups in two of the four resulting "types"—in addition to the two natural types, namely "characteristic two" or "even" type, and "characteristic not two" or "odd" type, we must also consider two more formal possibilities: "both" ("mixed") and "neither" ("degenerate").

DEFINITION 6.1. *Let G be an arbitrary group, p a prime.*

(1) *A* Sylow p-subgroup *is a maximal p-subgroup of G.*
(2) *A* Sylow° p-subgroup *is the connected component of a Sylow p-subgroup (not assumed definable).*

PROPOSITION 6.2 ([**52**, Theorem 6.21]). *Let G be a group of finite Morley rank, and P a 2-subgroup. Then P is nilpotent by finite, and is, in particular, locally finite.*

PROOF. We proceed by induction on the rank and degree of G. In particular, we may suppose that $G = d(P)$, and that G is connected. Then $Z(G) \cap P = Z(P)$, and we may factor out $Z(G)$, concluding by induction unless $Z(G)$ is finite.

By Lemma 3.9 we then have $G/Z(G)$ centerless, and since we may replace G by $G/Z(G)$ we may now assume that:

$$Z(G) = 1 .$$

In particular, for $x \in P^{\times}$, we have $C_G(x) < G$, and hence $C_P(x)$ is nilpotent by finite.

The family of subgroups of the form $C_P(a)$ or $C_P(a, b)$ are uniformly relatively definable, so we can apply the corresponding uniform chain condition of Lemma 3.13. Thus the set $\{C_P(a) : a \in P\}$ has maximal elements, and among all intersections of the form $C(a) \cap C(b)$ with $C(a)$, $C(b)$ maximal, $a, b \in P$, and $C(a) \neq C(b)$, there are maximal ones. So let $D = A \cap B$ be such a maximal intersection with $A = C(a)$, $B = C(b)$. As seen above, A and B are nilpotent by finite.

Let $H = N_G(D)$. Then $H \cap A, H \cap B > D$, by the normalizer condition, Lemma 1.22. Hence in $\bar{H} = H/D$ there are elements $\bar{a} \in \bar{A}$, $\bar{b} \in \bar{B}$ of order 2. Then $\langle \bar{a}, \bar{b} \rangle$ is a dihedral group contained in $\bar{P}$, hence in particular a finite 2-group, and $\langle a, b \rangle$ is therefore a nilpotent by finite 2-group. Hence $Z(\langle a, b \rangle) \neq 1$ (Lemma 1.22). Let $c \in Z(\langle a, b \rangle)^{\times}$, and let $C \leq P$ be a maximal centralizer of the form $C_P(c')$ containing $C_P(c)$. Then $A \cap C, B \cap C > D$, so by maximality $A = C = B$, a contradiction. □

6.2. Structure and conjugacy of 2-Sylow subgroups.

PROPOSITION 6.3. *Let G be a group of finite Morley rank, and let S be a p-Sylow° subgroup. Suppose S is nilpotent. Then S has the form*

$$U * T$$

with U unipotent and T a p-torus.

PROOF. The group $d(S)$ is definable, nilpotent, and connected. Let S_1 be a Sylow p-subgroup of $d(S)$ containing S. Then $S = S_1{}^\circ$. As $d(S)$ is nilpotent, $S_1 = d(S)_p$ (Lemma 1.17). In particular, $S \lhd d(S)$.

Thus $d(S) = \hat{U} * \hat{T}$ with $\hat{U}$ unipotent and $\hat{T}$ radicable (Proposition 5.8). Now $S \le S\hat{T} = \hat{T} * (S \cap \hat{U}) \le \hat{T} * \hat{U}_p$, and the latter is definable, so $d(S) = \hat{U}_p * \hat{T}$, hence $d(S) = U * \hat{T}$ with $U = \hat{U}_p{}^\circ$. Thus $S = U * T$ with $T = (S \cap \hat{T})^\circ$.

Here U is unipotent, and T is a connected p-subgroup of a radicable nilpotent group, hence contains only finitely much torsion of each order p^n, and is in particular abelian. It follows then from the connectivity that T is itself radicable. □

PROPOSITION 6.4. *Let G be a group of finite Morley rank, and S a 2-Sylow° subgroup. Then S has the form*

$$U * T$$

with U unipotent and T a 2-torus.

PROOF. By Proposition 6.2 and Lemma 3.16, S is nilpotent, and the preceding result applies. □

This leads to a fundamental division of the class of groups of finite Morley rank into four "types", as follows.

DEFINITION 6.5. *Let G be a group of finite Morley rank, and S a Sylow° 2-subgroup. Then G is:*

(1) *of* degenerate type *if $S^\circ = 1$ (Sylow 2-subgroups are finite);*
(2) *of* even type *if S° is nontrivial and 2-unipotent;*
(3) *of* odd type *if S° is a nontrivial 2-torus;*
(4) *of* mixed type *in the remaining case: $S = U * T$ with U unipotent, T a 2-torus, and both nontrivial.*

In Chevalley groups, only even and odd types occur, and the distinction corresponds to the characteristic of the base field being 2, or not; in this case lumping characteristic 0 under the "odd" case is an abuse of language. Mixed type groups can most easily be constructed as products of the two sorts. Nonsolvable connected groups of finite Morley rank of degenerate type are not known, and there is no consensus as to whether they should exist.

LEMMA 6.6. *Let G be a group of finite Morley rank whose Sylow 2-subgroups are finite. Then the Sylow 2-subgroups of G are conjugate.*

PROOF. Suppose P, Q are nonconjugate Sylow 2-subgroups of G, chosen so as to the maximize $|P \cap Q|$. Let $D = P \cap Q$, and $H = N_G(D)$. Then $H \cap P, H \cap Q > D$. Let $\bar{H} = H/D$ and take involutions $\bar{\imath}, \bar{\jmath}$ in $\overline{P \cap H}$, $\overline{Q \cap H}$.

If $\bar{\imath}, \bar{\jmath}$ are conjugate in $\bar{H}$, then there is an element $x \in H$ such that $j \in P^x \cap Q$. Hence $|P^x \cap Q| > |D|$ and by maximality P^x and Q are conjugate, hence P and Q are conjugate.

Suppose therefore that $\bar{i}, \bar{j}$ are nonconjugate in $\bar{H}$. Then they commute with an involution $\bar{k}$. Let R_1 be a Sylow 2-subgroup containing $\langle D, i, k \rangle$, and let R_2 be a Sylow 2-subgroup containing $\langle D, j, k \rangle$.

Then $P \cap R_1 > D$, $Q \cap R_2 > D$, and $R_1 \cap R_2 > D$, so by maximality P and R_1, Q and R_2, and R_1 and R_2 are conjugate, hence P and Q are conjugate as well. □

We now work toward the conjugacy of Sylow 2-groups in general, along lines similar to the preceding.

DEFINITION 6.7. *Let G be a group, p a prime. Then $O_p(G)$ denotes the maximal normal p-subgroup of G.*

The group $O_p(G)$ is typically not definable. We are interested in the condition $O_p(G) = 1$, meaning that there are no normal p-subgroups of G.

LEMMA 6.8. *Let G be a group of finite Morley rank, and suppose that G contains two nonconjugate Sylow 2-subgroups. Then there is a definable quotient $\bar{G}$ of G, also containing two nonconjugate Sylow 2-subgroups, with $O_2(\bar{G}) = 1$.*

PROOF. Let $O_2{}^\circ(G) = U * T$ with U 2-unipotent and T divisible. We may factor out the definable subgroup U, and suppose therefore that $O_2{}^\circ(G) = T$. Then every Sylow 2-subgroup of G contains, and therefore centralizes, T. Thus we may replace G by $C_G(T)$, and assume now that $T \leq Z(G)$, so $d(T) \leq Z(G)$. We factor out $d(T)$ and assume that our result applies in the quotient, so that any two Sylow subgroups P and Q may be assumed, after conjugation, to satisfy $Pd(T) = Qd(T)$. But then $P^\circ d(T) = Q^\circ d(T)$ is a nilpotent group, so $P^\circ = Q^\circ$ and after factoring out this group P and Q become finite, a case already treated. □

LEMMA 6.9. *Let G be group of finite Morley rank. Suppose that G contains no nontrivial 2-torus. Then all Sylow 2-subgroups of G are conjugate.*

PROOF. We proceed by induction on the rank and degree of G. By Lemma 6.8 we may suppose that $O_2(G) = 1$.

We show first that all maximal 2-unipotent subgroups of G are conjugate. Suppose toward a contradiction that U, V are maximal 2-unipotent subgroups of G, not conjugate, and chosen to maximize $\mathrm{rk}(U \cap V)$. Let $H = N_G(D)$. As $O_2(G) = 1$, we have $H < G$. By the normalizer condition of Proposition 5.3, we have $N_U{}^\circ(D), N_V{}^\circ(D) > D$. Let U_1, V_1 be maximal 2-unipotent subgroups of H containing $N_U{}^\circ(D)$ and $N_V{}^\circ(D)$ respectively. By induction, U_1 and V_1 are conjugate in H. Let U_2, V_2 be maximal 2-unipotent subgroups of G containing U_1 and V_1 respectively.

Then $\mathrm{rk}(U \cap U_2), \mathrm{rk}(V \cap V_2) > \mathrm{rk}(D)$, so U is conjugate to U_2 and V to V_2. As U_1 is conjugate to V_1 in H, there is a conjugate U_2^x of U_2 with $\mathrm{rk}(U_2^x \cap V_2) > \mathrm{rk}(D)$, and hence U_2^x, V_2 are also conjugate. So U and V are conjugate, a contradiction. □

LEMMA 6.10. *Let G be a group of finite Morley rank with $O_2(G) = 1$. Suppose P, Q are two 2-Sylow subgroups of G which are not conjugate. Then there are nonconjugate 2-Sylow subgroups P^*, Q^* of G such that $P^* \cap Q^* > P \cap Q$, and P, P^* are conjugate.*

PROOF. We proceed by induction on the rank and degree of G.

We deal first with the case $P \cap Q = 1$. Take $i \in P$, $j \in Q$ involutions. If they are conjugate, replace Q by a conjugate Q^* containing i and let $P^* = P$. If they are not conjugate, then by Lemma 2.20, there is an involution k commuting with both. Let R_i be a Sylow 2-subgroup containing i and k, and R_j a Sylow 2-subgroup containing j and k. If P is not conjugate to R_i, take $P^* = P$ and $Q^* = R_i$. If P is conjugate to R_i, and not conjugate to R_j, take $P^* = R_i$ and $Q^* = R_j$. Otherwise, take $P^* = R_j$ and $Q^* = Q$.

Now suppose $D = P \cap Q > 1$, $H = N_G(D)$, $\bar{H} = H/D$. Note that $H < G$ as $O_2(G) = 1$. Let $P_1 = P \cap H$, $Q_1 = Q \cap H$. By the normalizer condition (Lemma 1.22), $P_1, Q_1 > D$.

Let $i \in P \cap H$, $j \in Q \cap H$ represent involutions $\bar{i}, \bar{j}$ in $\bar{H}$. If $\bar{i}$ and $\bar{j}$ are conjugate in H, say $\bar{i}^x = \bar{j}$, then $P^x \cap Q > D$ and we take $P^* = P$, $Q^* = Q$. Assume therefore that $\bar{i}$ and $\bar{j}$ are nonconjugate.

By Lemma 2.20 there is an involution $\bar{k} \in \bar{H}$ commuting with both $\bar{i}$ and $\bar{j}$, represented in H by a 2-element k. Extend $\langle i, k, D\rangle$ and $\langle j, k, D\rangle$ to 2-Sylow subgroups R_i, R_j of G. We have $P \cap R_i$, $R_i \cap R_j$, and $R_j \cap Q$ all larger than D.

If P and R_i are nonconjugate, then take $P^* = P$, $Q^* = R_i$; if P and R_i are conjugate but R_i and R_j are not, then take $P^* = R_i$, $Q^* = R_j$, and otherwise take $P^* = R_j$, $Q^* = Q$. □

PROPOSITION 6.11. *Let G be a group of finite Morley rank. Then any two Sylow 2-subgroups of G are conjugate.*

PROOF. We proceed by induction on the rank and degree of G. By Lemma 6.8 we may suppose that $O_2(G) = 1$. By Lemma 6.9 we may suppose that G contains a nontrivial 2-torus.

Let P be a 2-Sylow subgroup of G containing a nontrivial 2-torus. We write $P^\circ = U_P * T_P$, the usual decomposition with U_P 2-unipotent and T_P a 2-torus. Let w be the bound afforded by Lemma 5.24, so that $[N_G(T_P) : C_G(T_P)] \le w$ for any such choice of P.

Let Q be a Sylow 2-subgroup assumed nonconjugate to P. By repeated application of Lemma 6.10, assume that $|P \cap Q| > w^2$ (the intersection may well be infinite). Then $|P \cap Q \cap C_G(T_P)| > w$.

If $T_Q \neq 1$, then $P \cap Q \cap C_G(T_P) \cap C_G(T_Q) \neq 1$, so we take $i \in P \cap Q$ centralizing both tori.

Then $H_i = C_G(i) < G$ since $O_2(G) = 1$. Applying the conjugacy of Sylow 2-subgroups in H_i, we may suppose that T_P and T_Q lie in a 2-group, hence commute. Now work in $H_P = C_G(T_P)$, which is definable by Corollary 2.9. Then $P^\circ \le H_P$, so P° is a Sylow° 2-subgroup of H_P, and by the conjugacy of Sylow 2-subgroups in H_P, T_Q can be conjugated into T_P in

H_P. So we may take $T_P \geq T_Q$, and arguing symmetrically we may take $T_P = T_Q$. Working in $N_G(T_P)$ (a finite extension of $C_G(T_P)$) we see that P and Q are conjugate, a contradiction.

Now suppose alternatively that Q° is unipotent. Take i an involution in $P \cap Q \cap C_G(T_P)$. Let $H_i = C_G(i)$. Note that $Z(Q) \leq H_i$. Applying the conjugacy of Sylow 2-subgroups in H_i, we may suppose that $Z(Q)$ commutes with T. Then working in $N_G(Z(Q))$ we see that Q can be conjugated to commute with T, which violates the assumption that Q is a Sylow 2-subgroup. □

This gives us the usual *Frattini argument*.

LEMMA 6.12 (Frattini). *Let G be a group of finite Morley rank, H a definable normal subgroup, and S a Sylow 2-subgroup of H. Then $G = H \cdot N_G(S)$.*

PROOF. For $g \in G$ we have $S^g = S^h$ with $h \in H$, hence $gh^{-1} \in N(S)$. □

6.3. Formal properties.

LEMMA 6.13. *Let G be a group of finite Morley rank. Then:*

(1) *The Sylow° 2-subgroups of G are conjugate.*
(2) *The maximal unipotent 2-subgroups of G are conjugate.*
(3) *The maximal 2-tori of G are conjugate.*

PROOF. Proposition 6.11. □

LEMMA 6.14. *Let G be a group of finite Morley rank, and P a Sylow 2-subgroup. Then $N_G(P^\circ)$ is definable.*

PROOF. Let $P^\circ = U * T$ with U 2-unipotent and T a 2-torus. Then $N_G(P^\circ) = N_G(U) \cap N_G(T)$ and both terms are definable: $N_G(U)$ because U is, and $N_G(T)$ because it is a finite extension of $C_G(T)$ (Lemma 5.24 and Corollary 2.9). □

LEMMA 6.15. *Let G be a group of finite Morley rank, P a Sylow 2-subgroup, H a normal subgroup. Then:*

(1) *$P \cap H$ is a Sylow 2-subgroup of H.*
(2) *If H is definable and $\bar{G} = G/H$, then $\bar{P}$ is a Sylow 2-subgroup of $\bar{G}$.*
(3) *Every Sylow 2-subgroup of $\bar{G}$ is the image of some Sylow 2-subgroup of G.*

PROOF.

Ad (1) Let Q be a Sylow 2-subgroup of G which contains a Sylow 2-subgroup of H. As P and Q are conjugate in G, and H is normal, the same applies to P.

Ad (2) We proceed by induction on the rank and degree of G.

Let $\bar{Q}$ be a Sylow 2-subgroup of $\bar{G}$ containing P, and suppose $\bar{Q} > \bar{P}$. Let $\bar{Q}_1 = N_{\bar{Q}}(\bar{P})$, with Q_1 its complete preimage in G. Then $Q_1 \leq N_G(PH) =$

$PH \cdot N_G(P) = H \cdot N_G(P) \leq H \cdot N_G(P^\circ)$. So $N_G(P^\circ)$ covers $\bar{Q}_1$, and as this group is definable, we may replace G by $N_G(P^\circ)$ and assume $P^\circ \lhd G$.

If $P^\circ = 1$, then P is finite and $Q_1 \leq HN_G(P)$ with $N_G(P)$ definable, so we may take P normal in G and factor it out as well; this reduces to the case $P = 1$, in which case Lemma 2.18 applies.

Assume therefore that $P^\circ \neq 1$ and let $K = d(P^\circ)$. Let $\tilde{G} = G/K$. Then by induction, the image $\tilde{P}$ of P in $\tilde{G}$ covers $\tilde{Q}_1$, that is, $Q_1 \leq PK$. As Q_1 contains P, this gives $Q_1 = P(Q_1 \cap K)$. Now the 2-torsion subgroup K_2 of K is connected (Lemma 5.12) and is normalized by P, hence coincides with P°. So $Q_1 \cap K \leq P$ and $Q_1 \leq P$.

Ad (3) By (2) and the conjugacy of Sylow 2-subgroups. □

One also uses the connected version of this.

LEMMA 6.16. *Let G be a group of finite Morley rank, P a Sylow° 2-subgroup, H a definable normal subgroup. Then the image $\bar{P}$ of P in $\bar{G} = G/H$ is a Sylow° 2-subgroup, and all Sylow° 2-subgroups of $\bar{G}$ are of this form.*

PROOF. Extend P to a Sylow 2-subgroup S of G. Then $\bar{S}$ is a Sylow 2-subgroup of $\bar{G}$ and $(\bar{S})^\circ = \bar{P}$.

The final point follows again by conjugacy. □

6.4. Control of fusion.

DEFINITION 6.17. *Let G be a group, H and K subgroups. We say that* K controls fusion *in H, if for any subsets X, Y of H with X, Y conjugate in G, there is $k \in K$ with $X^k = Y$.*

LEMMA 6.18. *Let G be a group of finite Morley rank, and P a Sylow° 2-subgroup with maximal 2-torus T. Then $N_G(T)$ controls fusion in P.*

PROOF. All we use here is the fact that T is central in P.

Take $X, Y \subseteq P$, $g \in G$ with $X^g = Y$. Then $Y \subseteq P \cap P^g \leq C_G(T, T^g)$. Hence T and T^g are maximal 2-tori in $C_G(Y)$, which is definable by Corollary 2.9, and hence T^g, T are conjugate by some $c \in C_G(Y)$: $T^{gc} = T$, $gc \in N_G(T)$, and $X^{gc} = Y^c = Y$. □

LEMMA 6.19. *Let G be a group of finite Morley rank. Suppose that G contains a nontrivial unipotent subgroup and a nontrivial 2-torus. Then G has at least two distinct conjugacy classes of involution.*

PROOF. Let P be a Sylow° 2-subgroup of G, $P = U * T$ with U 2-unipotent and T a 2-torus. By our hypotheses, $U, T \neq 1$. As T contains only finitely many involutions, we may take $i \in U$ an involution with $i \notin T$. Then i is not conjugate to any involution in T under the action of $N_G(T)$, and hence by control of fusion is not conjugate to any such involution under the action of G. □

7. Generalized Fitting subgroup

The fundamental properties of the Fitting subgroup in solvable groups are the following: it is normal; it has a simple structure (nilpotent); and it contains its centralizer. It was realized somewhat belatedly, as the classification of the finite simple groups got properly underway, that there is a *generalized Fitting subgroup* that plays the same role in general finite groups: it is normal and contains its centralizer, and it also has a very simple structure, namely a central product of a nilpotent group with normal quasisimple subgroups. The same theory goes through for groups of finite Morley rank, and when working with connected groups one may work with connected subgroups more or less throughout (though the centers of the quasisimple factors tend to be finite).

DEFINITION 7.1. *Let G be a group of finite Morley rank. Then $F^*(G) = \langle F(G), E(G)\rangle$.*

See Proposition 5.13 and Definition 1.7 for $F(G)$ and $E(G)$.

LEMMA 7.2. *Let G be a group of finite Morley rank. Then:*
(1) *$F^*(G) = F(G) * E(G)$.*
(2) *$E(G)$ is the central product of finitely many definable quasisimple subgroups.*
(3) *$F^*(G)$ is definable.*
(4) *Any subnormal quasisimple subgroup of G is normalized by G°.*
(5) *$E^\circ(G) = E(G^\circ)$ is the group generated by the subnormal quasisimple subgroups H of G with $H/Z(H)$ infinite.*

PROOF.
Ad (1) Lemma 1.10.
Ad (2) Lemmas 2.10, 1.9, 3.33.
Ad (3) By the first two points.
Ad (4) G acts on the set of its subnormal quasisimple subgroups by conjugation, and this set is finite, so Lemma 3.3 applies.
Ad (5) If $H/Z(H)$ is finite, then by Lemma 1.13 H' is finite. But then $H = H'$ is finite. As G° acts on H, G° centralizes H. Thus $E(G^\circ)$ is the product of the subnormal quasisimple factors H with $H/Z(H)$ infinite. Then $E(G^\circ)$ is connected, and of finite index in $E(G)$ and hence $E(G^\circ) = E(G)^\circ$ (denoted $E^\circ(G)$). □

PROPOSITION 7.3. *Let G be a group of finite Morley rank. Then:*
(1) *$C_G(F^*(G)) = Z(F(G))$.*
(2) *$C_G{}^\circ(F^{*\circ}(G)) \le F(G)$.*

PROOF.
Ad (1) Let $H = C_G(F^*(G))$. Then $E(H) \le E(G) \le F^*(G)$ so by definition of H, $E(H) \le Z(H)$, forcing $E(H) = 1$. Furthermore, $F(H) = F(G) \cap H = Z(F(G))$. In particular, $F(H) = Z(H)$.

It suffices to show that $H = Z(H)$. Supposing the contrary, let $\bar{H} = H/Z(H)$, and let $\bar{L}$ be minimal normal definable in $\bar{H}$, with full preimage L in H. Then $L_1 = L^\infty$ is definable (Lemma 3.34) and covers $\bar{L}$ as well, hence is nontrivial.

Now $L_1' = L_1$, $L_1 \lhd H$, and $L_1/Z(L_1)$ is semisimple. Hence L_1 is quasisemisimple and $L_1 \leq E(H) = 1$, a contradiction.

Ad (2) Let $H = C_G{}^\circ(F^{*\circ}(G))$. Then $E^\circ(H) \leq E^\circ(G) \leq F^{*\circ}(G)$. So by definition of H, $E^\circ(H) \leq Z(H)$, forcing $E^\circ(H) = 1$. Furthermore, $F^\circ(H) \leq F^\circ(G) \cap H \leq Z(F^\circ(H))$. In particular, $F^\circ(H) \leq Z(H)$. It suffices to show that $H = F^\circ(H)$.

Supposing the contrary, take a minimal normal connected subgroup L of H properly containing $F(H)$. As L is connected, $[L, F(H)] \leq F^\circ(H)$. If $a \in F(H)$, then commutation with a produces a homomorphism $\gamma_a : L \to F^\circ(H)$. Varying a over $F(H) \bmod F^\circ(H)$, we get a homomorphism γ from L to a product of finitely many copies of $F^\circ(H)$, with kernel $C_L(F(H))$. Let $L_0 = C_L(F(H))F(H)$. By minimality of L, either $L = L_0$ or $L_0/F(H)$ is finite. In the second case, since L/L_0 is nilpotent, and $[L, L_0] \leq F^\circ(H) \leq Z(H)$, it follows that L is nilpotent, hence $L \leq F(H)$, a contradiction.

So $L = L_0 = C_L(F(H))F(H)$. As L is connected, $L = C_L(F(H))Z(H) = C_L(F(H))$. Now L^∞ is quasisemisimple, so $L^\infty \leq E^\circ(H) = 1$, a contradiction. □

8. Solvable groups

We deal with the structure of solvable groups of finite Morley rank, with Schur-Zassenhaus splitting and conjugation theorems, the Hall theory, existence of the solvable radical, and some more specialized topics.

8.1. Structure. We begin with minimal modules in the sense of Definition 4.9.

LEMMA 8.1. *Let G be a connected group of finite Morley rank, and V a nontrivial G-minimal module (Definition 4.9). Then $\bar{V} = V/C_V(G)$ is definably irreducible, and $C_{\bar{V}}(G) = (0)$.*

PROOF. By assumption $C_V(G)$ is finite. Let $\bar{V} = V/C_V(G)$. If $\bar{W}$ is a proper definable submodule of $\bar{V}$ with preimage W in V, then by assumption W is finite, and by connectedness $W \leq C_V(G)$, so $\bar{W} = 0$. Thus $\bar{V}$ is irreducible.

If $C_{\bar{V}}(G) > 0$, then $C_{\bar{V}}(G) = \bar{V}$ and $[V, G] \leq C_V(G)$. So $[V, G]$ is connected and finite, hence trivial, a contradiction. □

LEMMA 8.2. *Let G be a connected solvable group and V a G-minimal module. Then G' acts trivially on V.*

PROOF. Proceed by induction on the solvability class of G. We may take G to act faithfully.

Let $U \leq V$ be G'-minimal. Then $V_1 = \langle U^G \rangle$ is G-invariant, hence $V_1 = V$. By induction, G'' acts trivially on U and its conjugates, hence on $V_1 = V$. In other words, $G'' = 1$ and G' is abelian.

We apply linearization, Proposition 4.11. Then V becomes a vector space over an algebraically closed field K, with G' a group of scalars and G linear; but the elements of G' have determinant 1 and hence the scalars involved are d-th roots of unity with $d = \dim V$. Thus G' is finite, but also connected; hence $G' = 1$, as claimed. □

LEMMA 8.3. *Let G be a connected solvable group of finite Morley rank. Then $G/F^\circ(G)$ is divisible abelian.*

PROOF. Form a G-invariant series $S : G = G_0 > G_1 > \cdots > G_n = 1$ with $A_i = G_i/G_{i+1}$ abelian and G-minimal, for example, by refining the series $(G^{(i)})$. Let $h_S : G \to \prod \mathrm{Aut}(A_i)$ be the map induced by the action of G on each factor.

By the foregoing, the action of G on each A_i is abelian, and the quotient $A_i/C_{A_i}(G)$ is irreducible, so G acts as a subgroup of the multiplicative group of a field. In particular, $G/\ker h_S$ is a connected subgroup of a divisible abelian group, hence is divisible abelian.

On the other hand, $\ker h_S$ stabilizes the chain S and hence has a nilpotent action on G, and in particular on itself; so $\ker h_S$ is nilpotent and normal in G. Thus $\ker h_S \leq F(G)$.

Thus $G/F(G)$ is divisible abelian. By Lemma 3.8, $G/F^\circ(G)$ is also divisible abelian. □

COROLLARY 8.4. *Let G be a solvable group of finite Morley rank, and U a unipotent subgroup (cf. Definition 5.4). Then $U \leq F^\circ(G)$.*

PROOF. We may take G to be connected. Then $U/F^\circ(G)$ is a unipotent subgroup of a divisible abelian group, hence finite. □

PROPOSITION 8.5. *Let Q and E be subgroups of a group of finite Morley rank such that Q is normal, solvable, definable and contains no π-unipotent subgroup, and E is a definable connected π-group of bounded exponent. Then $[Q, E]=1$.*

PROOF. Replacing E by $E/C_E(Q)$, we may suppose that the action of E on Q is faithful. Suppose that E is nontrivial. Then we may replace E by a connected abelian subgroup. So assume that E is abelian, and hence $Q \rtimes E$ is solvable. As E has bounded exponent, $E \leq F^\circ(Q \rtimes E)$. On the other hand, $F^\circ(Q \rtimes E) = U * T$ with U connected of bounded exponent, and T radicable. So $U = E \cdot (U \cap Q)^\circ$ and $(U \cap Q)^\circ = 1$ by hypothesis. Hence $E = U \lhd (Q \rtimes E)$ and $[Q, E] \leq Q \cap E = 1$. □

8.2. Hall subgroups.

DEFINITION 8.6. *Let G be a group, π a set of primes. A* Hall subgroup *is a maximal π-group contained in G; in particular, it is a torsion subgroup, and typically is undefinable.*

PROPOSITION 8.7. *Let G be solvable of finite Morley rank and let π be a set of primes. Then any two Hall π-subgroups of G are conjugate.*

PROOF. We prove the result for the wider class of solvable groups having a normal series as given in Lemma 3.36:

$$G = G_0 > G_1 > \cdots > G_n = 1$$

with G_i normal in G, such that each quotient G_i/G_{i+1} is abelian, and is either a π-group, or a π-divisible $\pi^\perp$-group which satisfies the DCC for G-centralizers.

We proceed by induction on the length n of such a series. We may suppose the result holds for $\bar{G} = G/G_{n-1}$. In particular, if G_{n-1} is a π-group the result follows at once for G.

Assume therefore that G_{n-1} is an abelian normal π-divisible $\pi^\perp$-subgroup of G with the DCC on centralizers in G. Let H_1, H_2 be maximal π-subgroups of G. Let $\bar{K}$ be a Hall π-subgroup of $\bar{G} = G/G_1$, and K its preimage in G. By induction on n, after conjugation we may assume that H_1 and H_2 are subgroups of K. By Proposition 1.28, K splits as $G_1 \rtimes L$ for some L, and after further conjugation in K we may assume that H_1 and H_2 are subgroups of L. Since L is a π-subgroup, we get $H_1 = H_2 = L$. □

COROLLARY 8.8. *Let G be a solvable group of finite Morley rank. Then any two maximal locally finite subgroups of G are conjugate.*

LEMMA 8.9. *Let G be a solvable group of finite Morley rank. Then:*

(1) *G has a Sylow basis.*
(2) *Any two such are conjugate.*

PROOF. Let U be a maximal locally finite subgroup of G. By Lemma 1.35, U belongs to $\mathfrak{U}$. By Proposition 1.37, U contains a Sylow basis for U, and any two such are conjugate in U. By Corollary 8.8, this Sylow basis is a Sylow basis for G, and any two Sylow bases for G are conjugate to those in U, hence to each other. □

LEMMA 8.10. *Let G be a solvable group of finite Morley rank, H a normal subgroup, π a set of primes, and U a Hall π-subgroup. Then $U \cap H$ is a Hall π-subgroup of H.*

PROOF. Let V be a Hall π-subgroup of G containing a Hall π-subgroup of H, and conjugate U to V. □

LEMMA 8.11. *Let G be a solvable group of finite Morley rank, π a set of primes, U a Hall π-subgroup, H a definable normal subgroup, and $\bar{G} = G/H$. Then $\bar{U}$ is a Hall π-subgroup of $\bar{G}$. Furthermore, all Hall π-subgroups of $\bar{G}$ are of this form.*

PROOF. We prove this more generally with G arbitrary, under the assumption that H has a characteristic series (H_i) as in Lemma 3.36: each quotient H_i/H_{i+1} is abelian, and is either a π-group, or a π-divisible $\pi^\perp$-group which satisfies the DCC for G-centralizers.

By induction on the length of this series, we suppose that a Hall π-subgroup of G/H_1 covers a Hall π-subgroup of G/H. This then reduces the problem to the case $H = H_1$, and if H is a π-group there is no problem. So suppose that H is an abelian π-divisible $\pi^{\perp}$-group with the DCC for G-centralizers. Let V be the pullback to G of a Hall π-subgroup $\bar{V}$ of $\bar{H}$. Then by Proposition 1.28, V splits over H_1, and the complement V_0 can be chosen to contain U. As V_0 is also a π-group, we have $U = V_0$ covers $\bar{V}$.

The last point then follows by conjugacy. □

PROPOSITION 8.12. *Let G be a connected solvable group of finite Morley rank, π a set of primes, and U a Hall π-subgroup. Then U is connected.*

PROOF. We proceed by induction on the rank of G. The group G contains a nontrivial definable connected abelian normal subgroup A, for example the last term of the commutator series $G^{(i)}$. Fix one such group A. Using Lemma 2.13, it is easy to see that the π-torsion subgroup A_π is connected. Furthermore, as $A_\pi \lhd G$, we have $A_\pi \leq U$.

Suppose that H is a definable subgroup of G such that $U \cap H$ has finite index in U. Then H contains A_π. Let $A_0 = d(A_\pi)$. If $A_0 = 1$, then U is definably isomorphic with its image in G/A, and we may conclude by induction. So suppose $A_0 > 1$.

Then we apply induction to G/A_0 and conclude that UA_0/A_0 is connected, hence $U \leq HA_0$. But $A_0 \leq H$, so $U \leq H$, as required. □

8.3. Splitting.

PROPOSITION 8.13. *Let G be a group of finite Morley rank and $H \lhd G$ definable and nilpotent, with G/H abelian. Suppose that for some $g \in G$ we have $C_H(g) = 1$. Then:*

(1) *$G = H \rtimes C_G(g)$.*
(2) *Any two complements of H in G are conjugate.*

PROOF.

Ad (1) Let $\bar{G} = G/Z(H)$, $\bar{H}_0 = C_{\bar{H}}(\bar{g})$, H_0 the preimage in G. Then commutation with g defines a homomorphism $\gamma_g : H_0 \to Z(H)$. By assumption this is injective and as H_0 contains $Z(H)$ we find that $H_0 = Z(H)$. Hence in $\bar{G}$ we have $C_{\bar{H}}(\bar{g}) = 1$.

Proceeding by induction on the nilpotency class of H, we may suppose that $\bar{G} = \bar{H} \rtimes C_{\bar{G}}(\bar{g})$ and hence $G = H \cdot C_G(g \bmod Z(H))$.

Let $G_1 = C_G(g \bmod Z(H))$. Then by our first remark, $G_1 \cap H = Z(H)$. It suffices therefore to show that G_1 splits as $Z(H) \rtimes C_{G_1}(g)$, or in other words, we may now take H to be abelian. Then commutation with g defines a homomorphism $\lambda_g : G \to H$ which is injective and hence surjective on H. So for $x \in G$ we have $[g, x] = [g, h]$ with $h \in H$ and hence $x \in hC_G(g) \subseteq H \rtimes C_G(g)$.

Ad (2) Again by induction on the nilpotency class, we may reduce to the case in which H is abelian.

Let T be a complement to H in G. Let $g = ht$, $h \in H$ and $t \in T$. Then $C_H(t) = C_H(g) = 1$, so $T = C_G(t)$. It suffices to show that g and t are conjugate. The equation $g = t^x$ may be rewritten as $h = [x, t^{-1}]$. So consider the commutation map $\gamma : H \to H$ given by $\gamma(x) = [x, t^{-1}]$. This is injective, hence surjective, on H, and hence the equation can be solved. □

PROPOSITION 8.14. *Let G be a connected group of finite Morley rank which is solvable of class 2 and has finite center, with G' being G-minimal. Then:*

(1) *G splits as $G' \rtimes T$ for some definable divisible abelian connected complement T containing $Z(G)$.*
(2) *Any two complements to G' in G are conjugate.*

PROOF. We show that $Z(G) \cap G' = 1$.

If G' is central in G, then G is nilpotent and has infinite center, a contradiction. Take $g \in G \setminus C_G(G')$. Consider the homomorphism $\gamma_g : G' \to G'$ induced by commutation with g. Its kernel is G-invariant and proper, hence finite.

Let $K = Z(G) \cap G'$. Note the G-equivariance of γ_g: for $x \in G'$, $\gamma_g(x)^y = [g^y, x^y] = [g, x^y] = \gamma_g(x^y)$. As K is normal in G, its inverse image under γ_g is also normal in G, hence central; hence K is trivial.

Now as observed, $C_{G'}(g)$ is finite and G-invariant, hence contained in $Z(G) \cap G' = 1$. Thus the hypotheses of Proposition 8.13 are satisfied. □

LEMMA 8.15. *Let G be a connected solvable nonnilpotent group of finite Morley rank. Then G has a definable quotient $\bar{G}$ with trivial center such that $\bar{G}'$ is abelian and G-irreducible.*

PROOF. Let H be maximal normal connected so that G/H is nonnilpotent. Replace G by G/H; then every quotient of G by a nontrivial definable connected normal subgroup is nilpotent. We may factor out the (finite) center of G and suppose that G is centerless (Lemma 3.9).

If $G'' \neq 1$, then G/G'' is nilpotent and hence G is nilpotent by Lemma 1.19, a contradiction. So $G'' = 1$ and G' is abelian.

The chain G^k eventually stabilizes at a nontrivial group, since G is nonnilpotent. Then G^k is G-irreducible, since a proper connected G-invariant definable subgroup would produce a nonnilpotent quotient, and a finite G-invariant subgroup would belong to $Z(G)$.

It suffices to show that $G' = G^k$. Supposing the contrary, let $\bar{G} = G/G^k$ and let $\bar{A} \leq \bar{G}'$ be $\bar{G}$-minimal, with preimage $A > G^k$. Note that $\bar{G}$ is nilpotent, so $\bar{A} \leq Z(\bar{G})$. For $g \in G$, as G' is abelian, $C_A(g)$ is normal in G. Hence $[G, C_A(g)] \lhd G$, while $[G, C_A(g)] \leq C_{G^k}(g)$. Taking g outside $C_G(G^k)$, we find that $[G, C_A(g)]$ is a proper G-invariant subgroup of G^k, hence trivial. Hence $C_A(g) \leq Z(G) = 1$, and commutation with g defines an injection $\gamma_g : A \to G^k$, a contradiction, since $A > G^k$. □

DEFINITION 8.16. *Let G be a group and H a subgroup. Then H is* abnormal *in G if $g \in \langle H, H^g \rangle$ for all $g \in G$.*

LEMMA 8.17. *Let G be a group, H a normal subgroup, and U a subgroup of G containing H. Then U is abnormal if and only if in $\bar{G} = G/H$, the image $\bar{U}$ is abnormal.*

PROOF. Immediate. □

PROPOSITION 8.18. *Let G be a solvable connected and nonnilpotent group of finite Morley rank. Then G has a proper abnormal subgroup.*

PROOF. By Lemma 8.15, G has a definable quotient $\bar{G}$ of class 2 with $\bar{G}'$ being G-irreducible and nontrivial. By the preceding lemma, it suffices to find an abnormal subgroup in $\bar{G}$, so we will take $G = \bar{G}$.

By Proposition 8.14 G splits as $G' \rtimes T$. We claim that T is abnormal. If $g \in G$, and $T_1 = \langle T, T^g \rangle$, then $T_1 = T \cdot (T_1 \cap G')$ with $T_1 \cap G'$ T-invariant, hence either trivial or equal to G'. Hence $g \in T_1$ in either case. □

LEMMA 8.19. *Let G be a connected solvable group of finite Morley rank. Then the following are equivalent:*

(1) *G is nilpotent;*
(2) *$G' \leq \Phi(G)$;*
(3) *$G/\Phi(G)$ is nilpotent.*

PROOF.
(1) $\implies$ (2): Let M be a maximal definable connected subgroup of G. By nilpotence, $M \lhd G$, and by maximality G/M has no proper infinite definable subgroups, hence is abelian by Lemma 3.10. Thus $G' \leq \Phi(G)$.

(3) $\implies$ (1):

Suppose that $\bar{G} = G/\Phi(G)$ is nilpotent, and G is not nilpotent. Then by Proposition 8.18, G has a proper abnormal subgroup H. Now $\bar{H}$ is abnormal in $\bar{G}$, a nilpotent group, so by the normalizer condition we have $\bar{H} = \bar{G}$, that is, $H\Phi(G) = G$, and hence $H = G$, a contradiction. □

8.4. Carter subgroups.

DEFINITION 8.20. *A* Carter subgroup *of a group G is a self-normalizing and nilpotent subgroup.*

LEMMA 8.21. *If G is a group of finite Morley rank and Q a Carter subgroup, then Q is definable.*

PROOF. The group $d(Q)$ is nilpotent; if $d(Q) > Q$, then the normalizer condition yields $N_{d(Q)}(Q) > Q$, contradicting the definition. □

LEMMA 8.22. *Let G be a connected solvable group of finite Morley rank and Q an abnormal subgroup. Then Q is definable and connected.*

PROOF. We proceed by induction on the rank of G. Let $A \lhd G$ be G-minimal, and $\bar{G} = G/A$. Then by induction $\bar{Q}$ is definable and connected, and the same applies to QA. Therefore, we may suppose that $G = QA$. In particular, $Q \cap A \lhd G$. If $Z(G)$ is infinite we take $A \leq Z(G)$ and hence $Q \lhd G$,

$Q = G$, and we are done. So we suppose $Z(G)$ is finite. We may factor it out and suppose $Z(G) = 1$ (Lemma 3.9). In particular, A is G-irreducible.

If $A \leq Q$, then $G = QA = Q$. So we may suppose that $A \cap Q < A$ and then $A \cap Q = 1$ by G-irreducibility. Thus Q is a complement to A in G.

We may suppose $A \leq G'$. Suppose first that $A = G'$. Then by Proposition 8.14, Q is definable and connected.

Now suppose $A < G'$, so $Q \cap G' \neq 1$. Then $Q \cap G'$ is normal in Q. As G' is nilpotent (Lemma 8.3) and A is G-irreducible, it follows that G' centralizes A, so $Q \cap G'$ is normal in G. As $Z(G)$ is trivial, $Q \cap G'$ contains a G-minimal (connected, definable) subgroup of G, which can be factored out to conclude by induction. □

PROPOSITION 8.23. *Let G be a connected solvable group of finite Morley rank, A a G-minimal abelian subgroup, and H a subgroup of G containing $Z^\circ(G)$ such that $G = C_G(A)H$ and HA is definable and connected. Then H is abnormal in HA.*

PROOF. If H contains HA, then the claim is vacuous.

Otherwise, A does not centralize G since $Z^\circ(G) \leq H$, and hence A does not centralize H. But $G = C_G(A)H$, so A is H-minimal. Again by Lemma 8.3, H' acts trivially and A is H/H'-irreducible, by Lemma 4.10. Thus $A \cap Z(G) = 1$, and $A \cap H = 1$.

Now suppose $g \in HA$, $g \notin H$. Let $H_1 = \langle H, H^g \rangle$. Then $H_1 = H \cdot (H_1 \cap A)$, and $H_1 \cap A$ is H-invariant. If $H_1 = A$ our claim is clear. Otherwise, by H-minimality, H_1 is finite, and by connectivity, central, hence trivial. So $g \in N(H)$. Then writing $g = ha$, we find $a \in N_A(H)$. Then $[a, H] \leq A \cap H = 1$, and hence $a \in C_A(H) \leq Z(G) \cap A = 1$. Thus $g \in H$. □

PROPOSITION 8.24. *Let G be a connected solvable group of finite Morley rank, H an abnormal subgroup of G, and K a subgroup of H which is abnormal in H. Then K is abnormal in G.*

PROOF. We know that K and H are connected and definable. Supposing the result fails, minimize the rank of G, and then maximize the rank of K.

Note that $G = G'H$ (by passage to G/G') and $H = H'K$, so $G = G'K$. Taking $A \lhd G$ abelian and G-minimal, we have, in particular, $G = C_G(A)K$ (via Lemma 8.3). Furthermore, $Z(G) \leq H$, and then $Z(G) \leq K$, by abnormality each time. So the Proposition 8.23 applies, and shows that K is abnormal in KA.

By induction on rank, KA/A is abnormal in G/A, so KA is abnormal in G.

We show that K is abnormal in G. Suppose that $g \in G$. Let $K_1 = \langle K, K^g \rangle$. By abnormality of KA in G, $g \in \langle KA, (KA)^g \rangle = AK_1$. Let $g = ak$ with $k \in K_1$, $a \in A$. Then $K_1^{g^{-1}} = K_1^{a^{-1}}$, so $K \leq K_1, K_1^{a^{-1}}$, and $K, K^a \leq K_1$. By abnormality of K in KA, $a \in \langle K, K^a \rangle \leq K_1$. So $g \in K_1$, as required. □

PROPOSITION 8.25. *Let G be a connected solvable group of finite Morley rank, and Q a minimal abnormal subgroup. Then Q is a Carter subgroup.*

PROOF. Certainly Q is self-normalizing. By Lemma 8.22, it is also definable and connected. We need to see that it is nilpotent.

If not, then by Proposition 8.18, Q has a proper subgroup K which is abnormal relative to Q. Then Proposition 8.24 contradicts the minimality of Q. □

PROPOSITION 8.26. *Let G be a connected solvable group of finite Morley rank. Then any two Carter subgroups of G are conjugate.*

PROOF. We proceed by induction on the rank of G.

We take Q_1 and Q_2 Carter subgroups in G. Then both contain $Z(G)$, so in the quotient $\bar{G} = G/Z(G)$, $\bar{Q}_1$ and $\bar{Q}_2$ are again Carter subgroups. If $Z(G)$ is infinite, we conclude by induction. So suppose $Z(G)$ is finite. We may still factor it out, to get $Z(G) = 1$.

Take A to be a G-minimal subgroup of G. We claim that in $\bar{G} = G/A$, the groups $\bar{Q}_1$ and $\bar{Q}_2$ are Carter subgroups, or, in other words, that Q_1A and Q_2A are self-normalizing in G. Consider for example Q_1A. If $Q_1A = G$ our claim certainly holds, and if $Q_1A < G$, then by induction we may suppose that the Carter subgroups of Q_1A are conjugate, and hence by a Frattini argument $N(Q_1A) \leq Q_1AN(Q_1) = Q_1A$. Thus the claim holds.

Now applying induction in G/A, we may suppose that $Q_1A = Q_2A$. Hence we may suppose that $G = Q_1A$. Since $C_{Q_1}(A)$ is normal in the nilpotent group Q_1, if this centralizer is nontrivial, then it meets $Z(Q_1)$, and hence meets $Z(Q_1A) = Z(G)$, a contradiction. So $C_{Q_1}(A) = 1$.

By minimality G' centralizes A (Lemma 8.3), so $G' = A$. Then Proposition 8.14 completes the argument. □

COROLLARY 8.27. *Let G be a connected solvable group of finite Morley rank, Q a Carter subgroup. Then Q is connected, definable, and abnormal.*

PROOF. Abnormality follows from Propositions 8.26 and 8.25, and implies connectedness and definability by Lemma 8.22. □

PROPOSITION 8.28. *Let G be a solvable group of finite Morley rank. Then $\Phi^\circ(G)$ is nilpotent; if the Sylow subgroups of G are all nilpotent, then $\Phi(G)$ is nilpotent.*

PROOF. Since $\Phi(G) = \Phi(G^\circ)$, we may suppose G is connected.

Let Q be a Carter subgroup of $\Phi^\circ(G)$. Then by the Frattini argument, $G = \Phi(G)N_G(Q)$. As G is connected we have $G = \Phi(G){N_G}^\circ(Q)$. Hence $G = {N_G}^\circ(Q)$, and in particular, $\Phi^\circ(G)$ normalizes Q. As Q is self-normalizing, we find $\Phi^\circ(G) = Q$ is nilpotent.

Similarly, for any prime p, if P is a Sylow p-subgroup of $\Phi(G)$, then $P \lhd G$ and hence $P \lhd \Phi(G)$.

Thus $F(\Phi(G))$ contains $\Phi^\circ(G)$, and if the Sylow subgroups are nilpotent, then $F(\Phi(G))$ contains all the Sylow p-subgroups of $\Phi(G)$, but the latter cover $\Phi(G)/\Phi^\circ(G)$, so in this case $\Phi(G)$ is nilpotent. □

PROPOSITION 8.29. *Let G be a connected solvable group of finite Morley rank, Q a Carter subgroup, K a normal subgroup, and $\bar{G} = G/K$. Then $\bar{Q}$ is a Carter subgroup of $\bar{G}$, and all Carter subgroups of $\bar{G}$ have this form.*

PROOF. By Corollary 8.27 Q is abnormal in G, and hence $\bar{Q}$ is abnormal in $\bar{G}$, and therefore self-normalizing. (One could also use the Frattini argument we just used above.)

The last statement follows by conjugacy. □

COROLLARY 8.30. *Let G be a connected solvable group of finite Morley rank and Q a Carter subgroup. Then $G'Q = G$.*

PROOF. The image of Q in G/G' is a Carter subgroup of an abelian group, hence Q covers the quotient. □

The more general form of this lemma in which G' is replaced by any normal subgroup with nilpotent quotient also has its uses, but for our purposes the foregoing is the important point.

LEMMA 8.31. *Let G be a connected solvable centerless group of class two and Q a Carter subgroup. Then $G = G' \rtimes Q$.*

PROOF. We have seen that $G = G'Q$ quite generally, so we need only consider $H = Q \cap G'$. By hypothesis G' is abelian, so $G' \leq C(H)$. As H is normal in the nilpotent group Q, if H is nontrivial, then it meets $Z(Q)$ nontrivially. But then $H \cap Z(Q)$ centralizes $G'Q = G$, producing a nontrivial center. □

8.5. The solvable radical.

PROPOSITION 8.32. *Let G be a group of finite Morley rank. Then there is a unique maximal normal solvable subgroup of G, and it is definable.*

PROOF. It is clear that there is a unique maximal definable connected normal solvable subgroup of G, and that we may factor it out, assuming therefore that every definable solvable normal subgroup of G is finite. In particular, $Z(G^\circ)$ is finite and after factoring this out, by Lemma 3.9 we may assume that $Z(G^\circ) = 1$. At this point, G° has no nontrivial solvable definable normal subgroup. In fact, by considering definable hulls, it follows that G° has no nontrivial solvable normal subgroup at all.

Now let $H \lhd G$ be solvable. Then $H \cap G^\circ = 1$. Therefore, there is a bound on the size of H, and hence a maximal such H, which is finite, hence definable. □

NOTATION 8.33. *Let G be a group of finite Morley rank. The largest normal solvable subgroup of G is called the* solvable radical, *and is denoted*

$$\sigma(G) .$$

8.6. The socle.

LEMMA 8.34. *Let G be a group of finite Morley rank and suppose that $\sigma(G) = 1$. Let* $\mathrm{Soc}(G)$ *be the subgroup of G generated by its minimal normal subgroups. Then* $\mathrm{Soc}(G) = E(G) = F^*(G)$ *is a direct sum of finitely many simple definable normal subgroups of G, and every nontrivial normal subgroup of G meets* $\mathrm{Soc}(G)$ *nontrivially.*

PROOF. Observe that G has no nontrivial abelian normal subgroup A, as otherwise $d(A) \leq \sigma(G)$. By Lemma 3.31 we have $\mathrm{Soc}(G) = E(G)$, and since $F(G) = 1$ also $E(G) = F^*(G)$. Furthermore, $Z(E(G)) \leq \sigma(G) = 1$ so $E(G)$ is a direct sum of finitely many simple definable normal subgroups of G.

For the last point, we can quote Proposition 7.3 or argue directly as follows. Let H be any nontrivial normal subgroup of G. Then H contains a minimal normal definable subgroup by the descending chain condition, and these are nonabelian by hypothesis, so by Lemma 3.31 they are minimal normal. □

In general, the appropriate socle to work with is the subgroup generated by minimal normal definable subgroups, but the groups $F^*(G)$ on the one hand, and $\sigma(G)$ on the other, play a more important role.

8.7. Unipotent radicals. Unipotent and p-unipotent groups were defined in Definition 5.4. We now consider the associated unipotent radicals.

NOTATION 8.35. *Let G be a group of finite Morley rank and π a set of primes. Then $U_\pi(G)$ is the subgroup of G generated by its unipotent π-subgroups.*

Observe that in a simple Chevalley group of characteristic p, $U_p(G)$ is equal to G.

LEMMA 8.36. *Let G be a solvable group of finite Morley rank, and p a prime. Then $U_\pi(G)$ is π-unipotent, and is contained in $F^\circ(G)$.*

PROOF. By Corollary 8.4, we have $U_\pi(G) \leq F^\circ(G)$. So by the structure of nilpotent groups (Proposition 5.8), $U_\pi(G)$ is π-unipotent. □

For the most part we take $\pi = \{p\}$ in practice. The previous lemma may be read as saying that the general notion is not, in fact, much more general.

9. Schur-Zassenhaus

9.1. A solvable factorization. We introduced U_π in the previous section, and now we introduce a complementary notion. We will get a rough analog of the structure theory for connected nilpotent groups in the solvable context.

DEFINITION 9.1. *Let G be a group of finite Morley rank, and π a set of primes. A definable subgroup H of G is a* π^*-group *if it is solvable and connected, and every definable connected abelian section of H is π-divisible.*

LEMMA 9.2. *Let G be a solvable group of finite Morley rank, π a set of primes. If $U = U_\pi(G)$, $H \lhd G$ is a π^*-subgroup, and $\bar{G} = G/H$, then $\bar{U} = U_\pi(\bar{G})$.*

PROOF. Proceeding inductively, we may suppose that H is G-minimal. Let V be the preimage in G of $U_\pi(\bar{G})$. Then V/H has bounded exponent, $H \le F^\circ(V)$, and $V/F^\circ(V)$ is divisible abelian, so V is nilpotent. As V/H is a π-group of bounded exponent, $V = U_\pi(V)H$; but $U_\pi(V) = U$. □

LEMMA 9.3. *Let π be a set of primes and let G be a solvable group of finite Morley rank.*

(1) *If $1 \to K \to G \to \bar{G} \to 1$ is a short exact sequence with K definable normal and connected in G, then G is a π^*-group if and only if K and $\bar{G}$ are.*
(2) *If G is a π^*-group and H is a normal Hall π-subgroup of G, then $H \le Z(G)$ is connected and divisible.*

PROOF.

Ad (1) It suffices to check that if K and $\bar{G}$ are π^*-groups, then G is. Let $H_1 \lhd H_2$ be connected definable subgroups of G with H_2/H_1 abelian. Then we have a short exact sequence

$$1 \to (H_2 \cap K)/(H_1 \cap K) \to H_2/H_1 \to H_2K/H_1K \to 1\ .$$

As this is in the category of abelian groups, and the connected components on both ends are π-divisible, with the middle term connected, the section H_2/H_1 is π-divisible.

Ad (2) Proceed by induction on the solvability class of G. Then $H \cap G'$ is a normal Hall π-subgroup of G', so $A = H \cap G' \le Z(G')$. As G' is a π^*-group, $A[n]$ is finite for each n. Hence $A \le Z(G)$. For $h \in H$, commutation with h defines a homomorphism $\gamma_h : G \to A$. As h is a π-element, the image is a π-group of bounded exponent; as a definable image of G, it is a π^*-group. Therefore $[h, G] = 1$ and $H \le Z(G)$.

By Proposition 8.12, H is connected. As $H \le Z(G)$ and G is a π^*-group, H is divisible. □

PROPOSITION 9.4. *Let G be a connected solvable group of finite Morley rank and π a set of primes. Then:*

(1) *Any two maximal π^*-subgroups of G are conjugate.*
(2) *If $H \le G$ is a maximal π^*-subgroup, then $G = U_\pi(G) \cdot H$.*

PROOF. We proceed by induction on the rank of G. We may suppose that G is not nilpotent. Let $U = U_\pi(G)$. Note that in the presence of condition (1), condition (2) reduces to:

(2′) $\qquad G = U_\pi(G) \cdot H$ for *some* π^*-subgroup.

Suppose first that G has a nontrivial normal π^*-subgroup K, and let $\bar{G} = G/K$. By Lemma 9.2, we have $\bar{U} = U_\pi(\bar{G})$, and hence by induction we have $G = UH$ with $\bar{H}$ a maximal π^*-group of $\bar{G}$, and H its preimage in G. Then H is a π^*-group, so we have (2′) in G. We also have conjugacy, by induction, since any two maximal π^*-subgroups of G will contain K. So we assume

(2) $\qquad$ G has no nontrivial normal π^*-subgroup .

In particular, $F^\circ(G) = U$.

Let H be a Carter subgroup of G. By Proposition 8.29, in $\tilde{G} = G/F^\circ(G)$, which is abelian, H covers a Carter subgroup, which must be $\tilde{G}$ itself. So $G = F^\circ(G)H = UH$. As H is nilpotent and connected (Lemma 8.22) there is a π^*-subgroup L of H for which $H = U_\pi(H)L$, hence $G = UL$.

Now let L_1 be any maximal π^*-subgroup of G, and $K = UL_1$. As $G = UL$, we have $K = U(K \cap L)^\circ$ and hence $(K \cap L)^\circ$ is a maximal π^*-group of K. If $K < G$, then after conjugation we may suppose that $L_1 = K \cap L$ and hence $L_1 = L$ by maximality, contradicting $K < G$. So we have $UL_1 = G$.

Now $L_1 \cap U$ is finite and normal in L_1, hence central, and $L_1/L_1 \cap U$ embeds in G/U, which is abelian. Thus $L_1/L_1 \cap U$ is abelian and connected, and $L_1 \cap U$ is finite, which implies that L_1 is abelian.

Let $H_1 = L_1 C_U(L_1)$. Then H_1 is nilpotent and L_1 is characteristic in H_1. Hence $N_G(H_1) \le N_G(L_1) = L_1 N_U(L_1)$. If $u \in N_U(L_1)$, then $[u, L_1]$ is a connected subgroup of the finite group $U \cap L_1$, so $[u, L_1] = 1$. Thus $N_G(H_1) \le H_1$ and H_1 is a Carter subgroup of G. We may therefore assume after conjugation that $H = H_1$, and it follows that $L = L_1$. □

9.2. Neoclassical Schur-Zassenhaus. The following purely algebraic statement is a useful variation on the classical Schur-Zassenhaus theorem.

FACT 9.5 ([**177**, Lemma 2.26]). *Let G be a group with a normal abelian subgroup A, and let H be a subgroup of G such that $A \le H$ and $[G : H] = m < \infty$. Assume that the map $a \mapsto a^m$ is bijective on A, and that H splits over A. Then G splits over A.*

9.3. Existence of complements.

PROPOSITION 9.6. *Let G be a solvable group of finite Morley rank, π a set of primes, and H a normal Hall π-subgroup of G. Then H has a complement in G.*

PROOF. We consider a hypothetical counterexample G of minimal rank and degree.

Recall that for any definable quotient $\bar{G}$ of G, the image $\bar{H}$ of H is a Hall subgroup of $\bar{G}$ (Lemma 8.11).

We show first that

(1) $\qquad$ G has no infinite, definable, normal $\pi^\perp$ subgroup A.

Assuming the contrary, applying induction to $\bar{G} = G/A$ one finds $G = HT$ with $A \leq T$, $H \cap T \leq H \cap A = 1$, and we have the desired complement,

$$(2) \qquad H \leq G^\circ \ .$$

Let $\tilde{G} = G/G^\circ$. Then $\tilde{H}$ is a normal π-subgroup of $\tilde{G}$ and hence $\tilde{G}$ splits as $\tilde{H} \rtimes \tilde{T}$ by the ordinary Schur-Zassenhaus theorem. Let T be the preimage of $\tilde{T}$ in G. If $H \not\leq G^\circ$, then $T < G$, and hence by induction T splits as $(H \cap T) \rtimes T_0$, and $G = HT = HT_0$ splits as well. So (2) follows.

Now let $U = U_\pi(G)$, and let K be a maximal π^*-subgroup of G. Then $G^\circ = UK$ by Proposition 9.4, and by the Frattini argument $G = UN_G(K)$. So if $N_G(K) < G$, then by induction

$$N_G(K) = (H \cap N_G(K)) \rtimes T \ ,$$

and this yields a splitting $G = H \rtimes T$. We conclude that K is normal in G. As $[U, K] \leq U \cap K$ is finite and connected, we find $[U, K] = 1$.

We claim:

$$(3) \qquad U = 1;\ G^\circ \text{ is a } \pi^*\text{-group} \ ,$$

G°/K is a nilpotent π-group, and G/G° is a finite π'-group. By Lemma 1.27, we have a splitting of $\bar{G} = G/K$ as $\bar{G}^\circ \rtimes \bar{T}$ with $\bar{T}$ finite. Let T be the preimage of $\bar{T}$, a finite extension of K. In particular, T is definable and we have $G = G^\circ T = UKT = UT$. If $U \neq 1$, then by induction we split T as $(H \cap T)T_0$ and get $G = HT_0$ split. So $U = 1$, and $G^\circ = K$ is a π^*-group.

In particular, by Lemma 9.3, we have $H \leq Z(G^\circ)$, and H is divisible. Let $L = (G^\circ)'$. Then L is a connected nilpotent π^*-group, and $L = L_1 * L_2$ with L_1 of bounded exponent and L_2 divisible. By (1), L_1 is a π-group, hence trivial. So L is radicable, and L splits further as $L_{\text{tor}} \times L_0$ with L_{tor} divisible abelian and L_0 torsion free (Proposition 5.11). As $L_{\text{tor}}[n]$ is finite for each n, $L_{\text{tor}} \leq Z(G)$. Furthermore, $L' = L_0'$ is torsion-free, connected, and definable, so by (1) we have $L' = 1$ and L is abelian.

We claim:

$$(4) \qquad L \leq Z(G^\circ) \ .$$

Take $g \in G^\circ$. Then $\tilde{L} = [L, g] = [L_0, g]$ is torsion free by Lemma 2.18. Hence $\langle \tilde{L}^G \rangle \leq L$ is also torsion-free, by Lemma 2.19. By (1) this group is trivial, so $[L, g] = 1$ and $L \leq Z(G^\circ)$.

It follows from (4) that G° is nilpotent and hence splits as a product $H \times T$ for some $\pi^\perp$-subgroup T. By Lemma 1.24 we may take T normal in G. Then by Lemma 1.26 we can split G/T, that is, we have $G = H\hat{T}$ with $\hat{T} \cap H = T \cap H = 1$. □

9.4. Conjugacy.

PROPOSITION 9.7. *Let G be a solvable group of finite Morley rank, π a set of primes, and H a normal Hall π-subgroup of G. Let T_1, T_2 be two definable complements to H in G. Then T_1 and T_2 are conjugate.*

PROOF. We proceed by induction on the rank and degree of G. Note that T_1 and T_2 provide complements to H in any subgroup of G containing H. We may suppose that G is infinite.

If G has an infinite definable normal $\pi^\perp$-subgroup, then we conclude easily by induction. It follows that G has a unique maximal definable $\pi^\perp$-subgroup, which is finite, and may be factored out. So we assume that

(1) G has no nontrivial definable normal $\pi^\perp$-subgroup .

Let $H_1 = T_1 \cap HC_G(H)$. Then $H_1/C_G(H)$ is a π-group and a $\pi^\perp$-group, hence trivial: $H_1 \le C_G(H)$. So $HC_G(H) = H \times H_1$. For $g \in G$, we have $H_1H_1^g \le HC_G(H)$ with $H_1H_1^g$ a $\pi^\perp$-group, and hence $H_1^g = H_1$, $H_1 \lhd G$, and by (1) $H_1 = 1$, $C_G(H) \le H$. In particular, H is infinite, as otherwise $G^\circ \le C_G(H) \le H$ and hence G is finite;

(2) H is G-minimal .

We show first that H contains no proper definable nontrivial G-invariant subgroup.

Suppose on the contrary $1 < A < H$ is definable and G-invariant. We may assume that $AT_1 = AT_2$. Furthermore, $AT_1 < G$, so we may conclude by induction.

On the other hand, $[G, H]$ is a G-invariant and nontrivial definable subgroup contained in H in view of Corollary 3.29. So $H = [G, H]$ is definable. Thus (2) follows.

In particular, H is connected and abelian. If $G > G^\circ$, then $G^\circ = H{T_1}^\circ = H{T_2}^\circ$ and by induction we may take ${T_1}^\circ = {T_2}^\circ$, so $T_1, T_2 \le N_G({T_1}^\circ)$. But by (1), $N_G({T_1}^\circ) < G$, and we can conclude by induction in this case. So we suppose G is connected.

As $C_G(H) \le H$ we find $[H, G] \ne 1$ and hence $H = [H, G] \le G'$. Hence $Z(G') \le H$ and as G' is connected, nilpotent, its center is infinite. Thus $Z(G') = H$ by minimality, and thus $G' \le H$; so $G' = H$.

Now Proposition 8.14 applies to conclude. □

PROPOSITION 9.8. *Let G be a solvable group of finite Morley rank, π a set of primes, and H a normal Hall π-subgroup of G. If H has bounded exponent, then the complements to H in G are definable, and conjugate, and such complements exist.*

PROOF. Existence of some complement was proved in Proposition 9.6. We will prove their definability, after which the conjugacy follows by Proposition 9.7.

Let $G = H \rtimes T$. Then $G^\circ = H^\circ T^\circ$. Accordingly, it suffices to treat the case in which G is connected. Then H is also connected.

If $K < H$ is infinite, definable, and G-invariant, then we conclude by induction. So we assume H is G-minimal, and, in particular, connected and abelian.

If $[G, H] = 1$, then as H has bounded exponent and T is π-radicable, T is definable. So suppose that $[G, H] = H$. Then $H \le G' < G$ and by induction we have $T' = T \cap G'$ definable. Now $T \leq N_G(T')$, so if $N_G(T') < G$, then by induction we have T definable, a contradiction.

So $T' \lhd G$. If $T' \neq 1$, then T' is infinite, and factoring it out, we may again conclude by induction. So $T' = 1$, T is abelian. Thus $T \leq Z(C_G(T))$, and the latter is definable. So we may suppose $T \leq Z(G)$. Then $[G, H] = 1$, contradicting our case assumption. □

9.5. Lifting centralizers.

PROPOSITION 9.9. *Let $G = H \rtimes T$ be a group of finite Morley rank, and $Q \lhd H$. Suppose that Q is a T-invariant solvable definable π-subgroup of bounded exponent, and T is a definable $\pi^\perp$ subgroup of G. Then*

$$C_H(T)Q/Q = C_{H/Q}(T).$$

PROOF. It suffices to show that $C_H(T \bmod Q) \leq C_H(T)Q$. Let $L = C_H(T \bmod Q)$. Then $[L, T] \leq Q$. We have $QT \leq LT \leq HT$. We will apply Proposition 9.7 to QT. For $x \in L$, $T^x \leq QT$. Therefore, by Proposition 9.7, we have $T^x = T^h$ for some $h \in Q$. This implies that $xh^{-1} \in N_L(T) = C_L(T)$, and therefore $x \in QC_L(T) \leq QC_H(T)$. □

COROLLARY 9.10. *Let $G = H \rtimes T$ be a solvable group of finite Morley rank with H and T definable. Assume that H is a π-group of bounded exponent and T is a $\pi^\perp$-group. Then $H = [H, T]C_H(T)$.*

COROLLARY 9.11. *In the situation of the previous corollary, if H is abelian and T is locally finite, then*

$$H = [H, T] \oplus C_H(T)$$

PROOF. Using additive notation, we just need to show that $C_H(T) \cap [H, T] = 0$ and as T is locally finite this reduces to the case in which T is finite. Then the endomorphism $E = \sum_T t$ satisfies $E([H, T]) = 0$ while on $C_H(T)$ it satisfies $E(x) = |T| \cdot x$. The claim follows. □

PROPOSITION 9.12. *Let $G = H \rtimes T$ be a group of finite Morley rank, with T a π-group of bounded exponent and $Q \lhd H$ a T-invariant $\pi^\perp$-subgroup. Suppose that Q and T are solvable, and definable in G. Then*

$$C_H(T)Q/Q = C_{H/Q}(T).$$

PROOF. It is enough to show that we have $C_H(T \bmod Q) \leq C_H(T)Q$.

Let $L = C_H(T \bmod Q)$. Then $[L, T] \leq Q$, so L normalizes QT. For $x \in L$, $T^x \leq QT$ is a Hall subgroup of QT. Therefore, by Proposition 8.7 $T^x = T^a$ for some $a \in Q$. This implies that $xa^{-1} \in N_L(T) = C_L(T)$ and therefore, $x \in QC_L(T) \leq QC_H(T)$. □

COROLLARY 9.13. *Let $G = H \rtimes T$ be a solvable group of finite Morley rank with H and T definable. Assume that H is a $\pi^\perp$-group and T is a π-group of bounded exponent. Then $H = [H, T]C_H(T)$.*

COROLLARY 9.14. *Let $G = H \rtimes T$ be a solvable group of finite Morley rank with H and T definable. Assume that H is an abelian $\pi^\perp$-group and T is a π-group of bounded exponent. Then*

$$H = [H,T] \oplus C_H(T) \ .$$

PROOF. As in Corollary 9.11. □

PROPOSITION 9.15. *Let $G = H \rtimes T$ be a group of finite Morley rank, with T a π-group of bounded exponent and $Q \lhd H$ a T-invariant definable π-divisible subgroup. Suppose that Q and T are solvable and H is connected. Then*

$$C_H(T)Q/Q = C_{H/Q}(T).$$

PROOF. Proceeding inductively we may suppose that Q has no proper, definable, π-divisible, definably characteristic subgroup. In particular, Q is abelian. If Q contains no π-torsion we apply Corollary 9.13. If Q does contain π-torsion, then by our initial reduction Q is the definable hull of its π-torsion subgroup Q_π. But Q_π has finitely many elements of each finite order and is therefore centralized by H. □

9.6. Generation.

PROPOSITION 9.16. *Let $Q \rtimes V$ be a group of finite Morley rank, with Q a definable, connected, solvable group with no nontrivial p-unipotent subgroup, and V a finite abelian p-group. Then $Q = \langle C_Q(V_0) : V/V_0 \text{ is cyclic}\rangle$.*

PROOF. We proceed by induction on the rank and degree of Q. Let $\mathcal{V} = \{V_0 \le V : V/V_0 \text{ is cyclic}\}$.

Suppose Q is not V-minimal, and $Q_0 < Q$ is infinite and V-invariant. Then by induction and Proposition 9.15, we have

$$Q/Q_0 = \langle C_{Q/Q_0}(V_0) : V_0 \in \mathcal{V}\rangle = \langle Q_0 C_Q(V_0) : V_0 \in \mathcal{V}\rangle$$

and again by induction, we have $Q_0 = \langle C_{Q_0}(V_0) : V_0 \in \mathcal{V}\rangle$, so $Q = \langle C_Q(V_0) : V_0 \in \mathcal{V}\rangle$ as well.

So we suppose that Q is V-minimal. In particular, $Q' = 1$, and Q is abelian. It follows that $C_Q(V_0)$ is normal in QV for $V_0 \le V$. Let $Q_1 = \langle C_Q(V_0) : V_0 \in \mathcal{V}\rangle$, and suppose $Q_1 < Q$. For $V_0 \in \mathcal{V}$, we have $C_{Q/Q_1}(V_0) = Q_1 C_Q(V_0) \le Q_1$ (Proposition 9.15), so replacing Q by Q/Q_1, we may suppose that $C_Q(V_0) = 1$ for all such V_0.

If Q is definably V-reducible, with $Q_0 < Q$ nontrivial and V-invariant, then Q_0 is finite and we may suppose it to be V-irreducible. Hence the natural map $V \to \mathrm{End}(Q_0)$ goes into a field, and its kernel contains a subgroup in $\mathcal{V}$, a contradiction.

If Q is definably V-irreducible, then the image of the natural map $V \to \mathrm{End}(Q)$ generates a subfield, and hence again the kernel contains a subgroup in $\mathcal{V}$. □

LEMMA 9.17. *Let G be a group of finite Morley rank, Q a definable, normal, connected, solvable $p^\perp$-subgroup, and T a p-torus of Prüfer rank n. Then*

$$Q = \langle C_Q(T_0) : T_0 \leq T, \text{ and the Prüfer rank of } T_0 \text{ is } n-1 \rangle \ .$$

PROOF. We may suppose that Q is a minimal definable T-invariant counterexample. Then $n \geq 2$.

Let $\mathcal{T} = \{T_0 \leq T : \text{the Prüfer rank of } T_0 \text{ is } n-1\}$.

Let $T_0 = C_T(Q)$. If T_0 is infinite, we may consider the action of T inside $d(T)/d(T_0)$, and reduce the Prüfer rank. Assume therefore that T_0 is finite, of exponent p^k. Let $V = T[p^{k+1}]$, $\mathcal{V} = \{V_0 \leq V : V/V_0 \text{ cyclic}\}$. Then $Q = \langle C_Q(V_0) : V_0 \in \mathcal{V}\rangle$, and it suffices to show that each subgroup $C_Q(V_0)$ with $V_0 \in \mathcal{V}$ is contained in $\langle C_Q(T_0) : T_0 \in \mathcal{T}\rangle$.

For $V_0 \in \mathcal{V}$ we have $V_0 \not\leq T[p^k]$ and thus $C_Q(V_0) < Q$. Hence by minimality of Q, $C_Q(V_0)$ is contained in $\langle C_Q(T_0) : T_0 \in \mathcal{T}\rangle$. □

9.7. A criterion for nilpotence.

PROPOSITION 9.18. *Let G be a connected solvable group of finite Morley rank containing no unipotent p-subgroup, p a prime, and E an elementary abelian p-group of p-rank at least 3 acting on G. Suppose that $C_G(a)$ is nilpotent for $a \in E^\times$. Then G is nilpotent.*

PROOF. We proceed by induction on the Morley rank of G. Observe that a p-Sylow subgroup of G is connected (Proposition 8.12), and hence is a p-torus by our hypothesis and Proposition 6.3. It follows easily that any connected abelian definable section of G is p-divisible.

Let A be a minimal definable connected E-invariant normal abelian subgroup of G contained in $Z^\circ(G')$. Then G/A is nilpotent by induction, and $[G, A]$ is either A or 1. It suffices to show that $[G, A] \neq A$. For this we may replace G by $A \rtimes (G/G')$ and hence assume that G/A is abelian.

Let $E_0 \leq E$ be elementary abelian of p-rank 2, and for $v \in E_0^\times$ let $G_v = C_G(v \bmod A)$. Then G is generated by the groups G_v for $v \in E_0^\times$ (Proposition 9.16), and these groups are normal in G. Thus it suffices to prove that the G_v are nilpotent. If $G_v < G$ this follows by induction on the Morley rank of G. Suppose therefore that $G_v = G$ for some $v \in E_0^\times$. Then by Proposition 9.15, we have $G = C_G(v)A$. Now $A = C_A(v) \times [A, v]$ (Proposition 9.15). Now $C_A(v)$ and $[A, v]$ are normalized by $C_G(v)$ and hence by G. If these groups are nontrivial, then by induction $G/C_A(v)$ and $G/[A, v]$ are nilpotent, and hence G, which embeds into $G/C_A(v) \times G/[A, v]$, is nilpotent.

Accordingly we may suppose $A = C_A(v)$ or $C_A(v) = 1$. In the first case $G = C(v)$ is nilpotent by hypothesis. Suppose therefore that $C_A(v) = 1$. Take $E_1 \leq E$ of p-rank 2, not containing the element v. Then by the foregoing, we may suppose that $E_1^\times$ also contains an element w centralizing G/A. Hence the subgroup $E_2 = \langle v, w\rangle$ of E centralizes G/A, and our analysis shows that $C_A(x) = 1$ for $x \in E_2^\times$. However, $A = \langle C_A(x) : x \in E_2 \rangle$ by Proposition 9.16, a contradiction. □

10. Automorphisms

10.1. Finite centralizers.

LEMMA 10.1. *Let G be a connected group of finite Morley rank and α a definable automorphism of G such that $C_G(\alpha)$ is finite. Then the set $[\alpha, G]$ is generic in G.*

PROOF. The function $\gamma : G \to G$ given by $\gamma(g) = g^{-1}\alpha(g)$ has finite fibers: if $g^{-1}\alpha(g) = h^{-1}\alpha(h)$, then $\alpha(gh^{-1}) = gh^{-1}$ and $g \in C_G(\alpha)h$. Hence the rank of the image of γ is the rank of G, □

We may rephrase this as follows.

LEMMA 10.2. *Let G be a group of finite Morley rank and $g \in G$ an element such that $C_{G^\circ}(g)$ is finite. Then the conjugacy class g^{G° is generic in gG°.*

PROOF. If $x \in G^\circ$, then $g^x = g[g, x]$. □

LEMMA 10.3. *Let G be a connected group of finite Morley rank, and α a definable involutory automorphism of G with $C_G(\alpha)$ finite. Then G is abelian and α inverts G.*

PROOF. Let $X = \{[\alpha, g] : g \in G\}$. Then α inverts X and $\mathrm{rk}(X) = \mathrm{rk}(G)$.
Fix $x \in X$ and let $Y = X \cap x^{-1}X$. Then for $y \in Y$ we have $y = \alpha(y^{-1}) = \alpha(xy)^{-1}\alpha(x) = xyx^{-1}$, and $y \in C(x)$. As the rank of Y is equal to the rank of G, we find that $C(x) = G$ and $X \subseteq Z(G)$. As the rank of X is equal to the rank of G, we find that $Z(G) = G$. □

LEMMA 10.4. *Let G be a group of finite Morley rank without involutions, and α a definable involutory automorphism of G. Then $G = C_G(\alpha)G^-$, where $G^- = \{g \in G : g^\alpha = g^{-1}\}$. Furthermore, the decomposition is unique, or, in other words, the multiplication map $C_G(\alpha) \times G^- \to G$ is a bijection.*

PROOF. First, every element of G has a unique square root. If $x \in G$ then $d(\langle x \rangle)$ is abelian and contains no involutions, hence is 2-divisible. So there is $y \in d(\langle x \rangle)$, $y^2 = x$. If also $z^2 = x$, then $x \in C(z)$ and hence $y \in C(z)$, so $(y^{-1}z)^2 = 1$ and $y = z$.
In particular, if $x \in G^-$ and $y^2 = x$, then $(y^\alpha)^2 = x^{-1} = (y^{-1})^2$ and thus $y \in G^-$, so G^- is also uniquely 2-divisible.
Suppose $cg_1 = g_2$ with $c \in C(\alpha)$ and $g_1, g_2 \in G^-$. Then $c = g_2g_1^{-1}$ and applying α, we have $g_2^{-1}g_1 = g_2g_1^{-1}$, or $g_1^2 = g_2^2$, and $g_1 = g_2$. So the multiplication map is injective. It suffices therefore to prove the first claim.
Let $x \in G$, and let $y^2 = [\alpha, x]$. Then $y \in G^-$ since $[\alpha, x] \in G^-$. Now $(xy^{-1})^\alpha = x^\alpha y = xy^{-2}y = xy^{-1}$, so $xy^{-1} \in C(\alpha)$ and $x \in C(\alpha)G^-$. □

LEMMA 10.5. *Let G be a connected group of finite Morley rank without involutions, and α a definable involutory automorphism of G. Then $C_G(\alpha)$ is connected.*

PROOF. By the preceding lemma, G is in a definable bijection with $C_G(\alpha) \times G^-$. As G has Morley degree one, each factor has Morley degree one. □

10.2. Relatively prime actions.

LEMMA 10.6. *Let G be a connected solvable $p^\perp$-group of finite Morley rank and P a finite p-group of definable automorphisms of G. Then $C_G(P)$ is connected.*

PROOF. Let A be a definably characteristic connected infinite abelian subgroup of G. We may suppose inductively that $C_{G/A}(P)$ is connected, and hence we may replace G by $C_G(P \bmod A)$. Thus according to Corollary 9.13, $G = AC_G(P)$. As G is connected, $G = AC_G{}^\circ(P)$ and $C_G(P) = C_A(P)C_G{}^\circ(P)$. But $A = [A, P] \oplus C_A(P)$ by Corollary 9.14, so $C_A(P)$ is also connected, and $C_G(P)$ is connected. □

PROPOSITION 10.7. *Let G be a group of finite Morley rank, π a set of primes, and let Q and X be definable subgroups with Q a solvable π-group of bounded exponent, X a $\pi^\perp$-group, and X acting on Q. Suppose that X acts trivially on the factors Q_i/Q_{i+1} of a definable normal series for Q. Then X acts trivially on Q.*

PROOF. We may deal with the elements of X individually, so replacing X by $d(x)$ for some $x \in X$, we may suppose X is abelian.

Then we may proceed by induction on the length of the series, and hence assume that X acts trivially on Q_1 as well as on Q/Q_1. By Proposition 9.9, we have $Q = C_{Q \bmod Q_1}(X) = Q_1 \cdot C_Q(X) = C_Q(X)$, as claimed. □

LEMMA 10.8. *Let A be an abelian p-group of finite exponent, Q a definable $p^\perp$-group of automorphisms of A acting trivially on $\Omega_1(A)$. Then Q centralizes A.*

PROOF. It will suffice to show that Q acts trivially on each quotient $V_i = \Omega_{i+1}(A)/\Omega_i(A)$. We have an embedding $\pi : V_i \to \Omega_1(A)$ induced by multiplication by p^i, and Q respects this map. The claim follows. □

PROPOSITION 10.9. *Let G be a group of finite Morley rank, π a set of primes, and let Q and X be definable subgroups with Q a solvable $\pi^\perp$-group, X a π-group of bounded exponent, and X acting on Q. Suppose that X acts trivially on the factors Q_i/Q_{i+1} of a definable normal series for Q. Then X acts trivially on Q.*

PROOF. We may proceed by induction on the length of the series, and hence assume that X acts trivially on Q_1 as well as on Q_0/Q_1. By Proposition 9.12, we have $Q = C_Q(Q \bmod Q_1) = Q_1 \cdot C_Q(X) = Q$, as claimed. □

10.3. Connected 2-group actions. Our goal here is to show that unipotent 2-groups can only act trivially on groups without unipotent 2-subgroups. In our analysis the minimal case will be a very special configuration known as a *strongly embedded subgroup*. This will be an important configuration for thorough analysis in Chapter VI. The appearance of the same configuration at this early stage seems to be merely a coincidence.

DEFINITION 10.10. *Let G be a group, M a proper subgroup. We say that M is* strongly embedded *in G if M contains an involution, $M < G$, and for all $g \in G \setminus M$, the intersection $M \cap M^g$ contains no involutions.*

LEMMA 10.11. *Let G be a group of finite Morley rank which contains an involution. Let M be a proper definable subgroup of G. Then the following are equivalent:*

(1) *M is strongly embedded in G.*
(2) *M contains the normalizer of a Sylow 2-subgroup S of G, and for any involution $i \in S$, we have $C_G(i) \leq M$.*
(3) *M contains an involution, and contains the normalizer of each of its nontrivial 2-subgroups.*

PROOF.

(1) $\Longrightarrow$ (3). Let $P \leq M$ be a nontrivial 2-subgroup, $g \in N_G(P)$. Then $P \leq M \cap M^g$, so $g \in M$.

Evidently (3) $\Longrightarrow$ (2).

(2) $\Longrightarrow$ (1). For any involution $i \in M$ there is $h \in M$ so that $i^h \in S$ (here the definability of M is invoked to allow application of Proposition 6.11, and hence $C_G(i^h) \leq M$. Conjugating by h, we find $C_G(i) \leq M$ for all involutions of M.

Now if $i \in M \cap M^g$, we may suppose $i^{g^{-1}} \in S$. Let $j \in Z(S)$ be an involution. Then $j^g \in C_G(i)$ and hence $j^g \in M$. As $S^g \leq C_G(j^g)$, we find $S^g \leq M$. Hence $S^g = S^h$, some $h \in M$, and $gh^{-1} \in N_G(S) \leq M$. Thus $g \in M$, as required. □

LEMMA 10.12. *Let G be a group of finite Morley rank with a strongly embedded subgroup M. Then all involutions in G are conjugate in G, and all involutions in M are conjugate in M.*

PROOF. We prove the first statement, since the second then follows by the definition of strong embedding.

By assumption, M contains an involution i, and as $M < G$ there is an involution j outside M: if $g \in G \setminus M$, then $j = i^g$ is one such. Hence it suffices to prove that for any involutions i, j with $i \in M$ and $j \notin M$, we have i conjugate to j.

If i and j are not conjugate, then by Lemma 2.20 we have an involution k commuting with both. Then $k \in C_G(i) \leq M$, and $j \in C_G(k) \leq M$, a contradiction. □

PROPOSITION 10.13. *Let G be a group of finite Morley rank, and H and U definable subgroups. Suppose that H contains no nontrivial unipotent 2-subgroup, and U is 2-unipotent and normalizes H. Then $[U,H]=1$.*

PROOF. We proceed by induction on the rank and degree of H. We may suppose $G=HU$. $U\cap H$ must be finite as H contains no 2-unipotent subgroups.

We may assume toward a contradiction that the action of U on H is faithful, and U is nontrivial. We may also suppose that U is elementary abelian:

$$N_H(U)=C_H(U)\ . \tag{1}$$

If $h\in N_H(U)$ then $[h,U]\le U\cap H$ so $[h,U]$ is connected and finite, hence trivial, and $h\in C_H(U)$.

In particular, if H normalizes U then H centralizes U, a contradiction.

Suppose that H contains a proper definable G-invariant subgroup H_1. By induction we may suppose that $[U,H_1]=1$. If H_1 is infinite, then after factoring it out induction applies, so we get $[U,H]\le H_1$. But for $h\in H$, commutation with h defines a homomorphism $\gamma_h:U\to H_1$, with unipotent image; so $[U,h]=1$ and $[U,H]=1$, a contradiction. Thus:

$$\text{Any definable } G\text{-invariant proper subgroup of } H \text{ is finite}\ . \tag{2}$$

Hence if H_1 is any definable U-invariant proper subgroup, then since U is connected we have $H_1\le C_H(U)$. Thus if we factor out $C_H(U)$, H will have no nontrivial definable proper U-invariant subgroup. If now the action of U becomes trivial, then $[U,H]\le C_H(U)$ and arguing as before we find $[U,H]=1$. Accordingly we may factor out $C_H(U)$, and assume

$$H \text{ has no nontrivial proper definable } G\text{-invariant subgroup}\ . \tag{2'}$$

In particular, H is connected.

If H is abelian, then $H\rtimes U$ is solvable, and by Corollary 8.4 we have $U\le F(H\rtimes U)$. Now U is maximal 2-unipotent in $H\rtimes U$, and hence H normalizes U, a contradiction. Thus, H is nonabelian and $Z(H)<H$; hence by (2),

$$Z(H)=1 \tag{3}$$

Suppose $Z(G)\ne 1$. Then as $Z(G)\cap H=1$, $Z(G)$ is a 2-group of bounded exponent. If $Z(G)$ is infinite, then $Z^\circ(G)\le U$, contradicting the faithfulness of the action. So $Z(G)$ is finite. Let $\bar G=G/Z(G)$. Then we claim that properties (1), (2′), (3) are preserved.

(1) Suppose that $h\in H$ and $U^h\le UZ(G)$. Then $U^h=(UZ(G))^\circ=U$, so $h\in C_H(U)$.

(2′) and (3) $H\to\bar H$ is a G-equivariant isomorphism.

Furthermore, the action of U on H is still faithful: if $[u,H]\le Z(G)$, then as H is connected, $[u,H]=1$. Now arguing as before, in $\bar G$, we conclude that $Z(\bar{G})$ is finite, hence pulls back to a finite normal subgroup of G. As G is connected, this shows $Z(\bar{G}) = 1$, and replacing G by $\bar{G}$ we now have

$$Z(G) = 1\,. \tag{4}$$

Let $M = N_G(U)$. We claim that M is strongly embedded in G.

Let S be a Sylow 2-subgroup of G containing U. Then U is the maximal 2-unipotent subgroup of S°, and hence $N_G(S)$ normalizes U.

In view of Lemma 10.11, it suffices now to check that $C_G(i) \leq C_G(U)$ for i an involution in S. As U is abelian and $N_H(U) = C_H(U)$, we have $i \in C(U)$.

As $i \in C_G(U)$, the group $C_H(i)$ is U-invariant. If $i \in C(H)$, then $i \in Z(G)$, a contradiction. So $C_H(i) < H$. By induction, therefore, we find that U centralizes $C_H(i)$, so $C_G(i) = C_H(i)U \leq C_G(U)$, as required.

Now as M is strongly embedded in G, it follows that the involutions of U are conjugate in M by Lemma 10.12; but M centralizes U. □

10.4. Automorphisms of p-tori.

DEFINITION 10.14. *Let A be a torus. The* Prüfer p-rank $\mathrm{Pr}_p(A)$ *is the p-rank of $A[p]$, that is, the dimension of $A[p]$ as a vector space over $\mathbb{F}_p$.*

The reason for the terminology is the following: the p-torsion subgroup A_p of A factors as a sum of quasicyclic abelian groups $\mathbb{Z}/p^\infty\mathbb{Z}$, called "Prüfer p-groups" (finitely many, in the case of finite Morley rank).

PROPOSITION 10.15. *Let A be a p-torus of Prüfer p-rank d (finite). Then* $\mathrm{Aut}(A) \simeq \mathrm{GL}(d, \mathbb{Z}_p)$*, naturally, where $\mathbb{Z}_p$ is the ring of p-adic integers.*

PROOF. A is the direct limit of the characteristic subgroups $A[p^n]$. These groups also form an inverse system with respect to multiplication maps rather than inclusions: $\mu_n : A[p^n] \to A[p^{n-1}]$ is given by multiplication by p. The inverse limit $\hat{A}$ is a free module of rank d over $\mathbb{Z}_p$. This gives a canonical embedding $\mathrm{Aut}(A) \to \mathrm{Aut}(\hat{A})$, and after choosing a basis we have $\mathrm{Aut}(\hat{A}) \simeq \mathrm{GL}(d, \mathbb{Z}_p)$. Conversely, any automorphism of $\hat{A}$ acts on the quotients $A[p^n]$, and commutes with multiplication by p, and hence acts on the direct limit A. So we have an identification of $\mathrm{Aut}(A)$ and $\mathrm{Aut}(\hat{A})$. □

This can also be extended to describe the endomorphism ring of a p-torus with the same proof (which implicitly goes via the associated Tate module).

This result can be used to make calculations of the following type.

LEMMA 10.16. *Let A be a p-torus of rank 2 and α an automorphism of A of order p, with finite centralizer in A. Then either $p = 2$ and p operates by inversion, or $p = 3$ and $C_A(\alpha)$ is cyclic of order 3.*

PROOF. We think of α as an element of $\mathrm{GL}(2, \mathbb{Q}_p)$. The eigenvalues of α are p-th roots of unity, and the cyclotomic polynomial ϕ_p is irreducible over

$\mathbb{Q}_p$, by Eisenstein's criterion applied to $((x+1)^p - 1)/x$. As the minimal polynomial for α has degree 2, this forces $p \leq 3$.

The eigenvalue 1 cannot occur, as it would give rise to a fixed vector in $\mathbb{Z}_p^2$ and then to an infinite fixed subgroup in the torus.

So if $p = 2$, then α has the form

$$\begin{bmatrix} -1 & 0 \\ 0 & -1 \end{bmatrix}$$

which is inversion.

If $p = 3$, then, up to conjugacy, α has the form

$$\begin{bmatrix} 0 & 1 \\ -1 & -1 \end{bmatrix}$$

whose centralizer in A (as opposed to $\hat{A}$) consists of all pairs $(a, b) \in A$ satisfying

$$b = a; -a - b = b$$

or $3a = 0; b = a$. This corresponds to a "diagonal" subgroup in A, with respect to the chosen decomposition of A, of order 3 (working modulo 3^n, the elements in question have the representation $i(3^{n-1}, 3^{n-1})$, $i = 0, 1, 2$, a representation which depends on the choice of n). $\square$

The one-dimensional case is also useful.

COROLLARY 10.17. *If Z is a quasicyclic p-group, then* Aut(Z) *is isomorphic to the group of units in the ring $\mathbb{Z}_p$.*

LEMMA 10.18. *Let Z be a quasicyclic p-group, with $p > 2$. Then Z has no nontrivial automorphism of order p.*

PROOF. By the last corollary, such an automorphism would represent an element of order p in the p-adic field $\mathbb{Q}_p$. But the usual proof of the irreducibility of the cyclotomic polynomial ϕ_p, namely by Eisenstein's criterion applied to $\phi_p(1 + x)$, works over $\mathbb{Q}_p$; so this polynomial is irreducible over $\mathbb{Q}_p$. $\square$

10.5. Continuously characteristic subgroups.

DEFINITION 10.19. *Let $H \leq K$ be a pair of groups interpreted in a third group G, which has finite Morley rank. Then H is* continuously characteristic *in K, relative to the ambient group G, if H is invariant under the action of every connected group of automorphisms of K which can be interpreted in G.*

The next lemma will be used at a couple of technical junctures in Chapter VI.

LEMMA 10.20. *Let G be a group of finite Morley rank, and $H \lhd K$ a pair of connected definable subgroups of G with K/H a good torus and H nontrivial. Then H contains a nontrivial connected, definable, continuously characteristic subgroup of K.*

PROOF. If the commutator subgroup K' is nontrivial, then as it is connected and contained in H, it will do. Assume therefore that K is abelian.

If K is not divisible abelian, then for some prime p the annihilator of p in K is infinite by Lemma 2.13, and the connected component of this group is contained in H, so this will do.

Now K contains a unique maximal definable good torus T, and T is normalized, and hence centralized, by any connected definable group of automorphisms of K. If H contains any nontrivial good torus, then it follows that any such is continuously characteristic in K.

So we may suppose that K is divisible abelian and that H contains no nontrivial good torus. Now we claim that any minimal nontrivial definable torus T_0 in H is torsion-free. If not, then T_0 must be the definable hull of its torsion, and being minimal is a good torus, which is a contradiction. Now there is a maximal torsion free definable torus T_1 in K, which is nontrivial as we have just seen. As K/H is a good torus and the image of T_1 in the quotient is torsion-free, it follows that $T_1 \leq H$, and we again have the desired subgroup. □

11. Modules

11.1. Composition series. For the most part we are concerned here with a definable action of a group G of finite Morley rank on an abelian group V. This means that we are working in a context in which both G and V are definable, and the action of G on V is definable, and all of this is interpreted into some group of finite Morley rank. In such cases one can always take the ambient group to be the semidirect product $V \rtimes G$, with predicates distinguishing its subgroups V and G present in the language.

LEMMA 11.1. *Let G be a connected group of finite Morley rank acting definably on an abelian group V. Let $W \leq V$ be an infinite nontrivial G-submodule, not necessarily definable. Then W contains an infinite definable G-submodule of V.*

PROOF. $[G, W] \leq W$ is connected and definable. □

LEMMA 11.2. *Let G be a connected group of finite Morley rank acting definably on an abelian group V. Then there is a series of G-invariant definable submodules*

$$V = V_0 > V_1 > \cdots > V_n = (0)$$

such that each quotient V_i/V_{i+1} is either G-irreducible or trivial.

PROOF. We proceed by induction on the rank and degree of V. Accordingly we may suppose that V is infinite and has no proper infinite definable G-invariant subgroup. In particular, V is connected.

We may suppose that G acts nontrivially on V. Therefore, $V_1 = C_V(G)$ is finite. Then V/V_1 is G-irreducible; otherwise, we have $V_0 > V_1$ G-invariant and finite, hence G acts trivially on V_1 as it is connected. □

When we have a group G acting definably on an abelian group V we must distinguish irreducibility of V as a G-module and *definable* irreducibility—the absence of nontrivial proper definable submodules. When G centralizes V these are very different notions. But otherwise, they tend to coincide.

LEMMA 11.3. *Let G be a connected group of finite Morley rank acting definably and nontrivially on an abelian group V. If V is definably irreducible as a G-module, then V is irreducible.*

PROOF. By Lemma 11.2 V has a definable nontrivial submodule V_0 on which the action is either trivial or irreducible, and by definable irreducibility $V = V_0$. □

DEFINITION 11.4. *Let G be a connected group of finite Morley rank acting on an abelian group V, and $V = V_0 > V_1 > \cdots > V_n = (0)$ a series with trivial and irreducible quotients, as above. Then the quotients V_i/V_{i+1} on which G acts nontrivially are called the* composition factors *with respect to the given series.*

LEMMA 11.5. *Let G be a connected group of finite Morley rank acting on an abelian group V. Let A_i $(i \in I)$ and B_j $(j \in J)$ be the set of nontrivial composition factors with respect to two series for V under the action of G. Then $|I| = |J|$ and after relabeling appropriately with $J = I$, we have $A_i \simeq B_i$.*

PROOF. Let the series be (V_i) and (W_j) respectively, with $A_i = V_i/V_{i+1}$ $(1 \leq i \leq m)$ and $B_j = W_j/W_{j+1}$ $(1 \leq j \leq n)$; then I and J are the corresponding sets of indices, for which the factors are nontrivial.

For any i, j we have

$$\frac{V_{i+1} + (V_i \cap W_j)}{V_{i+1} + (V_i \cap W_{j+1})} \simeq \frac{V_i \cap W_j}{(V_i \cap W_{j+1}) + (V_{i+1} \cap W_j)} \simeq \frac{W_{j+1} + (V_i \cap W_j)}{W_{j+1} + (V_{i+1} \cap W_j)}.$$

For $i \in I$, $j \leq n$, let $i \to j$ mean that j is maximal such that $V_i = V_{i+1} + V_i \cap W_j$. Then $V_i \cap W_{j+1} \leq V_{i+1}$ and

$$\frac{W_{j+1} + (V_i \cap W_j)}{W_{j+1} + (V_{i+1} \cap W_j)} \simeq \frac{V_{i+1} + (V_i \cap W_j)}{V_{i+1} + (V_i \cap W_{j+1})} = \frac{V_i}{V_{i+1}}$$

which is assumed irreducible.

On the other hand,

$$\frac{W_{j+1} + V_i \cap W_j}{W_{j+1} + V_{i+1} \cap W_j} \leq \frac{W_j}{W_{j+1} + V_{i+1} \cap W_j}$$

which must therefore be a nontrivial module, and which is also a quotient of W_j/W_{j+1}; so the latter is nontrivial and hence irreducible by construction. It follows that $W_j \cap V_{i+1} \leq W_{j+1}$, and by irreducibility $W_{j+1} + V_i \cap W_j = W_j$. Thus $j \to i$ in the reverse sense, and the foregoing shows also that corresponding quotients are isomorphic. □

11.2. Clifford Theory.

LEMMA 11.6. *Let G be a group of finite Morley rank acting definably and irreducibly on an abelian group V, and let $H \lhd G$. Suppose that H acts irreducibly on $U \le V$ with U infinite. Then V is completely reducible as an H-module, and in fact V is a finite direct sum of conjugates U^g of U under the action of G.*

PROOF. Let V_0 be a maximal H-submodule of V which is expressible as a direct sum of conjugates of U. Such a maximal submodule exists since $\mathrm{rk}(V) \ge \mathrm{rk}(V_0) = n\,\mathrm{rk}(U)$, where n is the number of conjugates involved.

For $g \in G$, U^g is again an irreducible H-module. Thus $V_0 = \langle U^g : g \in G\rangle$, since if $g \in G$ with U^g not contained in V_0, we find $U^g \cap V_0 = (0)$ and hence V_0 can be extended to $V_0 \oplus U^g$.

It follows that V_0 is a nontrivial G-module, so $V_0 = V$ by irreducibility. □

11.3. Faithful solvable actions.

PROPOSITION 11.7. *Let H be a connected solvable $\pi^\perp$-group of finite Morley rank acting faithfully on a nilpotent π-group V of bounded exponent. Then H is a good torus.*

PROOF. We work in $G = V \rtimes H$. Observe that $F(G) = V(F(G) \cap H)$ and that $F(G) \cap H$ centralizes V, hence is trivial by faithfulness. Thus $F(G) = V$ and $H \simeq G/F(G)$ is a torus.

Take a normal G-invariant series $V = V_0 > V_1 > \cdots > V_n = (0)$ with successive quotients finite or G-minimal. The stabilizer of this chain in H is trivial by Proposition 10.7. Thus by considering the action on all of the quotients $A_i = V_i/V_{i+1}$, we get a definable injection of H into $\prod_i \bar{H}_i$, where $\bar{H}_i$ is the image of H in $\mathrm{Aut}\, A_i$.

Now the $\bar{H}_i$ are subgroups of multiplicative groups of fields of finite characteristic, and hence are good tori by Proposition 4.20. So H is a good torus by Corollary 4.22. □

LEMMA 11.8. *Let $G = H \rtimes T$ be a group of finite Morley rank with H, T, and the action of T on H definable. Assume that H is a nilpotent π-group of bounded exponent and T is a connected solvable $\pi^\perp$-group. Then*

$$H = [H,T] \oplus C_H(T)\ .$$

PROOF. We know $H = [H,T] \cdot C_H(T)$ by Corollary 9.10, and the claim holds if T has bounded exponent by Corollary 9.14.

We may suppose that T acts faithfully on H, and then by Proposition 11.7 it follows that T is a good torus, and in particular abelian.

We will argue by induction on the rank and degree of H. Suppose $T \ne 1$ and pick a nontrivial torsion element $t \in T$. Then by the bounded exponent case we have $H = [H,t] \oplus C_H(t)$. Let $H_1 = C_H(t) < H$. By induction we have $H_1 = [H_1, T] \oplus C_{H_1}(T) = [H_1, T] \oplus C_H(T)$. Thus $H = ([H, t] \oplus [H_1, T]) \oplus C_H(T)$.

Now as $[H, t]$ and $[H_1, T]$ are T-invariant, we have $[H, T] = [[H, t], T] \oplus [H_1, T] \leq [H, t] \oplus [H_1, T]$ and hence $[H, T]$ centralizes $C_H(T)$ and is disjoint from it, as claimed. $\square$

12. Thompson $A \times B$

LEMMA 12.1 ([**177**], (1.13), page 8). *Let $G = HK$ be a group of finite Morley rank, where H is a definable normal π-subgroup of bounded exponent and K is a definable $\pi^\perp$-subgroup. Then $[[H, K], K] = [H, K]$.*

PROOF. By Lemma 3.30, $[H, K]$ and $[[H, K], K]$ are definable subgroups of G. Let $N = [H, K]K$. Then N is definable and normal in G. Furthermore, N is the smallest normal definable subgroup of G such that G/N is a π-group: if M is definable normal in G and $\bar{G} = G/M$ is a π-group, then $\bar{K}$ is a $\pi^\perp$-group hence $K \leq M$ (Lemma 2.18), hence also $[H, K] \leq M$. In particular, N is a definably characteristic subgroup of G.

Similarly, $N_1 = [[H, K], K]K$ is the smallest definable normal subgroup of N such that N/N_1 is a π-group, and N_1 is definably characteristic in N. Hence $N_1 \lhd G$. But $K \leq N_1$, hence $N_1 = N$, and $[[H, K], K] = [H, K]$. $\square$

We state the $A \times B$-lemma in two equivalent forms:

PROPOSITION 12.2 ([**177**, (1.15), (1.15)′]). *Let G be a group of finite Morley rank. The following conditions on G are equivalent:*

(1) *Let A be a definable $p^\perp$-subgroup of G. Let B be a definable p-subgroup of $C_G(A)$ of bounded exponent. Suppose $A \times B$ normalizes a definable subgroup p-subgroup P of bounded exponent. If A centralizes $C_P(B)$, then A centralizes P.*
(2) *Let Q be a definable p-subgroup of G of bounded exponent and U be a definable subgroup of Q such that $C_Q(U) \leq U$. Suppose A is a definable $p^\perp$-subgroup of G that normalizes Q and centralizes U. Then A centralizes Q.*

PROOF. We first prove (1) implies (2). Let U and Q be as in (2). The group $A \times U$ normalizes Q. By the assumption A centralizes $C_Q(U)$. Therefore, we can apply (1) with $B = U$ and $P = Q$ and conclude that A centralizes Q.

Now we prove (2) implies (1). By considering the semidirect product $P \rtimes (A \times B)$, we may assume that $B \cap P = 1$. In particular, $N_P(B) = C_P(B)$. Let $Q = BP$. Q is a definable p-group. If $U = N_P(B)B$, then A centralizes U and $C_Q(U) \leq C_Q(B) \leq N_Q(B) = N_P(B)B = U$. An application of (2) proves the result. $\square$

We will prove Proposition 12.2 (2). First a special case:

LEMMA 12.3 ([**177**, (1.16)]). *Let G be a group of finite Morley rank. Let X be a definable π-subgroup of bounded exponent of G and Y a normal* *definable subgroup of X such that $C_X(Y) \leq Y$. Suppose that A is a definable $\pi^{\perp}$-subgroup of G that normalizes X and centralizes Y. Then A centralizes X.*

PROOF. As $[A, Y] = 1$ and $Y \lhd X$, we have $[[X, Y], A] = 1$. Clearly, $[[Y, A], X] = 1$. By the three subgroups lemma, $[[A, X], Y] = 1$. Therefore, $[A, X] \leq C_X(Y) \leq Y$. Lemma 12.1 implies that $[A, X] = [[X, A], A] = 1$. □

Now we prove the $A \times B$-lemma:

PROPOSITION 12.4 ([**177**]). *Let G be a group of finite Morley rank whose Sylow p-subgroups are nilpotent-by-finite and of bounded exponent. Let Q be a definable p-subgroup of G and U be a definable subgroup of Q such that $C_Q(U) \leq U$. Suppose A is a definable $p^{\perp}$-subgroup of G that normalizes Q and centralizes U. Then A centralizes Q.*

PROOF. Let $Y = C_Q(A)$. Then $U \leq Y$ and we have $C_Q(Y) \leq C_Q(U) \leq U \leq Y$. Let $X = N_Q(Y)$. By Lemma 12.3, A centralizes X. Therefore, $X \leq Y$. But Q is a nilpotent-by-finite p-group and thus satisfies the normalizer condition. Therefore, $X = Y = Q$. □

13. Complex reflection groups

DEFINITION 13.1. *A linear transformation on a finite dimensional vector space is a (generalized, or complex)* reflection *if it is diagonalizable and has a fixed space of codimension exactly one. A* real *or* ordinary *reflection is a complex reflection of order two. Note that the identity is not considered to be a reflection.*

The finite groups generated by reflections were classified by Shephard and Todd [**168**]. Their numbering is referred to as the *Shephard-Todd* numbering. The table at the end of this section gives some of the properties of "sporadic" finite irreducible complex reflection groups in dimension at least two, organized according to the following scheme: *Shephard-Todd Number; dimension of the representation; Coxeter label (if applicable); Group order; Order of the center; Orders of reflections*, where the last item refers to the orders of the complex reflections occurring in the group. In groups defined over the real field these reflections must have order 2. There are also three infinite families: the first contains the standard representation of the symmetric group (Coxeter type A_n), the third consists of dihedral groups acting in dimension 2 and the second is a series $G(m, l, n)$ to which we will return below.

It will be observed that four of the groups listed are crystallographic Coxeter groups associated with exceptional Dynkin diagrams. Other than that, the most interesting group is probably the one with number 12, which crops up in various contexts such as singularity theory.

Series #2 in the Shephard-Todd classification is a family of groups denoted $G(m, \ell, n)$, where n is the dimension of the associated vector space, and m, ℓ are parameters with ℓ a divisor of m, which for $m = 2$ correspond to the Coxeter groups B_n (or C_n) and D_n. The groups $G(m, l, n)$ may be described explicitly as follows [**76**, p. 386]. Let $A(m, l, n)$ be the group of diagonal matrices D for which $D^m = 1$ and $\det(D)^{m/l} = 1$. Then $G(m, l, n)$ is the semidirect product $A(m, l, n) \rtimes \Pi_n$ with Π_n the group of permutation matrices.

We use the foregoing information to derive a criterion for a finite group to be isomorphic to an irreducible Coxeter group.

THEOREM 13.2. *Let W be a finite group, $I \subseteq W$ a subset, and n an integer, satisfying the following conditions.*

(1) *The set I generates W, consists of involutions, and is closed under conjugation in W.*
(2) *The graph Δ_I with vertices I and edges (i, j) for noncommuting pairs $i, j \in I$ is connected.*
(3) *For all sufficiently large prime numbers ℓ, W has a faithful representation V_ℓ over the finite field $\mathbb{F}_\ell$ in which the elements of I operate as complex reflections, with no common fixed vectors.*

Then one of the following occurs.

(a) *W is a dihedral group acting in dimension $n = 2$, or cyclic of order two.*
(b) *W is isomorphic to an irreducible crystallographic Coxeter group, that is, A_n, B_n, C_n, D_n ($n \geq 3$), E_n ($n = 6, 7,$ or 8), or F_n ($n = 4$),*
(c) *W is a semidirect product of a quaternion group of order* 8 *with the symmetric group* Sym_3, *acting naturally, represented in dimension* 2.

If, in addition, over some field, W has an irreducible representation of dimension at least 3, *in which the elements of I act as reflections, then case* (*b*) *applies.*

PROOF. Note that as W is generated by finitely many reflections, the dimensions of the representations V_ℓ are bounded. Let V be a nonprincipal ultraproduct of these representations, which is a representation of W over the field F obtained as the corresponding ultraproduct of the finite fields $\mathbb{F}_\ell$. Then the field F has characteristic zero and cardinality $2^{\aleph_0}$, and can be identified with a subfield of the complex field $\mathbb{C}$. Let $\tilde{V}$ be the complexification of V; we consider W with its complex representation $\tilde{V}$.

Then V and $\tilde{V}$ are finite dimensional as well, over their respective fields, and the elements of I operate as (ordinary) reflections on V and hence on $\tilde{V}$. We claim that the action of W on $\tilde{V}$ is irreducible. The action is completely reducible since W is finite and the characteristic is zero. If $\tilde{V}$ is reducible, then it factors as $V_1 \oplus V_2$ with V_1, V_2 nontrivial invariant subspaces. Then setting $I_i = \{w \in I : [w, \tilde{V}] \leq V_i\}$, it follows that (I_1, I_2) is a partition of I into commuting subsets, one of which must be empty. So we may suppose

$[I, \tilde{V}] \subseteq V_1$, so $[I, V] < V$; as V is an ultraproduct this yields $[I, V_\ell] < V_\ell$ for infinitely many ℓ, a contradiction.

We remark that the same argument shows that for ℓ not dividing the order of W, if the elements of I act as complex reflections on a vector space over $\mathbb{F}_\ell$ and have no common fixed vectors there, then the representation in question is irreducible.

Now returning to our complex representation, the classification of the irreducible complex reflection groups applies. Leaving aside the Coxeter groups, we have to deal with the groups numbered 4–27 or 29–34, as well as those of the form $G(m, l, n)$ with $m > 2$.

By a slight variation of Schur's lemma, we claim that the center of W acts via scalar matrices in every representation V_0 in which the generating set I acts via reflections. Take $z \in Z(W)$ and take $i \in I$. Then z preserves the one-dimensional space $[i, V_0]$ and hence has an eigenvalue α on this space. The α-eigenspace for z is W-invariant and hence equal to V.

Accordingly, the order of the center of W divides $\ell - 1$ for all sufficiently large primes ℓ. By Dirichlet's theorem, there are arbitrarily large primes congruent to -1 modulo $|Z(W)|$, and hence $|Z(W)|$ divides 2. But after leaving aside the crystallographic Coxeter groups, $|Z(W)| > 2$ with the exception of the groups numbered

$$4, 12, 23, 24, 30, 33$$

in the table following. As the last column in the table shows, group #4 contains no ordinary reflections, and may be excluded. Group #12 is referred to in case (c).

We claim that W cannot occur twice on our list. If $W \simeq G(m, \ell, n)$, then in any representation over $\mathbb{C}$, $A(m, \ell, n)$ is diagonalizable and its eigenspaces are permuted by W, so the representation is imprimitive. But the individually listed groups are primitive. So there is no overlap between the family $G(m, \ell, n)$ and the groups listed. As the Fitting subgroup of $G(m, \ell, n)$ is $A(m, \ell, n)$, it is easy to recover both m and n from the group $G(m, \ell, n)$; so any group $G(m, \ell, n)$ occurs at most once. The remaining groups on our list are of distinct orders. So the dimension n of the representation $\tilde{V}$ is independent of the nonprincipal ultrafilter chosen, and hence all but finitely many of the representations V_ℓ have dimension n.

For the groups numbered 23, 24, 30, 33 one works with the order, which must divide the order of $\mathrm{GL}_n(\ell)$ for almost all primes ℓ. We use the fact that the orders shown are divisible by the values 5, 7, 5^2, and 3^4 respectively, in dimensions 3, 3, 4, 5 respectively. For example in case 33 we may take ℓ congruent to 2 mod 3^4, so that $|\mathrm{GL}_5(\ell)|$ is congruent to $2^{10}(2^5 - 1)(2^4 - 1)(2^3 - 1)(2^2 - 1)(2 - 1)$, and the only factors divisible by 3 here are $2^4 - 1$, $2^2 - 1$ giving a factor of 3^2 but not 3^4, a contradiction.

It remains to consider the groups $G(m, l, n)$ with $m > 2$. We will work with particular elements of $G(m, l, n)$. Let ζ be a primitive m^{th} root of unity and let D_1, D_2 be the following diagonal matrices, considered as elements of W:

$$\operatorname{diag}(\zeta, \zeta^{-1}, \dots); \qquad \operatorname{diag}(\zeta, \zeta, \zeta^{-2}, \dots)$$

where diagonal entries not shown all equal 1. The coefficients are not necessarily in the base field F; this is the representation after complexification. However, the traces $\tau_1 = \zeta + \zeta^{-1}$ and $\tau_2 = 2\zeta + \zeta^{-2}$ are in the base field, and as this is an ultraproduct, with respect to whatever ultrafilter we like, it follows that we have similar elements τ_1, τ_2 in any field prime $\mathbb{F}_\ell$ with ℓ sufficiently large; that is, there is a primitive m-th root of unity ζ_ℓ in an extension of $\mathbb{F}_\ell$ for which the corresponding formulas hold.

Now one finds that $(\tau_1 - 2)\zeta = \tau_2 - \tau_1^2 + 1$, and over $\mathbb{F}_\ell$ this implies that either $\tau_1 = 2$ or $\zeta \in \mathbb{F}_\ell$; but when $\tau_1 = 2$ the equation $\zeta + \zeta^{-1} = 2$ yields $\zeta = 1$, and hence in either case $\zeta \in \mathbb{F}_\ell$. This means that m divides $\ell - 1$ for almost all ℓ, and hence $m \leq 2$, which corresponds to a Coxeter group.

This exhausts the treatment of all cases and proves that one of cases $(a - c)$ occurs.

Turning to the final point, if W has a faithful representation in which the elements of I act as reflections, in dimension $d \geq 3$, then it is certainly not dihedral. As far as the group listed as #12 is concerned (case (c)), this is generated by three reflections and hence has no suitable representation in dimension 4 or more. In dimension 3, since the commutator subgroup of W is the extension of a quaternion group Q by a cyclic group of order 3, and the center of Q is central in W, we find first that the central involution of Q is scalar, and second that it has no square root in SL_3, hence none in Q, and this is a contradiction. □

Number	Dim.	Name	$\lvert W\rvert$	$\lvert Z(W)\rvert$	$\lvert r\rvert$ (possible)
4	2	#4	$2^3 * 3$	2	[3]
5	2	#5	$2^3 * 3^2$	6	[3]
6	2	#6	$2^4 * 3$	4	[2, 3]
7	2	#7	$2^4 * 3^2$	12	[2, 3]
8	2	#8	$2^5 * 3$	4	[4]
9	2	#9	$2^6 * 3$	8	[2, 4]
10	2	#10	$2^5 * 3^2$	12	[3, 4]
11	2	#11	$2^6 * 3^2$	24	[2, 3, 4]
12	2	#12	$2^4 * 3$	2	[2]
13	2	#13	$2^5 * 3$	4	[2]
14	2	#14	$2^4 * 3^2$	6	[2, 3]
15	2	#15	$2^5 * 3^2$	12	[2, 3]
16	2	#16	$2^3 * 3 * 5^2$	10	[5]
17	2	#17	$2^4 * 3 * 5^2$	20	[2, 5]
18	2	#18	$2^3 * 3^2 * 5^2$	30	[3, 5]
19	2	#19	$2^4 * 3^2 * 5^2$	60	[2, 3, 5]
20	2	#20	$2^3 * 3^2 * 5$	6	[3]
21	2	#21	$2^4 * 3^2 * 5$	12	[2, 3]
22	2	#22	$2^4 * 3 * 5$	4	[2]
23	3	H_3	$2^3 * 3 * 5$	2	[2]
24	3	#24	$2^4 * 3 * 7$	2	[2]
25	3	#25	$2^3 * 3^4$	3	[3]
26	3	#26	$2^4 * 3^4$	6	[2, 3]
27	3	#27	$2^4 * 3^3 * 5$	6	[2]
28	4	F_4	$2^7 * 3^2$	2	[2]
29	4	#29	$2^9 * 3 * 5$	4	[2]
30	4	H_4	$2^6 * 3^2 * 5^2$	2	[2]
31	4	#31	$2^{10} * 3^2 * 5$	4	[2]
32	4	#32	$2^7 * 3^5 * 5$	6	[3]
33	5	#33	$2^7 * 3^4 * 5$	2	[2]
34	6	#34	$2^9 * 3^7 * 5 * 7$	6	[2]
35	6	E_6	$2^7 * 3^4 * 5$	1	[2]
36	7	E_7	$2^{10} * 3^4 * 5 * 7$	2	[2]
37	8	E_8	$2^{14} * 3^5 * 5^2 * 7$	2	[2]

TABLE 1. Sporadic complex reflection groups

14. Notes

We generally follow [**52**]. The main exceptions are the material on Carter subgroups, where we follow Frécon, and our treatment of two more specialized topics: the Thompson $A \times B$-theorem, which was treated in [**4**], and the material on complex reflection groups, for which we follow [**60**], which varies somewhat from the earlier [**34**].

The main focus of this set of tools can be seen to lie in the theory of solvable groups, and the closely related Schur-Zassenhaus material. A natural complement to the solvable theory for our purposes is furnished by the theory of Chevalley groups, covered in the next chapter. Putting the two together, we will get the theory of "K-groups".

§1. General group theory

Our first section is a miscellaneous collection of results from a variety of sources, many of them collected in [**52**]—frequently in the form of exercises.

One may notice a preoccupation with the commutator calculus and some aspects of the theory of groups with various finiteness conditions (locally finite, minimum condition on centralizers) and localization of finite notions (locally nilpotent, locally solvable), which go hand in hand. This furnishes very useful tools for the extension of results from finite group theory to our context. In the generalizations of these results to broader classes of interest in model theory, this aspect of the theory takes on a particular importance, as can be seen in the development of [**186**]. Here we get by with comparatively little in this vein, as our strong rank hypothesis simplifies some matters.

Whether the theory of $E(G)$ belongs here is uncertain, but it is closely tied to the three subgroups lemma, and a portion of this theory is completely general.

The versions of the Schur-Zassenhaus theorem given in this section are needed for the Hall theory in §8, and are due to Hartley.

The forms of Schur-Zassenhaus more commonly used in analyzing groups of finite Morley rank are given in §9.

§2. Rank

This section contains not only the key definitions but a number of powerful results from the early days of the subject.

We have reworded the rank axioms slightly. These were originally intended to serve as axiomatizations of Morley's rank notion "in a single model"—which, however, does not entirely obviate the need for an occasional compactness argument. Poizat devised the formulation given here (in a slight variant) and proved its equivalence to Morley's notion *in the context of groups*—outside that context, the equivalence fails. This analysis is fully developed in his [**156**], which is the essential reference for this point. From a practical point of view, the main point is that these axioms are true, in groups, whenever one has finite Morley rank—with additivity coming (to the model theorist) as an unexpected bonus.

As a result of these foundational matters being well in hand, the bulk of the later developments have been more group theoretic than model theoretic, with consequences of Wagner's results on fields of finite Morley rank in §4, coming as the major exceptions to this rule, to date.

We use the terms "definable" and "interpretable" interchangeably here; they refer, technically, to first order definability, with parameters, in G^{eq} (the extension of G by all equivalence classes for 0-definable equivalence relations). We sometimes say "definable" *over* rather than *in* to emphasize this point.

Lemma 2.7 would be considered fundamental from a model theoretic point of view. It gives the "stability" of the theory, and points toward generalizations of much of the theory. In our context (finite rank) it can usually, but not always, be replaced by more brutal, less combinatorial, considerations.

Lemma 2.8 is the chain condition of Baldwin and Saxl [**23**], one of the more subtle of the early results in the area.

The definable hull is an important tool in coming to terms with the need to work with undefinable groups, a line which has been considerably developed in Frécon's work on the theory of solvable groups, and which enters the theory early on, as soon as one takes up the Sylow theory.

The structure of abelian groups of finite Morley rank (Lemma 2.13) was given by Macintyre in [**138**] in preparation for his work on fields ([**137**], §4 below). This pair of articles were the first to suggest that stability assumptions had conventional algebraic content, and sparked a good deal of research, playing a role in the formulation of Zilber's conjectures (later, the Zariski structures of Hrushovski/Zilber) as well as the line of work represented here.

The Basic Fusion Lemma (Lemma 2.20) is a familiar group theoretic fact which fails badly in "superstable" groups. Its truth in our context, while not subtle, has a major impact on the theory. It is not out of the question that some of the structural analysis given here can be generalized substantially to contexts where this principle fails, but this would have to be reexamined from scratch, as everything we do with involutions here goes back to this principle.

§3. Connected groups

Lemma 3.6 is given in [**68**].

Lemma 3.10 is a version of an early result of Reineke.

The versatile Proposition 3.19 comes from [**195**], along with various applications given here. The simplest form—definability of groups generated by connected subgroups—will be used in a variety of further ways later on. It more or less guarantees the existence of a decent "connected" version of any subgroup in use in finite group theory.

A recurrent theme, and one we do not address systematically, is the behavior of all of our notions under elementary extension. We remark that if G^* is an elementary extension of G, then the natural relations hold:

$$(G^\circ)^* = (G^*)^\circ; \quad [G : G^\circ] = [G^* : G^\circ]$$

and are straightforward. One should be careful, however; although G° is definable, it does not follow from this that the function $H \to H^\circ$ is itself definable (this is meaningful whenever H varies over a uniformly definable family). Cf. Fact 3.17.

The simplicity of "definably simple" nonabelian groups (Proposition 3.25) is one of its less obvious consequences, also given by Zilber.

The full story on definability of commutators, including the troublesome "finite bits," is given in [**52**] as Corollary 5.31.

§4. Fields

The fundamental linearization results come from Zilber's abstract [**194**].

The result of Macintyre given as Proposition 4.2 was one of the earliest results relating to the meaning of stability in algebraic contexts, contemporaneous with his analysis of abelian groups in Lemma 2.13 [**138, 137**].

Lemma 4.3 is a quirk of the finite rank context, but useful. It can be strengthened considerably: there is no proper definable multiplicative subgroup containing a (not necessarily definable) infinite subfield.

Evidently, fields of finite Morley rank can have nontrivial automorphisms: powers of the Frobenius provide an example. These are the only examples known, and the problem of the existence of further examples is one of the outstanding open problems; Wagner's result, Proposition 4.18, casts some light on this.

Proposition 4.27 is due to Rose [**163**].

Bad fields made their appearance early, in Zilber's first field interpretation results, as emphasized here. The name came considerably later, as their relevance to the structure of simple groups of finite Morley rank became clearer. Wagner's result, Proposition 4.18 [**188**] has a number of practical consequences.

Some consequences for the theory of linear groups of finite Morley rank are given in [**159**], and applied to groups of even type with strongly embedded subgroups in [**11**].

The theory given here (in Parts B and, notably, C) was initially developed without the benefit of Wagner's result, but it has allowed us to considerably extend the scope of our results beyond our initial intentions.

For example, the material given in Chapter VI went through three stages of development. In the first instance the results were proved for even type K^*-groups not involving bad fields, beginning with [**1**]. After Jaligot's thesis [**125**] the "no bad field" hypothesis fell out of use in even type. At this stage more solvable group theory came into play, along with a much closer analysis of involutions and their fusion.

The difficulty with these approaches is that even if one is completely successful, in the end one is only dealing with groups having no simple degenerate sections. On the other hand, Proposition 10.13, from [**2**], suggests a "decoupling" of the degenerate type sections from the group theoretic analysis; but in following up this idea it turned out that Wagner's results, and the general theory of good tori to which they naturally lead, were essential [**12, 13, 14**].

When this idea came along quite a bit of material was already in print based on the K^* hypothesis, and we had to go back to the beginning and rework it in the L^* case. The material in Chapters VI and VII became considerably more delicate in the process, and everything afterward required some degree of adaptation, sometimes more, sometimes less, to fit the more general context. As a result of all this reworking, the publication of the K^* case was interrupted, and subsumed under the material in the present work.

§5. Nilpotent groups

The notion of nilpotence lends itself to considerable development outside the context of finite groups, and in the finite rank context is well behaved, and close to the abelian theory. Proposition 5.8 [**147**] was intended to be a direct analogue of Macintyre's analysis in the abelian case, and plays much the same role. The Fitting subgroup was introduced in [**146**] and in [**28**] (in the latter case, along with the generalized Fitting subgroup).

The Frattini subgroup has seen less use, but it begins to play more of a role in the work of Wagner and Frécon.

The results on generic equations in nilpotent groups are due to Jaber [**120**], in a broader context (stable groups).

The notion of unipotence has been extended to torsion free groups in [**60**] and has found substantial applications, notably in the theory of groups of odd type and the general theory of Carter subgroups (not necessarily in solvable groups).

§6. Sylow theory

The theory comes from [**53**], with the second point in Lemma 6.15 added in [**160**]. The type of classification it produces is fundamental for the approach taken in the present work, modeled on techniques of finite group theory. These can say very little about groups without involutions, as the techniques of the Feit-Thompson Theorem—and certainly the character theory—find no echo in our context. Our emphasis in the present work is on the more complete part of the theory, dealing with mixed type and even type. We will say something about the status of the other two cases, where considerable information is also available, at the end.

Recently there has been some improvement in our understanding of p-torsion for arbitrary primes p [**47, 64**]. Some of this is may be seen in Chapter IV.

§7. Generalized Fitting subgroup

The generalized Fitting subgroup plays a central role in the analysis of finite simple groups, one which was noticed comparatively late. Its generalization to our context, given the Fitting subgroup, is straightforward.

§8. Solvable groups

The exploitation of field interpretability results to elucidate the structure of solvable groups comes from Zilber's abstract [**194**], which includes the analog of the Lie-Kolchin theorem, rediscovered by Nesin ([**145**]).

The most incisive recent work on the general theory of solvable groups of finite Morley rank is in Frécon's thesis [**89**] and subsequent publications. There is now a very substantial body of theory which has considerable application in the structural analysis of groups of odd type, a subject which could take up another book the size of this one. See in particular [**95**], a review article by Frécon and Jaligot.

The Carter subgroup was treated first by Wagner in [**186**], and a full theory given by Frécon in a series of papers beginning with his thesis [**89**]. We have followed [**92**] here. His theory is considerably more general and more extensive than the one given here. We remark that the detailed theory of solvable groups is particularly relevant to the structural analysis of minimal simple connected groups, which in even type is not handled as a separate topic, but which comes into its own in the treatment of odd type, where problems are often reduced to the minimal simple connected case and then handled by close analysis there.

The theory of Carter subgroups is very powerful. To some extent it has given way in the study of even type groups to the closely related, but more specialized, theory of good tori. We will revisit this topic in §IV 1.

The final argument in the proof of 8.24 is given by D. R. Taunt [**118**, Theorem 11.17].

For the Hall theory we follow [**15**] This would actually follow from earlier results in the locally finite case, given the special case stated as Corollary 8.8. However, there seems to be no more direct route to that specific result.

For the Schur-Zassenhaus theory (§9) we follow the treatment of [**52**], based on papers of Borovik and Nesin.

§9. Schur-Zassenhaus

§9.1: See also Frécon's formation theory in [**92**]; cf. also [**95**].

§10. Automorphisms

This is largely a miscellany, consisting of results on automorphisms (and in particular, centralizers of elements) which do not fall neatly anywhere else. But these are important results. The result on actions of unipotent 2-groups on groups without unipotent 2-subgroups is fundamental, and many of the other topics covered here recur in our analysis.

For more on strong embedding see the notes to Chapter 5. Regarding the definition of strong embedding, we mention the following here.

REMARK 14.1. *Let G be group of finite Morley rank with a subgroup M satisfying any of the conditions of Lemma 10.11. If a Sylow 2-subgroup of G is infinite, then M is definable.*

Compare Lemma II 4.16.

EXAMPLE 14.2. *Let G be a generalized dihedral group $A \rtimes \langle i \rangle$ where i is an involution inverting the divisible abelian torsion-free group A. Let $M = A_0 \rtimes \langle i \rangle$ with $A_0 \leq A$ arbitrary. Then M is typically not definable, and satisfies conditions (2), (3) of Lemma 10.11. On the other hand, M is strongly embedded if and only if A_0 is 2-divisible.*

§11. Modules

This is all modeled tightly on the usual theory, but we cannot work with modules of finite length because of the possible presence of trivial composition factors (not necessarily submodules). The notion of G-irreducibility tends to give way to the related but distinct notions of definable G-irreducibility, and more particularly G-minimality. The connection between these definable irreducibility and minimality is given in Lemma 8.1, and the connection between ordinary and definable irreducibility is given in Lemma 11.3. The three notions do not differ by much.

§12. **Thompson** $A \times B$.

The proof of Thompson's $A \times B$-lemma that we give is a translation of the arguments in [**177**]. It was worked out in [**4**], in connection with "elimination of cores in $C(i)$".

§13. **Complex reflection groups**

The idea of bringing the theory of complex reflection groups to bear in the theory of groups of finite Morley rank as a way of recognizing Coxeter groups is introduced in [**34**].

The classification of finite complex reflection groups was given in [**168**] and again, modernized, in [**76**]. Theorem 13.2 is a weaker form of the criterion given in [**36**], and will be used in §III 10, following [**36**]. The analysis we give was presented in detail in [**60**, p. 98], from which we take also the table, in essentially the same form.

Remark. The Hall and Carter theories have useful generalizations to the "prime" $p = 0$, which turns out to split into infinitely many primes of the form 0_r with r a *rank*. This is very useful for the analysis of groups of odd type, but is not needed in even type. This theory should be part of any comprehensive treatment of the general theory of groups of finite Morley rank. The generalized Hall theory includes a Sylow theory for these "primes". We refer to [**62**] for this theory, and to [**66**] for an application.

CHAPTER II

K-groups and L-groups

... as we know, there are known knowns; there are things we know we know.
— D. Rumsfeld, 2002

Introduction

Having laid out the more general features of our theory, largely analogous to the finite case but with a number of quite distinctive features which will play a prominent role in the applications, we turn now to an examination of the properties of *K-groups*—that is, groups whose definable connected simple sections are Chevalley groups. We deal only with those properties, a motley crew, which we will actually need in our inductive analyses. Mainly, these can be expressed as properties of the simple sections, or occasionally their central extensions. So we collect some general facts about simple Chevalley groups, or from another point of view, simple algebraic groups. Another useful class of properties, and one which is not directly expressed in terms of properties of the simple or quasisimple factors, consists of the verifications of the conclusions of our various classification theorems from Part C, in the particular context of K-groups. This will provide the basis for our inductive treatment of these problems.

We must also go beyond the class of K-groups, to L-groups, which are permitted to have definable connected simple sections of degenerate type. In general, our results for L-groups follow quickly from the corresponding results for K-groups, using the general principle given in Lemma 6.3, which may be expressed loosely as saying that the *active portion* of an L-group is a K-group. This is an incarnation of Proposition I 10.13.

The first three sections concern Chevalley groups or algebraic groups. The availability of two different languages for discussing this material presents certain choices and complications. At some level these two languages become interchangeable, but they represent distinct points of view. What we aim at in our classification theorems is the recognition of our groups as Chevalley groups. The fact that they are then algebraic is really a theorem about Chevalley groups—though since it is a theorem that comes early in the subject, one tends to lose sight of this. At the same time, the language of algebraic groups is probably more convenient for discussing both the facts themselves and their proofs. So we adopt the policy that our primary language for discussing the subject is the language of algebraic groups, but when the focus is moving in the direction of generators and relations one may speak more of Chevalley groups.

We emphasize that we do not use the classification of the simple algebraic groups, but rather its converse. This is the natural flow of argument.

We begin with some general properties of algebraic groups, utilitarian and miscellaneous in character. This is supplemented by a discussion under the heading of Chevalley groups in the following section. We also need the theory of central extensions of simple Chevalley groups, and we remark that there are normally three different theories of central extensions: one in the category of algebraic groups, one in the category of abstract groups, but limited to extensions with finite kernel, and one in the category of abstract groups as such. We have a fourth category: groups of finite Morley rank. Using the abstract theory we show that our central extensions fall in the algebraic category. Again, this is the concatenation of two distinct results: one, that in our finite Morley rank category the universal central extension is the one given by Steinberg in the Chevalley context (one can hardly call this a category); and two, that this is also the universal extension in the algebraic category, which we simply quote from Steinberg.

Because a group of finite Morley rank which is a perfect central extension of a simple Chevalley group is in the algebraic category, we may conclude from the algebraic theory (or for that matter directly from the Chevalley theory) that the center is finite, and is contained in a maximal torus.

From §4 onward, we deal with K-groups and L-groups as a class. We deal first with the issue of how much "additional structure" a simple algebraic group over an algebraically closed field can carry, in an enriched language. By constructions of Hrushovski and others, a considerable amount of pathology is possible (the rank of the base field can be greater than 1, there can be more than one field structure on a given set, and so forth). But results of Wagner and Poizat limit the impact of this sort of pathology on the groups defined over such fields, particularly in positive characteristic.

The general structure of K-groups is very straightforward. We saw in the previous chapter that for a connected solvable group H of finite Morley rank, H modulo the Fitting subgroup is abelian. Similarly, for a connected K-group H, H modulo the solvable radical is a direct sum of finitely many simple algebraic groups. Using the theory of central extensions, one has a similar result modulo the connected part of the solvable radical, using quasisimple factors. Both versions are useful throughout.

At this point we consider of a more technical issue that is inspired by the techniques of finite group theory: weak embedding. As preparation for later analysis, we describe the K-groups with *weakly embedded* subgroups. In the last subsection we prove a small but useful result on the existence of Sylow 2-subgroups invariant under the action of a given group.

The discussion of K-groups continues at length in §5 with the focus shifting to groups of even type, reflecting the specific properties of algebraic groups in characteristic 2, viewed as abstract groups. In particular, we relate the Borel subgroups to the Sylow 2-subgroups, and discuss some other more or less abstract features of tori, O_2, and $O_{2\perp}$. Reductivity can be expressed by the condition $O_2(G) = 1$, and characterized also by the condition $G = E(G) * O(G)$. This is a point which occurs frequently in structural analyses, and sharpens the more general structural observations of the preceding section.

We discuss also the structure of K-groups of even type with abelian Sylow 2-subgroups and more generally strongly closed abelian 2-subgroups. It turns out that the latter topic is not required for our classification results (and as this section shows, once one has the classification results everything is known), but it is part of the finite group theoretic approach, and our treatment of abelian Sylow 2-subgroups in general in Part C can be arranged, with some additional effort, to include this case. This is in fact how it appears in the literature [**5**]. There are a number of other topics treated in this section, aimed at the specific needs of our classification results later on, and reflecting either the structure of algebraic groups or some ideas from finite group theory (or, frequently, both).

The next section moves this body of K-group theory into the framework of L-groups, beginning with the fundamental Lemma 6.3. Everything done up to this point has L-group versions, many coming easily from the fundamental lemma, while others require some further analysis.

One topic which would fall naturally into this chapter has been postponed to §III 5, namely various characterizations of the natural module for SL_2. In general, the theory of "pushing up" and the "amalgam method", which are interrelated, make use of some representation theory, which in recent times has become quite elaborate. But we have neither a representation theory nor any obvious prospects for developing one. Still, we can get just enough information to support these two approaches to group theoretic analysis, as we shall see.

Overview

In our first chapter we laid the foundations of our subject and then developed general group theory in this context. Here and there we encountered results more reminiscent of algebraic group theory than of group theory in general (structure of Sylow 2-subgroups, structure of connected solvable groups) and an occasional oddity having to do with "mixed characteristic" (actions of 2-unipotent groups on sections of degenerate type).

The present chapter deals with topics of a comparably general flavor, but controlled by the specific theory of Chevalley groups, and ultimately by the yoga of Coxeter groups and Dynkin diagrams. The reader looking for an overview of the theory will not linger long here—these topics will presumably be familiar. But there are some unexpected twists—simplifications, and also complications—which we will point out.

There are three levels to the theory: (1) Chevalley groups and algebraic groups; (2) K-groups (where the interesting sections are Chevalley groups); (3) L-groups (where the interesting sections are still Chevalley groups, but there are more uninteresting, or in any case annoying, sections).

Algebraic groups—specifically, linear algebraic groups—naturally constitute the framework one is most likely to adopt in order to think about these groups. A perennial question is the extent to which the notions in this subject go over to the context of groups of finite Morley rank. We have seen the notion of connectivity pass to the broader context, and come into use. The definition of "Borel subgroup" can be repeated, though some of the supporting machinery, such as the notion of complete variety, is lost, and the question of conjugacy becomes very difficult. We also lose the notion of tangent space (and for that reason, it will never be mentioned again), and with it, the Lie algebra. Unipotence and semisimplicity become problematic, but not entirely hopeless.

The material in §1, dealing with algebraic groups, is intended mainly to set the stage in a few respects. Some of the results given there are of use when applied later to definable sections of our ambient group, and they also may motivate some of the things we later do more broadly, but in this section we avoid dealing with the fine structure of the group, as encoded in the Dynkin diagram, which we think of as sitting in the context of *Chevalley groups*, to which we devote the second, far more utilitarian, section.

The section on Chevalley groups as such, §2, consists largely of standard fare. One exception is in the fourth subsection, a miscellany of remarks about the structure of Borel subgroups and the associated root systems that happen to be called on later, and would not otherwise be mentioned. But the main topics treated are the Chevalley commutator formula, the Bruhat decomposition, central extensions, and automorphisms, along with groups normalized by maximal tori and identification of groups in Lie rank 2 (also standard, though less standard than the others). All of this is enormously useful later on, and some of it calls for further comment now, namely: automorphisms, identification, and central extensions.

The description of the automorphism group of a Chevalley group is standard, but we are interested in definable automorphisms, and more specifically in definable groups of automorphisms, which as we might expect based on the previous chapter eliminates field automorphisms. The net result, since our base fields are algebraically closed, is that the automorphisms of interest are inner·graph. In particular connected groups of automorphisms are always inner, and definable groups of automorphisms of SL_2 (our favorite group, in practice) are always inner. So here we have a simplification to be exploited systematically.

As far as identification is concerned, there is nothing special in our case. But whereas the Curtis-Tits theorem says in Lie rank 3 or higher that a

Chevalley group is determined by its pattern of root SL_2-subgroups, which are encoded in the Dynkin diagram, matters are more difficult in the key case of Lie rank 2 (which is where our inductive analysis will really begin), and here a theorem of Tits fills the gap and says that it suffices to add in the structure of the normalizer of a maximal torus to have a defining amalgam determining the group. So we will be using this (in conjunction with the amalgam method) in Chapter IX.

Finally, the theory of central extensions is complicated in interesting ways by the fact that we work outside the algebraic category; so complicated, that we add an additional section to resolve the resulting difficulties. So in §2 we lay out the theory of central extensions of Chevalley groups as given by Steinberg, and in §3 we then work out the theory of central extensions of finite Morley rank. There are three categories of groups here, and three central extension theories: the category of algebraic groups, the category of groups of finite Morley rank (this is actually several categories, but let it pass), and the category of abstract groups. Chevalley groups live in all three categories of course, and have a central extension theory in each category. The algebraic and abstract central extension theories are very different; in fact, the abstract central extension theory depends heavily on the base field. The finite rank central extension theory happens to coincide with the algebraic, and this is what needs to be proved in §3. The method of proof is purely K-theoretic. We know from Steinberg that the difference between the algebraic and abstract central extension theories, working over the base field F, is measured by the K-group $K_2(F)$. There is also a "definable" K-group (representing K_2 in terms of "Steinberg symbols", and imposing a definability condition) and it is the definable K-theory that measures the gap between algebraic central extensions and central extensions interpretable in a given theory of finite Morley rank. So we need to prove that "definable $K_2 = 0$" and this uses the technology of §I 4, in particular the Newelski-Wagner genericity principle. The Steinberg symbols behave like bilinear maps, and are trivial on the algebraic closure of the prime field, so by treating them as linear maps, holding one variable fixed, one can apply the genericity principle to their kernels, and continue on in this vein. See the proof of Proposition 3.2.

With this out of the way, we can discuss systematically the structure of K-groups, and more particularly K-groups of even type, as well as the structure of L-groups. Let us give proper definitions first: A K-group is one whose *definable connected simple sections* are Chevalley groups; an L-group is one whose *definable connected simple sections of even or mixed type* are Chevalley groups. In particular, an L-group of even type is permitted to have simple sections of degenerate type (and an L-group of mixed type enjoys—if one may use this term—even greater liberty).

One can easily imagine the theory of K-groups, bootstrapped up from the theory of Chevalley groups. It is true that K-groups may have arbitrary *finite simple* sections, but this turns out not to be a complicating factor, and one hardly notices it.

The theory of L-groups is enormously simplified by the following circumstance. For H a group of finite Morley rank, we define the subgroup $U_2(H)$ as the group generated by all of its 2-unipotent subgroups. As these groups are connected, the subgroup $U_2(H)$ is definable, and as it contains all the 2-unipotent subgroups it contains, in a sense, the most active and interesting part of the group. But by Proposition 6.2, if H is an L-group of finite Morley rank, then $U_2(H)$ is a K-group. Notice that if G is a simple group of even or mixed type which is not algebraic, and if G is furthermore minimal with these properties, then its proper definable connected simple sections are L-groups, to which this result applies.

However, one should not swing from despair to euphoria; this result does not solve all our problems, just enough of them. Some of the K-group theory goes over to L-groups by quoting Proposition 6.2 to them. Others demand a little more analysis.

As far as the K-group theory is concerned, our main interest will be in even type. From the general theory, perhaps the outstanding result is the treatment of *weakly embedded subgroups*. For G a group of finite Morley rank with a nontrivial Sylow° 2-subgroup, and M a definable subgroup, we say that M is weakly embedded in G if it contains a Sylow° 2-subgroup of G, but $M \cap M^g$ has trivial Sylow° 2-subgroup whenever $g \notin M$. Then by Proposition 4.20, if G is a nonsolvable connected K-group with a weakly embedded subgroup M, then $G/O(G) \simeq (P)\,\mathrm{SL}_2(K)$, where K is an algebraically closed field. (The core $O(G)$ is the largest definable connected normal solvable subgroup without involutions.) This is not a subtle result, though it does rely on some preliminary structural analysis of K-groups.

Passing to K-groups of even type, there is a rich body of relevant results. Indeed, as far as preparation for the main work in the classification theorem is concerned, this is where that preparation is largely found. We go into some detail here. Fix H a K-group of finite Morley rank (sometimes it could even be a K^*-group, that is, a possibly simple group of unknown type whose proper definable connected simple sections are algebraic).

The Sylow 2-subgroups are definable and connected, and their normalizers are Borel subgroups; these are called *standard Borel subgroups*. This is a case in which the K^* case is important, as one then has very little control of the ambient group, or its other Borel subgroups.

Any torus, or more generally a connected definable subgroup without involutions, leaves invariant a Sylow° 2-subgroup.

A connected definable group without involutions which acts definably and faithfully on a semisimple K-group of even type is a good torus.

If H is connected, then $O_2(H)$ is connected and definable, and $O_{2^\perp}(H)$ is definable.

If $O_2(H) = 1$, then $H = E(H) * O(H)$ (this is a form of "reductivity").

We also give the structure of H when the Sylow 2-subgroups are abelian, or contain strongly closed abelian subgroups, or when there is a weakly embedded subgroup. Furthermore, there are versions of the Borel-Tits lemma and the L-balance property. All of this is familiar material in the finite case, and while being easily verified in the K-group context presents substantial challenges when one wishes to prove the same thing, or something similar, for an unknown simple group whose proper sections are K-groups (or for that matter, L-groups).

This list of special properties is still not complete, but nearly so, and so we will complete it, very briefly. We deal also with quasisimple subgroups normalized by Sylow 2-subgroups (useful in dealing with parabolic subgroups in §VIII 5), a generation (by $U_2(C(\cdot))$) property, and finally some characterizations of the natural module for SL_2 which feed into the amalgam method in Chapter IX and belong as much to Chapter I as the present one.

In the final section of the chapter, we move much of the K-group material over to L-groups, first proving the general Proposition 6.2. Some of the K-group material is only needed for $U_2(H)$ later on, and in such cases we do not attempt to generalize it, as the K-group result is fully adequate.

1. Algebraic groups

1.1. Algebraic groups. An algebraic group is an algebraic variety equipped with a group structure such that the multiplication map and the inversion map are morphisms of algebraic varieties. Similarly, an affine algebraic group is an algebraic group whose underlying variety is an affine algebraic variety.

We assume a great deal from the structure theory for algebraic groups, particularly simple algebraic groups. Another point of view is possible: what we really need is the explicit classification of algebraic groups, for example as Chevalley groups or via an explicit presentation (Curtis-Tits relations, for example), together with detailed knowledge of the latter class of groups, which can be very conveniently derived within the algebraic theory, but can also be obtained by direct calculation. As we have mentioned, since our methods in Parts B and C are inductive, and we identify the groups in question by one or another form of the Curtis-Tits relations, our groups are only recognized as algebraic groups after the fact, so that strictly speaking the theory of algebraic groups is not the one most tightly bound to our own approach. Still, that said, this theory provides a very convenient route to the required information about these simple sections.

For a fuller introduction to the subject we suggest [**167**], and for a historical overview see [**40**]. For the structure theory, see either [**115**] or [**171**]. The roots of the subject are found in [**74, 39**]. A particularly useful summary of the main facts, oriented toward the kind of material which is concretely useful in classification problems, is found in [**105**, Chapter A1] (30 pages), though without detailed references.

1.2. Linear and complete. In algebraic geometry the special classes of *affine* and *complete* varieties play an important role. An algebraic group is called affine or complete, respectively, if its underlying variety has the corresponding property. Affine groups are also called *linear*, on the basis of the following result.

FACT 1.1 ([**115**, 8.6]). *An algebraic group is affine if and only if it is isomorphic to a closed subgroup of some* GL_n*, or, in other words, has a finite dimensional linear algebraic representation.*

On the other hand, connected algebraic groups whose underlying variety is *complete* are called abelian varieties, and are commutative [**142**] (more generally, any morphism between abelian varieties is a translate of a homomorphism). Furthermore, a connected algebraic group has a maximal normal Zariski closed affine subgroup with complete quotient, a tricky result due independently to Chevalley and Barsotti [**75, 165, 77**]. In this connection, we should mention the following.

FACT 1.2 ([**115**]). *Let* G *be a linear algebraic group,* H *a Zariski closed subgroup. Then* G/H *is again a linear algebraic group.*

It follows from this discussion that simple algebraic groups are in the affine category, and for our purposes nothing would be lost in taking "algebraic" to mean "affine algebraic". As indicated at the outset, this type of information is not actually required for later sections of this book, but, on the other hand, it explains why our later definitions and results are consistent with the known special case of algebraic groups.

1.3. Borel subgroups. By definition, a Borel subgroup is a maximal connected solvable subgroup (such a group is automatically Zariski closed). The following basic fact about algebraic groups is not known in our more general setting, and its absence represents a serious obstacle to our classification project.

FACT 1.3 ([**115**]). *Let* G *be a connected nonsolvable algebraic group over an algebraically closed field, and* B *a Borel subgroup. Then* $Z(B) \leq Z(G)$.

LEMMA 1.4. *Let* G *be a connected nonsolvable algebraic group over an algebraically closed field, and* B *a Borel subgroup. Then* B *is nonnilpotent.*

PROOF. We reduce easily to the case $Z(G) = 1$, using Lemma I 3.9 if $Z(G)$ is originally finite. We conclude $Z(B) = 1$ by Fact 1.3. □

Another point worth noting is that Borel subgroups of algebraic groups are conjugate [**115**, 21.3]. Again, this is a point that does not go over easily into our more abstract setting.

1.4. Unipotent and semisimple. An element of a linear group is called *unipotent* if all of its eigenvalues are equal to 1, or, in other words, it can be written as $I+N$ with I the identity and N nilpotent. At the opposite extreme, it is *semisimple* if it is diagonalizable over the algebraic closure of the base field. These notions extend coherently to *affine* algebraic groups (that is to say, in a way independent of any particular linear representation) but not to the full category. A general element is a product of commuting factors, one unipotent and the other semisimple, but we will be mainly concerned with elements of one or the other special type.

As we will mix together various notions of algebraic group theory and finite group theory subsequently, we are particularly interested in the status of *involutions* (elements of order 2). Observe that in an affine algebraic group, involutions are *unipotent* if the characteristic is 2, and are *semisimple* otherwise. In the present work, devoted to analogs of algebraic groups in characteristic 2, one of our goals will be to show that involutions behave like unipotent elements (and then to exploit this fact).

DEFINITION 1.5.

(1) *An affine algebraic group is said to be* unipotent *if all of its elements are unipotent.*
(2) *The* unipotent radical *of an affine algebraic group is the maximal connected normal unipotent subgroup.*
(3) *An affine algebraic group is* reductive *if its unipotent radical is trivial.*
(4) *A* torus *is a connected affine abelian algebraic group all of whose elements are semisimple.*

We remark that unipotent groups are nilpotent [**115**, 17.5]. With this terminology, one can clarify the structure of a Borel subgroup.

FACT 1.6 ([**167**, p. 52], [**115**, 19.3], [**105**, 1.5.1]). *Let G be an affine algebraic group and B a Borel subgroup. Let T be a maximal torus of B and U the unipotent radical of B. Then T is a maximal torus of G, U is a maximal unipotent subgroup of G, and*

$$B = U \rtimes T \ .$$

One can also show that the maximal tori of G, like the Borel subgroups, are conjugate in G, or more precisely that the pairs (B, T) where T is a maximal torus contained in the Borel subgroup B are conjugate. In our more abstract context we will have better control over tori than over Borel subgroups, and a generalization of this conjugacy result will be given in Proposition IV 1.15.

The detailed structure of Borel subgroups in simple algebraic groups leads in the direction of generators and relations, and will be deferred to the next section, which goes more in that general direction.

We comment further on the situation in positive characteristic, having in view the case of characteristic 2.

LEMMA 1.7. *Let G be a connected affine algebraic group in positive characteristic p, realized over an algebraically closed field. Then the Borel subgroups are the groups of the form $N(S)$ with S a Sylow p-subgroup. In particular, Sylow p-subgroups of G are connected, and conjugate.*

PROOF. Let S be a Sylow p-subgroup of B, and U its connected component. Then U is a maximal connected unipotent subgroup of G, and thus $N^\circ(U) = B$ is a Borel subgroup. As B is self-normalizing, $N(U) = B$. Hence $S \leq B$ and $S = U$, $B = N(S)$. □

1.5. Reductive groups.

FACT 1.8 ([**167**, p. 52], [**115**, 22.3, 26.2]). *If G is reductive and T is a maximal torus of G, then $C(T) = T$. More generally, the centralizer of any torus in G is connected and reductive.*

In particular, in this context the action of a maximal torus T on a root subgroup U_0 is nontrivial, and hence$[U_0, T] = U_0$.

Another consequence is that the center of a reductive group consists of semisimple elements. This then leads to the following.

LEMMA 1.9. *Let G be a connected reductive algebraic group in characteristic p. Then the unipotent radical of G is its largest normal p-subgroup.*

PROOF. Let P be a normal p-subgroup of G. We claim P is contained in the unipotent radical.

We may suppose P is Zariski closed. Then its connected component is certainly contained in the unipotent radical U of G. Passing to $\bar{G} = G/U$, the image $\bar{P}$ is a finite normal p-subgroup of the connected reductive group $\bar{G}$. Thus $\bar{P}$ is central in $\bar{G}$ and hence consists of semisimple elements, forcing $\bar{P} = 1$. □

Thus in characteristic p, the unipotent radical can also be denoted $O_p(G)$ by analogy with finite group theory.

DEFINITION 1.10.

(1) *Let G be an affine algebraic group and T a torus in G, A* root subgroup *of G (with respect to T) is a minimal nontrivial T-invariant unipotent subgroup of G.*
(2) *A group H is* directly spanned *by an ordered sequence of subgroups $H_1, \ldots, H_n$ if every element of H has a unique representation as a product $h_1 \cdot \cdots \cdot h_n$ with $h_i \in G_i$.*

FACT 1.11 ([**115**, 28.1]). *If G is a reductive algebraic group, T is a maximal torus of G, and U is a unipotent T-invariant subgroup, then U is directly spanned by its root subgroups, in any order. In particular, $[T, U] = U$.*

Up to this point we have defined "root subgroups" without defining roots. The root subgroups afford 1-dimensional representations of a maximal torus T, which correspond to certain characters of T. These distinguished characters (finite in number) are the roots. The terminology arises from the study of the so-called *Weyl group* $W = N(T)/T$), which has a finite dimensional representation in which generators of W act as reflections in hyperplanes. In the theory of finite reflection groups, one associates to W a finite set of vectors perpendicular to the hyperplanes in question, and invariant under the action of W, and these vectors, also called roots, are correlated with the roots in our present sense. This comes into play in the next statement, in which the language of root systems is used.

We will return to this subject more fully in the next section.

FACT 1.12 ([**115**, Theorem 26.3]). *Let G be a reductive algebraic group, T a fixed maximal torus and $\Phi = \Phi(G,T)$. Let Δ be a base of Φ. Let Z_α denote $C_G(T_\alpha)$ where $T_\alpha = (\ker\alpha)^\circ$ and $\alpha \in \Phi$. Then G is generated by the Z_α ($\alpha \in \Delta$), or equivalently by T along with all U_α ($\pm\alpha \in \Delta$).*

1.6. Centralizers of semisimple elements.

DEFINITION 1.13.

(1) *A connected algebraic group is said to be* simple *(as an algebraic group) if it contains no nontrivial proper connected closed normal subgroup; equivalently, the center is finite, and modulo the center, the group is simple. We will, however, refer to such a group as* quasisimple, *using a terminology which is compatible with the terminology for abstract groups. When dealing with algebraic groups which are simple in the abstract sense, we tend to emphasize this point with the phrase "abstractly simple".*
(2) *A connected algebraic group is said to be* semisimple *if it has no nontrivial connected normal abelian subgroup (or, equivalently, no nontrivial connected normal solvable subgroup).*

The motivation for the term "semisimple" is a structural result; modulo the (finite) center, the group is a direct product of abstractly simple Zariski closed subgroups. Compare Proposition 4.8 in the abstract setting.

We have referred above to the structural significance of the "unipotent/semisimple" distinction, and now we arrive at a concrete statement of one of the central structural points.

FACT 1.14 ([**172**, Corollary 4.6]). *Let G^* be a semisimple algebraic group and x a semisimple element of G^* of prime order p. Let $\pi : \tilde{G} \to G^*$ be the canonical map from the simply connected cover. If p does not divide $|\ker\pi|$, then $C_{G^*}(x)$ is connected.*

FACT 1.15 ([**172**, 3.19]). *Let G^* be a semisimple algebraic group and y any semisimple element. Then $C_{G^*}(y)$ is reductive.*

Combining these two facts, we get:

COROLLARY 1.16. *With the hypotheses and notation of Fact* 1.14, $C_{G^*}(x)$ *is connected and reductive. In particular, if* G^* *is one of the groups* SL_3, Sp_4, *or* G_2 *over an algebraically closed field of characteristic* 2 *and* x *is a semisimple element of prime order* $p > 3$, *then* $C_{G^*}(x)$ *is a torus or the product of a torus with* SL_2.

PROOF. $C_{G^*}(x)$ is reductive of Lie rank 2, and contains a central element of order greater than 3. The claim follows. □

In a similar vein we have the following.

FACT 1.17 ([**67**]). *Let* L *be a quasisimple algebraic group and* α *a semisimple automorphism of* L. *Then* $C_L(\alpha)$ *is reductive.*

FACT 1.18 ([**115**, Section 27.5]). *Let* G *be a perfect algebraic group such that* $G/Z(G)$ *is a simple algebraic group. Then* G *is a simple algebraic group.*

1.7. Field definability implies group definability.

FACT 1.19 ([**156**], Corollaire 4.16). *In a simple algebraic group over an algebraically closed field, definability from the field and definability from the pure group coincide.*

PROPOSITION 1.20. *In a quasisimple algebraic group over an algebraically closed field, definability from the field and definability from the pure group coincide.*

PROOF. Let $G = G(F)$ be a quasisimple algebraic group over an algebraically closed field F. We will use $\bar{\ }$-notation to denote quotients by $Z(G)$.

If B is a Borel subgroup of $\bar{G}$, then by Lemma 1.4, B is nonnilpotent. Therefore, we can interpret an algebraically closed field K in $\bar{G}$ using a Borel subgroup. As K is interpretable in F, these two fields are definably isomorphic in F by Fact I 4.30. We denote this isomorphism by θ.

Let $G(K)$ be a linear algebraic group over K isomorphic to $G(F)$ by an isomorphism ψ induced by θ. We claim that the isomorphisms $\psi_\alpha : X_\alpha(F) \longrightarrow X_\alpha(K)$ defined by $\psi_\alpha(x_\alpha(t)) = x_\alpha(\theta(t))$ are definable in G. The isomorphism ψ_α can be written as the composition of the following maps: the isomorphism $X_\alpha(F) \longrightarrow \overline{X_\alpha(F)}$ induced by the canonical homomorphism $G \longrightarrow \overline{G}$, the isomorphism $\overline{X_\alpha(F)} \longrightarrow \overline{X_\alpha(K)}$ given by

$$\overline{x_\alpha(t)} \longmapsto \overline{x_\alpha(\theta(t))},$$

and the isomorphism $\overline{X_\alpha(K)} \to X_\alpha(K)$ which is inverse to the map induced by $G(K) \to \overline{G(K)}$. Here the map $\overline{x_\alpha(t)} \longmapsto \overline{x_\alpha(\theta(t))}$ is F-definable in $\bar{G}$ and hence $\bar{G}$-definable in $\bar{G}$ by Fact 1.19. So the maps ψ_α are G-definable.

The isomorphisms ψ_α are induced by the isomorphism $\psi : G(F) \longrightarrow G(K)$, and conversely ψ is definable from the collection (ψ_α), since every element of G is a product of a bounded number of elements from the root groups X_α. Therefore, ψ is G-definable.

Now, let A be a subset of $G(F)^n$ definable from F. Then $\psi(A)$ is definable in $G(K)^n$ from K and hence is definable in G. But ψ and K are definable in G. Hence, $A = \psi^{-1}(\psi(A))$ is definable in G. □

1.8. Constructible groups.

FACT 1.21 ([**156**], Théorème 4.13]). *Let G be a group which is definable over an algebraically closed field. Then G is isomorphic to an algebraic group.*

1.9. Borel-Tits.

DEFINITION 1.22. *Let G be an algebraic group. A* parabolic subgroup *is a connected algebraic subgroup containing a Borel subgroup.*

Actually, any subgroup containing a Borel subgroup will be algebraic and connected.

FACT 1.23 ([**115**, §30.3]). *Let G be an algebraic group, P a proper parabolic subgroup, and U the unipotent radical of P. Then $C_P(U) \leq U$.*

FACT 1.24 ([**41**], cf. [**115**, Corollary 30.3 A]). *Let G be a reductive algebraic group and let U be a Zariski closed unipotent subgroup of G. Then $N_G(U)$ is contained in a parabolic subgroup $\mathcal{P}(U)$ of G such that $U \leq R_U(\mathcal{P}(U))$, where R_U denotes the unipotent radical.*

We will need the next result in characteristic 2.

PROPOSITION 1.25. *Let G be a simple algebraic group over a field of positive characteristic, let U be a nontrivial Zariski closed unipotent subgroup of G, and let $H = N_G(U)$. Then $C_H(R_U(H)) \leq R_U(H)$, where R_U denotes the unipotent radical.*

Another way to express the conclusion here is as follows:

$$F^*(H) = O_p(H) \ .$$

PROOF. Embed H into a parabolic subgroup $\hat{H} = \mathcal{P}(U)$ as in Fact 1.24. Let $Q = R_U(H)$ and $V = R_U(\hat{H})$. We claim $C(Q) \leq Q$.

Observe that $C(Q) \leq C(U) \leq H$. In particular $C_V(Q) \leq V \cap H \leq Q$ since $V \cap H$ is a normal p-subgroup of H.

First we show that $C(Q)$ is a p-group. Otherwise, we can find a definable $p^{\perp}$-group A contained in $C(Q)$. Now we apply Proposition I 12.4 with G, Q, V, A here playing the role of G, Q, U, A there. So Proposition I 12.4 applies and yields $A \leq C_{\hat{H}}(V) = V$, a contradiction.

So $C(Q) \lhd H$ is a p-group and thus $C^\circ(Q) \leq R_U(H) = Q$. Hence $C(Q)/C^\circ(Q)$ is a finite normal p-subgroup of H/Q; but the latter is reductive in characteristic p and hence contains no central p-elements. Thus $C(Q)/C^\circ(Q)$ is trivial and $C(Q) \leq Q$. □

LEMMA 1.26. *Let G be a quasisimple algebraic group over an algebraically closed field of characteristic 2, and P a definable connected 2-subgroup of G (in an expanded language, in which G has finite Morley rank). Then $O(N_G(P)) = 1$.*

PROOF. Let $\hat{P}$ be the Zariski closure of P in G. By Lemma I 3.22 we have $O(N_G(P)) = O(C_G(P))$, and the same applies to $\hat{P}$. Since $C_G(P) = C_G(\hat{P})$, we may suppose that P is Zariski closed.

Let Q be the unipotent radical of $N_G(P)$. Then Q is a 2-group, and $O(N_G(P))$ commutes with Q. By Proposition 1.25 we have $O(N_G(P)) \leq Q$ and hence $O(N_G(P)) = 1$. □

FACT 1.27 (L-balance). *Let L be a quasisimple algebraic group in characteristic 2 and $U \leq L$ a nontrivial 2-subgroup of L. Then $E(C_L(U)) = 1$.*

PROOF. This follows from Fact 1.24, as explained in [**102**, 13-4] or [**103**, §3]. □

2. Chevalley groups

2.1. Origins (existence). Remarkably enough, most finite simple groups are forms of groups associated with simple Lie algebras over $\mathbb{C}$. The classification of these Lie algebras was initiated by Killing and completed by Cartan. Dynkin later introduced the very simple "Dynkin diagram" description of the classification which encodes the essential information in a small finite graph; these graphs were introduced independently a little earlier by Coxeter in the study of crystallographic groups. It was realized early on that the so-called "classical matrix groups" (general linear, symplectic, orthogonal, unitary) had analogs over finite fields, though it was apparently not realized until 1937 that the underlying geometry could be treated coherently in that context (a point on which Leonard Dickson had expressed skepticism).

In three papers in the period 1901–1908 Dickson, who came under the direct influence of both Lie and Jordan, gave constructions of the finite simple groups corresponding (from the modern point of view) to Lie algebras of type G_2, and E_6. This point went largely undeveloped until Chevalley's article in 1955 [**73**], which introduced the full correspondence (enriched rapidly thereafter by the so-called "twisted" versions). In this approach, one begins by showing the highly nontrivial fact that finite dimensional simple complex Lie algebras $\mathfrak{L}$ have bases with respect to which the "structure constants" (giving the Lie bracket) are rational integers (this involves a substantial refinement of Cartan's analysis). The resulting basis is not uniquely determined, but is rigidly constrained. At this point one may consider the Lie algebra $\mathfrak{L}_{\mathbb{Z}}$ spanned by this basis over $\mathbb{Z}$, and its tensor product with any field K,

$$\mathfrak{L}_K = \mathfrak{L}_{\mathbb{Z}} \otimes K$$

which is a Lie algebra over K. The Chevalley groups are constructed as subgroups of the automorphism group of $\mathfrak{L}_{\mathfrak{K}}$, which, however, contains $\mathrm{Aut}(K)$ and other unwanted elements. To overcome this, one first defines "root groups" over K as groups of automorphisms of $\mathfrak{L}_K$, and one then takes as the corresponding Chevalley groups the subgroup of $\mathrm{Aut}(\mathfrak{L}_K)$ generated by these root groups. In particular, in making computations one has, initially, only unipotent elements available, and the initial phase of analysis is heavily computational in this approach.

It remains to say a word about root groups. One has in mind subgroups of $\mathrm{Aut}(\mathfrak{L}_K)$ parametrized explicitly by the additive group of K. These may be constructed first for the original algebra $\mathfrak{L}$, where the notation used is $x_r(\zeta)$, with ζ running over $\mathbb{C}$ and r a root associated with $\mathfrak{L}$. The matrix representation of these elements, with respect to a Chevalley basis, has entries of the form $m\zeta^d$ with m and d integers, $d \geq 0$. These are replaced in an ad hoc manner by entries of the form mt^d, with t varying over K, and it is then checked that these matrices do indeed represent automorphisms of $\mathfrak{L}_K$ (for this point, see [**67**, p. 63]). So one may consider the subgroup of $\mathrm{Aut}(\mathfrak{L}_K)$ generated by these automorphims, and consider it as the group $G_{(L)}(K)$ associated with $\mathfrak{L}$ over K.

For our purposes, this construction and the subsequent calculations justify the existence of these groups (though existence is not, strictly speaking, our concern here) as well as the fact that they have a structure parallel to the structure of simple algebraic groups over algebraically closed fields, a point with a wide variety of practical consequences. The first phase of analysis produces generators and relations for the groups, and from these everything else which is needed can be derived.

The discussion above concerns only the *adjoint forms* of Chevalley groups. A more general construction which uses lattices in representation spaces for $\mathfrak{L}$ rather than $\mathfrak{L}$ itself produces certain perfect central extensions, including the so-called "universal Chevalley group", which, though is not the universal central extension in the category of abstract groups, turns out to be universal among the perfect central extensions with finite center, when working over an algebraically closed field. For the more general construction, see [**173**] or the very accessible account in [**117**, Chap. 27].

The algebraicity (and of course linearity) of all forms of Chevalley groups follows directly from the construction. By definition, they are generated inside some $\mathrm{GL}(V)$ by the root groups, represented as automorphisms of a lattice. Since these root groups are isomorphic to the additive group $\mathbb{G}_a$, as is seen early on, the group they generate inside $\mathrm{GL}(V)$ is also algebraic [**173**, Theorem 6].

2.2. Coxeter groups and Dynkin diagrams. We will touch on the theory of Coxeter groups in very broad terms here, as this provides the standard framework for classifying and analyzing both simple complex Lie algebras and the associated Chevalley groups. While this is a central topic as far as structure theory is concerned, we refer to [**67**] or [**116**] for a satisfactory account of the matter.

A *Coxeter group* is a finite group generated by Euclidean reflections. A Euclidean reflection is a reflection in a hyperplane in a vector space over $\mathbb{R}$. More geometrically, one may allow reflections in an arbitrary Euclidean hyperplane, but as the group is finite it follows that it fixes a point, and taking a fixed point as origin reduces the situation from affine space to a vector space.

In more vivid terms, a Coxeter group corresponds to a finite set of mirrors such that in each mirror one sees only finitely many images of the other mirrors.

If the group in question also preserves a lattice in the space, it is said to satisfy the *crystallographic condition.* The Coxeter groups relevant to the theory of algebraic groups all satisfy the crystallographic condition.

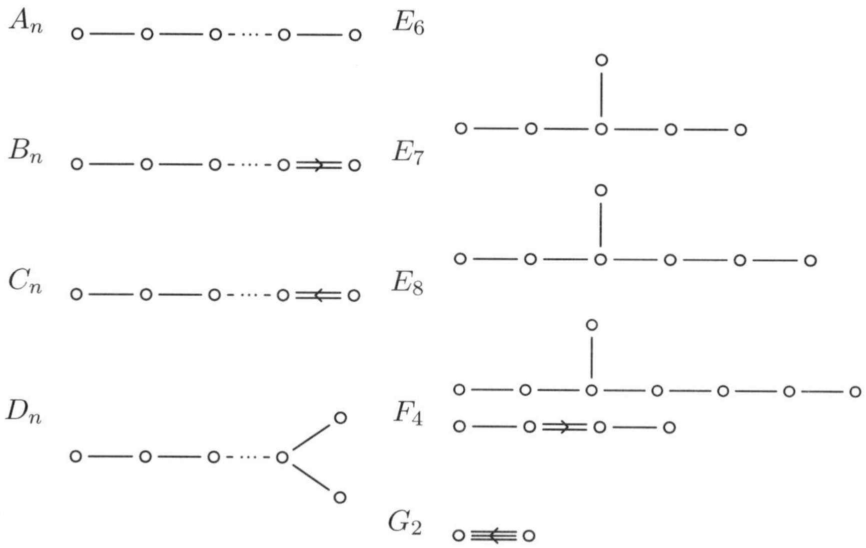

TABLE 1. Dynkin diagrams

The explicit classification of the Coxeter groups involves the following notions. A *root system* in a vector space is a finite collection of vectors, called *roots*, which spans the space and is invariant under the reflections in the hyperplanes determined by these vectors (their orthogonal complements). A *basis* for a root system is a basis for the space such that every root is either a positive integral combination of the roots or a negative integral combination of the roots. Bases are described by the angles between pairs of roots, which are very limited, and even more so under the crystallographic condition. With a basis fixed, its elements are called *simple* roots, and in general most pairs of simple roots are orthogonal. If one forms the so-called *Dynkin diagram*, a graph with vertices the simple roots and with edges between nonorthogonal pairs, this will be a tree, and in fact a tree which is not much more elaborate than a path. In addition, the edges are generally labeled by data indicating the precise angle involved, and if adjacent roots are of different lengths, then an arrow is added pointing from the longer to the shorter. With these additions, the diagram completely determines the structure of the root system and the associated group.

The following fact is fundamental to this point of view.

FACT 2.1 ([**115**, Lemma 10.1]). *Let G be a Chevalley group, and let Φ be a root system for G. If Δ is a base of Φ, then $(\alpha, \beta) \leq 0$ for $\alpha \neq \beta$ in Δ, and $\alpha - \beta$ is not a root.*

The root system is *indecomposable* if its Dynkin diagram is connected; so the general root system is an orthogonal sum of indecomposable root systems, and the corresponding Coxeter groups are direct products of the factors corresponding to the indecomposable constituents. In addition, the Coxeter group has a presentation in which the reflections associated to the simple roots are the generators, and the relations give the order of these reflections (2) as well as the orders of all products of two simple reflections (2 if they are orthogonal, and more otherwise).

As an example, the symmetric group Sym_n on $\{1, \ldots, n\}$ is a Coxeter group, whose reflection representation is given by restricting the natural representation on $\mathbb{R}^n$ to the hyperplane orthogonal to the fixed vector $(1, \ldots, 1)^T$. Relative to the standard basis (e_i) of $\mathbb{R}^n$, the simple roots may be taken to be the $n-1$ vectors $v_i = e_i - e_{i+1}$ and the corresponding reflections correspond to the elementary transpositions $(i, i+1)$. Adjacent roots v_i, v_j are those with adjacent indices i, j, with the angle in question always $120°$. All roots have the same length. For an example with distinct root lengths, take the symmetries of the square, with roots pointing to the midpoints of the edges and the corners; or in three dimensions, in the case of a cube, include the centers of the faces. These two examples are labeled A_{n-1}, B_n where the subscript indicates the dimension of the representation space and the letter goes from A to G according to the form of the root system (with various more or less severe restrictions on n); or indeed from A to H if one includes the non-crystallographic cases. More precisely, types $A - D$ classify the classical groups, GL_n and the symplectic and orthogonal groups, and the exceptional diagrams E_6, E_7, E_8, F_4, G_2 are associated with non classical Chevalley groups; in the case of G_2, the associated finite groups were constructed explicitly by Dickson in analogy with the complex case, while the construction of the others was achieved only by Chevalley in the context of a general theory.

This is all generalized by the theory of complex reflection groups, which we summarized in §I 13, though the Coxeter groups are scattered randomly throughout that classification, which contains a large number of sporadic additions.

The point to emphasize is that the Coxeter group, occurring as the Weyl group $N(T)/T$ in a Chevalley or simple algebraic group, is a finite combinatorial object which together with the base field completely determines the structure of the associated group. As a result, computations in Chevalley groups frequently become computations in Coxeter groups, and these become computations with roots. An example of this is the Chevalley commutator formula itself.

At this juncture, we note that there is an *extended Dynkin diagram* which also comes into play in some computations. In this diagram, one adjoins the negative of the sum of the simple roots to the set of simple roots and forms the associated diagram. So in the case of A_n the additional vector is $e_n - e_1$, which completes the cycle and gives rise to a graph which is no longer a tree, but in fact a circuit.

2.3. The Chevalley commutator formula.

FACT 2.2 ([**173**, Corollary to Lemma 15]). *Let α, β be roots with $\alpha+\beta \neq 0$. Let k be a field. Then*

$$(x_\alpha(t), x_\beta(t)) = \prod x_{i\alpha+j\beta}(c_{ij}t^iu^j)$$

where $(g,h) = ghg^{-1}h^{-1}$, and the product on the right is taken over all roots $i\alpha+j\beta$ $(i,j \in \mathbb{Z})$ arranged in some fixed order, and where the c_{ij} are integers depending on α, β and the chosen ordering, but not on t and u.

It is useful to fix a reasonable ordering and give the constants explicitly, particularly since they have to be interpreted as elements of the base field, and may therefore degenerate in some characteristics. Ordering the roots by height, the computation may be made as follows.

The roots of the form $\alpha + i\beta$ form a so-called *chain* $\alpha - p\beta, \dots, \alpha + q\beta$ where $0 \leq p, q$ and $p+q \leq 3$, and similarly there is a chain $\beta - p'\alpha, \dots, \beta + q'\alpha$. We then have

$$\begin{aligned} |c_{1j}| &= \binom{p'+j}{j} \\ |c_{i1}| &= \binom{p+i}{i} \\ |c_{32}| &= 1 \\ |c_{23}| &= 2 \end{aligned}$$

We will mainly be interested in characteristic 2, in which case these coefficients are equal to 1 (when defined), with the following exceptions:

$$(*) \qquad \begin{array}{l} c_{i1} = 0 \text{ if } p = 1, i = 1 \text{ (root system of type } B_2 \text{ or } G_2\text{), and} \\ \text{similarly for } c_{1j}; \\ c_{23\alpha\beta} = 0\ . \end{array}$$

Bear in mind that $p + i \leq 3$.

The next fact will be needed once, in §VIII 5. While certainly suggested by the general formulas, it does require some verification, particularly in view of the possibility of some degeneracy in the coefficients in certain characteristics.

LEMMA 2.3. *Let G be a Chevalley group, and U a maximal unipotent subgroup corresponding to the set of positive roots Φ. Then $[U, U]$ is the product of the root groups S_α, with α running over the roots of Φ which are not simple.*

PROOF. It is best to treat G_2 as a separate case (especially in characteristic 2), which we leave to the reader. In the other cases we proceed as follows. We claim that for any positive root γ which is not a simple root, the root subgroup S_γ is generated by commutators of the form $[S_\alpha, S_\beta]$, with α simple and β positive (but not necessarily with $\alpha + \beta = \gamma$). Once one has this result, it can be applied repeatedly to prove our claim, since the group is nilpotent and there is therefore a limit to the number of times it can be iterated.

So fix a positive root γ which is not simple. We will prove by *downward* induction on the height of γ that S_γ lies in the group spanned by $[S_\alpha, S_\beta]$, with α varying over simple roots and β varying over positive roots. Now γ can be expressed as $\alpha + \beta$ with α simple and β positive. Now unless the coefficient $p_{\alpha,\beta}$ is 1 (which means that the chain associated with α, β begins with $\alpha - \beta$), we will have $[x_\alpha(t), x_\beta(u)] = x_\gamma(tu) \cdot$ higher root elements. In this case, by our induction hypothesis, we can remove the higher order terms.

It remains to consider the case in which $p_{\alpha\beta} = 1$, and the first root in the (α, β)-chain is $\alpha - \beta$. Then we need to consider the commutator formula for $[S_{\alpha'}, S_\beta]$ with $\alpha' = \alpha - \beta$ and to show that $S_{\alpha+\beta}$ is contained in this commutator, leaving aside the case of a root system of type G_2. In this case we may take α' and β to be a basis for a root system of type B_2, and the commutator formula has two terms, involving the commuting groups S_α and S_γ, of the form $x_1(tu)x_2(tu^2)$, with distinct exponents. So one may easily extract S_γ from a combination of such expressions. □

We give another simple result in the same vein.

LEMMA 2.4. *Let G be a simple Chevalley group of Lie rank 2, α, β a fundamental system of roots, and S_α, S_β the corresponding root subgroups. Then S_α, S_β generate a maximal unipotent subgroup.*

PROOF. As the exponents i, j occurring in the commutator formula vary with the root $i\alpha + j\beta$, and all of these root groups together generate a maximal unipotent subgroup, if all of the coefficients c_{ij} are nonzero the result follows.

In this context, we have $p = p' = 0$ and thus $c_{1i} = c_{j1} = 1$, which cannot degenerate regardless of the characteristic. This covers all relevant coefficients except in the case of G_2. But here, if we take α short and β long, then we need only consider the further coefficient $c_{32} = 1$. □

We remark that the Chevalley commutator formula gives a very clear picture of the structure of a Borel subgroup; it gives an explicit description of the unipotent radical U, which is the product of the root subgroups relative to a choice of *positive roots*; and the full Borel subgroup has the form $U \rtimes T$ with T a maximal torus contained in B, acting on the root groups via characters which in the algebraic theory are natural representatives of the positive roots. We touch here on the point of transition between the algebraic theory and the Chevalley theory, which belongs within the algebraic theory, and is covered in [**115**]. This correlation is crucial, in the sense that it allows us to continue our discussion from the point at which it left off in the previous section, rather than redefining everything from scratch.

From this point on, it seems reasonable to mix the language of algebraic groups freely with the language of Chevalley groups.

In particular, Fact 1.11, concerning the structure of unipotent groups invariant under a maximal torus, can be seen as belonging to the present context.

2.4. Structure. We expand on the structure of Borel subgroups, with more detail concerning the action of tori. We first repeat our closing remarks from the last subsection in a more explicit form. Throughout this subsection, one can consult [**115, 171**] for the theory from an algebraic point of view, or [**67**] for the same results, in some instances with different terminology, from a computational point of view. Note that we are working toward results which are helpful when the groups in question are considered as abstract groups.

NOTATION 2.5. *Let G be a Chevalley group, Φ the associated root system, X_α the root group attached to α, and Δ a set of positive roots. Set $S_\Delta = \langle X_\alpha : \alpha \in \Delta \rangle$.*

FACT 2.6. *Let G be a Chevalley group, Φ the root system associated with G, and a maximal torus T, Δ a system of positive roots. Then $N(S_\Delta)$ is a Borel subgroup of G.*

FACT 2.7. *Let G be a Chevalley group, Φ the root system associated with G and a maximal torus T, and Δ a system of positive roots. Then $N(S_\Delta) = S_\Delta \rtimes T$.*

FACT 2.8. *Let G be a Chevalley group, B a Borel subgroup of G, and $T \leq B$ a maximal torus of G. Then there is an involution $w \in N(T)$ such that $B \cap B^w = T$. For such an involution, $G = \langle B, B^w \rangle$.*

One refers to B, B^w in the foregoing as a pair of "opposite" Borel subgroups.

LEMMA 2.9. *Let G be a Chevalley group over a field of characteristic p, Φ the associated root system, Δ a system of positive roots, and $S_\Delta = \langle X_\alpha : \alpha \in \Delta \rangle$. Then S_Δ is a Sylow p-subgroup of G.*

PROOF. Evidently S_Δ is a p-group. If it is not a Sylow p-group, then $N(S_\Delta)$ contains a larger p-group, and hence the torus T of Fact 2.7 contains a nontrivial p-group, a contradiction. $\square$

LEMMA 2.10. *Let H be a quasisimple Chevalley group, B a Borel subgroup, $S = U(B)$. If $S = S_1S_2$ with $[S_1, S_2] = 1$ and $S_1, S_2 \lhd B$, then $S = S_1$ or S_2.*

PROOF. S_1 and S_2 are products of root groups, more precisely they are *directly spanned* by root groups in the sense of Definition 1.10, by [**115**, Proposition 28.1]. Furthermore, we claim that every root group lies in one or the other of these two groups: if an element s of a root group is written as s_1s_2 with $s_1 \in S_1$, $s_2 \in S_2$, where s_1 and s_2 are expressed as products of elements of distinct root groups (so that in particular no root group makes a nontrivial contribution to both s_1 and s_2), then necessarily one of these elements is s and the other is trivial.

In particular the set of fundamental roots is expressed as the union of two sets A and B of roots, so that if $r \in A \setminus B$ and $s \in B$, then $r+s$ is not a root, and similarly with A and B reversed. Since the Dynkin diagram for H is connected, this forces one of these two sets to contain all the fundamental roots (even in characteristic 2 the coefficients in the Chevalley commutator formula do not degenerate sufficiently to affect this, though one has to check the various cases, such as G_2, individually). □

The foregoing extends to central products of quasisimple groups as a decomposition lemma. This is another result needed in the context of §VIII 5, and only there.

LEMMA 2.11. *Let H be a central product of quasisimple algebraic groups, B a Borel subgroup of H, and S the unipotent radical of B. Suppose that $S = S_1 * S_2$, a central product, with $S_1, S_2 \lhd B$. Then $H = H_1 * H_2$, a central product, where for $i = 1, 2$ the group $S_i \cap H_i$ is a maximal unipotent subgroup of H_i. If in addition $S = S_1 \times S_2$ is a direct product, then S_i is a maximal unipotent subgroup of H_i, and $H_1 = E(C_H(S_2))$, $H_2 = E(C_H(S_1))$.*

PROOF. For the first claim, we may factor out the center of H and suppose that it is a direct product of simple factors. Let L be one of the simple components of H, and $U = S \cap L$. Then $U = U_1 * U_2$ with U_i is the projection of S_i into L. As S_1, S_2 are invariant under the action of a maximal torus T of L, we have $U_i = [T, U_i] = [T, S_i] \leq S_i$ and thus $U_i = S_i \cap L$. Now by the preceding lemma $U = U_i$ for at least one $i = 1$ or 2. Thus if we let H_1 be the product of the factors L for which $S_1 \cap L$ is a maximal unipotent subgroup, and H_2 the product of the remaining factors, we have our first claim.

The second claim then follows directly from the first. □

2.5. The Bruhat decomposition. The *Weyl group* associated to a Chevalley group is the group $N(T)/C(T)$, with T a maximal torus, viewed as a group of automorphisms of T. In view of Fact 1.8 this may be written more simply as $N(T)/T$.

A fundamental fact about the Weyl group is that it is the *Coxeter group* associated with the root system of G. The information in the Dynkin diagram, the Coxeter group, or the root system are all interchangeable; the Dynkin diagram encodes the geometry of a system of fundamental roots and is in this sense a minimal amount of information.

The Weyl group parametrizes a double coset decomposition of G with respect to a Borel subgroup containing the torus T. This is referred to as the *Bruhat decomposition.*

FACT 2.12. *G is the disjoint union of the double cosets BwB ($w \in W$).*

Observe that w is an element of $N(T)/T$ rather than G, but the notation BwB is unambiguous as $T \leq B$.

At this point we have arrived at a chapter of the subject conveniently handled in the context of (B, N)-pairs, a matter on which we will enlarge subsequently. This material is discussed from that precise point of view in [**115**].

A *parabolic subgroup* of a Chevalley group is one which contains a Borel subgroup. Generally one fixes the Borel subgroup B in advance and considers the parabolic subgroups containing that fixed group. These are correlated with subgroups of the Weyl group generated by some of the simple reflections; so there are few such groups, and their structure is clear. For a brief account see [**167**, §3.2], and for details see [**115**, §29.3]. One can also get rather good control over the class of groups containing a maximal torus, which we discuss in the next subsection.

Our identification procedure for simple groups of finite Morley rank relies heavily on the properties of parabolic subgroups. More precisely, these properties motivate our approach, and are occasionally useful in a more concrete way as properties holding inductively in simple sections. For an elaboration on this point, see the notes at the end of the chapter.

FACT 2.13 ([**115**, 30.2]). *Let G be a Chevalley group and P a proper parabolic subgroup. Then the unipotent radical of P is nontrivial.*

FACT 2.14. *Let G be a Chevalley group, and B a Borel subgroup. Then G is generated by the minimal parabolic subgroups properly containing B.*

FACT 2.15 ([**115**, Corollary 32.3]). *A semisimple Chevalley group of* Lie rank 1 *is isomorphic to* $\mathrm{SL}_2(K)$ *or* $\mathrm{PSL}_2(K)$.

2.6. Groups normalized by a maximal torus.

DEFINITION 2.16. *If G is a Chevalley group and T is a maximal torus of G, then a* subsystem *subgroup of G (relative to T) is a central product of Zariski closed quasisimple subgroups normalized by T.*

FACT 2.17 ([**167**, Prop. 3.1], [**166**, 2.5]). *Let G be a Chevalley group over an algebraically closed field, T a maximal torus of G, and X a Zariski closed subgroup of G which contains T. Let U be the unipotent radical of*

X. Then $X = DZU$ with D a subsystem subgroup and Z a subtorus of T commuting with D.

DEFINITION 2.18. *The* root SL_2-*subgroup associated to a root α is the group $\langle U_\alpha, U_{-\alpha}\rangle$ generated by the root subgroups corresponding to α and its opposite.*

The name is a bit loose, as these groups may well be of the form PSL_2.

FACT 2.19. *A root* SL_2*-subgroup is isomorphic to either* SL_2 *or* PSL_2.

LEMMA 2.20. *If K and L are root* SL_2 *subgroups of the Chevalley group A with respect to a fixed maximal torus, and K and L do not commute, then $\langle K, L\rangle$ is a Lie rank 2 Chevalley group.*

PROOF. Let $M = \langle K, L\rangle$. We have $M = DZU$ with U the unipotent radical of M, by Fact 2.17. There is an automorphism of the root system for A carrying a root to its opposite, and this gives rise to an automorphism ϕ of A normalizing T and switching root subgroups with their opposites, hence leaving K and L, and also M, invariant. If U is nontrivial then as it is T-invariant U contains a root subgroup, and applying ϕ we see that U also contains the opposite root subgroup, a contradiction. So $U = 1$.

Now in view of the Chevalley commutator formula the semisimple group M, which is generated by two pairs of opposite root subgroups, has a root system of rank 2. Everything follows. □

2.7. Central extensions. The theory of central extensions of simple algebraic groups has a number of different aspects, depending on the category in which one works, and we will have to revisit this topic at length in the next section, in the broad context of groups of finite Morley rank. But here we are concerned mainly with the structure of quasisimple groups already known to be algebraic groups or Chevalley groups.

Mainly, we need the following.

LEMMA 2.21. *Let G be a quasisimple group with $G/Z(G)$ a simple Chevalley group. Then the following are equivalent:*

(1) *G is algebraic,*
(2) *$Z(G)$ finite.*

There are two distinct theorems here, one from the general theory of algebraic groups (cf. [**115**]) and the other from the theory of Chevalley groups, working heavily with generators and relations (and leading eventually to K-theory); see [**173**].

DEFINITION 2.22. *A quasisimple algebraic group is said to be* simply connected *if it has no proper algebraic perfect central extension.*

For the following minor fact see again [**115**] or [**67**].

LEMMA 2.23. PSL_2 *is simply connected in characteristic* 2.

In particular, there is no point, or little point, in distinguishing PSL_2 and SL_2 in characteristic 2, a point considerably more evident than the full lemma. One of the less pleasant consequences of this is that when working exclusively in characteristic 2 one is faced with the choice of using a single not always appropriate notation, or varying the notation according to the context, either of which produces some peculiar effects. The groups in question are not the same as algebraic groups, but we are actually working with their realizations over particular fields.

FACT 2.24 ([**173**, §7]). *If G is a Chevalley group over a field of characteristic p, then p does not divide $|Z(G)|$.*

In fact the center of the universal Chevalley group lies in a torus.

We remark also that the Chevalley commutator formula is essentially a presentation of the universal Chevalley group, which corresponds to the universal central extension of the corresponding simple group, taken in the algebraic category. One must of course add the basic relations giving the structure of the root groups themselves:

$$x_r(s+t) = x_r(s)x_r(t) \ ,$$

and in rank 1 it is also necessary to give relations determining the structure of a torus; these relations are also taken as part of the definition in higher rank, but have been shown to be superfluous.

We will return to this in a more explicit way in the next section, when we deal with central extensions in a broader category.

2.8. Automorphisms.

FACT 2.25. *Let G be a group of finite Morley rank, and H a definable group of automorphisms acting (faithfully) on G. Assume that G is an infinite simple Chevalley group over an algebraically closed field F. Then $H \leq G \rtimes \Gamma$, where G is identified with the group of inner automorphisms of G, and Γ is the group of graph automorphisms of G relative to a choice of maximal torus and Borel subgroup.*

PROOF. By Theorem 27.4 of [**115**], $\mathrm{Aut}(G) = G\Gamma\,\mathrm{Aut}(F)$ where $\mathrm{Aut}(F)$ acts naturally. The subgroup $G\Gamma$ consists of algebraic automorphisms, and G is of finite index, so it is definable. If H is not contained in $G\Gamma$, then $HG\Gamma$ contains a definable automorphism induced by $\mathrm{Aut}(F)$. As F can be interpreted in a Borel B and the action of $\mathrm{Aut}(F)$ on F can be interpreted via its action on B, $H \cap \mathrm{Aut}(F)$ defines a definable group of automorphisms on F, which is trivial by Lemma I 4.5.

So $H \leq G\Gamma$. □

COROLLARY 2.26. *Let G be a group of finite Morley rank, and L a normal subgroup isomorphic to a quasisimple algebraic group over an algebraically closed field K. Suppose that G is connected, or has no graph automorphisms. Then $G = L * C_G(L)$.*

PROOF. Suppose first that L is centerless. Then G acts by inner automorphisms by the previous fact, and the claim follows.

Now in general, the center $Z(L)$ is finite, and the result applies modulo $Z(L)$. So we get $G = L \cdot H$ with $[H, L] \leq Z(L)$. But since H is connected, so is $[H, L]$ (Corollary I 3.29), and as $[H, L]$ is finite it is trivial, and $G = L * C_G(L)$. □

LEMMA 2.27. *Let G be a quasisimple Chevalley group of finite Morley rank, and α a definable involutory automorphism of G such that $C_G{}^\circ(\alpha)$ is solvable. Then α is inner.*

PROOF. We may take G to be simple. We use the detailed information given in [**21**, §§8,19]. This deals with groups over arbitrary finite fields of characteristic 2, and goes over easily to the algebraically closed case.

We know that either α is inner or α lies in the coset of an involutory graph automorphism modulo inner automorphisms, and in the latter case the associated Dynkin diagram is of type A_n $(n \geq 2)$, D_n $(n \geq 4)$, or E_6. If α is itself conjugate to a graph automorphism, then the result is standard (and of course included in [**21**] as well), as the centralizer is the associated twisted group. But other cases arise.

Type D_n is dealt with in §8 of [**21**]. According to (8.10) given there, the relevant involutions are those of "type b" in the notation of that section (and the previous section). The centralizer for this type of involution is discussed in (8.7).

For types A_n and E_6 we turn to §19 of [**21**], beginning with the bottom of p. 78. The classification of the relevant involutions is given in (19.8); up to conjugacy we are dealing either with a graph automorphism or a product σt with σ a graph automorphism and t a 2-central involution commuting with σ. The centralizers in all cases are described in (19.9) and for those of the form $C(\sigma t)$ we have the following.

(1) In type A_n: the centralizer of a transvection in $\mathrm{PSp}(n)$.
(2) In type E_6: the centralizer of a 2-central involution in F_4.

Descriptions of these centralizers are given in (7.10) and (13.2), though the latter is given more explicitly just after the statement of (13.1); it contains a factor of the form $\mathrm{Sp}(6)$. □

2.9. Identification in Lie rank 2. The following result of Tits lays out the information needed to identify a Chevalley group of Lie rank 2.

FACT 2.28. *Let G be a Chevalley group of Lie rank 2 and let P_1, P_2 be minimal parabolic subgroups containing a common Borel subgroup B. Let N be the normalizer of a Cartan subgroup of B. Then G is the universal closure of the triple amalgam of P_1, P_2, and N.*

An elegant proof of this has been given by Bennett and Shpectorov [**31**]. The idea of the proof is to adjoin to the natural point/line geometry associated with G a third kind of object, the set of apartments, where an apartment is incident with its elements. This has the effect of making the geometry simply connected, and a very general result of Tits [**184**] on groups acting flag-transitively on simply connected geometries then applies.

The analogous statement in Lie rank greater than 2 does not require the additional group N. This gives a very flexible criterion for the final identification of simple groups of this type.

FACT 2.29 (Curtis-Tits, [**105**, Theorem 2.9.3],[**182**]). *Let Σ be an indecomposable root system of rank at least 2 with a fundamental system Π, and let F be a field. Let G be the universal Chevalley group constructed from Σ and F. For each $r \in \Sigma$ denote by X_r the root subgroup $\{x_r(t) : t \in F\}$ and for $J \subseteq \Pi$ let G_J be the subgroup of G generated by all X_r for $\pm r \in J$. Then G is the universal completion of the amalgam $(G_J : J \subseteq \Pi, |J| \leq 2)$.*

This includes the fact that G is generated by the minimal parabolic subgroups containing a fixed Borel subgroup (which may also be considered as an instance of the classification of parabolic subgroups).

3. Central extensions

In this section we show that groups of finite Morley rank which are perfect central extensions of quasisimple Chevalley groups are themselves Chevalley groups.

PROPOSITION 3.1. *Let G be a perfect group of finite Morley rank such that $G/Z(G)$ is a quasisimple Chevalley group. Then G is a Chevalley group over the same field. In particular, $Z(G)$ is finite.*

Once the first point is proved, the second point follows from Fact 1.18.

The center of the universal extension of a simple Chevalley group, as an abstract group, is not finite in general; this only holds in the algebraic category. Thus the finite Morley rank assumption cannot be omitted.

We will prove this proposition in the following more technical form, to be explained below:

PROPOSITION 3.2. *Let G be a perfect group of finite Morley rank and let $C_\circ$ be a definable central subgroup of G such that $G/C_\circ$ is a universal Chevalley group over an algebraically closed field; that is, G is a perfect central extension of finite Morley rank of a universal Chevalley group. Then $C_\circ = 1$.*

The main model theoretic ingredient in the argument is Lemma I 4.16, which eliminates certain "very bad fields" from consideration.

The second ingredient in the proof is the theory of central extensions of linear algebraic groups as explained in [**173**], blended with a dose of model theory needed for definability arguments. This theory runs as follows. We begin with a fuller statement of the relations holding in Chevalley groups and their central extensions, elaborating on Fact 2.2.

NOTATION 3.3. *Let F be a field and Φ a root system. Consider the following relations over a set of formal symbols $\{x_\alpha(t) : \alpha \in \Phi, t \in F\}$.*

(A) *$x_\alpha(t)$ is additive.*

(B) *If α, β are roots and $\alpha + \beta \neq 0$, then*

$$(x_\alpha(t), x_\beta(u)) = \prod x_{i\alpha+j\beta}(c_{ij}t^i u^j) ,$$

where i and j are positive integers and the c_{ij} are integers depending on α, β, and the chosen ordering of the roots, but not on t or u. Here $(g,h) = ghg^{-1}h^{-1}$.

(B′) *$w_\alpha(t)x_\alpha(u)w_\alpha(-t) = x_{-\alpha}(-t^{-2}u)$ for $t \in k^*$, where*

$$w_\alpha(t) = x_\alpha(t)x_{-\alpha}(-t^{-1})x_\alpha(t)$$

for $t \in k^*$.

(C) *$h_\alpha(t)$ is multiplicative in t, where $h_\alpha(t) = w_\alpha(t)w_\alpha(-1)$ for $t \in k^*$.*

(1) *We define the group X_u as follows:*
   (a) *X_u is the group presented by relations* (A) *and* (B) *if the rank of Φ is greater than 1;*
   (b) *X_u is the group presented by the relations* (A) *and* (B') *if the rank of Φ is equal to 1.*

(2) *X is the quotient of X_u obtained by adding the relation* (C)*; this is called the* universal Chevalley group ([**173**]) *of type Φ over F.*

FACT 3.4 ([**173**, Lemma 39, p. 70]). *Let α be a root and X_u be as above. In X_u, set $f(t,u) = h_\alpha(t)h_\alpha(u)h_\alpha(tu)^{-1}$. Then:*

- *$f(t, u^2v) = f(t,u^2)f(t,v)$.*
- *If t, u generate a cyclic subgroup of k^*, then $f(t,u) = f(u,t)$.*
- *If $f(t,u) = f(u,t)$, then $f(t,u^2) = 1$.*
- *If $t, u \neq 0$ and $t + u = 1$, then $f(t,u) = 1$.*

FACT 3.5 ([**173**, Theorem 9, p. 72]). *Assume that Φ is indecomposable and that F is an algebraic extension of a finite field. Then the relations* (A) *and* (B) (*or* (B′) *if rank $\Phi = 1$) suffice to define the corresponding universal Chevalley group, i.e. they imply the relations* (C).

FACT 3.6 ([**173**, Theorem 10, p. 78]). *Let Φ be an indecomposable root system and k a field such that $|k| > 4$, and if rank $\Phi = 1$, assume further that $|k| \neq 9$. If X is the corresponding universal Chevalley group (abstractly defined by the relations* (A), (B), (B′), (C) *above), if X_u is the group defined by the relations* (A), (B), (B′) ((B′) *is used only if rank $\Phi = 1$), and if π is the natural homomorphism from X_u to X, then (π, X_u) is a universal central extension of X.*

FACT 3.7 ([**173**, Corollary 2, p. 82]). *X_u is centrally closed. Each of its central extensions splits, i.e. its Schur multiplier is trivial. It yields the universal central extension of all the Chevalley groups of the given type.*

FACT 3.8 ([**173**, Theorem 12 (Matsumoto, Moore)]). *Assume that Φ is an indecomposable root system and k a field with $|k| > 4$. If X is the universal Chevalley group based on Φ and k, if X_u is the group defined by* (A), (B), (B′), *and if π is the natural map from X_u to X with $C = \ker\pi$, the Schur multiplier of X, then C is isomorphic to the abstract group generated by the the symbols $\{t, u\}$ $(t, u \in k^*)$ subject to the relations:*

(a) $\{t,u\}\{tu,v\} = \{t,uv\}\{u,v\}$; $\{1,u\} = \{u,1\} = 1$,
(b) $\{t,u\}\{t,-u^{-1}\} = \{t,-1\}$,
(c) $\{t,u\} = \{u^{-1},t\}$,
(d) $\{t,u\} = \{t,-tu\}$,
(e) $\{t,u\} = \{t,(1-t)u\}$, *and in the case Φ is not of the type C_n $(n \geq 1)$ the additional relation*
(ab′) $\{\ ,\ \}$ *is bimultiplicative. In this case relations* (a)-(e) *may be replaced by* (ab′) *and*
(c′) $\{\ ,\ \}$ *is skew,*
(d′) $\{t,-t\} = 1$,
(e′) $\{t, 1-t\} = 1$.

The isomorphism is given by $\phi : \{t,u\} \longmapsto h_\alpha(t)h_\alpha(u)h_\alpha(tu)^{-1}$, α a fixed long root.

We now consider a perfect central extension G of finite Morley rank of a universal linear algebraic group X over an algebraically closed field K. Let (π, X_u) be the universal covering extension of X, $C = \ker\pi$ and $C_\circ = \ker\psi$ where ψ is the covering map from G onto X. By the universality of (π, X_u), there exists a map θ from X_u into G such that $\psi\theta = \pi$.

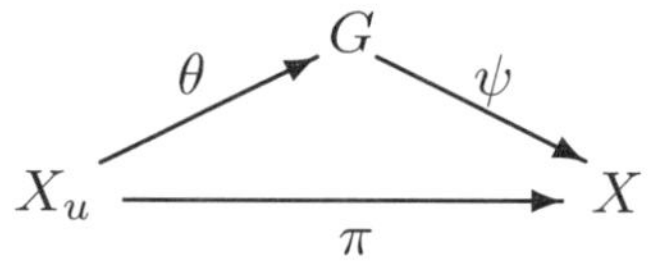

As G is a perfect group, one can show that θ is surjective and $C = \theta^{-1}(C_\circ)$. By Fact 3.8 C is generated by $f(t,u)$ where f is as in Fact 3.4. We must prove the interpretability in G of the function $\hat{f} = \theta \circ f$ in order to understand the structure of $C_\circ$.

PROPOSITION 3.9 ([**3**, Proposition 4.12]). *Let G be a group of finite Morley rank. Assume that G is a perfect central extension of a universal Chevalley group X, such that the kernel of the covering map from G onto X is a definable central subgroup of G. If X_u is the universal central extension of X and $\theta : X_u \to G$ is the unique induced map, and $f : K \times K \to X_u$ is the function defined in Fact 3.4, then the function $\hat{f} = \theta \circ f$ is interpretable in G.*

PROOF. In order to prove Fact 3.6, Steinberg proves that the relations (A), (B) and (B′) can be lifted from a universal linear algebraic group X

to any of its central extensions. To do so he starts with a central extension (ψ, G) of X and he constructs a map ϕ from the root subgroups of X into G. We will make use of this map in order to show that the function $\theta \circ f$ is interpretable in G.

The first step in the proof is to show that ϕ is interpretable in G. To do so, we need to look at the definition of ϕ. First, an element a of K^* is chosen so that $c = a^2 - 1 \neq 0$. In G/C, $(h_\alpha(a), x_\alpha(t)) = x_\alpha(ct)$ for all $\alpha \in \Sigma$, $t \in K$. Then $\phi(x_\alpha(t))$ is defined so that:

(i) $\psi(\phi(x_\alpha(t))) = x_\alpha(t)$,

(ii) $(\phi(h_\alpha(a)), \phi(x_\alpha(t))) = \phi(x_\alpha(ct))$.

Steinberg observes that this determines ϕ as a map from the root group $X_\alpha = \langle x_\alpha(t) : t \in K \rangle$ into G. The x_α are definable from the field over which X is defined. Therefore, by Proposition 1.20, they are definable from the pure group G. On the other hand, the following formula defines ϕ:

$$\phi(x) = y$$

if and only if

$$\exists x_1, y_1 (\psi(x_1) = y_1 \ \&\ (g_\circ, y_1) = y \ \&\ \exists t (x_1 = x_\alpha(t) \ \&\ x = x_\alpha(ct))),$$

where $g_\circ$ is the group element defined by $h_\alpha(a)$. As a result, we conclude that ϕ is an interpretable map from X_α into G. One can do the same thing for all roots and get a map ϕ which lifts (interpretably in G) the X_α from X to G.

Now we define the following functions from K into G:

$$\begin{aligned} \overline{w_\alpha}(t) &= \phi(x_\alpha(t))\phi(x_{-\alpha}(-t^{-1}))\phi(x_\alpha(t)), \\ \overline{h_\alpha}(t) &= \overline{w_\alpha}(t)\overline{w_\alpha}(-1). \end{aligned}$$

As ϕ and the x_α are interpretable in G, so is $\overline{w_\alpha}$ and therefore, $\overline{h_\alpha}$. Hence, using Proposition 1.20, the following function also is interpretable in G:

$$\begin{aligned} \bar{f} \ :\ K^* \times K^* &\rightarrow G \\ (t, u) &\longmapsto \overline{h_\alpha}(t)\overline{h_\alpha}(u)\overline{h_\alpha}(tu)^{-1} . \end{aligned}$$

But $\bar{f} = \hat{f}$ since (i) and (ii) hold in X_u for $x_\alpha(t)$ and are preserved by homomorphisms. This finishes the proof. □

Before we start the proof of Proposition 3.2 we need one last ingredient, coming from K-theory. The kernel of the covering map π is known in K-theory as $K_2(K)$, where K is the field over which the Chevalley group is defined. Facts 3.4 and 3.8 describe how the group $K_2(K)$ is presented. The definition of K_2 can be generalized to commutative rings, and K_2 is actually a functor from commutative rings to abelian groups. We will make use of some properties of the functor K_2 to show that in our case the algebraically closed field K contains an infinite subfield over which the generators f are trivial.

The characteristic p of the field K plays an important role. If $p \neq 0$, then Fact 3.5 proves that over the algebraic closure of the prime field, the generators $f(t, u)$ are all equal to 1. This will provide the necessary infinite subfield. When $p = 0$, the following two results from K-theory imply that f is trivial on $\mathbb{Q} \times \mathbb{Q}$:

FACT 3.10 ([**164**, Proof of Theorem 4.4.9, p. 225]). *The group* $K_2(\mathbb{Q})$ *is a direct limit of finite abelian groups.*

FACT 3.11 ([**24**]). *If* F *is an algebraically closed field, then* $K_2(F)$ *is a divisible torsion-free group.*

LEMMA 3.12. *The function* f *is trivial on* $\mathbb{Q} \times \mathbb{Q}$ *(where we are viewing* $\mathbb{Q}$ *as a subfield of an algebraically closed field* K*).*

PROOF. The functor K_2 is covariant, that is, the values $f(a, b)$ for $a, b \in \mathbb{Q}$ can be interpreted as the images in $K_2(K)$ of the corresponding symbols in $K_2(\mathbb{Q})$. As these symbols can be taken to lie in a finite group and their image lies in a torsion-free group, the natural homomorphism must kill them. □

Now we can prove our main result.

PROOF OF PROPOSITION 3.2. The arguments above show that in all characteristics K has an infinite subfield $K_\circ$ such that for t, $u \in K_\circ$, $f(t, u) = 1$. Let $t \in K^*$. We define $B_t = \{u \in K^* : f(t, u) = 1\}$. As K is an algebraically closed field, B_t is a subgroup of K^* by Fact 3.4 (a). We will show that for any $t \in K^* \setminus \{1\}$, $B_t = K^*$. This will prove the proposition. First, suppose $t \in K_\circ$. Since $B_t \geq K_\circ^*$, Lemma I 4.16 implies that B_t is generic in K^*. But K^* is connected (Corollary I 3.7), therefore $B_t = K^*$. Now choose t to be any element of $K^* \setminus \{1\}$. For any $u \in K \setminus \{1\}$, $f(t, u) = f(u^{-1}, t)$ by Fact 3.8 (c). But if $u \in K_\circ^*$ then $f(u^{-1}, t) = 1$ by the first part of the argument. Hence, $B_t \geq K_\circ^*$ and we conclude again by Lemma I 4.16 that $B_t = K^*$. □

PROOF OF PROPOSITION 3.1. We start with a perfect group G of finite Morley rank such that $G/Z(G)$ is a quasisimple algebraic group. Let X be the universal group of the same type as $G/Z(G)$. Then we have the following diagram:

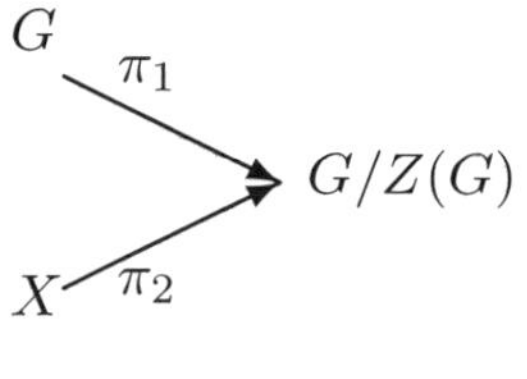

We form the pullback of this diagram:

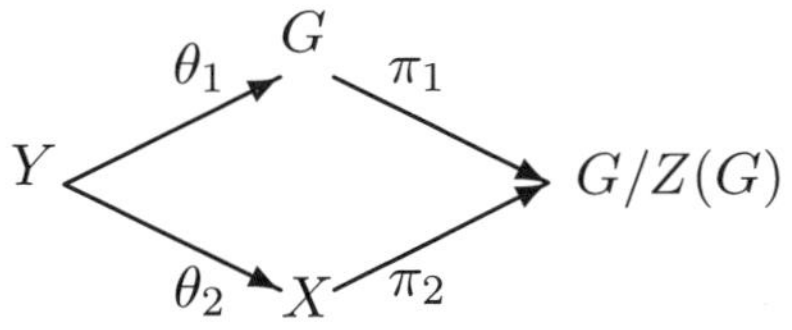

Then $Y \simeq \{(g,x) \in G \times X : \pi_1(g) = \pi_2(x)\}$. In this diagram, π_1 and $G/Z(G)$ are interpretable in G. On the other hand, as X is an algebraic group, it is interpretable in $G/Z(G)$ and hence in G. Moreover, the triple $(X, G/Z(G), \pi_2)$ is algebraic and hence interpretable in G, say as $(X^*, \overline{G}^*, \pi^*)$, where $\overline{G}^* \cong G/Z(G)$ definably; hence we may take $\overline{G}^* = G/Z(G)$ and $\pi^* : X^* \to G/Z(G)$.

The pullback $\tilde{Y}$ of π_1 and π^* is interpretable in G. Hence it is of finite Morley rank. Since $\tilde{Y} \simeq Y$, Y also has finite Morley rank. Moreover, Y is a definable central extension of X. Therefore, we can apply Proposition 3.2 to Y and X and conclude, using [**173**, (iii), p. 75] that $Y \simeq X * A$, where A is abelian. Note that $\theta_1(Y') = (\theta_1(Y))' = G' = G$; but $Y' \simeq X$ is an algebraic group. Therefore, G is a quotient of an algebraic group by a finite group. We conclude that G also is an algebraic group. □

PROPOSITION 3.13. *Let G be a connected group of finite Morley rank, and suppose that $G/Z(G)$ is a quasisimple Chevalley group, and $L = G^{(\infty)}$. $G = Z(G) * L$ is a central product and L is a quasisimple Chevalley group.*

PROOF. Certainly $G = Z(G) * L$. Now $L/Z(L)$ is quasisimple and perfect, so L is a quasisimple Chevalley group. □

We have worked in the context of Chevalley groups rather than algebraic groups throughout, and with good reason as we depend on [**173**]. However, there is considerable value in knowing that we have not left the category of algebraic groups, and we record this fact, which will be applied without explicit mention.

FACT 3.14 ([**173**, §5]). *A universal Chevalley group is an algebraic group.*

Indeed, we will generally read Proposition 3.1 as saying that the group in question is algebraic.

4. Structure of K-groups

4.1. K-groups and K^*-groups.

DEFINITION 4.1.

(1) *A K-group is a group G of finite Morley rank such that every definable connected simple section of G is isomorphic to a Chevalley group over an algebraically closed field.*

(2) *A K^*-group is a group G of finite Morley rank such that every proper definable connected simple section of G is isomorphic to a Chevalley group over an algebraically closed field.*

PROPOSITION 4.2. *If G is a simple algebraic group, with no further structure, then G is a K-group.*

PROOF. Fact 1.21. □

Our discussion in this subsection is directed not so much toward methods to be used in the classification phase, as toward a general discussion of some results we find quite striking that bear on the Algebraicity Conjecture in the context of locally finite or linear groups.

The following result is striking in itself, and leads via Wagner's results to results of considerably greater generality, if one opens up the proof a bit (the statement as we give it omits the uniformities in the proof, which are useful).

FACT 4.3 ([**178**]). *Let G be a locally finite simple group of finite Morley rank. Then G is isomorphic to a Chevalley group over an algebraically closed field of positive characteristic.*

The proof of this theorem involves a reduction to properties of finite simple groups, and requires the classification of the finite simple groups. Using the material in Parts B and C of the present book, and taking pains never to invoke this result one can get a reasonably self-contained proof which needs nothing beyond the Feit-Thompson theorem—but only because it runs a miniaturized version of the classification of the finite simple groups through to the end. In order to carry this through one has to eliminate the uses of Proposition 4.4 below, as they involve results on locally finite simple groups which in turn depend on the classification of the finite simple groups. This is done by restricting oneself to the class of K^*-groups throughout, which suffices for the proof of Fact 4.3. In that context it turns out that Proposition 4.4 is not needed. In the locally finite case it provides a mechanism for verifying the algebraicity of sections of degenerate type, which by Theorem IV 4.1 contain no involutions, which requires at worst the Feit-Thompson Theorem—and in particular configurations, somewhat less.

Here we find it useful to apply the "change of group" idea alluded to in §I 2.

PROPOSITION 4.4 ([**159**]). *Let G be a simple group of finite Morley rank with a definable finite dimensional representation over a definable field F of characteristic $p > 0$. Then G is definably isomorphic to an algebraic group over the same field.*

PROOF. (Sketch) By considering the Zariski closure of G we easily eliminate the possibility that G would have bounded exponent. Then replacing G by an elementary extension, we may suppose it contains an element g of infinite order. As we are in finite characteristic, replacing g by a suitable power we may suppose that g is semisimple. Let $T = d(g)$. By Proposition I 3.19 the group G is generated (in finitely many steps) by a finite number of conjugates $T_i = T^{g_i}$ of T. Now T sits in an algebraic torus of $\mathrm{GL}_n(F)$ for some n, which we identify with $(F^\times)^n$. We view (F, T) as a structure in its own right, F being a field and T a multiplicative subgroup of $(F^\times)^n$, and we note that G is parametrically definable in (F, T).

Now we invoke Proposition I 4.15; the structure (F, T) has an elementary substructure $(F_{\mathrm{alg}}, T \cap F_{\mathrm{alg}})$ where $F_{\mathrm{alg}} = F \cap \mathrm{acl}(\emptyset)$ consists of the elements of F which in (F, T) are model theoretically algebraic. As the Frobenius $x \mapsto x^p$ is an automorphism of (F, T), any 0-definable subset is invariant under the Frobenius and hence algebraic in the field theoretic sense. So the structure $(F_{\mathrm{alg}}, T \cap F_{\mathrm{alg}})$ is locally finite, as is any group defined in it. Thus the group G satisfies any sentence belonging to the theory of all locally finite groups.

For the rest, one enters into the classification of the locally finite groups of finite Morley rank [**178**]. Here what is needed is a slight refinement; after fixing some parameters associated with the group G (notably the length of chains of centralizers), the methods of [**178**] force any locally finite group into one of a finite number of families, in each of which a field can be interpreted and the structure of the group completely elucidated, in a way that can be encoded by a single rather elaborate first-order statement. This statement then goes over to G by "transfer", as it is something true of all locally finite groups: if their basic parameters are sufficiently restricted, they are of the stated form. One then finds that the underlying field must be algebraically closed. To show that the field in question is the one with which we began then requires some further inspection of the situation within $\mathrm{GL}_n(F)$, and tends not to play a role in applications in any case.

This completes our sketch. □

PROPOSITION 4.5. *Let G be a group of finite Morley rank with a definable finite dimensional representation over a definable field F of characteristic $p > 0$. Then G is a K-group.*

PROOF. We work in the group $\mathrm{GL}(n, F)$ with F an algebraically closed field of characteristic $p > 0$. We assume that the pair $(\mathrm{GL}(n, F), G)$, with whatever structure G carries, has finite Morley rank. What we need to see is that an arbitrary connected simple section of $\mathrm{GL}(n, F)$ which is definable in the extended language is isomorphic to a Chevalley group.

Let H/N be the simple section in question, and let R be the Zariski closure of $\sigma(H)$. Then R is solvable, so $H \cap R = \sigma(H)$. Now $N(R)$ is a linear algebraic group, so $N(R)/R$ is also a linear algebraic group.

Working in the linear algebraic group $N(R)/R$, the group H is replaced by $H/\sigma(H)$. Thus we may suppose $\sigma(H) = 1$. As H is connected, it follows that $F^*(H)$ is a finite product of infinite simple groups normalized by H (Lemma I 7.2). Furthermore, $F^*(H)$ embeds in $N(R)/R$, hence by Poizat's result (Proposition 4.4) each factor is algebraic. Therefore, by Fact 2.25, for each simple factor L of $F^*(H)$, we have $H = L \cdot C_H(L)$. It follows that $H = F^*(H)C_H(F^*(H)) = F^*(H)$ by Proposition I 7.3. Accordingly, H/N must be isomorphic to one of the factors of $F^*(H)$. □

The following variation does not require the assumption of positive characteristic. The proof involves a subtle calculation with traces which seems to have no natural extension to more general groups.

FACT 4.6. [**159**] *If K is a field of finite Morley rank, every definable subgroup of $\mathrm{GL}_2(K)$ is either solvable-by-finite or contains $\mathrm{SL}_2(K)$.*

PROPOSITION 4.7. *Let G be a group of finite Morley rank which is definable over a bad field of positive characteristic. Then G is a K-group.*

PROOF. Let F be the field involved. Varying the parameters of the definition, G belongs to a uniformly definable family $\mathcal{G}$ of groups of finite Morley rank. Arguing as in Proposition I 3.25, one can express the property that G is generated by each of its nontrivial conjugacy classes (in a bounded number of steps). Hence we may require that property for each group in the family, forcing them all to be simple.

By Proposition I 4.18, F_{alg} is an elementary substructure of F. Over F_{alg}, the groups in the family $\mathcal{G}$ are locally finite, hence are Chevalley groups by Fact 4.3. One can then bound their dimensions by using the uniform bound on lengths of chains of centralizers, and find a first order sentence that expresses the fact that they are Chevalley groups; this then passes to the field F. □

4.2. Structure.

PROPOSITION 4.8. [**1**] *Let G be a connected nonsolvable K-group. Then $G/\sigma(G)$ is isomorphic to a direct sum of finitely many simple algebraic groups over algebraically closed fields. In particular, the definable connected $2^\perp$-sections of G are solvable.*

PROOF. We may suppose that $\sigma(G) = 1$. Then $F^*(G)$ is a finite product of infinite simple groups, each of which is a Chevalley group over an algebraically closed field (Lemma I 7.2).

G normalizes the factors of $F^*(G)$, and arguing as in the proof of Proposition 4.5 we see that $G = F^*(G)C_G(F^*(G)) = F^*(G)$. □

COROLLARY 4.9. *Let G be a connected group of finite Morley rank with no definable connected simple sections. Then G is solvable.*

PROPOSITION 4.10. *Let G be a connected nonsolvable K-group. Then $G/\sigma^\circ(G)$ is isomorphic to a central product of finitely many quasisimple algebraic groups over algebraically closed fields.*

PROOF. We may assume that $\sigma^\circ(G) = 1$ and hence that $\sigma(G) = Z(G)$ as G is connected. Furthermore, $\bar{G} = G/Z(G)$ is a direct product of simple algebraic groups. Let $H = G^{(\infty)}$. Then $\bar{H} = \bar{G}$. By Lemma I 3.38, H is a central product of quasisimple groups, which by Proposition 3.1 are Chevalley groups. Now $G = H \cdot Z(G)$ and as G is connected and $Z(G)$ is finite, we have $G = H$. □

LEMMA 4.11. *Let G be a connected perfect K-group of finite Morley rank, with $\sigma^\circ(G) = Z^\circ(G)$. Then G is a finite central product of quasisimple Chevalley groups. In particular, $\sigma^\circ(G) = 1$.*

PROOF. $G/Z^\circ(G)$ is a finite central product of quasisimple Chevalley groups, each of which is covered by a perfect normal subgroup H of G which has the property that $H/Z(H)$ is a quasisimple Chevalley group. In other words, we may suppose that $G/Z(G)$ is a quasisimple Chevalley group. In particular, $\sigma(G)/Z(G)$ is finite, and $[G, \sigma(G)] \leq Z(G)$, so by the three subgroups lemma we have $\sigma(G) \leq Z(G)$ and G is a perfect central extension of a simple Chevalley group, to which the theory of central extensions applies (Proposition 3.1). □

LEMMA 4.12. *Let G be a connected K-group of finite Morley rank and H a product of quasisimple factors of $E(G)$ (the main cases are: $H = E(G)$; $H = L$ a single such factor). Then $G = H * C_G{}^\circ(H)$.*

PROOF. The quasisimple factors of H are quasismple Chevalley groups normal in G. If L is one such, then, by Fact 2.25, G acts on L via inner automorphisms. Hence G acts on H via inner automorphisms. This yields $G = H * C_G(H)$ and as G is connected our claim follows. □

PROPOSITION 4.13. *Let G be a connected K-group. Then $C_G(F^\circ(G)) \leq F^*(G)$.*

PROOF. We have $G = E(G) \cdot C_G{}^\circ(E(G))$ and $E(G) \leq C_G(F^\circ(G))$. Hence

$$C_G(F^\circ(G)) = E(G) \cdot C_G{}^\circ(F^\circ(G)E(G)) ;$$

but $F^\circ(G)E(G) = F^{*\circ}(G)$, and we have $C_G{}^\circ(F^{*\circ}(G) \leq F(G)$ by Proposition I 7.3, so

$$C_G(F^\circ(G)) \leq E(G)F(G) = F^*(G)$$

□

4.3. Weak embedding.

DEFINITION 4.14. *Let G be a group and M a proper definable subgroup. We say that M is* weakly embedded *in G if M contains an infinite 2-subgroup, and for $g \in G \backslash M$, $M \cap M^g$ does not contain an infinite 2-subgroup.*

PROPOSITION 4.15. *Let G be a group of finite Morley rank of even type, M a proper definable subgroup of G, and S a Sylow 2-subgroup of M. Then the following are equivalent:*

I *M is weakly embedded in G.*

II (i) *M has infinite Sylow 2-subgroups.*

(ii) *For any infinite Sylow 2-subgroup P of M, we have $N_G(P) \leq M$.*

III (i) *M has infinite Sylow 2-subgroups.*
(ii) *For any unipotent 2-group U of M, $N_G(U) \leq M$.*
(iii) *For any 2-torus T of M, $N_G(T) \leq M$.*

IV (i) *S is infinite.*
(ii) *For any unipotent 2-group U of S, $N_G(U) \leq M$.*
(iii) *For any 2-torus T of S, $N_G(T) \leq M$.*

PROOF. (II $\implies$ I): Suppose $M \cap M^g$ has an infinite Sylow 2-subgroup Q, and let $R \leq G$ be a Sylow 2-subgroup containing Q. Then by hypothesis $N_G(Q) \leq M$ and similarly, conjugating, $N_G(Q) \leq M^g$, so $N_R(Q) = Q$. By the normalizer condition, $Q = R$, and it is a Sylow 2-subgroup of G. Hence Q and $Q^{g^{-1}}$ are Sylow 2-subgroups of M. Now for some $x \in M$ we have $Q^x = Q^{g^{-1}}$, so $xg \in N(Q) \leq M$ and $g \in M$.

(III $\implies$ II): Clear, since P will contain a definably characteristic subgroup which is either 2-unipotent or a 2-torus.

(IV $\implies$ III): With $Q = U$ or T we have $Q^g \leq S$ for some $g \in M$ and hence $N_G(Q) = [N_G(Q^g)]^{g^{-1}} \leq M$.

(I $\implies$ IV): With $Q = U$ or T, if $g \in N(Q)$, then $Q \leq M \cap M^g$ and hence $g \in M$. □

The definability hypothesis is superfluous here.

LEMMA 4.16. *Let G be a group of finite Morley rank, M a subgroup satisfying any of the conditions of Proposition 4.15, where, however, we allow "unipotent" subgroups to be intersections of definable subgroups of G with M, rather than insisting that they be definable in G. Then M is definable in G.*

PROOF. With S a Sylow 2-subgroup of M, our hypotheses imply that $d(S^\circ) \leq M$ and thus M contains nontrivial G-definable connected subgroups. Let M_0 be the subgroup of M generated by its G-definable connected subgroups. Then M_0 contains S° and by a Frattini argument, $N(M_0) \leq M_0 N(S_0) \leq M$. Thus $M = N(M_0)$ is definable. □

COROLLARY 4.17. *Let G be a group of finite Morley rank, M a weakly embedded subgroup, and H a definable proper subgroup of G containing M. Then H is weakly embedded in G.*

LEMMA 4.18. *Let G be a group of finite Morley rank and M a weakly embedded subgroup. Then M is definable.*

PROOF. Let $M_0 = \langle N_G{}^\circ(S) : S \text{ a Sylow 2-subgroup of } M \rangle$. Then $M_0 \leq M$ is definable by Lemma I 6.14, and $M \leq N(M_0)$. It will suffice to show that $M = N(M_0)$.

If $g \in N(M_0)$ and S is any Sylow 2-subgroup of M, then $S, S^g \leq M_0$ and $S^g = S^x$ for some $x \in M_0$. Hence $gx^{-1} \in N(S) \leq M$, and $g \in M$. □

PROPOSITION 4.19. *Let G be a group of finite Morley rank, H a definable normal subgroup, $\bar{G} = G/H$, and M a weakly embedded subgroup of G. If $\overline{M} < \bar{G}$ and $\overline{M}$ has an infinite Sylow 2-subgroup, then $\overline{M}$ is weakly embedded in G.*

PROOF. We take $\bar{P} \leq \overline{M}$ as an infinite 2-subgroup, and P as a Sylow 2-subgroup of its preimage in G. Then P covers $\bar{P}$ (Lemma I 6.15). Let $\bar{K} = N_{\bar{G}}(\bar{P})$, with preimage K in G. Then $K = N_G(HP)$. For $k \in K$ we have $P^k = P^h$ with $h \in H$, hence $K \leq N_G(P)H \leq MH$, and $\bar{K} \leq \overline{M}$. □

PROPOSITION 4.20. *If G is a nonsolvable connected K-group with a weakly embedded subgroup M, then $G/O(G) \simeq (\mathrm{P})\mathrm{SL}_2(K)$, where K is an algebraically closed field.*

PROOF. We know that $G = G/\sigma^\circ(G)$ is a central product of quasisimple Chevalley groups. We will show that $\sigma^\circ(G) = O(G)$, and that $\bar{G}$ has Lie rank 1, and apply Fact 2.15.

Suppose first that $O(G) = 1$. If $O_2(G)$ is infinite, then $O_2(G) \leq M$ and $G = N(O_2(G)) \leq M$, a contradiction. As the Sylow 2-subgroup of $F^\circ(G)$ is connected by Proposition I 8.12, we find that it is trivial and $F^\circ(G) \leq O(G) = 1$. Accordingly $\sigma^\circ(G) = \sigma^\circ(G)/F^\circ(G)$ is abelian and hence contained in $F^\circ(G) = 1$. So by Proposition 4.10, G is a central product of finitely many quasisimple Chevalley groups over algebraically closed fields.

In this case, weak embedding of M forces G to consist of a single quasisimple Chevalley group. Suppose G has Lie rank greater than 1. In characteristic 2, M contains a Borel subgroup $N_G(S)$ with S a Sylow 2-subgroup, and every minimal parabolic containing B normalizes a connected subgroup of S, so M contains these minimal parabolics, which generate G, a contradiction. If the characteristic is not 2, then the Sylow° 2-subgroups of G are contained in maximal tori. Let T_α and Z_α be as in Fact 1.12. As the Lie rank is greater than 1, we have $T_\alpha \neq 1$ for any $\alpha \in \Delta$. Moreover, $Z_\alpha \leq M$. This forces $M = G$, a contradiction. So in all cases we arrive at a contradiction, and the Lie rank of G must be 1. This proves the result in the case $O(G) = 1$.

Next, suppose

$$O(G) \leq M\ .$$

Let $\bar{G} = G/O(G)$. Then $\overline{M}$ is a weakly embedded subgroup of $\bar{G}$, and $O(\bar{G}) = 1$, so the case just treated applies.

Accordingly we now suppose

$$O(G) \text{ is not contained in } M.$$

Suppose M contains a nontrivial unipotent subgroup U. Then U centralizes $O(G)$ by Proposition I 10.13. Hence by weak embedding $O(G) \leq M$, a contradiction.

So a Sylow° 2-subgroup S of M is a 2-torus. If the Prüfer rank is greater than 1, then $O(G)$ is generated by centralizers of nontrivial 2-tori contained in S, by Lemma I 9.17, forcing $O(G) \leq M$ by weak embedding, a contradiction. So S is a 2-torus of Prüfer rank 1.

Suppose $\sigma^\circ(G) > O(G)$. Then $S \cap \sigma^\circ(G)$ is connected by Proposition I 8.12, and is nontrivial. As S has Prüfer rank 1, this gives $S \leq \sigma^\circ(G)$, and hence $G/\sigma^\circ(G)$ has a finite Sylow 2-subgroup. This contradicts Proposition 4.10. We conclude that $\sigma^\circ(G) = O(G)$.

Thus $G/O(G)$ is a central product of simple Chevalley groups, and has Prüfer rank 1, as claimed. □

5. K-Groups of even type

According to the classification presented in §I 6, a group G of finite Morley rank is of *even type* if its Sylow 2-subgroups are infinite and of bounded exponent. We will be concerned with the structure of K-groups of even type in this section.

What we really need to understand is the structure of L-groups, but in most cases this follows readily from the K-group analysis. We will make this extension in the next section.

5.1. Borel subgroups.

LEMMA 5.1. *If G is of even type, then its Sylow 2-subgroups are definable.*

PROOF. Let S be a Sylow 2-subgroup. By Proposition I 6.4, S° has the form $U * T$ with U unipotent and T a 2-torus; as S has bounded exponent, $T = 1$ and $S^\circ = U$ is definable. Hence S is definable. □

PROPOSITION 5.2. *Let G be a K^*-group of finite Morley rank and of even type, and S a Sylow$^\circ$ 2-subgroup. Then*

(1) *$N^\circ(S)$ is a Borel subgroup of G, and the only one which contains S.*
(2) *$N^\circ(S)$ splits as $S \rtimes T$ with T a definable complement.*

PROOF.

Ad (1) Let $H = N^\circ(S)/S$. Then H is connected, and has a finite Sylow 2-subgroup by Lemma I 6.15. The same applies to any definable section of H, and as H is a K-group, it has no definable connected simple sections, hence is solvable by Corollary 4.9. So $N^\circ(S)$ is solvable.

Now let H be any connected solvable definable group containing S. As S is unipotent, we have $S \leq F(H)$ by Corollary I 8.4, and hence S is the unique maximal 2-unipotent subgroup of $F(H)$, hence normal in H.

So $N^\circ(S)$ is a Borel subgroup, and the only one containing S.

Ad (2) Any solvable connected group H containing S will split definably by Propositions I 9.6 and I 9.8. □

5.2. Tori.

PROPOSITION 5.3 ([**3**, Proposition 3.4]). *Let G be a connected K-group of even type, and T a connected $2^\perp$-group acting definably on G. Then T leaves invariant a Sylow° 2-subgroup of G.*

PROOF. Suppose first that G is a simple algebraic group over an algebraically closed field of characteristic 2. Then by Fact 2.25, T acts by inner automorphisms, and its image in G is again a connected $2^\perp$-group (Lemma I 2.18), As G is a K-group, Corollary 4.9 implies that the image of T in G is solvable. Hence so is its Zariski closure $\hat{T}$. The connected component $\hat{T}_0$ of $\hat{T}$ in the sense of algebraic groups is definable in G by Fact 1.19. As the image of T is connected, it is contained in $\hat{T}_0$, and hence in a Borel subgroup of G. So T acts as a subgroup of a Borel subgroup, and therefore leaves invariant a Sylow° 2-subgroup.

More generally, if

$$\sigma(G) = 1,$$

then by Proposition 4.8 G is a product of simple algebraic groups over algebraically closed fields of characteristic 2, and the first case applies in each factor.

We now deal with the general case. Applying the previous case to $G/\sigma(G)$, we may suppose that $G/\sigma(G)$ is a unipotent 2-group. In particular, G is solvable. Then by Proposition 5.2, $S = U_2(G)$ is T-invariant. □

PROPOSITION 5.4. *Let G be a connected K-group of even type, and T a connected $2^\perp$-group acting definably on G. If U is a T-invariant unipotent 2-subgroup of G, then U is contained in a T-invariant Sylow° 2-subgroup of G.*

PROOF. We may suppose that U is a maximal T-invariant and unipotent s-subgroup of G. Applying Proposition 5.3 to $N_G{}^\circ(U)$, it follows that U is a Sylow° 2-subgroup of $N_G{}^\circ(U)$, and hence of G. □

LEMMA 5.5. *Let G be a connected K-group of even type, and T a connected $2^\perp$-group acting definably on G and faithfully on $G/\sigma(G)$. Then T is a good torus.*

PROOF. This reduces at once to the case $\sigma(G) = 1$, so that G is a direct sum of simple Chevalley groups on which T acts. By Corollary I 4.22 it suffices to treat a single simple factor, replacing T by its image in that factor. So we may suppose that G is a simple algebraic group over an algebraically closed field of characteristic 2, and T is a definable subgroup of G containing no involutions.

By Proposition 5.3, T normalizes a Sylow° 2-subgroup of G, and hence lies in a Borel subgroup B of G. As $B = U \rtimes T_0$ with T_0 a maximal torus, and $T \cap U = 1$, there is a definable embedding $T \to T_0$. Now T_0 is a good torus by Proposition I 4.20 and Corollary I 4.22. □

5.3. O_2 is connected. The notation $O_p(G)$ may be problematic; it stands for the largest normal p-subgroup of G, but it is perhaps better to restrict its use to cases where this group is definable. In the most commonly occurring case there is no difficulty, in view of the following.

LEMMA 5.6. *Let G be a connected K-group of even type. Then $O_2(G)$ is connected and definable.*

PROOF. By Proposition 4.10, $G/\sigma^\circ(G)$ is a central product of quasisimple algebraic groups L_i over algebraically closed fields of characteristic 2. The image of $O_2(G)$ in $\prod_i L_i$ is a central 2-subgroup; but $Z(L_i)$ consists of semisimple elements by Fact 1.8, so $O_2(G) \le \sigma^\circ(G)$. Thus the problem reduces to the case in which G is solvable and connected.

In this case, the Sylow 2-subgroup S of G is connected by Proposition I 8.12, and is unipotent as G is of even type. Hence $S = O_2(G)$ by Corollary I 8.4. □

5.4. $O_{2^\perp}$.

DEFINITION 5.7. *Let G be a group of finite Morley rank, p a prime. Then $O_{p^\perp}(G)$ is the group generated by the definable normal $p^\perp$-subgroups of G.*

LEMMA 5.8. *Let G be a group of finite Morley rank and of even type. Then $O_{2^\perp}(G)$ is definable.*

PROOF. Passing to a quotient, we may suppose that G has no nontrivial connected definable normal $2^\perp$-subgroup. We may also suppose that G is connected. Then the definable normal $2^\perp$-subgroups of G are finite and central, so $O_{2^\perp}(G) = O_{2^\perp}(Z(G))$, and we may suppose that G is abelian.

By Lemma I 2.13, $G = A + B$ with A a 2-group of bounded order and B 2-divisible, and both factors definable. As G is of even type, B is a $2^\perp$-group, and hence $O_{2^\perp}(G) = B$ is definable (and indeed $B = 1$ at this point, in view of earlier reductions). □

5.5. Reductive groups.

DEFINITION 5.9. *A group G of finite Morley rank and even type is said to be* reductive *if $O_2{}^\circ(G) = 1$.*

PROPOSITION 5.10. *Let G be a connected reductive K-group of finite Morley rank of even type. Then $G = E(G) * O(G)$.*

PROOF. We have $G = E(G) * C_G{}^\circ(E(G))$ and it suffices to show that $C_G{}^\circ(E(G)) = O(G)$.

Let $G_1 = C_G{}^\circ(E(G))$. Then $O_2(G_1) = 1$ by reductivity, and $E(G_1) = 1$. We claim that $G_1 = O(G_1)$. It suffices to show that $U_2(G_1) = 1$.

Let $H = U_2(G_1)$. Then $O_2{}^\circ(H) = E(H) = 1$ and $U_2(H) = H$. By Proposition I 10.13, H centralizes $O(H)$. Furthermore $\sigma^\circ(H) = O(H) =$

$Z^\circ(H)$ and thus $H/Z^\circ(H)$ is a finite product of quasisimple Chevalley groups over algebraically closed fields of characteristic 2.

For each quasisimple factor $\bar{L}$ of $H/Z^\circ(H)$, with preimage L in G_1, we have $L = Z(L) * L_1$ with L_1 quasisimple by Proposition 3.13. Thus $H = Z^\circ(H) * E(H) = Z^\circ(H)$. Then as $H = U_2(H)$ and $O_2{}^\circ(H) = 1$ we find $H = 1$, as claimed. □

COROLLARY 5.11. *Let G be a connected K-group of even type. Then the Sylow 2-subgroups of G are connected.*

PROOF. This holds in $G/O_2{}^\circ(G)$ by the previous lemma and inspection, which gives the result in general. □

We will take advantage of this corollary to lighten the notation a little, but the connectedness must then be borne in mind as it is not recalled by the notation. But in a K^* context we must be careful to observe the distinction.

5.6. Perfect groups.

LEMMA 5.12. *Let H be a K-group of finite Morley rank of even type, generated by its 2-unipotent subgroups. Then $H^{(\infty)}$ is generated by its connected $2^\perp$-subgroups.*

PROOF. Let $K = H^{(\infty)}$. As the group $H/O_2(H)$ is generated by 2-unipotent groups, it is perfect, by Proposition 5.10. Hence $H = KO_2(H)$ and $K/O_2(K) \simeq H/O_2(H)$. In particular, $K/O_2(K)$ is a central product of algebraic groups, and it suffices to show that the subgroup K_1 of K generated by its connected $2^\perp$ groups covers $K/O_2(K)$, since it then follows that K/K_1 is both solvable and perfect, hence trivial.

Now $K/O_2(K)$ is generated by its maximal tori. Consider a maximal torus $\bar{T}$ in a quasisimple factor of $K/O_2(K)$, and its preimage T in K. Then $T/O_2(K) = \bar{T}$ and T splits by Schur–Zassenhaus as $O_2(K) \cdot T_0$, with T_0 the desired connected $2^\perp$-group. □

5.7. Abelian Sylow 2-subgroups.

PROPOSITION 5.13. *Let G be a connected K-group of finite Morley rank and of even type with abelian Sylow 2-subgroups. Then*

$$G = L_1 \times \cdots \times L_n \times \sigma(G)$$

is a direct product, with $L_i \simeq \mathrm{SL}_2(F_i)$ for suitable algebraically closed fields F_i of characteristic 2.

PROOF. Let $\bar{G} = G/O_2(G)$. Then by Proposition 5.10, we have $\bar{G} = O(\bar{G})E(\bar{G}) = \overline{\sigma^\circ(G)}\bar{H}$ where H is the preimage in G of $E(\bar{G})$. Hence we may suppose $G = H$, or in other words $G/O_2(G)$ is a central product of quasisimple Chevalley groups with abelian Sylow 2-subgroups of bounded exponent (Lemma I 6.15), each of the form $L_i = \mathrm{SL}_2(F_i)$, with F_i algebraically closed of characteristic 2 (Lemma 2.23). In particular, $G = U_2(G)$.

As Sylow 2-subgroups are abelian and $O_2(G)$ is connected (Lemma 5.6), we find that $O_2(G) \leq Z(G)$, and hence for each quasisimple factor $\bar{L}$ of $\bar{G}$, its preimage L in G is of the form $Z(G) * L_1$ with L_1 quasisimple. Thus $G = Z(G) * E(G)$ and our claim is proved. □

5.8. Strongly closed abelian subgroups.

DEFINITION 5.14. *Let G be a group, and $A \leq H \leq G$ subgroups. We say that A is* strongly closed *in H, relative to G, if every element of H which is conjugate, in G, to an element of A, is in A.*

LEMMA 5.15. *Let G be a group with an abelian 2-subgroup A which is strongly closed in some Sylow° 2-subgroup S containing it. Then A is strongly closed in any Sylow 2-subgroup containing it.*

PROOF. Let S_1 be a Sylow° 2-subgroup containing A. Then $S_1 = S^g$ for some $g \in G$ and $B = A^g$ is strongly closed in S_1. Now $B^{g^{-1}} = A \leq S_1$, so by strong closure of B in S_1 we have $A = B$. Thus A is strongly closed in S_1. □

LEMMA 5.16. *Let G be a group of finite Morley rank having a Sylow° 2-subgroup S with a strongly closed abelian subgroup A. and let $B \leq A$ be $N(A)$-invariant. Then B is strongly closed in S.*

PROOF. Suppose that $b^g \in S$, with $b \in B$ and $g \in G$. Then $b^g \in A$ and $A, A^g \leq C(b)$, so conjugating further in $C(b^g)$ we may suppose that $A^g = A$, $g \in N(A)$, and then by assumption $b^g \in B$. □

LEMMA 5.17. *Let G be a group of finite Morley rank, $A \leq S$ where S is a Sylow° 2-subgroup, and $H \lhd G$. Suppose that A is strongly closed in S. Then in $\bar{G} = G/H$, we have A strongly closed in $\bar{S}$.*

PROOF. We take $a \in A$ and $g \in G$ such that $\overline{a^g} \in \bar{S}$. We claim $\overline{a^g} \in \bar{A}$.

We have $a^g \in SH$. We may conjugate a^g by an element h of H into S, which is a Sylow 2-subgroup of SH. So $\overline{a^g} = \overline{a^{gh}} \in \bar{A}$. □

LEMMA 5.18. *Let G be a connected K-group of finite Morley rank and of even type with a nontrivial abelian 2-subgroup A which is strongly closed in a Sylow 2-subgroup of G. Then G is a finite direct product $G = L_1 \times \cdots \times L_n \times H$ with the following properties.*

(1) *$L_1, \ldots, L_n$ are groups of type* PSL_2 *in characteristic* 2.
(2) *$A \cap L_i$ is a Sylow 2-subgroup of L_i for each i.*
(3) *$A \cap H \lhd H$.*
(4) *$A = (A \cap L_1) \times \cdots \times (A \cap L_n) \times (A \cap H)$.*

PROOF. Let G_1 be the subgroup of G generated by the conjugates of A. Then all of our hypotheses apply to G_1, and $G_1 \lhd G$. If G_1 has the specified structure, then the same picture lifts easily to G. In particular, we will have the $L_i \lhd G$ since they are permuted by G and G is connected, and our results on definable groups of automorphisms show then that $G = L \times C_G(L)$ with

$L = L_1 \times \cdots \times L_n$. The last point is that $C_A(L)$ should be normal in G. Indeed, $C_{G_1}(L) \lhd G$ so by a Frattini argument, if S is a Sylow 2-subgroup of $C_{G_1}(L)$, then $G \le C_{G_1}(L)N(S) \le N(C_A(L))$ as $A \cap S = C_A(L)$. So we may make the additional hypothesis:

G is generated by conjugates of A.

Now supposing the center of G is infinite, we make a further reduction. The preceding lemma allows us to pass to $\bar{G} = G/Z(G)$ and do induction on rank. This gives us a decomposition (not a direct product) of G into connected groups $L_1 \cdot \cdots \cdot L_n \cdot H$ with $\bar{L}_i \simeq \mathrm{PSL}_2$ in characteristic 2 and $\bar{A}$ normal in $\bar{H}$. Then $L_i = K_i Z(G)$ with K_i a perfect and quasisimple, hence a covering group of PSL_2 in characteristic 2, hence K_i is itself of type PSL_2. Furthermore, $AZ(G) \cap L_i$ is a Sylow 2-subgroup of L_i and there is a maximal torus T of K_i such that $\bar{T}$ normalizes $\bar{A}$, and so T normalizes $AZ(G)$. For $t \in T$ we have $A^t \le AZ(G)$ commuting with A, so AA^t belongs to a Sylow° 2-subgroup, and by strong closure $A^t = A$, also $[t, A] \le A$. It follows that A meets K_i in a Sylow 2-subgroup of K_i, and the rest of the structure of $\bar{G}$ easily pulls down to G in similar fashion.

So let us suppose that $Z(G)$ is finite. We claim then that $O_2{}^\circ(G) = 1$. Suppose first that $A \cap O_2{}^\circ(G)$ is infinite. By strong closure it then follows that $A \cap O_2{}^\circ(G) \lhd G$. So any conjugate of A centralizes $A \cap O_2{}^\circ(G)$, and as these conjugates generate G this group is central in G, a contradiction. So $A \cap O_2{}^\circ(G)$ is finite. On the other hand, $AO_2{}^\circ(G)$ is contained in a Sylow° 2-subgroup of G, so $O_2{}^\circ(G)$ normalizes A and hence $[A, O_2{}^\circ(G)] \le A \cap O_2{}^\circ(G)$ is finite; being connected, it is trivial, and now we find $O_2{}^\circ(G) \le Z(G)$. So if $Z(G)$ is finite then $O_2{}^\circ(G)$ is trivial.

By Proposition 5.10 we now have $G = O(G) * E(G)$; as G is generated by conjugates of A, we have

$$G = E(G) .$$

There is a maximal torus in $E(G)$ normalizing a Sylow 2-subgroup of G containing A, and by strong closure also normalizing A. Consequently by the structure of such groups, A is a product of its intersections with the quasisimple components L of $E(G)$. Now we can use the classification of the quasisimple Chevalley groups. Invariance under the action of a maximal torus implies that A is a product of root groups, and strong closure, together with the irreducibility of the root system, implies that A is a full Sylow 2-subgroup. So we are dealing with groups with abelian Sylow 2-subgroups, and the factors are of type PSL_2. □

5.9. Quasisimple subgroups normalized by a 2-Sylow subgroup.

PROPOSITION 5.19. *Let L be a K-group of even type with $L = L_1 \times \cdots \times L_t$, where the L_i are simple algebraic groups. If K is a definable simple subgroup of L normalized by a Sylow 2-subgroup of L, then $K = L_i$ for some i.*

PROOF. Let U be a Sylow 2-subgroup of L which normalizes K. Then $U = U_1 \times \cdots \times U_t$ where each U_i is a Sylow 2-subgroup of L_i. For some i we have $1 \neq [K, U_i] \leq K \cap L_i \lhd K$. Therefore $K \leq L_i$, and we may assume $L = L_i$ is simple.

We claim that $U \leq K$. Since U normalizes K, $U \cap K$ is a Sylow 2-subgroup of K. Now U acts on K by inner automorphisms by Corollary 2.26, so U acts on K as $U \cap K$ does, and $U = C_U(K) \times (U \cap K)$.

Let $V = C_U(K)$ and consider the subgroup $H = N_L(V)$. Note that $K, U \leq H$. By Proposition 1.25, $C_H(O_2(H)) \leq O_2(H)$. But $V \leq O_2(H) \leq U$ so $O_2(H) = V \times (O_2(H) \cap K) = V$. Thus $K \leq C(O_2(H)) = O_2(H)$, a contradiction. This shows $V = 1$. So $U \leq K \leq L$.

Let $w \in K$ be an involution with $U \cap U^w = 1$ (that is, taking the Borel subgroup of K containing U to its opposite). Then $\langle U, U^w \rangle = L$ and thus $K = L$. □

PROPOSITION 5.20. *Let H be a connected K-group of finite Morley rank and even type, and L a definable quasisimple subgroup of H such that $N^\circ(L)$ contains a Sylow 2-subgroup of H. Then $L \lhd H$.*

PROOF. Let $S \leq N^\circ(L)$ be a Sylow 2-subgroup of H. Let $\bar{H} = H/O_2(H)$. By Proposition 5.10 $\bar{H} = E(\bar{H}) * O(\bar{H})$, and by Proposition 5.19, $\bar{L}$ is normal in $E(\bar{H})$ and hence in $\bar{H}$. In terms of H we have $LO_2(H) \lhd H$. But $O_2(H) \leq S$ so $[L, O_2(H)] \leq L \cap O_2(H) \leq Z(L)$, and as L is quasisimple $[L, O_2(H)] = 1$ by the three subgroups lemma. Thus $L = E(LO_2(H)) \lhd H$. □

5.10. Borel-Tits.

LEMMA 5.21. *Let G be a K-group of even type and H a definable connected subgroup of G such that $H = N_G{}^\circ(O_2(H))$. Then H contains a Sylow$^\circ$ 2-subgroup of G.*

PROOF. We may suppose that G is connected. We set $Q = O_2(H)$. By Lemma 5.6, Q and $O_2(G)$ are connected. The subgroup $QO_2(G)$ is a connected 2-subgroup, hence nilpotent. Thus $N_{O_2(G)}{}^\circ(Q)$ is nontrivial. As this group is contained in H and normalized by H, it is a subgroup of Q. Hence, $N_{QO_2(G)}(Q) = Q$ and thus $O_2(G) \leq Q$. Thus we may factor out $O_2(G)$ and assume that $O_2(G) = 1$. By Proposition 5.10, $G = E(G) * O(G)$. We may therefore assume that $G = E(G)$.

As $Q = O_2(QZ(G))$, $N_G(QZ(G)) = H$ and hence we may pass to $\bar{G} = G/Z(G)$, a direct product of simple algebraic groups over algebraically closed fields of characteristic 2. This is almost the situation to which Fact 1.24 applies, though as the base fields of the factors may vary one cannot say that this is literally so. While it would suffice to apply that result to each factor, we may argue more directly as follows.

Let G^* be an elementary extension of G in which each direct summand is uncountable, and of fixed cardinality. Then the base fields of the factors may be identified and G^* becomes an algebraic group over an algebraically closed field of characteristic 2. Thus after replacing G by G^* we may suppose that G is itself algebraic. Then the condition on H implies that Q is Zariski closed and hence by Fact 1.24 H is contained in a parabolic subgroup P of G whose unipotent radical U contains Q. Then $N_U(Q) \leq O_2(H) = Q$ so $U = Q$ and H is a parabolic subgroup of G. □

5.11. *L*-Balance. The next result is referred to as the *L-balance* property:

PROPOSITION 5.22. *Let H be a K-group of finite Morley rank of even type, and U a 2-subgroup of H. Then $E^\circ(C(U)) \leq E^\circ(H)$.*

PROOF. Let T be a torus contained in a component of $E^\circ(C(U))$. Let P be $O_2(H)$. Now $C_P(U) \leq O_2(C(U))$, so T commutes with $C_P(U)$. By the Thompson $A \times B$-lemma, Proposition I 12.4, with $A = T$, T commutes with $O_2(H)$. As such tori generate $E^\circ(C(U))$, $E^\circ(C(U))$ centralizes $O_2(H)$. On the other hand, $E^\circ(C(U))$ also centralizes $O(H)$ since $E^\circ(C(U))$ is generated by unipotent 2-subgroups (Proposition I 10.13). Thus $E^\circ(C(U))$ centralizes $F^\circ(H) = O_2{}^\circ(H) * O(F(H))$; but the connected component of the centralizer of $F^\circ(H)$ in H is $Z^\circ(F^\circ(H)) * E^\circ(H)$ by Proposition 4.13, so $E^\circ(C(U)) \leq E^\circ(H)$. □

PROPOSITION 5.23. *Let H be a connected K-group of finite Morley rank, of even type, and let U be a 2-subgroup of H. Then $E^\circ(C_H(U)) \lhd E(H)$.*

PROOF. By Proposition 3.1, $Z(E(H))$ is finite, and $E(H)$ is a central product of quasisimple Chevalley groups. As H is connected, it acts by inner automorphisms on $E(H)$. Hence so does U.

By Proposition 5.22, $E^\circ(C(U)) \leq E(H)$, so $E^\circ(C(U)) = E(C_{E(H)}{}^\circ(U))$. As U acts by inner automorphisms, $C_{E(H)}(U)$ is the central product of $C_L(U)$ as L varies over the factors of $E(H)$, and $E^\circ(C(U))$ is correspondingly the central product of the groups $E^\circ(C_L(U))$.

For any factor L of $E(H)$, U acts on L as a 2-subgroup $\bar{U}$ of L. If this group is trivial, then $E(C_L(U)) = L$, and otherwise $E(C_L(U)) = E(C_L(\bar{U})) = 1$ by Fact 1.27. □

PROPOSITION 5.24. *Let H be a connected K-group of finite Morley rank and even type. Let α be a definable automorphism of G of order an odd prime p. Then $O_2(C_H(\alpha)) \leq O_2(H)$.*

PROOF. We may suppose that $O_2(H) = 1$. Then by Proposition 5.10 $H = O(H) * E(H)$. So it suffices to consider the action of α on $E(H)$, and more particularly the action of α on the group $\langle L^{\langle\alpha\rangle}\rangle$, with L a single component of $E(H)$. This is either L itself, or a central product of p groups cycled by α. In the latter case the centralizer covers a diagonal subgroup modulo the center. This contributes nothing to $O_2(C(\alpha))$.

So suppose α normalizes the component L. Then Fact 1.17 gives the desired result. □

5.12. Weak embedding. We apply the weak embedding classification to the case of even type.

LEMMA 5.25. *If G is a nonsolvable connected K-group of even or mixed type with a weakly embedded subgroup M, then $G \simeq \mathrm{PSL}_2(K) \times O(G)$, where K is an algebraically closed field of characteristic* 2.

PROOF. By Proposition 4.20, $G/O(G)$ has the stated structure. By Proposition I 10.13, G centralizes $O(G)$. Furthermore, since $O_2(G) = 1$, by Proposition 5.10 we have $G = O(G) * L$ with $L = E(G)$, which in the present case means $L = \mathrm{SL}_2(K)$. □

5.13. Generation.

LEMMA 5.26. *Let A be an elementary abelian p-group with $m_p(A) \geq 2$, contained in a p-torus $\hat{A}$, and let H be a K-group of even type which is generated by its 2-unipotent subgroups, with $\hat{A}$ acting definably on H. Then*

$$H = \langle U_2(C^\circ(x)) : x \in A^\times \rangle \ .$$

PROOF. We may suppose toward a contradiction that H is a counterexample of minimal Morley rank. Let $H_0 = \langle U_2(C^\circ(x)) : x \in A^\times \rangle$

By Lemma 5.6, $O_2(H)$ is connected. By Proposition I 9.16, $O_2(H)$ is contained in H_0. If $O_2(H)$ is nontrivial, then induction applies to $\bar{H} = H/O_2(H)$ and hence $\bar{H} = \langle U_2(C_{\bar{H}}(x)) : x \in A^\times \rangle$. But by Proposition I 9.12, $C_{\bar{H}}(x) = C_H(x)O_2(H)/O_2(H)$. So in this case our claim follows for H.

There remains only the case $O_2(H) = 1$. Then by Proposition 5.10 we have $H = E(H) * O(H)$ and by our hypotheses we have $H = E(H)$. The p-torus $\hat{A}$ must normalize the quasisimple components of H, so by our minimality hypothesis $H = L$ is a quasisimple algebraic group in characteristic 2.

The p-torus $\hat{A}$ induces inner automorphisms of L, and acts faithfully (otherwise our claim would be immediate). So we may take A to be a subgroup of a maximal torus T of L. In particular, $\hat{A}$ normalizes two opposite Borel subgroups of L. In particular, $O_2(B) \leq H_0$ for each such Borel subgroup B. As these two groups generate L, our claim follows. □

5.14. Properties of SL_2**.** The present subsection contains a number of characterizations of the natural module for SL_2, beginning with Lemma 5.31, which are needed for the *amalgam method* beginning in §III 5.

We begin with a general lemma of Timmesfeld, which may be applied in its original form.

DEFINITION 5.27. *Let G be a group and V an elementary abelian 2-group on which G acts, and A a subgroup of G. The action of A on V is said to be* quadratic *if $[V, A, A] = 0$.*

FACT 5.28 ([**180**, Proposition 2.7]). *Let V be a $\mathbb{Z}X$-module where $X \simeq \mathrm{SL}_2(K)$ with K a field. Suppose the following:*

(1) *$C_V(X) = 0$ and $[V, X] = V$,*
(2) *$[V, A, A] = 0$, where A is a maximal unipotent subgroup of X.*

Then for some field action on $\langle v^X \rangle$, the vector space $\langle v^X \rangle$ is a natural module for each $v \in C_V(A)^\times$.

LEMMA 5.29. *Let G be group of finite Morley rank which is isomorphic to $\mathrm{SL}_2(K)$ as an abstract group with K an algebraically closed field. Suppose A is an infinite definable unipotent subgroup of G. Then for some conjugate B of A, $\langle A, B \rangle = G$.*

PROOF. Let A be such that we may suppose A is connected. Let $U = C(A)$ be the maximal unipotent subgroup of G containing A, and let B be a conjugate of A which does not normalize U. Let $H = \langle A, B \rangle$. Then H is a definable connected subgroup of G by Corollary I 3.29. If H is solvable then H is contained in a Borel subgroup of G, and this must be $N(U)$, contradicting the choice of B. Thus Fact 4.6 applies and $H = G$. □

LEMMA 5.30. *Let G be a group of finite Morley rank which is isomorphic to $\mathrm{SL}_2(K)$ as an abstract group with K an algebraically closed field of characteristic 2. Let $S \rtimes R$ be a Borel subgroup of G, with S a Sylow 2-subgroup of G and R a maximal torus. Then the following hold:*

(1) *G is generated by S together with any involution i not in S.*
(2) *Let V be an elementary abelian 2-group on which G acts faithfully so that the structure (G, V) has finite Morley rank, and set $f = \mathrm{rk}(K)$. Then*
   (a) *If $C_V(G) = 0$, then for some $v \in V$, $\mathrm{rk}(C_G(v)) \leq f$;*
   (b) *$\mathrm{rk}(V) \geq 2f$.*

PROOF.

Ad (1) This follows from Fact 4.6 applied to $\langle S, S^i \rangle$.

Ad (2a) Let V be as stated.

Assume toward a contradiction that $C_V(G) = 0$, and $\mathrm{rk}(C_G(v)) > f$ for all nontrivial $v \in V$.

Fix $v \in C_V(S)^\times$. As $\mathrm{rk}(C_G(v)) > f$, we have $C_G{}^\circ(v) > S$ and thus $C_G{}^\circ(v)$ has the form $S \rtimes R_0$ with R_0 a nontrivial torus, which is not necessarily algebraic. Let w be an involution that inverts R_0 and set $v_1 = v + v^w$. Note that $v_1 \neq 0$; in fact, if $w \in C_G(v)$, then by Lemma 5.29, we have $C_G(v) = G$, contradicting our hypothesis.

Now, $\langle w, R_0 \rangle \leq C_G(v_1)$. As $\mathrm{rk}(C_G(v_1)) > f$, $C_G{}^\circ(v_1)$ has a nontrivial Sylow° 2-subgroup Q, which is normal in $C^\circ(v_1)$ and in particular is normalized by R_0 and by w; but there is no such 2-group Q in G since the only Sylow 2-subgroups normalized by R_0 are S and S^w.

Ad (2b) We may suppose that V is irreducible and faithful, and apply (2*a*). □

LEMMA 5.31. *Let G be a group of finite Morley rank which is isomorphic to $\mathrm{SL}_2(K)$ as an abstract group with K an algebraically closed field of* *characteristic* 2. *Let* V *be an elementary abelian 2-group on which* G *acts definably and faithfully. Let* $f = \mathrm{rk}(K)$, *and suppose* $\mathrm{rk}(V) = 2f$. *Then* V *is a natural module for* G.

PROOF. We use Fact 5.28. As $\mathrm{rk}(V) = 2f$, and the action is faithful, it follows from Lemma 5.30 (2) that V is irreducible, and thus $[V, G] = V$. The only point that needs to be checked is the quadratic action $[V, S, S] = 0$ where S is a Sylow 2-subgroup of G.

Let $V_0 = C_V(S)$. and $X = \bigcup\{V_0^g) : g \in G\}$. Distinct conjugates of V_0 intersect trivially since $C_V(G) = 0$. Thus X has rank $\mathrm{rk}(V_0) + f$, and it follows that $\mathrm{rk}(V_0) \leq f$. We claim

$$(*) . \qquad \mathrm{rk}(V_0) = f$$

Suppose the contrary. Let O_1 be an orbit of maximal rank for T in V_0, and $T_1 = C_T{}^\circ(O_1)$. Since $\mathrm{rk}(O_1) < f$, the group T_1 is nontrivial as well as connected. Furthermore, $V_1 = C_{V_0}{}^\circ(T_1)$ is nontrivial; V_1 contains O_1, and if $\mathrm{rk}(O_1) = 0$, then the orbits of T on V_0 are all finite, in which case $T_1 = T$ and $V_1 = V_0$.

Now take an involution w inverting T_1. For $v \in V_1^\times$, the element $v + v^w$ is centralized by T_1 and by w, and is nontrivial since $\langle S, w\rangle = G$. Now the group $C_G{}^\circ(v + v^w)$ is solvable since it is not G, and its unipotent radical is w-invariant and normalized by T_1, hence trivial. Thus $C_G{}^\circ(v + v^w)$ is a torus and hence has rank at most f. Since the rank of the orbit of $v + v^w$ under G is at most $2f$, it follows that $C_G{}^\circ(v + v^w)$ is a maximal torus; since it contains T_1 it is T, and $C_G(v + v^w) = \langle T, w\rangle$. In particular, the orbit of $v + v^w$ is generic in V.

Now if $v_1, v_2 \in V_1$ are distinct and nontrivial, then the elements $v_1 + v_1^w$ and $v_2 + v_2^w$ both have centralizer $\langle T, w\rangle$ and a generic orbit in V, hence are conjugate under the action of G. This conjugation must preserve the centralizer, and normalize T; but as v_1 and v_2 are centralized by $N_G(T)$, we conclude $v_1 = v_2$, contradicting our choice. This contradiction proves $(*)$.

Thus $X = \bigcup V_0^G$ is generic in V, or in other words, a generic element of V is fixed by a Sylow subgroup of G.

We claim that every element $v \in C_V(S)^\times$ has $C_G{}^\circ(v) = S$. Suppose the contrary. We suppose $v \in C_V(S)^\times$ is centralized by a nontrivial torus R and we take w inverting R. Consider $v_1 = v + v^w$. Then as above, $C_G(v_1)$ must be a torus. In particular, $\mathrm{rk}(C_G(v_1)) = f$ and thus v_1^G is also generic in V. But this contradicts the result of the previous paragraph.

Let T be a maximal torus in $N_G(S)$. For $v \in C_V^\circ(S)^\times$ as $C_G^\circ(v) = S$, the orbit v^T is generic in $C_V^\circ(S)$ and as $C_V^\circ(S)$ is connected, $C_V^\circ(S)^\times$ is a single orbit under T. But if $S_1 \neq S$ is a conjugate of S normalized by T then $V = C_V(S) \oplus C_V(S_1)$ as a T-module and thus $\bar{V}^\times = (V/C_V(S))^\times$ is also a single orbit under T. Since $C_{\bar{V}}(S) \neq 1$, it follows that $C_{\bar{V}}(S) = \bar{V}$, or in other words, $[V, S] \leq C_V(S)$, and $[V, S, S] = 0$. □

PROPOSITION 5.32. *Let G be a group of finite Morley rank which, as an abstract group, is isomorphic to $\mathrm{SL}_2(K)$ with K an algebraically closed field of characteristic 2. Let A be an infinite definable 2-subgroup of G, and V a connected elementary abelian 2-group which is a G-module such that (G,V) has finite Morley rank. Suppose $C_V(G)=0$. Then:*

(1) $\mathrm{rk}(A) \le \mathrm{rk}(V/C_V(A))$.
(2) *Equality holds only if A is a Sylow 2-subgroup of G, and V is a natural G-module.*

PROOF. Let $f = \mathrm{rk}(K)$. By Lemma 5.29, we have $G = \langle A, B\rangle$ with B some conjugate of A. As $C_V(G)=0$, the natural map

$$V \longrightarrow [V/C_V(A)] \times [V/C_V(B)]$$

is injective, and thus $\mathrm{rk}(V) \le 2\,\mathrm{rk}(V/C_V(A))$. By Lemma 5.30 (2),

$$\mathrm{rk}(V/C_V(A)) \ge f \ge \mathrm{rk}(A)\ .$$

This proves the first point.

Now suppose $\mathrm{rk}(A) = \mathrm{rk}(V/C_V(A))$. Then $\mathrm{rk}(A) = f$ and A is a Sylow 2-subgroup of G. Furthermore $\mathrm{rk}(V/C_V(A)) = f$ so $\mathrm{rk}(V) \le 2f$ and by Lemma 5.30, $\mathrm{rk}(V) = 2f$. So by Lemma 5.31, V is a natural module. □

The following corollary is an analog of a result given in [**175**] and [**176**].

PROPOSITION 5.33 ([**176**], [**175**, (2.1)]). *Let G be a group of finite Morley rank which is isomorphic to $\mathrm{SL}_2(K)$ with K an algebraically closed field of characteristic 2. Let V be a faithful $\mathbb{F}_2G$-module. Let S be a Sylow 2-subgroup of G. Assume that $T \le S$ is definable and nontrivial, and:*

(i) $[V,T,T]=1$,
(ii) $\mathrm{rk}(V/C_V(T)) \le \mathrm{rk}(T)$.

Then the following hold:

(a) $\mathrm{rk}(T) = \mathrm{rk}(V/C_V(T))$,
(b) $T = S$,
(c) $V/C_V{}^\circ(G)$ *is a natural $\mathbb{F}_2$-module for G,*
(d) $C_V(S) = [V,S]C_V(G)$.

PROOF. Point (d) is a special case of (c). We have proved (a-c) under the assumption that $C_V(G)=0$. All that we need to prove now is that $C_{V/C_V{}^\circ(G)}(G) = 0$.

Let $V_0/C_V{}^\circ(G) = C_{V/C_V{}^\circ(G)}(G)$. Then $[V_0,G,G]=1$, so by the Three Subgroups Lemma $[V_0,G]=1$, as claimed. □

6. L-Groups and L^*-groups

6.1. L-groups and L^*-groups. The notion of K-group must be extended as follows, given that groups of odd and degenerate type are poorly understood.

DEFINITION 6.1. *Let G be a group of finite Morley rank.*

(1) *G is an* L-group *if every definable connected simple section of G of even type is isomorphic to a Chevalley group over an algebraically closed field of characteristic* 2.
(2) *G is an* L^*-group *if every proper definable connected simple section of G of even type is isomorphic to a Chevalley group over an algebraically closed field of characteristic* 2.
(3) *G is a* U_2-type group *if G is generated by its* 2*-unipotent subgroups.*

In the case of L-groups of even type, the only doubtful sections are those of degenerate type. In the case of L-groups of mixed type, a priori the situation could be fairly wild, but in Part B we will show that simple mixed type groups of finite Morley rank do not exist, and after that we will know that only odd and degenerate type sections are problematic. For that matter, when we prove the nonexistence of simple groups of finite Morley rank of mixed type, we will consider a hypothetical minimal counterexample, so even in that context we will know that there can be no proper simple definable sections of mixed type. So we now prepare some tools for the study of L-groups without proper definable simple sections of mixed type.

PROPOSITION 6.2. *Let G be an L-group of finite Morley rank, with no definable simple sections of mixed type. Then $U_2(G)$ is a K-group.*

PROOF. We may suppose $G = U_2(G)$ and $\sigma(G) = 1$.

By Lemma I 8.34, $\mathrm{Soc}(G)$ is a direct sum of definable simple subgroups. As G is an L-group, these factors are either algebraic groups, or odd type, or degenerate, and as G is connected they are normal in G. Suppose there is a degenerate or odd type factor K. Then by Proposition I 10.13, the unipotent subgroups of G centralize K, and as $G = U_2(G)$, we find that K is central in G, a contradiction. Thus $\mathrm{Soc}(G)$ is a finite sum of simple algebraic groups (of even type), and in view of Fact 2.25, as G is connected it follows that $G = \mathrm{Soc}(G)C_G(\mathrm{Soc}(G))$. If $C_G(\mathrm{Soc}(G)) \neq 1$, then $Z(G) = C_G(\mathrm{Soc}(G)) \cap \mathrm{Soc}(G) \neq 1$ by Lemma I 8.34, hence $\sigma(G) \neq 1$, a contradiction. So $G = \mathrm{Soc}(G)$. □

LEMMA 6.3. *Let G be an L-group of finite Morley rank with no definable simple sections of mixed type. Then $U_2(G)$ is a K-group of even type.*

PROOF. We may suppose that $G = U_2(G)$, and by the preceding, G is a K-group. We may factor out $U_2(\sigma(G))$ and suppose that $U_2(\sigma(G)) = 1$. Then G centralizes $\sigma(G)$ by Proposition I 10.13. Let $H = G^{(\infty)}$. Then H is a perfect central extension of $H/\sigma(H)$, hence H is a finite central product of quasisimple algebraic groups. Then $G = H * C_G(H)$ and $C_G(H)/Z(H)$ is a solvable group of U_2-type, hence 2-unipotent; but then $U_2(C_G(H)) \leq U_2(\sigma(G)) = 1$, so $C_G(H) = Z(H)$ and $G = H$. In particular, H is also of U_2-type and hence, as a central product of quasisimple Chevalley groups, is of even type. □

The following is a typical consequence, following from the K-group analog.

COROLLARY 6.4. *Let G be a connected L-group of even type. Then $O_2(U_2(G))$ is connected and definable.*

This can be improved; cf. Lemma IV 4.19.

6.2. Weak embedding.

LEMMA 6.5. *Let H be a connected L-group of even type with a weakly embedded subgroup M. Then*

$$H = L \times D .$$

where $L = U_2(H) \simeq \mathrm{SL}_2(K)$ with K algebraically closed of characteristic 2, and $D = C_H(L)$ a subgroup of degenerate type. Furthermore,

$$M^\circ \cap L \text{ is a Borel subgroup of } L;\ D \leq M .$$

PROOF. We define $L = U_2(H)$ and $D = C_H(L)$. Let S be a Sylow° 2-subgroup of M. Then $S \leq L$, and by a Frattini argument we have $H \leq L \cdot N(S)$. So if we had $L \leq M$, we would get $H \leq M$, a contradiction.

It follows that $M \cap L$ is weakly embedded in L, which is a K-group of even type by Lemma 6.3. So $L \simeq \mathrm{PSL}_2(K)$ for some algebraically closed field K of characteristic 2, by Lemma 5.25. It follows easily that $M \cap L$ is a Borel subgroup of L.

Now by Fact 2.25 we have $H = LD$, and since L is simple we have $L \cap D = 1$. It follows then that D is of degenerate type. Furthermore $D \leq C(L) \leq C(S) \leq M$, and all of our claims are proven. □

6.3. Strongly closed abelian subgroups. In much the same vein as the preceding, we may state the following, derived from the K-group case.

LEMMA 6.6. *Let G be a connected L-group of finite Morley rank and of even type, with a nontrivial abelian 2-subgroup A which is strongly closed in a Sylow° 2-subgroup of G. Then G is a finite direct product $G = L_1 \times \cdots \times L_n \times H$ with the following properties.*

(1) *$L_1, \ldots, L_n$ are groups of type PSL_2 in characteristic 2.*
(2) *$A \cap L_i$ is a Sylow 2-subgroup of L_i for each i.*
(3) *$A \cap H \lhd H$.*
(4) *$A = (A \cap L_1) \times \cdots \times (A \cap L_n) \times (A \cap H)$.*

PROOF. $G_0 = U_2(G)$ is a K-group, and thus by Lemma 5.18 the stated result holds for G_0 in place of G, say $G_0 = L \times H_0$ with $L = \prod_i L_i$. Then easily $G = L \times H$ with $H = C_G(L)$ and $H_0 \lhd H$. So with S a Sylow° 2-subgroup of H_0 containing $A \cap H_0$, we have $H = H_0 \cdot N_H(S) \leq N(A)$ and everything follows. □

In particular, a connected simple L-group of even type with an abelian Sylow° 2-subgroup is of type SL_2, and one of our tasks later on will be to extend this to the L^* case, in Chapter VII.

6.4. Borel-Tits. We give the L-group version of the Borel-Tits theorem.

LEMMA 6.7. *Let G be an L-group of even type and H a definable connected subgroup of G such that $H = N_G{}^\circ(O_2(H))$. Then H contains a Sylow° 2-subgroup of G.*

PROOF. Let $G_0 = U_2(G)$ and $H_0 = (H \cap G_0)^\circ$. Then these are K-groups. Furthermore, $O_2(H) \leq O_2(H_0) \leq O_2(U_2(H)) \leq O_2(H)$, so these groups all coincide and $H_0 = N_{G_0}{}^\circ(O_2(H_0))$.

So by the K-group version, Lemma 5.21, H_0 contains a Sylow° 2-subgroup of G_0, which is a Sylow° 2-subgroup of G. □

We also give the L-groups version of Proposition 5.20.

PROPOSITION 6.8. *Let H be a connected L-group of finite Morley rank and even type, and L a definable quasisimple subgroup of H such that $N^\circ(L)$ contains a Sylow° 2-subgroup of H. Then $L \triangleleft H$.*

PROOF. $L \leq U_2(H)$ and $L \triangleleft U_2(H)$ by Proposition 5.20. But H normalizes the quasisimple components of $U_2(H)$. □

6.5. Reductivity.

NOTATION 6.9. *Let G be a group of finite Morley rank. We write $\hat{O}(G)$ for the largest connected normal definable subgroup of G of degenerate type.*

In a K-group, the group $\hat{O}(G)$ coincides with $O(G)$.

LEMMA 6.10. *Let H be a reductive L-group of even type. Then $U_2(H) = E(U_2(H))$, and $H = U_2(H) * \hat{O}(H)$.*

PROOF. As $U_2(H)$ is a connected K-group with O_2 trivial, the structure of $U_2(H)$ is given by Proposition 5.10, together with the definition of U_2. Writing $L = U_2(H)$, Fact 2.25 yields $H = L * C_H(L)$, with finite intersection, and it is clear then that $C_H(L) = \hat{O}(H)$. □

7. Notes

§1. Algebraic Groups

We use the theory of algebraic groups as a framework for understanding the structure of Chevalley groups. Hence for our purposes, the point is not that a simple algebraic group is a Chevalley group, but the converse.

§2. Chevalley Groups

The Borel-Tits theorem plays an important role. The original form is in [**41**], and its relevance to the analysis of even type groups (or their finite analogues) is discussed in [**102, 103**].

We have already mentioned the central role played by parabolic subgroups in our classification results, where we follow primarily a *unipotent* rather than a *semisimple* strategy. As we will be treating the even type case, which is parallel to the case of characteristic 2 in algebraic group theory, our involutions behave as unipotent elements, and it is natural to work directly with parabolic subgroups. This is not the method used in finite group theory, which aims at a more rapid identification of semisimple elements, and focuses ultimately on semisimple elements in all characteristics; but it is the central idea of the so-called "third generation" approach in the finite case. (For more on the "semisimple" strategy see [**170**], notably pp. 338-9 and 344-5.

The usual notion of Borel subgroup is available in groups of finite Morley rank: maximal definable connected solvable subgroups (one can omit "definable" here), and thus the notion of parabolic subgroup is also available. However, with this definition, the theory of Borel subgroups is less satisfactory in our context than in the algebraic context, and there are various ways to build in a bit more of the theory into the definition. In characteristic 2, a Borel subgroup is also the normalizer of a Sylow 2-subgroup, and if we replace the normalizer by its connected component, we get another notion of Borel subgroup which is well adapted to our abstract setting. We call these "standard" Borel subgroups. It is not obvious a priori that these groups are even solvable. It turns out that on one hand the solvability can be proved, eventually, and on the other hand it is not absolutely necessary for our arguments; to bypass the solvability argument it would be sufficient to build appropriate subgroups of our standard Borel subgroups consisting of the extension of the Sylow 2-subgroup in question by a suitable maximal torus. Lining up the tori involved in different Borel subgroups then requires some effort, but we remark that this kind of modification of Borel subgroups is already a standard part of the amalgam method in the finite case.

In any case, our primary notion of parabolic subgroup is really *standard parabolic*, that is, containing a fixed *standard Borel* subgroup, (however, the latter term is defined). We will not get any real control over these groups until §VIII 5; everything up to that point can be viewed as very extensive preparation, and it is precisely at this point that the proof of the classification in even type really gets underway.

We are indebted to Richard Lyons and Franz Timmesfeld for discussions of the Curtis-Tits theorem. Each has pointed out in his own way that the matter involves certain subtleties, in its general formulation.

§3. Central extensions

Our treatment follows [**10**]. A model theoretic definability result of Wagner and Newelski [**187, 148**] plays a major role.

§4. Structure of K-groups

Sometimes a K^*-group is defined as one such that every proper definable subgroup is a K-group. Our K^*-groups have a stronger property, but as far as simple groups go the definitions are equivalent. With our definitions, a K^*-group is either a K-group, or else is a simple group which violates the Algebraicity Conjecture.

The structural information in Proposition 4.8 and its later elaborations shows that connected K-groups have a very limited structure, unlike their finite counterparts.

At this point we begin to prepare the technical K-group results needed for various phases of the classification argument later. Most of these are in the next section, since we concentrate on even type groups in the present section.

Proposition 4.4 comes from [**159**], where the proof is given in more detail, though one must still consult [**178**] to complete the argument.

In connection with Proposition 4.7, one may wonder if there is a theory of algebraic geometry over bad fields which would allow a full development of the theory of algebraic groups over bad fields. This seems unlikely. It is more probable that there are no bad fields in positive characteristic, which would trivialize Proposition 4.7. Whether bad fields in characteristic 0 (with no additional structure) have the potential to produce new simple groups of finite Morley rank remains completely unclear. If one has a bad group, there are general methods for combining the theory of a strongly minimal subset with the theory of an arbitrary algebraically closed field to produce a field with enriched structure over which the group is now definable, but the additional structure will have nothing to do with the field structure.

The notion of weak embedding, and the criteria given in Proposition 4.15, appears first in [**3**]. See the discussion of strong and weak embedding in the notes to Chapter V.

§5. K-groups of even type

The first four sections elaborate on the structure of K-groups in the even type case, and continue the analysis of the preceding section. The remainder prepare technical points which serve as the basis for the inductive analysis of K^*-groups.

The L-balance theorem of Gorenstein and Walter reads as follows.

FACT 7.1. *Let G be a finite group. For any 2-local subgroup H of G we have*

$$L_{2'}(H) \leq L_{2'}(G) \ .$$

Here the subgroup $L_{2'}(G)$ is defined as the subgroup of the full preimage of $G/O(G)$ generated by 2-elements.

The material in sections 5.7–5.9 comes from [**5**]. The study of strongly closed abelian subgroups in the context of groups of finite Morley rank began in the concluding section of [**4**].

For our purposes, the use of the classification theorem for Chevalley groups in the proof of Lemma 5.18 is somewhat beside the point, since in our inductive framework the only sections that can occur at any point are those on our list of known groups. So if we chose to redefine K^*-groups and L^*-groups in terms of the explicitly known groups, we would actually wind up deriving this classification as a byproduct. (The same is true in the finite case.)

The treatment of natural modules for SL_2 is essential preparation for applications of the amalgam method in §§III 5 and VIII 2. The use of Fact 4.6 could be seen as overkill, as that result is considerably more subtle than the special cases actually needed, but at the same time it seems like the natural approach.

§6. L-groups and L^*-groups

The notion of L-group was introduced in [**2**] as a proposed inductive framework allowing one to prove results on groups of even type without necessarily disposing of the degenerate case first. One of our main goals in the present text is to bring this line of thought to completion.

We will show in Part C that simple L^*-groups of even type are Chevalley groups, which implies that simple groups of even type are Chevalley groups. The material in Chapter VI becomes distinctly more difficult in this extended setting, relative to the K^*-case. The material in Chapter VII also would seem to pose some challenges, but these relate mainly to the theory of pseudoreflection groups and the necessary adjustments will be made in the preparatory material, notably §IV 5, after which the analysis itself does not vary widely from the K^*-case. Later sections tend to run quite parallel to their K^*-versions, with a fairly systematic alteration in some key definitions and a reliance on Lemma 6.3 (and, frequently, a Frattini argument). We have already seen examples of this kind of bootstrapping in this section.

Also, in Part B, we show that mixed type simple L^*-groups are algebraic, so that once the work of Part C is complete the induction terminates, and we have our main results: simple groups of finite Morley rank with nontrivial unipotent 2-subgroups are algebraic, in characteristic 2. The analogous results limited to K^*-groups are easier, but to pass from those results to a general conclusion would then require a separate analysis of the problematic degenerate type groups, which is not presently in view.

From a technical point of view, *everything* we do depends on this section. A major open problem is to develop a parallel theory in odd type groups, at a comparable level of generality.

We note that the implementation of this line of analysis is also facilitated by the results of §IV 4, which, however, will not available when it was first worked out. The structure of the argument is not noticeably affected by this, but there are fewer distractions along the way as a result.

CHAPTER III

Specialized Topics

I remember that during the whole of that memorable day he lost himself in a monograph which he had undertaken upon the Polyphonic Motets of Lassus.
— A. Conan Doyle,
The Adventure of the Bruce-Partington Plans

Introduction

In this chapter we prepare a number of special topics, which in their own more limited spheres are as useful as the Schur-Zassenhaus splitting and conjugacy theorems or K-group facts are in a broader sphere. This includes a number of topics with a distinctly geometric flavor: the classification of groups generated by pseudoreflection groups, with the flavor of linear algebra, the theory of Zassenhaus groups in permutation group theory, (B, N)-pairs, buildings, and more particularly Moufang polygons, and Niles' Theorem. In addition we have some topics well known in finite group theory: the theory of Suzuki 2-groups, the analysis of a another special 2-group configuration of Landrock and Solomon in which one can pin down very tightly the structure of a Sylow 2-subgroup, a theorem of Baumann proved by the amalgam method following Stellmacher, and signalizer functor theory. Several of these ingredients will be combined in the present chapter to provide a "generic identification theorem" which can serve as an alternative to the use of the theory of buildings in the generic case, while still leaving the theory of Moufang polygons as the natural way to complete the analysis in the exceptional cases where the amalgam method applies.

It is in this chapter, more than any other, that our subject makes contact with a range of mathematical subjects that yield conclusive information in various particularly well structured configurations that arise—or that can be made to arise—in the analysis of more general cases. That is, the key configurations are all treated here, each by its own method, and the rest of the work in Parts B and C is largely a matter of induction. But before we reach that point, we will need some other, more model theoretic and global methods, given in Chapter IV.

Overview

We pick up where we left off at the end of Chapter I, developing group theoretic topics which require a more extended analysis than those of Chapter I, and which are generally less standard fare, but which can be obtained with a similar mix of model theoretic and group theoretic ideas, and which are for the most part topics of group theory adapted to our category of groups.

0.1. Pseudoreflection groups. The first of these topics is the most difficult to place: the classification of groups generated by pseudoreflection groups. It has the feeling of a "connected" version of the theory of complex reflection groups, but very little in common with that theory at a technical level; and the only group resulting is GL_n. This theory is absent from finite group theory. We use it to classify groups with abelian Sylow° 2-subgroups, and we could use it similarly to classify groups whose Sylow° 2-subgroups contain strongly closed abelian subgroups (cf. [**5**]), though we will take a different tack here. These are results obtained by a very different line of argument in the finite case, and incidentally they present one of the major challenges in passing from K^*-groups to L^*-groups.

Pseudoreflection groups should be groups which act trivially on a codimension one subspace and preserve a complementary subspace; but we need to define them without the crutch of an auxiliary vector space structure. The setting in which they arise involves the action of a connected group on an elementary abelian group (for us, a 2-group), and a subgroup resembling a 1-dimensional torus both in structure and action—coming from an actual copy of SL_2 found in the ambient group. We define a pseudoreflection group as a divisible abelian group T acting on an elementary abelian 2-group A in such a way that $A = C_A(T) \oplus [A,T]$, with T acting faithfully on $[A,T]$ and transitively on its nontrivial elements. This last condition is both more and less than 1-dimensionality; in any case it is both very strong, and satisfied in practice.

We show that in a K-group context, if a group H acts faithfully and irreducibly on an elementary abelian 2-group A, and contains a nontrivial group of pseudoreflections, then A carries a vector space structure and H is $\mathrm{GL}(A)$ (Theorem 1.5). The weight of the argument is borne by the K-group assumption, and the thrust of the argument is: What else could it be?

But this is not enough. We will need this result in the L-group context, and the magic wand of Proposition II 6.2, wave it as one might, is but hand-waving. So we will return to this again in §IV 5 after preparing some further model theoretic techniques, and settle it in the L-group case as well.

0.2. Zassenhaus, Suzuki, Landrock, and Solomon. The next three topics, Zassenhaus groups, Suzuki 2-groups, and the Landrock-Solomon configuration, deal with three extreme configurations in which either the group or the Sylow 2-subgroup can be pinned down. In the context of Zassenhaus groups one can give the recognition theorem for SL_2 to which all our subsequent recognition theorems reduce, and the only one in which SL_2 is recognized as such. This result was also given in [**52**] but we give it a full treatment here. The Suzuki 2-group configuration is somewhat different, but equally extreme, and in the so-called "free Suzuki" case reduces to homocyclic abelian groups (largely because our base fields are algebraically closed, and in particular quadratically closed). The Landrock-Solomon configuration arises, for us, in the context of weakly embedded groups which are not strongly embedded, by considering centralizers of appropriate involutions. Here the conclusion is that the Sylow° 2-subgroup is either homocyclic abelian or of a very particular and explicitly determined type, to be eliminated from configurations subsequently by special pleading. The treatment of Zassenhaus groups and Suzuki 2-groups is not very close to the treatment in the finite case. The Landrock-Solomon analysis on the other hand is quite reminiscent of its model.

A Zassenhaus group is a doubly transitive group acting on a set with at least three points, where the stabilizer of any three points is trivial. Suppose G is an infinite Zassenhaus group of finite Morley rank with 1-point stabilizer B and 2-point stabilizer T, and suppose in addition B splits as $U \rtimes T$, and that

$$Z(U) \text{ contains an involution.}$$

Under these hypotheses Theorem 2.2 shows that the group G either acts sharply transitively, or can be identified with PSL_2 acting naturally, in characteristic 2.

We will not say much about the proof. One considers the subgroup U_0 of U generated by the involutions of U, which lies in $Z(U)$. One may show first that the pair (U_0, T), with the action of T on U_0, can be identified with a pair $(K_+, K^\times)$ with K a field of characteristic 2, and the natural action, and one can check fairly directly that if $U_0 = U$, then the whole group embeds into, and then coincides with PSL_2 in its customary action on the projective line. One is left desiring the relation

$$U_0 = U$$

at which point an inordinate use is made of elements of order three, and not for the last time. This seems to be an occupational hazard associated with PSL_2. These arguments are found in the proofs of Lemmas 2.10 and 2.11.

A *Suzuki 2-group* is defined as a nilpotent 2-group S on which an abelian group T acts in such a way that all involutions of S form a single orbit. A *free Suzuki 2-group* is one for which the action is free, or in other words, regular. The important theorem for our purposes is that a free infinite Suzuki group of finite Morley rank is abelian and homocyclic.

A good deal of induction is possible in this context, with a little care. One disposes of various extreme cases by careful inspection and calculation of the behavior of commutators. Typical of the analysis in these cases is the following result.

LEMMA 3.10. *Let K, F be fields of characteristic 2, with $K \subseteq F$ as sets and with compatible multiplication operations, but possibly different addition operations. Let $\gamma : K \to F$ be an additive map. Then the following are equivalent:*

(1) $\gamma(x^{-1}) = x^{-2}\gamma(x)$ *for all* $x \in K$,
(2) γ *is linear over the subfield* K^2.

However, while this disposes of a number of critical cases it does not dispose of the problem. The key argument is as follows. Let A be a maximal normal abelian T-invariant subgroup of S (automatically definable since $A = C(A)$). If $\Omega_2(S) \leq A$, we arrive at a previously treated special case, essentially by induction. So we can fix an element g of order 4, lying outside of A.

Also by induction, we may suppose $S = N(A)$ and $\bar{S} = S/A$ is T-irreducible. So the subring of $\mathrm{End}(\bar{S})$ generated by T is a field K, and a general lemma shows that T must be the multiplicative group of K in this case.

By computation with the functional equation one shows

$$S' \leq 2A$$

(using additive notation); one works with one commutation map at a time. Accordingly, $V = S/2A$ can be viewed as a T-module. Everything comes down to an analysis of the subring R of $\mathrm{End}(V)$ generated by T. In view of the two-layer structure of V (V/A, $A/2A$) it is not surprising that this is a local ring whose maximal ideal $\mathfrak{m}$ satisfies $\mathfrak{m}^2 = 0$. One then succeeds in embedding the field K into R, giving V a vector space structure, and hence one can find a proper T-invariant subgroup $\hat{A}$ of S not contained in A. By induction we may also suppose $\hat{A}$ is abelian, and this configuration is one of the "special" configurations treated at the outset, computationally.

The thrust of the argument is the following: One can just about envision the canonical counterexample to the theorem, but a close consideration of the endomorphism ring in this case leads to a contradiction. This general line of analysis is useful elsewhere, and is quite separate from the use of functional equations to analyze minimal cases.

The *Landrock-Solomon configuration* looks as follows. We have a unipotent 2-group S, a divisible abelian group T acting on S, and an elementary abelian T-invariant 2-group $A \leq S$ such that the pair (A, T), with its action, can be identified with a pair $(K_+, K^\times)$, associated with a field K of characteristic 2. Finally, we have a definable involutory automorphism α of $S \times T$ whose centralizer is $A \rtimes T$, and the whole configuration, of course, has finite Morley rank.

One finds that S is either homocyclic abelian or of a specific form (in particular of exponent 4) which is subsequently eliminated in applications. One may note a certain similarity to the Suzuki 2-group situation.

The main point here, we feel, is not the analysis of this configuration, but its isolation as a distinct configuration. This comes up quite naturally in the case of groups with weakly embedded but not strongly embedded subgroups, as will be copiously visible in Chapter VI. The order of events is as follows: the pair (A, T) is extracted from a group of type SL_2 (AT is a Borel subgroup), and then S is chosen as a Sylow° 2-subgroup invariant under T (and with some attention paid to α, which begins its life as an involution in the ambient group).

The next few sections survive transplantation from abstract categories to the context of finite Morley rank with little disarray.

0.3. Baumann. We consider the following situation: a group G of finite Morley rank and even type, a definable connected subgroup of G of minimal parabolic type (Definition 5.1), and the following condition holds:

(P) No nontrivial definable connected subgroup of S is normalized by both M and $N_G(S)$.

Under these hypotheses, the structure of M is closely determined (and the matter is taken up again in §VIII 2).

Here we follow the theory of finite groups very closely indeed, both in the content of the result, and the proof. Still there is a good deal to be said.

First, we consider P as an obstacle to further analysis. If we have a group X normal in M and also in $N_G(S)$, then its normalizer is a group we are looking for: it contains M, and it also contains $N_G(S)$, and in particular contains $N_G{}^\circ(S)$, which we have encountered before as a *standard Borel subgroup.* In other words, we are trying to push M into a (standard) parabolic subgroup—and failing, at the moment.

Conversely, if M does sit in a parabolic subgroup, then the theory of algebraic groups would lead one to expect that group to be 2-local (as we work in characteristic 2) and thus the group X referred to ought to exist.

So much for the context. As regards the proof, we follow closely the method of Stellmacher, who showed that the amalgam method is effective here. In particular, this constitutes an introduction to the amalgam method, which will be applied on a much broader scale (again, following Stellmacher) in Chapter IX. This amalgam method is remarkably indifferent to the category in which one works—in essence one works in the category of abstract groups throughout much of the analysis. In adapting it to our context, we have to pay some attention to connected components, and some attention to issues of definability, so we cannot just carry it over, but nonetheless it is very robust.

0.4. Generalized polygons, buildings, and BN-pairs. Critical for our recognition of groups of Lie rank 2, once the lengthy amalgam analysis of

Chapter IX is over, is the classification of Moufang polygons of finite Morley rank; this is the Tits rank case version of the classification of Moufang buildings.

This is based on the full classification of Moufang polygons, and can be found in [**130**]. We will give details, modulo the results of [**185**].

The higher dimensional theory, the theory of buildings, is also useful for recognition purposes, though we will offer an alternative approach; the last two sections of this chapter are in fact devoted to setting up the machinery for this approach.

We will just add a few words here about the notions involved. A typical example of a simple Chevalley group is the projective special linear group PSL_n, which acts naturally on the projective space of dimension $n-1$ whose points are the lines of a vector space of dimension n. In the case $n = 3$ we have a Lie rank 2 group acting on a projective plane, also known in the modern terminology as a *generalized triangle* in view of the generalization due to Tits, to generalized polygons. An "ordinary triangle" in this terminology would be an extremely degenerate geometry, consisting of three points, any pair of which forms a line. One notes that this satisfies all of the axioms of projective geometry except the axiom that lines should contain more than two points (since the order of a line is the order of the base field plus one, this is, morally speaking the same thing as $0 \neq 1$).

Of course, not all projective planes can be coordinatized by fields, or even division rings. A nice class of projective planes, now called *Moufang planes* was introduced in [**141**] in connection with the little Desargues theorem. These turn out to be classified by alternative division rings, which are not necessarily associative, and are also characterized by a certain richness of their automorphism groups; it is this latter condition which Tits adopted as the definition of Moufang polygon, and more generally Moufang building, rather than some intrinsic definition from synthetic geometry.

We showed in Proposition I 4.27 that Moufang projective planes of finite Morley rank are coordinatized by fields, and this is typical of the way the theory collapses in general. In fact, if one looks at the classification of *finite* Moufang polygons as derived in [**185**], one will see that the same line of argument works in the finite Morley rank case.

0.5. Niles' theorem. This is a transposition to the case of finite Morley rank of a theorem in the finite case. For "generic" groups of even type it produces a definable spherical BN-pair in the group, after which one can invoke the classification of the latter from [**130**], or, alternatively, make a direct reduction to an identification theorem of Curtis-Tits type, using mainly the fact that Niles' Theorem brings the Weyl group and the associated Dynkin diagram under control.

This alternative argument, bypassing the classification of buildings (in Tits rank at least 3), is prepared by the next two sections.

0.6. Signalizer functors and a generic identification theorem. These are two entirely different subjects, but they go together nonetheless. For us the signalizer functor theory is a way of killing the core, or much of it, in a 2-local subgroup.

The generic identification theorem gives a reduction to the Curtis-Tits theorem of a somewhat more general configuration which can be briefly described as "generation by reductive centralizers". Here reductivity is essentially the killing, or control, of the core in the centralizer of an element of odd prime order.

As we have suggested, one can proceed directly to the theory of buildings, or pass through this more elementary approach. We give the details for the latter.

1. Pseudoreflection groups

The present section does not have any clear antecedents in the context of finite groups. It is to a degree a semisimple analog of the theory of groups generated by root subgroups. It seems likely that one can give a finite analog of this theory using the classification of the finite simple groups; whether one can turn that around as we do here, and develop the theory also in the finite setting as a tool for use in classification, is unclear, and motivates some of the discussion in our last chapter.

1.1. Definitions.

DEFINITION 1.1. *If A is an elementary abelian 2-group, then a torus T acting on A is called a* group of pseudoreflections *on A if $A = C_A(T) \times [A, T]$ and T acts faithfully on the second factor, and transitively on its nonzero elements.*

Intuitively, the main idea is that T should act trivially on a subspace of codimension 1. We assume both less than this (in that we have no useful vector space structure) and substantially more, in that transitivity is a very restrictive condition. In the end, we will show that the "more" wins out over the "less", and that under moderate hypotheses the only interesting group generated by groups of this type is the group $\mathrm{GL}(V)$ with V, indeed, a vector space.

Our analysis of groups generated by pseudoreflections depends on a theorem of McLaughlin, involving groups of *root type* in the following sense.

DEFINITION 1.2. *Let V be a vector space of dimension $n \geq 2$ over a field K.*

(1) *For $U \leq V$ a K-subspace of codimension 1, and $L \leq V$ a K-subspace of dimension 1, $X(L, U)$ is the group of linear transformations of the form $I + T$ where $T = 0$ on U, and $T[V] \leq L$.*
(2) *A subgroup of $\mathrm{GL}(V)$ is said to be of* root type *if it has the form $X(L, U)$ for suitable L and U.*

Note that a group of root type is in fact contained in $\mathrm{SL}(V)$.

The theorem we need is the following.

THEOREM 1.3 ([**136**, Theorem]). *Let V be a vector space of dimension at least 2, over a field with more than two elements, and G a subgroup of $\mathrm{SL}(V)$ which is generated by subgroups of $\mathrm{SL}(V)$ of root type. Suppose furthermore that G has no nontrivial normal unipotent subgroup. Then there is a decomposition of V as a direct sum $\oplus_i V_i$ of G-invariant subspaces, and a corresponding decomposition of G as a direct product $\prod_i G_i$, such that*

(1) *each G_i acts trivially on V_j for $j \neq i$;*
(2) *G_i induces either $\mathrm{SL}(V_i)$ or $\mathrm{Sp}(V_i)$ on V_i.*

We also make use of the following.

LEMMA 1.4. *Let V be a vector space of dimension greater than 2 over a field K. Then*

(1) *the normalizer of $\mathrm{Sp}(V)$ in $\mathrm{GL}(V)$ consists of the group generated by $\mathrm{Sp}(V)$ and the scalars;*
(2) *this group contains no pseudoreflection subgroups.*

PROOF.

Ad (1) Certainly the normalizer G of $\mathrm{Sp}(V)$ contains the group generated by $\mathrm{Sp}(V)$ and the scalars. Conversely, for $g \in G$, if $V_0 \leq V$ is maximal isotropic then it is easy to see that $g[V_0]$ is also maximal isotropic, and after adjusting by an element of $\mathrm{Sp}(V)$ we may suppose g leaves V_0 invariant. We may then consider the subgroup H of $\mathrm{Sp}(V)$ which leaves V_0 invariant. This induces $\mathrm{GL}(V_0)$ on V_0 and so after a further adjustment we may suppose $g = 1$ on V_0. Then $[g, H] \leq \mathrm{Sp}(V)$ acts trivially on V_0 and thus $[g, H] = 1$. Let V_1 be a maximal isotropic subspace complementary to V_0. Let H_0 be the subgroup of H leaving V_0 and V_1 invariant. Then g commutes with H_0 and hence leaves it invariant, and thus g leaves V_1 invariant. Since g commutes with the action of H_0 on V_1, again g is a scalar on V_1. Since V_1 is any maximal isotropic subspace complementary to V_0, and the dimension of V is greater than 2, it follows easily that this scalar is 1.

Ad (2) In this linear context, a nontrivial pseudoreflection r will have two eigenspaces, of dimension 1 and $n-1$, with n the ambient dimension, and the same applies to a product $r\alpha$ with α scalar. We wish to see that an element of this explicit form cannot belong to the symplectic group. Indeed, taking v nontrivial in the 1-dimensional eigenspace and v' a nonorthogonal element in the complementary eigenspace, consider the restriction of $r\alpha$ to $W = \langle v, v' \rangle^{\perp}$. As $r\alpha$ leaves W invariant, and W is disjoint from the 1-dimensional eigenspace for r, r acts trivially on W and thus $r\alpha$ acts both as a scalar and as a symplectic transformation on W; so $\alpha = 1$, and r is symplectic. Then as r has a trivial eigenspace of dimension $n-1$, this would force r to be trivial. □

1.2. ***K*-groups.** Our main result is the following classification theorem, which will be extended to the L-group case afterward (in §IV 5).

THEOREM 1.5. *Let $A \rtimes H$ be a connected K-group of finite Morley rank and of even type, in which A is an elementary abelian definable 2-subgroup and H acts irreducibly and faithfully on A. Assume that H contains a group T of pseudoreflections on A. Then A can be given a vector space structure over an algebraically closed field K in such a way that $H \simeq \mathrm{GL}(A)$ acting naturally.*

PROOF. Observe that since $A \rtimes H$ is infinite and connected, and H acts faithfully, the group A is also infinite and connected.

Furthermore, $O_2(H) = 1$, since $O_2(H)$ centralizes a nontrivial subgroup B of A, and by irreducibility we have $B = A$. Therefore $H = E(H) * O(H)$ by Proposition II 5.10.

It follows from Lemma I 8.2 that $O(H)$ is abelian; indeed, if $B \leq A$ is minimal nontrivial $O(H)$-invariant, then $O(H)'$ centralizes B and hence by irreducibility $C_A(O(H)') = A$, and $O(H)' = 1$. Thus $O(H) = Z^\circ(H)$.

The question as to whether the group $Z^\circ(H)$ is trivial or not provides a major case division. Assume first that

$$(*) \qquad Z^\circ(H) > 1 \; .$$

Then by Proposition I 4.11, A has a natural vector space structure over an algebraically closed field K, with $Z^\circ(H)$ acting via scalars and H acting linearly. We assume $\dim A > 1$, as otherwise there is nothing to prove.

Now T has some eigenspace $L \leq A$ on which T does not act trivially (Proposition I 10.7), and as T is a group of pseudoreflections, T must act transitively on $L \setminus (0)$, and hence L is 1-dimensional, and T acts via scalars on L. Thus the elements of T are pseudoreflections also from a linear point of view.

Let H_1 be the subgroup of H generated by pseudoreflection subgroups. As H acts irreducibly and $H_1 \lhd H$, the action of H_1 on A is completely reducible (Lemma I 11.6). Write $A = A_1 \oplus \cdots \oplus A_n$ as a sum of irreducible H_1-submodules. Each pseudoreflection subgroup acts nontrivially on exactly one factor A_i. Hence H_1 is the direct product of subgroups $H_1^{(i)}$, where $H_1^{(i)}$ acts trivially on all factors A_j for $j \neq i$; A_i is an irreducible $H_1^{(i)}$-module. In particular, the A_i are *all* the irreducible H_1-submodules of A, and these factors are therefore permuted by H, which is connected and irreducible. Accordingly there is only one such factor, and A is irreducible as an H_1-module.

In particular, there are two pseudoreflection subgroups T_1, T_2 of H which do not commute. The group $\langle T_1, T_2 \rangle$ fixes a subspace of codimension 2 and acts on a complementary space as a subgroup of $\mathrm{GL}_2(K)$. It follows by inspection that this group contains a subgroup of root type in the sense of Definition 1.2.

Let H_0 be the subgroup of H generated by subgroups of root type. Consider an irreducible H_0-submodule B of A. Note that $\dim B > 1$, as otherwise H_0 acts trivially on B and hence on A. By McLaughlin's theorem,

Theorem 1.3, H_0 induces $\mathrm{SL}(B)$ or $\mathrm{Sp}(B)$ on B. If T is a pseudoreflection subgroup of H, then T fixes a subspace of codimension 1, and hence fixes a nonzero vector in B. Therefore B is T invariant: for $t \in T$, B^t is an irreducible H_0-submodule meeting B nontrivially. Thus B is H_1-invariant, and thus $A = B$ is H_0-irreducible. Now $\mathrm{SL}(A)$ or $\mathrm{Sp}(A)$ is normal in H. In the former case we have $H = \mathrm{GL}(A)$ as claimed, since H also contains a pseudoreflection group and in the latter case H is an extension of $\mathrm{Sp}(A)$ by the scalars, which does not in fact contain a pseudoreflection group except in dimension 2, where in any case $\mathrm{Sp}(A) = \mathrm{SL}(A)$ (Lemma 1.4).

Now suppose that

$$(\neg *) \qquad Z^\circ(H) = 1 \, .$$

In other words, $H = E(H)$. In this case we will arrive eventually at a contradiction. Note that we can no longer view A as a finite dimensional vector space.

We show first that H is simple. Let T be a pseudoreflection subgroup of H and let H_1 be a simple factor of H not commuting with T. Using Lemma I 1.5, we have $[T, H_1] = H_1$. Then T normalizes H_1 and acts by inner automorphisms, so T normalizes a pair of opposite Sylow 2-subgroups (maximal unipotent subgroups) S^+, S^- of H_1. Set $A^\pm = C_A(S^\pm)$. Then $A^+ \cap A^- = 0$, in additive notation, since H_1 is generated by $S^+ \cup S^-$ and has no fixed points on A. As the groups $A^\pm$ are T-invariant, T acts trivially on at least one of them, say A^+. Let $B \le A$ be H_1-irreducible. Then B meets A^+ and thus T stabilizes B. If $B = A$, then evidently $H = H_1$. Suppose $B < A$. As A is completely reducible as an H_1-module and T acts as a pseudoreflection group, we may suppose that the action of T on B is trivial. Then $H_1 = [T, H_1]$ acts trivially on B, a contradiction.

Thus H is simple. Let P be a maximal parabolic subgroup corresponding to deletion of a terminal node in (a component of) the Dynkin diagram, and L the associated Levi factor. Now L contains a maximal torus of H and hence contains a pseudoreflection group T.

Suppose that V is a composition factor for A as an L-module, and that T acts trivially on V. As L' is simple, either L' acts trivially on V, or $[T, L'] = 1$, in which case $[T, L] = 1$.

We may exclude the case $T \le Z(L)$ as follows. If $T \le Z(L)$, then T is a root torus in H. Hence T and some conjugate T^g generate a subgroup $L_1 \simeq \mathrm{SL}_2$ in H such that $A/C_A(L_1)$ has Morley rank $2t$, where $t = \mathrm{rk}\, T$. We consider the action of L_1 on $A_1 = A/C_A(L_1)$. T is a torus of L_1 and normalizes two "opposite" Sylow 2-subgroups S^+, S^- in L_1, each of which centralizes a nontrivial T-invariant subgroup of A_1; and the two subgroups involved are disjoint as S^+, S^- generate L_1. Now the Weyl group stabilizes $C_{A_1}(T)$ and interchanges the centralizers of S^+ and S^-, so T acts nontrivially on each of these two subgroups. However, as T is a pseudoreflection group this is not possible.

Our conclusion is that L' acts trivially on any composition factor on which T acts trivially. However, L' cannot act trivially on all the factors of a composition series for A, as the $2^\perp$ elements of L' would then act trivially on A itself (Proposition I 10.7). Accordingly, let V be a composition factor of A on which L' acts nontrivially. By the above, T also acts nontrivially on V, and therefore acts as a pseudoreflection group on V. By induction on $\mathrm{rk}(H)$ we may suppose therefore that $L \simeq \mathrm{GL}(V)$ acting naturally on V. In particular, $Z(L)$ acts as an algebraically closed field K on V.

We may suppose that $V = A_1/A_0$ where L normalizes A_0 and A_1, and L' acts trivially on A_0. As A_0 is T-invariant and T acts nontrivially on V, T acts trivially on A_0. Thus L acts trivially on A_0. Let $T_1 = Z(L)$ and let $a \in T_1^\times$. Then commutation with a induces an isomorphism $\gamma : V \to [a, A_1]$ which is an isomorphism of L-modules. Thus we may suppose that V is a subgroup of A. Furthermore T acts trivially on every composition factor of A/V and hence by the above L acts trivially on every such composition factor, forcing L to act trivially on A/V since it is generated by $2^\perp$-elements.

In particular if $\hat{T}$ is a maximal torus of H contained in L, then $V = [\hat{T}, A]$ and thus the Weyl group W of H acts on V. Let $w \in W$ invert T_1. Then for $v \in V^\times$ and α a scalar we find $(\alpha v)^w = \alpha^{-1} v^w$ and on considering $((\alpha+\beta)v)^w$ this yields $(\alpha+\beta)^{-1} = \alpha^{-1} + \beta^{-1}$, a contradiction. □

2. Zassenhaus groups

DEFINITION 2.1.

(1) *A permutation group (G, X) is a structure consisting of a group G acting on a set X, with the action included as part of the structure; a permutation group has finite Morley rank if and only if this structure does.*

(2) *A permutation group is a* Zassenhaus group *if it is doubly transitive, the underlying set contains at least three points, and the stabilizer of any three points is trivial.*

A Zassenhaus group G (really, (G, X)) is said to be *split* if for a pair of points $x, y \in X$ the point stabilizer G_x splits as $U \rtimes G_{xy}$ for some subgroup U. By double transitivity, if this holds for one pair x, y, then it holds for all such pairs. We need the classification by DeBonis and Nesin of a particular class of split Zassenhaus groups, as follows.

THEOREM 2.2. *Let G be an infinite Zassenhaus group of finite Morley rank with 1-point and 2-point stabilizers B and T respectively. Assume that B splits as $U \rtimes T$ where $Z(U)$ contains an involution. Then either*

- *G is sharply 2-transitive, or*
- *G is of the form $\mathrm{PSL}_2(K)$, for some algebraically closed field K of characteristic 2.*

This is the fundamental recognition theorem as far as the various *uniqueness cases* of our even type analysis (Part C) are concerned, and we will give the proof. We will see in a moment that under our hypotheses, the stabilizer of a point in G is a *strongly embedded* subgroup of G (§I 10.3). In Chapter VI we will reduce the classification of groups with a strongly embedded subgroup to this particular configuration, and even extend this considerably to the case of *weak embedding* (§II 5.12).

We begin with a definability result which can be proved in considerably greater generality [**83**].

LEMMA 2.3. *Let G be a group of finite Morley rank which has a faithful representation as a split Zassenhaus group with associated subgroups B, T, U, where $Z(U)$ contains an involution i. Then $C_G(i) = U$, the groups B, T, and U are definable, and the action of G on X is interpretable in G.*

PROOF. The fixed point set of i is the point $x \in X$ associated with B. Hence $C_G(i)$ fixes x, that is $C_G(i) \leq B$. On the other hand, for $t \in T^\times$, the fixed point set of t is the set $\{x, y\}$ associated with T (and no more, as G is a Zassenhaus group). Hence $C_G(t)$ leaves $\{x, y\}$ invariant and thus $C_B(t) \leq T$. Taken all together then we have $U \leq C_G(i) \leq B$ and $C_T(i) = 1$, hence $C_G(i) = U$ and U is definable.

Now the fixed point set of U is $\{x\}$ and hence $N(U) \leq B$; so $B = N(U)$ is definable. Hence the action of G on X is interpretable in G, and thus also T is interpretable in G. □

For the remainder of this section, assume the following.

(∗)
- G is a split Zassenhaus group of finite Morley rank with associated subgroups B, T, U;
- $Z(U)$ contains an involution.

By the preceding lemma, the groups B, T, U and the action of G on X may all be considered within the same ranked universe as G.

LEMMA 2.4. *With hypotheses and notation as in (∗), we have the following.*

(1) *The involutions of B are conjugate in B, and lie in $Z(U)$; the involutions of G are conjugate in G.*
(2) *$N(T)$ contains an involution w inverting T, and T is abelian.*
(3) *$G = B \sqcup UwTU$, with uniqueness of representation for $g \in UwTU$; that is, $g = u_1 w t u_2$ with u_1, t, u_2 unique.*

PROOF. Suppose there is an involution $t \in T$. Then letting t act on the torsion subgroup of $Z(U)$ of exponent 2, t commutes with an involution $i \in U$. But the fixed point set of t is $\{x, y\}$ and hence i leaves this invariant, while fixing x. So i fixes y and we have a contradiction.

Therefore every involution in B fixes a single point. It follows easily that B is strongly embedded in G, and hence all involutions of B are conjugate in B (Lemma I 10.12). In particular, they all lie in $Z(U)$. By the same token, all involutions of G are conjugate in G. This proves the first point.

For the second point, we have $T = G_{x,y}$ and we consider an element w switching x and y. Then $w \in N(T) - T$, and $w^2 \in T$. We may take w to be a 2-element (Lemma I 2.18). As T contains no involutions, by the first point, it follows that w is an involution. Hence w has exactly one fixed point z and since w leaves $\{x, y\}$ invariant, we have $z \neq x, y$. Now $C_T(w)$ fixes z and hence is trivial. It follows that w inverts T, and that T is abelian.

As G is doubly transitive, and $w \notin B$, it follows that $G = B \sqcup BwB$ (for $g \in G - B$, take $b \in B$ taking x^g to y; then $x^{gbw} = x$ and $gbw \in B$). We have $BwB = UTwTU = UwTU$, and it remains to deal with uniqueness.

As T normalizes U, a failure of uniqueness will lead to

$$u_1^w = tu_2$$

with $u_1 \neq 1$, $u_1, u_2 \in U$, $t \in T$; but then $u_1 \in B \cap B^w$ fixes $\{x, y\}$ and as $u_1 \in U$ we find $u_1 = 1$. □

From this point on we put aside the sharply 2-transitive case, that is we assume

$$T > 1 \ .$$

We will also assume that

$$G \text{ is infinite} \ .$$

As a consequence of the uniqueness of representation, we find

$$\mathrm{rk}(G) = 2\,\mathrm{rk}(U) + \mathrm{rk}(T) \ .$$

LEMMA 2.5. *With the hypotheses and notation of* (∗), *and assuming* $T > 1$ *and* G *infinite, the groups* G, B, U, *and* T *are connected.*

PROOF. If G° fixes a point, then as the action of G is transitive, G° is trivial, a contradiction. So $G^\circ \not\leq B$.

Fix some $g \in G^\circ - B \subseteq BwB$, which after conjugation under B may be taken to be of the form

$$g = bw$$

with $b \in B$.

It follows from the uniqueness of representation that the subset $X = U^\circ wU^\circ gT^\circ = U^\circ wU^\circ bT^\circ \cdot w$ has rank equal to $\mathrm{rk}(G)$ and hence is generic in G°. Hence for $u \in U$ we have a nontrivial intersection $X \cap uX$, and by uniqueness of the representation this means uU° meets U°, hence $u \in U^\circ$. Thus the group U is connected.

Now $U \leq G^\circ$ acts transitively on the set $X \setminus \{x\}$, and G° is not contained in B, so the action of G° on X is doubly transitive, and therefore G° is a Zassenhaus group, satisfying the same hypotheses as G. In particular, we may take $w \in G^\circ$.

So we have $G^\circ = B_0 \sqcup UwUT_0$ with $B_0 = B \cap G^\circ$ and $T_0 = T \cap G^\circ$. By the uniqueness of the representation, T_0 is connected and $T_0 = T^\circ$.

We have shown that T contains no involutions. So T is 2-divisible. Let $t = s^2$ with $s \in T$. Then $t = [w, s] \in G^\circ$ since $w \in G^\circ$, and hence $t \in T_0 = T^\circ$, and T is connected. Thus B is also connected, and as

$$G = B \sqcup UwUT$$

with uniqueness of representation, the Morley degree of G is one and G is connected. □

LEMMA 2.6. *Let $U_0 = \langle I(U) \rangle$. Then the pair (U_0, T), with T acting on U_0, is equivalent to a pair of the form $(K_+, K^\times)$ with K an algebraically closed field of characteristic 2, where the multiplicative group acts naturally.*

PROOF. The involutions of U are conjugate under the action of B and hence lie in $Z(U)$, so U_0 is an elementary abelian 2-group and T acts regularly on $U_0^\times$. Also T is abelian. So Proposition I 4.11 applies. □

LEMMA 2.7. *Suppose $U = U_0$. Then $G \simeq \mathrm{PSL}_2(K)$ for some algebraically closed field of characteristic 2.*

PROOF. By Lemma 2.6 the action of B on X can be identified with the natural action of a 1-dimensional affine group on the projective line, where the element $x \in X$ corresponds to ∞, and where y corresponds to 0. The involution w has a unique fixed point z which we identify with the point 1; as w inverts T, its action on X corresponds to inversion. Accordingly, the group G is embedded naturally into $\mathrm{PGL}_2(K)$ with its usual action on the projective line, where as we have seen the field K is algebraically closed. But $\mathrm{PGL}_2(K) \simeq \mathrm{PSL}_2(K)$ is sharply 3-transitive, and G is 3-transitive since T is transitive on $U^\times$, so $G = \mathrm{PSL}_2(K)$ under this identification. □

So the missing ingredient in all of this is the relation

$$U = U_0 \,.$$

For this, the only known approach is extensive calculation with elements of order 3.

We make the following observation.

LEMMA 2.8. *If $t \in T$ has order 3 then the set*

$$\{u \in U : uu^t u^{t^2} = 1\}$$

is generic in U.

PROOF. We have $C_U(t) = 1$ since $t \in T^\times$, so $[t, U]$ is generic in U and this identity holds on $[t, U]$. □

The following lemma simplifies calculations.

LEMMA 2.9. *Every element of $G - B$ is conjugate under T to an element of UwU.*

PROOF. It suffices to treat elements of wTU. So let $g = wtu$ with $u \in U$, $t \in T$. As T is 2-divisible we may take $s \in T$ with $s^2t = 1$. Then

$$g^s = wu \ .$$

□

LEMMA 2.10.

(1) *There is an involution* $i \in U$ *with* $(wi)^3 = 1$.
(2) *If* $u \in U$ *and* $(wu)^3 = 1$, *then* u *is an involution.*

PROOF.

Ad (1) Let $t \in T$ have order 3. Then $w^t w = t$ has order 3. Let $j = w^t$. Note that j and w lie in distinct conjugates of B, as otherwise they commute. If j, w lie in the conjugates B_j, B_w of B, then we may conjugate the pair (B_j, B_w) to the pair (B, B_w) by double transitivity, replacing j by an involution $j' \in U$ and w by an involution $w' \in B_w$. Now $B \cap B_w$ is a 2-point stabilizer and hence acts transitively on the involutions of B_w. Hence after a further conjugation, we replace j' by another involution i of U, and w' by w.

Ad (2) Suppose $(wu)^3 = 1$ and fix $i \in I(U)$ with $(wi)^3 = 1$.

Then $(wu)^{iwu} = u^{-1}wiwuiwu = u^{-1}iwiuiwu = u^{-1}iwuwu = u^{-1}iu^{-1}w = u^{-2}iw$, and hence $u^{-2}iw$ has order 3 as well.

Then

$$\begin{aligned}
(iu)^{wu^{-2}w} &= wu^2wiuwu^{-2}w &&= wu(uwu)iwu^{-2}w \\
&= wuwu^{-1}wiwu^{-2}w &&= wuwu^{-1}iwiu^{-2}w \\
&= u^{-1}wu^{-2}iwu^{-2}iw &&= u^{-1}w(u^{-2}iw)^2 \\
&= u^{-1}iu^2 = iu
\end{aligned}$$

and hence $(u^{-2})^w$ commutes with iu, forcing either $iu = 1$ or $u^{-2} = 1$ and in either case $u^2 = 1$. □

LEMMA 2.11. $U = U_0$.

PROOF. Fix an element $t \in T$ of order 3. We define a function

$$f : U^\times \to U$$

as follows. For $u \in U^\times$, write $u^w = u_1wu_2s$ with $u_1, u_2 \in U$ and $s \in T$. Set $f(u) = u_1^{-t}u_2^{-st^{-1}}$. In principle one expects $f(u) = u^{-1}$, which may explain why the function f has useful properties.

We show first that

$$f(u^x) = f(u)^{x^{-1}} \quad (u \in U, x \in T) \ .$$

With $u^w = u_1wu_2s$ we find

$$(u^x)^w = (u^w)^{x^{-1}} = u_1^{x^{-1}}w^{x^{-1}}u_2^{x^{-1}}s \ .$$

As $w^{x^{-1}}u_2^{x^{-1}} = wu_2^x x^{-2}$ this gives $f(u^x) = (u_1^{-tx^{-1}})(u_2^{x\cdot(-sx^{-2}t^{-1})}) = f(u)^{x^{-1}}$.

Now let X be the (generic) set of $u \in U^\times$ such that

$$uu^t u^{t^2} = 1 \ .$$

For $u \in X$, we claim that $(wf(u))^3 = 1$. Conjugating by T, we may suppose that $u^w = u_1 w u_2$, and we compute:

$$\begin{aligned} f(u)^w &= u_1^{-wt^{-1}} u_2^{-wt} = (u_2 w u^{-1})^{t^{-1}} (u^{-1} w u_1)^t \\ &= u_2^{t^{-1}} w u^{-t} u^{-1} w t u_1^t = u_2^{t^{-1}} [u^{t^2}]^w t u_1^t \\ &= u_2^{t^{-1}} (u_1 w u_2)^t t u_1^t = u_2^{t^{-1}} u_1^t w u_2^{t^{-1}} u_1^t = f(u)^{-1} w f(u)^{-1} \ . \end{aligned}$$

So by the preceding lemma, $f[X] \subseteq I(U)$. We claim that this implies that $X \subseteq I(U)$, and as X is generic therefore $U = U_0$.

Suppose then that $u \in X$. Again, conjugating by T, we may suppose that $u^w = u_1 w u_2$. As $f(u) \in Z(U)$, we have also

$$f(u) = f(u)^{u_1^{-t}} = u_2^{-t^{-1}} u_1^{-t} \ .$$

and we can compute $f(u)^w$ a second time:

$$\begin{aligned} f(u)^w = &\ u_2^{-wt} u_1^{-wt^{-1}} = u^{-t} w u^{t^{-1}} u_2 w u t^{-1} = u^{-t} [f(u)]^{-tw} u t^{-1} \\ = &\ u^{-t} (f(u) w f(u))^{t^{-1}} u t^{-1} = u^{-t} f(u)^{t^{-1}} w f(u)^t u^{t^2} \ . \end{aligned}$$

So by uniqueness of representation we find

$$f(u)^{-1} = u^{-t} f(u)^{t^{-1}} \ .$$

and as $f(u) \in I(U)$ it follows that $u \in I(U)$. □

This completes the proof of Theorem 2.2.

3. Suzuki groups

DEFINITION 3.1.

(1) *A* Suzuki 2-group *is a nilpotent* 2*-group* S *which is equipped with the action of an abelian group* T *such that the involutions of* S *lie in a single orbit under the action of* T.

(2) *A Suzuki* 2*-group is said to be* free *if the action is free, that is, the nontrivial elements of* T *have no nontrivial fixed points in* S.

The next result will play an important role in the treatment of groups with strongly embedded subgroups in §VI 1, and also in the analysis of groups with standard components of type SL_2 in Chapter VII.

THEOREM 3.2. *Let* (S,T) *be an infinite free Suzuki group of finite Morley rank. Then* S *is abelian and homocyclic.*

The following principles will be applied freely. Recall the notation $\Omega_i(S)$ (Notation I 1.1), which is very useful in this context.

LEMMA 3.3. *Let* S *be a Suzuki* 2*-group with operator group* T.

(1) *S is free if and only if the action of T on the involutions of S is regular.*
(2) *$Z(S) = \Omega_i(S)$ for some i.*
(3) *Setting $A = \Omega_1(S)$, the action of $T/C_T(A)$ on A is regular, and A may be identified with the additive group of a field (algebraically closed if A is infinite) in such a way that $T/C_T(A)$ is identified with $K^\times$ acting naturally.*
(4) *Let S be a Suzuki 2-group of exponent greater than 2, with $\Omega_2(S)$ abelian. Then $\bar{S} = S/\Omega_1(S)$ is a Suzuki 2-group with the same group of operators T, and the T-modules $\Omega_1(S)$ and $\Omega_1(\bar{S})$ are naturally isomorphic. In particular, if S is a free Suzuki 2-group, then $\bar{S}$ is as well.*

PROOF. The first point is immediate.

For the second point, suppose that $g \in Z(S)$ and $h \in S$ both have order 2^i. Then $h^{2^{i-1}}$ is an involution and in view of the action of T, there is an element $g' \in Z(S)$ such that $g'^{2^{i-1}} = h^{2^{i-1}}$ and thus $(g'h^{-1})^2 = 1$. Now the action of T shows that $g'h^{-1} \in Z(S)$ and thus $h \in Z(S)$.

The third point is essentially Schur's Lemma. Let K be the additive subgroup of $\mathrm{End}(A)$ generated by the image of T. Then K is a commutative ring and for each element of K, its kernel and range are T-invariant. Thus K is a field and as $\bar{T} = T/C_T(A)$ is transitive on $A^\times$, $\bar{T}$ is its multiplicative group. If A is infinite, then Proposition I 4.2 applies.

For the last point, observe first that $\Omega_1(\bar{S})$ is $\Omega_2(S)/\Omega_1(S)$. There is a natural map $\Omega_2(S) \to \Omega_1(S)$ given by squaring, which is surjective in view of the action of T. This induces the desired isomorphism. □

By the field associated with a Suzuki 2-group S, we mean the field K described above, with its action on $\Omega_1(S)$. If the Suzuki group is not free, then this field need not act naturally on S.

This result is helpful also when $\Omega_1(S)$ is finite, as it implies that the involutions are permuted cyclically, and as this was required by Higman's original definition of Suzuki 2-group in the finite case, this means that the finite free Suzuki 2-groups are actually the Suzuki 2-groups classified by Higman. In particular, we have the following, following from an explicit classification.

FACT 3.4 ([**112**]). *A finite nonabelian freeSuzuki 2-group has exponent 4.*

The last point in Lemma 3.3 can be iterated, and yields the following.

COROLLARY 3.5. *Let A be an abelian Suzuki 2-group with operator group T, and H a proper T-invariant subgroup of A. Then the following hold.*

(1) *$H = \Omega_i(A)$ for some i.*
(2) *$\bar{A} = A/H$ is a Suzuki 2-group with respect to the same group T.*
(3) *There is a natural isomorphism between the T-modules $\Omega_1(\bar{A})$ and $\Omega_1(A)$.*

LEMMA 3.6. *Let S be a free Suzuki 2-group with operator group T, and $H \lhd S$ a T-invariant abelian subgroup. Then the action of T on S/H is free.*

PROOF. Suppose that $t \in T$, $g \in S$, and $[t,g] \in H$, with $t \neq 1$. We claim then that $g \in H$.

By Corollary 3.5 and induction it is easy to see that $[t,H] = H$. Hence $[t,g] = [t,h]$ for some $h \in H$ and then t commutes with gh^{-1}, and $g = h \in H$. □

3.1. Special cases. Now we take up the topic of Theorem 3.2. We will focus initially on cases in which $S' \leq \Omega_1(S)$.

First, we consider a functional equation which arises in this context.

LEMMA 3.7. *Let K be a field of characteristic 2 and let $f : K \to K$ be a function which satisfies the following condition for $x \neq 0, 1$:*

$$(*) \qquad xf(y/x) + (1+x)f(y/(1+x)) = f(x+y) + f(x) \ .$$

Then $f(x^2) = xf(1)$ for all $x \in K$.

PROOF. Taking $y = 0$ we get $f(0) = 0$.

Now in $(*)$ the left hand side is invariant under the substitution $x \mapsto 1+x$, and hence the right hand side is as well:

$$f(x+y) + f(x) = f(1+x+y) + f(1+x) \ .$$

Taking $y = x$ this gives the additive law

$$f(1+x) = f(x) + f(1) \ .$$

Now it suffices to take $y = x(1+x)$ in $(*)$ to get

$$xf(1+x) + (1+x)f(x) = f(x^2) + f(x)$$

and then simplifying by the additive law we have

$$f(x^2) = xf(1)$$

as claimed. □

This yields the following group theoretic result which will also be useful in §4.

PROPOSITION 3.8. *Let S be a unipotent 2-group of exponent at most 4. Assume that*

$$0 \to Z \to S \to S/Z \to 0$$

is an exact sequence, where Z is central and both Z and S/Z are isomorphic to K_+, K a perfect field of characteristic 2. Assume also that $T \simeq K^*$ acts on S, inducing the natural action on both Z and S/Z. Then S is abelian, and it is either homocyclic or else is elementary abelian of the form $S = S_1 \oplus S_2$, splitting as a T-module. In the case where S is homocyclic, it is isomorphic to the group on $K \times K$ with the operation

$$(x,y) \cdot (x',y') = (x+x', y+y'+\sqrt{xx'}) \ .$$

PROOF. We identify $T\cup\{0\}$ with K and assume K acts on S with $g^0 = 1$ for $g \in S$, where the action on S/Z and on Z gives each the structure of a 1-dimensional vector space.

Fix two elements $g \in S\backslash Z$ and $h \in Z^\times$, and coordinatize S by associating the pair $s, t \in K$ with $g^s h^t$.

For $x \in K$ we have

$$gg^x = g^{1+x}h^{f(x)}$$

with some function $f : K \to K$. Writing out the law $(gg^x)g^y = g(gg^{x^{-1}y})^x$ in terms of f yields

$$f(x) + (1+x)f(y/(1+x)) = xf(y/x) + f(x+y) \ .$$

This is equivalent to the functional equation of Lemma 3.7, so $f(x^2) = xf(1)$, or since K is perfect we may write:

$$f(x) = f(1)\sqrt{x} \ .$$

To work out the full product in terms of coordinates we compute

$$(g^s h^t)(g^{s'} h^{t'}) = g^s g^{s'} h^{t+t'}$$

and then

$$g^s g^{s'} = (gg^{s^{-1}s'})^s = g^{s+s'} h^{s(1+f(s^{-1}s'))} \ .$$

Now as far as the function f is concerned there are really only two cases. If S has exponent 2, then $f(1) = 0$ and thus $f(x) = 0$, and S splits as a T-module.

If S has exponent 4, then we can choose g to have order 4 and then choose $h = g^2$, getting $f(1) = 1$ and $f(x) = \sqrt{x}$, and again everything is clear. □

In particular we arrive at the following critical case of Suzuki 2-groups in exponent 4. Whereas in the previous lemma we did not need to assume a priori that our group S is abelian, here we need that assumption to recover the situation of the foregoing lemma.

COROLLARY 3.9. *Let S be an abelian free Suzuki 2-group of exponent 4 over a perfect field K. Then S is isomorphic to the group with underlying set $K \times K$, and multiplication defined as follows:*

$$(x, y) \cdot (x', y') = (x + x', y + y' + \sqrt{xx'}) \ ,$$

and with $K^\times$ acting by multiplication, componentwise.

The next result will show that when K is quadratically closed, it does not allow a free Suzuki 2-group of one critical type. In the course of the analysis the following functional equation will arise.

LEMMA 3.10. *Let K, F be fields of characteristic 2, with $K \subseteq F$ as sets and with compatible multiplication operations, but possibly different addition operations. Let $\gamma : K \to F$ be an additive map. Then the following are equivalent:*

(1) $\gamma(x^{-1}) = x^{-2}\gamma(x)$ *for all* $x \in K$,
(2) γ *is linear over the subfield* K^2.

PROOF. Since $x^{-1} = x^{-2} \cdot x$, the second hypothesis implies the first. For the converse, we use the identity

$$(x + x^2 z)^{-1} = x^{-1} + (x + z^{-1})^{-1} .$$

Applying the functional equation for γ and multiplying through by $(x+x^2z)^2$ we find

$$\gamma(x + x^2 z) = (1 + xz)^2\gamma(x) + x^2z^2\gamma(x + z^{-1}) = \gamma(x) + x^2z^2\gamma(z^{-1})$$

which simplifies to

$$\gamma(x^2 z) = x^2\gamma(z),$$

and this is our claim. □

LEMMA 3.11. *Let S be a free Suzuki 2-group over a perfect field K, and suppose that S contains two distinct K-invariant abelian subgroups M and N of exponent 4, with M normalizing N. Then MN can be coordinatized by the base field K so that the underlying set is identified with $K \times K \times K$, with T acting by multiplication coordinatewise, and with multiplication given by*

$$(*) \qquad (\mathbf{u}, z) \cdot (\mathbf{u}', z') = (\mathbf{u} + \mathbf{u}', z + z' + \sqrt{B(\mathbf{u}, \mathbf{u}')})$$

with B a bilinear form over K for which the associated quadratic form $B(\mathbf{u}, \mathbf{u})$ is nonisotropic.

PROOF. We may suppose that $S = MN$ and N is normal in S. Observe that MN is nonabelian by Corollary 3.5. Then $S' = [N, M] < N$ and as N has exponent 4, that same corollary shows that $S' \leq \Omega_1(S)$. Hence $S' = \Omega_1(S)$. Similarly, $M \cap N = \Omega_1(S)$, and, in particular, M is also normal in S.

The structure of M and of N is given by Corollary 3.9. We may now coordinatize as follows. Fix $g \in M$, $h \in N$ of order 4 with $g^2 = h^2$, and set $a = g^2$. Every element of S may be expressed uniquely as $g^x h^y a^z$ with $x, y, z \in K$, and will be identified with the corresponding triple (x, y, z). Then K acts coordinatewise, and it remains to work out the multiplication in S, which has the form $(*)$, but with B at the moment simply some bi-additive function of $\mathbf{u}$ and $\mathbf{u}'$. We need to show that B is K-linear in each variable and that the associated form Q is nonisotropic.

Note that $(\mathbf{u}, z)^2 = (0, Q(\mathbf{u}))$ and as all involutions have $\mathbf{u} = \mathbf{0}$ it follows that Q is anisotropic, so it is only the K-linearity which is in question. Furthermore, this is known already separately on M and on N, so it suffices to consider the commutation map $M \times N \to M \cap N$, with respect to a fixed element of M or N. Without loss of generality, we may restrict our attention to $B(g, x)$ with x varying over N. In other words, we consider the function $f : K \to K$ defined by

$$[g, h^x] = a^{f(x)} .$$

This is an additive map. We have the additional relations

$$[g^x, h] = [g, h^{x^{-1}}]^x; \qquad [g^{x+y}, h] = [g^x, h][g^y, h],$$

and these combine to give the functional equation

$$xf(x^{-1}) + yf(y^{-1}) = (x+y)f((x+y)^{-1}) .$$

Replacing x and y by 1 and $(1+x)^{-1}$ respectively, and multiplying through by $(1+x)$, this gives

$$f(x^{-1}) = xf(x) .$$

Since the field is perfect, the function is K-linear by Lemma 3.10. □

In the previous case S is "short and fat". The next one is more Laurelian.

LEMMA 3.12. *Let S be a Suzuki group over a perfect field K. If $\Omega_2(S)$ is abelian and $S' \le \Omega_1(S)$, then S is abelian.*

PROOF. We suppose the contrary. Let $\bar{S} = S/\Omega_1(S)$, an abelian group. By Lemma 3.3 the group $\bar{S}$ is a Suzuki 2-group, with the same associated field K.

Now $\Omega_1(S) \le Z(S) < S$, and by Lemma 3.5 the groups $S/Z(S)$ and $\bar{S}$ also have the same associated field K. Furthermore $S/Z(S)$ is elementary abelian since $S' \le \Omega_1(S) \le Z(S)$. So the K-module $S/Z(S)$ is naturally isomorphic with $V = \Omega_1(S)$.

Fix nontrivial elements $\bar{g} \in S/Z(S)$ and $i \in I(S)$. We consider the commutation map $\gamma : S/Z(S) \to \Omega_1(S)$ given by $\gamma(x) = [g, x]$, using "coordinates" from K. That is, we define a map $f : K \to K$ as follows:

$$[g, g^t] = i^{f(t)} .$$

We claim that the function $f(t)$ has the following properties:

$$f \text{ is additive; } f(1) = 0; \ f(t^{-1}) = tf(t^{-1}) .$$

The first is immediate as the commutation map $S/S' \times S/S' \to S' \le Z(S)$ is bilinear, and the second is equally clear. For the "functional equation" given by the last clause, we make the following computation:

$$[g, g^{t^{-1}}] = [g^t, g]^{t^{-1}} = [g, g^t]^{t^{-1}}$$

which indeed becomes $f(t^{-1}) = f(t)t^{-1}$.

Observe then that the function $\gamma(x) = f(x)^2$ satisfies the functional equation of Lemma 3.10, and as K is perfect, it is a linear function of one variable, vanishing at $x = 1$, and hence identically 0. This means that g lies in $Z(G)$ and yields a contradiction. □

Combining these two yields a generalization of our first case.

LEMMA 3.13. *Let S be a free Suzuki 2-group over a quadratically closed field K. Then the T-invariant abelian subgroups of S are linearly ordered by inclusion.*

This can also be expressed as follows: If A is maximal abelian T-invariant, then all other abelian T-invariant subgroups are of the form $\Omega_i(A)$ for some i.

PROOF. Suppose that M and N are abelian T-invariant subgroups of S. We claim that one is contained in the other. We may suppose $S = MN$. Then $M \cap N$ is central in S and hence is $\Omega_i(S)$ for some i.

If $i > 1$, then $\Omega_2(S)$ is abelian. Since $\Omega_1(S) \leq M \cap N$, passage to $S/\Omega_1(S)$ allows us to conclude by induction.

So we may suppose that $M \cap N = \Omega_1(S)$. Hence also $S' = [M, N] = \Omega_1(S)$. If one of M or N is $\Omega_1(S)$ our claim is clear, while if neither one is $\Omega_1(S)$, then the groups $\Omega_2(M)$ and $\Omega_2(N)$ are incomparable, and we aim at a contradiction. We may suppose therefore that M and N have exponent 4, and as $S = MN$ with $S' \leq \Omega_1(S)$ it follows easily that S has exponent 4. Now Lemma 3.11 applies and the structure of S is determined. As it involves a nonisotropic quadratric form, we contradict the hypothesis that the base field is quadratically closed. □

Now we arrive at a broader case, but under model theoretic restrictions. We first make a small remark which will be useful again a little further on.

LEMMA 3.14. *Let (K, L) be a structure of finite Morley rank with K and L infinite fields of positive characteristic p. Suppose that their multiplicative groups are related by $K^\times \leq L^\times$, without any restriction on their respective additive groups. Then $K = L$ as sets.*

PROOF. Both K and L are algebraically closed, and have the same torsion elements (multiplicatively). Hence as a set, the algebraic closure $\tilde{\mathbb{F}}_p$ of $\mathbb{F}_p$ in K or L is the same.

By Lemma I 4.16 the set $K^\times$ is generic in $L^\times$, and as these are connected groups it follows that $K = L$ as sets. □

LEMMA 3.15. *Let S be an infinite free Suzuki 2-group of finite Morley rank. If $S' \leq \Omega_1(S)$, then S is abelian.*

PROOF. Let us first deal with the case in which $\Omega_1(S)$ is finite. Then $(S^\circ)'$ is a connected subgroup of $\Omega_1(S)$ and hence S° is abelian. As S° is infinite it follows that S° contains elements of order 2^n for all n, and hence $S = S^\circ$. So the claim holds in this case. For the remainder of the argument we assume $\Omega_1(S)$ is infinite.

If $\Omega_2(S)$ is abelian, we apply Lemma 3.12. So we may suppose that $S = \Omega_2(S)$ and thus S has exponent 4. We assume S is nonabelian. Thus $Z(S) = \Omega_1(S) = S'$.

In particular, $\bar{S} = S/\Omega_1(S)$ is a nontrivial elementary abelian 2-group, which we consider as a K-module. We may suppose that every proper definable T-invariant subgroup of S is abelian.

Let A be a maximal T-invariant abelian subgroup of S. Then $A = C(A)$ is definable.

Take $v \in \bar{S}$, $v \notin \bar{A}$, and let $V = \langle [T, v] \rangle$.

Let $v = \bar{g}$ with $g \notin A$. We show first that $[T, \langle g \rangle] = S$.

If $[T, \langle g \rangle] < S$, then $[T, \langle g \rangle]$ is abelian, and by Lemma 3.13 we find $[T, \langle g \rangle] \leq A$, and in particular $[T, g] \leq A$. Take any $t \in T^\times$. Then t acts freely on A and hence $[t, A] = A$. Thus we have $[t, g] = [t, x]$ for some $x \in A$, and t fixes gx^{-1}, forcing $g = x$ and $g \in A$, a contradiction. So $[T, \langle g \rangle] = S$.

On the other hand, $[T, \langle g \rangle] \leq \langle [T, g] \rangle S'$. So $\langle [T, g] \rangle S' = S$ and it follows from Lemmas I 5.19 and I 5.20 that $\langle [T, g] \rangle = S$. In particular, $V = \bar{S}$.

Consider the additive subgroup L of $\mathrm{End}(\bar{S})$ generated by the image of the field K associated with S. What we have shown is that L acts transitively on $\bar{S}$. Furthermore, by our choice of S, $\bar{S}$ is definably irreducible, and thus L is a field by Schur's Lemma. By Lemma I 4.8 K is definable.

We have an inclusion of multiplicative groups $K^\times \leq L^\times$, though the additions operations may be incompatible. By Lemma 3.14 we have $K = L$ as sets. Since the addition operations are not necessarily related, it is nonetheless important to distinguish the two fields.

Now fix elements $g, a \in S$ of order 4 and two respectively, and define a function $f : L \to K$ by

$$[g, g^x] = a^{f(x)} .$$

As usual f is additive, and from $[g^x, g] = [g, g^{x^{-1}}]^x$ we derive $f(x) = xf(x^{-1})$. So applying Lemma 3.10 to $\hat{f}(x) = f(x)^2$, we find that $f(x)^2$ is L-linear. As $f(1) = 0$, it follows that $f \equiv 0$. So $\langle g^T \rangle$ is commutative. In particular, $S = \langle [T, g] \rangle$ is commutative, a contradiction. □

3.2. The general case. We can now treat the general case. If we wish to treat a Suzuki group as an abstract group (that is, without assuming that all elementarily equivalent groups are Suzuki groups), then we cannot immediately assume it has bounded exponent. But the following remark eliminates any cause for concern on this point.

LEMMA 3.16. *Let S be an infinite nonabelian free Suzuki group. Then there is a definable section S_1 of S of bounded exponent which is also an infinite nonabelian free Suzuki group, with the same operator group.*

PROOF. Suppose first that S is not nilpotent of class 2. Then we may factor out $Z(S)$ unless $Z(S)$ has finite index in S. On the other hand, if $Z(S)$ has finite index in S, then either $Z(S)$ has bounded exponent and then so does S, or $Z(S)$ has unbounded exponent and then $S = Z(S)$ is abelian.

Iterating this argument, we may suppose S is nilpotent of class 2. Then $\Omega_i(S)$ has exponent at most 2^{i+1} for each i, since for $x, y \in \Omega_i(S)$ we have $(xy)^2 = x^2y^2[x, y]$ with $[x, y]$ central and of order at most 2^i. So it suffices to choose i so that $\Omega_i(S)$ is nonabelian, and of exponent at least 8, so that by Fact 3.4 it is again an infinite group. □

PROOF OF THEOREM 3.2. We suppose toward a contradiction that S is an infinite nonabelian free Suzuki 2-group of finite Morley rank, with operator group T identified with $F^\times$, where F is a field interpreted in S. As just seen, we may suppose that S has bounded exponent. It follows easily that $\Omega_1(S)$ is infinite.

Let A be a maximal normal abelian T-invariant subgroup of S. Then $A = C(A)$ is definable in S. If $\Omega_2(S) \le A$, then by Lemma 3.3, induction applies to $\bar{S} = S/\Omega_1(S)$, which is therefore abelian. Thus $S' \le \Omega_1(S)$ and now Lemma 3.12 applies to give a contradiction. So we may fix an element $g \in S \setminus A$ of order 4.

We may suppose that any proper definable T-invariant subgroup of S is abelian, and hence by Lemma 3.13 is contained in A. In particular, $S = N_S(A)$, that is $A \lhd S$, and furthermore S/A is T-minimal. By Lemma I 4.10, $\bar{S} = S/A$ is T-irreducible and hence by Schur's Lemma the subring of $\mathrm{End}(\bar{S})$ generated by T is a field K. The action of T on S/A is free by Lemma 3.6, so $T \le K^\times$. As in the proof of Lemma 3.15, K is interpretable in S, and by Lemma 3.14 we have $K = F$ as sets.

Now we can begin to compute.

Let $B = 2A$ (in additive notation). We show first that

$$(*) \qquad S' \le B\,.$$

We know $S' \le A$ and $[S, A] \le B$.

Fix $g \in S$. Commutation with g gives a function $S \to [g, S] \le A$, and the induced map $S \to A/B$ is a homomorphism since $[g, xy] = [g, y][g, x]^y$ and $[y, A] \le B$. Furthermore, $[g, S'] \le [g, A] \le B$, so there is an induced homomorphism

$$\gamma : S/S' \to A/B\,.$$

We claim that this map is trivial, that is, $[g, S] \le B$, which will prove $(*)$.

For the sake of coordinatization, fix an element $\bar{a} \in (A/B)^\times$. Define an additive function $f : K \to F$ by

$$\gamma(g^t) = \bar{a}^{f(t)}\,.$$

As usual by considering $[g^t, g]$ we find $f(t^{-1}) = t^{-1} f(t)$ and then Lemma 3.10 shows that f is K-linear. Since $f(1) = 0$ we find $f \equiv 0$ and $[g, S] \le B$.

We may now consider the T-module $V = S/B$ and the submodule $U = A/B$. Note that $S = A\langle\Omega_2(S)\rangle$ and hence V is an elementary abelian 2-group. Our assumptions at this point are that $V > U$ and that any proper definable submodule of V is contained in U.

Let R be the subring of $\mathrm{End}(V)$ generated by T. Let $R \to K$ be the map induced by restriction to U, and $\mathfrak{m}$ the kernel of this map, the annihilator of U.

Let $r \in R \setminus \mathfrak{m}$. We claim that r acts freely on V. Consider $V[r] = \{v \in V : rv = 0\}$. Then $V[r]$ is T-invariant and proper in V, hence contained in U; but then $V[r] = U[r] = 0$.

Now we claim that $\mathfrak{m}V \le U$. Let $v \in V$ and $a \in \mathfrak{m}$. Consider the submodule $\langle Tav\rangle$ of V. This is definable. If $\langle Tav\rangle = V$, then we have $rv = v$ for some r which is in the additive closure of Ta, and hence lies in

$\mathfrak{m}$. Thus $(1-r)v = 0$ and $1 - r \in \mathfrak{m}$, and thus $1 \in \mathfrak{m}$, a contradiction. So $\langle Tav \rangle < V$ and thus $\langle Tav \rangle \leq U$, and in particular $av \in U$. Thus $\mathfrak{m}V \leq U$. As $\mathfrak{m}U = 0$, we find

$$\mathfrak{m}^2 = 0 \ .$$

Extend the inclusion $T \to R$ to a map $\iota : K \to R$ by $\iota(0) = 0$. We claim that this is an embedding of the field into R. It suffices to check additivity for pairs $s, t \in T$ with $s +_K t \neq 0$, with $+_K$ computed in K. In R, we have $\sqrt{s} + \sqrt{t} \in \sqrt{s +_K t} + \mathfrak{m}$, and after squaring this yields $s + t = s +_K t$. So we have a natural embedding of K into R, and V becomes an K-vector space in a way compatible with the K-structure of U. Now take a complement U_1 to U in V as an K-space, which lifts to a proper T-invariant subgroup of S not contained in A. This gives a contradiction. □

4. Landrock-Solomon

4.1. The Landrock-Solomon configuration. The configuration dealt with in the following theorem arises in the course of the analysis of groups with weakly embedded subgroups which are not strongly embedded.

THEOREM 4.1. *Let $H = S \rtimes T$ be a group of finite Morley rank, where S is a definable, connected 2-group of bounded exponent, and T is definable. Assume:*

1. *S has a definable subgroup A such that $A \rtimes T \cong K_+ \rtimes K^\times$ for some algebraically closed field K of characteristic 2, with the multiplicative group acting naturally on the additive group.*
2. *There is a definable involutory automorphism α of H such that $C_H{}^\circ(\alpha) = A \rtimes T$.*

Then S is isomorphic to one of the following groups.

- (A) *If S is abelian, then either*
    - (1) *S is homocyclic with $I(S) = A^\times$, or*
    - (2) *$S = E \oplus E^\alpha$, with E an elementary abelian group isomorphic to K_+.*

    In the second case, $A = \{xx^\alpha : x \in E\}$.
- (B) *If S is nonabelian, then S is an algebraic group over K whose underlying set is $K \times K \times K$ and the group multiplication is as follows:*

$$(*)\qquad \begin{aligned}(a_1, b_1, c_1) \cdot (a_2, b_2, c_2) \\ = (a_1 + a_2, \quad b_1 + b_2, \quad c_1 + c_2 + \epsilon\sqrt{a_1 a_2} + \sqrt{b_1 b_2} + \sqrt{b_1 a_2})\end{aligned}$$

for $a_1, b_1, c_1, a_2, b_2, c_2 \in K$, where ϵ is either 0 or 1.

In this case, α acts by $(a, b, c)^\alpha = (a, a + b, a + b + c + \sqrt{ab})$ and $[\alpha, S] = \{(0, b, c) : b, c \in K\}$.

In particular, if S is nonabelian, then S has exponent 4.

4.2. Around Suzuki 2-groups. It is not surprising that §3 has some relevance here, since in the conclusion of Theorem 4.1, in the important case (A.1), the pair (S,T) constitutes a free Suzuki 2-group of finite Morley rank. In particular, Lemma 3.8 gives the analysis of an important special case.

We mention here a cohomological splitting argument which enters into the proof of Theorem 4.1, and is of use elsewhere. For the moment the one-dimensional case is of interest.

LEMMA 4.2. *Suppose T is a divisible abelian group of finite Morley rank acting definably on each term of the short exact sequence*

$$0 \to M_\circ \to M \to M_1 \to 0$$

of T-modules, where $M_\circ$ is finite, and the action of T on M_1 is that of the multiplicative group of a field K on a vector space. Then the sequence splits, definably.

PROOF. We fix $x \in M \setminus M_\circ$ (in other words, the lift to M of a nontrivial element of M_1), and we write $x^{g+h} = x^g + x^h + a(g,h)$ (taking $x^0 = 0$). Then a is a definable function from $(T \cup \{0\}) \times (T \cup \{0\})$ into $M_\circ$ which satisfies the standard cocycle condition for every $g,h,k \in T \cup \{0\}$:

$$a(g,h) + a(g+h,k) = a(g,h+k) + a(h,k) \; .$$

Since $M_\circ$ is finite, for fixed g the function $a(g,\_)$ generically takes on a fixed value a_g. This yields a definable function from T into $M_\circ$ which associates to every $g \in T$ the corresponding generic value a_g. Applying the cocycle condition with an arbitrary pair g,h of elements from T, and with $k \in T$ independent from g and h, one obtains $a(g,h) + a_{g+h} = a_g + a_h$. This means that $a(g,h)$ is a definable coboundary, and hence the extension splits definably. □

4.3. Action of T. In the present subsection we will show that T acts freely on S, and we derive some consequences.

LEMMA 4.3. *$C_S(t)$ is finite and equal to $C_S(T)$ for every $t \in T^\times$.*

PROOF. Fix $t \in T$. Since $C_S(t)$ is T-invariant and T is connected, it suffices to show the first point, namely that $C_S(t)$ is finite.

By assumption, T acts freely on A and $A = C_S{}^\circ(\alpha)$. Let $Q = C_S(t)$. Then Q is an $(\langle\alpha\rangle \times T)$-invariant 2-group. Now $C_Q{}^\circ(\alpha) \le C_A(t) = 1$, so $C_Q(\alpha)$ is finite and hence Q is finite. □

LEMMA 4.4. *Let S, T and α satisfy the hypotheses of Theorem 4.1. Let R be a nontrivial definable connected $\langle\alpha,T\rangle$-invariant subgroup of S. Let R_1 be a maximal proper definable connected normal $\langle\alpha,T\rangle$-invariant subgroup of R which contains R'. Then the following hold:*

(a) *R/R_1 is an elementary abelian group.*
(b) *$C_{R/R_1}(\alpha) = R/R_1$, or in other words $[\alpha,R] \le R_1$.*
(c) $\mathrm{rk}(R) = \mathrm{rk}(R_1) + \mathrm{rk}(A)$.

PROOF. (a) As $\Omega_1{}^\circ(R/R_1)$ is a nontrivial definable connected subgroup of R/R_1, R/R_1 is an elementary abelian 2-group.

(b) By Lemma I 10.3, $C_{R/R_1}{}^\circ(\alpha)$ is a nontrivial definable connected subgroup of R/R_1. Thus, $C_{R/R_1}(\alpha) = R/R_1$.

(c) We will use $\overline{}$-notation to denote quotients by R_1. By Lemma 4.3 we can find $x \in R \setminus C_R(T)R_1$. By Proposition I 9.9 and Lemma 4.3, $C_T(\bar{x}) = 1$, so $\mathrm{rk}(\bar{x}^T) = \mathrm{rk}(T) = \mathrm{rk}(A)$. We have $\bar{x}^T \subseteq R/R_1$ and thus $\mathrm{rk}(\bar{x}^T) \leq \mathrm{rk}(R) - \mathrm{rk}(R_1)$, so $\mathrm{rk}(R) \geq \mathrm{rk}(R_1) + \mathrm{rk}(A)$. Since we also have $\mathrm{rk}(R) = \mathrm{rk}(C_R(\alpha)) + \mathrm{rk}(\alpha^R)$ and $\alpha^R \subseteq \alpha R_1$ by (b), we conclude $\mathrm{rk}(R) \leq \mathrm{rk}(A) + \mathrm{rk}(R_1)$. □

PROPOSITION 4.5. *Let S and T be as in the the statement of Theorem 4.1. Then for every $t \in T^\times$, $C_S(t) = 1$.*

PROOF. Let $Q = C_S(t)$. By Lemma 4.3, $Q = C_S(T)$ and Q is finite. We suppose toward a contradiction that $Q \neq 1$. Let R be a minimal definable connected $\langle \alpha, T \rangle$-invariant subgroup of S which contains Q. Let R_1 be a maximal proper definable connected normal $\langle \alpha, T \rangle$-invariant subgroup of R which contains R'. By Lemma 4.4 (a) and (c), R/R_1 is a connected elementary abelian 2-group of rank $\mathrm{rk}(T)$. We therefore have the following short exact sequence

$$0 \to M_\circ \to M \to M_1 \to 0$$

where $M = R/R_1$, M_1 is the natural T-module K_+ (by Lemma I 4.10, T acts on M_1 by scalar multiplication) and $M_\circ = QR_1/R_1$ is the kernel of the action of T on M. The choice of R and R_1 implies that $M_\circ \neq (0)$.

By Lemma 4.2, this sequence splits definably, contradicting the connectedness of R, and we conclude that $Q = 1$. □

COROLLARY 4.6. *Let S, T, and α satisfy the hypotheses of Theorem 4.1. Then the following hold.*

(1) *If X is a definable normal T-invariant subgroup of S, then for any element t of $T^\times$, $C_{S/X}(t) = 1$.*
(2) *Any definable normal T-invariant subgroup X of S is connected.*
(3) $C_S(\alpha) = A$.

PROOF.

Ad (1) Let T_1 be the definable hull of $\langle t \rangle$. Then T_1 is a definable $2^\perp$-group and $C_{S/X}(t) = C_{S/X}(T_1) = C_S(T_1)X/X$ by Proposition I 9.9. By Proposition 4.5 this is trivial.

Ad (2) As T is connected, it centralizes X/X°. By the preceding corollary, we get $X = X^\circ$.

Finally, the third point is an instance of the second, with $X = C_S(\alpha)$. □

4.4. The abelian case.

LEMMA 4.7. *If $A < S$, then $\mathrm{rk}(C_{S/A}(\alpha)) = \mathrm{rk}(A)$ and $C_{S/A}(\alpha)$ is an elementary abelian group. Furthermore, $C_{S/A}(\alpha)$ is isomorphic as a T-module with A.*

PROOF. Let $X/A = C_{S/A}(\alpha)$, which is nontrivial by Lemma I 10.3. Commutation with α induces an isomorphism of X/A with A. It is surjective because the image is nontrivial and T-invariant. □

The next proposition classifies the abelian 2-groups which satisfy the conditions of Theorem 4.1:

PROPOSITION 4.8. *Let S be an abelian 2-group satisfying the conditions of Theorem 4.1. Then either*

(1) *S is homocyclic with $I(S) = A^\times$, or*
(2) *$S = E \oplus E^\alpha$, where E is a T-invariant elementary abelian group.*

In the second case, $A = \{xx^\alpha : x \in E\}$ and both E and E^α are T-modules.

PROOF. Note that the assumption that S is abelian implies that for $x \in S$, xx^α is centralized by α, and thus $xx^\alpha \in A$. As a result α inverts S/A.

If $I(S) = A^*$, by Lemma I 1.16, S is a direct sum of cyclic groups. Since $\mathrm{Aut}(S)$ is transitive on $I(S)$, it follows that S is homocyclic, as claimed. So we will assume on the contrary that $I(S \setminus A) \neq \emptyset$.

Let $S_0 = \Omega_1(S)$ be an $(\langle\alpha\rangle \times T)$-invariant definable subgroup of S. By Corollary 4.6, S_0 is connected. As S/A is inverted by α, $S_0/A \leq C_{S/A}(\alpha)$. Lemma 4.7 and the connectedness of definable normal T-invariant subgroups of S imply that $S_0/A = C_{S/A}(\alpha)$. In particular, S_0 and T satisfy the conditions of Proposition 3.8. Thus, S_0 is an elementary abelian subgroup which is the direct sum of two T-modules E_1 and E_2. As the actions of α and T commute, we may assume that $E_2 = E_1^\alpha$.

So it suffices now to show that $S = S_0$. Let $R = \Omega_2(S)$. For $x \in R$, as $xx^\alpha \in A$, we have $x^2(x^2)^\alpha = (xx^\alpha)^2 = 1$, which implies that x^2 is an involution inverted by α, hence $x^2 \in A$. Therefore R is a T-invariant subgroup of S such that $R/A \leq C_{S/A}(\alpha)$. But from the previous paragraph we know that $C_{S/A}(\alpha) = E/A$. Thus $R = E$ and hence $S = E$. □

4.5. Exponent 4. With the abelian case disposed of, there remains a considerable amount of analysis to be carried out to complete the proof of Theorem 4.1. In essence the point is to show that in exponent 4 the nonabelian examples are as stated, and then to use this to show that there are no others.

For the remainder of the proof of Theorem 4.1 we assume that S is nonabelian, unless the contrary is explicitly noted. We choose S_1 to be a maximal definable proper normal $(\langle\alpha\rangle \times T)$-invariant subgroup of S containing S'. Note that, by Corollary 4.6, S_1 is connected. Lemma 4.4 applies to S and S_1. We note the resulting conclusions, which will be used later:

- S/S_1 is an elementary abelian group.
- $C_{S/S_1}(\alpha) = S/S_1$, or in other words $[\alpha, S] \leq S_1$.
- $\mathrm{rk}(S) = \mathrm{rk}(S_1) + \mathrm{rk}(A)$.

We will show first that $S_1 = [\alpha, S]$ which implies that this group is uniquely determined.

PROPOSITION 4.9. *$S_1 = [\alpha, S]$ is abelian, and α inverts S_1.*

PROOF. Let $X = \{[\alpha, x] : x \in S\}$ be a subset of S_1 inverted by α and of rank $\text{rk}(S) - \text{rk}(A) = \text{rk}(S_1)$. Thus X is generic in S_1. As S_1 is connected, we find $S_1 = \langle X \rangle$ and furthermore $X \cap gX$ is generic in S_1 for any $g \in S_1$. If $g, h, gh \in X$, then α inverts all three elements and hence $[g, h] = 1$. Thus for $g \in X$, $C_{S_1}(g)$ contains the generic subset $X \cap g^{-1}X$, and hence $C_{S_1}(g) = S_1$, $X \subseteq Z(S_1)$. As X is generic we conclude that S_1 is abelian and then $S_1 = [\alpha, S]$.

As S_1 is abelian, the subset of S_1 inverted by α is a subgroup, and as this set contains the generic set X, S_1 must be inverted by α. □

In particular, α centralizes the involutions of S_1, hence:

COROLLARY 4.10. *$\Omega_1(S_1) = A$, and thus $A \triangleleft S$.*

COROLLARY 4.11. *$A \leq Z(S)$.*

PROOF. As A is normal in S, $A \cap Z(S) \neq 1$. But $A \cap Z(S)$ is T-invariant and T acts on A transitively. Therefore, $A \leq Z(S)$. □

Next we prove a special case of Theorem 4.1. The proof makes use of computations very similar to those of §3.

PROPOSITION 4.12. *Let α, S, A and T be as in Theorem 4.1, with S being of exponent 4. Then the conclusions of Theorem 4.1 hold.*

PROOF. We have assumed that S is nonabelian. If $S_1 = A$, then S, T and the actions of T on A and S/A are as described in Proposition 3.8, which forces S to be abelian. Thus $S_1 > A$. As $\Omega_1(S_1) = A$, S_1 is of exponent 4. Moreover, the action of T on A and S_1/A is as described in the assumptions of Proposition 3.8; note that commutation with α (i.e., squaring) gives a T-module isomorphism of S_1/A with A. Hence S_1 is homocyclic of exponent 4.

For $x \in S$, commutation with x gives an endomorphism h_x of S_1. Since $A = \Omega_1(S_1)$ lies in the kernel of h_x, the image is elementary abelian. In other words, $[S, S_1] \leq A$.

We will now see that the map

$$\begin{array}{rcccc} \text{ad}_\alpha & : & S/S_1 & \longrightarrow & S_1/A \\ & & xS_1 & \longmapsto & [\alpha, x]A \end{array}$$

is a well-defined T-module isomorphism.

As $[S, S_1] \leq A$ the map from S to S_1/A induced by commutation with α is a homomorphism: $[\alpha, xy] = [\alpha, y][\alpha, x]^y \equiv [\alpha, x][\alpha, y]$ modulo A. The kernel of this map contains S_1, so we have an induced homomorphism ad_α, which is surjective by Proposition 4.9. As S/S_1 and S_1/A have the same rank as A, the kernel of ad_α is finite; since it is also T-invariant, it is trivial by Corollary 4.6. As α commutes with T, it also respects the T-module structure.

As S/S_1 is elementary abelian and S is of exponent 4, for any $x \in S$ we have $x^2 \in \Omega_1(S_1) = A$. Thus S/A is elementary abelian. Combining the T-module isomorphism given by ad_α with Proposition 3.8, we find that S/A splits as a T-module: $S/A = S_0/A \oplus S_1/A$.

We can now completely coordinatize S in terms of the base field K. Fix $x_0 \in S_0 \setminus A$, and set $x_1 = [\alpha, x_0]$, $x_2 = x_1^2$. Now $A \rtimes T \cong K_+ \rtimes K^\times$ for some algebraically closed field K of characteristic 2. We identify K_+ with $T \cup \{0\}$, and then we identify S as a set with $K_+ \times K_+ \times K_+$; (a, b, c) corresponds to $x_0^a x_1^b x_2^c$, where elements of T act by conjugation and $x_i^0 = 1$.

For $a_1, b_1, c_1, a_2, b_2, c_2 \in K$, we have

$$x_0^{a_1} x_1^{b_1} x_2^{c_1} x_0^{a_2} x_1^{b_2} x_2^{c_2} = (x_0^{a_1} x_0^{a_2})(x_1^{b_1} x_1^{b_2})([x_1^{b_1}, x_0^{a_2}] x_2^{c_1} x_2^{c_2}).$$

If we let $[x_1, x_0^t] = x_2^{g(t)}$ and apply Proposition 3.8 to S_0 and S_1, we get the following formula:

$$(a_1, b_1, c_1)(a_2, b_2, c_2) = (a_1+a_2, b_1+b_2, c_1+c_2+\epsilon\sqrt{a_1a_2}+\sqrt{b_1b_2}+b_1g(b_1^{-1}a_2))$$

where ϵ is either 0 or 1 depending on whether S_0 is elementary abelian or homocyclic respectively. Note that g is an additive map.

The associativity of the group law implies

$$(b + c)g((b + c)^{-1}a) = cg(c^{-1}a) + bg(b^{-1}a) .$$

Letting $a = (b + c)x$ implies

$$bg(b^{-1}cx) = cg(c^{-1}bx),$$

hence,

$$g(yx) = yg(y^{-1}x) .$$

In particular, if $x = y$ then $g(x^2) = xg(1)$. By taking square roots we conclude

$$g(x) = \sqrt{x}g(1).$$

We will show finally that $g(1) = 1$; in other words, that $[x_0, x_1] = x_2$. We have

$$x_0^2 = (x_0^{-2})^\alpha = [(x_0^{-1})^\alpha]^2 = (x_1 x_0^{-1})^2$$

and as S has exponent 4, this becomes $1 = x_1 x_0^{-1} x_1 x_0 = x_1^2 [x_0, x_1]^{-1}$, and our claim follows.

This shows that the structure of S is determined by the structure of S_0 and finishes the proof of the theorem apart from the calculation of α, which may be done directly, using $x_0^\alpha = x_0[x_0, \alpha] = x_0 x_1^{-1}$:

$$\begin{aligned}(x_0^a x_1^b x_2^c)^\alpha &= (x_0^\alpha)^a (x_1^\alpha)^b (x_2^\alpha)^c \\ &= (x_0 x_1^{-1})^a (x_1^{-1})^b x_2^c \\ &= (x_0)^a (x_1 x_2)^a (x_1 x_2)^b x_2^c \\ &= (a,a,a)(0,b,b+c) \\ &= (a, a+b, a+b+c+\sqrt{ab})\end{aligned}$$ □

4.6. The nonabelian case. Having handled the minimal nonabelian case, we now assume that S is a counterexample of minimal rank to the statement of Theorem 4.1.

PROPOSITION 4.13. *$2S_1 \leq Z(S)$; equivalently, $[S, S_1] \leq A$.*

PROOF. The equivalence of the two conditions is straightforward, as $A = \Omega_1(S_1)$ and S_1 is abelian.

We may suppose inductively that Theorem 4.1 holds in S/A, since by the assumption on the action of T on definable quotients of S and Lemma 4.7, the group S/A together with T and the map induced by α satisfy the assumptions of Theorem 4.1. By induction, we have the following three possibilities for S/A:

(i) S/A is abelian:

In this case certainly $[S, S_1] \leq A$.

(ii) S/A is nonabelian and in part (B) of Theorem 4.1, $\epsilon = 1$:

The analysis in Theorem 4.12 shows that $S/A = S_0/A \cdot S_1/A$ where S_0/A is homocyclic of exponent 4, $2S_0/A = 2S_1/A$.

Fix $s_1 \in S_1$ and choose $s_0 \in S_0$ such that $s_0^2 = s_1^2$. Then $[S_1, s_1^2] = 1$ and $[S_0, s_1^2] = [S_0, s_0^2]$. As S_0/A is abelian, $[S_0, s_0] \leq A$ and $[S_0, s_0^2] = 1$; thus s_1^2 commutes with $S_0 S_1 = S$.

(iii) S/A is nonabelian and in part (B) of Theorem 4.1, $\epsilon = 0$:

In this case we will obtain a contradiction.

The analysis in Proposition 4.12 shows that $S/A = S_0/A \cdot S_1/A$ where S_0/A is elementary abelian. Let $B/A = Z(S/A)$. B is a homocyclic group of exponent 4. We will show that every element of $S_0 \setminus A$ inverts B, which yields a contradiction by considering a triple x, y, xy of elements in $S_0 \setminus A$.

The image $\bar{x}$ of x in S/A acts on S_1. Let $X = \{[\bar{x}, s] : s \in S_1\}$. Then $XA/A = B/A$, by inspection in S/A, and the involution $\bar{x}$ inverts the elements of X, as well as the elements of A. Thus $\bar{x}$ inverts $XA = B$, as claimed. □

COROLLARY 4.14. *S/A is abelian.*

PROOF. As in the preceding proof, if S/A is not abelian then by induction we have $S/A = S_0/A \cdot S_1/A$, where S_0/A and S_1/A are abelian. Furthermore, by the preceding proposition, these two factors commute. □

PROPOSITION 4.15 ([**80**]). *If S is not abelian then $S \setminus S_1$ contains an involution.*

PROOF. If S has exponent 4 then our claim follows from Proposition 4.12. Assume that S has exponent greater than 4 and $I(S) = A^\times$. We will show that this implies S is abelian.

$S/Z(S)$ is an elementary abelian 2-group: if x, y are in S, then as S/A is abelian (Corollary 4.14) we compute $[x, y^2] = [x, y][x, y]^y = [x, y]^2 = 1$.

We claim that $S_1 = Z(S)$. If S_1 has exponent 4, then as S has exponent greater than 4, there is an element of S whose square lies in $S_1 - A$. Hence $S_1 \setminus A$ meets $Z(S)$ and as $(S_1/A)^\times$ is a single T-orbit, and $A \leq Z(S)$, we have $S_1 = Z(S)$ in this case. Now suppose the exponent of S_1 is greater than 4 and $Z(S) < S_1$, hence as $Z(S)$ is T-invariant, $Z(S) = 2S_1$. For $x \in S \setminus S_1$ we can solve $x^2 = s^2$ with $s \in S_1$, hence xs^{-1} has order at most 4 and lies in $S \setminus S_1$. Then $(xs^{-1})^2 = s_1^2$ with $s_1 \in 2S_1$, so $xs^{-1}s_1^{-1}$ is an involution, which is a contradiction. Thus in all cases we get $S_1 = Z(S)$.

As $(S/S_1)^\times$ is a single T-orbit, it will suffice to show now that for $x \in S \setminus S_1$ the conjugates of x under T commute with each other.

Fix $x \in S \setminus Z(S)$ and $i \in A^\times$. Define $g : K \to K$ by $[x, x^t] = i^{g(t)}$ following the line of §3. Note that g depends only on xS_1, and therefore g is additive (the action of T on S/S_1 is by multiplication on K_+). Furthermore, $g(1) = 0$. Working modulo S_1, the equations $[\overline{x}, \overline{x}^t] = [\overline{x}^t, \overline{x}] = [\overline{x}, \overline{x}^{t^{-1}}]^t = i^{g(t^{-1})t}$ imply $g(t^{-1}) = t^{-1}g(t)$. Then g^2 satisfies the functional equation of Lemma 3.10 and hence g^2 is linear. As $g(1) = 0$, we conclude $g \equiv 0$, and x commutes with its T-conjugates, as claimed. □

PROPOSITION 4.16. *For every involution $x \in S \setminus S_1$, $[x, x^\alpha] \neq 1$.*

PROOF. Let $x^\alpha = xs$ with $s = [x, \alpha] \in S_1$. If x and x^α commute, then s is an involution, hence $s \in A$. Accordingly there is $s_1 \in S_1$ with $[s_1, \alpha] = s$ and thus $[xs_1, \alpha] = s^2 = 1$, $xs_1 \in A$, $x \in S_1$, a contradiction. □

PROPOSITION 4.17. *The exponent of $Z(S)$ is 2.*

PROOF. Suppose toward a contradiction that the exponent of $Z(S)$ is at least 4. As $Z(S)$ is a T-invariant subgroup of S_1, it is homocyclic.

Let y be an involution in $S \setminus S_1$. Then $y^\alpha = yx$, where $x \in S_1$. As $[S, S_1] \leq A$, $[y, y^\alpha] = [y, x] \in A$. Let $s \in Z(S)$ be such that $s^2 = [y, x]$.

Now $1 = (y^\alpha)^2 = (yx)^2 = y^2x^2[x, y] = x^2s^{-2} = (xs^{-1})^2$. So $xs^{-1} \in A$ and $x \in Z(S)$. Thus $[y, y^\alpha] = 1$ and this contradicts the preceding proposition. □

COROLLARY 4.18. *The exponent of S is 4.*

PROOF. We know $S' \leq A$ (Corollary 4.14), so $S^2 \leq Z(S)$. □

PROOF OF THEOREM 4.1. Proposition 4.8 proves the theorem if S is abelian. The nonabelian case is handled by Corollary 4.18 and Proposition 4.12. □

It is also useful to have the formula for commutation in terms of coordinates. This does not depend on the value of ϵ.

COROLLARY 4.19. *If S is nonabelian, then in the notation of Theorem 4.1 we have the following commutation formula:*

$$[(a,b,c),(a',b',c')] = (0,0,\sqrt{ab'+a'b}) \ .$$

We record some observations about the triple (A, S, α) which are useful in practice.

COROLLARY 4.20. *With the notation and hypotheses of Theorem 4.1, we have the following.*

(1) *If the action of α on S is induced by an inner automorphism, then $S = A$.*
(2) $\Omega_1([\alpha, S]) \leq A$.
(3) *If S is nonabelian, then any $\langle\alpha\rangle \times T$-invariant elementary abelian subgroup of S is contained in A.*
(4) *Any proper nontrivial $(\langle\alpha\rangle \times T)$-invariant subgroup S_0 of S is abelian homocyclic, with $A = \Omega_1(S_0)$.*
(5) *$[\alpha, S]$ is the unique maximal proper $(\langle\alpha\rangle \times T)$-invariant subgroup of S.*
(6) $\alpha \notin S$.

5. A theorem of Baumann

5.1. The theorem. The main result of the present section is Theorem 5.3 following, analogous to a result of Baumann in the finite case [**27**]. For the proof we follow [**175**] closely. For us this plays the role of a strong and abstract form of the Borel-Tits Lemma II 6.7, aimed at putting certain subgroups into parabolic subgroups, as will become clear in Proposition VIII 2.2.

We need a little preliminary terminology.

DEFINITION 5.1.

(1) *Let M be a connected group of finite Morley rank and of even type. We will say that M is of* minimal parabolic type *if $F^*(M) = O_2(M)$ and $M/O_2(M) \simeq \mathrm{SL}_2(K)$ for some algebraically closed field of characteristic 2.*
(2) *For any group G of finite Morley rank, $O^2(G)$ denotes the smallest definable normal subgroup K for which G/K is a 2-group, or in other words, the definable hull of the subgroup generated by $2^\perp$-subgroups.*

There are several equivalent notations one can use here, and for the sake of clarity we now compare them.

LEMMA 5.2. *If M is of minimal parabolic type, then $O^2(M) = M^{(\infty)}$, and this is also the minimal normal subgroup H of M for which $M = H \cdot O_2(M)$.*

PROOF. For a normal subgroup H of M, M/H is a 2-group if and only if $M/(HO_2(M))$ is a 2-group, and as the latter is a quotient of $\mathrm{SL}_2(K)$, this means $M = HO_2(M)$. Thus $O^2(M)$ is the minimal normal subgroup H for which $M = H \cdot O_2(M)$.

Evidently $M^{(\infty)}$ covers $M/O_2(M)$, so $O^2(M) \le M^{(\infty)}$. Conversely, $M^{(\infty)}/O^2(M)$ is a perfect 2-group, hence trivial. □

The present section will be devoted to the proof of the following theorem, parallel to the main result (and proof) of [**175**] in the finite case. For the present we retain a relatively technical formulation, deferring a more intuitive version to a late stage of our classification project, in §VIII 2.

THEOREM 5.3. *Let G be a group of finite Morley rank of even type. Let M be a definable connected subgroup of G of minimal parabolic type. Assume that for S a Sylow 2-subgroup of M:*

(P) *No nontrivial definable connected subgroup of S is normalized by both M and $N_G(S)$.*

Set $Q = O_2(M)$, $L_0 = O^2(M)$, $V = [L_0, Q]$, and $D = C_Q{}^\circ(L_0)$.
Then the following hold:

1. *V is an elementary abelian 2-group central in Q.*
2. *$V/V \cap Z(M)$ is a natural $\mathbb{F}_2(\overline{M})$-module.*
3. *$Q = DV$.*
4. *$S/\Omega_1{}^\circ(Z(S))$ is an elementary abelian 2-group.*
5. *$Z^\circ(Q)$ is an elementary abelian 2-subgroup.*

Observe that condition (P) above is the "bad" case. If $X \le S$ were a nontrivial definable connected subgroup of S normalized by both M and $N_G(S)$, then $N^\circ(X)$ would be a parabolic subgroup in the sense that it contains $N_G{}^\circ(S)$, and M would be a subgroup of this parabolic subgroup. Note that in condition (P), we consider not only connected X, but the full subgroup $N_G(S)$; in both cases, we have taken the *weak* form of the condition, thereby strengthening the result.

5.2. The associated graph. We begin the proof of Theorem 5.3, which will occupy us to the end of this section.

Suppose M, S, and G are as in the statement of the theorem. By Lemma II 5.6 S is a connected group. We let $H = N_G(S)$ and $G_0 = \langle M, H\rangle$. By Lemma I 3.23, G_0 is definable in G, and we may replace G by G_0. So we assume $G = \langle M, H\rangle$. Set $B = M \cap H$. Note that B is a Borel subgroup of M. We keep most of this notation fixed; namely M, G, G_0, and H are as above. We do not need to retain the notation S for a particular Sylow 2-subgroup and in fact we will prefer to vary the Sylow 2-subgroups under consideration.

We consider the bipartite coset graph Γ associated with the free product $M *_B H$ corresponding to the pair of subgroups M and H. The two types of vertices will be the cosets of M and H in G. The edges are the cosets of B in G. An edge Bx has as its vertices the cosets Mx and Hx. The natural action of G on Γ is definable. One may wish to look at §IX 1.2 for a discussion of definability issues.

We will refer to a coset of M as a vertex of *type* M.

The following properties given in [**99**] apply here.

LEMMA 5.4 ([**175**, 1.1]).

(1) Γ *is connected and bipartite.*
(2) G *is edge but not vertex transitive on* Γ.
(3) *The vertex stabilizers in* G *are conjugate to* M *or* H.
(4) *The edge stabilizers in* G *are conjugate to* $M \cap H = B$.
(5) *For* $\lambda \in \Gamma$, *the vertex-stabilizer* G_λ *is transitive on the set of vertices adjacent to* λ.

LEMMA 5.5 ([**175**, 1.2]). *No nontrivial definable connected subgroup of* G *is normal in the stabilizer of two adjacent vertices. The kernel of the action of* G *on* Γ *is a finite subgroup of* $O_2(Z(G))$.

PROOF. If K is a definable subgroup of G which is normal in the stabilizers of two adjacent vertices, then by edge transitivity we may suppose that these vertices are M and H. Then $K \lhd M$ and $K \leq B$, so $K \leq O_2(M)$ and condition (P) applies. Hence K cannot be nontrivial and connected.

In particular, if K is the kernel of the action of G on Γ, then $K \leq O_2(M)$ and $K^\circ = 1$. As G is connected, $K \leq Z(G)$ as well. □

Since we prefer to work with a faithful action, we will factor out the kernel of the action of G on Γ, which will not affect our hypotheses. We will also have to check the validity of our conclusions in the original context, at some point; this is actually done in the proof of Corollary 5.13. Thus we will generally suppose:

$$(*) \qquad G \text{ acts faithfully on } \Gamma.$$

5.3. The module Z_α. What follows is typical of the setup in the amalgam method, though the definitions that follow, of the module Z_α and the parameter b, can be varied considerably with similar results.

NOTATION 5.6. *Let* α, α' *be vertices of* Γ.

(1) $d(\alpha, \alpha')$ *will denote their distance in* Γ.
(2) $G_\alpha^{(1)}$ *is the intersection of the vertex stabilizers* G_β, *where* β *varies over vertices for which* $d(\alpha, \beta) \leq 1$.
(3) $Q_\alpha = O_2(G_\alpha)$.
(4) $Z_\alpha = \langle \Omega_1{}^\circ(Z(S)) : S$ *a Sylow* 2*-subgroup of* $(G_\alpha)\rangle$.
(5) $b_\alpha = \min\{d(\alpha, \beta) : \beta \in \Gamma, Z_\alpha \not\leq G_\beta^{(1)}\}$. *Let* $b = b_\delta$ *with* δ *of type* M.
(6) (α, α') *is a* critical pair *for* Γ *if* α *is of type* M, $d(\alpha, \alpha') = b$, *and* $Z_\alpha \not\leq G_{\alpha'}^{(1)}$.

REMARK 5.7.

(1) *Q_α and Z_α are of interest only when α is of type M; otherwise, Q_α is the unique Sylow 2-subgroup of G_α, and Z_α is $\Omega_1{}^\circ(Z(Q_\alpha))$.*
(2) *For α of type M, Z_α is the critical object of study. We will see momentarily that this is an elementary abelian 2-group which affords a nontrivial representation of $G_\alpha/Q_\alpha \simeq \mathrm{SL}_2(K)$, which will essentially be the natural representation.*
(3) *The parameter b_α is well defined (finite) since Z_α is nontrivial, Γ is connected, and the action of G on Γ is faithful. Furthermore, b_α evidently depends only on the type of α, so b is also well defined. Large values of b lead quickly to implausible (and contradictory) configurations; our main concern will be with the possibilities $b = 2$ and $b = 4$.*
(4) *The definition of a critical pair implies that $Z_\alpha \leq G_{\alpha'}$.*

LEMMA 5.8 ([**175**, 1.3, 3.1]). *Let $\alpha \in \Gamma$ be of type M. Then:*

(1) *$Q_\alpha = O_2(G_\alpha^{(1)})$ is a Sylow 2-subgroup of $G_\alpha^{(1)}$.*
(2) *For S a Sylow 2-subgroup of G_α, $Z_\alpha > \Omega_1{}^\circ(Z(S))$.*
(3) *$Z_\alpha \leq \Omega_1{}^\circ(Z(Q_\alpha))$ and $C_{G_\alpha}(Z_\alpha) = Q_\alpha$.*
(4) *$b \geq 2$ is even.*

In particular, G_α/Q_α acts on Z_α, and the action is nontrivial.

PROOF.

Ad (1) We may suppose that $\alpha = M$. For S a Sylow subgroup of M, the vertex $\beta = N^\circ(S)$ is a neighbor of α and hence $G_\alpha^{(1)} \leq N^\circ(S)$. Hence a Sylow 2-subgroup of $G_\alpha^{(1)}$ is contained in $O_2(M) = Q_\alpha$. On the other hand, M acts transitively on its neighbors, by edge transitivity, so they are of the form $N^\circ(S)$ with S a Sylow 2-subgroup of M. Thus $Q_\alpha \leq G_\alpha^{(1)}$ is a Sylow 2-subgroup of $G_\alpha^{(1)}$.

Ad (2) If $Z_\alpha = \Omega_1{}^\circ(Z(S))$, we contradict Lemma 5.5

Ad (3) Again we suppose $\alpha = M$. Let S be a Sylow 2-subgroup of M. Then $Z^\circ(S) \leq C_M{}^\circ(Q_\alpha) \leq Q_\alpha$ as $F^*(M) = O_2(M)$, so $Z^\circ(S) \leq Z(Q_\alpha)$. Hence $Z_\alpha \leq \Omega_1{}^\circ(Z(Q_\alpha))$ and $C_{G_\alpha}(Z_\alpha) \geq Q_\alpha$. But G_α/Q_α is simple so by point (2), $C_{G_\alpha}(Z_\alpha) = Q_\alpha$.

Ad (4) As $Z_\alpha \leq Q_\alpha \leq G_\alpha^{(1)}$, we have $b \geq 1$. It suffices now to check that b is even, or in other words, taking (α, α') to be a critical pair, we claim that α' is of type M. If this is not the case, then $O_2(G_{\alpha'})$, which is the Sylow 2-subgroup of $G_{\alpha'}$, is contained in $G_{\alpha'}^{(1)}$. Since $Z_\alpha \leq G_{\alpha'}$ by the definition of a critical pair, we have $Z_\alpha \leq O_2(G_{\alpha'} \leq G_{\alpha'}^{(1)}$, a contradiction. □

LEMMA 5.9 ([**175**, 1.4]). *Let (α, α') be a critical pair. Then:*

(1) *$1 \neq [Z_\alpha, Z_{\alpha'}] \leq Z_\alpha \cap Z_{\alpha'}$.*
(2) *$[Z_\alpha, Z_{\alpha'}, Z_{\alpha'}] = 1 = [Z_{\alpha'}, Z_\alpha, Z_\alpha]$.*
(3) *(α', α) is a critical pair.*

PROOF. By the minimality of b we have $Z_\alpha \le G_{\alpha'}$ and thus Z_α normalizes $Z_{\alpha'}$. As this is a critical pair, however, $Z_\alpha \not\le Q_{\alpha'}$ and thus $[Z_\alpha, Z'_\alpha] \ne 1$ (Lemma 5.8). So (1) holds.

In particular, $Z_{\alpha'} \not\le Q_\alpha$ and thus the pair (α', α) is also critical. This gives (3).

Now applying Lemma 5.8 and (1) to both (α, α') and (α', α) we get (2). □

LEMMA 5.10 ([**175**, 2.2]). *Let (α, α') be a critical pair for Γ and set $\overline{G}_\alpha = G_\alpha / Q_\alpha$. Then:*

(1) *$Z_\alpha / Z_\alpha \cap Z(G_\alpha)$ is a natural module for $\overline{G}_\alpha$.*
(2) *$Z_{\alpha'} Q_\alpha$ is a Sylow 2-subgroup of G_α.*
(3) *Setting $S = Z_{\alpha'} Q_\alpha$, $\Omega_1{}^\circ(Z(S)) = [Z_\alpha, Z_{\alpha'}](Z_\alpha \cap Z(G_\alpha))^\circ$*

PROOF. As both (α, α') and (α', α) are critical pairs, we will first suppose that for the pair under consideration we have:

$$\text{(1)} \qquad \mathrm{rk}(Z_{\alpha'} / Z_{\alpha'} \cap Q_\alpha) \ge \mathrm{rk}(Z_\alpha / Z_\alpha \cap Q_{\alpha'}) \ .$$

We may also assume $G_\alpha = M$.

We apply Proposition II 5.33 to $\bar{G}_\alpha$ and its subgroup $T = \bar{Z}_{\alpha'}$, acting on the module $V = Z_\alpha$. With this notation, the hypotheses of the corollary are that Z_α is a faithful module (Lemma 5.8), that $[Z_\alpha, Z_{\alpha'}, Z_{\alpha'}] = 1$ (Lemma 5.9), and that

$$\mathrm{rk}(Z_\alpha / C_{Z_\alpha}(\bar{Z}_{\alpha'})) \le \mathrm{rk}(\bar{Z}_{\alpha'})$$

which decodes to condition (1).

Proposition II 5.33 then yields the following four conditions:

(1) $\mathrm{rk}(\bar{Z}_{\alpha'}) = \mathrm{rk}(Z_\alpha / C_{Z_\alpha}(\bar{Z}_{\alpha'}))$, and thus our results apply equally to (α, α') or (α', α);
(2) $\bar{Z}_{\alpha'}$ is a Sylow 2-subgroup of $\bar{G}_\alpha$, which was our second point;
(3) $Z_\alpha / C_{Z_\alpha}(\bar{G}_\alpha)$ is indeed a natural module;
(4) $C_{Z_\alpha}(Z_{\alpha'}) = [Z_\alpha, Z_{\alpha'}] C_{Z_\alpha}(G_\alpha)$; this is our final claim, taking into account $C^\circ_{Z_\alpha}(Z_{\alpha'}) = \Omega_1{}^\circ(Z(Z_{\alpha'} Q_\alpha))$.

□

5.4. The case $b = 2$. We know that $b \ge 2$ is even. In this subsection we show that the case $b = 2$ leads to the configuration described in the theorem, always assuming that G acts faithfully on Γ. Subsequently we will show that the case $b > 2$ leads to a contradiction.

Since the present case does not lead to a contradiction, but rather to the desired conclusions about the structure of G, at the end of our analysis we will give the argument to pass from the special case of a faithful action to the general case.

We recall the notation involved in analyzing the structure of M:

$$Q = O_2(M), \qquad L_0 = O^2(M),$$
$$V = [L_0, Q], \qquad D = C_Q{}^\circ(L_0).$$

The following lemma will be useful in this subsection as well as in the following.

LEMMA 5.11. *If α, β are vertices of type M in Γ with $d(\alpha,\beta)=2$, then $G_\alpha \cap G_\beta$ contains a unique Sylow subgroup of G_α and G_β.*

PROOF. There is a vertex γ of the form $N^\circ(S)$ adjacent to both α and β, with S a Sylow 2-subgroup of G_α and G_β. If the intersection contained another Sylow 2-subgroup of G_α, then by Lemma II 5.30 the two together would generate G_α. □

PROPOSITION 5.12 ([**175**, 3.2]). *Assume that $b=2$ and that the action of G on Γ is faithful. Then the following hold:*

(1) *$Q = DV$, and V is an elementary abelian 2-group central in Q.*
(2) *For S a Sylow 2-subgroup of M, $S/\Omega_1{}^\circ(Z(S))$ is an elementary abelian group.*
(3) *$Z_\alpha = Z^\circ(Q_\alpha)$. In particular, $Z^\circ(Q)$ is an elementary abelian group.*

PROOF. Let (α,α') be a critical pair for Γ with $\alpha = M$. Then the subgroups Q, L_0, D, V lie in G_α and in particular $Q = Q_\alpha$.

As $b=2$, Lemma 5.11 implies that $G_\alpha \cap G_{\alpha'}$ contains a unique Sylow 2-subgroup S of G_α.

(1) $S = Z_\alpha Q_{\alpha'}$.

By Lemma 5.10 $Z_\alpha Q_{\alpha'}$ is a Sylow 2-subgroup of $G_{\alpha'}$. Since it is contained in G_α as well, it coincides with S. The same applies to $Z_{\alpha'}Q_\alpha$.

(2) $Q_\alpha = Z_\alpha(Q_\alpha \cap Q_{\alpha'})$.

$Z_\alpha \le Q_\alpha \le S = Z_\alpha Q_{\alpha'}$. Thus (2) holds.

Now we introduce some additional notation. We fix $g \in G_\alpha$ so that $G_\alpha = \langle Z_{\alpha'}, Z_{\alpha'}^g\rangle Q_\alpha$, which is possible since $Z_{\alpha'}$ covers a Sylow 2-subgroup of $G_\alpha/Q_\alpha \simeq \mathrm{SL}_2(K)$. Set $F = \langle Z_{\alpha'}, Z_{\alpha'}^g\rangle$.

(3) $Z_\alpha \le [S, Z_\alpha][S^g, Z_\alpha]Z(G_\alpha)$.

We work in the natural module $\bar{Z}_\alpha = Z_\alpha/C_{Z_\alpha}(G_\alpha)$. Then $[S,\bar{Z}_\alpha]$ is a 1-dimensional subspace of $\bar{Z}_\alpha$, as is $[S^g,\bar{Z}_\alpha]$. On the other hand $[S,Z_\alpha] = [Z_{\alpha'}, Z_\alpha] \le Z_{\alpha'} \cap Z_\alpha \le C_{Z_\alpha}(Z_{\alpha'})$, so $[S,\bar{Z}_\alpha] \le C_{\bar{Z}_\alpha}(Z_{\alpha'})$ and $[S,\bar{Z}_\alpha] \cap [S,\bar{Z}_{\alpha'}] \le C_{\bar{Z}_\alpha}(F) = C_{\bar{Z}_\alpha}(G_\alpha) = 1$.

Thus $\bar{Z}_\alpha = [S,\bar{Z}_\alpha] \oplus [S^g,\bar{Z}_\alpha]$ and (3) follows.

(4) $Q_\alpha = Z_\alpha(Q_\alpha \cap Q_{\alpha'} \cap Q_{\alpha'}^g)$.

By (1) $[S, Z_\alpha] \le Q_{\alpha'}$ and $[S^g, Z_\alpha] \le Q_{\alpha'}^g$, so by (3) $Z_\alpha \le (Z_\alpha \cap Q_\alpha \cap Q_{\alpha'})Q_{\alpha'}^g$. Now $Q_\alpha \cap Q_{\alpha'} \le S^g = Z_\alpha Q_{\alpha'}^g \le (Z_\alpha \cap Q_\alpha \cap Q_{\alpha'})Q_{\alpha'}^g$ so $Q_\alpha \cap Q_{\alpha'} \le (Z_\alpha \cap Q_\alpha \cap Q_{\alpha'})(Q_\alpha \cap Q_\alpha' \cap Q_{\alpha'}^g) \le Z_\alpha(Q_\alpha \cap Q_\alpha' \cap Q_{\alpha'}^g)$, and this combines with (2) to give (4).

(5) $Q_\alpha = C_{Q_\alpha}{}^\circ(F)Z_\alpha$.

Evidently $Q_\alpha \cap Q_{\alpha'} \cap Q_{\alpha'}^g \le C_{Q_\alpha}(F)$ and thus (5) follows from (4).

(6) $FZ_\alpha \lhd G_\alpha$.

We have $G_\alpha = FQ_\alpha$. Now $[Q_\alpha, FZ_\alpha] = [C_{Q_\alpha}(F)Z_\alpha, FZ_\alpha] \le Z_\alpha$, and $[F, FZ_\alpha] \le FZ_\alpha$, so $[G_\alpha, FZ_\alpha] \le FZ_\alpha$.

(7) $Z_\alpha \le F$; in particular $F \lhd G_\alpha$ and $L_0 \le F$.

By (3) $Z_\alpha = [Z_{\alpha'}, Z_\alpha][Z^g_{\alpha'}, Z_\alpha]C_{Z_\alpha}(G_\alpha) = [F, Z_\alpha]\Omega_1{}^\circ(Z(G_\alpha))$. Consider the factors. We have $\Omega_1{}^\circ(Z(G_\alpha)) \le Z_{\alpha'} \le F$. Also $[Z_{\alpha'}, Z_\alpha] \le Z_{\alpha'} \le F$ and $[Z^g_{\alpha'}, Z_\alpha] \le Z^g_{\alpha'} \le F$. Thus $[F, Z_\alpha] \le F$.

Thus $Z_\alpha \le F$ and $F = FZ_\alpha \lhd G_\alpha$. As $G_\alpha = FQ_\alpha$, the quotient G_α/F is a 2-group and $L_0 \le F$.

(8) $Q_\alpha = DV$.

We apply (5). $C^\circ_{Q_\alpha}(F) \le D$ by (7). As

$$Z_\alpha/C_{Z_\alpha}(G_\alpha) = [G_\alpha/Q_\alpha, Z_\alpha/C_{Z_\alpha}(G_\alpha)],$$

we have $Z_\alpha \le [L_0, Z_\alpha]C_{Z_\alpha}(G_\alpha) \le VD$. Thus (8) follows.

(9) $\Phi(S) \le D\Omega_1(Z(S))$.

As $[Z_\alpha, Z_{\alpha'}]$ centralizes Q_α and $Z_{\alpha'}$, and $S = Z_{\alpha'}Q_\alpha$, we find $[Z_\alpha, Z_{\alpha'}] \le \Omega_1(Z(S))$. Now $S = Z_{\alpha'}Q_\alpha = Z_{\alpha'}DV$. As $V = [Q_\alpha, L_0] \le [Q_\alpha, F] = [Z_\alpha, F] \le Z_\alpha$, we find $S = DZ_\alpha Z_{\alpha'}$.

Let $\hat{S} = S/D\Omega_1(Z(S))$. Then $\hat{S} = \langle \hat{Z}_\alpha, \hat{Z}_{\alpha'}\rangle$. Furthermore, $[Z_\alpha, Z_{\alpha'}] \le \Omega_1(Z(S))$ and thus $[\hat{Z}_\alpha, \hat{Z}_{\alpha'}] = 1$. Hence $\hat{S}$ is elementary abelian and (9) follows.

(10) $S/\Omega_1{}^\circ(Z(S))$ is elementary abelian.

The groups $[\Phi(S), S]$ and $\mho^1(\Phi(S))$ are contained in D by (9), and are normal in S. Hence they are normalized by $L_0S = G_\alpha$. But as they are characteristic in S, they are normal in $N^\circ(S)$ as well. As these groups are also connected, by our basic assumption (P), this forces them to be trivial. Thus $\Phi(S) \le \Omega_1(Z(S))$, and as $\Phi(S)$ is connected, (10) follows.

(11) $Z^\circ(S)$ is elementary abelian.

$\mho^1(Z^\circ(S))$ is connected, definable, and characteristic in S, and is contained in $C_{Z(Q_\alpha)}(F)$ which is contained in D. Thus $\mho^1(Z^\circ(S))$ is normalized by $L_0S = G_\alpha$ and by $N^\circ(S)$, which by our main assumption (P) implies (11).

(12) $Z_\alpha = Z^\circ(Q_\alpha)$.

By Lemma 5.8(2), $Z_\alpha \le Z(Q_\alpha)$. As $Q_\alpha = C_{Q_\alpha}(F)Z_\alpha$, we have $Z(Q_\alpha) = C_{Z(Q_\alpha)}(F)Z_\alpha$. We have $C_{Z(Q_\alpha)}(F) \le Z(S)$ so $Z^\circ(Q_\alpha) \le Z^\circ(S)Z_\alpha = Z_\alpha$ by (11).

This proves all parts of the theorem. □

COROLLARY 5.13. *Assume that $b = 2$. Then the following hold:*

(1) *$Q = DV$, and V is an elementary abelian 2-group central in Q.*
(2) *For S a Sylow 2-subgroup of M, $S/\Omega_1{}^\circ(Z(S))$ is an elementary abelian group.*
(3) *$Z_\alpha = Z^\circ(Q)$. In particular, $Z^\circ(Q)$ is an elementary abelian group.*

PROOF. This is the same statement as the previous one, without the proviso that G act faithfully on Γ. So let K be the kernel of the action of G on Γ, a finite central 2-group, and let G_1, M_1, S_1 be the quotients of G, M, S by K. Set

$$Q_1 = O_2(M_1), \qquad L_1 = O^2(M_1),$$
$$V_1 = [Q_1, L_1], \qquad D_1 = C_{Q_1}{}^{\circ}(L_1).$$

By the previous proposition our three claims hold for these groups. Note that $Q_1 = Q/K$ and $L_1 = L_0K/K$. Thus $V_1 = VK/K$. We will check also that $D_1 = DK/K$. Certainly $DK/K \leq D_1$. Conversely, let $\hat{D}$ be the preimage of D_1 in G. Then $[\hat{D}, L_0] \leq K$, so by Corollary I 3.29, $[\hat{D}^{\circ}, L_0] = 1$ and $\hat{D}^{\circ} \leq D$. As $\hat{D}^{\circ}$ covers D, $\hat{D} \leq \hat{D}^{\circ}K \leq DK$.

Ad (1) From $Q_1 = D_1V_1$ it follows that $Q \leq DVK$. Since Q is connected we conclude that $Q = DV$.

Ad (2) Let S_0 be the preimage of $\mho^{1\circ}(S_1)$ in S. Then $[S, S_0] \leq K$. As S is connected and K is finite, by Corollary I 3.29 we find $S_0 \leq Z(S)$. Further S_0/K is elementary abelian and $\Phi(S_0{}^{\circ})$ is connected, so $S_0{}^{\circ}$ is elementary abelian. Thus $S_0{}^{\circ} \leq \Omega_1(Z(S))$. Now $\Phi(S) \leq S_0$ and $\Phi(S)$ is connected so $\Phi(S) \leq S_0{}^{\circ} \leq \Omega_1{}^{\circ}(Z(S))$.

Ad (3) Let $Z_{1\alpha}$ be Z_α computed in G_1. It suffices to check that Z_α covers $Z_{1\alpha}$ and that $Z^{\circ}(Q)$ covers $Z^{\circ}(Q_1)$. Let A be the preimage in G of $Z^{\circ}(Q_1)$. Then $[A, Q] \leq K$. As Q is connected, $A \leq Z(Q)$. Thus $Z^{\circ}(Q)$ covers $Z^{\circ}(Q_1)$. The argument for Z_α is similar. □

5.5. The case $b > 2$. In this final subsection we eliminate the case $b > 2$. As b is even, we have $b \geq 4$. The case $b \geq 6$ leads more quickly to a contradiction, while the case $b = 4$ takes a closer analysis.

NOTATION 5.14. *Let (α, α') be a critical pair in Γ. A path of length b from α to α' is fixed, and its vertices are denoted by $(\alpha, \alpha+1, \ldots, \alpha+b)$ or, counting from the other end, $(\alpha'-b, \ldots, \alpha'-1, \alpha')$.*

In the next lemma we discuss the prolongation of a path linking a critical pair "to the left" in a natural way.

LEMMA 5.15 ([**175**, 2.3]). *Let (α, α') be a critical pair in Γ. Then there is a vertex β such that $d(\alpha, \beta) = 2$ and:*

(a) $Z_\beta \not\leq G_{\alpha'}$,

With such a choice of β we have:

(b) $\langle O_2(G_\beta \cap G_\alpha), Z_{\alpha'}\rangle = G_\alpha$.
(c) *$(\beta, \alpha'-2)$ is a critical pair,*
(d) *If $b > 2$ then $[Z_\beta, Z_{\alpha'-2}] \leq Z(G_\alpha)$.*

PROOF. Suppose first that β has been found satisfying (a) with $d(\alpha, \beta) = 2$. Note that $d(\beta, \alpha') = b + 2$ as a consequence of condition (a). Let λ be adjacent to α, β, and let $S = O_2(G_\lambda) = O_2(G_\alpha \cap G_\beta)$ by Lemma 5.11. As $\lambda \neq \alpha + 1$, S is distinct from $O_2(G_{\alpha+1}) = Z_{\alpha'}Q_\alpha$. Thus $\langle S, Z_{\alpha'}\rangle$

covers G_α/Q_α and hence $\langle S, Z_{\alpha'}\rangle = G_\alpha$. This is condition (b). For (c), note that $d(\beta, \alpha'-2) \le b$ while $Z_\beta \not\le G^{(1)}_{\alpha'-2}$ as otherwise we would find $Z_\beta \le O_2(G_{\alpha'-1}) \le G_{\alpha'}$. Thus $(\beta, \alpha'-2)$ is a critical pair. Thus (b) and (c) both hold.

If $b > 2$ then $[Z_{\alpha'-2}, Z_{\alpha'}] = 1$. As $(\beta, \alpha'-2)$ is a critical pair $[Z_\beta, Z_{\alpha'-2}] \le Z_{\alpha'-2} \cap Z_\beta$. Thus the group $[Z_\beta, Z_{\alpha'-2}]$ is centralized by $Z_{\alpha'}$ and also by S as $S = Z_{\alpha'-2}Q_\beta$. Now (b) implies (d).

Accordingly, we turn our attention to condition (a). Let $\lambda \ne \alpha+1$ be any other neighbor of α. Then as seen above, while checking (b), we have $\langle O_2(G_\lambda), Z_{\alpha'}\rangle = G_\alpha$. We will find β adjacent to λ so that $Z_\beta \not\le G_{\alpha'}$. Then as $\beta \ne \alpha$, we have $d(\alpha, \beta) = 2$.

Suppose toward a contradiction that $Z_\beta \le G_{\alpha'}$ for every neighbor β of λ, so that in fact $Z_\beta \le G_{\alpha'} \cap G_{\alpha'-2}$ for each such β. Let $T = O_2(G_{\alpha'} \cap G_{\alpha'-2}) = O_2(G_{\alpha'-1}) = Z_\alpha Q_{\alpha'}$ and set $V_\lambda = \langle Z_\beta : d(\lambda, \beta) = 1\rangle$. Then our hypothesis amounts to $V_\lambda \le T$. As $T = Z_\alpha Q_{\alpha'}$ this yields $[V_\lambda, Z_{\alpha'}] \le [Z_\alpha, Z_{\alpha'}] \le Z_\alpha \le V_\lambda$, and hence V_λ is normalized by $Z_{\alpha'}$.

As V_λ is normal in G_λ and $\langle O_2(G_\lambda), Z_{\alpha'}\rangle = G_\alpha$, we find that V_λ is normalized by G_α as well. This contradicts Lemma 5.5. □

As a matter of notation, when we apply the foregoing lemma, we will call the vertex β which is selected "$\alpha - 2$". Formally, this has no special meaning, but it serves as an aide-mémoire.

PROPOSITION 5.16 ([**175**, 2.4]). *$b < 6$.*

PROOF. Suppose toward a contradiction that $b \ge 6$. Fix a vertex $\beta = \alpha - 2$ as afforded by Lemma 5.15, and let a common neighbor of α and $\alpha - 2$ be called $\alpha - 1$. We consider the following groups:

$$V_\alpha = \langle Z_{\alpha-2}^{G_\alpha}\rangle Z_\alpha, \qquad V_{\alpha-2} = \langle Z_\alpha^{G_{\alpha-2}}\rangle Z_{\alpha-2}\ .$$

Then $V_\alpha \lhd G_\alpha$ and $V_{\alpha-2} \lhd G_{\alpha-2}$. As $b > 2$ we have $V_\alpha \le Q_\alpha$ and $V_{\alpha-2} \le Q_{\alpha-2}$.

(1) $[Q_\alpha, V_\alpha] \le Z(G_\alpha)$.

It suffices to check that $[Q_\alpha, Z_{\alpha-2}] \le Z(G_\alpha)$. As $Z_{\alpha'-2}Q_{\alpha-2}$ is a Sylow 2-subgroup of G_α, we have $[Q_\alpha, Z_{\alpha-2}] \le [Z_{\alpha'-2}Q_{\alpha-2}, Z_{\alpha-2}] = [Z_{\alpha'-2}, Z_{\alpha-2}]$ and condition (d) of Lemma 5.15 applies.

The idea now is to "reflect" the "path" $(\alpha - 2, \dots, \alpha')$ around $\alpha - 2$ and to consider the view from within the resulting long "path".

As $(\alpha - 2, \alpha' - 2)$ is a critical path, $Z_{\alpha'-2}$ covers a Sylow 2-subgroup of $G_{\alpha-2}/Q_{\alpha-2}$ and thus we may choose an element $t \in G_{\alpha-2}$ such that $G_{\alpha-2} = \langle Z_{\alpha'-2}, Z^t_{\alpha'-2}\rangle Q_{\alpha-2}$. We consider the sequence of vertices

$$((\alpha'-2)^t, (\alpha'-4)^t, \dots, \alpha^t, \alpha-2, \alpha, \dots, \alpha'-2)$$

in which $\alpha - 2$ is the central point, and only the even terms, as indicated, play any real role.

(2) $V_\alpha \le G_{(\alpha'-2)^t}$.

We check first that

$$V_\alpha \le G_{(\alpha'-6)^t} .$$

For $g \in G_\alpha$ we have $d(\alpha, (\alpha-2)^g) \le 2$, and $d(\alpha-2, (\alpha'-6)^t) = d(\alpha-2, \alpha'-6) \le b-4$. Thus $d((\alpha-2)^g, (\alpha'-6)^t) \le b$. So $Z_{(\alpha-2)^g} \le G_{(\alpha'-6)^t}$ and $V_\alpha \le G_{(\alpha'-6)^t}$.

Now suppose toward a contradiction that $V_\alpha \not\le G_{(\alpha'-2)^t}$. Then $V_\alpha \not\le Q_{(\alpha'-4)^t}$. Thus we may fix i, $i = 4$ or 6, so that $V_\alpha \le G_{(\alpha'-i)^t}$ while $V_\alpha \not\le Q_{(\alpha'-i)^t}$. The two possibilities can be analyzed to some extent simultaneously.

We fix $\beta \in (\alpha-2)^{G_\alpha} \cup \{\alpha\}$ such that $Z_\beta \not\le Q_{(\alpha'-i)^t}$, and we take $\beta = \alpha$ if possible. Set $R = [Z_\beta, Z_{(\alpha'-i)^t}]$.

As $Z_\beta \le G_{(\alpha'-i)^t}$, we have $R \le Z_{(\alpha'-i)^t}$. As $d((\alpha'-i)^t, (\alpha'-2)^t) \le 4 < b$ we have $[R, Z_{(\alpha'-2)^t}] \le [Z_{(\alpha'-i)^t}, Z_{(\alpha'-2)^t}] = 1$. Thus R centralizes $Z_{(\alpha'-2)^t}$.

Now $d((\alpha'-i)^t, \alpha) \le (b-i)+4 \le b$ so $Z_{(\alpha'-i)^t} \le G_\alpha$ and thus $Z_{(\alpha'-i)^t} \le O_2(G_{\alpha-1})$. In particular, $R \le O_2(G_{\alpha-1})$.

We now consider two cases separately:

(Case 1) $\qquad Z_{(\alpha'-i)^t} \le Q_\alpha.$

Then $R = [Z_\beta, Z_{(\alpha'-i)^t}] \le [V_\alpha, Q_\alpha] \le Z(G_\alpha)$ by (1). By the choice of t, $G_{\alpha-2} = \langle G_{\alpha-2} \cap G_\alpha, Z_{(\alpha'-2)^t}\rangle$ and thus $R \le Z(G_{\alpha-2})$ as well. As $\beta \in (\alpha-2)^{G_\alpha} \cup \{\alpha\}$, we have $R \le Z(G_\beta)$.

On the other hand, we have $Z_{(\alpha'-i)^t} \le Q_\alpha \le G_\beta$ acting nontrivially on Z_β. As $\bar{Z}_\beta = Z_\beta / C_{Z_\beta}(G_\beta)$ is a natural module for $\bar{G}_\beta = G_\beta/Q_\beta$, the commutator R is nontrivial in $\bar{Z}_\beta$, and thus $R \not\le Z(G_\beta)$, a contradiction.

Now suppose:

(Case 2) $\qquad Z_{(\alpha'-i)^t} \not\le Q_\alpha.$

As $d((\alpha'-i)^t, \alpha) \le (b-i)+4$, we conclude that $i = 4$ and that $(\alpha, (\alpha'-i)^t)$ is a critical pair. Hence $\beta = \alpha$.

Now $R = [Z_\alpha, Z_{(\alpha'-i)^t}] \le [V_{\alpha-2}, Q_{\alpha-2}] \le Z(G_{\alpha-2})$ by (1).

We have $G_\alpha \le \langle O_2(G_{\alpha-1}), Z_{\alpha'}\rangle$ and hence $G_{\alpha^t} \le \langle O_2(G_{\alpha-1})^t, Z_{(\alpha')^t}\rangle$; but R centralizes $G_{\alpha-2}$, hence $O_2(G_{\alpha-1})^t$, and $d((\alpha'-4)^t, \alpha'^t) = 4 < b$, so $R \le Z_{(\alpha'-4)^t} \le Q_{\alpha'^t}$ and $[R, Z_{\alpha'^t}] = 1$. Thus R centralizes G_{α^t} and as t centralizes R, we have $R \le Z(G_\alpha)$ as well; but $\bar{Z}_\alpha = Z_\alpha / C_{Z_\alpha}(G_\alpha)$ is a natural module for G_α/Q_α, and $\bar{R} = [\bar{Z}_\alpha, Z_{(\alpha'-i)^t}]$ with $Z_{(\alpha'-i)^t}$ acting nontrivially, a contradiction.

(3) V_α, $Z_{\alpha-2}Z_\alpha$, and $Q_\alpha \cap Q_{\alpha-2}$ are normal in $G_{\alpha-2}$.

$Z_{\alpha-2}Q_{(\alpha'-2)^t}$ is a Sylow 2-subgroup of $G_{(\alpha'-2)^t}$ as $(\alpha-2, \alpha'-2)$ is a critical pair; but this is a subgroup of $V_\alpha Q_{(\alpha'-2)^t}$ which is a 2-group by point (2). Hence $V_\alpha \le Z_{\alpha-2}Q_{(\alpha'-2)^t}$.

$G_{\alpha-2}$ is generated by $G_{\alpha-2} \cap G_\alpha$ and $Z_{(\alpha'-2)^t}$.

Now G_α normalizes V_α and $[V_\alpha, Z_{(\alpha'-2)^t}] \le [Z_{\alpha-2}Q_{(\alpha'-2)^t}, Z_{(\alpha'-2)^t}] \le Z_{\alpha-2} \le V_\alpha$. Thus V_α is normal in $G_{\alpha-2}$.

Again, $G_{\alpha-2} \cap G_\alpha$ normalizes $Z_{\alpha-2}Z_\alpha$ and by the calculation of the previous paragraph $[Z_{(\alpha'-2)^t}, Z_\alpha Z_{\alpha-2}] \leq [Z_{(\alpha'-2)^t}, V_\alpha] \leq Z_{\alpha-2}$ so $Z_{(\alpha'-2)^t}$ also normalizes $Z_{\alpha-2}Z_\alpha$. Thus $Z_{\alpha-2}Z_\alpha$ is normal in $G_{\alpha-2}$.

Finally, $Q_{\alpha-2} \cap Q_\alpha = C_{G_{\alpha-2}}(Z_\alpha Z_{\alpha-2})$.

(4) $Q_\alpha \cap Q_{\alpha-2} \lhd G_\alpha$.

Let X be the normal closure of $Q_{\alpha-2} \cap Q_\alpha$ in G_α. Then $X \leq Q_\alpha$ and our claim is that $X \leq Q_{\alpha-2}$.

Let $Y = [V_\alpha, Q_\alpha \cap Q_{\alpha-2}]$. By (1) Y is central in G_α and thus $Y = [V_\alpha, X]$ as well.

Since Y is central in G_α it centralizes a Sylow 2-subgroup of $G_{\alpha-2}$. But Y is normal in $G_{\alpha-2}$ by (3), so Y is central in $G_{\alpha-2}$. Thus $[Z_{\alpha-2}, X] \leq [V_\alpha, X] \leq Z(G_{\alpha-2})$. As $\bar{Z}_{\alpha=2} = Z_{\alpha-2}/C_{Z_{\alpha-2}}(G_{\alpha-2})$ is a natural module and $[\bar{Z}_{\alpha-2}, X] = 0$, we find $X \leq Q_{\alpha-2}$ as claimed.

The final contradiction is derived as follows. As $\alpha - 1$ is conjugate under G_α to $\alpha + 1$, $(\alpha - 2)$ is conjugate under G_α to a neighbor λ of $\alpha + 1$. Suppose $\lambda = (\alpha - 2)^g$ with $g \in G_\alpha$. As $d(\lambda, \alpha' - 2) < b$, we have $Z_{\alpha'-2} \leq Q_\alpha \cap Q_\lambda = (Q_\alpha \cap Q_{\alpha-2})^g = Q_\alpha \cap Q_{\alpha-2}$ by (4). Then $[Z_{\alpha'-2}, Z_{\alpha-2}] = 1$, while $(\alpha - 2, \alpha' - 2)$ is a critical pair, a contradiction. □

PROPOSITION 5.17 ([**175**, 3.3]). $b \neq 4$.

PROOF. Suppose toward a contradiction that $b = 4$. Fix a critical pair (α, α'). Choose $\alpha - 2$, and then $\alpha - 4$, in accordance with Lemma 5.15 so that $d(\alpha, \alpha - 2) = d(\alpha - 2, \alpha - 4) = 2$ and $(\alpha - 4, \alpha)$ and $(\alpha - 2, \alpha + 2)$ are critical pairs.

(1) $Z_\alpha = (Z_\alpha \cap Z_{\alpha+2})[Z_\alpha, Z_{\alpha-4}]$.

This reflects the fact that the module $\bar{Z}_\alpha = Z_\alpha/C_{Z_\alpha}(G_\alpha)$ is a natural module. $Z_{\alpha-4}$ covers a Sylow 2-subgroup of G_α/Q_α so $[Z_{\alpha-4}, \bar{Z}_\alpha]$ is a 1-dimensional subspace of this module, and similarly $[Z_{\alpha'}, \bar{Z}_\alpha]$ is a 1-dimensional subspace. As $Z_{\alpha'}$ and $Z_{\alpha-4}$ generate G_α modulo Q_α, by Lemma 5.15(b), we find $\bar{Z}_\alpha = [Z_{\alpha-4}, \bar{Z}_\alpha] \oplus [Z_{\alpha'}, \bar{Z}_\alpha]$.

As the commutator $[Z_{\alpha'}, Z_\alpha] \leq Z_{\alpha'} \cap Z_\alpha$ centralizes $Z_{\alpha'}Q_\alpha = O_2(G_{\alpha+1})$, we have $[Z_{\alpha'}, Z_\alpha] \leq Z_\alpha \cap Z_{\alpha+2}$, and (1) follows.

We introduce the following additional notation:

$$U = Z_\alpha Z_{\alpha-2} Z_{\alpha+2}; \qquad \tilde{D} = Q_{\alpha-4} \cap Q_{\alpha-2} \cap Q_\alpha \cap Q_{\alpha+2} \cap Q_{\alpha'}\ .$$

We observe that U is a subgroup with $U' = [Z_{\alpha-2}, Z_{\alpha+2}] \neq 1$, as Z_α centralizes all three factors and $[Z_{\alpha-2}, Z_{\alpha+2}] \leq Z_{\alpha-2} \cap Z_{\alpha+2}$, since $b = 4$.

(2) $Q_\alpha = \tilde{D} * U$.

This is similar to the proof of point (4) in Proposition 5.12. As $Z_{\alpha-2} \leq Q_\alpha \leq O_2(G_{\alpha+1}) = Z_{\alpha-2}Q_{\alpha+2}$ we find $Q_\alpha = Z_{\alpha-2}(Q_\alpha \cap Q_{\alpha+2})$. Similarly, using successively $Q_\alpha \cap Q_{\alpha+2} \leq O_2(G_{\alpha+3}) = Z_\alpha Q_{\alpha'}$ and $Q_\alpha \cap Q_{\alpha+2} \cap Q_{\alpha'} \leq O_2(G_{\alpha-1}) = Z_{\alpha+2}Q_{\alpha-2}$ we find $Q_\alpha \leq U \cdot (Q_{\alpha-2} \cap Q_\alpha \cap Q_{\alpha+2} \cap Q_{\alpha'})$.

For the final step, $Q_{\alpha-2} \cap Q_\alpha \cap Q_{\alpha+2} \cap Q_{\alpha'} \le O_2(G_{\alpha-3}) = Z_\alpha Q_{\alpha-4} = (Z_\alpha \cap Z_{\alpha+2})Q_{\alpha-4}$, using (1), and as $Z_\alpha \cap Z_{\alpha+2} \le Q_{\alpha-2} \cap Q_\alpha \cap Q_{\alpha+2} \cap Q_{\alpha'}$, we find $Q_\alpha = U \cdot \tilde{D}$, and the two factors evidently commute.

(3) $UZ(G_\alpha) \lhd G_\alpha$.

Set

$$F = \langle Z_{\alpha-4}, Z_{\alpha'} \rangle \ .$$

By Lemma 5.15(b), $G_\alpha = FQ_\alpha$. By (2) $[U, Q_\alpha] \le U$ so it remains to be seen that $[F, U] \le UZ(G_\alpha)$.

Let $U_0 = U[U, F]$. Then $U_0 = U(U_0 \cap \tilde{D})$. Now $U_0 \cap \tilde{D}$ commutes with $\tilde{D}$, since U_0 does, and with U, since $\tilde{D}$ does, and thus with Q_α. Furthermore, $U_0 \cap \tilde{D}$ commutes with F, since $\tilde{D}$ does. So $U_0 \cap \tilde{D} \le Z(G_\alpha)$, and $[F, U] \le U_0 \le UZ(G_\alpha)$, as desired. Thus (3) holds.

(4) $\bar{U} = UZ^\circ(G_\alpha)/Z_\alpha Z^\circ(G_\alpha)$ *is a nontrivial* G_α/Q_α*-module.*

Point (3) implies that G_α/Q_α acts on $\bar{U}$. It remains to show that this action is nontrivial.

If $(\alpha-1)^g = \alpha+1$, we will show that g acts nontrivially. Let $\lambda = (\alpha-2)^g$. Then $d(\lambda, \alpha+2) \le 2$ and hence $[Z_\lambda, Z_{\alpha+2}] = 1$. As $[Z_{\alpha-2}, Z_{\alpha+2}] \ne 1$, it follows easily that the action of g on $\bar{U}$ is nontrivial.

(5) $Z_{\alpha'}$ acts quadratically on $\bar{U}$.

$\bar{U} = \bar{Z}_{\alpha-2}\bar{Z}_{\alpha+2}$ where the bar refers to factoring out $Z_\alpha Z^\circ(G_\alpha)$. As $Z_{\alpha'}$ centralizes $Z_{\alpha+2}$, it suffices to consider the action on $\bar{Z}_{\alpha-2}$.

Now $[Z_{\alpha-2}, Z_{\alpha'}] \le Q_{\alpha+2}$. As $Q_{\alpha+2} \le O_2(G_{\alpha'-1}) = Z_\alpha Q_{\alpha'}$, we have $[Q_{\alpha+2}, Z_{\alpha'}] \le [Z_\alpha, Z_{\alpha'}] \le Z_\alpha$. Thus $[Z_{\alpha-2}, Z_{\alpha'}, Z_{\alpha'}] \le Z_\alpha$ and (5) follows.

(6) $\bar{U}$ is a natural module for $F/C_F(\bar{U})$.

Here $F = \langle Z_{\alpha-4}, Z_{\alpha'} \rangle$ as in (3). As Q_α acts trivially on $\bar{U}$ and $FQ_\alpha = G_\alpha$, $F/C_F(\bar{U}) \simeq G_\alpha/Q_\alpha$ is of type SL_2.

We apply Proposition II 5.33 with $G = F/C_F(\bar{U})$ and $T = Z_{\alpha'}$. In view of point (5), we need only check that $\mathrm{rk}(\bar{U}/C_{\bar{U}}(Z_{\alpha'})) \le \mathrm{rk}(Z_{\alpha'})$, which is clear, to conclude that $\bar{U}/C_{\bar{U}}(F)$ is a natural module. But $\mathrm{rk}(\bar{U}) \le 2f$ where f is the rank of the base field, so $\bar{U}$ must itself be a natural module.

(7) $Z^\circ(G_\alpha) \le Z_\alpha$.

We know $Z^\circ(G_\alpha) \le Q_\alpha$. We need to show that $Z^\circ(G_\alpha)$ is elementary abelian. Let $S = O_2(G_{\alpha+1}) = Z_{\alpha'}Q_\alpha$. It suffices to show that $Z^\circ(S)$ is elementary abelian.

$Z(S) \le Q_\alpha$ and $U \cap \tilde{D} \le Z(G_\alpha)$ so $Z(S) = [Z(S) \cap U] * [Z(S) \cap \tilde{D}]$. By (6) $\mathrm{rk}(\bar{U}) = \mathrm{rk}(\bar{Z}_{\alpha-2}) + \mathrm{rk}(\bar{Z}_{\alpha+2})$, so $[U \cap Z(G_\alpha)]^\circ \le Z_\alpha$. Hence $\Phi(Z^\circ(S)) \le Z(\tilde{D}) \le Z(G_\alpha)$. Our original hypothesis (P) forces $\Phi(Z^\circ(S)) = 1$ and (7) follows.

After these preparations we reach a contradiction as follows. By (7) the action of G_α on $\bar{U}$ is induced by an action on U. By (6) this action is transitive on $(\bar{U})^\times$. If $u \in U \setminus Z_\alpha$ is an involution, then the class uZ_α

consists entirely of involutions. By transitivity of the action, U is elementary abelian; but $U' \neq 1$.

This contradiction shows that $b \neq 4$. □

PROOF OF THEOREM 5.3. Lemma 5.8, Corollary 5.13, and Propositions 5.16 and 5.17 yield the result. □

6. Generalized n-gons

In the present section we take up the theory of generalized n-gons insofar as it relates to the identification of simple groups of finite Morley rank. In the next section we will pass to the more general context of Tits' *buildings*, and the related topic of (B, N)-pairs.

Tits' work on the classification of buildings provides a powerful tool for the identification of simple groups. Buildings are very symmetrical combinatorial geometries which are intimately connected with the theory of Coxeter groups, and every algebraic group acts on an associated building which encapsulates the relations among its parabolic subgroups. So a natural strategy in group theoretic classification problems is to reconstruct the building from the group theoretic data, and then use Tits' classification to conclude.

Tits' classification includes, and is modeled on, the classification of projective spaces via coordinatization. As is well known, this works well in dimension three or greater, where Desargues' theorem can be proved from the combinatorial axioms, and is meaningless in dimension 1, where the geometry is just a set of points. In dimension two the matter is more subtle, and the semi-classical notion of a "Moufang plane" has been generalized by Tits to "Moufang buildings" generally. He in fact classifies the Moufang buildings with a finite Coxeter group of rank at least two, and along the way proves the fundamental fact that buildings with a finite Coxeter group of rank at least three are always Moufang (a result of considerable depth, proved in [**183**] and again in [**193**]).

In the case of rank two, Tits' buildings are essentially the same thing as *generalized polygons* (up to duality), and can be described by axioms closely parallel to the case of projective planes (which are generalized triangles in Tits' sense). This case is of considerable importance to us, as these buildings correspond to groups of Lie rank two, which are parallel to the *quasithin* groups in the finite simple case, a class which tends to require special methods for classification, being a bit too small for the generic methods, and at the same time having some nontrivial internal structure that requires analysis in its own right.

We will eventually show, using the "amalgam method", that our quasithin groups are associated with Moufang generalized polygons, so that the known classification theorems apply. Note, however, that this method of identifying groups, while it ends in some sense rapidly, in fact invokes a very large body of material, found in [**185**]. As has been remarked in the finite case, by the time one gets within reach of such general classification results one has accumulated a very considerable body of additional information, from which it may be possible in principle to obtain generators and relations for the original group by a direct analysis. On the other hand, we are interested only in the classification of Moufang polygons of finite Morley rank, and the classification in this special case would not be so very long, though it would necessarily follow the general lines of the full classification.

6.1. Definitions. A (rank 2) *combinatorial geometry* is a structure $(P, L; I)$ where P and L are disjoint sets and $I \subseteq P \times L$ is called the *incidence relation.* As a rule, the elements of P are called *points*, and the elements of L are called *lines* or *blocks.* The associated *incidence graph* is the graph on the set of vertices $P \cup L$ in which edges represent incident pairs. There is at least one difference between a combinatorial geometry and its incidence graph: interchanging P and L gives a new geometry, the *dual* of the original, but leaves the incidence graph unaffected.

A *generalized n-gon* may be defined as a combinatorial geometry for which the associated incidence graph is connected of diameter n and girth $2n$ and every vertex has at least three neighbors. Here the *girth* is the length of the shortest cycle. An *ordinary n-gon* is a combinatorial geometry whose associated incidence graph is a circuit of length $2n$. It may be observed that such a geometry does indeed have the essential properties of an n-gon, consisting of n distinct vertices connected cyclically by n distinct lines.

In the case $n = 3$, the axioms for a generalized triangle decode to give the axioms for projective planes. As the incidence graph is bipartite, if the diameter is three, then two distinct elements of the same kind lie at distance two precisely, or in geometrical terms two points lie on a line, two lines contain a common point. The absence of cycles of length 4 signifies that the line or point in question is unique. (We should also require a minimum of three points per line to have a projective plane in the usual sense—otherwise what we have is an *ordinary* triangle.)

The axioms for generalized n-gons may be decoded similarly, and may be expressed in a variety of ways (cf. [**140**, §1.2]).

As is well known in the case of generalized triangles (projective planes), there is an extensive coordinatization theory, and the class of projective planes associated with division rings is characterized by the *Desargues axiom* concerning triangles in perspective, while commutativity corresponds to the *Pappus* axiom; cf. [**114**] for a thorough account of the coordinatization theory in general, or [**140**] for a review of the essentials special cases relevant here.

A broader class of projective planes is relevant here: the so-called *Moufang planes.* These turn out to be the planes coordinatized by alternative division rings; they may be characterized in geometrical terms as those satisfying the *Little Desargues* theorem, and this is the essential content of [**141**].

(See [**185**, p. 176] for a more precise account of the history, worked out by H. van Maldeghem.)

A third characterization is more important here, because it furnishes a notion useful for generalized n-gons for any n. This goes as follows.

6.2. Moufang generalized n-gons. If Γ is a generalized n-gon, and v a point or line of Γ, then $\Delta(v)$ (or $\Delta_\Gamma(v)$, if we need to be more explicit) denotes the set of neighbors of v in the incidence graph of Γ. More generally let $\Delta(X) = \bigcup_{v \in X} \Delta(v)$.

DEFINITION 6.1. *Let $\gamma = (x_0, \ldots, x_n)$ be a path of length n in the generalized n-gon Γ, with $n \geq 3$.*

(1) *An automorphism of Γ is called a* γ-elation *if it fixes the neighbors of each x_i for $1 \leq i \leq n-1$ pointwise (in particular, it fixes γ pointwise).*
(2) *U_γ denotes the group of γ-elations; this group is called the* root group *associated with γ.*
(3) *The generalized n-gon Γ is said to be* Moufang *if for every such path γ, the root group U_γ acts transitively on $\Delta(x_n) \setminus \{x_{n-1}\}$ (hence also, by reversing the path, a similar condition applies at x_0).*

The transitivity condition (3) may be expressed in another way. An *apartment* in a generalized n-gon Γ is simply an ordinary n-gon contained in Γ. It can be shown that every path of length $n+1$ lies in a unique apartment, and hence the transitivity condition can be expressed also as follows: if $\gamma' = (x_0, \ldots, x_n)$ and $\gamma = (x_1, \ldots, x_{n-1})$, then $U(\gamma)$ acts transitively on the set of apartments containing $\gamma' = (x_0, x_1, \ldots, x_n)$.

We will see later that in using the amalgam method (Chapter IX) we arrive naturally at the Moufang condition (what is much harder to achieve in that context is the verification of the basic axioms for generalized n-gons), with the relevant root group actually being a root group in a copy of the group SL_2.

The monumental classification of all Moufang generalized n-gons is given in [**185**], with the principle difficulties arising for $n = 4$.

From this point on, to lighten the terminology, we use the term "n-gon" or "polygon" in the sense of "generalized n-gon". Since ordinary n-gons are not n-gons in this sense, it is preferable to refer to them as "apartments".

6.3. Moufang n-gons. Our objective is the following.

THEOREM 6.2 ([**130**]). *If Γ is an infinite Moufang polygon of finite Morley rank, then Γ is either the projective plane, the symplectic quadrangle, or the split Cayley hexagon over an algebraically closed field.*

We describe these three examples (or rather, specific representations of them) briefly. For the projective plane, as usual one takes a 3-dimensional vector space V, and one considers the subspaces of V of dimensions one and two respectively as the points and lines of a geometry, with incidence being containment. For the symplectic quadrangle one begins with a 4-dimensional space V equipped with a nondegenerate symplectic form ($(x, x) = 0$ identically); points are again 1-dimensional subspaces and lines are now totally isotropic planes, that is, subspaces of dimension two on which the induced form is trivial. The axioms for a quadrangle (4-gon) are readily verified.

The split Cayley hexagon is more subtle and less familiar. One begins with an 8-dimensional space V carrying a nondegenerate quadratic form q of Witt index 4. The associated *quadric hypersurface* defined by $q(v) = 0$ can be viewed as living in the associated 7-dimensional projective space $P(V)$, and by assumption contains 3-dimensional linear subspaces. One may then introduce a certain *trilinear* form $T : V \times V \times V \to K$ (with K the base field) related to the quadratic form q, in such a way that for fixed $v \in V - \{0\}$ the set of those $w \in V$ for which the form $T(v, w, x)$ vanishes identically in x is a 4-dimensional totally isotropic q-space, or in other words, represents a 3-dimensional projective subspace of the associated quadric. The same condition—$T(v, w, x)$ vanishing—provides an incidence relation which allows us to consider the points of $P(V)$ as representing both points and lines in an associated combinatorial geometry, which turns out to be a generalized hexagon. For details, see [**140**, §2.4.6]. The description in [**185**, Example 15.20, $E = F$] is less geometrical.

The projective planes, symplectic quadrangles, and split Cayley hexagons over arbitrary (but commutative) fields are referred to collectively as *Pappian polygons* in [**140**]. These Pappian n-gons all have the property that the root groups can be identified naturally with the base field, and, in particular, in the finite Morley rank context the root groups all have the same Morley rank. The latter property will in fact already be known before the classification theorem is applied in Chapter IX, and could be used to reduce considerably the number of cases which need to be considered in the proof.

Other simplifications are also possible in the specific context of Chapter IX. We will also know a priori that the value of n is 3, 4, or 6; this information is obtained prior to the actual construction of the associated n-gon we consider, and is used in that construction. Furthermore, all of our polygons will be interpreted in groups, and in these groups the field structure on the root groups will already be visible, so issues of coordinatization or interpretability also trivialize.

We now discuss the proof of Theorem 6.2. One begins with the full classification of Moufang n-gons. Moufang n-gons exist only for $n = 3$, 4, 6, or 8 by a theorem of Tits; this can be proved by passing to the universal cover of the incidence graph, which is a tree, and studying the situation there ([**192**], [**140**, 5.3.3]).

In general, it can be shown directly that the root groups U_γ, together with their actions on the polygon, are uniformly interpretable in the polygon itself, from the parameter γ; cf. [**130**, 3.2]. The group generated by all root subgroups is called the *little projective group*. For each d, the subset G_d of elements expressible as a product of at most d elations, together with its action on the polygon, is interpretable in the polygon.

Using the classification, one proceeds to consider Moufang projective planes, quadrangles, hexagons, and octagons individually.

For the case of *projective planes* one uses the coordinatization by alternative division rings [**114**], and Proposition I 4.27. The relationship between the older geometric notion of Moufang plane and the current definition in terms of root groups is covered by [**114**]; this point goes back to [**150**].

Quadrangles are more complicated, and will be left to the next subsection.

In the case of Moufang hexagons, one studies the commutation relation on a fixed sequence of six successive root subgroups, as described in [**140**, 5.5.13]. In particular, it is shown that one of these root subgroups carries the structure of a field K, and another that of a vector space V over K, equipped with a quadratic mapping from V to $\mathrm{GL}(V)$, all interpretable directly in terms of the actions of the root subgroups (or somewhat less: the structure of G_d with d small). This is shown to give rise to a quadratic Jordan algebra structure on V. There are six cases [**140**, p. 222], [**185**, p. 148]; we will follow the latter listing, where, however, the base field is called F and K denotes a possible extension field; but we continue to refer to the base field here as K.

One possibility is $V = K$ (which falls within type $1/F$), corresponding to the split Cayley hexagon, as expected. Other possibilities included under type $1/F$ involve nontrivial algebraic extensions of K, hence are excluded. The remaining constructions, all given explicitly in [**185**], involve proper finite dimensional field or division ring extensions of K, and hence are excluded since K is algebraically closed.

We note that it seems hard to avoid this line of argument even in the context of Chapter IX, as the various cases envisaged appear to be consistent with the group information obtained there, until the fact that K is algebraically closed is invoked.

Moufang octagons are less numerous and are associated with Suzuki groups. In particular, they involve a field K of characteristic two and a field automorphism whose square is the Frobenius. This structure is visible in the root groups (see [**140**, 5.5.18] or [**185**, 16.9]) and hence the field in question must be definable, hence algebraically closed, and no such automorphism exists.

6.4. The case of quadrangles. We note that the analysis in Chapter IX which makes use of the present section involves quadrangles in which, among other things, the ranks of the two classes of root groups are equal, and all root groups are abelian. Restricting to this case would eliminate most of what follows.

In any case, the full classification of Moufang quadrangles in [**185**] is completely explicit, always involving fields or division rings and some related subgroups and/or vector spaces, and relying on algebraic extensions, imperfection, or anisotropic forms (or pseudoforms, see below). None of these objects exist over algebraically closed fields, and over finite fields they are finite (though one must check the definitions to verify this latter point, as not everything is assumed to be finite dimensional). However, this observation does not suffice to read off the classification of Moufang quadrangles of finite Morley rank, because the constructions give an interpretation of the Moufang polygon in the associated structure, but not (immediately) the converse. Implicitly, the classification theorem itself amounts, among other things, to an interpretation of the auxiliary structures in the polygons, but it would be tedious in the extreme to follow this through (and in case 16.4 of [**185**], it is not literally true; only approximations to the ambient field K are definable in general).

In fact the situation can be handled very efficiently. Each Moufang polygon is characterized by commutation relations involving the root groups, and from this one can read off an explicit structure bi-interpretable with the Moufang polygon. In some cases, however, the root groups themselves have a composite structure (see the examples below), and hence the structure associated naturally with the commutation relations does not contain the field structure in an explicit form, and it is necessary to extract it. This can be done easily in all cases, and we follow this route below. There is a certain inefficiency in this procedure, because in reality it suffices to treat three of the six classes of Moufang quadrangles; the three more complicated classes all contain Moufang quadrangles of a simpler kind as definable subpolygons, and hence the field interpretation can be derived indirectly without close examination of the data.

The six types of Moufang polygon are listed in [**185**, p. 165, Fig. 3] and are described in paragraphs 16.2-16.7 of that source.

In case 16.2 of [**185**], the quadrangle is constructed from an involutory set (K, K_0, σ) (cf. §I 4.5) in which K and K_0 can be identified with adjacent root groups, and the commutator relation between these two root groups includes a component involving the multiplication map from $K_0 \times K$ to K (living in three different root groups, which can however be definably identified with subgroups of one copy of K). Hence we can interpret into the polygon the structure consisting of the additive group of K together with the subgroup K_0 and the restriction of the multiplication map to $K_0 \times K$. In this case Lemma I 4.29 shows that σ is trivial and K is an algebraically closed field interpretable in the polygon. But in an involutory set, K_0 contains $a^\sigma a$ for $a \in K$, hence in the present case $K_0 = K$ and we have the desired class of quadrangles.

In case 16.3 the construction involves a quadratic space (K, V, q) where q is an anisotropic quadratic K-form defined on V, where V is not assumed to be finite dimensional. Using the commutator relations on the root groups we may interpret this structure in our polygon, where we have the two sets K and V, the quadratic form, and the action of K on V, as well as the additive structure on each set. But from the action of K on V we at once also have the field structure on K, which is then algebraically closed. As the form is anisotropic, it then follows that V is 1-dimensional. At this point the definitions degenerate and we have the same quadrangles as in the previous case.

In case 16.4, the construction involves a field K of characteristic two, together with two additive subgroups K_0 and L_0, satisfying

(1) $K_0^2 L_0 \subseteq L_0$, $K_0 L_0 \subseteq K_0$,
(2) K_0 generates K as a ring.

Here the root groups correspond to K_0 and L_0, and the commutator relations allow us to interpret the two multiplication maps

$$K_0 \times L_0 \to K_0, K_0^2 \times L_0 \to L_0$$

in the polygon. When this structure has finite Morley rank, it follows readily that the ranks and degrees of K_0, K_0^2, and L_0 all coincide, and that the indicated multiplication maps are bijections. In particular K_0 and L_0 both contain the element 1, and hence $L_0 = K_0$ and K_0 is a ring. Hence $K_0 = K$. We note here that the definability of K depends strongly on our model theoretic assumption.

At this point one can either treat the remaining cases, 16.5, 16.6, and 16.7, or use the fact that such examples involve examples of the kind already analyzed as definable subpolygons. We indicate both lines of argument. We note that the last two types are considerably more complicated to construct, though not much more trouble to deconstruct (that is, to recover the underlying field structure definably).

We begin with the analysis involving a reduction to the cases already considered. We use the more explicit form of the classification theorem given at the beginning of Chapter 21 of [**185**]. The Moufang quadrangles are of three kinds, called *indifferent, reduced*, and *wide* [**185**, 21.2]; we have considered the indifferent and reduced above, and in the wide case, every such quadrangle Γ is an extension of a reduced polygon Ω in the following sense (or rather, in a stronger sense).

(1) Ω contains an apartment of Γ.
(2) The root subgroups of Ω are definable subgroups of the root subgroups of Γ (in one case coinciding with the original root subgroups), and they act faithfully on Ω.

Note then that the associated data we use (root subgroups, commutator relations) are all interpretable in the original polygon. Hence the analysis above forces these reduced quadrangles to be orthogonal or symplectic over an algebraically closed field.

The relevant results are given as 21.8-21.12 in [**185**]. The reduced quadrangles fall under cases 16.2 and 16.3, where in the former case we may suppose σ is nontrivial. The indifferent quadrangles fall under case 16.4, and as noted the wide cases are extensions of reduced cases.

As we have treated cases 16.2 to 16.4, and the cases of 16.2 with $\sigma \neq 1$ do not arise, this leaves us with extensions of quadrangles of type 16.3 where the quadratic space involved is 1-dimensional. There are three cases [**185**, 21.12], all involving quadratic spaces of dimension greater than one: either V carries a division ring structure, and is a proper extension of K of finite dimension, or V is a quadratic space associated with quadrangles of exceptional types (E_6, E_7, E_8 or R_4). In the E_i cases, the dimensions involved are respectively 6, 8, and 12, and not in any case 1. In the F_4 case, the dimension is at least 5, by definition [**185**, 14.1]. So this eliminates all cases.

We now take note of a second approach in each of the remaining cases, continuing directly along the lines of those used for cases 16.2, 16.3, and 16.4.

In case 16.5, the construction involves an anisotropic *pseudo-quadratic space* (K, K_0, V, σ, q). For the most part it is not necessary to go into the details, as it will suffice to exploit the structure of the field K.

In the associated polygon, one family of root groups may be identified with the additive group of the field K, and the other with a slightly more complicated group whose underlying set is $T = V \times K$. The structure of T is irrelevant here; the commutation conditions include a function from $T \times K$ to K corresponding to multiplication in K:

$$(v, a) \times b \mapsto ab \ .$$

With $b = 1$ this allows us to treat the set K as a definable quotient of T, and then to recover the multiplication on K. Hence K has finite Morley rank. In particular, K is a field, either finite or algebraically closed, and σ becomes an automorphism of K of order at most two; either σ is trivial or K is finite. To complete the analysis one needs to look into the definitions, notably the notion of an anisotropic pseudoquadratic form. So we now give the rest of the description of this case.

Here (K, K_0, σ) is an involutory set, V is a right vector space over K, and q is a *pseudoquadratic form* on V with respect to K and σ. The latter condition means the following.

(1) There is a σ-skew hermitian bilinear form (a, b) on V such that $q(u + v) - q(u) - q(v) \in (u, v) + K_0$ for all $u, v \in V$.
(2) $q(vt) \in t^\sigma q(v)t + K_0$ for $t \in K$, $v \in V$.

The form q is also *anisotropic* in the sense that $q(v) \notin K_0$ for $v \neq 0$.

When σ is trivial, then the definition of involutory set implies that K_0 contains K^2 and hence in all relevant cases $K_0 = K$ and there are no anisotropic forms. When σ is nontrivial, the field K is finite and K_0 is its fixed field. Then again as q is anisotropic, V is finite as well.

In case 16.6, the construction involves a triple (K, V, q) where K is a field, V a vector space over K of dimension 6, 8, or 12, and q an anisotropic quadratic form of a very specific type on V. One family of root groups can be identified with the additive group of V, and the other has a more complicated structure, whose underlying set $S = X_0 \times K$ involves an auxiliary finite dimensional vector space X_0 whose structure is inessential at this point. Among the maps recoverable from the commutator relations there is one of the form

$$f : V \times V \to \{0\} \times K \leq S$$

which corresponds to the bilinear form $V \times V \to K$ associated to q. In particular, the image of this map yields a copy of the additive group of K. As noted in [**185**, 12.12], the definitions (which we have omitted) force the associated bilinear form to be nondegenerate, so we may easily restrict f to a pair of 1-dimensional spaces which are nonorthogonal. At this point we can recover the multiplication on K from f, and it follows that either K and V are finite, or K is algebraically closed. However, there are no examples for K algebraically closed (V would be at most 1-dimensional).

Essentially the same argument works in case 16.7, though in this case neither root group is entirely straightforward. The construction involves a highly degenerate quadratic form (K, V, q), on a not necessarily finite dimensional space. The root groups are parametrized by the sets $A = X_0 \times K$ and $B = W_0 \times F$ with $F = q(V)$ a subfield of K. Here W_0 is a vector space of dimension 4 over K equipped with a specific quadratic form arising from a norm on a quadratic extension of K. Among the data recoverable from the commutation relations is a map

$$\phi : B \times B \to \{0\} \times K$$

corresponding to the (nondegenerate) bilinear form on W_0 associated with its quadratic form $f(w, w')$, that is,

$$\phi((w, a), (w', a')) = (0, f(w, w')) \ .$$

From this one recovers W_0 as a definable quotient of B and then the multiplication on K.

6.5. Groups acting on n-gons.

PROPOSITION 6.3 ([**130**]). *Let G^* be a group of finite Morley rank acting faithfully and definably as a group of automorphisms on a Moufang polygon Γ, and containing the little projective group of* $\mathrm{Aut}(\Gamma)$. *Then G^* is a Chevalley group of Lie rank two over some algebraically closed field.*

PROOF. The little projective group G is by definition the group of automorphisms generated by the root subgroups, and in these classical cases is a Chevalley group of Lie rank two; the full automorphism group of the n-gon is the extension of this group by $\mathrm{Aut}(K)$, where K is the underlying field ([**183**, 5.10]). As these groups are finite products of root subgroups, the group G is a definable subgroup of G^*. As G acts transitively on the points, its centralizer in G^* is trivial and thus G^* induces a group of definable automorphisms of G; by Fact II 2.25 this is an extension of G by at most a graph automorphism. But a graph automorphism interchanges a point and line stabilizer, so it cannot act on Γ. □

6.6. Groups acting on graphs. In conjunction with the amalgam method in Chapter IX, we will need a result from [**84**] concerning the universal cover of a generalized n-gon on which a group acts subject to certain conditions. This will be needed in §IX 9, and will be more transparent in that context.

Our setting is as follows (cf. [**84**, p. 73]).

HYPOTHESIS 6.1. *Γ is a tree, and G is a group of automorphisms of Γ with the following properties.*

(1) *G operates faithfully and edge transitively, but not vertex transitively, on Γ.*
(2) *The vertex stabilizers G_δ have finite Morley rank for each δ.*
(3) *$C_{G_\delta}(O_2(G_\delta)) \leq O_2(G_\delta)$ for each vertex δ.*
(4) *For α, β adjacent vertices, $G_{\alpha,\beta}$ contains a Sylow 2-subgroup of G_α and G_β.*

Of course, Delgado and Stellmacher assume the vertex stabilizers are actually finite. There is a difference in the two settings, and we will explain this before resuming our account of their theory.

NOTATION 6.4. *With the hypotheses as above, let δ be a vertex of Γ, and $k \geq 0$.*

(1) *$\Delta^k(\delta)$ is the set of vertices within distance k of δ.*
(2) *$G_k(\delta)$ is the* set *of elements of G which are products of at most k elements of G, with each element stabilizing some vertex in $\Delta^k(\delta)$.*

Now if the vertex stabilizers are finite, then the sets $G_k(\delta)$ are finite, and this happens in particular if Γ is obtained as the universal cover of a graph associated with a finite group. The analog for finite Morley rank is slightly different, and requires a further definition.

DEFINITION 6.5. *With G, Γ as above, we say that G is* locally of finite Morley rank *if $G_k(\delta)$ has finite Morley rank for all vertices δ and all k, where this structure carries a partial group operation as well as a partial action on $\Delta^k(\delta)$.*

We do not claim that this condition follows from the corresponding condition on vertex stabilizers (the case $k = 0$), but it is this condition which will matter ultimately.

After these preliminaries we return to the algebraic development.

DEFINITION 6.6. *Let α, β be adjacent vertices in Γ.*

(1) *A* Cartan *subgroup of $G_\alpha \cap G_\beta$ is a subgroup K whose normalizer contains elements t_α, t_β not in $(G_\alpha \cap G_\beta)$ with $t_\alpha^2, t_\beta^2 \in K$.*
(2) *We associate to a Cartan subgroup the* Weyl group *W defined as $\langle t_\alpha, t_\beta \rangle K/K \leq N(K)/K$.*

This definition is very weak (but adequate); it does not force K to be nontrivial, and there is no conjugacy theorem.

DEFINITION 6.7. *Suppose that α, β are adjacent vertices in Γ and K is a Cartan subgroup of $G_{\alpha,\beta}$ with associated Weyl group W. Then the graph induced on the set $\alpha^W \cup \beta^W$ is called the associated* apartment.

The following is elementary, but clearly depends on the hypothesis that Γ is a tree.

LEMMA 6.8 ([**84**, 3.5, p. 75]). *If T is the apartment associated to a pair α, β and a Cartan subgroup K as above, then T is a 2-way infinite path whose vertices are fixed by K. Furthermore, for every vertex δ of T, the reflection of T about δ is induced by some element of W.*

DEFINITION 6.9. *Let T be an apartment in Γ, and s a positive integer.*

(1) *T satisfies the* uniqueness condition *if every path of length s in Γ is contained in a unique G-conjugate of T.*
(2) *T satisfies the* exchange condition *at $\delta \in T$ if for every pair of paths γ^-, γ^+ contained in T of length $s-1$, with right and left endpoint respectively equal to δ, and for every $x \in G_{\gamma^+}$, there is $y \in G_{\gamma^-}$ so that xy carries γ^- to γ^+.*
(3) *T satisfies the* exchange condition *if it satisfies this condition with respect to each of its vertices.*

Here finally is the key result.

PROPOSITION 6.10 ([**84**, 3.6, p. 77]). *With G and Γ satisfying the hypotheses above, let T be an apartment and suppose that it satisfies the uniqueness and exchange conditions. Then there is a G-invariant equivalence relation on T such that the corresponding quotient $\tilde{\Gamma}$ of Γ is a generalized n-gon, and such that the kernel of the action of G on $\tilde{\Gamma}$ is disjoint from the vertex stabilizers G_δ in Γ. Furthermore, if G is locally of finite Morley rank on Γ, then the induced automorphism group $G/G_{\tilde{\Gamma}}$ has finite Morley rank.*

The last statement of course varies considerably from that in [**84**], and has a slightly different proof. They show that if the vertex stabilizers are finite, the quotient group is finite. The corresponding statement for finite Morley rank requires slightly more attention, as it involves issues of definability. We will give some elements of the proof, but omit the issues treated explicitly in [**84**].

PROOF. Write $\delta_1 \sim \delta_2$ if the vertices in question lie at distance $2(s-1)$, and the path between them lies in a conjugate of T. Let $\approx$ be the equivalence relation generated by the relation $\sim$. This is G-invariant, and in the quotient $\tilde{\Gamma}$ we take two equivalence classes to be adjacent if they have adjacent representatives.

For the (efficient and rapid) verification of the axioms for a generalized n-gon we refer to [**84**]. We turn to the last statement.

The point here is that the graph $\tilde{\Gamma}$ has finite diameter, and even more: for δ any vertex of Γ, the extended neighborhood $\Delta^{s-1}(\delta)$ contains a set of representatives for the quotient. This is an explicit and important part of the proof in [**84**], and when the vertex stabilizers are finite it follows that $\Delta^{s-1}(\delta)$ is finite and hence the induced group is finite.

Now consider the case in which the group G is locally of finite Morley rank. Fix a vertex δ of Γ. For $g \in G$, the vertex δ^g is represented by a vertex δ_1 whose distance from δ is at most $s-1$, and even, since G respects a bipartition of Γ. If δ_0 is the midpoint of the path from δ to δ_1, then there is a sequence of at most $(s-1)/2$ elements from vertex stabilizers of intermediate vertices that carries the path from δ_0 to δ_1 to the path from δ_0 to δ. After adjusting g by this short sequence, the product fixes $\delta/\approx$; so the action of g on $\tilde{\Gamma}$ is expressible as a product, of length at most $(s-1)/2+1$, of elements of $G_{(s-1)/2}(\delta)$. Accordingly, $G/G_{\tilde{\Gamma}}$ has a set of representatives of finite Morley rank. Furthermore, two such representatives are equivalent if and only if their actions on $\Delta^{s-1}(\delta)$ induce the same action on $\tilde{\Gamma}$; we need to see that this last relation, which is an equivalence relation, is definable from the structure induced on $G_{s-1}(\delta)$. For this we need to check that the equivalence relation $\approx$ on $\Delta^{s-1}(\delta)$ is definable. Since the quotient is a generalized $(s-1)$-gon, it follows from our hypotheses that the relation $\approx$, on $\Delta^{s-1}(\delta)$, may be characterized as follows:

($\approx$) $\alpha \approx \beta$ iff $d(\alpha,\beta) = 2(s-1)$ and the path from α to β is conjugate under G to a subpath of T.

This is not quite the desired definition, because of the presence of the full group G and the full apartment T here.

Obviously the apartment T can be replaced by a fixed path of length $2(s-1)$ which may be supposed to lie in $\Delta^{s-1}(\delta)$ and have endpoints in the appropriate orbit under G. So it suffices to define the conjugacy relation for short paths in $\Delta^{s-1}(g)$ using only local data, and this can easily be done using the fact that vertex stabilizers act transitively on their neighbors. Namely, we can assume that the two paths have the same initial point, and then the question becomes one of determining whether the vertex stabilizer of an initial segment of one path carries one neighbor of an endpoint to another, and this is part of the local structure. □

We require one further definability result in the same vein, in order to verify that the graphs of interest to us carry automorphism groups of locally finite Morley rank. This is a corollary of the following.

LEMMA 6.11. *Let $\mathcal{P} = (P, Q, B)$ be a structure consisting of two groups P, Q. Let $G = P *_B Q$ be the free product with amalgamation and let Γ be the associated tree of cosets, on which G acts naturally. Then the structure $\mathcal{G} = (G, \Gamma)$ consisting of G acting on Γ is locally interpretable in $\mathcal{P}$ in the following sense: for any vertex $\delta \in V(\Gamma)$ and any $k \geq 0$ the graph $\Delta_k(\delta)$,* *the partial group $G_k(\delta)$, and the partial action of $G_k(\delta)$ on $\Delta_k(\delta)$ are all interpretable in $\mathcal{P}$.*

PROOF. Let $X = P \cup Q$ and let $R_k(x_1, \ldots, x_k)$ be the relation on X defined by: "$x_1 \cdots x_k = 1$ in G". Everything comes down to the definability of this relation in $\mathcal{P}$, which is proved by induction based on the following property of free products with amalgamation: if $x_1, \ldots, x_k$ are alternately from $P \setminus B$ and $Q \setminus B$, then the product is nontrivial. In the remaining cases, either the product can be shortened, and induction applies, or else $k = 1$. Bearing in mind that the natural maps of P and Q into G are embeddings, the claim follows. □

COROLLARY 6.12. *Under the stated hypotheses, if $\mathcal{P}$ has finite Morley rank, then (G, Γ) is locally of finite Morley rank.*

7. Buildings and (B, N)-pairs

7.1. Buildings. We record the definition, or in any case one of the definitions, of Tits' buildings, and some facts about them. This section can be skipped since we will give the same information in the language of BN-pairs in the next subsection, but as the more geometric language of buildings is actually used to carry out classification arguments, it seems worth giving here, even very briefly.

DEFINITION 7.1.

(1) *A* chamber system *with index set I is a set Δ, whose elements are called chambers, together with a system of equivalence relations $\sim_i$ for $i \in I$.*
(2) *Two chambers in a chamber system are i-*adjacent *if they are i-equivalent and distinct. A* gallery *between two chambers x, y is a sequence of chambers $x = x_0, x_1, \ldots, x_n = y$ such that $x_j \neq x_{j+1}$ for $i < n$, and x_j is i_j-adjacent to x_{j+1} for some $i_j \in I$; the sequence $(i_0, \ldots, i_{n-1})$ is the* type *of the gallery. The* distance *from x to y is the length of the shortest gallery (or ∞).*
(3) *Let W be a Coxeter group with Dynkin diagram Π and associated generating set I (corresponding to a fundamental system of roots, or to the vertices of Π). A* building *of type Π (or type W) is a chamber system of type I together with a "W-metric" δ, namely a function*

$$\delta : \Delta \times \Delta \to W$$

such that the following two conditions are satisfied:
   (a) *Every i-equivalence class is nontrivial (contains at least two elements).*
   (b) *If $w \in W$ and $\hat{w}$ is a reduced word in the generators I representing w, then for $x, y \in \Delta$ we have $\delta(x, y) = w$ if and only if there is there is a gallery of type $\hat{w}$ from x to y.*

One can show that the building (taken as a chamber complex) uniquely determines the associated Coxeter group W, which is called the *Weyl group*, and the W-metric δ. By definition, the *rank* of the building is the rank of the Coxeter group, which is just the cardinality of the index set I. A building is said to be *spherical* if the Weyl group is finite (and hence acts naturally on a sphere in the associated reflection representation). The building is called *irreducible* if the group W is irreducible.

EXAMPLE 7.2. *The building $\Delta(W)$ (more properly, $\Delta(W, I)$) associated to a Coxeter group W with distinguished generators I has as its set of chambers the elements of W, with w_1, w_2 r-equivalent, for $r \in I$, if and only if $w_2 \in w_1\langle r\rangle$; the W-metric is given by $\delta(w_1, w_2) = w_1^{-1}w_2$.*

A building is called *thick* if every i-equivalence class contains at least three chambers; and *thin* if every i-equivalence class contains exactly two chambers; the buildings $\Delta(W)$ are thin.

DEFINITION 7.3. *Let Δ be a building of type* Π. *An* apartment *in Δ is any isomorphic copy of $\Delta(\Pi)$.*

Let us connect this with Moufang polygons. The *flags* of a Moufang polygon Γ are the pairs (v, e) with v a vertex and e an edge incident with v. These flags make up a chamber system of rank 2 with the two natural equivalence relations whose classes are the vertices and edges of Γ, respectively. An adjacent pair of flags corresponds either to two vertices linked by an edge, or to two edges with a common vertex. The Weyl group W is the associated dihedral group D_{2n}, and its associated building $\Delta(W)$ is an ordinary n-gon.

Now we define roots and Moufang buildings in general. We begin with roots. Recall that a root in an n-gon is a half-apartment, or a path of length n.

DEFINITION 7.4. *Let Δ be a building of type* Π, *with Weyl group W.*

(1) *A* root *of $\Delta(W)$ is a subset of the form*

$$\{w : d(w, x) < d(w, y)\}$$

where (x, y) is an ordered pair of adjacent chambers, and d is the distance in $\Delta(W)$.

(2) *A* root *of Δ is a root of any apartment of Δ.*

(3) *The* interior *α° of a root α is the set of all i-equivalence classes in Δ (for any $i \in I$) which contain two chambers of α.*

Note that in an apartment, each such equivalence class contains two chambers, or none.

Now we may define root groups and Moufang buildings; as always, comparison with the case of polygons may be useful.

DEFINITION 7.5. *Let Δ be a building.*

(1) *If α is a root of Δ, then define the "root group" U_α as*

$$\{g \in \mathrm{Aut}(\Delta) : g \textit{ acts trivially on each } i\textit{-equivalence class in } \alpha^\circ\} .$$

(2) *A building is called* Moufang *if it is thick, irreducible, of Tits rank at least 2, and for each root α of Δ the root group U_α acts transitively on the set of apartments containing α.*

We have piled up quite a few definitions here; but essentially, all that has happened is that the dihedral group has been replaced by an arbitrary Coxeter group in the definitions of the preceding section.

We will give the classification of Moufang buildings in the notation of [**185**], referring to that source not only for the proof but also for a complete elucidation of the notation, which however is a mixture of standard notation for Coxeter groups together with some terminology we have largely seen in the case of Moufang polygons.

In order to state this classification one must fix an apartment Σ and a chamber $c \in \Sigma$, and associate a system of roots α_i $(i \in I)$ relative to the fixed pair (c, Σ); namely, α_i is the root of Σ which contains c and does not contain the unique chamber of Σ which is i-adjacent to c. With these conventions, the main result goes as follows (rather schematically):

FACT 7.6 ([**185**, 40.22]). *Let Δ be a Moufang spherical building of Coxeter type Π, and I the set of vertices of Π and rank $\ell \geq 3$. Let Σ be an apartment of Δ and c a chamber of Σ. Then there is a parameter system (algebraic structure) Ξ and a root group labeling of Π of a standard form which is isomorphic to the induced root group labeling and satisfies one of the following:*

(1) $\Pi = A_\ell$: *Ξ is a division ring;*
(2) $\Pi = B_\ell$: *Ξ is an anisotropic quadratic space;*
(3) $\Pi = C_\ell$: *Ξ is an involutory set, either proper, quadratic, or (for $\ell = 3$ only) honorary;*
(4) $\Pi = BC_\ell$: *Ξ is a proper anisotropic pseudoquadratic space;*
(5) $\Pi = D_\ell$ $(\ell \geq 4)$, *or* E_ℓ $(\ell = 6, 7, 8)$: *Ξ is a field;*
(6) $\Pi = F_4$: *Ξ is a quadratic or honorary involutory set.*

The precise root group labelings are given in [**185**, 40.25, 40.50, 40.51]. This is considerably more precise than one needs in the finite Morley rank context to conclude that the underlying parameter set is just an algebraically closed field in that context. We note that the proof given in [**185**] is efficient and inductive, and when specialized to the case of finite Morley rank, since most of the candidate n-gons disappear, the analysis is even shorter. However, to bring all this to bear one more result is crucial.

FACT 7.7. *Every thick irreducible spherical building of Tits rank at least 3 is Moufang.*

For this one needs [**183**, 4.16] (see [**185**, 40.3]). A recent exposition is given in [**193**].

One may deduce from this the following.

PROPOSITION 7.8. *Let Δ be an irreducible spherical building of Tits rank at least* 3 *and finite Morley rank. Then the subgroup of* $\mathrm{Aut}(\Delta)$ *generated by the root groups is an algebraic group over an algebraically closed field.*

Compare [**185**, 41.16]; one may also argue more abstractly that the group in question is definable over an algebraically closed field, and apply Fact II 1.21. Cf. [**130**, Theorems 5.1 and 5.3] (Fact 7.11 below).

7.2. BN-pairs. The notion of BN-pair is an abstraction of the situation arising in algebraic groups when one considers a Borel subgroup B, a maximal torus T contained in B, and the normalizer $N = N(T)$ in the ambient group.

DEFINITION 7.9. *Let G be a group, and B, N subgroups of G; set $T = B \cap N$.*

(1) *We say that the pair (B, N) is a BN-*pair *for G if the following conditions hold, with respect to some subset X of N/T:*
 BN 1. $G = \langle B, N \rangle$.
 BN 2. $B \cap N \triangleleft N$. *We write $T = B \cap N$ and $W = N/T$.*
 BN 3. *X is a set of involutions generating the group W.*
 BN 4. *$B^s \neq B$, for all $s \in X$.*
 BN 5. *$nBs \subseteq BnB \cup BnsB$ for all $n \in N$ and all $s \in X$.*
(2) *If (B, N) is a BN-pair for G, the pair is said to be* definable *if B is.*
(3) *A BN-pair (B, N) is* spherical *if $W = N/T$ is finite.*
(4) *If X is finite, its cardinality is called the (*Tits*)* rank *of the BN-pair (B, N).*

Note that a BN-pair is more properly speaking a triple (B, N, S). One can reproduce a good deal of the structure theory of algebraic groups in this context.

FACT 7.10 ([**115**]). *Let G be a group with a BN-pair. Then*

(1) *W is a Coxeter group with distinguished generators X;*
(2) *the subgroups containing B are those of the form $P_I = \langle B, I \rangle$ where $I \subseteq X$;*
(3) *G is the disjoint union of the double cosets BwB $(w \in W)$.*

This includes the case of algebraic groups, with B a Borel subgroup, and the terminology is extended to the present more combinatorial settings; in particular, groups containing B (and their conjugates) are called *parabolic*.

One may convert BN-pairs into buildings as follows, bearing in mind that we have defined buildings as chamber complexes. The chambers are the conjugates of B; the set of types I may be identified with the fixed set X

of generators for W, or with the *minimal parabolic subgroups* P containing B; two chambers are P-adjacent, for such a subgroup P, if they lie jointly in a conjugate of P.

When buildings are defined as complexes rather than as chamber complexes, then the associated building becomes the complex of parabolic subgroups (that is, those containing a conjugate of B).

As a matter of convention we may transfer all of the building terminology to BN-pairs. With those conventions, we have the following.

FACT 7.11 ([**130**, Theorems 5.1 and 5.3]). *Let G^* be an infinite simple group of finite Morley rank with a definable spherical BN-pair of Tits rank at least* 3. *Then G^* is a Chevalley group over an algebraically closed field.*

FACT 7.12 ([**130**]). *Let G^* be an infinite simple group of finite Morley rank with a spherical Moufang BN-pair of Tits rank* 2. *Then* $G^* \simeq \mathrm{PSL}_3(F)$, $\mathrm{PSp}_4(F)$, *or* $G_2(F)$ *for some field F.*

In the above, the field in question must be algebraically closed as it will also have finite Morley rank.

7.3. Pairwise BN-pairs. We will need a weakening of the BN-pair condition, easier to verify in practice.

DEFINITION 7.13. *Let G be a group, B and N subgroups of G, $T = B \cap N$, and X a subset of N/T. Then we say that (B,N,X) forms a* pairwise BN-pair *if the following conditions hold.*

BN 1. $G = \langle B, N\rangle$.
BN 2. $B \cap N \lhd N$.
BN 3. *X generates N/T and $s^2 = 1$ for all $s \in X$.*
BN 4. *$sBs \neq B$ for all $s \in X$.*
BN 5_2. *For all $u, v \in X$, $n \in \langle u, v\rangle$ and $s \in \{u, v\}$. $nBs \subseteq BnB \cup BnsB$.*

NOTATION 7.14.

(1) *If $R \subseteq X$, denote $N_R = \langle B \cap N, R\rangle$ and $W_R = N_R/B \cap N$. It is well-known (and easily follows from BN 5) that $P_R = BN_RB$ is a subgroup (known as a* parabolic subgroup*).*
(2) *In a spherical definable BN-pair, every parabolic subgroup P_R is a finite union*
$$P_R = \bigcup_{w \in W_R} BwB$$
and therefore is definable.
(3) *We set $B_R = \bigcap_{r \in \langle R\rangle} B^r$ for $R \subseteq X$, writing, in particular, B_s for $B_{\{s\}}$.*

The next result is a purely algebraic fact, proved by Niles in the form we use, which applies to groups which may be finite or infinite. We impose no model theoretic hypotheses.

FACT 7.15 ([**149**, Theorem A]). *Let G be a group with a pairwise BN-pair (B, N, S). Assume the following condition:*

$$(*)\qquad \begin{array}{l} \text{If } X \text{ is a } T\text{-invariant subgroup of } B \\ \text{such that } B = XB_s \text{ and } B = XB_t, \\ \text{then } B = XB_{s,t}. \end{array}$$

Then (B, N, S) forms a BN-pair for G.

Hidden within this fact is a similar characterization of Coxeter groups, due to Goldschmidt.

8. A theorem of Niles

8.1. The theorem. Our principal tool for identifying the "generic" group of finite Morley rank will be the following.

THEOREM 8.1. *Let G be a group of finite Morley rank and even type. Let S be a Sylow° 2-subgroup of G, and $B \leq N_G(S)$ a connected solvable group. Assume that G contains definable connected subgroups $P_1, \ldots, P_r$ which satisfy the following conditions:*

(A) $G = \langle P_1, P_2, \ldots, P_r \rangle$.

(B) $N_{P_i^\circ}(S) = B$ *for all* $i = 1, \ldots, r$.

(C) *Setting* $L_i := U_2(P_i)$ *and* $\bar{L}_i = L_i/O_2{}^\circ(P_i)$, *we have* $L_i \simeq \mathrm{SL}_2(K_i)$, *with K_i some algebraically closed field of characteristic 2, for* $i = 1, \ldots, r$.

(D) *Setting* $L_{ij} := \langle L_i, L_j \rangle$, *and* $\bar{L}_{ij} = L_{ij}/O_2{}^\circ(L_{ij})$, *the group* $\bar{L}_{ij}$ *is either a quasisimple Chevalley group of Lie rank 2 over an algebraically closed field of characteristic 2:*

$$\mathrm{PSL}_3(K), \mathrm{SL}_3(K), \mathrm{Sp}_4(K), G_2(K)$$

or a direct product of two Chevalley groups of Lie rank 1:

$$\mathrm{SL}_2(K_1) \times \mathrm{SL}_2(K_2)$$

with K_1 and K_2 two algebraically closed fields of characteristic 2.

Let $G_0 = \langle L_1, L_2, \ldots, L_r \rangle$. Then

(1) *G_0 is normal in G, and*

(2) *G_0 has a definable spherical BN-pair of Tits rank r.*

COROLLARY 8.2. *Assume that G is a simple group which satisfies all the requirements of Theorem 8.1. Then G is a simple Chevalley group over an algebraically closed field K of characteristic 2.*

PROOF. By Theorem 8.1, G has a definable BN-pair of Tits rank r. Notice that if $r \leq 2$, then $G/O_2{}^\circ(G)$ has the desired form by the assumptions of Theorem 8.1. Therefore we can assume that $r \geq 3$. Now we are in a position to apply Fact 7.11 which states that, in this situation, G is a Chevalley group over an algebraically closed field. □

We take up the proof of Theorem 8.1. We can deal quickly with the issue of normality.

LEMMA 8.3. *Under the hypotheses of Theorem* 8.1, *setting* $P_{ij} = \langle P_i, P_j \rangle$, *we have the following:*

(1) $P_i = L_i B$,
(2) $L_{ij} = U_2(P_{ij})$,
(3) G_0 *is normal in* G.

PROOF.
(1) By the Frattini argument and connectedness, $P_i = L_i N_{P_i}{}^\circ(S) = L_i B$.
(2) $P_{ij} = \langle L_i B, L_j B \rangle = \langle L_{ij}, B \rangle$. Furthermore, B normalizes L_i, L_j, and L_{ij}, so $P_{ij} = L_{ij} \cdot B$. As $U_2(B) = S \leq L_{ij}$, it follows that $U_2(P_{ij}) = L_{ij}$.
(3) By the second point, P_{ij} normalizes L_{ij}, and in particular P_i normalizes all L_{ij} for $j = 1, \ldots, r$. Hence P_i normalizes G_0, for all i, and thus G normalizes G_0. □

8.2. The proof. From now on, we work under the hypotheses and notation of Theorem 8.1. We disposed of normality above, so we are concerned with the construction of a spherical BN-pair associated with G. We have a group B furnished by the hypotheses of the theorem. We will look for an appropriate group N and a set X of generators to accompany it.

We may factor B as $S \rtimes T$, by Proposition I 9.6. For $i = 1, \ldots, r$, let $N_i = T N_{L_i}(T)$, and set $N = \langle N_i : i = 1, \ldots, r \rangle$. Recall that $L_i = U_2(P_i)$ and $\bar{L}_i = L_i / O_2(L_i) \simeq \mathrm{SL}_2(K_i)$ for some field K_i depending on i.

LEMMA 8.4. *With the notations and hypotheses as above, and letting* $T_i = T \cap L_i$, *we have the following.*

(1) $\bar{T}_i$ *is a maximal torus of* $\bar{L}_i$.
(2) $\overline{N_i \cap L_i} = N_{\bar{L}_i}(\bar{T}_i)$.

PROOF.
(1) As $B \leq P_i$, B normalizes L_i. Then by Proposition I 9.4 T contains a complement to S in $N_{L_i}{}^\circ(S)$, which covers a maximal torus of $\bar{L}_i$.
(2) We work in $\bar{P}_i = P_i / O_2{}^\circ(P_i)$. We have $N_i \cap L_i = N_{L_i}(T)$. Let $\bar{s} \in N_{\bar{L}_i}(\bar{T})$. Then lifting $\bar{s}$ to $s \in L_i$, we have $T^s \leq O_2(P_i)^\circ T$. Hence there is an element $x \in O_2{}^\circ(P_i)$ conjugating T^s to T, again by Proposition I 9.4. So $sx \in N_{L_i}(T)$ and sx represents $\bar{s}$ in $\bar{L}_i$. □

NOTATION 8.5. *Let* $r_i \in N_i$ *represent the involution of* $N_{\bar{L}_i}(\bar{T}_i)$. *Let* $X = \{r_i : i = 1, \ldots, r\}$.

LEMMA 8.6. $G_0 = \langle B, N \rangle$.

PROOF. We have $L_i = \langle S T_i, r_i \rangle$ for all i, $1 \leq i \leq r$. Therefore $\langle L_1, \ldots, L_r \rangle = \langle S, T_1, \ldots, T_r, N \rangle \leq \langle B, N \rangle$. □

LEMMA 8.7. $B \cap N \lhd N$

PROOF. By our construction $B \cap N \leq N_B(T) = N_S(T) \cdot T = C_S(T) \times T \leq B \cap N$. Now for each i, $C_S(T) \leq C_S(T_i) \leq O_2(L_i)$. So $C_S(T) = C_{O_2(L_i)}(T)$ for each i, and $C_S(T)$ is normalized by each N_i, hence by N. Thus $B \cap N \lhd N$. □

LEMMA 8.8. *The triple (B, N, X) provides a pairwise BN-pair for G_0.*

PROOF. The first two conditions, $G_0 = \langle B, N \rangle$ and $B \cap N \lhd N$, have just been verified. By construction, X generates $W = N/(B \cap N)$. Now $r^2 = 1$ in W for $r \in X$, since this holds in $\bar{L}_i$, and $O_2(L_i) \leq B$. Similarly, $B^{r_i} \neq B$ by inspection in $\bar{L}_i$.

Finally, the main condition, (BN 5_2), can be verified by inspection in $P_{ij} = L_{ij}T$, as the structure of L_{ij} is known. □

LEMMA 8.9. *Assume that L is isomorphic to one of the following:*

$$\mathrm{PSL}_3(K), \mathrm{SL}_3(K), \mathrm{Sp}_4(K), \mathrm{G}_2(K), \mathrm{SL}_2(K_1) \times \mathrm{SL}_2(K_2)$$

where K, K_1, K_2 are algebraically closed fields of characteristic 2. Let B and N be subgroups of L which form a BN-pair for L, with B a Borel subgroup. Let S be a Sylow 2-subgroup of B, and let P_1, P_2 be two parabolic subgroups containing B. Set $U_i := O_2{}^{\circ}(P_i)$ for $i = 1, 2$ and let T be a complement to U in B. Then there is no proper T-invariant subgroup S_0 of S such that $S = S_0U_1$ and $S = S_0U_2$.

PROOF. In each case the Weyl group has two generators, say $S = \{s_1, s_2\}$, such that $P_i = B\langle s_i \rangle B$ for $i = 1, 2$. Assume that S_0 is a T-invariant subgroup of S satisfying $S = S_0U_1$ and $S = S_0U_2$.

Now by Fact II 1.11, S_0 is a product of root subgroups, and every element of S_0 can be written as a product of root elements in a unique way. Now $S = S_0U_1$ implies that S_0 contains the root subgroup S_{s_2} corresponding to s_2. Similarly, S_0 contains the root subgroup S_{s_1} corresponding to s_1; but $S = \langle S_{s_1}, S_{s_2} \rangle$ by Lemma II 2.4. Hence $S_0 = S$, which completes the proof. □

LEMMA 8.10. *G_0 has a BN-pair of Tits rank r.*

PROOF. We have constructed a pairwise BN-pair (B, N, X) satisfying Niles' condition (Lemma 8.9), so by Fact 7.15 this is a BN-pair.

Now as $\bar{L}_i = \langle \bar{S}, \bar{s}_i \rangle$, the elements s_i are distinct, and $|X| = r$. □

To conclude the proof of Theorem 8.1, it suffices to check that our definable BN-pair is of spherical type. We give this in a more general form.

LEMMA 8.11. *Let G be a group of finite Morley rank and (B, N, X) a definable BN-pair in G, with B connected. Then the BN-pair (B, N, X) is spherical.*

PROOF. Let $W = N/(B \cap N)$. The minimal parabolic subgroups $M_s = B \cup BsB$, $s \in X$ are definable. We claim

(1) M_s is connected.

Note that $B < M_s{}^\circ$ since $B^s \neq B$. Thus BsB is a generic subset of M_s. On the other hand the multiplication map

$$B \times B \to BsB$$

has fibers of constant rank, namely $\mathrm{rk}(B \cap B^s)$. Since B is connected, it follows that the image BsB has Morley degree 1. Thus M_s has Morley degree 1, and is connected.

Now G is generated by the groups M_s, and hence is the setwise product of finitely many of them in view of Proposition I 3.19. Then by expanding the representation of G as $M_1 \cdot M_2 \cdot \ldots \cdot M_n$ for some n, where $M_i = B \cup (Bs_iB)$ for some $s_i \in X$, using axiom BN 5, we find that G is the union of finitely many double cosets of the form BwB, $w \in W$. On the other hand, the decomposition

$$G = \bigcup_{w \in W} BwB$$

is a disjoint union (Fact 7.10), so W is finite. □

Now Lemmas 8.3, 8.10, and 8.11 prove Theorem 8.1.

9. Signalizer functors

In the next section we give a generic identification theorem which can be used in place of the theory of buildings to complete the identification of the generic simple group of even type. The present section prepares an ingredient needed in order to verify a hypothesis of that generic identification theorem in practice.

9.1. Definitions.

NOTATION 9.1. *Let G be a group of finite Morley rank, and p a prime. Then $O_{p^\perp}(G)$ is the largest definable connected normal solvable $p^\perp$-subgroup of G. For example, $O_{2^\perp}(G) = O(G)$.*

DEFINITION 9.2. *Let G be a group of finite Morley rank, p a prime, and $E \leq G$ an elementary abelian p-subgroup of G.*

(1) *An E-*signalizer functor *is a function θ with domain $E^\times$, satisfying the following conditions.*
   (a) *For $a \in E^\times$, $\theta(a)$ is an E-invariant connected subgroup of $O_{p^\perp}(C(a))$.*
   (b) *For $a, b \in E^\times$, we have*

(Balance) $$\theta(a) \cap C(b) \leq \theta(b)\ .$$

(2) *An E-signalizer functor θ on X is* nilpotent *if $\theta(a)$ is nilpotent for all $a \in E$.*

(3) *If θ is an E-signalizer functor, then*
   (a) *$\theta(E) = \langle \theta(a) : a \in E^\times \rangle$.*

(b) θ *is* E-complete *if* $\theta(E)$ *is a solvable* $p^{\perp}$*-group and for* $a \in E^{\times}$*,* $\theta(E) \cap O_{p^{\perp}}(C(a)) = \theta(a)$.

Our goal in this section is to prove that nilpotent E-signalizer functors on groups of finite Morley rank are E-complete, for E of p-rank at least 3.

PROPOSITION 9.3. *Let G be a group of finite Morley rank, p a prime, E an elementary abelian p-subgroup of G of rank at least 3, and θ a nilpotent E-signalizer functor on G. Then θ is E-complete.*

PROOF. We will proceed by induction on the rank of G.

A subgroup Q of G will be called a *θ-subgroup* if Q is a connected, definable, solvable $p^{\perp}$-subgroup of G normalized by E, and for all $a \in E^{\times}$ we have

$$Q \cap \theta(a) = C_Q(a) .$$

Note that here $C_Q(a)$ is connected (Lemma I 10.6). Furthermore, a θ-subgroup Q will be nilpotent since $C_Q(a)$ is nilpotent for $a \in E^{\times}$ and Proposition I 9.18 applies.

In view of the balance condition, each of the groups $\theta(a)$ is a θ-group for $a \in E^{\times}$. Furthermore, any connected definable E-invariant subgroup of a θ-subgroup is a θ-subgroup. To prove completeness of θ it suffices to show that the group $\langle \theta(a) : a \in E^{\times} \rangle$ is a θ-subgroup. This follows at once from the following claim:

(1) There is a unique maximal θ-subgroup in G.

Indeed, if Q is the unique maximal θ-subgroup, then Q contains $\langle \theta(a) : a \in E^{\times} \rangle$ which is therefore also a θ-subgroup. So we will prove (1).

Suppose first that the intersection of any two distinct maximal θ-subgroups is finite. Let P be a maximal θ-subgroup. Then P contains any θ-subgroup whose intersection with P is infinite. By Proposition I 9.16, $P = \langle C_P(E_0) : E_0 \le E, [E : E_0] = p \rangle$. In particular, $C_P(E_0)$ is nontrivial for some $E_0 \le E$ with $[E : E_0] = p$. If Q is another maximal θ-subgroup, then similarly $C_Q(E_1)$ is nontrivial for some $E_1 \le E$ with $[E : E_1] = p$. In particular, if $a \in E_0 \cap E_1$ is nontrivial, then $C_P(a) = P \cap \theta(a)$, $C_Q(a) = Q \cap \theta(a)$ are nontrivial; but these groups are connected by Lemma I 10.6, hence infinite, and hence by our case assumption $\theta(a) \subseteq P \cap Q$, forcing $P = Q$. Thus in this case (1) holds.

Now suppose that there are distinct maximal θ-subgroups with infinite intersection, and let P, Q be two such, with the rank of $P \cap Q$ maximal. Let $R = (P \cap Q)^{\circ}$, $H = N^{\circ}(R)$, $\bar{H} = H/R$. Then θ induces a function $\bar{\theta}$ on H by

$$\bar{\theta}(\bar{a}) = \overline{\theta(a)},$$

which however depends on a choice of representatives for $\bar{E}$ in E. In any case, $\bar{\theta}$ is a nilpotent $\bar{E}$-signalizer functor on $\bar{H}$, because in the balance condition we may apply Proposition I 9.12, and we find

$$\bar{\theta}(\bar{a}) \cap C_G(\bar{b}) = C_{\bar{\theta}(\bar{a})}(\bar{b}) = C_{\overline{\theta(a)}}(\bar{b}) = \overline{C_{\theta(a)}(b)} ,$$

and here the original balance condition applies.

Now R is infinite so the rank of $\bar{H}$ is less than that of G, and by induction the nilpotent $\bar{E}$-signalizer functor $\bar{\theta}$ is E-complete. In particular, $\langle \bar{P}, \bar{Q} \rangle$ is a $\bar{\theta}$-subgroup of $\bar{H}$. We claim that $S = \langle P, Q \rangle$ is a θ-subgroup of H, which contradicts the maximality of P and Q. What needs to be checked is the condition $C_S(a) \leq \theta(a)$, and indeed, $\overline{C_S(a)} \leq \bar{\theta}(\bar{a})$, so $C_S(a) \leq R\theta(a)$, and thus $C_S(a) \leq C_{R\theta(a)}(a) = \theta(a)C_R(a) = \theta(a)$. Thus S is a θ-subgroup and we have a final contradiction. □

PROPOSITION 9.4. *Let G be a reductive K^*-group of finite Morley rank and of even type, $p > 2$ prime, and T a p-torus in G of Prüfer p-rank 3. Suppose that*

$$(*) \qquad G = \langle U_2(C^\circ(a)) : a \in \Omega_1(T)^\times \rangle \ .$$

Then for $a \in T$ of order p, $C_G{}^\circ(a)$ is reductive.

PROOF. Let $E = \Omega_1(T)$ be an elementary abelian p-group of p-rank 3. For $a \in E^\times$ let $\theta(a) = O_2(C_G{}^\circ(a))$, which is connected by Lemma II 5.6. We claim that θ is a nilpotent E-signalizer functor on G; it is the balance condition that needs to be checked. For $a, b \in E$, we have $\theta(a) \cap C(b) = C_{\theta(a)}(b)$ connected by Lemma I 10.6. Thus writing $H = C^\circ(b)$, we have $\theta(a) \cap C(b) \leq O_2(C_H(a)) \leq O_2(H) \leq \theta(b)$, using Proposition II 5.24,

Now by Proposition 9.3 θ is E-complete, and $Q = \langle O_2(C(a)) : a \in E^\times \rangle$ is a nilpotent subgroup of G, and in particular is a proper 2-subgroup.

Now suppose toward a contradiction that Q is nontrivial. Then $M = N_G(Q) < G$, by reductivity. Let $A \leq T_p$ be elementary abelian of rank at least 2. Then $Q = \langle C_Q{}^\circ(a) : a \in A^\times \rangle$ by Proposition I 9.16. In particular $N(A)$ normalizes Q, that is $N(A) \leq M$.

Now we will show that $U_2(C^\circ(a)) \leq M$ for all $a \in \Omega_1(T)$, contradicting our hypothesis $(*)$.

For $x \in \Omega_1(T)$ choose $A \leq \Omega_1(T)$ elementary abelian of rank 2 and not containing x. Let $H = U_2(C^\circ(x))$. Then by Lemma II 5.26, $H = \langle C_H{}^\circ(a) : a \in A^\times \rangle = \langle C_H(\langle x, a \rangle) : a \in A^\times \rangle \leq M$ since each of the groups $A_a = \langle x, a \rangle$ involved is elementary abelian of rank 2. This contradicts our hypothesis $(*)$. □

10. Generic identification

The material in the present section gives a method for identifying the "generic" group of even type by a direct reduction to standard generators and relations. In our treatment of even type we can avoid this analysis by invoking Tits' powerful classification of buildings of spherical type and Tits rank at least three. The approach given in the present section is less general but more efficient in our particular case. The main result will be Theorem 10.2, but it reduces in turn to a theorem of Curtis-Tits-Phan type, with which we begin.

10.1. A theorem of Curtis-Tits-Phan type. Our first result is purely algebraic, and does not involve any model theoretic hypotheses.

PROPOSITION 10.1. *Let I be a connected Dynkin diagram of Tits rank at least three, let F be an algebraically closed field, and let G^* be the simply connected quasisimple algebraic group of type I over the field F. Let G be a group, and suppose that for each vertex i of I there is an associated subgroup L_i of G isomorphic to $\mathrm{SL}_2(F)$ or $\mathrm{PSL}_2(F)$, and a fixed maximal torus T_i of L_i, for which the following conditions hold.*

R1. *G is generated by the groups L_i for $i \in I$.*

R2. *The groups T_i commute.*

R3. *For any pair of vertices i, j of I, the structure of the group $L_{ij} = \langle L_i, L_j \rangle \leq G$ is compatible with the diagram I in the following sense:*

R3.0. *If there is no edge (i, j) in I, then L_i, L_j commute.*

R3.1. *If there is a simple edge (i, j) in I, then L_{ij} is isomorphic to $\mathrm{SL}_3(F)$ or $\mathrm{PSL}_3(F)$*

R3.2. *If there is a double edge (i, j) in I, then L_{ij} is isomorphic to $\mathrm{Sp}_4(F)$ or $\mathrm{PSp}_4(F)$; if the characteristic is not 2, there is a suitable isomorphism in which L_i corresponds to a root group for a long or short root according to whether the vertex i in I corresponds to a long or short root respectively, as indicated by the orientation of the corresponding edge in the Dynkin diagram.*

Then there is a homomorphism from G^* onto G which carries the root SL_2 subgroup of G^* corresponding to the vertex i onto the group L_i in G.

PROOF. In the simply connected group G^* we fix a maximal torus T and the associated root SL_2 groups as indexed by the diagram I, which we may denote L_i^*, and we consider the groups $L_{ij}^* = \langle L_i^*, L_j^* \rangle$ corresponding to pairs of vertices in I, which are the simply connected ([**173**, p. 47, §6]) versions of the Lie rank 2 groups encoded by the corresponding rank 2 Dynkin diagrams induced by I (including the reducible case, $A_1 + A_1$ corresponding to a product). We also consider the tori $T_i^* = T \cap L_i^*$ and $T_{ij} = T \cap L_{ij}^*$.

Our hypotheses amount to a family of surjective homomorphisms $h_{ij} : L_{ij}^* \to L_{ij}$, not necessarily respecting any further structure, other than the condition imposed in the case $L_{ij}^* \simeq \mathrm{Sp}_4(F)$, with characteristic not 2. We first need to adjust this family so that each homomorphism h_{ij} carries the quadruple $(L_i^*, L_j^*, T_i^*, T_j^*)$ onto the corresponding quadruple (L_i, L_j, T_i, T_j), so that furthermore for each i the homomorphisms $h_i : L_i^* \to L_i$ arising by restriction are independent of j.

To begin with, we adjust the $h_{i,j}$ separately, for those pairs for which there is an edge (i, j) in the Dynkin diagram, without regard for the compatibility condition. By an inner automorphism of $L_{i,j}$ we carry the image of $T_{i,j}^*$ to $T_{i,j}$. Then the images $\tilde{L}_i, \tilde{L}_j$ of L_i^* and L_j^* are root SL_2 subgroups for $T_{i,j}$, each one corresponding to a pair of opposite roots. If $L_{i,j} \simeq \mathrm{SL}_3$ or PSL_3, then the pair $(\tilde{L}_i, \tilde{L}_j)$ can be moved to (L_i, L_j) by a combination of a Weyl group element and graph automorphism (possibly trivial), which leave $T_{i,j}$ invariant. If $L_{i,j} \simeq \mathrm{Sp}_4$ or PSp_4, then by hypothesis if the characteristic is not 2, and using an "extra" automorphism available if the characteristic is 2, we may suppose that $\tilde{L}_i$, $\tilde{L}_j$ correspond to roots of the appropriate lengths, and use the Weyl group to align $(\tilde{L}_i, \tilde{L}_j)$ with (L_i, L_j).

Now we begin again with our system $L_{i,j}$ (for pairs (i,j) corresponding to edges in the Dynkin diagram). As the Dynkin diagram is a tree, the vertices may be ordered so that the graph induced on any initial segment is connected. Inductively, we will adjust the maps further, so that for $i_0 < i < j$ with (i_0, i) and (i,j) edges, the homomorphisms $h_{i_0,i}$ and $h_{i,j}$ agree on L_i^*. For the first two vertices we need do nothing, and otherwise at stage j we have a unique $i < j$ for which (i,j) is an edge of the Dynkin diagram, and we need to adjust the single homomorphism $h_{i,j}$ to match a given homomorphism $h_i : L_i^* \to L_i$ obtained by restricting $h_{i_0,i}$ for some neighbor i_0 of i; by induction, this homomorphism h_i is well defined. If h_i has a kernel, we may factor it out; to lighten the notation we will suppose that h_i is already an isomorphism. Then the map h_i' induced by $h_{i,j}$ is also an isomorphism, and we need to "untwist" $h_{i,j}$ by an automorphism of $L_{i,j}$ which should preserve the data (L_i, L_j, T_i, T_j) and operate on L_i as $\iota_i = h_i \circ h_i'^{-1}$. Now ι_i preserves T_i and is a combination of an inner automorphism induced by an element of T_i with a field automorphism. Both of these extend to $L_{i,j}$ preserving the data, so we succeed.

At this stage, we have a family $h_{i,j}$ of homomorphisms, for (i,j) an edge of the Dynkin diagram, compatible on their overlap. We may then extend this to a family $h_{i,j}$ defined for all pairs (i,j). This gives us a *map of amalgams* from the data presenting G^* to G, and applying Fact II 2.29, we have a homomorphism from G^* onto G, as stated. □

10.2. A generic identification theorem. Now we arrive at the main result, which gives sufficient conditions for a reduction to the Curtis-Tits-Phan type theorem above.

THEOREM 10.2. *Let G be a simple L^*-group of finite Morley rank and even type, and p an odd prime. Let T_0 be a maximal p-torus in G, supposed to be of Prüfer rank at least 3. Assume the following* generation *and* reductivity *hypotheses.*

(G) *G is generated by the subgroups*

$$U_2(C_G^\circ(x)) \text{ for } x \in T_0 \text{ of order } p.$$

(R) *For every element x of order p in T_0 the following conditions are satisfied:*

(R.1) *$U_2(C_G^\circ(x))$ contains no nontrivial p-unipotent subgroup;*

(R.2) *$U_2(C_{G^\circ}(x)) = F^*(U_2(C_{G^\circ}(x)))$.*

Then G is a Chevalley group over an algebraically closed field of characteristic 2.

The "genericity" to which we refer is the assumption of Prüfer rank at least three. Assumption (G) is a weak form of the generation of a Chevalley group by "root SL_2" subgroups associated with a fixed maximal torus, and conversely when assumptions (G) and (R) are combined one is in a position to reconstruct this pattern of root SL_2 subgroups. By itself, assumption (R) expresses the reductivity of centralizers of semisimple elements.

The proof makes use of the theory of complex reflection groups, and aims at the reconstruction of the group G as generated by appropriate "root SL_2" subgroups. We will first present our general strategy for making this reduction.

Fix the notation and hypotheses of Theorem 10.2 relating to G, p, and T_0. In addition, we introduce the set Σ consisting of all definable subgroups L of G with the following properties.

(1) T_0 normalizes L.

(2) L is isomorphic to SL_2 or PSL_2 over some algebraically closed field.

These groups are intended to serve as root SL_2-subgroups in our abstract group G. We aim to show that with a suitable labeling, these groups can in fact serve as the groups L_i of Theorem 10.1.

For $L \in \Sigma$, let $T_L = C_L(T_0)$. Then T_L will be a maximal torus of L, and its p-torsion belongs to T_0. But T_L may be larger than the definable hull of $T_0 \cap L$.

In the first place, we should show that we have enough "root groups".

LEMMA 10.3. *With the hypotheses and notation as above, G is generated by the groups K for $K \in \Sigma$.*

PROOF. Let $G_0 = \langle K : K \in \Sigma \rangle$. Let $\hat{T} = C_G(T_0)$ (we make no special claim about the structure of $\hat{T}$).

We have by hypothesis $G = \langle U_2(C^\circ(x)) : x \in T_0 \text{ of order } p \rangle$. We claim for $x \in T_0$ of order p we have the following:

$$(*) \qquad U_2(C^\circ(x)) \leq G_0\hat{T} \ .$$

This follows from our hypothesis on the structure of $U_2(C^\circ(x))$. We have $U_2(C^\circ(x)) = F(U_2(C^\circ(x)))E(U_2(C^\circ(x)))$, with

$$F(U_2(C^\circ(x))) \leq \hat{T},$$

since this nilpotent group centralizes any p-torus contained in it, as well as $E(U_2(C^\circ(x)))$. As far as $E(U_2(C^\circ(x)))$ is concerned, it is easy to see that T_0 acts on each component of $E(U_2(C^\circ(x)))$ like the p-torsion in a maximal torus. As Chevalley groups are generated by root SL_2-subgroups relative to a maximal torus, it follows that $E(U_2(C^\circ(x))) \leq G_0$. So $(*)$ holds.

Now applying our generation hypothesis we conclude that $G = G_0\hat{T}$. On the other hand, $\hat{T}$ normalizes G_0 and hence $G_0 \lhd G$. Since $G \neq \hat{T}$, we have $G_0 > 1$, and thus $G_0 = G$, as claimed. □

The second point is to get some control over the groups generated by pairs of "root SL_2-subgroups".

LEMMA 10.4. *Let $K, L \in \Sigma$. Then $C_{T_0}(\langle K, L\rangle) > 1$.*

PROOF. K and L are normalized by T_0, which has Prüfer rank at least three. The definable hull of T_0 acts on K and L by inner automorphisms, and hence the same applies to T_0. The image of T_0 in $\mathrm{Aut}(K) \times \mathrm{Aut}(L)$ has Prüfer rank at most 2, so the kernel is nontrivial. □

Note that in the foregoing lemma, as K and L are connected and definable the group $\langle K, L\rangle$ is also definable.

LEMMA 10.5. *Let K, L in Σ be noncommuting. Then $\langle K, L\rangle$ is isomorphic with a Lie rank 2 Chevalley group.*

PROOF. Let $M = \langle K, L\rangle$. We have $M \le U_2(C^\circ(x))$ for some $x \in T_0^\times$ of order p, by Lemma 10.4. By hypothesis (R.2), we have $U_2(C^\circ(x)) = F^*(C^\circ(x))$ and hence $K, L \le E(U_2(C^\circ(x)))$. Note that T_0 is a maximal p-torus of $C^\circ(x)$ and hence contains a maximal p-torus of $E(U_2(C^\circ(x)))$. As this normalizes K and L, it follows easily that each of K and L is contained in a single component of $E(U_2(C^\circ(x)))$; as K and L do not commute, these two components coincide and $M \le A$ with A a quasisimple component of $E(U_2(C^\circ(x)))$. Let $T_A = C_A(T_0)$, a maximal torus of A. It will suffice to show that K and L are root SL_2-subgroups of A, with respect to T_A, and then apply Lemma II 2.20.

Let U be a root subgroup of K, and $\hat{U}$ its Zariski closure in A. Then $\hat{U}$ is abelian and $T_0 \le N(\hat{U})$, hence $T_A \le N(\hat{U})$. Also $U \le [T_A, \hat{U}]$ and hence $\hat{U} = [T_A, \hat{U}]$ is unipotent, and a product of root subgroups for T_A. Let T_1 be the kernel of the action of T_0 on U. Then T_1 acts trivially on $\hat{U}$ and hence the Zariski closure $\hat{T}_1$ of T_1 acts trivially on $\hat{U}$. As T_0/T_1 has Prüfer rank 1, $\hat{T}_1$ is a torus of codimension 1 in T_A. Thus $\hat{U}$ is a root subgroup for T_A, say $\hat{U} = U_\alpha$. Then the Zariski closure of the opposite root subgroup in K must be $U_{-\alpha}$, since the pair generates a non nilpotent subgroup. So the Zariski closure $\hat{K}$ of K is a root SL_2-subgroup of A. Now consideration of the embedding $K \to \hat{K}$ produces a definable inclusion between the base fields, and hence must be the identity.

So K, and similarly L, is a root SL_2 subgroup of A. □

As a consequence we have the following, which is not essential to the argument, but is certainly welcome at this stage.

LEMMA 10.6. *The base fields of the groups $K \in \Sigma$ are definably isomorphic.*

PROOF. Let Δ be the graph whose vertex set is Σ, with edges between noncommuting pairs. As G is generated by the groups in Σ, the subgroup generated by those groups lying in one connected component of Δ is normal in G. Thus Δ is connected. On the other hand, for two noncommuting groups $K, L \in \Sigma$ it is easy to find a definable isomorphism between their base fields, in view of the structure of $\langle K, L\rangle$. □

Now we need to improve on the foregoing by finding a labeling of the root subgroups by the vertices of a Dynkin diagram in the manner of Theorem 10.1. We note that in practice, when we apply this result, we may well have the necessary data already in hand, but we wish to work at a substantial level of generality. So we will argue that our present hypotheses allow us to find an appropriate Coxeter group in G, and to extract the Dynkin diagram, and the relevant information, from the Coxeter group. For this we will take a detour through the theory of complex reflection groups, for which the background is found in §I 13.

10.3. The Coxeter group. Set $T = \langle T_L : L \in \Sigma\rangle$. In view of the structure of the groups $\langle K, L\rangle$ for $K, L \in \Sigma$, the tori T_L commute for $L \in \Sigma$ and T is a definable divisible abelian group with $T \leq C(T_0)$. One expects $T = C(T_0)$, but this would be deduced after the identification has been made. In the meantime we work with T as a well-behaved torus, not embedded as yet in an algebraic group.

For each group $L \in \Sigma$ we have a Weyl group $N(T_L)/T_L$ of order 2, and we choose an element $r_L \in L$ representing a generator. Then r_L acts as an involution on T, and the action is independent of the choice of r_L. So r_L represents an element w_L of the "big Weyl group" $W = N_G(T)/C_G(T)$. We will not work with W, but rather with the subgroup $W_0 = \langle w_L : L \in \Sigma\rangle$ generated by our distinguished involutions. We need to show that W_0 is a Coxeter group, that is a finite group generated by real reflections.

By the criterion of Theorem I 13.2, it suffices to show the following points, where the group in question is W_0, I is the distinguished set of involutions generating W_0, and n is the Prüfer p-rank of T:

(1) The set I is closed under conjugation in W_0.
(2) The graph Δ_I on the vertex set I in which edges correspond to noncommuting pairs of involutions is connected.
(3) For all sufficiently large prime numbers ℓ, W_0 has a faithful irreducible representation over $\mathbb{F}_\ell$ in which the elements of I act as generalized reflections.
(4) In the action of W_0 on T_0, the elements of I act as reflections of order 2, and have no common fixed points.
(5) W_0 has an irreducible representation of dimension at least three over some field.
(6) W_0 is finite.

Now W_0 acts on the set Σ of distinguished "root SL_2" subgroups, hence preserves I. In view of the structure of the groups $\langle K, L\rangle$ for $K, L \in \Sigma$, if K and L do not commute then w_K and w_L do not commute, so the graph Δ_I is connected. This disposes of the first two points. For the rest, we must examine the action of W_0 on T, and specifically on the subgroup

$T_\ell = T[\ell]$ consisting of the torsion of exponent ℓ. We claim that all of these W_0-modules are faithful and irreducible, with the generators r_L acting as reflections of order 2. As the Prüfer p-rank is at least three, the module T_p is at least 3-dimensional over $\mathbb{F}_p$. Furthermore, as these representations are finite, if they are faithful, then W_0 is finite. So this will suffice.

As far as the action of r_L on T_p is concerned, we have $T = T_L C_T(L)$ and thus r_L acts as a reflection of order 2.

For the irreducibility, since the representations are generated by reflections and the graph Δ is connected, it suffices to check that the r_L have no common centralizer in T. But an element of T which centralizes r_L must centralize L and hence $C_T(W_0)$ centralizes the subgroup generated by all $L \in \Sigma$, which is G (Lemma 10.3).

So it remains only to check that these representations are faithful. Let $N = N_G(T)$. We claim that more generally the action of $N/C_G(T)$ on each T_ℓ is faithful (for ℓ odd), or in other words that $C_N(T_\ell)$ centralizes T.

So consider $x \in N(T)$ centralizing T_ℓ for some prime ℓ. Then x acts on the set Σ. If $L \in \Sigma$, then $L \cap L^x$ contains $T_\ell \cap L$. If $L \neq L^x$, then $|L \cap L^x| \leq Z(L)$ has order at most 2, a contradiction. So for each $L \in \Sigma$, x acts on L and centralizes $T_\ell \cap L$. As x normalizes $T \cap L$ and acts as an inner automorphism of L, it either inverts or centralizes $T \cap L$; since it centralizes $T_\ell \cap L$, x centralizes T_L. Since this holds for all L, x centralizes T.

So we have

$$W_0 \text{ is a crystallographic Coxeter group.}$$

By the proof of that result, the generators r_L correspond to reflections in W_0 (that is, elements of W_0 which act as reflections in the usual real representation of W_0). We claim:

$$\text{All reflections of } W_0 \text{ are of the form } r_L \ (L \in \Sigma) .$$

Since the set of generators r_L is closed under conjugation, and since reflections corresponding to roots of fixed length are conjugate, there are only two possibilities: either the reflections r_L exhaust all reflections in W_0, or else there are two root lengths, and the r_L vary over roots of one length. But in the latter case the group generated by the r_L is associated to the root system consisting of roots of that fixed length, and is not the group W_0. So this proves our claim.

Now the group W_0 largely determines the associated Dynkin diagram, apart from an indication of root lengths. So let I_0 be the Dynkin diagram without the root length information, correlated with a set r_i $(i \in I_0)$ of reflections generating W_0. Here $r_i = r_{L_i}$ for some $L_i \in \Sigma$.

We claim the groups L_i $(i \in I_0)$ generate G. Let $L \in \Sigma$. We claim that $\langle L_i : i \in I_0 \rangle$ contains L. Since r_L is conjugate to some r_i, $i \in I_0$ under the action of W_0, and W_0 is generated by the r_i, with $r_i \in L_i$, we may suppose that $r_L = r_i$ for some $i \in I_0$. On the other hand, if L and L_i are distinct, we have already determined the structure of $\langle L, L_i \rangle$, a Chevalley group of

Lie rank 2, and if $L \neq L_i$, it follows that $r \neq r_i$. So we have $L = L_i$ in this case, and our claim holds.

At this point we have the data needed for the application of Proposition 10.1. As we are in characteristic 2 the condition R3.2 falls away. So we conclude by applying that proposition.

11. Notes

§1. Pseudoreflection groups

This material is applied in Chapter VII and represents a major departure from the approach taken in the finite case.

§2. Zassenhaus groups

This material is used in Chapter VI.

The classification theorem of DeBonis and Nesin is given in [**82**], and makes use of a body of material from [**83**]; it is also treated completely in [**52**, 11.90, p. 245]. Its hypotheses are the goal to which all of the analysis in Chapter VI is directed. The theorem is applied in §VI 4; the remainder of that chapter is devoted to obtaining contradictions in configurations that diverge from this one.

Of course, the extensive calculations with elements of order three are modeled on DeBonis/Nesin. Calculations with a similar flavor occur also in the finite theory.

§3. Suzuki groups

This material deals with an extreme configuration which arises in Chapter VI, specifically in §§VI 1,VI 7, and again in the last section of Chapter VII. Some of its methods are also used in §4.

In the finite case Suzuki 2-groups were both named and classified by G. Higman [**112, 119**].

The topic was treated in the context of groups of finite Morley rank by Davis and Nesin [**80**], whose treatment we follow with minor adjustments. They also deal in part with Suzuki 2-groups without assuming freeness, a matter which we generally leave aside.

Here we required Suzuki groups to be nilpotent. It suffices for the arguments given to have them nilpotent by finite, which is no restriction at all in the context of groups of finite Morley rank, since we deal with 2-groups. We do not know whether the main result holds for quadratically closed fields, in the absence of a model theoretic hypothesis. One might also wish to examine the broader framework in which locally nilpotent groups are allowed.

We note that Nesin's work on Zassenhaus groups and free Suzuki groups dealt very early on with two essential minimal configurations on which the entire classification of groups of even type is based. It took quite some time before the theory was sufficiently developed to make good use of these results.

§4. Landrock-Solomon

This result was given in [**4**, §4]. It corresponds closely in content to the result of Landrock and Solomon in [**132**]. The method of proof used here relies on connectedness.

We noted the relationship of Proposition 3.8 with the analysis of Suzuki 2-groups in [**80**]. There is a close connection between these two topics. This proposition finds further application in a different situation arising in the detailed analysis of a configuration we encounter late in Chapter VII, specifically in the proof of Lemma VII 1.7.

§5. A theorem of Baumann

In this section we carry over the results in [**175**] to the context of groups of finite Morley rank of even type. The adaptation to groups of finite Morley rank was given in [**6**, Appendix], following closely the notation and the arguments of [**175**], which uses the amalgam method.

While that method is not intrinsically tied to the theory of finite groups, and lends itself to much broader application, still there are some deviations here from the arguments as given in [**175**], and that for a number of reasons: on the one hand the representation theory of $\mathrm{SL}_2(K)$ over the field of 2 elements now involves infinite dimensional representations, and in particular ranks are used rather than dimensions to compare sizes; second, rather than passing to a free product with amalgamation (or in graph theoretical terms, a tree) we work here in the context of an ambient group of finite Morley rank. However, in Chapter IX, which concerns "quasithin" groups, we will work with the amalgam method in another way, leaving the category of groups of finite Morley rank and returning to it at the end, much as occurs in the case of finite groups.

There is also an important deviation is in the statement of the theorem itself. Our condition (P) is weaker than the most natural analog of Stellmacher's condition, and thus the result is stronger. It is essential for applications that the proof goes through with this slightly weaker assumption.

Our main goal is the simpler statement given later as Proposition VIII 2.2,

§6. Generalized n-gons

The classification in the case of finite Morley rank (Theorem 6.2) is from [**130**], and relies on a major classification theorem announced by Tits, and proved by Tits and Weiss in a modified form. The proof of the classification theorem is outlined in [**140**], and given in full in [**185**], along with existence proofs and a great deal of clarifying material. They reduce the three complex cases to simpler cases. This was prior to the publication of [**185**], and in that context was by far the easiest approach to document. The potential nondefinability of the ambient field in case 16.4 of [**185**] is noted in [**130**, 3.6].

§7. Buildings and (B, N)-pairs

The classification is from [**130**].

The criterion in terms of pairwise BN-pairs is from [**149**], and is the basis of the proof of Theorem 8.1 in the following section.

We noted that Fact 7.15 incorporates within it a characterization of Coxeter groups due to Goldschmidt. One of the essential points in any classification project like ours is to pin down the Coxeter group at some point. We have two different approaches to this; one uses the theory of complex reflection groups, and the other goes via this result of Goldschmidt, but tacitly, when Fact 7.15 is invoked.

§8. A theorem of Niles

The finite version is in [**149**], and the adaptation to the case of finite Morley rank is in [**35**].

§9. Signalizer functors

The nilpotent signalizer functor theorem generalizes a result given in [**52**]. In the finite case one has also a solvable signalizer functor theorem, lacking in general in the context of finite Morley rank because of difficulties with torsion-free groups. In odd type groups, Burdges has shown that one can extract a nilpotent signalizer functor from a solvable one.

§10. Generic identification

This material comes from [**37**], with variations incorporated from [**60**]. The general approach was suggested to us by Lyons, and is similar to [**105**, Cor. 2.9.6]. Compare [**182**].

While this material allows us to circumvent the classification of buildings in Tits rank at least three, we still rely on the classification of Moufang polygons in our treatment of quasithin groups, in conjunction with the amalgam method. The mass of material needed could be reduced by taking into account the full configuration on hand, but it could not be easily done away with completely. The axioms for Moufang polygons capture much of the essence of the information obtained by the amalgam method, so in sense the logical continuation of the argument at that point amounts to the classification of at some classes of Moufang polygons, though from a point substantially farther along than at the beginning of that theory.

There is no particular reason to restrict the argument to characteristic two, but in that case one would work with K^*-groups rather than with L^*-groups.

CHAPTER IV

Generic Covering and Conjugacy Theorems

If you already have group coverage, the experts recommend hanging onto it no matter what ...
— Teresa McUsic,
Heal (Summer 2007)

Introduction

The present chapter will complete our development of general tools useful for the classification problems to be dealt with in Parts B and C. It is complementary to the preceding chapter, dealing in topics in which global model theoretic ideas take precedence over the close consideration of specific configurations, with a leading role being played by considerations of genericity and generic conjugacy. We are particularly fond of this chapter, as it represents an approach which is both geometrical and model theoretic, and whose DNA carries markers associated with the theory of algebraic groups. This has a noticeable impact on the flavor of the theory as a whole, which is not simply the transposition to our domain of the very powerful methods of finite group theory.

In the first section we will again encounter good tori and begin to make good use of them. In particular, we will see in Proposition 1.15 that maximal good tori are conjugate. The following section introduces generic covering arguments, based on very direct rank computations. These lead in certain cases to the conclusion that the conjugates of a certain subgroup, or coset, are generic in the ambient group, and this type of fact may be used sometimes as a substitute for a conjugacy theorem, or in other cases as a step on the way toward a conjugacy theorem (as is already the case in §1 with the conjugacy of maximal good tori). There are further results which arise when one has a generic covering by good tori, and these are considered separately in §3.

In the last two sections of the chapter we arrive at considerably more concrete results.

In §4 we prove the striking Theorem 4.1, which states that a connected group of degenerate type contains no involutions. Here the model theoretic ideas of the present chapter are combined with techniques we have not seen otherwise, coming from computational group theory and specifically the theory of "black box groups". This result is a very convenient one to have available when working with L-groups, as it gives a measure of control over precisely those degenerate sections about which we have assumed nothing. We would like to note, however, that this result was not available when the material of Parts B and C was first worked out, and its absence did not in fact lead to major complications. Naturally, with that result in hand, we take advantage of it wherever possible. Combining this result with the methods that we use in Parts B and C to obtain classification results, we can also answer some questions put forward by Poizat in the general theory of groups of finite Morley rank. These were intended to cast some light on the difficulties associated with the classification project, so it seems appropriate that we can now use the techniques developed for classification to settle some of these problems, relating to generic equations and to the nontriviality of connected centralizers.

In §5 we come back to the further consideration of pseudoreflection groups, extending the existing K-group theory to the L-group context. This has been deferred from its normal position in Chapter III to allow us to make use of the model theoretic ideas of the present chapter.

One of the central techniques of this chapter is based on Lemma I 3.6, the irreducibility of connected groups. The underlying philosophy is as follows: one way to approximate a counterexample to the Algebraicity Conjecture is by considering a disconnected group whose connected component is algebraic; elements lying outside the connected component may then exhibit pathological properties, if one thinks of them as elements of the connected component. Configurations often arise in practice which can be interpreted in this way. The problem is then to reach a contradiction by showing that the group in question is disconnected, and this is often done by constructing two disjoint generic sets. One can say a little bit more about the origin of these generic sets. One of them would normally consist of the generic elements of the connected component; another would consist of the pathological elements (and their immediate relatives), making up a generic subset of a distinct coset of the connected component. Still, sometimes one cannot get the first generic subset at all, but one may compensate by getting two generic, disjoint, pathological subsets. This happens in particular in the analysis of groups of degenerate type.

The basic mechanism for constructing generic sets is the following: if an almost self-normalizing subgroup is generically disjoint from its conjugates (Def. 1.1), then the union of its conjugates is generic in the ambient group. This principle also applies to cosets of almost self-normalizing subgroups (Lemma 1.8).

Overview

The main line of argument in the present chapter is the following. Let G be a connected group of finite Morley rank, and H a definable connected subgroup which is almost self-normalizing in the sense that $H = N_G{}^\circ(H)$. Then heuristically one might expect the union $\bigcup H^G$ of its conjugates to be generic, based on the expectation that

$$\mathrm{rk}(\bigcup H^G) = \mathrm{rk}(\{H^g : g \in G\}) + \mathrm{rk}(H) = \mathrm{rk}(G) - \mathrm{rk}(N(H)) + \mathrm{rk}(H) = \mathrm{rk}(G)\,.$$

Only the first equation is doubtful here; it will hold in the very common case in which the conjugates of H are *generically disjoint* in the following sense:

$$\mathrm{rk}(H \cap \bigcup\{H^g : g \notin N(H)\}) < \mathrm{rk}(H)$$

or more generally, if the set of elements of H lying in *infinitely many* conjugates of H is non-generic in H.

Now the intersection of two generic subsets of G will itself be generic, so the significance of this kind of result emerges when one either has another generic subset already in view, or when the result can be applied to two distinct groups.

For example, Proposition 1.15 implies the following.

Conjugacy of Maximal Good Tori. *Let G be a group of finite Morley rank. Then any two maximal good tori of G are conjugate.*

Here a good torus is one for which every definable subgroup is the definable hull of its torsion subgroup. The result is proved more generally, for decent tori (where the group in question is the definable hull of its torsion subgroup), but this is a technical variation. Specific properties of good tori are essential for the proof in either version.

Let us unpack this a little. Let T_1 and T_2 be two maximal good tori, and consider the groups $H_i = N^\circ(T_i)$ for $i = 1, 2$.

The first point is that with $H = H_1$ or H_2, the theory alluded to above applies to H: H is almost self-normalizing and generically disjoint from its conjugates. The first of these two claims is straightforward, the second considerably more subtle. Both depend on *rigidity properties* of good tori, three in number:

- $N^\circ(T) = C^\circ(T)$.
- Any uniformly definable family of subgroups of T is finite.
- Any uniformly definable family of homomorphisms $H \to T$ (with H and T both fixed) is finite.

The proofs of these rigidity principles are straightforward: the first holds for decent tori by the results of Chapter I, and the other two are proved by working in a highly saturated model (that is, a large elementary extension, analogous to a universal domain) and observing that as the sets in question are controlled by torsion elements, they have at most $2^{\aleph_0}$ elements, and as this is absolutely bounded regardless of the model considered, these sets must be finite. The underlying idea here is captured quite well by the following instance: if the cardinality of the set of rational points on a variety is bounded, independently of the base field, then the variety is finite.

So suppose we know that $\bigcup H_i^G$ is generic for $i = 1$ and 2; then we may suppose in particular that $H_1 \cap H_2 > 1$, and consider an element $a \neq 1$ in the intersection. By the first rigidity principle, we have $T_1, T_2 \leq C^\circ(a)$. If $C^\circ(a) < G$, then we can argue inductively. The other possibility is that $Z(G) > 1$, in which case either $Z(G)$ is infinite and one argues inductively again, after passing to a quotient, or else, finally, $Z(G)$ is finite and after factoring it out, one reduces to the case $Z(G) = 1$.

This line of argument illustrates quite well most of the themes of the present chapter. There is one other point of an entirely different, and essential, character, coming from black box group theory, used in §4 in the proof that connected degenerate groups contain no involutions. Here there are two quite distinct lines of argument. Supposing there are involutions (and taking some pains to make sure they are noncentral) one considers a fixed conjugacy class I_0 of involutions, and one asks the following question: for a generic pair (i, j) of involutions in I_0, does $d(\langle ij \rangle)$ contain a third involution? If the answer is *No*, black box group theory leads to the conclusion that $C(i)$ is *connected*, which is absurd, or rather becomes absurd if one works in a minimal counterexample to our claim: a connected group of degenerate type containing involutions, and so that *no proper connected definable subgroup* contains involutions. If, on the other hand, the answer is *yes*, we argue in a very different way, using a genericity computation of the sort sketched above.

Let us examine this second genericity argument. Taking G once more connected, of degenerate type, and containing (noncentral) involutions, and *minimal* in the sense that no proper definable connected subgroup contains involutions, we attach to an involution i the group $H_i = N^\circ(\dots N^\circ(C^\circ(i))\dots)$ obtained from $C^\circ(i)$ by taking normalizers until the sequence stabilizes. As one easily reduces to the case in which the ambient group is simple, one has $H_i < G$ and by hypothesis $i \notin H_i$, while of course $i \in N(H_i)$. Now a close examination of the coset iH_i reveals that H_i is recoverable from any element of this coset, and thus no element can belong to two distinct conjugates of this coset. So we have the following: H_i is almost self-normalizing, and the coset iH_i is disjoint from its conjugates. Then a minor variation of the argument sketched above shows that the union $\bigcup(iH_i)^G$ is a generic subset of G. Then arguing somewhat as in the case of maximal good tori, it follows that the Sylow 2-subgroup of G is *elementary abelian*; the reason for this is that we can work with elements of order four much as we work with involutions, and arrive at two disjoint generic subsets of the same general form if there are any such elements.

We have seen at this point that very simple genericity arguments give a great deal of control over the Sylow 2-subgroup. Let us now sketch the final contradiction toward which this all tends. Fix a Sylow 2-subgroup S. Consider a generic pair of involutions (i, j) (they are all conjugate now, as a consequence of our genericity arguments). Then (j, i) is another generic pair, and has the same type over S as (i, j) does; this is an instance of the Fubini principle (or symmetry) of Chapter I. Suppose $d(\langle ij \rangle)$ contains a third involution k. Then of course $d(\langle ji \rangle)$ contains this same involution.

From the pair i, j we define the set

$$X_{ij} = \{(a, b) \in S : \exists g (i, k)^g = (a, b)\}$$

and then with this definition we have

$$X_{ji} = \{(a, b) \in S : \exists g (j, k)^g = (a, b)\} .$$

But as the pairs i, j and j, i have the same type over S, these sets coincide, and hence the pairs (i, k) and (j, k) are conjugate, or in plainer language: i is conjugate to j under the action of $C(k)$. But by computation in $d(\langle i, j \rangle)$, j is conjugate to ik under the action of $C(k)$, and thus finally i is conjugate to ik under this action, at which point we can replace the conjugating element by a 2-element and conclude that the Sylow 2-subgroup is nonabelian, a final contradiction.

Of course we simply invoked the hypothesis that for the generic pair (i, j) above, we do find an involution in $d(\langle ij \rangle)$, which is not particularly plausible. To eliminate the more plausible alternative we have recourse to a black box group technique for which the reader can consult §4.4

Now let us mention some other results which play a leading role in later chapters. The first repackages Wagner's results on fields of finite Morley rank in positive characteristic along with the conjugacy argument.

LEMMA 1.6. *Let $A \rtimes H$ be a group of finite Morley rank in which A, H, and the action are definable. Assume that A is 2-unipotent and H is connected of degenerate type, and acts faithfully on A. Then the Borel subgroups of H are good tori, and are conjugate in H.*

The results in §2 are proved by genericity arguments involving cosets, like the one we gave above, but in the simpler case when the group H under consideration is also generically disjoint from its conjugates, so that the set $\bigcup H^G$ is already generic, and any other generic set must meet it (generically).

In §3 we encounter a covering lemma which seems rather technical—and certainly the proof is rather technical—but proves to be very handy. In this result, we assume the ambient group has finite Morley rank and is also sufficiently saturated. Then, specializing the statement slightly, we have the following.

THEOREM 3.1. *Let $\mathcal{F}$ be a uniformly definable family of good tori in G such that $\bigcup \mathcal{F}$ is generic in G. Then there is a maximal good torus in $\mathcal{F}$.*

Let us consider just the simplest case to see why this is plausible. Suppose that G is itself a good torus. In that case, the family $\mathcal{F}$ is finite by a rigidity principle. So then one of the groups in it must be G. Of course, the statement as we give it is considerably more powerful, and involves some extended rank computations, but in the end it comes down to the same principle. But we have not given it in its most general form, and this is important for applications. We can extend the theory of good tori to *rigid abelian groups*, with a similar definition but omitting connectedness. For example, a definable subgroup of a good torus is a rigid abelian group, and this allows for greater flexibility. We need the version of Theorem 3.1 corresponding to this broader setting. So with the same hypotheses on G, we may state the following.

THEOREM 3.1. *Let $\mathcal{F}$ be a uniformly definable family of rigid abelian subgroups of G such that $\bigcup \mathcal{F}$ is generic in G. Then there is a group $A \in \mathcal{F}$ such that A° is a maximal good torus in $C(A)$.*

To see that this is really a stronger statement, imagine for a moment that every group in $\mathcal{F}$ is finite. Then the orders of the groups in $\mathcal{F}$ are bounded, and the elements of G are generically of finite order, a situation that may be called "Poizat's nightmare". In that case, the conclusion is that G contains no good tori, a result which is not entirely trivial.

The proof of Theorem 3.1 is rather unpleasant, largely because of unavoidable definability issues that lead to a very heavy notation. But the trivial instance we have given above really does capture the underlying idea.

At the end of this chapter, we return to the consideration of pseudoreflection groups and finally show that the classification extends to the L-group context. Here the pseudoreflection groups themselves tend to play the role of maximal good tori, and the line of analysis is at points very much in the same vein as the preceding.

1. Borel subgroups

1.1. Almost self-normalizing good tori. A good deal of what we do here will be extended in the next subsection, notably by replacing almost self-normalizing good tori by maximal good tori. But the methods here are more direct and the special configurations considered are of particular importance. Furthermore, the essential rank computations are the same. Still, the more powerful version in the following subsection also plays a central role.

DEFINITION 1.1. *Let G be a group of finite Morley rank and H a definable subgroup of G.*

(1) *H is* almost self-normalizing *if H is of finite index in $N(H)$, or in other words, $N^\circ(H) \subseteq H$.*
(2) *H is* generically disjoint from its conjugates *in G if*

$$H \setminus \textstyle\bigcup_{g \in G \setminus N(H)} (H \cap H^g) \text{ is generic in } H$$

or, in other words,

$$\operatorname{rk}(H \setminus \bigcup_{g \in G \setminus N(H)} (H \cap H^g)) = \operatorname{rk}(H) .$$

In this definition, it is not necessary to take H to be connected.

LEMMA 1.2. *Let G be a connected group of finite Morley rank and H a definable subgroup of G which is almost self-normalizing and generically* *disjoint from its conjugates in G. Then the union of the conjugates of H in G is generic in G.*

PROOF. Let $X = H \setminus \bigcup_{g\in G\setminus N(H)}(H \cap H^g))$. We have assumed $\mathrm{rk}(X) = \mathrm{rk}(H)$. Observe that X is invariant under $N(H)$, and hence we can speak of X^γ for $\gamma \in N(H)\backslash G$. Thus $\bigcup X^G = \bigcup X^{(N(H)\backslash G)}$ (here $N(H)\backslash G$ is a set of cosets). Furthermore, if X^g meets $X^{g'}$, then $N(H)g = N(H)g'$ and thus the latter union is disjoint: $\mathrm{rk}(X^G) = \mathrm{rk}(X) + \mathrm{rk}(N(H)\backslash G) = \mathrm{rk}(H) + \mathrm{rk}(G) - \mathrm{rk}(N(H)) = \mathrm{rk}(G)$ by hypothesis, and of course $\bigcup H^G \supseteq \bigcup X^G$. □

The condition of generic covering by conjugates is of fundamental importance. In [**122**] a group such that the union of its conjugates is generic is called *generous*. In reductive algebraic groups, these are the groups containing maximal tori.

LEMMA 1.3. *Let G be a connected group of finite Morley rank, and let K_1, H_1, H_2 be definable subgroups of G with $K_1 \le H_1$. Suppose in addition:*

(1) *K_1 is almost self-normalizing in G;*
(2) *K_1 is generically disjoint from its conjugates in G*
(3) *$\bigcup_G H_2^G$ is generic in G.*

Then the conjugates of H_2 in G generically cover H_1, in the sense that $H_1 \cap \bigcup H_2^G$ is generic in H_1.

PROOF. Let $X = K_1 \setminus \bigcup_{g\in G\setminus N(K_1)} K_1^g$. Then X is $N(K_1)$-invariant, and $N(K_1)$ is the full stabilizer of X in G, while distinct conjugates of X are pairwise disjoint.

By hypothesis, $\mathrm{rk}(X) = \mathrm{rk}(K_1)$. For any $Y \subseteq X$ which is $N(K_1)$-invariant, and for any group H containing K_1, we find:

$$\mathrm{rk}(\bigcup Y^H) = \mathrm{rk}(Y) + \mathrm{rk}(N_H(K_1)\backslash H) = \mathrm{rk}(Y) + \mathrm{rk}(H) - \mathrm{rk}(K_1)$$

and hence $\bigcup Y^H$ is generic in H if and only if Y is generic in X.

Taking $H = H_1$ here, we see that it suffices to show that $Y = X \cap \bigcup H_2^G$ is generic in K_1.

Let $Z = \bigcup H_2^G$. If $X \setminus Z$ is generic in K_1, then $\bigcup(X \setminus Z)^G$ is generic in G, by our first remark with $H = G$; but for any g, $(X \setminus Z)^g \le G \setminus Z$, and Z is generic in G by hypothesis, which, as G is connected, is a contradiction. So $X \setminus Z$ is nongeneric in K_1, and hence $X \cap Z$ is generic in K_1, as required. □

LEMMA 1.4. *Let G be a connected group of finite Morley rank and suppose that T is a good torus in G. Then T is generically disjoint from its conjugates in G.*

PROOF. The family $\mathcal{F} = \{T \cap T^g : g \in G\}$ is a uniformly definable family of subgroups of T. Hence by Lemma I 4.23, this family is finite. It follows that the union $\bigcup_{g\in G\setminus N(T)} T \cap T^g$ is a finite union, hence nongeneric in G. □

The following is a special case of the principle that maximal good tori are conjugate, which will be given below.

LEMMA 1.5. *Let G be a connected group of finite Morley rank, and T_1, T_2 almost self-normalizing good tori in G. Then T_1 and T_2 are conjugate in G.*

PROOF. We apply Lemma 1.3 with $K_1 = H_1 = T_1$ and $H_2 = T_2$. As both are good tori, they are generically disjoint from their conjugates and hence the union of their conjugates is generic in G. The hypotheses of the lemma apply, and T_1 is generically covered by the uniformly definable family $\{T_1 \cap T_2^g : g \in G\}$, which must be a finite family since T_1 is a good torus. So we have $T_1 \subseteq T_2^g$ for some g, and as T_1 is almost self-normalizing we find $T_1 = T_2^g$. □

LEMMA 1.6. *Let $A \rtimes H$ be a group of finite Morley rank in which A, H, and the action are definable. Assume that A is 2-unipotent and H is connected of degenerate type, and acts faithfully on A. Then the Borel subgroups of H are good tori, and are conjugate in H.*

PROOF. The Borel subgroups of H are $2^{\perp}$-groups and hence are good tori by Proposition I 11.7. Borel subgroups are always almost self-normalizing, so the preceding corollary applies to show that the Borel subgroups are conjugate. □

COROLLARY 1.7. *Let $A \rtimes H$ be a group of finite Morley rank in which A, H, and the action are definable. Assume that A is 2-unipotent, H is of degenerate type, and the kernel of the action is a solvable subgroup of H. Then the Borel subgroups of H are conjugate.*

The following variant of Lemma 1.2 is occasionally useful. The proof is essentially the same.

LEMMA 1.8. *Let G be a connected group of finite Morley rank, H a definable subgroup of G which is almost self-normalizing, and H_1 a coset of H in G which is generically disjoint from its conjugates in G. Then the union of the conjugates of H_1 in G is generic in G.*

PROOF. The only point requiring attention is the notion of generic disjointness. Here one considers $H_1 \setminus \bigcup_{g \in G \setminus N(H)} H_1^g$. A point to note is that for $g \in N(H)$ we have either $H_1^g = H$ or $H_1 \cap H_1^g = \emptyset$, so we may replace $N(H)$ here by $N(H_1)$, if we extend the notation so that the latter is in fact defined. With these conventions one may then use the same proof. □

We will pursue this important theme further in Lemma 2.1. What we have so far is sufficient for a number of applications.

1.2. Maximal good tori and maximal decent tori.

LEMMA 1.9. *Let H be a connected group of finite Morley rank, T a good torus of H. Suppose T is central in H. Let $\mathcal{F}$ be a uniformly definable family of subgroups of H, none of which contain T.*

Then the union $\bigcup \mathcal{F}$ is not generic in H.

Proof. There are only finitely many intersections of the form $X \cap T$ for $X \in \mathcal{F}$ (Lemma I 4.23). So passing to a subfamily of $\mathcal{F}$ we may suppose that the intersections $X \cap T$ for $X \in \mathcal{F}$ are independent of X. After taking a quotient we may even suppose that $X \cap T = 1$ for $X \in \mathcal{F}$.

Now suppose $\bigcup \mathcal{F}$ is generic in H. Then by the Fubini principle (Lemma I 2.1 (5)), there is a coset V of T in H such that $V \cap \bigcup \mathcal{F}$ is generic in V. We may suppose that $g \in V \cap \bigcup \mathcal{F}$ is chosen to minimize the rank and degree of the definable hull $d(g)$.

Then for $gt \in V \cap \bigcup \mathcal{F}$, we have $d(gt) \leq d(g) \times T$, and $d(gt) \cap T = 1$. Hence the projection $\pi_1 : d(gt) \to d(g)$ is injective, and by the choice of g, also surjective. It follows that the group $d(gt)$ is the graph of a homomorphism $h_t : d(g) \to T$. Furthermore, if $gt \in X \in \mathcal{F}$, and we set $\tilde{X} = X \cap (d(g) \times T)$, then as $d(gt) \leq \tilde{X}$ and $\tilde{X} \cap T = 1$, the same considerations show that $d(gt) = \tilde{X}$. Thus the family of homomorphisms $\{h_t : gt \in \bigcup \mathcal{F}\}$ is uniformly definable, and hence finite (Lemma I 4.23).

On the other hand, for $X \in \mathcal{F}$, we have $|V \cap X| \leq 1$, and hence $V \cap \bigcup \mathcal{F}$ is finite, contradicting the genericity. □

It is sometimes useful to extend the range of applicability of this result a little, at the price of a few additional lemmas.

Definition 1.10. *A divisible abelian group T of finite Morley rank is called a* decent torus *if it is the definable hull of its torsion subgroup.*

There are a number of alternative ways to formulate this notion.

Lemma 1.11. *Let T be a divisible abelian group of finite Morley rank. Then the following conditions are equivalent.*

(1) *T is the definable hull of its torsion subgroup.*
(2) *Any nontrivial quotient of T by a definable subgroup contains torsion.*
(3) *Any quotient of T by a maximal proper connected definable subgroup of T is a good torus.*
(4) *$T/\Phi(T)$ is a good torus.*

Proof. (1) $\Longrightarrow$ (2). If K is a proper definable subgroup of T, then by (1) the torsion of T is not contained in K, and thus T/K contains torsion.

(2) $\Longrightarrow$ (3). Let $H < T$ be a maximal proper connected subgroup of T. If the torsion subgroup of T/H is finite, then after factoring it out we contradict (2). So the torsion subgroup of T/H is infinite and by minimality of T/H, the latter is a good torus.

(3) $\Longrightarrow$ (4). As $\Phi(T)$ is a finite intersection of maximal proper definable connected subgroups of T, this implication follows from the closure properties of the class of good tori with respect to extensions.

(4) $\Longrightarrow$ (1). let T_0 be the definable hull of the torsion subgroup of T. If $T_0 < T$, then $T_0\phi(T) < T$ by Lemma I 5.19 and thus $T/T_0\phi(T)$ is a

nontrivial good torus. In particular, this quotient contains torsion, and as that torsion lifts to T (Lemma I 2.18), we contradict the definition of T_0. □

We need some formal properties as well.

LEMMA 1.12.

(1) *If T is a decent torus and $\bar{T}$ is a definable quotient of T, then $\bar{T}$ is a decent torus.*
(2) *Let $1 \to T_0 \to T \to T_1 \to 1$ be a short exact sequence with $T_0{}^\circ$ and T_1 decent tori and T divisible abelian. Then T is a decent torus.*

PROOF. We make use of criterion (2) for decency. Then the first point is immediate. As for the second point, any quotient of T is the middle term of a similar short exact sequence, and by criterion (2) it suffices to show that T contains some torsion. This holds by lifting (Lemma I 2.18) if T_1 is nontrivial, and otherwise $T = T_0$. □

LEMMA 1.13. *Let H be a connected group of finite Morley rank, T a decent torus in H. Suppose T is central in H. Let $\mathcal{F}$ be a uniformly definable family of subgroups of H, none of which contain T.*

Then the union $\bigcup \mathcal{F}$ is not generic in H.

PROOF. We consider $\bar{H} = H/\Phi(T)$ and the corresponding family $\bar{\mathcal{F}}$. If $\bar{X} \in \bar{\mathcal{F}}$ contains $\bar{T}$, then $T \leq X\Phi(T)$ and $T = \Phi(T)(X \cap T)^\circ$, and thus $T \leq X$ by Lemma I 5.19, contradicting our hypotheses. So Lemma 1.9 applies to $\bar{H}$, $\bar{T}$, and $\bar{\mathcal{F}}$ and thus the union $\bigcup \bar{\mathcal{F}}$ is nongeneric in $\bar{H}$, and its preimage $\bigcup \mathcal{F}$ is nongeneric in H. □

LEMMA 1.14. *Let G be a connected group of finite Morley rank, T a decent torus in G, and $H = C^\circ(T)$. Then:*

(1) *H is generically disjoint from its conjugates in H.*
(2) *$\bigcup H^G$ is generic in G.*

PROOF. Let T_0 be the maximal decent torus contained in $Z(H)$. Then $N^\circ(H) \leq N^\circ(T_0) = C^\circ(T_0) \leq C^\circ(T) = H$. Hence H is almost self-normalizing in G. Furthermore we may suppose $T = T_0$.

In view of Lemma 1.2, the second point will follow from the first. So we focus on the first point.

Let $\mathcal{F} = \{H \cap H^g : g \in G \setminus N(H)\}$. We claim that $\bigcup \mathcal{F}$ is not generic in H. Suppose $T \leq H \cap H^g$. Then as T^g is central in H^g and is a maximal decent torus of H^g, we have $T \leq T^g$, hence $T = T^g$ and $H = H^g$. So the previous lemma applies to $\mathcal{F}$. □

PROPOSITION 1.15. *Let G be a group of finite Morley rank. Then any two maximal decent tori of G are conjugate, and the same holds for maximal good tori.*

PROOF. As the maximal good tori are contained in maximal decent tori, it suffices to treat the case of decent tori.

We proceed by induction on the rank of G. We may suppose that G is connected. Let T_1, T_2 be two maximal decent tori of G.

Suppose first that G is centerless. Let $H_i = C^\circ(T_i)$. As $\bigcup H_i^G$ is generic in G for $i = 1, 2$, we may suppose $H_1 \cap H_2 \neq 1$. Let $h \in (H_1 \cap H_2)^\times$. Then $T_1, T_2 \leq C(h)$ and we may conclude by induction.

If G has an infinite center, then by induction we may suppose that $T_2 \leq Z(G)T_1$. But $Z(G)T_1$ is nilpotent, and from the structure of nilpotent groups of finite Morley rank it follows that T_1 is central in $Z(G)T_1$, so this is an abelian group, and by maximality $T_2 = T_1$.

Now suppose that G has a finite center. Then $G/Z(G)$ is centerless [**52**, elementary], and the first case applies. So after conjugating we may suppose $T_2 \leq (T_1 \cdot Z(G))^\circ = T_1$, and thus $T_2 = T_1$. □

We have the following corollary, where now we must return to the case of good tori.

COROLLARY 1.16. *Let G be a group of finite Morley rank, and $\mathcal{F}$ a uniformly definable family of good tori in G. Then the tori in $\mathcal{F}$ fall into finitely many conjugacy classes under the action of G.*

PROOF. Let T be a maximal good torus in G. Every torus in $\mathcal{F}$ is conjugate to a subtorus of T, so we may as well suppose that every torus in $\mathcal{F}$ is a subtorus of T. In that case $\mathcal{F}$ is finite by the second Rigidity Lemma (Lemma I 4.23, part 2). □

1.3. Rigid abelian groups. The theory of good tori has another mild extension, to the not necessarily connected case, which is well worth recording.

DEFINITION 1.17. *A* rigid abelian *group is a group of finite Morley rank which is abelian, whose connected component is a good torus.*

An equivalent and highly relevant condition is the following: a rigid abelian group is a good abelian group A such that the annihilator $A[n] = \{a \in A : na = 0\}$ is finite for all n.

These have much the same properties as good tori, notably the finiteness of families of uniformly definable subgroups, and in carrying out certain genericity arguments we will find we need to deal directly with disconnected groups (in some cases, one may even need to deal seriously with finite groups).

REMARK 1.18. *A definable subgroup of a rigid abelian group is a rigid abelian group.*

We can prove generic disjointness from the conjugates for these groups, extending the result for good tori. This is based on the following useful principle.

LEMMA 1.19. *Let H be a group of finite Morley rank such that H/H° is cyclic. Then the complement in H of a finite union of proper definable subgroups has full rank.*

PROOF. This is little more than a combination of the result for the cases of connected groups and finite cyclic groups.

In general, there is a coset C of H° in H which generates H. If $H_1, \ldots, H_n < H$ are proper definable subgroups, we consider their intersections with C. Evidently these are proper as well. In particular, any groups H_i which contain H° are disjoint from the coset C, and the remainder meet C in a set of lower rank. □

LEMMA 1.20. *Let G be a group of finite Morley rank and A a definable rigid abelian subgroup with A/A° cyclic. Then A is generically disjoint from its conjugates.*

PROOF. The proof is the same as for good tori, using the previous lemma. □

1.4. Carter° subgroups. In §I 8.4 we gave a good deal of the general theory of Carter subgroups of connected solvable groups of finite Morley rank. There is also a very interesting theory for groups which are not necessarily solvable, but in that case it is appropriate to loosen the definition a little.

DEFINITION 1.21. *Let G be a group of finite Morley rank. A* Carter° *subgroup of G is a connected nilpotent subgroup of finite index in its normalizer.*

EXAMPLE 1.22. *Let A be a connected abelian group of finite Morley rank and H an extension of A by an involution i acting on A by inversion. Then A is a Carter° subgroup. If A is a $2^\perp$-group, then $\langle i \rangle$ is a Carter subgroup. If A is a divisible abelian 2-group, then H has no Carter subgroup.*

As this example shows, the two notions are quite distinct. Experience suggests that it is the notion of Carter° subgroup which is the useful one. In fact, the recent literature uses the term "Carter subgroup" in place of our "Carter° subgroup". We avoid this usage only because we include some older material in Chapter I.

The existence of Carter° subgroups in arbitrary connected groups of finite Morley rank has also been shown [**94**], but we do not have space here to develop the relevant machinery.

It can be shown that for G connected solvable of Morley finite rank, the two notions of Carter subgroup coincide [**93**]. We remark that by the definition, a Carter° subgroup of G or of G° is the same thing.

There is a conjugacy result which falls very much in the line of the present chapter, which we will now give. The following terminology is a recent and we think useful innovation by Jaligot.

DEFINITION 1.23. *Let G be a group of finite Morley rank and X a definable subset. We say that X is* generous *in G if $\bigcup X^G$ is generic in G.*

There are a number of related open problems concerning Carter° subgroups, of a fundamental character. One is the conjugacy of Carter subgroups, which seems to us out of reach at present. Another is the existence of a generous Carter° subgroup. Linking these two problems is the following noteworthy structural fact.

THEOREM 1.24. *Let G be a connected group of finite Morley rank. Then any two generous Carter° subgroups of G are conjugate.*

This is another "conjugacy of tori" result. While it has a different character from the conjugacy of maximal good tori, there is a certain affinity not only in content but in the mechanism of the proof.

For the proof, we begin with yet another version of Lemma 1.2.

LEMMA 1.25. *Let G be a connected group of finite Morley rank and H a definable, connected, and almost self-normalizing subgroup of G. Let $\mathcal{F}$ be the family of all conjugates of H in G. Then the following are equivalent.*

(1) *H is generous in G.*
(2) *The definable set*
$$H_0 = \{h \in H : \{X \in \mathcal{F} : h \in X\} \text{ is finite}\}$$
is generic in H.
(3) *The definable set*
$$G_0 = \{g \in \bigcup H^G : \{X \in \mathcal{F} : g \in X\} \text{ is finite}\}$$
is generic in G.

PROOF. We make a rudimentary geometry in which the points are the elements of $\bigcup H^G$, and the lines are elements of the family $\mathcal{F}$ of conjugates of H in G, with incidence given by membership.

Observe that $\text{rk}(\mathcal{F}) = \text{rk}(G/N(H)) = \text{rk}(G) - \text{rk}(H)$ since H is almost self-normalizing. So considering the incidence relation I as a subset of $\bigcup H^G \times \mathcal{F}$, and projecting onto $\mathcal{F}$, as the fiber above $X \in \mathcal{F}$ is the set $X \subseteq \bigcup H^G$, whose rank is $\text{rk}(H)$, we get
$$\text{rk}(I) = \text{rk}(\mathcal{F}) + \text{rk}(H) = \text{rk}(G) \ .$$

On the other hand, projecting in the other direction, onto $\bigcup H^G$, the fibers will be of variable rank, but one value, say r, will occur generically, and then we have
$$\text{rk}(I) \geq \text{rk}(\bigcup H^G) + r \ .$$

Now H is generous if and only if $\text{rk}(\bigcup H^G) = \text{rk}(G)$, and by the preceding equations this implies
$$r = 0 \ .$$

The fiber above $g \in \bigcup H^G$ is the set $\{X \in \mathcal{F} : g \in X\}$ referred to in condition (3). So we see that (1) implies (3).

Now observe that $G_0 = \bigcup H_0^G$ since the fiber ranks are invariant under conjugation. If H_0 is not generic in H, then by comparing the rank of the union making up G_0 to the rank of the corresponding disjoint union, we find

$$\mathrm{rk}(G_0) \leq \mathrm{rk}(H_0) + \mathrm{rk}(G/N(H)) < \mathrm{rk}(G)$$

contradicting (3). So (3) implies (2).

Finally, supposing (2) holds we will prove (1). We make another rudimentary geometry in which the points are the elements of $\bigcup H_0^G$ and the lines come from the family $\mathcal{F}_0$ of conjugates of H_0 in G, with an incidence relation $I_0 \subseteq \bigcup H_0^G \times \mathcal{F}_0$ given by membership.

Observe that the normalizer of the set H_0 coincides with $N(H)$, since H is connected and therefore H_0 generates H. Thus $\mathrm{rk}(\mathcal{F}_0) = \mathrm{rk}(G/N(H)) = \mathrm{rk}(G) - \mathrm{rk}(H_0)$. Now the projection of I_0 onto the second factor has fibers in $\mathcal{F}_0$, all of rank $\mathrm{rk}(H_0)$, so $\mathrm{rk}(I_0) = \mathrm{rk}(\mathcal{F}_0) + \mathrm{rk}(H_0) = \mathrm{rk}(G)$.

Now observe that the projection of I_0 onto the first factor has finite fibers, since each conjugate of H contains a unique conjugate of H_0. Thus $\mathrm{rk}(\bigcup H_0^G) = \mathrm{rk}(I_0) = \mathrm{rk}(G)$. So (1) follows. □

Now the following very pretty result will give the theorem immediately.

LEMMA 1.26. *Let G be a connected group of finite Morley rank, and $\mathcal{F}$ a uniformly definable family of Carter° subgroups of G, closed under conjugation. If $g \in G$ belongs to finitely many elements of $\mathcal{F}$, of Q, then g belongs to at most one element of $\mathcal{F}$.*

PROOF. Let $\mathcal{F}_g$ be the set of elements of $\mathcal{F}$ which contain g, and $X = \bigcap \mathcal{F}_g$ its intersection. Then $\mathcal{F}_g$ is also the set of elements of $\mathcal{F}$ which contain X. Hence $N(X)$ acts on the finite set $\mathcal{F}_g$, by conjugation, and thus $N^\circ(X)$ normalizes each group in $\mathcal{F}_g$. But these are Carter° subgroups, and it follows that $N^\circ(X)$ is contained in each group in $\mathcal{F}_g$, and thus $N^\circ(X) \leq X$.

On the other hand, if $X \leq Q \in \mathcal{F}_g$, then as Q is connected and nilpotent and $N_Q(X) \leq X$, we find $X = Q$. Hence $\mathcal{F}_g$ contains at most one element. □

PROOF OF THEOREM 1.24. Let Q_1 and Q_2 be generous Carter° subgroups of G, and $\mathcal{F}$ the family consisting of all their conjugates. By Lemma 1.25, a generic element of G belongs to finitely many elements of $\mathcal{F}$, and hence by Lemma 1.26, to just one. On the other hand, a generic element of G belongs to both a conjugate of Q_1 and a conjugate of Q_2; so these groups are conjugate. □

We make a further observation concerning the notion of generosity.

LEMMA 1.27. *Let G be a group of finite Morley rank and H a definable subgroup of G. If H is generous, then H is almost self-normalizing.*

PROOF. Let X be the disjoint union of the conjugates of H, suitably interpreted in G. Then

$$\mathrm{rk}(X) = \mathrm{rk}(H) + \mathrm{rk}(N(H)\backslash G) = \mathrm{rk}(G) - (\mathrm{rk}(N(H)/H)) .$$

Now X maps definably onto $\bigcup H^G$, so if H is generous, then $\mathrm{rk}(X) \geq \mathrm{rk}(G)$. Hence $\mathrm{rk}(N(H)/H) = 0$. □

2. Generic cosets

We deal here with variations on a standard genericity argument, continuing in the line of Lemmas 1.2, 1.8, 1.25, but with more emphasis on cosets and less on subgroups. The underlying idea runs roughly as follows: if the conjugates of a Borel subgroup B generically cover the connected group G, and if $N(B) > B$, then one might expect that the conjugates of a coset of B in $N(B)$ would give a disjoint generic subset, and a contradiction. The matter is not so simple as that: for example, in simple algebraic groups, the conjugates of a maximal torus cover the group generically, but the normalizer of the torus is strictly larger, and this is the source of the Weyl group. Admittedly, the underlying geometry that allows this to happen is an interesting one in this case.

The basic principle is the following, which is more manageable in practice than might immediately appear.

LEMMA 2.1. *Let G be a connected group of finite Morley rank and H a proper definable almost self-normalizing subgroup of G such that $\bigcup H^G$ is generic in G. Let $x \in N(H) \setminus H$. Then the set*

$$X = \{x_1 \in xH : x_1 \in (\langle x\rangle H)^g \text{ for some } g \in G \setminus N(H)\}$$

is generic in xH.

PROOF. Suppose the contrary, and let $Y = xH \setminus X$. Then $Y \subseteq \langle x\rangle H$ and $\mathrm{rk}(Y) = \mathrm{rk}(H)$. For $g \in G \setminus N(H)$ we have $Y \cap Y^g = \emptyset$ since $Y \cap Y^g \subseteq Y \cap (xH)^g = \emptyset$. So the usual rank computation shows

$$\mathrm{rk}(\bigcup Y^G) = \mathrm{rk}(\bigcup Y^{N(H)\backslash G}) = \mathrm{rk}(Y) + \mathrm{rk}(G) - \mathrm{rk}(H) = \mathrm{rk}(G)$$

and $\bigcup Y^G$ is generic in G. As G is connected, it follows that $\bigcup Y^G$ meets $\bigcup H^G$, and thus Y meets some conjugate H^g of H. Since $Y \cap H = \emptyset$, it follows that $g \notin N(H)$, but then $Y \cap H^g = \emptyset$, a contradiction.

So Y is not generic in xH, and consequently X is generic in xH. □

As we will see, this result combines well with results about torsion elements such as the following.

LEMMA 2.2. *Let H be a group of finite Morley rank with H° abelian, and let $x \in H \setminus H^\circ$ be an element so that for some fixed integer n, the elements of xH° are generically of order n. Then every element of the coset xH° is of order n.*

PROOF. We may suppose $x^n = 1$. Then there is an element r in the integral group ring $\mathbb{Z}[x]$, such that for $h \in H^\circ$ the condition

$$(xh)^n = 1$$

can be expressed (additively) in the form $rh = 0$. Our assumption is that the annihilator of r in H° is generic in H°, and hence, as it is a subgroup, is equal to H°. □

The following is a specialized variant of this type of argument, needed in Chapter VI.

LEMMA 2.3. *Let G be a connected group of finite Morley rank and suppose that there is a definable divisible abelian subgroup T of G which is almost self-normalizing and such that the intersection of any two distinct conjugates of T is finite. Then for any $x \in N(T) \setminus T$, the centralizer $C_T(x)$ is finite.*

PROOF. By our hypothesis, the set $\{T \cap T^g : g \in G \setminus N(T)\}$ is a definable family of finite sets; hence they are of bounded order, and as T is divisible abelian, their union is finite. Accordingly, T is generically disjoint from its conjugates. Thus the union of the conjugates of T is generic in G.

We are now set up for a genericity argument. The set

$$X = \{x_1 \in XT : x_1 \in (\langle x \rangle T)^g, \text{ for some } g \in G \setminus N(T)\}$$

is generic in xT. For each element x_1 of X, if $n = [N(T) : T]$, then we have $x_1^n \in T \cap T^g$ for some $g \in G \setminus N(T)$, a group of bounded order. So there is an absolute exponent N such that $x_1^N = 1$ for all $x_1 \in xT$. As this holds generically, it holds over the whole coset xT, by the preceding lemma. Now for $t \in C_T(x)$, the conditions $x^N = (xt)^N = 1$ imply $t^N = 1$; as T is divisible abelian, the subgroup defined by this condition is finite. □

3. Generic covering

We now take up the subject of generic covering from a different point of view. We aim to show that a generic covering of a connected group by good tori, and more generally by rigid abelian subgroups, necessarily involves at least one "maximal" subgroup in an appropriate sense. A good torus is maximal if it is maximal within the class of good tori. A rigid abelian group A, on the other hand, will be considered maximal if A° is maximal among good tori centralizing A. The proof of the next result is long and technical, dealing mainly with issues of definability (if one chooses to ignore those issues, only the last few paragraphs are relevant). We will also make a saturation hypothesis here, which can be avoided at the cost of additional combinatorics ([**69**, §3]). For our intended applications such a hypothesis is harmless. We will call G "sufficiently saturated" if it is $|\mathcal{L}|^+$-saturated, with $\mathcal{L}$ the underlying first order language.

THEOREM 3.1. *Let G be a sufficiently saturated group of finite Morley rank, and $\mathcal{F}$ a uniformly definable family of rigid abelian subgroups of G such that $\bigcup \mathcal{F}$ is generic in G. Then there is a group $A \in \mathcal{F}$ such that A° is a maximal good torus in $C(A)$.*

An equivalent way to formulate the condition on A would be the following:

$$\text{If } A \leq \tilde{A} \text{ with } \tilde{A} \text{ rigid abelian, then } [\tilde{A} : A] < \infty.$$

And it is equally clear in this formulation that this is a notion of maximality.

PROOF. The family $\mathcal{F}$ is a definable subset of G^{eq}, possibly involving parameters. We may treat these parameters as constants and assume that the family is 0-definable.

We will first make some adjustments to the family $\mathcal{F}$. By Fact I 3.17, there is a finite bound $m = m_{\mathcal{F}}$ on the indices $[A : A^\circ]$ for $A \in \mathcal{F}$. Therefore, $m! \cdot A = A^\circ$ for all $A \in \mathcal{F}$, and A° is uniformly definable from A. Accordingly, the modified family $\hat{\mathcal{F}} = \{A^\circ\langle a \rangle : A \in \mathcal{F}, a \in A\}$ is another uniformly definable family of rigid abelian subgroups, covering the same subset of G, and with the added condition that A/A° is cyclic for all $A \in \hat{\mathcal{F}}$.

Now we will prove our claim for the family $\hat{\mathcal{F}}$. Since each $A \in \hat{\mathcal{F}}$ is contained in some $B \in \mathcal{F}$ our result then follows easily for $\mathcal{F}$. So we may again write $\mathcal{F}$ for $\hat{\mathcal{F}}$, and assume that the groups in $\mathcal{F}$ are finite cyclic extensions of good tori.

We may now begin. Suppose toward a contradiction that for every $A \in \mathcal{F}$ there is a definable rigid abelian subgroup $\tilde{A}$ of G such that $A \leq \tilde{A}$ and $[\tilde{A} : A] = \infty$. We may suppose further that the groups in question have the form $\tilde{A} = A \cdot \tilde{A}^\circ$. We can put $\tilde{A}$ into some 0-definable family of subgroups of G, which may not necessarily all be rigid. We will find it useful to make this precise, aiming at condition $(*)$ below.

With A and $\tilde{A}$ fixed, there is a definition $\phi_{\tilde{A}}(x, \bar{a})$ for $\tilde{A}$ involving parameters $\bar{a}$ from G. We may associate to any formula $\phi(x, \bar{y})$ the uniformly definable family

$$\mathcal{C}_\phi = \{\phi[G, \bar{g}] : \phi[G, \bar{g}] \text{ is an abelian group}\}$$

where $\bar{g}$ varies over G. As the set

$$\{\bar{g} \in G^{l(\bar{y})} : \phi[G, \bar{g}] \text{ is an abelian group}\}$$

is 0-definable, the family $\mathcal{C}_\phi$ is 0-definable, as a subset of G^{eq}. In particular, the group $\tilde{A}$ belongs to $\mathcal{C}_{\phi_{\tilde{A}}}$, but this family need not consist exclusively of *rigid* abelian groups. We need to refine the construction further to get some approximation to rigidity.

Fix m_ϕ so that the indices $[A : A^\circ]$ are bounded by m_ϕ for $A \in \mathcal{C}_\phi$. We introduce the abbreviation "$B \ll_\varphi A$" to stand for the condition

$$A = B \cdot m_\varphi! A \ \& \ [A : B] > m_\phi .$$

For $A \in \mathcal{C}_\phi$ and $B \in \mathcal{F}$ this implies: $A = B \cdot A^\circ$, hence A/A° is cyclic, and $A^\circ > B^\circ$. The converse also holds when $A \in \mathcal{C}_\phi$ and $B \in \mathcal{F}$, if A is a rigid abelian group (in which case $m_\phi!A = A^\circ$ is a torus).

Let $\mathcal{C}^*_\varphi$ be

$$\{A \in \mathcal{C}_\varphi : \exists B \in \mathcal{F} \quad B \ll_\varphi A\} .$$

Then for $A \in \mathcal{C}^*_\varphi$, we have $A^\circ = m_\varphi!A$, and the quotient A/A° is cyclic. In particular, degree(A) is uniformly definable from A for $A \in \mathcal{C}^*_\varphi$.

Since $\tilde{A}$ is rigid, the set of intersections $\{\tilde{A} \cap A : A \in \mathcal{C}^*_{\varphi_{\tilde{A}}} \cup \mathcal{F}\}$ is finite (Lemma I 4.23), of size $k_{\tilde{A}}$, say. For any finite k and any formula $\varphi(x, \overline{y})$, we may consider the family

$$\mathcal{C}^*_{\varphi,k} = \{A \in \mathcal{C}^*_\varphi : |\{A \cap B : B \in \mathcal{C}^*_\varphi \cup \mathcal{F}\}| \leq k\} .$$

The family $\mathcal{C}^*_{\varphi,k}$ is uniformly definable over $\emptyset$ (i.e., 0-definable as a subset of G^{eq}) since $\mathcal{F} \cup \mathcal{C}^*_\varphi$ is, and k is fixed.

By our choice of $k_{\tilde{A}}$, we have $\tilde{A} \in \mathcal{C}^*_{\varphi_{\tilde{A}},k_{\tilde{A}}}$.

The preceding discussion may be summarized as follows:

> $(*)$ For every $A \in \mathcal{F}$ there exists a finite number k, a formula φ, and some $A \in \mathcal{C}^*_{\varphi,k}$ such that $A \ll_\varphi A$.

We claim next that condition $(*)$ holds uniformly: there exist finitely many pairs of the form $(\varphi_1, k_1), \ldots, (\varphi_n, k_n)$, consisting of formulas φ_i and natural numbers k_i as in $(*)$, such that for any $A \in \mathcal{F}$ the pair (φ, k) in $(*)$ can be taken to be one of the (φ_i, k_i).

Indeed, consider the following 1-type $p(S)$ in G^{eq} , where φ varies over all formulas defined over $\emptyset$ and k varies over all natural numbers, and S is a variable of a suitable sort for representing elements of $\mathcal{F}$,

$$S \in \mathcal{F}; \qquad \neg\exists X \in \mathcal{C}^*_{\varphi,k}\,(S \ll_\varphi X) .$$

Observe that the cardinality of this 1-type is at most the cardinality of the language $|\mathcal{L}|$.

By condition $(*)$, the type $p(S)$ is not realized in G^{eq} . As we take G to be $|\mathcal{L}|^+$-saturated, we conclude that $p(S)$ is inconsistent. Hence there are finitely many formulas φ_i and natural numbers k_i $(i \leq n)$ such that

$$S \in \mathcal{F} \implies \exists i \leq n\, \exists X \in \mathcal{C}^*_{\varphi_i,k_i}(S \ll_{\varphi_i} X) .$$

This is the desired uniformity.

Let $\mathcal{C}^*_i = \mathcal{C}^*_{\varphi_i,k_i}$. Before proceeding, it will be convenient to modify this choice of the $\mathcal{C}^*_i$. We would like the rank and degree rk(A), degree(A) for $A \in \mathcal{C}^*_i$ to be constant; this is achieved by partitioning $\mathcal{C}^*_i$ into finitely many subsets on which the rank and degree are constant—and the defining formula φ_i is altered accordingly, while n, the number of formulas, increases. Let $r_i = \text{rk}(A)$ for $A \in \mathcal{C}^*_i$ (a constant), and similarly $d_i = \text{degree}(A)$ for $A \in \mathcal{C}^*_i$. We will write $\ll_i$ for $\ll_{\varphi_i}$.

Now with φ_i, k_i $(1 \le i \le n)$ as described, let $\mathcal{C}_i^* = \mathcal{C}_{\varphi_i,k_i}^*$ and set

$$\mathcal{F}^i = \{A \in \mathcal{F} : \exists X \in \mathcal{C}_i^* \, (A \ll_i X)\} \ .$$

Then $\mathcal{F}^i$ is a uniformly definable family, over $\emptyset$.

We now pass to rank computations. We have $\bigcup \mathcal{F}$ generic in G, and $\mathcal{F}$ is the union of the $\mathcal{F}^i$, so for some i the union $\bigcup \mathcal{F}^i$ is generic in G. To reach a contradiction it suffices to show that $\mathrm{rk}\left(\bigcup \mathcal{F}^i\right) < \mathrm{rk}\left(\bigcup \mathcal{C}_i^*\right)$.

For $A \in \mathcal{C}_i^*$, let $X_A = \bigcup\{B \in \mathcal{F} : B \ll_i A\}$ and let $Y_A = \bigcup\{A \cap B : B \in \mathcal{C}_i, B \neq A\}$. Note that if $A \neq B$ with $A, B \in \mathcal{C}_i$, then $A \cap B < A$, as $\mathrm{rk}(A) = \mathrm{rk}(B)$ and $\mathrm{degree}(A) = \mathrm{degree}(B)$. Thus X_A and Y_A are unions of proper subgroups of A, and by the definition of the classes $\mathcal{C}_{\varphi,k}$ only finitely many subgroups are involved. We consider these two sets in more detail.

The subgroups making up X_A have infinite index in A, so their union has rank less than r_i. Furthermore, $Z_A = A \setminus Y_A$ has rank r_i by Lemma 1.19.

As $Z_A \cap Z_B = \emptyset$ and $\mathrm{rk}(Z_A) = r_i$ for $A \neq B$ in $\mathcal{C}_i^*$, we have $\mathrm{rk}\left(\bigcup \mathcal{C}_i^*\right) \geq r_i + \mathrm{rk}(\mathcal{C}_i^*)$. On the other hand, $\mathrm{rk}\left(\bigcup \mathcal{F}^i\right) \leq \mathrm{rk}\left(\bigcup\{X_A : A \in \mathcal{C}_i^*\}\right) < r_i + \mathrm{rk}(\mathcal{C}_i^*)$. Thus $\mathrm{rk}\left(\bigcup \mathcal{F}^i\right) < \mathrm{rk}\left(\bigcup \mathcal{C}_i^*\right)$, as claimed. This contradicts the supposed genericity of $\mathcal{F}^i$ and proves the theorem. □

LEMMA 3.2. *Let G be a sufficiently saturated group of finite Morley rank, T a good torus in G, and $K = C_G{}^\circ(T)$. Suppose that K is generically covered by a family of intersections $\{K \cap K^g\}$ where g varies over some definable subset of G, and where each of these intersections is a rigid abelian group.*

Then K contains a good torus which is almost self-normalizing in G.

PROOF. By Theorem 3.1, there is at least one intersection $B = K \cap K^g$ in the specified family which is maximal in the following sense: B° is the unique maximal good torus of $C_K(B)$. We will show that B° is almost self-normalizing in G.

As T is central in K, we have $T \leq B$. Thus $C^\circ(B^\circ) \leq C^\circ(T) = K$. Similarly, $T^g \leq C^\circ(B) \leq K$. Since $T^g \leq C_K(B)$, the maximality condition implies $T^g \leq B$. So again, $C^\circ(B^\circ) \leq C^\circ(T^g) = K^g$.

So $N^\circ(B^\circ) = C^\circ(B^\circ) \leq K \cap K^g = B$, and B° is almost self-normalizing. □

4. Degenerate type groups

In this section we will prove the following result about groups of degenerate type. The proof is given in §4.5

THEOREM 4.1. *Let G be a connected group of finite Morley rank and degenerate type. Then G contains no involutions.*

This result is of considerable use in treating groups which may have sections of degenerate type, and yet it is not actually essential to any of the applications given in Parts B or C, though it has other applications which we will give in the present section.

4.1. The function $d(g)$. In a group G of finite Morley rank, we consider the function $d(g)$ which associates to an element g its definable hull in G. This function d, with G as its domain and range among the definable subgroups of G, is not usually a definable function. Indeed, for g of finite order, $d(g)$ is nothing but the cyclic group generated by g, and thus typically the values of d include finite sets of unbounded size, something which is not possible for a definable function.

On the other hand, it is convenient to use the function d to define other definable functions, and it is useful to systematize this process. One way to proceed is by introducing a definable approximation $\hat{d}$ to the function d which has analogous properties. In principle such an approximation can be refined arbitrarily, and there is no canonical choice. However, the following will cover all of our needs (and if one replaces 2 by an arbitrary prime, it appears to cover every application of the method currently envisaged).

LEMMA 4.2. *let G be a group of finite Morley rank with Sylow 2-subgroups of bounded exponent. Then there is a definable function $\hat{d}(a)$, from elements of G to definable subgroups of G, with the following properties:*

(1) $d(a) \leq \hat{d}(a)$;
(2) *If $d(a) = d(b)$, then $\hat{d}(a) = \hat{d}(b)$;*
(3) *For $g \in G$, we have $\hat{d}(a^g) = \hat{d}(a)^g$;*
(4) *$\hat{d}(a)$ is abelian;*
(5) *The groups $d(a)$ and $\hat{d}(a)$ have the same Sylow 2-subgroup;*
(6) *If $x \in G$ conjugates a to its inverse, then x normalizes $\hat{d}(a)$ and acts on it by inversion.*

PROOF. Consider the following two functions.

- $d_1(a) = Z(C(a))$.
- $d_2(a) = d_1(a)^q\langle a\rangle$ where q is a bound on the order of the 2-torsion in G.

One sees easily that d_1 is definable and satisfies our first four conditions. As $[d_2(a) : d_1(a^q)] \leq q$, it follows easily that d_2 is also a definable function, and it also satisfies condition (1), and inherits conditions $(2-4)$ from d_1.

Furthermore, $d_2(a)$ also satisfies the fifth condition, since $d_1(a)$ is abelian and $d_1(a)^q$ is 2-torsion free.

Now to achieve the final point, let $d_3(a)$ be the subgroup of $d_2(a)$ consisting of elements inverted by every element that inverts a. □

4.2. Minimization. We examine the structure of a minimal counterexample to Theorem 4.1.

LEMMA 4.3. *Let G be a connected group of finite Morley rank and degenerate type, containing an involution, and minimal among all such groups. Then $\bar{G} = G/Z(G)$ is simple and contains an involution, while no proper definable connected subgroup of $\bar{G}$ contains an involution.*

PROOF. By our minimality hypothesis no proper connected subgroup of G contains involutions. If $H < G$ is a nontrivial definable connected normal subgroup, then passing to G/H we contradict the minimality of G. So $Z(G)$ is finite and $G/Z(G)$ is simple. It suffices to show that $G/Z(G)$ contains involutions.

Supposing the contrary, after passing to a quotient of G we may suppose that $Z(G)$ is a 2-group. We now introduce a function

$$\eta : G \to Z(G)$$

which though not necessarily a homomorphism will be covariant with respect to the action of $Z(G)$. This is defined as follows.

For $g \in G$, we consider the subgroup $\hat{d}(g)$, which splits as $\hat{d}(g)^q \times S_g$, with q the exponent of $Z(G)$ and $S_g \leq Z(G)$ the Sylow 2-subgroup of $\hat{d}(g)$ (or of $d(g)$). So the projection $\pi_2 : \hat{d}(g) \to S_g$ is well defined, and we may set $\eta(g) = \pi_2(g) \in Z(G)$.

The desired covariance property is the following.

$$\eta(zg) = z\eta(g) \text{ for } z \in Z(G),\ g \in G\ .$$

Writing $g = g_0 s$ with $g_0 \in d(g)^q$ and $s \in S_g$, we have $g_0^q \in d(zg)$ and as $d(g_0^q)$ is 2-divisible, and $d(g_0)$ is uniquely 2-divisible, we have $g_0 \in d(zg)$; but $zg = g_0 zs$ and hence $zs \in d(zg) \leq \hat{d}(zg)$ as well, and our claim follows.

Now in view of the covariance of the map η, its fibers have constant rank. Thus G is partitioned by the fibers of η into finitely many sets of equal rank, and as G is connected this yields a contradiction (Lemma I 3.6). □

We will use a similar argument again in §4.5.

4.3. Genericity. Let us now suppose the following.

(†) G is a simple group of finite Morley rank and degenerate type, containing an involution, while no proper connected definable subgroup of G contains an involution.

Let q be a the exponent of a Sylow 2-subgroup of G.

We will show that the generic elements of G lie outside every proper connected subgroup of G, and we will pin down their location with sufficient precision to give useful structural information. In particular, we will show that the Sylow 2-subgroup of G is elementary abelian.

DEFINITION 4.4.

(1) *Let $i \in G$ be an involution. Set*

$$H_i = N^\circ(\ldots(N^\circ(C^\circ(i)))\ldots)$$

where the operator N° is applied sufficiently many times to ensure that it stabilizes.

(2) *Let $a \in G$, and suppose $d(a)$ contains an involution i. Set $H_a = H_i$.*

Observe that H_a is well defined in clause (2), as the Sylow 2-subgroup of $d(a)$ is cyclic, and that notations (1) and (2) are compatible. We have the following formal properties whenever the groups in question are defined:

(1) H_a is a proper definable connected subgroup of G;
(2) $H_a = N^\circ(H_a)$ ("almost self-normalizing");
(3) $a \in N(H_a) \setminus H_a$;
(4) $H_{a^g} = H_a^g$.

We check the first and third points. Let $i \in d(a)$ be an involution. For the first point, note that $C^\circ(i) > 1$ in view of Lemma I 10.3. For the third point, we have $i \notin H_i$ by our hypothesis on G, and hence $a \notin H_i$.

The following property lies deeper.

LEMMA 4.5. *Let $a \in G$, and suppose $d(a)$ contains an involution. Then the following hold.*

(1) *For $c \in aH_a$ we have $H_a = H_c$, and hence $aH_a = cH_c$.*
(2) *The union $\bigcup_{g\in G}(aH_a)^g$ is generic in G.*

PROOF. For any $c \in aH_a$, there is some $c' \in d(c) \cap iH_a$ and hence some involution $j \in d(c') \cap iH_a$. In particular, $j \in d(c)$ and $H_c = H_j$.

Now the group $H_a\langle i\rangle = H_a\langle j\rangle$ has a Sylow 2-subgroup of order 2, and hence i and j are conjugate under the action of H_a. But $H_i = H_a$ and hence $H_j = H_a$, so $H_c = H_a$. So the first claim holds.

For the second point, we show first that aH_a is disjoint from its conjugates $(aH_a)^g$ for $g \notin N(H_a)$. So suppose we have an element $c \in (aH_a) \cap (aH_a)^g$. Then $H_c = H_a$ by the preceding lemma and similarly $H_c = H_{a^g} = H_a^g$. So $H_a = H_a^g$, and $g \in N(H_a)$.

Now we may apply Lemma 1.8 to conclude that $\bigcup_{g\in G}(aH_a)^g$ is generic in G. □

This has the following structural consequence.

LEMMA 4.6. *Under the hypothesis (†), the Sylow 2-subgroup of G is elementary abelian.*

PROOF. We claim there is no element of order 4. If on the contrary a is an element of order 4, then the cosets aH_a and a^2H_a consist of elements c such that modulo H_c the element c is of order 4 or 2, respectively. But $\bigcup(aH_a)^G$ and $\bigcup(a^2H_a)^G$ are both generic, and this is a contradiction. □

At this point, we shift gears and invoke quite different methods, inherited from "black box" group theory.

4.4. Black box methods. We retain the hypothesis (†) of the preceding subsection.

For the moment we need not invoke the results of our genericity arguments. Rather we introduce a crucial case division, dispose of one case using so-called "black box" group theoretic methods, and then return to handle the other case using the information from the last subsection.

We turn to the case division. Fix a conjugacy class of involutions $\mathcal{C}$, and note that as this set can be identified definably with $G/C(i)$ for any fixed $i \in \mathcal{C}$, it has Morley degree 1. Thus the notion of "generic element" or pair of elements in $\mathcal{C}$ is robust. The first case we will treat is the following one:

(Case I) For generic and independent $i, j \in \mathcal{C}$ the group $d(ij)$ contains no involution.

The following two facts are elementary but important, and are used in combination.

LEMMA 4.7. *Let G be a group of finite Morley rank and a an element of G. Suppose that the Sylow 2-subgroups of G have bounded exponent. Then the following conditions on the group $d(a)$ are equivalent.*

(1) *$d(a)$ contains no involutions.*
(2) *$d(a)$ is 2-divisible.*
(3) *$d(a)$ is uniquely 2-divisible.*

On the other hand, if $d(a)$ does contain an involution, then that involution is unique.

LEMMA 4.8. *Let G be a group of finite Morley rank with Sylow 2-subgroups of bounded exponent, and i, j involutions of G. Let $a = ij$. Then $d(i, j) = d(a) \rtimes \langle i \rangle$, where i acts by inversion on $d(a)$. Furthermore, i and j are conjugate under the action of $d(a)$ if and only if $d(a)$ contains no involution.*

PROOF. Since i inverts a, i inverts $d(a)$. If $i \in d(a)$, then $d(i, j) = d(a)$ is abelian and we arrive quickly at a contradiction. So $i \notin d(a)$ and the structure of $d(i, j)$ is clear.

For $b \in d(a)$ we have $i^b = b^{-1}ib = ib^2$ and thus $i^b = j$ if and only if $b^2 = a$. If $d(a)$ contains no involution, then we can find $b \in d(a)$ with $b^2 = a$ and $i^b = j$. Conversely, if $d(a)$ contains an involution, then as the Sylow 2-subgroups have bounded exponent, we have

$$d(a) = A \times C$$

with A $2^{\perp}$ and C a cyclic 2-group. Then $a = a_1 c$ with $a_1 \in A$ and c a generator of C. Evidently the equation $b^2 = a$ has no solution in $d(a)$. □

This leads to consideration of the following partial functions from the group G to $C(i)$, for any fixed involution i, under the hypothesis that the Sylow 2-subgroups of G have bounded exponent:

(1) $\zeta_0(g)$ is the unique involution in $d(i \cdot i^g)$, if $d(i \cdot i^g)$ contains an involution;
(2) $\zeta_1(g)$ is the unique element in $gd(i \cdot i^g) \cap C(i)$, if $d(i \cdot i^g)$ contains no involution

Indeed, under our hypothesis ζ_0 is well defined on its domain. So suppose $d(i \cdot i^g)$ contains no involution. Then this group is 2-divisible, and there is an element $x \in d(i \cdot i^g)$ conjugating i to i^g, so gx^{-1} belongs to $C(i) \cap gd(i \cdot i^g)$.

As far as uniqueness is concerned, if $x, y \in C(i) \cap gd(i \cdot i^g)$, then $x^{-1}y \in C(i) \cap d(i \cdot i^g)$ is both centralized and inverted by i, hence is an involution or trivial, and as there is no involution in $d(i \cdot i^g)$ we conclude $x = y$.

One could also compute more directly that $x^{-1} = x^i = x[x,i] = xi^g i$ and thus $x^2 = ii^g$, so that x is uniquely determined within $d(i \cdot i^g)$, symbolically $x = \sqrt{ii^g}$, with the square root operation restricted to $d(i \cdot i^g)$, though this extra precision is useful mainly as a way of verifying the existence of x.

The functions ζ_0 and ζ_1 are definable, because we can replace d by $\hat{d}$ everywhere in their definitions, and the Sylow 2-subgroups remain the same. The uniqueness argument also relies on the properties of $\hat{d}$ given at the outset, which mimic the properties of d.

LEMMA 4.9. *Case I does not occur.*

PROOF. Assume we are in Case I. We again use a covariant function in the manner of Lemma 4.3.

We fix $i \in \mathcal{C}$ and consider the definable partial function $\zeta_1 : G \to C(i)$ discussed above, which is defined on a generic subset of G, namely

$$\zeta_1(g) \in C(i) \cap gd(ii^g) \ .$$

It follows by inspection of the definition that we have the covariance property

$$\zeta_1(cg) = c\zeta_1(g)$$

for g in the domain of ζ_1 and $c \in C(i)$. This implies that the fibers of ζ_1 are of constant rank, say f, and hence that any subset of $C(i)$ of rank r lifts under ζ_1 to a subset of G of rank $r + f$. Now since $i \in C(i) \setminus C^\circ(i)$, the group $C(i)$ is disconnected and hence has disjoint subsets of full rank, and these lift under ζ_1 to disjoint generic subsets of G, which contradicts the connectivity of G. □

4.5. Proof of Theorem 4.1. In order to prove that a connected group of degenerate type contains no involutions, it suffices now to focus on the remaining Case II:

(Case II) For generic and independent $i, j \in \mathcal{C}$ the group $d(ij)$ contains a unique involution.

Supposing G to be a counterexample, then after applying Lemma 4.3 we arrive at the following minimal configuration.

- G is simple.
- No proper definable connected subgroup of G contains an involution.

Now fix a conjugacy class of involutions $\mathcal{C}$ in G. In view of Lemma 4.9 the following will hold.

By Lemma 4.6 the Sylow 2-subgroups of G are elementary abelian. This then yields the following.

LEMMA 4.10. *If i, j are involutions and $k \in d(ij)$ is an involution, then i and j are not conjugate under the action of $C(k)$.*

PROOF. We show first that i and ik are not conjugate in $C(k)$. Suppose on the contrary $i^u = ik$ with $u \in C(k)$. Then $i^{u^2} = i$ and u acts on the group $\langle i, k \rangle$ as a nontrivial automorphism of order 2. It follows that $d(u)$ contains a 2-element with the same action, and as G has abelian Sylow 2-subgroups this is impossible.

On the other hand, as in the case of ordinary dihedral groups one may see that the group $d(i, j)$ has two conjugacy classes of noncentral involutions, represented by i and j, and in particular j is conjugate to ik under the action of $d(ij)$, and in particular under $C(k)$. If i is conjugate to j under $C(k)$, then i is conjugate to ik under $C(k)$ and we have a contradiction. □

Now we may conclude our proof by a model theoretic argument.

Fix a Sylow 2-subgroup S of G and consider a pair i, j of involutions in $\mathcal{C}$ which are independent and generic over S, that is to say with the elements of S treated as constants. Define a subset $S_{i,j} \subseteq S \times S$ as follows:

$$\{(s, t) \in S \times S : (i, k) \sim (s, t)\} .$$

Here k is the unique involution in $d(ij)$, and "$\sim$" refers to conjugacy under the action of G. As i and k commute, the set $S_{i,j}$ is nonempty.

Now the pair (i, j) and the pair (j, i) have the same type over S, so $S_{i,j} = S_{j,i}$. As the involution k is also the unique involution in $d(ji)$, this means that (i, k) and (j, k) are conjugate to the same pairs in $S \times S$, and hence to each other. But to conjugate (i, k) to (j, k) in G means that i is conjugated to j in $C(k)$. This contradicts the preceding lemma, and completes the proof of Theorem 4.1.

4.6. No decent tori. Our results on degenerate type groups have an interesting application to some very general problems put forward by Bruno Poizat. He stressed their relationship to the classification project, and indeed it turns out that the methods of the classification project are sufficiently developed to shed some light on these problems, as we will see in the next subsection. But we can avoid an appeal to the classification of even type groups from Part C by treating the following special case.

THEOREM 4.11. *Let G be a connected group of finite Morley rank containing no nontrivial decent torus. Then $G/O_2{}^\circ(G)$ has degenerate type.*

This is a reformulation of the content of the Algebraicity Conjecture in this limited case. We leave it as an exercise for the interested reader to derive this result directly from the Algebraicity Conjecture (or Theorem, as we deal with even type).

Before taking up the proof of Theorem 4.11, we will examine both its main hypothesis and its conclusion more carefully. The following clarifies the hypothesis.

LEMMA 4.12. *Let G be a group of finite Morley rank. Then the following are equivalent.*

(1) *G contains no nontrivial decent torus.*
(2) *There is no prime p for which G contains a nontrivial p-torus.*
(3) *No definable section of G is a good torus.*

These conditions are inherited by passage to definable sections or elementary extensions.

PROOF. Let us show first that condition (2) passes to elementary extensions. So suppose that in an elementary extension G^* of G we have a nontrivial p-torus T_0, and consider $A = d(T_0)$. Then A is definable and p-divisible, and contains p-torsion; this property passes to the elementary substructure G and contradicts (2). So (2) is preserved by passage to elementary extensions.

As the third condition is inherited by definable sections, it suffices now to check the stated equivalences.

The equivalence of the first two is clear.

The third condition implies the first, since every nontrivial decent torus has a nontrivial good torus as a definable quotient (Lemma 1.11).

Now, it suffices to check that (2) $\implies$ (3), and since condition (2) is inherited by definable subgroups, we need only concern ourselves with definable quotients. So suppose that $\bar{G}$ is a definable quotient of G and that $\bar{G}$ contains a good torus, and in particular contains a p-torus $\bar{T}_0$ for some p. Each element $\bar{t}$ of $\bar{T}_0$ lifts to a p-element t of G, and hence the groups $Z(C(t))$ as t varies over G contain p-subgroups of unbounded order. It follows easily that an elementary extension of G contains a p-torus, and thus as we have seen G also contains a p-torus, contradicting (2). So (2) $\implies$ (3). □

The conclusion of Theorem 4.11 also deserves further analysis.

LEMMA 4.13. *Let G be a connected group of finite Morley rank containing no decent torus and $U = O_2{}^\circ(G)$. Suppose that G/U is of degenerate type. Then $G = U \cdot C_G(U)$.*

PROOF. Let G be a counterexample of minimal rank. Let $A = Z^\circ(U)$. Then $G/C_G(A)$ is a group of degenerate type acting faithfully on A. By Lemma 1.6, the Borel subgroups of $G/C_G(A)$ are good tori, hence trivial, and thus G centralizes A.

On the other hand, by the minimality of G, the group $G/A = (U/A) \cdot C_{G/A}(U/A)$. Let $H/A = C_{G/A}(U/A)$. Then $G = UH$ and $[H, U] \le A$.

For any $2^\perp$-subgroup X of H we have $[X, U] = 1$. It follows that $H/C_H(U)$ is a 2-group, and thus $H \le UC_H(U)$, and $G = UC_G(U)$. □

Let us now take up the structure of a minimal counterexample to the theorem.

LEMMA 4.14. *Let G be a group of finite Morley rank containing no nontrivial decent torus, and such that $O_2{}^\circ(G)$ is not a Sylow° 2-subgroup of G.*

Suppose that G is of minimal rank among such groups. Then the following hold.

(1) *$Z(G)$ is finite and $G/Z(G)$ is simple.*
(2) *For $U \leq G$ a nontrivial definable 2-subgroup, not contained in $Z(G)$, setting $H = N^\circ(U)$ and $V = O_2{}^\circ(H)$, we have $H = VC_H{}^\circ(V)$ and H/V is of degenerate type.*

PROOF.
Ad (1) It follows from the minimality hypothesis and Lemma 4.12 that $O_2{}^\circ(G) = 1$ and that $U_2(G) = G$. We must show that G contains no nontrivial proper definable connected normal subgroup H. Supposing the contrary, with $\bar{G} = G/H$ our minimality hypothesis implies that $\bar{G}/O_2{}^\circ(\bar{G})$ is of degenerate type; since also $\bar{G} = U_2(\bar{G})$ we find that $\bar{G}$ is a 2-group. Let S be a Sylow° 2-subgroup of G. Then $G = HS$, and by Proposition I 10.13 we have $S \leq C(H)$. Thus $S \leq O_2{}^\circ(G)$, contradicting our hypothesis. This proves the first point.

Ad (2) With U, H, and V as specified, observe that $H = N^\circ(U) < G$. Thus by minimality H/V is of degenerate type, and by Lemma 4.13 we have $H = VC_H(V)$. □

Now let us fix our notation in accordance with the preceding lemma. We may take G to be a group of finite Morley rank and even type, containing no decent torus, and of minimal rank among such groups, and we may factor out $Z(G)$. We then have the following conditions.

(∗) G is simple, and for $U \leq G$ a nontrivial definable 2-subgroup, setting $H = N^\circ(U)$ and $V = O_2{}^\circ(H)$, we have $H = VC_H{}^\circ(V)$, with H/V of degenerate type.

Now, using the notion of strong embedding, we can arrive quickly at a proof of Theorem 4.11.

LEMMA 4.15. *Let G be a group of finite Morley rank and even type satisfying the conditions (∗) above. Let S be a Sylow° 2-subgroup of G. Then $N(S)$ is strongly embedded in G.*

PROOF. We show first that

$$N(S) \text{ is weakly embedded in } G\,.$$

For this, it suffices to show that for $U \leq S$ nontrivial, connected, and definable, we have $N(U) \leq N(S)$.

Supposing the contrary, take $U \leq S$ maximal connected definable such that $H = N(U)$ is not contained in $N(S)$. Evidently $U < S$. Let $V = O_2{}^\circ(N(U))$. As V is a Sylow° 2-subgroup of H, it follows that $U < V$ and thus $N(V) \leq N(S)$. Now $N(U) \leq N(V) \leq N(S)$, a contradiction. This proves that $N(S)$ is weakly embedded in G.

Now it suffices to prove that $C(i) \leq N(S)$ for any involution $i \in N(S)$. Let $U = C_S{}^\circ(i)$, a nontrivial connected definable 2-group. Then $N(U) \leq$

$N(S)$. Let $V = O_2{}^\circ(C(i))$. Then V is a Sylow° 2-subgroup of $C(i)$ and thus contains U. Furthermore, $N_V(U) \leq N(S)$ so $N_V{}^\circ(U) \leq C_S{}^\circ(i) = U$. It follows that $V = U$ and $N(V) \leq N(S)$. In particular, $C(i) \leq N(S)$. □

PROOF OF THEOREM 4.11. Assuming the theorem fails, we find a group G of finite Morley rank and even type satisfying the hypothesis $(*)$ above. Then with S a Sylow° 2-subgroup of G we find that $N(S)$ is strongly embedded in G. Setting $H = N^\circ(S)$, we have $H = S \cdot C_H(S)$.

Now all involutions of $N(S)$ are conjugate under the action of $N(S)$ (Lemma I 10.12). In particular, they all lie in $Z(S)$; but H acts trivially on $Z(S)$ and hence S contains only finitely many involutions, contradicting our assumptions. □

4.7. An application. Poizat has pointed out that the Algebraicity Conjecture has some very general consequences, reflecting the presence of a natural topology in the algebraic case; most notably, an equation which holds generically in a connected group should hold identically. The case of a generic equation of the simple form $x^n = 1$ is particularly striking, and apart from low values of n (2 or 3) or special contexts (solvable groups), this problem has long resisted in analysis. We will prove the following, which is a substantial step in a more general direction.

PROPOSITION 4.16. *Let G be a connected group of finite Morley rank which generically satisfies the equation*

$$x^n = 1$$

for some fixed n. Then $O_2(G)$ is unipotent and $G/O_2(G)$ contains no involutions, and generically satisfies the equation

$$x^{n_0} = 1$$

where n_0 is the odd part of n.

This has the following noteworthy consequences.

COROLLARY 4.17. *Let G be a connected group of finite Morley rank which generically satisfies the equation*

$$x^{2^k} = 1\ .$$

Then G satisfies this equation identically.

Indeed, Proposition 4.16 reduces this to the case in which G is 2-unipotent, and hence nilpotent. We may conclude via Jaber's Proposition I 5.32.

COROLLARY 4.18. *Let G be a connected group of finite Morley rank, and $a \in G$ an arbitrary element. Then $C^\circ(a)$ is infinite.*

PROOF. If $a \in G$ has a finite centralizer, then a has finite order and its conjugacy class a^G is generic in G, as is the conjugacy class of ${a^{-1}}^G$. Thus a and a^{-1} are conjugate in G. By Proposition 4.16, the quotient group $\bar{G} = G/O_2(G)$ contains no involutions. However, the images $\bar{a}$ and $\bar{a}^{-1}$

are conjugate, and as $\bar{G}$ contains no involutions it follows that $\bar{a} = \bar{a}^{-1}$. Thus $a \in O_2(G)$ is a 2-element and by Corollary 4.17 G is 2-unipotent. Then passing to a nontrivial abelian quotient of G, the image of a^G is both generic and trivial, and we have a contradiction. □

PROOF OF PROPOSITION 4.16. Let G be a connected group of finite Morley rank of generic exponent n. Let U be a Sylow° 2-subgroup of G.

We will proceed by induction on the rank of G. We claim:

G contains no decent torus.

If G contains a nontrivial decent torus and if T is a maximal such, then $H = N^\circ(T)$ is generically disjoint from its conjugates, that is $H \setminus \bigcup_{g \notin N(H)} H^g$ is nongeneric in H, and $\bigcup H^G$ is generic in G. Again, as H is generically disjoint from its conjugates, the subset of H consisting of elements of order n must itself be generic in H. As $H = C^\circ(T)$, some coset of $hd(T)$ in H must also satisfy $x^n = 1$ generically (Fubini), and hence h commutes with $d(T)$, we conclude that $d(T)$ is itself of bounded exponent, a contradiction. So there is no nontrivial p-torus in G for any p. In particular, taking $p = 2$, we find that U is 2-unipotent.

As G contains no decent torus we may apply Theorems 4.11 and 4.13. Hence U is normal in G and $G = U \cdot C_G(U)$.

Consider the quotient G/U, which again satisfies $(*)$ generically. As G/U is of degenerate type it contains no involutions, and thus we may replace n by its odd part. □

We also can improve upon Lemmas II 6.4 and II 5.11 at this point.

LEMMA 4.19. *Let G be a connected L-group of even type, and S a Sylow 2-subgroup of G. Then $O_2(G)$ and S are connected and definable.*

PROOF. The quotient $G/U_2(G)$ is of degenerate type, hence contains no involutions. Thus $O_2(G) = O_2(U_2(G))$ and $S \leq U_2(G)$, and Lemmas II 6.4 and II 5.11 apply. □

5. Pseudoreflection L-groups

It is now convenient to take up the theory of pseudoreflection groups once more, continuing on from §III 1. We extend the theory from K-groups to L-groups, making use of Lemma 1.6 along the way. We recall the main definition.

DEFINITION 5.1. *If A is an elementary abelian group, then a torus T acting on A is called a* group of pseudoreflections *on A if $A = C_A(T) \times [A, T]$ and T acts faithfully on the second factor, and transitively on its nonzero elements.*

We will now pass to the L-group classification, with the same outcome as in the K-group case. This depends on the treatment of the following additional special case, where hypothetical degenerate type groups intervene with a vengeance. We use Theorem 4.1 to simplify the analysis very slightly; the real difficulties lie elsewhere.

PROPOSITION 5.2. *Let H be a group of finite Morley rank of degenerate type, acting faithfully and definably on a definable elementary abelian group A. Suppose that H is generated by pseudoreflection subgroups. Then H is abelian.*

PROOF. By Theorem 4.1, the group H contains no involutions.

We proceed by induction on the rank of the group AH, in other words, on $\mathrm{rk}(A) + \mathrm{rk}(H)$. As H is generated by connected definable subgroups, it is connected. Thus $[A, H] \leq A^\circ$, and it follows easily that H acts faithfully on A°. So we may suppose that A is connected. We may also suppose that $C_A(H) = 1$.

We suppose toward a contradiction that H is nonabelian, and we fix two pseudoreflection subgroups T_1, T_2 of H which do not commute. By our inductive hypothesis, we have $H = \langle T_1, T_2 \rangle$.

Our first claim is the following:

(1) $$A \text{ is irreducible under the action of } H.$$

By Lemma I 11.3 it suffices to check definable irreducibility.

Suppose the contrary, $1 < A_0 < A$ with A_0 H-invariant. Let V be either of the H-modules A_0 or A/A_0. Then $H/C_H(V)$ is still generated by pseudoreflection groups (one of which may become trivial) and by induction $H/C_H(V)$ is abelian, so if $K = C_H(A_0) \cap C_H(A/A_0)$, then H/K is also abelian; but K is a $2^\perp$-group so K acts trivially on A (Proposition I 10.7) and thus $K = 1$, H is abelian.

Now we can limit the structure of H:

(2) $$\sigma^\circ(H) = 1 \ .$$

Assuming the contrary, H will have a normal infinite abelian definable subgroup, and by Proposition I 4.11 we may give A the structure of a vector space over an infinite, definable field of characteristic 2, on which the group H acts linearly. Then by Fact II 4.5, H is a K-group, and being both connected and of degenerate type, is solvable. Then by Lemma I 8.2, H is abelian after all, a contradiction.

From this point on, we no longer have any direct possibility of imposing a vector space structure on A. However, we may retain the intuition that $[T, A]$ should be in some sense 1-dimensional. We will now fix just one pseudoreflection subgroup T of H, and we set $f = \mathrm{rk}([T, A])$, where "f" stands, with pardonable optimism, for "field". One might reasonably hope that the rank of A would be a multiple of f, but at this point we know nothing of the kind. However, our next claim is very suggestive:

(3) $$\mathrm{rk}(A) \leq 2f \text{ and } \mathrm{rk}(C_A(T)) \leq f \ .$$

Let $H_0 = \langle T^h : h \in H \rangle$. As H_0 is connected and normal in H, it is nonsolvable by (2). In particular, some conjugate T^h of T does not commute with T. So by minimality $H = \langle T, T^h \rangle$. Then as A is irreducible, and the sum

$$[T, A] + [T^h, A]$$

is H-invariant (being T-invariant and T^h-invariant), we have

$$A = [T, A] + [T^h, A]$$

and $\mathrm{rk}(A) \leq 2f$. Again, $A = C_A(T) \oplus [T, A]$ and thus $\mathrm{rk}(C_A(T)) \leq f$ as well.

(4) The Borel subgroups of H are good tori, and are conjugate.

This is Lemma 1.6. In our situation this is very strong. It follows, in particular, that every Borel subgroup of H contains some pseudoreflection subgroup of H, which it centralizes.

We next eliminate the extreme case in which $C_A(T) = 1$:

(5) $$C_A(T) > 1 \ .$$

Suppose the contrary. Then the pseudoreflection subgroup T acts transitively on $A^\times$. As H acts faithfully on A, $H = T \cdot C_H(a)$ for $a \in A^\times$. Now if $C_H(a)$ is finite, then we find $H = T$ as H is connected, and this is a contradiction as we assume H nonabelian. So $C_H(a)$ is infinite and contains some infinite connected definable abelian subgroup H_0. But then H_0 is also contained in a Borel subgroup, and hence commutes with a conjugate of T; but H_0 fixes a nontrivial point, and T acts acting transitively on $A^\times$, so the fixed point set of H_0 must be A, and we have a contradiction. This proves (5).

(6) If $T_1, T_2 \leq H$ are pseudoreflection groups with $[T_1, A] = [T_2, A]$, then $T_1 = T_2$.

If T_1 and T_2 commute, then $[T_2, C_A(T_1)] \leq C_A(T_1) \cap [T_1, A] = 1$, and thus $C_A(T_1) = C_A(T_2) = C_A(T_1T_2)$. It follows easily that T_1T_2 acts freely on $[T_1, A]$ and thus $T_1 = T_2$ in this case.

If T_1 and T_2 do not commute, then by minimality they generate H, and hence $[T_1, A] = [T_2, A]$ is H-invariant, forcing $[T_1, A] = A$ and contradicting point (5). So (6) holds in either case.

Now we make a case division, depending on whether or not we can find two commuting pseudoreflection subgroups in H.

(**Case I**) There is a pair of commuting pseudoreflection subgroups of H.

For the duration of this case analysis, we fix such a commuting pair T_1, T_2, and we set $B = T_1T_2$.

(I.1) B is a Borel subgroup of H.

Without loss of generality, the pseudoreflection group T considered previously is T_1, and $f = \mathrm{rk}(T_1) \le \mathrm{rk}(T_2)$. As T_2 commutes with T_1 it respects the decomposition $A = C_A(T_1) \oplus [T_1, A]$, and as T_2 is a pseudoreflection group, the subgroup $[T_2, A]$ is contained in one factor or the other. But if $[T_2, A] \le [T_1, A]$, then $[T_1, A] = [T_2, A]$, forcing $T_1 = T_2$, which is not the case. So $[T_2, A] \le C_A(T_1)$, and then by rank considerations we have $[T_2, A] = C_A(T_1)$ and $f = \mathrm{rk}(T_2)$. Thus the situation is symmetrical and the decomposition of A can be written in a number of equivalent ways, e.g., $A = [T_1, A] \oplus [T_2, A]$.

Now let $B \le B_1$ with B_1 a Borel subgroup of H. Then B_1 is a good torus and in particular is contained in $C(T_1T_2)$. But it is easily seen from the decomposition of A and the action of T_1 and T_2, that $C_H{}^\circ(T_1T_2) = T_1T_2$. So $B_1 = B$.

(I.2) The intersection of distinct Borel subgroups of H is finite.

Suppose on the contrary that $(B \cap B_1)^\circ = X > 1$ with B_1 another Borel subgroup of H. As $Z(H)$ is finite, we have $C^\circ(X) < H$; but $C^\circ(X)$ contains B and B_1, each of which is a product of pseudoreflection groups, so by induction we may suppose that the subgroup generated by B and B_1 is commutative. Since B and B_1 are Borel subgroups, they then coincide.

Now we examine the conjugates of $A_1 = C_A(T_1)$.

(I.3) For $g \in H \setminus N(T_1)$ we have $A_1^g \cap A_1 = 1$.

Suppose on the contrary $a \ne 1$ belongs to $C_A(T_1) \cap C_A(T_1^g)$. If T_1 and T_1^g do not commute, then by minimality they generate H, and $a \in C_A(H)$, contradicting our initial setup. So T_1 and T_1^g commute, in which case we could take $T_2 = T_1^g$ and then as we have seen above $C_A(T_1^g) = [T_1, A]$, so the intersection is trivial.

Now we can arrive at a contradiction in this first case. We have $\mathrm{rk}(A_1) = f$, $\mathrm{rk}(A) = 2f$, and the conjugates of $A_1^\times$ under the action of H are pairwise disjoint, while $N_H{}^\circ(A_1) = N_H{}^\circ(T_1) = N_H{}^\circ(B) = B$. Thus $\mathrm{rk}(H/B) \le f$. On the other hand, for any $g \in H \setminus N(B)$ we have $B \cap B^g$ finite, and hence $\mathrm{rk}(H/B) \ge \mathrm{rk}(B) = 2f$. So this is a contradiction.

We pass to the second case.

(Case II) No two distinct pseudoreflection subgroups of H commute.

As the Borel subgroups are abelian and conjugate, this produces the following:

(II.1) Every Borel subgroup of H contains a unique pseudoreflection subgroup of H.

Let us fix a Borel subgroup B of H, and the corresponding pseudoreflection subgroup T. Let $H_T = N_H{}^\circ(T) = C_H{}^\circ(T)$.

(II.2) H_T is almost self-normalizing.

Observe that by our case hypothesis, T is the only pseudoreflection subgroup of H_T. Thus $N_H(H_T) \le N_H(T)$ and our claim follows.

(II.3) For $g \in H \setminus H_T$, we have $H_T \cap H_T^g \le Z(H)$.

Suppose on the contrary that $X = H_T \cap H_T^g$ is not central in H. Then $T, T^g \le C^\circ(X) < H$, and by minimality of H, T and T^g must commute. Then by our case assumption, $g \in N(T) = N(H_T)$.

As $Z(H)$ is finite, we have the following as well:

(II.3) The union of the conjugates of H_T in H is generic in H.

(II.4) For $g \in H \setminus N(T)$, the rank of $H_T g H_T$ is $2\,\mathrm{rk}(H_T)$.

We consider the natural map $H_T \times H_T \to H_T g H_T$ and we claim it has finite fibers. So suppose

$$ugv = g$$

with $u, v \in H_T$. That is, $u^g = v^{-1} \in H_T^g \cap H_T \le Z(H)$. It follows that $u = v^{-1} \in Z(H)$, and this is a finite set.

Now we may follow the line of argument of the previous case. Let $A_T = C_A(T)$.

(II.5) For $g \in H \setminus N(T)$ we have $A_T \cap A_T^g = 1$.

By our case assumption T and T^g do not commute, and so by minimality of H we have $H = \langle T, T^g \rangle$; but then H centralizes the intersection $A_T \cap A_T^g$, which must be trivial.

(II.6) $\mathrm{rk}(H/H_T) \le f = \mathrm{rk}(T)$.

We have $\mathrm{rk}(H_T) = \mathrm{rk}(N(T))$, $\mathrm{rk}(A) = \mathrm{rk}(A_T) + f$, and thus our claim follows from the previous point.

Now combining (II.4) and (II.6), we will arrive at a contradiction. We have $f \ge \mathrm{rk}(H/H_T) \ge \mathrm{rk}(H_T)$. It follows that $H_T = T$ and $\mathrm{rk}(H) = 2f$. In particular, applying (II.4) to both g and g^{-1} with $g \in H \setminus N(T)$, we find that TgT and $Tg^{-1}T$ are generic subsets of H, hence equal,

$$t_1 g t_2 = g^{-1}$$

for suitable $t_1, t_2 \in T$. Then $(gt_1)^2 \in T$, $gt_1 \notin T$, and this produces an involution, a contradiction. □

THEOREM 5.3. *Let $A \rtimes H$ be a connected L-group of finite Morley rank and of even type, in which A is an elementary abelian definable 2-subgroup and H acts irreducibly and faithfully on A. Assume that H contains a group T of pseudoreflections on A. Then A can be given a vector space structure over an algebraically closed field K in such a way that $H \simeq \mathrm{GL}(A)$ acting naturally.*

PROOF. We analyze the structure of H:

$$O_2(H) = 1 \; . \tag{1}$$

Indeed, $C_A(O_2(H))$ is nontrivial and H-invariant, hence equal to A, and $O_2(H) = 1$. By Lemma II 6.10, we then have $U_2(H) = E(U_2(H))$, and $H = U_2(H) * \hat{O}(H)$:

$$O(H) = 1 \; . \tag{2}$$

In the contrary case, our aim is to show that $\hat{O}(H) = O(H)$, which puts us in the case of a K-group, treated in Theorem III 1.5.

We consider a series of definable $\hat{O}(H)$-invariant subgroups

$$A = A_0 > A_1 > \cdots > A_n = (0)$$

with successive quotients $\hat{O}(H)$-minimal.

If $O(H)$ acts trivially on some quotient $V_i = A_i/A_{i+1}$, then $C_{A_i}(O(H))$ covers V_i, by Proposition I 9.9. In particular, $C_A(O(H))$ is nontrivial in this case; but this group is H-invariant and must then be A, forcing $O(H) = 1$, as desired.

So we suppose that $O(H)$ acts nontrivially on every quotient V_i and therefore by Proposition I 4.11 we get a linear action of $\hat{O}(H)$ on V_i with respect to some infinite definable field, varying with i. We apply Proposition II 4.5 to conclude that the quotient $\hat{O}(H)/C_{\hat{O}(H)}(V_i)$ is a K-group, and hence solvable, for each i. Now consider $H_0 = \bigcap_i C_{\hat{O}(H)}(V_i)$, the kernel of all these actions. Then $H_0{}^\circ$ is a group of degenerate type, and Proposition I 10.7 applies, showing $H_0{}^\circ = 1$. So H_0 is finite, hence central in $\hat{O}(H)$, hence solvable after all, and we find $\hat{O}(H) = O(H)$, and H is a K-group, to which our earlier result applies. We therefore have (2).

Again, leaving aside the K-group case already treated, we may suppose

$$\hat{O}(H) > 1 \; . \tag{3}$$

Now we invoke the special case treated just above to reduce to the following case:

$$U_2(H) > 1 \; . \tag{4}$$

If $H = \hat{O}(H)$ is of degenerate type, we invoke Proposition 5.2. Then H is abelian and it follows easily that $H = \mathrm{GL}(A)$ with A viewed as 1-dimensional over a suitable field. Leaving this case aside, we have (4).

Now we may conclude. We fix a nontrivial torus R in $U_2(H)$, and a definable simple normal subgroup K of $\hat{O}(H)$, using Lemma I 8.34, bearing in mind that $\hat{O}(H)$ is connected and hence normalizes the simple components of its socle. We consider the $(R \times K)$-module $V = A/C_A(R)$, on which the torus R acts without fixed points. If we pass to a series of definable (RK)-invariant subgroups

$$V = V_1 > V_2 > \cdots > V_n = (0)$$

with successive quotients (RK)-minimal, then the action of R on each quotient is nontrivial, by Proposition I 9.9. Accordingly, the action of K on each factor is linear, by Proposition I 4.11, and thus as K is simple and of degenerate type its action is trivial on each factor. Then again by Lemma I 10.7 we find that K is trivial, a contradiction. □

6. Notes

The material in the present chapter has continued to evolve considerably as this book was in preparation. Most of the more recent developments will have more of an impact on the analysis of groups of odd type, but they combine neatly with our results on even and mixed type to give some information about connected groups of finite Morley rank in general, something that was not in view when we first began work on the present text.

§1. **Borel subgroups** The construction of Carter° subgroups in general is found in [**94**]. It requires machinery we have no space for here, machinery which has numerous applications to the study of groups of odd and degenerate type.

The theory of Carter° subgroups has implications for classification problems, but is not involved in those aspects of the theory dealt with here, in Parts B and C. At the present time it also has the flavor of an alternative to the classification project, giving structural information of a geometric kind very reminiscent of the behavior of tori in reductive algebraic groups, without passing through an inductive analysis. The subject has interacted strongly with classification projects in odd type groups and has potential for application in the very difficult degenerate type context as well.

Our "Carter°" subgroups are usually called simply Carter subgroups, and this is justified by the fact that they coincide with the usual Carter subgroups in the connected solvable case. Our terminology is more cumbersome, but then we do not actually use the more general theory here.

Maximal good tori are discussed in [**69**], which was inspired by arguments found in early drafts of [**14**].

§2. Generic cosets

The genericity argument used here is one of the earliest arguments developed in the study of bad groups, and naturally depends on arguments in the style of §1. It has been developed further for various purposes, and was used particularly heavily in [**70**].

§3. Generic covering

The covering theorem, Theorem 3.1, first appears in [**14**], and will be used toward the end of Chapter VI.

§4. Degenerate type groups

The results on degenerate type groups are taken from [**47**], which contains additional information about torsion in degenerate type groups, also for odd primes. This section was not part of the first draft of the present book, but it eliminates a host of minor complications which were treated originally by ad hoc modifications of the main line of argument. But we stress that this plays an inessential role; as one sees also in the case of finite groups, the main lines of argument are very robust and can easily be adapted to handle a variety of complications, a point to which we will return in the last chapter. Results which may seem to be essential when one has them available for use often turn out to have been merely convenient, when one is forced to work without them (see [**30**, p. 1] for a more forceful illustration of this point).

The main line of argument, via the covariant map $\zeta_1 : G \to C(i)$, has an unusual source: black box group theory [**127**], the study of efficient computation in large groups for which random elements can be efficiently generated and (in most cases) combined. Its direct antecedents may be found in [**55, 17, 44**]. This is a very "global" line of argument, very different from what we do elsewhere, and not particularly tied to any inductive framework. It has something in common with techniques of finite group theory which are typically not available in our settings, notably the transfer map, though as we use it here, the method is very tightly tied to connectivity, which is of course not the case in the original black box setting, where the global feature of interest is equidistribution.

This work was carried out at the Newton Institute, Cambridge, during the month devoted to groups of finite Morley rank within the larger semester program on model theory and its applications, in Spring 2005.

§5. Pseudoreflection L-groups

The treatment of pseudoreflection subgroups in the L-group context involves some substantial issues relative to the K-group case, and in fact is the last point in the adaptation of K-group methods to the L-group context that requires new ideas. This material has not previously appeared, though the K-group version was one of the central ingredients of [**5**]. With the pseudoreflection subgroup theory in hand, the adaptation of the remainder of [**5**] is routine, given the technology already developed at that point. This will be carried out in Chapter VII; the relevant L-group techniques have already been presented in Chapter II. They were developed in [**2**] and a series of subsequent works aimed at the material we will give in Chapter VI. There are some points in later chapters where the divergence between the K-group theory and the L-group theory can be felt, but this is a matter of adaptation. The most delicate point among these later developments is that "standard Borel" subgroups continue to be solvable (Lemma VIII 5.7).

The proof of Lemma 5.2 relies on Poizat's linearization result for the degenerate case, which goes through the theory of locally finite groups and ultimately depends on the Feit-Thompson Odd Order Theorem.

The proof of point (II.2) within the proof of Proposition 5.2, short as it is, is superfluous because pseudoreflection subgroups are good tori. This suggests a point of contact between this result and earlier considerations in the chapter.

Part B

Mixed Type Groups

CHAPTER V

Mixed Type

Leges meas custodite
iumenta tua non facies coire
cum alterius in generis animantibus
agrum non seres diverso semine
veste quae ex duobus texta est non indueris
— Leviticus 19:19

Introduction

In this part, consisting of a single short chapter, we will prove the following.

MIXED TYPE THEOREM. *If every simple group of finite Morley rank of even type is algebraic, then there is no simple group of finite Morley rank of mixed type.*

In Part C we will prove that every simple group of finite Morley rank of even type is indeed algebraic. Hence we can eliminate simple groups of finite Morley rank of mixed type *absolutely*. In an inductive approach, it may be natural that we need to treat even type groups in order to treat mixed type groups. But then we are very fortunate that we do not need to treat odd type or degenerate type groups, which could also appear as sections of our group.

We may sharpen the Mixed Type Theorem slightly, as follows. Recall that a group of finite Morley rank is an *L^*-group* if every proper definable connected infinite simple section of even type is a Chevalley group.

MIXED TYPE L^* THEOREM. *There is no simple L^*-group of finite Morley rank of mixed type.*

Another way to express the main result of Part C is that *every* group of finite Morley rank is an L^*-group. One should observe that there is a strong asymmetry in the L^*-hypothesis, as we make no assumption at all about the structure of definable sections of *odd type.*

The proof of the Mixed Type L^* Theorem is motivated by the following considerations. If the group G in question were an L-group, then it would follow that the subgroups $U_2(G)$ and $T_2(G)$ must commute, where $U_2(G)$ is generated by its 2-*unipotent subgroups*, and $T_2(G)$ is generated by the *definable hulls of its* 2-*tori.* This point will be established in §2. If one could establish this result for the ambient group G as well, this would furnish two nontrivial proper normal subgroups, in contradiction to the assumed simplicity. Of course since G is not assumed to be an L-group we have to reach some such conclusion by internal analysis.

Pursuing this line of thought, as the groups $U_2(G)$ and $T_2(G)$ are expected to commute, if we then fix a unipotent subgroup U or a 2-torus T, we find that $T_2(C_G(U))$ should be $T_2(G)$, and $U_2(C_G(T))$ should be $U_2(G)$. Notice however that the groups $T_2(C_G(U))$ and $U_2(C_G(T))$ are necessarily proper subgroups of the simple group G, so we have much better control over them. We can therefore shift our goal slightly, and aim at proving that, for example $T_2(C_G(U))$ is *independent* of the choice of U, in which case it furnishes a nontrivial normal subgroup of G, and a contradiction reminiscent of our original idea.

Now this second line of argument actually succeeds *generically*, namely in the case where a certain graph $\mathcal{U}$ introduced below is connected, and most of our actual analysis will be devoted to "mopping up" the specific configurations that arise when the generic argument fails. This is of course a common strategy in finite group theory, where however departure from the main line often occurs at the outset, while the consideration of pathological cases may then occupy hundreds of pages. Here the situation is more balanced; the exceptional configuration in which the main line fails requires as much attention as the main line, but not much more. The critical configuration off the main line will be presented in §3 and then exploited, twice, once in that section and once in the final §5. The two configurations involved have much in common. As is generally the case in such circumstances, we are mainly concerned with the smallest possible group, $\mathrm{SL}_2(K)$.

To implement our overall strategy involves making use of the notions of strong and weak embedding, which we now recall.

DEFINITION. *Let G be a group of finite Morley rank and M a subgroup.*

(1) *M is* strongly embedded *in G if for all $g \in G$ we have*

$$M \cap M^g \textit{ contains an involution iff } g \in M.$$

(2) *M is* weakly embedded *in G if for all $g \in G$ we have*

$$M \cap M^g \textit{ has an infinite Sylow subgroup iff } g \in M.$$

We gave other criteria for strong and weak embedding in §§I 10.3 and II 4.3.

These two notions are notions of largeness relative to G (or in more graphic terminology, black hole properties relative to G). They become interesting when $M < G$. Strong or weak embedding turns out to be a rare occurrence in reality, but a common enough occurrence in configurations which need to be driven to a contradiction—for example, when M is a well understood subgroup of G which one hopes is in fact equal to G!

Strategy

We will proceed as follows.

For $U \leq G$ 2-unipotent and nontrivial, we consider the group

$$U^{\perp} = T_2(C_G(U)) .$$

We show that if U_1, U_2 are two nontrivial 2-unipotent subgroups of G which commute, then $U_1^{\perp} = U_2^{\perp}$. At this point, one considers the graph $\mathcal{U}(G)$ whose *vertices* are the nontrivial 2-unipotent subgroups of G, with *edges* between all commuting pairs. What we have just said is that $U^{\perp}$ depends, not on U itself, but rather on the connected component of the graph $\mathcal{U}(G)$ containing the vertex U. So if the graph $\mathcal{U}(G)$ is connected, one sees that the group $U^{\perp}$ is independent of U, and then normal in G, as well as proper and nontrivial, and we have the desired contradiction.

How does this go wrong—in other words, where does the case $\mathcal{U}(G)$ disconnected take us? If G is a simple K-group with $\mathcal{U}(G)$ disconnected, this forces $G \simeq \mathrm{SL}_2$ by Proposition II 4.20. So the methods we use will be related to methods used to identify SL_2 in the even type context. There are several such methods, and we will become acquainted with a variety of them in Part C, where all of Chapters VI and VII and much of Chapter VIII are devoted to this single issue. The method we need now is the first of these methods, corresponding to Chapter VI of Part C: strong and weak embedding.

The way to exploit the disconnectedness of $\mathcal{U}(G)$ is by considering the setwise stabilizer M of a connected component of $\mathcal{U}(G)$. In SL_2, the graph $\mathcal{U}(G)$ consists of isolated points, and M is a Borel subgroup, which is indeed strongly embedded. In our context, if we can prove that M is strongly embedded we will have an immediate contradiction, as this implies that G has only one class of involutions, by Lemma I 10.12, but in fact it has more than one conjugacy class by control of fusion, Lemma I 6.18. So in a mixed type context strong embedding is extremely powerful.

Our strategy from this point on is clear. We must first show that M is weakly embedded, and then improve this to strong embedding. In each case, failure of the desired conclusion produces a definite configuration, and it is fortunate that the two configurations that arise can be killed by a single method. We will enter somewhat into the details. It will be helpful to bear in mind what our nonexistent group G "really" is, or should be.

Since G is of mixed type, one should imagine that G is not actually SL_2 over a field of characteristic 2, but rather $T \times \mathrm{SL}_2(K)$ with T the multiplicative group of a field of odd characteristic, and K of characteristic 2. This is a mixed type group in which the graph $\mathcal{U}(G)$ again consisting of isolated vertices, and with M of the form $T \times B$ with B a Borel subgroup of SL_2. Here M is not actually weakly embedded, but satisfies a weaker condition: the normalizer of any 2-unipotent subgroup of M is contained in M. However, the normalizer of T is all of G. This configuration, and a close relative, will reappear in the course of our general analysis, and will become the focus of our attention.

To recapitulate, suppose that G is a simple L^*-group of mixed type, $\mathcal{U}(G)$ the associated graph on the nontrivial 2-unipotent subgroups, $\mathcal{C}$ a connected component of the graph $\mathcal{U}(G)$, and M the setwise stabilizer of $\mathcal{C}$ under the natural action of G on $\mathcal{C}$. After showing that $\mathcal{U}(G)$ is disconnected, we will argue that M is weakly embedded in G, encountering along the way a configuration involving SL_2 which will need to be eliminated by technical considerations. Thinking in terms of the case $G = T \times \mathrm{SL}_2$, one may observe two more or less "local" properties that could be used to contradict the hypothesis that G is simple without actually pinning down its global structure: (1) the involutions in T and in SL_2 form two commuting conjugacy classes; (2) any involution in a conjugate of T (i.e., in the case at hand, in T itself) lies in M. Both of these properties will in fact make an explicit appearance in the course of our analysis, and suffice to dispose of the general case.

In the case that M is not weakly embedded, we find a 2-torus T in M such that $N_G(T)$ is not contained in M, and the definition of M leads quickly to the identification of $L = U_2(N_G(T))$ as a group of type SL_2 over an algebraically closed field of characteristic 2. With some additional adjustment we may take T to be a maximal 2-torus in G. This is one of the key configurations which must be treated separately. We dispose of this particular configuration in advance, in §3.

Once we have M weakly embedded, then since M cannot be strongly embedded, the criterion for strong embedding shows that the subgroup M must contain so-called *offending* involutions in M; these are involutions $\alpha \in M$ such that $C_G(\alpha)$ is not contained in M. One can quickly pin down the structure of the connected centralizer $C_G{}^\circ(\alpha)$ of any offending involution α, showing in particular that it contains a normal subgroup of the form $L = \mathrm{SL}_2(K)$ with K an algebraically closed field of characteristic 2. The picture begins to resemble the special case eliminated previously; the details are different, but the key technical lemma, Lemma 3.3, is formulated so as to apply in both cases

There is an underlying idea behind the last part of this analysis, which is not actually visible as one goes through it: use of the *Thompson rank formula*. This formula provides a powerful general technique for analyzing groups with more than one conjugacy class of involutions, if those conjugacy classes are reasonably well understood, and if, more particularly, the structure of the centralizers of the various involutions is also well understood. It will become clear in the course of our analysis here that one rapidly reaches a point at which information of the appropriate type is available. Jaligot found that as one then proceeds to pin down the relationships among the involutions a little more closely, along the lines needed for a Thompson rank computation, the technical Lemma 3.3 leads to a "premature" contradiction, somewhat before the point at which one could actually make the relevant computation. Later we will apply the full Thompson rank method, with no such premature contradiction, in our analysis of groups of even type, specifically in Chapter VII.

The organization of this chapter is as follows. In the first two sections we prepare the ground by discussing criteria for weak embedding, and the general theory of $U_2(H)$ and $T_2(H)$ for H an L-group, on which our argument depends. We then present the particular configuration involving SL_2 which was identified as crucial by Jaligot, and prove the relevant technical lemma about the behavior of involutions in that context, as well as the application that leads quickly to the elimination of the configuration critical to the proof of weak embedding.

The proof of the Mixed Type L^* theorem occupies the last 2 sections. We use the U_2/T_2 theory and the criterion of §4.5 to produce a weakly embedded subgroup, making use of Lemma 3.3 to verify the weak embedding criterion, and in the last section we show that the failure of strong embedding leads to a similar configuration to which Lemma 3.3 again applies.

1. Weak embedding

The following is part of Proposition II 4.15.

PROPOSITION II 4.15. *Let G be a group of finite Morley rank, M a definable subgroup having an infinite Sylow 2-subgroup S. Then the following are equivalent:*

(1) *M is weakly embedded in G.*
(2) *For every nontrivial subgroup Q of S° which is either*
    (a) *2-unipotent or,*
    (b) *a 2-torus,*
    we have $N_G(Q) \subseteq M$.

We will focus initially on the condition that M contains the normalizer of any nontrivial definable 2-unipotent subgroup of M. In groups of even type this is equivalent to weak embedding, and in groups of mixed type it will lead to a weakly embedded subgroup under suitable inductive hypothesis.

The following definition is fundamental.

DEFINITION 1.1. *For G a group of finite Morley rank, the graph $\mathcal{U}(G)$ is the graph whose vertex set is the set of nontrivial 2-unipotent subgroups of G, and whose edges are pairs of vertices U, V which commute: $[U, V] = 1$.*

Note that for any nontrivial unipotent subgroup U of G and any maximal unipotent 2-subgroup V of G containing U, the vertex V of $\mathcal{U}(G)$ is in the connected component containing the vertex U, and indeed at distance at most 2, with a connecting path going through $Z^\circ(V)$.

PROPOSITION 1.2. *Let G be a group of finite Morley rank and $\mathcal{C}$ a connected component of the graph $\mathcal{U}(G)$. Let M be the setwise stabilizer of $\mathcal{C}$ under the action of G by conjugation. Then we have the following.*

(1) *M is the normalizer of the group generated by $\bigcup\mathcal{C}$.*
(2) *M is definable.*
(3) *$\mathcal{U}(M) = \mathcal{C}$.*
(4) *For $U \le M$ nontrivial and unipotent, $N_G(U) \le M$.*
(5) *If $\mathcal{C} \ne \mathcal{U}(G)$, then $M < G$.*

PROOF. The group G acts on the graph $\mathcal{U}$ by conjugation. Furthermore, as every component of $\mathcal{U}$ contains the connected component of a Sylow 2-subgroup, G conjugates the connected components of $\mathcal{U}$ transitively. Let $M_1 = N_G(\langle\bigcup\mathcal{C}\rangle)$, and let $S \in \mathcal{C}$ be the connected component of a Sylow 2-subgroup. Observe that M_1 contains the stabilizer M of $\mathcal{C}$ in G. In fact, M_1 coincides with M, since by the Frattini argument $M_1 \le \langle\bigcup\mathcal{C}\rangle N(S)$, which evidently stabilizes $\mathcal{C}$.

Thus we have proved our first claim, $M_1 = M$, and it follows that M is definable. Now as the maximal unipotent 2-subgroups of M are conjugate under the action of M, they lie in $\mathcal{C}$, and the same applies to any definable 2-unipotent subgroup of G. Thus the third claim holds, and this implies the fourth and fifth claims. □

LEMMA 1.3. *Let G be a simple group of finite Morley rank of even type and H a proper definable subgroup with infinite Sylow 2-subgroups, which contains the connected component of the normalizer of any nontrivial unipotent 2-subgroup of H. Let S be a Sylow° 2-subgroup of H. Then there is a definable subgroup H_1 of H containing S, for which $N(H_1)$ is weakly embedded in G.*

PROOF. Let $\mathcal{C}$ be the connected component of $\mathcal{U}(G)$ which contains the unipotent part of S. Let $H_1 = \langle\bigcup\mathcal{C}\rangle$. We claim that $\mathcal{C}$ is not all of $\mathcal{U}(G)$.

Observe that any $U \in \mathcal{C}$ is contained in H, by the assumption on H. So $H_1 \le H$, and if $\mathcal{C} = \mathcal{U}(G)$, then $H_1 \lhd G$, a contradiction.

So $\mathcal{C} \ne \mathcal{U}(G)$, and by Proposition 1.2 $N(H_1)$ is weakly embedded in G. □

We apply this to complete a circle of ideas around weak embedding developed in earlier sections in the K-group context.

LEMMA 1.4. *Let G be an L-group of even type with $\mathcal{U}(G)$ disconnected. Then $U_2(G) \simeq \mathrm{SL}_2(K)$ for some algebraically closed field K of characteristic 2.*

PROOF. We may suppose $G = U_2(G)$. So G is an L-group of U_2-type, and by Lemma II 6.3 it is a K-group of even type. The hypothesis that $\mathcal{U}(G)$ is disconnected produces a weakly embedded subgroup M as the stabilizer of any connected component of $\mathcal{U}(G)$, by Lemma 1.3. Hence $G = O(G) \times L$ with $L \simeq \mathrm{SL}_2(K)$ by Proposition II 5.25. Now as $G = U_2(G)$, we find $G = L$. □

2. Subgroups of type U_2 or T_2

We defined the graph $\mathcal{U}(G)$ in §1 and the related subgroup $U_2(G)$ as early as §I 3. We will also make use of the parallel notions for 2-tori in place of unipotent subgroups.

DEFINITION 2.1. *Let G be a group of finite Morley rank.*

(1) *$\mathcal{T}(G)$ the set of 2-tori in G, construed as a graph with edge relation $\sim$ given by centralization: $T_1 \sim T_2$ if $[T_1, T_2] = 1$.*
(2) *$T_2(G) = \langle d(T) : T \in \mathcal{T}(G)\rangle$.*
(3) *G is of T_2-type if $G = T_2(G)$.*

Note that for any group G of finite Morley rank, the groups $U_2(G)$ and $T_2(G)$ are definable and connected by Proposition I 3.20.

We can also set up a Galois connection linking these notions, of which the following is a special case. We will return to this point at the end of the present section.

NOTATION 2.2. *Let G be a group of finite Morley rank, U a unipotent 2-subgroup, and T a 2-torus.*

(1) *$U^\perp = T_2(C_G(U))$.*
(2) *${}^\perp T = U_2(C_G(T))$.*

We are getting ahead of ourselves. The basic result which makes all of this reasonable is the following.

LEMMA 2.3. *If H is an L-group with no definable simple section of mixed type, then $U_2(H)$ and $T_2(H)$ commute.*

PROOF. By Lemma II 6.3, $K = U_2(H)$ is a K-group.

Let $\hat{T}$ be the definable hull of a 2-torus T, and let U be a unipotent 2-subgroup of H. We claim that $\hat{T}$ centralizes U.

By Proposition II 4.8, $K/\sigma(K)$ is a direct product of finitely many simple algebraic groups, over fields of characteristic 2. So by Corollary II 2.26 $\hat{T}$ acts trivially on $K/\sigma(K)$. In particular $\hat{T}$ normalizes $U\sigma(K)$, a solvable group. By Corollary I 8.4, $U \le U_2(F(K))$, a $\hat{T}$-invariant unipotent 2-group. So we may suppose that $\hat{T}$ normalizes U. Then UT is a Sylow 2-subgroup of $U\hat{T}$ and thus U and T commute, hence U and $\hat{T}$ commute. □

PROPOSITION 2.4. *Let G be a centerless L^*-group with no proper definable simple section of mixed type, and $U_1, U_2 \in \mathcal{U}(G)$. If $U_1 \sim U_2$, then $U_1^\perp = U_2^\perp$.*

PROOF. Let $H = T_2(C_G(U_2))$. The group U_1 normalizes H, and $N(H)$ is an L-group with no proper definable simple section of mixed type. So by Lemma 2.3, U_1 and H commute. Thus $H \le U_1^\perp$ and by symmetry we get equality. □

COROLLARY 2.5. *Let G be a centerless L^*-group with no proper definable simple section of mixed type, and $U_1, U_2 \in \mathcal{U}(G)$. If U_1 and U_2 are in the same connected component of $\mathcal{U}(G)$, then $U_1^\perp = U_2^\perp$.*

COROLLARY 2.6. *Let G be a centerless L^*-group with no proper definable simple section of mixed type. If $\mathcal{U}(G)$ is a connected graph, then $U^\perp \lhd G$, for any unipotent 2-subgroup U of G.*

PROPOSITION 2.7. *Let G be a centerless L^*-group of finite Morley rank with no proper definable simple section of mixed type. Let $T_1, T_2 \in \mathcal{T}(G)$ such that $T_1 \sim T_2$. Then $U_2(C_G(T_1)) = U_2(C_G(T_2))$.*

PROOF. Let T_1 and T_2 be as in the statement of the proposition. Then T_1 normalizes $U_2(C_G(T_2))$. By Lemma 2.3, T_1 centralizes $U_2(C_G(T_2))$. This forces $U_2(C_G(T_2)) \leq U_2(C_G(T_1))$. We get equality by symmetry. □

COROLLARY 2.8. *Let G be a centerless L^*-group of finite Morley rank with no proper definable simple section of mixed type. If T_1 and T_2 are in the same connected component of $\mathcal{T}(G)$, then $U_2(C_G(T_1)) = U_2(C_G(T_2))$.*

COROLLARY 2.9. *Let G be a centerless L^*-group of finite Morley rank with no proper definable simple section of mixed type. If $\mathcal{T}(G)$ is connected, then $U_2(C_G(T)) \lhd G$ for every $T \in \mathcal{T}(G)$.*

The next lemma is an analog of Proposition 1.2.

LEMMA 2.10. *Let G be a group of finite Morley rank. Let $\mathcal{C}$ denote a connected component of the graph $\mathcal{T}(G)$, and set $M = N_G(\langle\mathcal{C}\rangle)$. Then $\mathcal{T}(M) = \mathcal{C}$.*

PROOF. Let T_1 and T_2 be two 2-tori in M such that $T_1 \in \mathcal{C}$. We will show that $T_2 \in \mathcal{C}$. We may assume that T_1 and T_2 are maximal since any 2-torus is in the same connected component of $\mathcal{T}(M)$ as a maximal 2-torus which contains it. By the Sylow theorem, there exists $g \in M$ such that $T_1^g = T_2$. Therefore, $T_2 \leq \langle\mathcal{C}\rangle$ and g can be taken to be in $\langle\mathcal{C}\rangle$. It suffices to treat the case in which $g \in T \in \mathcal{C}$. In this case T_1, $T \in \mathcal{C}$ and therefore their conjugates $T_2 = T_1^g$ and $T^g = T$ are in $\mathcal{C}$. □

COROLLARY 2.11. *Let G be a group of finite Morley rank. Let $\mathcal{C}$ be a connected component of $\mathcal{T}(G)$. Then $N_G(\langle\mathcal{C}\rangle) = \mathrm{Stab}(\mathcal{C})$, the setwise stabilizer of $\mathcal{C}$. In particular, $\mathrm{Stab}(\mathcal{C})$ is definable.*

To conclude this section, we give the actual Galois connection between U_2-type and T_2-type subgroups of an L^*-group of finite Morley rank with no proper definable simple section of mixed type. Let G be a group of finite Morley rank. Let $\mathcal{U}$ and $\mathcal{T}$ denote the posets of U_2-type and T_2-type subgroups of G respectively, ordered by inclusion. We define the following mappings:

$$
\begin{array}{rcccl} {}^\perp\cdot : & \mathcal{T} & \longrightarrow & \mathcal{U} \\ & Y & \longmapsto & {}^\perp Y = U_2(C_G(Y)) \end{array}
\qquad
\begin{array}{rcccl} \cdot^\perp : & \mathcal{U} & \longrightarrow & \mathcal{T} \\ & X & \longmapsto & X^\perp = T_2(C_G(X)) \end{array}
$$

These mappings define a *Galois connection* (see [**38**]) between $\mathcal{U}$ and $\mathcal{T}$. That is, they satisfy the following properties.

PROPOSITION 2.12.

- *The mappings* $\cdot^{\perp}$ *and* ${}^{\perp}\cdot$ *are order-reversing.*
- *If* $X \in \mathcal{U}$, *then* $X \leq^{\perp} (X^{\perp})$, *and if* $Y \in \mathcal{T}$, *then* $Y \leq ({}^{\perp}Y)^{\perp}$.

The following properties of these operations (or of any Galois connection) can be checked easily:

PROPOSITION 2.13. *Let* $X \in \mathcal{U}$ *and* $Y \in \mathcal{T}$. *Then* $X^{\perp} = ({}^{\perp}(X^{\perp}))^{\perp}$ *and* ${}^{\perp}Y =^{\perp} (({}^{\perp}Y)^{\perp})$.

COROLLARY 2.14. *The operations* ${}^{\perp}(\cdot^{\perp})$ *and* $({}^{\perp}\cdot)^{\perp}$ *are* closure operations *on* $\mathcal{T}$ *and* $\mathcal{U}$ *respectively, i.e. setting* $\overline{X} = {}^{\perp}(X^{\perp})$ *and* $\overline{Y} = ({}^{\perp}Y)^{\perp}$ *for* $X \in \mathcal{U}$ *and* $Y \in \mathcal{T}$, *we have the following.*

- *For* $X \in \mathcal{U}$, $Y \in \mathcal{T}$, *we have* $X \leq \overline{X} = \overline{\overline{X}}$ *and* $Y \leq \overline{Y} = \overline{\overline{Y}}$.
- *Monotonicity: for* $X_1 \subseteq X_2$ *in* $\mathcal{U}$ *and* $Y_1 \subseteq Y_2$ *in* $\mathcal{T}$, *we have* $\overline{X}_1 \subseteq \overline{X}_2$ *and* $\overline{Y}_1 \subseteq \overline{Y}_2$.

The closed elements of $\mathcal{U}$ *and* $\mathcal{T}$ *are of the form* $\overline{X}$ *and* $\overline{Y}$ *respectively.*

3. The SL_2 configuration

Our aim is to deal with the following special case of the Mixed Type L^* Theorem.

PROPOSITION 3.1. *Let* G *be a simple group of mixed type, with no proper definable simple section of mixed type, and let* T *be a maximal* 2*-torus in* G. *Then* ${}^{\perp}T = U_2(C_G(T))$ *cannot be of the form* $\mathrm{SL}_2(K)$ *with* K *an algebraically closed field of characteristic* 2.

This is exactly the configuration which will arise in our proof of weak embedding in the next section. Furthermore, the technical result on which it rests is the key to both of the major steps of the proof. That technical result, which occurs as Lemma 4.1 in [**123**], is given below as Lemma 3.3.

3.1. Sketch of the proof.

We first sketch the broad outlines of the proof of Proposition 3.1. Assume therefore that G, T, and K are as described, with ${}^{\perp}T \simeq \mathrm{SL}_2(K)$. Set $L = {}^{\perp}T$.

DEFINITION 3.2. *An involution which lies in a* 2*-torus will be called* toral, *and one which lies in a nontrivial unipotent group will be called* unipotent; *we may also say these involutions are of toral or unipotent* type. *In principle an involution can be of both types. Of particular importance are the* properly unipotent *involutions, that is those which are of unipotent type and not of toral type.*

We will argue first that each unipotent involution belongs to a unique maximal 2-unipotent subgroup U_i of G, essentially because this holds in L, and that whenever a toral involution normalizes a group of the form U_i, it commutes with it.

Now use the simplicity of G to find a pair i, j of involutions, one of unipotent type and the other toral, which do not commute (this is the point at which it is clear that G is not a "simple version" of the group $T \times \mathrm{SL}_2(K)$). By the fusion principle, Lemma I 2.20, there is an involution k in $d(\langle ij \rangle)$ commuting with both i and j. Now one argues that $L = U_2(C(k))$ is of the form SL_2 (more specifically, generated by U_i and U_i^j, which are distinct by our choice of i, j).

Now a close examination of the situation reveals that $jk \in L$, and in particular the involution jk is of unipotent type; this is Lemma 3.3 below. So we can now consider $U_{jk} \leq L$; as both T and U_{jk} lie in $C(j)$, one can conclude in succession that T centralizes the following: U_{jk}; k; L. So now both j and k centralize L, while jk lies in L, a contradiction.

3.2. The technical lemma. We now turn to the detailed proof of Proposition 3.1. We first state the technical lemma on which it all depends, in a generalized form which will be applied again in §5; in the present case, we will take $k = k'$ to recapture the situation described above.

LEMMA 3.3. *Let G be a group of finite Morley rank containing involutions i, j, k, and k', and set $L = U_2(C_G(k))$. Suppose the following five conditions are satisfied.*

(1) *The involutions i and j are not conjugate.*
(2) *The group L is of the form $\mathrm{SL}_2(K)$ with K an algebraically closed field of characteristic 2.*
(3) *The involution k' is the unique involution in $d(\langle ij \rangle)$.*
(4) *$i \in L$; $j \in C_G(k)$.*
(5) *$k' \notin L$.*

Then jk' is in L.

Under the hypotheses of this lemma it follows easily that jk' is an involution commuting with k; the question dealt with here is whether this involution belongs to the subgroup $L = U_2(C_G(k))$ of $C_G(k)$, or lies outside of it.

PROOF. Let $a = ij$, and $A = d(\langle a \rangle)$. By hypothesis, i and j are in $C(k)$, and $i \in L \lhd C(k)$. Thus $a^2 = ii^j \in A \cap L$.

If $a \in L$, then as $i \in L$, also $j \in L \simeq \mathrm{SL}_2(K)$ and hence i and j are conjugate, a contradiction. Thus a represents an involution modulo $A_0 = A \cap L$, and it follows that the coset aA_0 contains a nontrivial 2-element a_1 by Lemma I 8.11. We claim that $a_1 = k'$. Now $a_1^{2^n}$ is an involution for some $n \geq 0$, and this involution can only be k' (hypothesis 3). If $n > 0$, then we find $k' \in L$, a contradiction. So $a_1 = k'$. This means $ijk' \in L$.

Since $i \in L$, we find $jk' \in L$, as claimed. □

3.3. Local structure of G. Now let us examine the situation of Proposition 3.1. That is, we suppose the following.

(*)
G is a simple group of mixed type and of finite Morley rank.
T a maximal 2-torus of G.
$L = {}^{\perp}T$ is of type $SL_2(K)$ with K algebraically closed of characteristic 2.

We keep this notation fixed for the remainder of the section.

The next four lemmas show that in many ways the structure of G resembles that of $L \times H$ for some group H of odd type.

LEMMA 3.4. *If U is a Sylow 2-subgroup of L, then $U \times T$ is a Sylow° 2-subgroup of G.*

PROOF. Since T commutes with L, we have $T \cap U = 1$. Now $U \times T$ is contained in a Sylow° 2-subgroup S of G, and $S = T_1 * U_1$ with T_1 the maximal 2-torus of S and U_1 the maximal 2-unipotent subgroup of S. So $T = T_1$ by maximality, and $U_1 \le U_2(C(T)) = L$; thus $U_1 = U$, and $S = U \times T$. □

LEMMA 3.5. *Any 2-unipotent subgroup in G is abelian, of exponent 2, and is conjugate to a subgroup of L.*

PROOF. By the previous lemma, any 2-unipotent subgroup is conjugate to a subgroup of L, and is therefore abelian of exponent 2. □

LEMMA 3.6. *Let i be an involution of G which belongs to a nontrivial unipotent subgroup of G. Then the following hold:*

(1) *i is of properly unipotent type;*
(2) *i belongs to a unique maximal unipotent 2-subgroup U_i of G;*
(3) *$U_i = U_2(C(i))$.*

PROOF. If an involution i belongs both to a nontrivial unipotent subgroup U and to a nontrivial 2-torus T_0, then after conjugating in $C(i)$, we may suppose that UT_0 lies in some Sylow° 2-subgroup of G, which contradicts the structure of that subgroup (Lemma 3.4). So i is of properly unipotent type, proving (1).

Now if i belongs to the unipotent subgroup U, then $U \le U_2(C(i))$ and thus it suffices to prove that $U_2(C(i))$ is unipotent.

We may suppose $T \le C(i)$; then $U_2(C(i))$ commutes with $T_2(C(i))$, which contains T, so $U_2(C(i)) \le L$. On the other hand, $i \in U_2(C(i))$, so $i \in L$ and $U_2(C(i)) \le C_L(i)$, which is a unipotent 2-group. □

LEMMA 3.7. *Let j be a toral involution of G, U a unipotent 2-subgroup normalized by j, and T_0 a 2-torus containing j. Then T_0 centralizes U.*

PROOF. $C_U{}^\circ(j)$ is nontrivial by Corollary I 5.2 and Lemma I 1.23. For i any involution of $C_U{}^\circ(j)$, we have $U \le U_2(C(i))$ and $T_0 \le T_2(C(i))$, since $U_2(C(j))$ and $T_2(C(j))$ commute. So U and T commute. □

LEMMA 3.8. *Let i and j be commuting involutions of unipotent type. Then $U_i = U_j$.*

PROOF. Since j normalizes U_i, $V = C_{U_i}{}^{\circ}(j)$ is a nontrivial unipotent subgroup of U_i. But $V \leq U_2(C(j)) = U_j$, so $U_i \cap U_j$ is nontrivial and for k an involution in the intersection we have $U_i = U_k = U_j$ by Lemma 3.6. □

Now we begin a finer analysis. Since G is simple there must be some involution i of unipotent type and some involution j of toral type such that i and j do not commute. By the fusion principle, Lemma I 2.20, i and j must commute with a third involution in $d(\langle ij \rangle)$, and we ask ourselves how this one behaves. We fix the notation.

NOTATION 3.9. *Fix i and j involutions with i unipotent, j toral, and $[i, j] \neq 1$.*

LEMMA 3.10. *The group $A = d(\langle ij \rangle)$ contains a unique involution.*

PROOF. By Lemma I 2.20, there is at least one involution in A.

By Lemma I 2.16, A is the direct sum of a finite cyclic subgroup and a divisible group. We claim that in the case at hand, A contains no 2-torus, and hence contains a unique involution.

Suppose on the contrary that T_0 is a nontrivial 2-torus contained in A, and hence inverted by i. Then i centralizes an involution $j' \in T_0$, and hence by Lemma 3.7, T_0 commutes with i, which is a contradiction. □

NOTATION 3.11. *The unique involution of $d(\langle ij \rangle)$ is denoted k.*

We are approaching the situation of Lemma 3.3.

LEMMA 3.12. *$U_2(C(k))$ is of the form SL_2 over an algebraically closed field of characteristic 2.*

PROOF. We claim first that $U_i \cap U_i^j = 1$. If not, then $U_i = U_i^j$ and by Lemma 3.7 we find that j, being toral, commutes with U_i, and hence with i, a contradiction.

The involution k acts on $U_2(C(i))$, and the latter group is U_i by Lemma 3.8. So $P_i = C_{U_i}{}^{\circ}(k)$ is nontrivial. Also $P_i^j \leq U_2(C(k))$. Since distinct maximal unipotent subgroups of G intersect trivially, the same applies to $U_2(C(k))$, and as there are at least two such, the associated graph $\mathcal{U}(U_2(C(k)))$ is disconnected. By Lemma 1.4, $U_2(C(k)) \simeq \mathrm{PSL}_2(K)$ for some algebraically closed field K of characteristic 2. □

And here is the last ingredient.

LEMMA 3.13. *The involution i belongs to $U_2(C(k))$.*

PROOF. The involution i acts on $L_1 = U_2(C(k))$ and centralizes $U_i \cap L_1$. So i acts on L_1 like some element $t \in U_i \cap L_1$, and $it \in U_i$ centralizes U_i^j. Hence $it = 1$ and $i = t \in L_1$. □

PROOF OF PROPOSITION 3.1. We have i an involution of unipotent type, j a toral involution, which we may suppose is in T, with i and j not commuting, and k the unique involution in $d(\langle ij\rangle)$, with $L_1 = U_2(C(k))$ of type SL_2 in even characteristic.

We apply Lemma 3.3 to the involutions i, j, and k, taking $k' = k$ as well. As i is of unipotent type and j is of toral type, they are not conjugate. The remaining conditions of the lemma are all clear at this point. The conclusion is

$$jk \in U_2(C(k)) .$$

In particular, jk is of unipotent type.

Now j is a toral involution commuting with jk, and $j \in T$, so by Lemma 3.7, the torus T commutes with U_{jk}, and in particular with jk, hence with k. So we have $T \leq C(k)$ and hence T commutes with $U_2(C(k)) = L_1$. In particular, the element $j \in T$ commutes with element $i \in L_1$, contradicting our initial choice. □

4. A weakly embedded subgroup

In this and the next section we will carry out the proof of the Mixed Type L^* Theorem.

THEOREM 4.1. *There is no simple L^*-group of finite Morley rank of mixed type.*

Until the completion of this proof, we fix the following notation. We will arrive at a contradiction in the next section.

NOTATION 4.2.

(1) *G is a simple L^*-group of mixed type and of minimal rank. Accordingly, all simple sections of G are either of degenerate type, of odd type, or algebraic.*
(2) *$\mathcal{U} = \mathcal{U}(G)$ is the graph whose vertices are the nontrivial 2-unipotent subgroups of G, and whose edges consist of pairs of vertices which commute.*
(3) *S is a fixed Sylow° 2-subgroup of G.*
(4) *$S = U * T$ with U 2-unipotent, and T a 2-torus.*

Our goal in the present section is to show, first, that the graph $\mathcal{U}$ is disconnected, and secondly, that the stabilizer of any connected connected component of $\mathcal{U}$, under the natural action of G, is a weakly embedded subgroup of G. The resulting configuration will be analyzed in the next section.

LEMMA 4.3. *The graph $\mathcal{U}$ is not connected.*

PROOF. This is a consequence of Corollary 2.6. □

NOTATION 4.4. *$\mathcal{C}$ is some connected component of the graph $\mathcal{U}$, and M is the setwise stabilizer of $\mathcal{C}$ with respect to the natural action of G on $\mathcal{U}$.*

The subgroup M is definable, and $\mathcal{U}(M) = \mathcal{C}$ (Lemma 1.2).

PROPOSITION 4.5. *The subgroup M is weakly embedded in G.*

PROOF. As $\mathcal{U}(M) \neq \emptyset$, M has an infinite Sylow 2-subgroup S. By the criterion of Proposition II 4.15, it suffices to check that the normalizer in G of any subgroup of M which is either 2-unipotent, or a 2-torus, is contained in M.

(1) For U a nontrivial 2-unipotent subgroup of M, we have $N_G(U) \leq M$.

Let U be such a subgroup. Then $U \in \mathcal{U}(M) = \mathcal{C}$, so $N(U)$ must stabilize $\mathcal{C}$ setwise, and $N(U) \leq M$.

(2) For T a nontrivial 2-torus in M, we have $N_G(T) \leq M$.

This is more substantial. Let T be such a subgroup, and let $H = N_G(T)$. Suppose H is not contained in M. We claim that $\mathcal{U}(H)$ is disconnected.

Now T is contained in a Sylow° 2-subgroup S of M, and hence commutes with the 2-unipotent factor U of S°; so $U \in \mathcal{U}(M) = \mathcal{C}$ and $U \in \mathcal{U}(H)$. On the other hand, as H is not contained in M and M is the stabilizer of $\mathcal{C}$, there is $h \in H$ for which $\mathcal{C}^h \neq \mathcal{C}$. Now $\mathcal{U}(H)$ meets both $\mathcal{C}$ and $\mathcal{C}^h$, and $\mathcal{U}(H)$ is simply the restriction of the graph $\mathcal{U}(G)$ to its vertex set. So the graph $\mathcal{U}(H)$ is disconnected.

Consider $L = U_2(H)$. Then L is a K-group by Lemma II 6.3, and $\mathcal{U}(L)$ is disconnected since $\mathcal{U}(L) = \mathcal{U}(H)$. Then by Proposition II 5.25, L must be of the form $\mathrm{SL}_2(K)$ with K algebraically closed of characteristic 2. Furthermore, the group L is not contained in M, since $\mathcal{U}(M) = \mathcal{C}$.

The next step is to trade in T for a maximal 2-torus. Let T_1 be a maximal 2-torus of G containing T. Then T_1 normalizes L. By Lemma 2.3, T_1 commutes with L. Since $L \cap M$ contains a nontrivial 2-unipotent subgroup, it follows from (1) that $T_1 \leq M$, and since $L \leq N_G(T_1)$, T_1 can play the role of T. So replacing T by T_1 and L by $U_2(N_G(T_1))$, we will still have L of the form SL_2, and now T is a maximal 2-torus of G. This is the configuration we eliminated in §3. This contradiction proves (2), and the proposition follows. □

5. Offending involutions

We continue the proof of the Mixed Type L^* Theorem, Theorem 4.1.

NOTATION 5.1.

(1) *G is a simple L^*-group of mixed type. All simple sections of G are either of degenerate type, or algebraic.*
(2) *M is a weakly embedded subgroup in G.*
(3) *S_0 is a fixed Sylow° 2-subgroup of G.*
(4) *$S_0 = U_0 * T_0$ with U_0 2-unipotent, and T_0 a 2-torus.*

An *offending involution* is an involution α in M whose centralizer is not contained in M.

LEMMA 5.2. *There is an offending involution in M.*

PROOF. Otherwise, by the criterion for strong embedding of Lemma I 10.11, we would have M strongly embedded in G, and hence by Lemma I 10.12 all involutions in G would be conjugate. But $N_G(T_0)$ controls fusion in $S_0{}^\circ$, so involutions in T_0 and $U_0 \setminus T_0$ cannot be conjugate. □

Our task now is to analyze the structure of $C^\circ(\alpha)$ for α an offending involution, which will produce a group of type SL_2 which we can work with in the manner of the preceding section. Indeed, there is a strong similarity between our current offending involution and the similarly "offending" toral involutions of the previous section.

LEMMA 5.3. *Let $\alpha \in M$ be an offending involution. Then $C^\circ(\alpha)$ has the form $L \times H$, with the following properties.*

(1) *L is a group of type $\mathrm{SL}_2(K)$ with K algebraically closed of characteristic 2.*
(2) *H is a group of degenerate type.*
(3) *$L \cap M$ is a Borel subgroup of L.*

The involution α does not belong to any Sylow° 2-subgroup, and in particular, it is neither toral nor unipotent.

PROOF. Let $X = C_G(\alpha)$. We define $L = U_2(X)$, and $H = C_X{}^\circ(L)$. We show first that L is not contained in M. Let U be a Sylow 2-subgroup of L. By the Frattini argument, if $L \leq M$, then $X \leq N(L) \leq L \cdot N(U) \leq M$, a contradiction.

Now α belongs to a Sylow 2-subgroup S of M, and hence centralizes a nontrivial 2-unipotent subgroup of M. So $L \cap M$ contains a nontrivial unipotent subgroup, and since L is not contained in M, L also contains a unipotent subgroup not contained in M. So $\mathcal{U}(L)$ is disconnected. Now by Lemma 1.4, L is of type SL_2 in characteristic 2. Then the decomposition $X^\circ = L \times H$ follows from Corollary II 2.26.

Now we have seen that L contains a nontrivial unipotent subgroup U_0 of M, and if U_0 is chosen maximal unipotent in L, and then extended to a Sylow 2-subgroup of L, the strong embedding hypothesis says that $N_U{}^\circ(U_0) = U_0$ and hence that $U = U_0$. So $L \cap M$ contains a Sylow 2-subgroup of L, which is to say the unipotent radical U of a Borel subgroup B of L. As $B \leq N(U)$, and M is weakly embedded, we find $B \leq M$. As B is a maximal subgroup of L and L is not contained in M, we find that $B = L \cap M$.

We claim that H is of degenerate type. If there is a nontrivial 2-torus $T \leq H$, then $T \leq M$ by weak embedding, and then $L \leq M$ by weak embedding since T commutes with L. As this is a contradiction, H is of degenerate type.

Finally, if α belongs to a Sylow° 2-subgroup, then $C(\alpha)$ contains a nontrivial torus, contradicting the structure of $C^\circ(\alpha)$. □

NOTATION 5.4. *For α an offending involution of M, set $L_\alpha = U_2(C(\alpha))$, and $A_\alpha = O_2(L_\alpha \cap M)$, a Sylow 2-subgroup of L_α.*

Once more, we prepare for an application of Lemma 3.3. First, we need to clarify the nature of unipotent involutions.

LEMMA 5.5.

(1) *Any nontrivial unipotent group contains a properly unipotent involution.*
(2) *If α is an offending involution, then any involution in L_α is properly unipotent.*

PROOF. Let U be nontrivial and unipotent, and let S be a Sylow° 2-subgroup containing U. Then $S = U * T$ with T a 2-torus, and with $U \cap T$ finite. Furthermore, $N(T)$ controls fusion in S (Lemma I 6.18), so involutions in $U \setminus T$ are not conjugate to involutions in T, and hence are not toral.

For the second point, every involution in L_α belongs to a nontrivial unipotent subgroup $U \leq L_\alpha$, and these involutions are all conjugate; as some of them are not toral, none of them are. □

LEMMA 5.6. *Let t be a toral involution not in M, α an offending involution in M, and $i \in A_\alpha$. Then $d(\langle it \rangle)$ contains a unique involution β, and β is an offending involution of M.*

PROOF. The involution i is properly unipotent, hence not conjugate to t, and also not offending. Thus there is some involution β in $A = d(\langle it \rangle)$ by the basic fusion principle.

As A is the direct sum of a finite cyclic group and a divisible group, for the uniqueness claim it suffices to show that A contains no nontrivial 2-torus. Suppose on the contrary that $T \leq A$ is a nontrivial 2-torus and let s be an involution of T. Then $s \in C(i) \leq M$, and as s is toral it is not offending. But then $t \in C(s) \leq M$, a contradiction. So β is unique.

Now β commutes with both i and t. As i is not offending, β is in M; as t is not in M, β is offending. □

Now we can complete our analysis by "absorbing" toral involutions into M, contradicting the simplicity of G.

PROOF OF THE MIXED TYPE L^* THEOREM. Assuming the contrary, our mixed type group G contains a weakly but not strongly embedded subgroup M, and M contains an offending involution α_0. As G is simple, it contains a toral involution t outside M. If $i_0 \in A_{\alpha_0}$, then by the previous lemma, there is a unique involution $\alpha \in d(\langle i_0 t \rangle)$, and it is again offending. So we now have an offending involution (in M) that commutes with t.

Take an involution $i \in L_\alpha$. Then again there is a unique involution $\alpha' \in d(\langle it \rangle)$.

We can now consider the involutions i, t, α, and α', which play the roles of i, j, k, and k' in Lemma 3.3. In particular, L_α plays the role of L. All of the conditions of that lemma hold; for example, $\alpha' \notin L_\alpha$ because α' is an offending involution.

The conclusion is that $t\alpha' \in L_\alpha$. Now t centralizes α and hence acts on L_α. The action is by an inner automorphism, in view of Corollary II 2.26. Since t centralizes $t\alpha'$, it centralizes $A_{t\alpha'}$, and thus $t\alpha' \in U_2(C(t))$. Now if T is a maximal torus containing t, then T is also in $C(t)$ and hence centralizes $U_2(C(t))$, and in particular, $t\alpha'$; in other words, T centralizes α', and $T_2(C(\alpha'))$ is nontrivial, a contradiction as α' is an offending involution. □

With this, the proof of the Mixed Type Theorem for L^*-groups is complete, and the analysis of mixed type groups therefore depends entirely on the analysis of even type groups, to be carried out in the remainder of this volume.

6. Notes

Groups of mixed type were first eliminated in [**3**], under the two additional hypotheses that the group in question is a K^*-group and that it involves no bad field. The second hypothesis was eliminated in Jaligot's thesis [**125, 123**]. The eliminability of the K^*-hypothesis became clear around the time of Altınel's habilitation [**2**], where the idea of L^*-groups was first pursued systematically, and was discussed in the first published paper on L^*-groups [**11**, §7], in the general form given here. The overall structure of the argument is the same in all three cases.

Strong and weak embedding are typical *"uniqueness conditions"* on M in the sense of finite group theory; strong embedding is the oldest of these uniqueness conditions, and was introduced by Bender [**29**]. Weak embedding first appears in [**3**]. While weak embedding has no direct analog in finite group theory, it has the general flavor of "proper 2-generated" k-core for various values of k, notably $k = 2$. For some discussion of the finite case, see [**103**, §I.1.20, p. 52]. In the case of mixed or even type groups, weak embedding is much easier to achieve than strong embedding, and in the long run equally powerful.

Once one has traced through the general line of argument of this chapter, it becomes clear that the crucial configuration, corresponding to a potential failure of the main line of argument, is the SL_2 configuration treated in §3. Earlier versions of the arguments given here differ in their treatment of this case. The key to eliminating the hypothesis of "no bad fields" is Lemma 3.3, taken from Jaligot's thesis. As we remarked in the introduction, the underlying idea is an application of the *Thompson rank formula*, a powerful method which over time has been somewhat supplanted in our treatment by other fusion-theoretic arguments, but which still plays an important role, as we will see in Chapter VII. We will expand on this point here.

Previous analyses strongly suggested that the configurations arising throughout this chapter could be treated using the Thompson rank formula. Such an approach involves first getting a good understanding of the conjugacy classes of involutions in G, and then (in order to pin down the parameter f) also the structure of their centralizers. A close analysis of this problem led precisely to the configuration described in Lemma 3.3, which instead of producing the input for a contradictory calculation, produced the desired contradiction directly. So while the Thompson rank formula plays no role in the proof, we still consider this argument, to a degree, as a successful application of that method. Lemma 3.3 can be construed as an "invisible" use of the Thompson rank formula.

Part C

Even Type Groups

CHAPTER VI

Strong Embedding and Weak Embedding

Ya tutarsa?
— *Nasreddin Hoca*

Introduction

We now take up our main subject, the inductive analysis of simple groups of finite Morley rank of even type. In this long chapter, we deal with the *weak embedding* classification (the notion was introduced in §II 5.12). The main result is the following.

THEOREM 10.12. *Let G be a simple L^*-group of finite Morley rank and even type with a definable weakly embedded subgroup. Then G is of the form* $\mathrm{PSL}(2, K)$ *for some algebraically closed field K of characteristic* 2.

This is a very flexible result, which ostensibly only characterizes one very small group but in fact will be one of the main engines of our analysis in general. The proof is long, and greatly complicated by the possible presence of simple degenerate sections in G. The proof we give follows closely the structure of the argument in the K^*-case, deviating more toward the end when the older argument begins to rely in a substantial way on the structure of solvable groups, as well as conjugacy theorems holding in solvable groups.

The proof takes place in three steps. The first point is the following: the proof varies somewhat depending on whether one is in the strongly embedded case or not.

PROPOSITIONS 1.1 AND 2.1. *Let G be a simple L^*-group of finite Morley rank with a definable weakly embedded subgroup M. Then $U_2(M)$ is 2-unipotent.*

This can be rephrased in a number of ways, e.g., $U_2(M) = O_2{}^\circ(M)$. Observe that $M/U_2(M)$ is of degenerate type; so another way to express this is that M contains a normal 2-unipotent subgroup with degenerate quotient. In the K^*-context, an equivalent statement would be that M° is solvable (or M, if one invokes Feit-Thompson) and the foregoing result provides the best approximation to solvability one can expect at this point in the analysis.

One must distinguish between the cases of strong embedding (introduced in §I 10.3) and weak embedding here. When one has a subgroup which is weakly embedded and *not* strongly embedded, a very particular configuration arises, which was analyzed in the finite case by Landrock and Solomon;

and the parallel analysis has been given in §III 4. This analysis produces a list of possible isomorphism types for a Sylow° 2-subgroup S, from which one argues in the case at hand that S is abelian. After that, the two cases, strong and weak embedding, run more or less in parallel.

The remainder of the analysis leading to the identification of G splits in two. One considers the following condition, which will eventually be proved.

(∗) Whenever A_1, A_2 are two distinct conjugates of $\Omega_1(O_2{}^\circ(M))$, the group $C_G(\langle A_1, A_2\rangle)$ is finite.

This is the "expected" case, since we aim at showing $G \simeq \mathrm{PSL}_2(K)$. When the condition (∗) holds, we do get the desired result in a reasonably natural way, ultimately by understanding the permutation representation of G on the coset space G/M, showing that we have a *split Zassenhaus* group.

But this favorable case must be further subdivided. Assuming first that M is *strongly embedded*, we head directly for the identification of G in terms of the action of G on G/M. If, on the other hand, M is weakly embedded but not strongly embedded, then we again pass through the Landrock–Solomon analysis of §III 4, which gives us a short list of configurations, some quite distant from the desired one. In this case we first eliminate the possibilities for nonabelian S, and shortly thereafter we return to the main line and identify G as a Zassenhaus group. Since at this point we have assumed M is not strongly embedded, and we have also used that hypothesis to the fullest, the ultimate conclusion reached in this line of argument is a contradiction, rather than an identification. All of this yields the following.

PROPOSITION 6.20 AND THEOREM 4.13. *Let G be a simple L^*-group of finite Morley rank and even type with a definable weakly embedded subgroup, and assume that the condition (∗) above holds. Then G is of the form* $\mathrm{PSL}(2, K)$ *for some algebraically closed field K of characteristic* 2.

Finally, we must prove that the condition (∗) does in fact hold, and this portion of the analysis is as long as all the rest together. Assuming that (∗) fails, it is immediate that G contains proper subgroups of the form $\mathrm{PSL}_2(K)$, and in particular contains algebraic tori. We study these subgroups and their tori in particular detail, arriving very belatedly at a contradiction.

In the first part of the analysis, we show that the tori in question can all be conjugated into M, and that those which are contained in M break up into *finitely* many conjugacy classes under the action of M. While not difficult, this is a powerful conclusion, unpromising as it may seem at first. It quickly allows us to show that for at least one such conjugacy class of tori T, we have $C(T) \subseteq M$, and we can control the rank of G accurately. After this, we quickly get detailed structural information for G, reminiscent of $\mathrm{PSL}_2(K)$, or perhaps $\mathrm{PSL}_2(K) \times O$ with O some degenerate type group.

This corresponds to a situation which arises late in Jaligot's treatment of the K^*-case [**125, 124**], where the internal information is largely self-consistent but the group resists identification, and in fact satisfies a condition incompatible with the desired identification. At this point one switches over to a careful examination of elements of order 3, and eventually a contradiction ensues. We will have information reminiscent of the (B, N)-pair description of $\mathrm{PSL}_2(K)$, meaning that while we lack explicit relations among the generators we can compute a good deal, and apply the associative law directly to an appropriate triple of elements closely related to the subgroup M. For this, it seems to be necessary to work with certain elements of order 3 which have a relatively clear description in these terms.

In this roundabout way we arrive at the following.

PROPOSITION 10.1. *Let G be a simple L^*-group of finite Morley rank and even type with a definable weakly embedded subgroup. Then the condition* $(*)$ *above holds.*

The main theorem then follows. In the proof of this last proposition, we do not need to make any real distinction between weak and strong embedding, or to invoke the Landrock–Solomon analysis. Most of the analysis is sufficiently robust to go through in the case of weak embedding, and then at a very late stage we prove finally that the subgroup in question is strongly embedded, just before embarking on the final analysis leading to a contradiction.

As the present chapter is a long one, the reader would be well advised to omit some sections on a first reading. If the sections relating to the case of weak but not strong embedding are omitted, then the line of analysis should be quite clear. The last four sections, which have to be taken as a single unit, seem to us the most interesting, and use increasingly geometrical lines of reasoning to compensate for our inability to invoke the theory of solvable groups in a useful way, given the weakness of our inductive hypothesis. One may wonder if the whole thing might be put on a more geometrical footing. We doubt it.

The treatment of the "weak but not strong" case may be viewed as a perturbation in which some good control is provided at the outset by an "offending involution", which is then used to compensate for the lack of control in other directions. As a result, many of the arguments proceed along similar lines; that is, similar facts are proved (always reminiscent of SL_2), but in a different order, and their relative difficulty is not at all preserved in passing from one case to the other.

One final point. It will take us quite some time to reach the point at which the underlying strategy for the proof of Theorem VIII 6.1 (the classification of groups of finite Morley rank of even type) is ready to be laid out—until §VIII 6, in fact. But there is nothing to prevent the reader from beginning the proof with that section. Everything from this point forward, up to that section, aims at setting up the argument outlined there.

Once we are in a position to lay out the global strategy, all that will remain to implement that strategy is one, admittedly lengthy, amalgam argument (Chapter IX). A curious coincidence, if indeed it is a coincidence, is that the result which makes that strategy viable in the first place—namely the $C(G,T)$-theorem—also disposes of one of the three cases that need to be treated.

1. Weak solvability: Strong embedding

Our goal in the present section is the following.

PROPOSITION 1.1 ([**1, 3**]). *If G is a simple L^*-group of even type of finite Morley rank with a definable strongly embedded subgroup M, then $U_2(M)$ is 2-unipotent.*

We will prove the corresponding result for the case of weakly, not strongly, embedded subgroups in the next section. In the case of strong embedding, all involutions are conjugate in M, and there are additional simplifying properties, which are not only useful in arguments but tend to lighten the notation as well. When we turn to the weakly but not strongly embedded case, we will have to revisit our notation. On the other hand, Theorem IV 4.1 will serve to lighten the notation in that case.

We can phrase our goal as follows: *$U_2(M)$ is solvable.* This immediately implies that $U_2(M)$ is 2-unipotent (Lemma I 8.36). We approach this by setting $M_1 = (U_2(M))^{(\infty)}$ and considering whether M_1 is trivial or not. This group is in any case a K-group, and we will eventually show that it is quasisimple, hence has finite center, by the theory of central extensions, while on the other hand we will show early on that its involutions are central, and conclude that it is trivial.

We first show that the involutions of M are central in $U_2(M)$. We then consider the subgroup M_1 and the group $O_2^\circ(M)$ which may be defined as the largest connected normal 2-unipotent subgroup of M, and which equals $O_2(U_2(M))$ (Proposition II 6.2). The next step is to show that $[M_1, O_2^\circ(M)] = 1$, and then we may show that M_1 is a perfect quasisimple K-group, and invoke the theory of central extensions to show that $M_1 = 1$.

1.1. $\Omega_1(\mathbf{U_2(M)})$.

Our first goal is to show that $I(M) \subseteq Z(U_2(M))$, and in particular the involutions of M generate an elementary abelian subgroup. The following is both basic and powerful.

LEMMA 1.2. *Let G be a group of finite Morley rank with a definable strongly embedded subgroup M. Let $i, a \in M$ and $j \in G$ with with i, j involutions, such that i centralizes a, j inverts a, and $a \neq 1$. Then $j \in M$.*

PROOF. Let $H = N(d(\langle a \rangle))$. If $H \leq M$, then our claim follows since $j \in H$. And if not, then since $H \cap M$ is a proper subgroup of H containing the involution i, it follows that $H \cap M$ is strongly embedded in H, and hence the involutions of H are conjugate in H (Lemma I 10.12). So in this case i and j are conjugate in H, say by $h \in H$. Then $a^h = (a^i)^h = (a^h)^{i^h} = (a^h)^j = (a^h)^{-1}$. So a^h is an involution, and a is an involution, and j commutes with a. Then by strong embedding, $j \in M$. □

LEMMA 1.3. *Let G be a connected group of finite Morley rank, and M a definable strongly embedded subgroup. Then there is a definable subgroup K of M° normalized by an involution $w \in G \setminus M$, such that for any involution $i \in M$ we have $M^\circ = C^\circ(i) \cdot K^\circ$.*

PROOF. Note first that as G is connected and contains an involution, it has an infinite Sylow 2-subgroup by Theorem IV 4.1. As M contains a Sylow 2-subgroup of G, there is an involution in M°, and as all involutions of M are conjugate in M (Lemma I 10.12), all involutions in M lie in M°.

Now all involutions in G are conjugate (Lemma I 10.12), so the rank $c = \mathrm{rk}(C(i))$ is constant for $i \in I(G)$, and $\mathrm{rk}(I(G)) = \mathrm{rk}(i^G) = \mathrm{rk}(G/C(i)) = g - c$ where $g = \mathrm{rk}(G)$. Similarly, $\mathrm{rk}(I(M)) = \mathrm{rk}(M) - c < \mathrm{rk}(I(G))$. Consider the function

$$\pi : I(G) \setminus I(M) \to G/M^\circ$$

defined by $\pi(w) = wM^\circ$. This map takes a set of rank $g - c$ into a set of rank $g - \mathrm{rk}(M)$, and therefore it has a fiber of rank at least $\mathrm{rk}(M) - c$. Such a fiber amounts to a set $X \subseteq I(G)$ included in a single coset of M° outside M; fixing one element $w \in X$, the set $Y = wX$ is a subset of M° which is inverted by w, and again is of rank at least $\mathrm{rk}(M) - c$. Let $K = d(Y)$. Then K is a subgroup of M° normalized by w, and $w \notin M$.

Now we claim that for any involution $i \in M$, $\mathrm{rk}(K/C(i)) \geq \mathrm{rk}(Y)$, or more concretely that the elements of Y lie in distinct left cosets of $C(i)$. Suppose on the contrary that $y, y' \in Y$ with $y^{-1}y' \in C(i)$. Write $y = wj$, $y' = wj'$ with j, j' involutions in X; then $y^{-1}y' = jj'$ is inverted by j and centralized by i. So we have a nontrivial element $y^{-1}y'$ of M, centralizing an involution in M and inverted by $j \in wM$; hence $j \in M$, a contradiction.

It follows that $\mathrm{rk}(K/C(i)) \geq \mathrm{rk}(M) - c$ and hence $\mathrm{rk}(K^\circ/C(i)) \geq \mathrm{rk}(M) - c$ as well. Hence for $i \in I(M^\circ)$ we have $\mathrm{rk}(i^{K^\circ}) \geq \mathrm{rk}(M) - c = \mathrm{rk}(i^M)$, and it follows that $\mathrm{rk}(i^{K^\circ}) = \mathrm{rk}(i^M) = \mathrm{rk}(i^{M^\circ})$. On the other hand, i^{M° is in bijective correspondence with the set of cosets $M^\circ/C_{M^\circ}(i)$, which has Morley degree 1. As the sets i^{K° are pairwise equal or disjoint as i varies, it follows that $i^{K^\circ} = i^{M^\circ}$ for $i \in I(M^\circ)$, and this translates into $M^\circ = C_{M^\circ}(i) \cdot K^\circ$.

On the other hand, $C_{M^\circ}(i) \cdot K^\circ$ is a finite union of double cosets $C^\circ(i)aK^\circ$ with $a \in M^\circ$. Furthermore, each of these double cosets is of rank $\mathrm{rk}(M^\circ)$ because $\mathrm{rk}(K^\circ/C^\circ(i)^a) = \mathrm{rk}(K^\circ/C^\circ(i^a)) \geq \mathrm{rk}(M) - c$. So these double cosets must all coincide as M° is connected. □

REMARK 1.4. *As $K \leq M \cap M^w$, K contains no involutions.*

Let us retain a little more from the proof, namely the provenance of the involution w.

COROLLARY 1.5. *Let G be a connected group of finite Morley rank, and M a definable strongly embedded subgroup. Then there is an involution $w \in G \setminus M$ such that $\mathrm{rk}(I(wM^\circ)) \geq \mathrm{rk}(I(M))$, and for any such involution w, there is a definable subgroup K of M° normalized by w such that for any involution $i \in M$ we have $M^\circ = C^\circ(i) \cdot K^\circ$.*

PROPOSITION 1.6. *Let G be a simple L^*-group of even type of finite Morley rank with a definable strongly embedded subgroup M. Then the involutions of M are central in $U_2(M)$.*

PROOF. We observed above that the involutions of M belong to M°. Let us fix a definable subgroup K of M, normalized by an involution w outside M, such that $M^\circ \leq C(i)K^\circ$ for each $i \in I(M)$, as given by Lemma 1.3. Then K contains no involutions. Recall that $U_2(M)$ is a K-group (Lemma II 6.3).

In this situation, Lemma II 5.3 applies, and K° normalizes a Sylow 2-subgroup U of $U_2(M)$. Let $i \in I(Z(U))$. Note that as $M^\circ = C^\circ(i)K^\circ$, also $M^\circ = K^\circ C^\circ(i)$. So

$$U_2(M) = U_2(M^\circ) = \langle U^{M^\circ} \rangle = \langle U^{K^\circ C(i)} \rangle = \langle U^{C^\circ(i)} \rangle \leq C(i),$$

that is, i is central in $U_2(M)$. Since all the involutions in M are conjugate, they are all central in $U_2(M)$. □

This has the following important consequences.

LEMMA 1.7. *Let G be a simple L^*-group of even type of finite Morley rank with a definable strongly embedded subgroup M. Then the set of involutions of M has Morley degree equal to 1, and they are all conjugate in M°.*

PROOF. We know now that the involutions of M generate a normal elementary abelian subgroup A, and A is infinite. As all involutions are conjugate in M, they belong to A°, and thus $A = A^\circ$ and $I(M) = (A^\circ)^\times$. This set has Morley degree 1, proving the first claim.

Consider any conjugacy class i^{M° of involutions in M°. The rank of this set is

$$\mathrm{rk}(M^\circ) - \mathrm{rk}(C_{M^\circ}(i)) = \mathrm{rk}(M) \setminus \mathrm{rk}(C_M(i)) = \mathrm{rk}(i^M) = \mathrm{rk}(I(M)) \ .$$

So i^{M° is a set of full rank in $I(M)$. As $I(M)$ has Morley degree 1, and distinct conjugacy classes under the action of M° are disjoint, it follows that $I(M)$ is a single conjugacy class under the action of M°. □

Recall that strongly real elements are those inverted by involutions (Remark I 1.43).

LEMMA 1.8. *Let G be a group of finite Morley rank with a definable strongly embedded subgroup M such that the involutions of M generate an elementary abelian subgroup. Then for each involution $i \in G$, any strongly real element in the centralizer of i is an involution.*

PROOF. Let $A = \langle I(M) \rangle = I(M) \cup \{1\}$. We may suppose (conjugating) that $i \in A$. If $a \in C(i)$ is strongly real, inverted by j, then by Lemma 1.2 we have $j \in M$. Now j and ja are involutions in M, hence are both in A, and commute; so a is also an involution. □

1.2. M_1, K°, and $O_2(M)$.

LEMMA 1.9. *Let G be a simple L^*-group of even type of finite Morley rank with a definable strongly embedded subgroup M. Suppose that $U_2(M)$ is nonsolvable, and $K \leq M$ is w-invariant for some involution $w \notin M$. Then K° is abelian.*

PROOF. We let $M_1 = (U_2(M))^{(\infty)}$, and we suppose M_1 is nontrivial. We let $\overline{M}_1 = M_1/O_2(M_1)$. By Lemma II 6.3 and Proposition II 5.10, M_1 is a central product of quasisimple Chevalley groups.

The group K contains no involutions as $w \notin M$. We consider the action of K° on $\overline{M}_1$, with kernel $K_0 = C_{K^\circ}(\overline{M}_1)$. As K° acts on $\overline{M}_1$ by inner automorphisms, and contains no involutions (Lemma II 5.5), K°/K_0 is abelian. We phrase this as follows: $(K^\circ)' \leq K_0$.

Now we claim

$$(K^\circ)' = 1 \ .$$

Suppose $(K^\circ)' \neq 1$. Now w acts on $(K^\circ)'$, and either centralizes a nontrivial element of this group, or inverts them all (Lemma I 10.3). So take $x \in (K^\circ)'$ nontrivial, and either centralized or inverted by w.

We show that $C_{M_1}(x)$ covers $\overline{M}_1$. This follows from Proposition I 9.9, taking $T = d(\langle x \rangle)$, a $2^\perp$-group acting trivially on $M_1/O_2(M_1)$; as T centralizes $M_1/O_2(M_1)$, $C_{M_1}(x) = C_{M_1}(T)$ covers $\overline{M}_1$ as claimed. In particular, $C_{M_1}(x)$ contains an infinite Sylow 2-subgroup S. Observe that $S^w \cap M = 1$ since $M^w \cap M$ contains no involutions. So $C^\circ(x)$ is not contained in M, but contains S°, and thus $M \cap C^\circ(x)$ is a strongly embedded subgroup of $C^\circ(x)$.

This gives us the structure of $C^\circ(x)$: $C^\circ(x) = L \times D$, with L of PSL_2 type, and D degenerate (Lemma II 6.5). Now $C^\circ(x) \cap M$ is supposed to contain a subgroup covering $\overline{M}_1$, but this is a subgroup of $B \times D$ with B solvable and D degenerate, and this is impossible. So we have a contradiction, and $(K^\circ)' = 1$. □

LEMMA 1.10. *Let G be a simple L^*-group of even type of finite Morley rank with a definable strongly embedded subgroup M. Suppose that $U_2(M)$ is nonsolvable. Then $O_2(M)$ is abelian.*

PROOF. We set $M_1 = U_2(M)^{(\infty)}$ and we let K be a group furnished by Lemma 1.3, to which the preceding lemma will apply; so K° is abelian. We let K_1 be the subgroup of K° inverted by w.

We consider the structure $(O_2(M), K_1)$, where the second group is acting on the first. Each element of K_1 has trivial centralizer in $O_2(M)$ (Lemma 1.2). We wish to conclude by applying Nesin's result on free Suzuki groups, Theorem III 3.2. We have seen that the action is free, and we claim also:

$$K_1 \text{ acts transitively on the involutions of } M.$$

By Lemma I 10.4 applied to the action of w on K and on K° we get: $\mathrm{rk}(K) = \mathrm{rk}(C_K(w)) + \mathrm{rk}(K^-)$; $\mathrm{rk}(K^\circ) = \mathrm{rk}(C_{K^\circ}(w)) + \mathrm{rk}(K_1)$. Now $\mathrm{rk}(K) = \mathrm{rk}(K^\circ)$ and $\mathrm{rk}(C_K(w) = \mathrm{rk}(C_{K^\circ}(w))$, so by cancellation $\mathrm{rk}(K^-) = \mathrm{rk}(K_1)$. Recall that by the choice of w, $\mathrm{rk}(K^-) \geq \mathrm{rk}(I(M))$. So this gives us the following:

$$\mathrm{rk}(K_1) \geq \mathrm{rk}(I(M)) \ .$$

Furthermore, $K_1 \cap C(i) = 1$ for any $i \in I(K)$ since the elements of K_1 are inverted by w (Lemma 1.2). So $\mathrm{rk}(i^{K_1}) = \mathrm{rk}(I(M))$ for all $i \in I(M)$. Since the conjugacy classes with respect to K_1 are pairwise disjoint or equal, and of full rank in $I(M)$, while $I(M)$ has Morley degree 1, therefore $I(M)$ consists of a single conjugacy class under the action of K_1. So by Theorem III 3.2, $O_2(M)$ is abelian. □

LEMMA 1.11. *Let G be a simple L^*-group of even type of finite Morley rank with a definable strongly embedded subgroup M. Let $M_1 = (U_2(M))^{(\infty)}$. If $O_2{}^\circ(M)$ is abelian, then $[O_2(M), M_1] = 1$.*

PROOF. Let X be any definable subgroup of M_1 containing no involutions. Let $P = O_2{}^\circ(M)$. As $\Omega_1(P) \leq Z(U_2(M))$, we have $[X, \Omega_1(P)] = 1$, so $[X, P] = 1$ (Lemma I 10.8).

As these groups generate M_1 (Lemma II 5.12), we find $[M_1, P] = 1$, as claimed. □

1.3. Weak solvability.

LEMMA 1.12. *Let G be an L^*-group of finite Morley rank of even type with a weakly embedded subgroup M, with $O_2{}^\circ(M)$ nontrivial, and suppose that*

$$[O_2{}^\circ(M), (U_2(M))^{(\infty)}] = 1 \ .$$

Then $\sigma^\circ((U_2(M))^{(\infty)}) = 1$.

PROOF. We let $M_1 = (U_2(M))^{(\infty)}$.

We consider $H = \sigma^\circ(M_1)$. We have $U_2(H) \leq O_2{}^\circ(M) \cap M_1 \leq Z(M_1)$, and H splits as a semidirect product $U_2(H) \cdot T$ (Proposition I 9.6), hence as a direct product in this case. Here $T = O(M_1) \leq O(U_2(M))$, and $U_2(M)$ centralizes T (Proposition I 10.13). Thus $\sigma^\circ(M_1)$ is central in M_1. But M_1 is perfect, so by Proposition II 4.11 it follows that $\sigma^\circ(M_1) = 1$. □

PROOF OF PROPOSITION 1.1. G is a simple L^*-group of even type of finite Morley rank with a definable strongly embedded subgroup M. Let $M_1 = (U_2(M))^{(\infty)}$.

If $M_1 = 1$ our claim is proved. If not, then by the sequence of lemmas above $O_2(M)$ is abelian and commutes with M_1, and is nontrivial. So by Lemma 1.11, $\sigma^\circ(M_1) = 1$ and M_1 is a central product of quasisimple Chevalley groups in characteristic 2. However, the involutions in M are central in $U_2(M)$ by Proposition 1.6, while the center of any quasisimple component of M_1 consists of semisimple elements by Fact II 1.8, and this is a contradiction. □

2. Weak solvability: Weak embedding

Our goal in the present section is the following.

PROPOSITION 2.1 ([**1, 3**]). *If G is a simple L^*-group of even type of finite Morley rank with a definable weakly embedded subgroup M which is not strongly embedded, then $U_2(M)$ is 2-unipotent.*

DEFINITION 2.2. *Let G be a group of finite Morley rank with a weakly but not strongly embedded definable subgroup M. An* offending involution *in M is an involution $\alpha \in I(M)$ for which $C_G(\alpha) \not\leq M$.*

It is the presence of offending involutions which is characteristic for the case we are now considering, and our analysis revolves around the structure of $C^\circ(\alpha)$ with α offending; this will have the familiar form $L \times D$ with L of type PSL_2 and D degenerate, and with L meeting M in a Borel subgroup, and in particular with M containing a unique Sylow 2-subgroup A of L, a subgroup to which we will devote considerable attention. After invoking the Landrock-Solomon analysis of §III 4 to conclude that the Sylow° 2-subgroup of M is either abelian or of exponent 4, we show that $O_2{}^\circ(M)$ is nontrivial, that the subgroup A is central in $U_2(M)$, and, as in the previous section, that $O_2(M)$ commutes with $M_1 = (U_2(M))^{(\infty)}$. From this, as we showed in the previous section, the structure of M_1 is determined, which leads to a rapid contradiction unless $M_1 = 1$, and the result follows.

2.1. Offending involutions.

LEMMA 2.3. *Let G be a simple L^*-group of finite Morley rank and even type, and M a definable subgroup which is weakly but not strongly embedded. Then for any offending involution, writing C_α for $C_G(\alpha)$, we have the following:*

(1) *$M \cap C_\alpha{}^\circ$ is strongly embedded in $C_\alpha{}^\circ$;*
(2) *$C_\alpha{}^\circ = L \times D$ with $L \simeq \mathrm{PSL}_2(K)$, K algebraically closed of characteristic 2, and D of degenerate type;*
(3) *$M \cap C_\alpha{}^\circ = B \times D$ with B a Borel subgroup of L;*
(4) *$\alpha \notin C_\alpha{}^\circ$.*

PROOF. The main point here is that ${C_\alpha}^\circ$ is not contained in M, as otherwise taking S_α to be Sylow subgroup of ${C_\alpha}^\circ$, a Frattini argument gives

$$C_\alpha \le {C_\alpha}^\circ \cdot N(S_\alpha) \le M$$

and α is not offending.

Furthermore, α lies in a Sylow 2-subgroup S of M and $Z^\circ(S) \le {C_\alpha}^\circ$. So ${C_\alpha}^\circ$ is an L-group with a weakly embedded subgroup, namely $M \cap {C_\alpha}^\circ$. Now by Lemma II 6.5 the structure of ${C_\alpha}^\circ$ is determined, as well as the intersection with M.

It follows that $\alpha \notin {C_\alpha}^\circ$, as the factor D of degenerate type contains no involutions. □

In this situation, we will adopt the notation of the Landrock–Solomon theorem, Theorem III 4.1. This requires an additional lemma.

LEMMA 2.4. *Let G be a simple L^*-group of finite Morley rank and even type, M a definable subgroup which is weakly but not strongly embedded, α an offending involution, $L = U_2(C(\alpha))$, $B = M \cap L$ a Borel of L, $A = U_2(B)$, and T a complement to A in B (a maximal torus of L). Then there is an $\langle\alpha\rangle \times T$-invariant Sylow° 2-subgroup of M.*

PROOF. Let Q be a maximal 2-unipotent subgroup of M normalized by $\langle\alpha\rangle \times T$, and S a Sylow 2-subgroup of M containing $Q\langle\alpha\rangle$. We claim that $Q = S^\circ$.

Supposing the contrary, let $R \le S$ be the preimage in S of $Z^\circ(N_S(Q)/Q)$. Then $R > Q$ (Corollary I 5.2) and since $\alpha \in N_S(Q)$ we have $[\alpha, R] \le Q$. Let K/Q be $C_{U_2(N_G(Q)/Q)}(\alpha)$. Any subgroup of K containing Q is α-invariant, and a Sylow° 2-subgroup of K properly contains Q. As K is T-invariant, and is a K-group (Proposition II 6.2), there is a T-invariant Sylow° 2-subgroup of K (Proposition II 5.4). As this properly contains Q, we contradict the maximality of Q.

So $Q = S^\circ$, as claimed. □

So consider $A = U_2(M \cap C_\alpha)$, T is a torus of $U_2({C_\alpha}^\circ)$ normalizing A, and S a Sylow° 2-subgroup of M normalized by $\langle\alpha\rangle \times T$ and containing A. With these choices, the hypotheses of Theorem III 4.1 are fulfilled.

2.2. M_1, A, $O_2(M)$.

LEMMA 2.5. *Let G be a simple L^*-group of finite Morley rank and even type, and M a definable subgroup which is weakly but not strongly embedded. Then ${O_2}^\circ(M) \ne 1$.*

PROOF. Supposing the contrary, we consider $M_1 = (U_2(M))^{(\infty)}$. Note that $U_2(M)$ is a K-group (Proposition II 6.2).

Now $\sigma^\circ(M)$ is a $2^\perp$-group since ${O_2}^\circ(M) = 1$. Hence by Lemma I 10.13 we have $[\sigma^\circ(M), U_2(M)] = 1$. Thus M_1 is a central product of quasisimple algebraic groups (Lemma II 4.11).

The group M_1 covers $U_2(M)/\sigma(U_2(M))$ and thus $U_2(M)/M_1$ is solvable and contains no 2-unipotent subgroup. Thus $U_2(M) = M_1$ is a central product of quasisimple Chevalley groups.

We take an offending involution $\alpha \in M$ and we consider its action on $U_2(M)$. As $C_M{}^\circ(\alpha)$ has the form $B \times D$ with B solvable and D degenerate, α must normalize each quasisimple component of $U_2(M)$. Hence $A = U_2(C_M(\alpha))$ meets each such component. At the same time the connected group T normalizes each factor, and acts transitively on $A^\times$, and so $M_1 = U_2(M)$ must consist of a single quasisimple Chevalley group.

Now $C_{M_1}{}^\circ(\alpha)$ is the product of a solvable and a degenerate factor, both connected, and the degenerate factor must also be solvable (Proposition II 4.5). So by Lemma II 2.27, α acts on M_1 as an inner automorphism.

Now α normalizes a Sylow° 2-subgroup S of M whose structure is given by the Landrock-Solomon analysis, Theorem III 4.1. As α acts as an inner automorphism it acts on S like an involution in S. By Fact III 4.20, this forces $S = A$.

So A is a Sylow 2-subgroup of M. In particular, M_1 has abelian Sylow 2-subgroups and hence is of the form $\mathrm{PSL}_2(K)$. Now $\langle\alpha\rangle \times T$ acts on M_1 by inner automorphisms, and as M_1 contains no such subgroup the action cannot be faithful. On the other hand, the kernel of the action on A is just $\langle\alpha\rangle$, and α acts faithfully on M_1, so we arrive at a contradiction. □

LEMMA 2.6. *Let G be a simple L^*-group of finite Morley rank and even type, and M a definable subgroup which is weakly but not strongly embedded. Let $\alpha \in I(M)$ be an offending involution, and $A = U_2(C_M(\alpha))$. Then $A \leq Z(U_2(M))$.*

PROOF. By Lemma 2.5 $Z^\circ(O_2(M))$ is nontrivial, so $C_{Z^\circ(O_2(M))}{}^\circ(\alpha)$ is nontrivial, and contained in A. But there is a subgroup T of $C_M(\alpha)$ acting transitively on A, so $A \leq Z(O_2{}^\circ(M))$.

On the other hand, consider $B = \Omega_1(Z(O_2{}^\circ(M)))$. We have $A \leq B \leq S$ where S is an $(\langle\alpha\rangle \times T)$-invariant Sylow° 2-subgroup of M which contains A, whose structure is given by the Landrock-Solomon Theorem III 4.1.

Suppose S is nonabelian. Then as B is an $(\langle\alpha\rangle \times T)$-invariant elementary abelian subgroup of S, B must be contained in A by Corollary III 4.20, and thus $B = A$. Again, if S is abelian and not elementary abelian, then $B \leq \Omega_1(S) = A$. If S is elementary abelian, then $B = A$ or $B = S$.

If $B = A$, then A is normal in M and hence is central in each conjugate of S, and hence also in $U_2(M)$.

If $B = S$, then $U_2(M) = S$. So in either case our claim follows. □

LEMMA 2.7. *Let G be a simple L^*-group of finite Morley rank and even type, and M a definable subgroup which is weakly but not strongly embedded. Then $[O_2{}^\circ(M), (U_2(M))^{(\infty)}] = 1$.*

PROOF. We fix an offending involution α of M and the usual associated notation A, S. We set $M_1 = (U_2(M))^{(\infty)}$.

If $O_2{}^\circ(M) = S$, then $U_2(M) = S$ and $M_1 = 1$, and our claim holds.

So $O_2{}^\circ(M)$ is a proper $(\langle\alpha\rangle \times T)$-invariant subgroup of S containing A. In particular, $O_2{}^\circ(M)$ is abelian and homocyclic, with $A = \Omega_1(O_2{}^\circ(M))$, (Corollary III 4.20).

Now M_1 is generated by its definable connected $2^\perp$-subgroups in view of Lemma II 5.12. If X is one such, then X centralizes A in view of the preceding lemma, and hence X centralizes $O_2{}^\circ(M)$ in view of Lemma I 10.8. Thus M_1 centralizes $O_2{}^\circ(M)$, as claimed. □

Proof of Proposition 2.1. By Lemma 1.12 and the preceding results, $M_1 = (U_2(M))^{(\infty)}$ is a central product of quasisimple Chevalley groups. But α normalizes M_1, so if M_1 is nontrivial, then α centralizes a nontrivial 2-unipotent subgroup of M_1, and thus $A \cap M_1$ is infinite; but $A \leq O_2{}^\circ(M)$, so this is a contradiction. □

Lemma 2.8. *If S is a Sylow° 2-subgroup of M and $i \in S$, then $C(i) \leq M$.*

Proof. Suppose on the contrary that i is an offending involution. By Theorem 2.1, $S = U_2(M)$. Thus the Landrock-Solomon Theorem applies and in particular Corollary III 4.6 gives $C_S(i) = A$.

But then $i \in A$ and hence $i \in C^\circ(i)$, contradicting Lemma 2.3. □

3. Recognition: Strong embedding, I

We will now begin the "recognition" phase of our classification theorem in the strong embedding case. In other words, we will assume the following favorable hypothesis, and identify the group, in several stages.

(∗) Whenever A_1, A_2 are two distinct conjugates of $\Omega_1(O_2{}^\circ(M))$, the group $C_G(\langle A_1, A_2\rangle)$ is finite.

Later, we will show that the *failure* of this hypothesis leads eventually to a contradiction.

The main result of this section concerns intersections of the form $M \cap M^w$ with M strongly embedded and w a suitable involution.

Notation 3.1.

(1) *For $w \in I(G)$, let $T(w) = \{x \in M^\circ : x^w = w^{-1}\}$.*
(2) *Let $I_M^+ = \{w \in I(G) \setminus M : \mathrm{rk}(T(w)) \geq \mathrm{rk}(I(M))\}$.*

We observe that $T(w) \subseteq M \cap M^w$ but that $T(w)$ is a group only if its elements commute, and in particular there is no obvious reason to suppose that $T(w)$ will coincide with the intersection. In PSL_2, however, it does, and the goal of this section is to prove a version of this more generally. This will then give us the point of departure for a close analysis of the ambient group.

THEOREM 3.2. *Let G be a simple L^*-group of even type and finite Morley rank with a definable strongly embedded subgroup M, and let $A = \Omega_1(O_2{}^\circ(M))$. Assume the hypothesis $(*)$ above. Let w be an involution in I_M^+. Then w inverts the group $T = (M \cap M^w)^\circ$.*

We must first settle the general framework more precisely. We will keep our basic notation fixed throughout this and the next section. So throughout, G is a simple L^*-group of finite Morley rank and even type, and M is a definable strongly embedded subgroup of G. We set $A = \Omega_1(O_2{}^\circ(M))$, and hypothesis $(*)$ is assumed to hold.

Recalling that all involutions in M are conjugate in M, and that $U_2(M)$ is 2-unipotent (§1), it follows that $I(M) \subseteq Z(U_2(M))$ and, in particular the group A is elementary abelian, and $A = \langle I(M) \rangle$. Later, in dealing with the weakly embedded case, we will have to treat the group A more circumspectly, but here it may be viewed in a wide variety of ways.

In particular, $\mathrm{rk}(I(M)) = \mathrm{rk}(A)$, and thus we will express the characteristic property of involutions in I_M^+ as follows: $\mathrm{rk}(T(w)) \geq \mathrm{rk}(A)$.

Observe also that

$$I_M^+ \neq \emptyset\ .$$

This was argued earlier, in the proof of Lemma 1.3, and noted as a corollary to that lemma. We run over the argument again. In the first place, $T(w)$ is equal to the set $\{wi : i \in I(wM^\circ)\}$, so the condition $\mathrm{rk}(T(w)) \geq \mathrm{rk}(I(M))$ can also be expressed as $\mathrm{rk}(I(wM^\circ)) \geq \mathrm{rk}(I(M))$. Second, by a direct computation, the set of all involutions in G has rank equal to the rank of $I(M)$ plus the corank of M° in G (i.e., $\mathrm{rk}(G) - \mathrm{rk}(M)$), and hence some coset other than M° must contain at least $\mathrm{rk}(I(M))$ involutions.

Lemma 1.3 states that for each involution $w \in I_M^+$, there is a connected subgroup K of M such that $M^\circ = C^\circ(i)K$ for all involutions $i \in M$, with $K \leq M \cap M^w$.

3.1. Preparation. Notation and hypotheses were fixed above; G, M, A are fixed throughout. We prepare for the proof of Theorem 3.2. The next result will be applied mainly with $H = (M \cap M^w)^\circ$, and occasionally to other subgroups of $(M \cap M^w)^\circ$.

LEMMA 3.3. *With hypotheses and notation as above, if $w \in I(G) \setminus M$, then for any connected subgroup H contained in $M \cap M^w$, the Borel subgroups of H are good tori, and are conjugate in H.*

PROOF. We have natural actions of H on A and on A^w, hence on $A \times A^w$, and the kernel of this combined action is finite, according to hypothesis $(*)$, hence central in H°. By Corollary IV 1.7, the Borel subgroups of H are conjugate, and are good tori after factoring out the kernel of the action. In other words, if B is a Borel subgroup of H, then B/B_0 is a good torus for some finite normal subgroup B_0. Then B' is both connected and finite, so B is abelian, and hence B is a good torus by Lemma I 4.21. □

NOTATION 3.4. *We fix an involution $w \in I_M^+$ and set $T = (M \cap M^w)^\circ$.*

We will need to make a case distinction eventually depending on whether $C_T{}^\circ(A)$ is trivial or not, and we prepare now for the more problematic case, in which it is trivial.

LEMMA 3.5. *With the hypotheses and notation above, if $C_T{}^\circ(A) = 1$, then the intersection of any two distinct Borel subgroups of T is finite.*

PROOF. Suppose that B_1, B_2 are distinct Borel subgroups of T, and $B_0 = (B_1 \cap B_2)^\circ > 1$. The Borel subgroups are good tori, and are, in particular, abelian. As $C_T{}^\circ(A) = 1$, applying w we also have $C_T{}^\circ(A^w) = 1$, and thus $C_{B_0}{}^\circ(A) = C_{B_0}{}^\circ(A^w) = 1$.

Let $X = C_T{}^\circ(B_0)$. Then X contains the two Borel subgroups B_1 and B_2, so is nonsolvable. Now $[A, B_0] \neq 1$. Let A_1 be an X-minimal subgroup of $[A, B_0]$. As $A = [A, B_0] \oplus C_A(B_0)$ by Lemma I 11.8, it follows that B_0 acts freely on A_1. Let K_1 be the kernel in X of this action.

If the action of X on A_1 is not irreducible, then there must be a finite X-invariant subgroup of A_1 which is then centralized by X, and in particular by B_0, contradicting our choice of A_1. So A_1 is irreducible as an X-module and thus by Proposition I 4.11 the quotient X/K_1 has a linear representation over the field generated by B_0 in $\mathrm{End}(A_1)$; this is a field of characteristic 2. So by Proposition II 4.5 X/K_1 is a K-group, and as X contains no involutions, neither does the quotient, and X/K_1 is solvable. As X is nonsolvable, K_1 is nonsolvable. In particular, K_1 is infinite, as otherwise it would be central in X. In the remainder of the argument we will show that K_1 is solvable, getting a contradiction.

We claim the following:

(1) For any connected definable nonsolvable subgroup K of K_1, $C_{A^w}(K)$ is finite.

Suppose on the contrary that $K \leq K_1$ is connected, definable, and nonsolvable, and $C_{A^w}(K)$ is infinite. Now $[K, A_1] = 1$, so $C_A(K)$ is infinite. Since $C_{A^w}(K)$ is also infinite, the group $L = U_2(C(K))$ is of type PSL_2 by Lemma II 6.5.

Let S_1 and S_2 be the two Sylow 2-subgroups of L contained in A and A^w respectively. There is a maximal torus T_1 of L normalizing S_1 and S_2, so $T_1 \leq M \cap M^w$, that is $T_1 \leq T$. As T_1 is inverted by an involution outside M, no element of T_1 commutes with an element of A, by Lemma 1.2.

Now T_1 commutes with K. Let $A = V_0 > V_1 > \cdots > V_n = (0)$ be a KT_1-invariant series with KT_1-minimal quotients. As T_1 acts freely on A, also T_1 acts freely on the quotients, by Proposition I 9.9. So again the image of KT_1 in $\mathrm{Aut}(V_i/V_{i+1})$ is linear over the field generated by the image of T_1, and hence, as it contains no involutions, also solvable, for each i.

On the other hand, the kernel of the map from K into the product of all $\mathrm{Aut}(V_i/V_{i+1})$ is finite, by assumption. So K is solvable, a contradiction; this proves (1).

Now consider a definable $K_1{}^\circ B_0$-minimal subgroup V of $[A^w, B_0]$. Again, by Lemma I 11.8, B_0 acts freely on V. So letting K_2 be the kernel of the action of $K_1{}^\circ B_0$ on V, we find as usual that $K_1{}^\circ B_0/K_2$ is linear and solvable, and hence that $K_1/(K_1 \cap K_2)^\circ$ is solvable. So the subgroup $(K_1 \cap K_2)^\circ$ is nonsolvable. But this contradicts (1). With this final contradiction, the proof is complete. □

LEMMA 3.6. *With the hypotheses and notation above, if* $C_T{}^\circ(A) = 1$, *then for any Borel subgroup* B *of* T *we have* $[w, B] \neq 1$.

PROOF. We suppose $[w, B] = 1$ for some Borel subgroup of T. Then B contains no strongly real element, by Lemma 1.8.

Let y be an element of T which is inverted by w. Let $K = C_T{}^\circ(y)$. Note that by Lemma I 3.11 the group K is nontrivial. Now w normalizes K. By Lemma 3.3, the Borel subgroups of K are conjugate. Fix a Borel subgroup B_1 of K. We will find an involution in the normalizer of B_1 which, like w, inverts y. By the Frattini argument $N(K) = K \cdot N(B_1)$, so $w = kw_1$ with $k \in K$, $w_1 \in N(B_1)$. Then $w_1^2 \in K$. Let $X = d(w_1^2)$. Then $X \leq N_K(B_1) \cap C(w_1)$ and X contains no involutions, so X is 2-divisible. Let $x \in X$, $x^2 = w_1^2$. Then $w' = w_1 x^{-1}$ is an involution normalizing B_1. Furthermore, $w' \in wK$, so w inverts y.

Extend this Borel subgroup B_1 of K to a Borel subgroup C of T. As disjoint Borel subgroups of T have finite intersection, B_1 determines C, and hence C is normalized by w' and by y. Since C is conjugate to B in T, C contains no strongly real elements. Now w' and $w'y$ are two involutions that normalize C, and as we have just seen they invert no elements of C, and hence centralize C. So y also centralizes C. On the other hand, y is strongly real, so y is not in C.

Now Lemma IV 2.3 says that this configuration cannot occur. We have a connected group of finite Morley rank (namely T) whose Borel subgroups are divisible abelian and almost self-normalizing, and intersect pairwise in finite groups, and an element y normalizing one of these groups (namely C), and lying outside it; but according to that lemma, this configuration cannot occur. (We recall that the contradiction arises from a genericity argument based on consideration of the set $\bigcup (yC)^G$.) □

This concludes our preparations.

3.2. Analysis of T. We retain the notation and hypotheses fixed above: G, M, A, w and T are all fixed, and the condition $(*)$ is included among our hypotheses. We aim to prove that w inverts T.

Assuming the contrary, $C_T(w)$ is nontrivial by Lemma I 10.4, and is connected by Lemma I 10.5. So we work throughout with this hypothesis:

(Hyp) $C_T(w)$ is connected and nontrivial.

We consider separately the cases in which $C_T(A)$ is finite, or infinite.

(Case I) $C_T(A)$ is finite.

Extend a Borel subgroup of $C_T(w)$ to a Borel subgroup B of T. By Lemma 3.5, w normalizes B.

We have an action of T on A. By Lemma 1.7, $A^\times = I(M)$ is a single conjugacy class in M°. By Corollary 1.5 we have a connected group K normalized by w for which $M^\circ = C^\circ(i)K$, for $i \in I(M)$, and thus $A^\times = i^K$. Of course, $K \leq T$, and thus T acts transitively on $I(M)$.

Fix $w \in I_M^+$. Then $\mathrm{rk}(T(w)) \geq \mathrm{rk}(A)$ so $\mathrm{rk}(T) = \mathrm{rk}(T(w)) + \mathrm{rk}(C_T(w)) > \mathrm{rk}(A)$. Since T acts on A, we conclude that $C_T(u)$ is infinite for $u \in A$. This will be the point of departure leading eventually to a contradiction.

We claim that $u \in A^\times$ can be chosen so that $C_B(u)$ is infinite. Let $u \in A^\times$, and let B_0 be a Borel subgroup of $C_T(u)$. Extending B_0 to a Borel subgroup of T, and using the fact that the Borel subgroups of T are conjugate in T, we may conjugate u so as to have $B_0 \leq B$. Then $B_0 = C_B{}^\circ(u)$.

We claim that $C_A(B_0)$ is infinite. We have $B_0 \leq B$, with B abelian, so B normalizes $C_A(B_0)$. If $C_A(B_0)$ is finite, then B centralizes $C_A(B_0)$. On the other hand, w normalizes B, and by Lemma 3.6 w does not centralize B, hence inverts an element $y \in B$. By Lemma 1.8, y centralizes no involution in A, which would give a contradiction. So $C_A(B_0)$ is infinite, and B acts nontrivially on $C_A(B_0)$. So far we have the following:

$$B_0 \leq B \text{ is nontrivial, connected; } A_0 = C_A{}^\circ(B_0) \text{ is nontrivial.}$$

Furthermore, $A_0 < A$ by our case hypothesis, namely $C_T(A)$ is finite.

In the next part of the argument, we will show that $N_T(B) \not\leq B$, and choose an element σ of $N_T(B)$ of prime order p modulo B, showing eventually that $p = 3$, and then that σ has order 3, after which we make a precise analysis of the resulting configuration, or at least the part of it involving elements of order 3 and associated involutions.

Now B_0 and B are good tori. Fix a prime ℓ such that

$$\mathrm{Pr}_\ell(B_0) > 0 \ .$$

We claim

$$\mathrm{Pr}_\ell(B) > \mathrm{Pr}_\ell(B_0) \ .$$

Supposing the contrary, then as $B_0^w \leq B$ we find that $B_1 = (B_0 \cap B_0^w)^\circ$ is nontrivial. As $A_0, A_0^w \leq C(B_1)$, letting $L = U_2(C(B_1))$ it follows by Lemma II 6.5 that L is of type PSL_2. In particular, L contains a maximal torus T_0 which normalizes the Sylow 2-subgroups of L containing A_0 and A_0^w, and hence lies in T. As T_0 commutes with B_1, we have $T_0 \leq B$. However, T_0 is a full algebraic split torus, and hence $\mathrm{Pr}_\ell(T_0) = 1$. Furthermore, T_0 acts freely on A and hence $T_0 \cap B_0 = 1$. So $\mathrm{Pr}_\ell(B) > \mathrm{Pr}_\ell(B_0)$, a contradiction.

Now we can examine $N_T(B)$. Our first claim is as follows.

(1) $$N_T(B) \not\leq B \ .$$

Fix a definable B-invariant subgroup A_1 of A containing A_0, with $\bar{A}_1 = A_1/A_0$ B-minimal. Then $C_{\bar{A}_1}(B)$ is trivial; by Proposition I 9.9 this centralizer is covered by $C_{A_1}(B)$, which as we have noticed is trivial. By Proposition I 4.11, B acts on $\bar{A}_1$ like a multiplicative subgroup of a field, and as $\mathrm{Pr}_\ell(B) > 1$ it follows that $B_2 = C_B{}^\circ(\bar{A}_1)$ is infinite. Then $A_2 = C_{A_1}(B_2)$ covers $\bar{A}_1$. Let $B_3 = C_B{}^\circ(A_2)$. Then $B_2 \leq B_3$, and A_2 is not contained in A_0.

Now take $u \in A_0^\times$ and $t \in T$ such that $u^{t^{-1}} \in A_2 \setminus A_0$. Then $B_3^t \leq C^\circ(u)$. Now the Borel subgroups of $C^\circ{}_T(u)$ are conjugate, so there is an element $t_1 \in C^\circ{}_T(u)$ carrying B_3^t into a Borel subgroup of $C^\circ(u)$ containing B_0; in particular, $B_3^{tt_1} \leq B$, and hence $tt_1 \in N_T(B)$.

On the other hand, $u^{(tt_1)^{-1}} = u^{t^{-1}} \notin A_0$, so $tt_1 \notin B$. This proves (1):

Now $[N_T(B) : B]$ is finite. For the rest of our analysis, we fix the following notation.

$$\sigma \in N_T(B) \text{ is an element of order } p\text{, a prime, modulo } B.$$

We may adjust the element σ in the course of our analysis.

We will use a number theoretic argument to show that $p = 3$, first showing that for each odd prime the Prüfer rank of B is either 1 or 2. For this we need first to analyze the structure of B in considerable detail.

We have $B = B^+ \oplus B^-$ where $B^+ = C_B(w)$ and B^- is the subgroup inverted by w (Lemma I 10.4).

Let A_w be the conjugate of A containing w, and let $V = C_{A_w}(B^+)$. Then $w \in V$. We claim that V is infinite. Supposing the contrary, there is a conjugate V^* of V contained in A, and the corresponding conjugate R of B^+ then lies in M°, with $C_A(R)$ finite and nontrivial. Let $\overline{M}^\circ = M^\circ/C_{M^\circ}(A)$. Since $U_2(M) \leq C(A)$, the group $\overline{M}^\circ$ is of degenerate type. Hence the Borel subgroups of M° are abelian and conjugate, by Lemma IV 1.6. Conjugating R again, we may suppose that $\bar{R}$ and $\overline{B}^-$ lie in a Borel subgroup of $\overline{M}^\circ$. Then B^- normalizes the finite group $C_A(R)$, and hence centralizes it. But as w inverts B^-, this is a contradiction to Lemma 1.8.

So $C_{A_w}(B^+)$ is infinite. On the other hand, $B^- \leq C(B^+)$ and B^- is not contained in $M_w = N(A_w)$, so $M_w \cap C(B^+)$ is a strongly embedded subgroup of $C(B^+)$. Let $L = U_2(C(B^+))$. Then L is of type PSL_2 by Lemma II 6.5, and we have $w \in V \leq L$.

We use L as a tool to analyze B. As B centralizes B^+, B normalizes L, and thus $B \leq L \times C(L)$, by Corollary II 2.26. In particular, elements of B^- inverted by w have the form ti with $t \in L$, $i \in I(C(L))$, so $x^2 \in L$ for $x \in B^-$. As B^- is 2-divisible, it follows that $B^- \leq L$. It follows easily that

$$B^- \text{ is contained in a maximal torus } T_0 \text{ of } L.$$

Later we will show that B^- is exactly a maximal torus of L.

At this point, we know that for each prime the Prüfer rank of B^- is at most 1. We make the same claim for B^+:

(2) For each prime, the Prüfer rank of B^+ is at most 1.

Let U be a Sylow 2-subgroup of L normalized by B, and let $\hat{A}$ be the conjugate of A containing U. By our hypothesis (*), we have $U < \hat{A}$. Let $U < U_1 \leq \hat{A}$ with U_1 definable, B^+-invariant, and B^+-minimal. The action of B^+ on U_1/U is nontrivial, as usual, since $C_{\hat{A}}(B^+)$ covers $C_{U_1/U}(B^+)$. Let $U_2/U = C_{U_1/U}(B^+)$, a finite quotient, and $\bar{U}_1 = U_1/U_2$. If condition (2) fails for some prime, then $R_0 = C_{B^+}(\bar{U}_1)$ is infinite, and $\bar{U}_1$ is covered by $C_{U_1}(R_0)$. But $U \leq C(R_0)$ as well, so $U_1/C_{U_1}(R_0)$ is finite and hence U_1 centralizes R_0. Thus $U_2(C(R_0)) > U_2(C(B^+))$. This is impossible for a number of reasons. For example, B^+ centralizes R_0 and hence normalizes $L_1 = U_2(C(R_0))$, while centralizing some involutions in L_1; hence B^+ centralizes L_1 and $L_1 \leq U_2(C(B^+))$ after all. This contradiction proves (2).

From the analysis so far, it follows that for each prime, the corresponding Prüfer rank of B is at most 2. We wish to show that for odd primes, the Prüfer rank is at least 1. For this it will suffice to show that B^- is a maximal torus of L; but some preliminary analysis is needed, involving the p-torsion of B, where p is the order of σ modulo B.

By Lemma I 3.11, the group $C_T(\sigma)$ is infinite. Let H_0 be a Borel subgroup of $C_T(\sigma)$, and extend H_0 to a Borel subgroup H of T. Then σ normalizes H, by Lemma 3.5. If σ is not in H, then this is exactly the configuration ruled out by Lemma IV 2.3 (taking $G = T$, and $T = H$). So $\sigma \in H$. On the other hand, σ can be taken to be a p-element; so H contains p-torsion, and as the Borel subgroups of T are conjugate, the same applies to B. So the Prüfer p-rank of B is at least 1.

Let B_p be the p-torsion subgroup of B. By Lemma IV 2.3 again, it follows that $C_B(\sigma)$ is finite. But σ acts on B_p, and if B has Prüfer rank 1 then this action is trivial, by Lemma I 10.18.

So the Prüfer p-rank of B is at least 2, and by our analysis above it is exactly 2, with each of B^+ and B^- having Prüfer p-rank 1.

Now we claim:

(3) The conjugates of B^+ under the action of $\langle\sigma\rangle$ intersect trivially.

Suppose on the contrary that $x \in B^+ \cap (B^+)^\tau$ is nontrivial, where $\tau \in \langle\sigma\rangle$ is nontrivial. Then $U_2(C(x))$ contains L and L^τ. It follows easily via Lemma II 6.5 that $L = U_2(C(x)) = L^\tau$. On the other hand, $B \cap L$ contains B^- and is disjoint from B^+, so this group, which is invariant under τ, has Prüfer p-rank 1. Thus τ acts trivially on B^- (Lemma I 10.18). But $C_B(\sigma)$ is finite, a contradiction.

It follows that $\Omega_1(B_p)$, a group of order p^2, contains at least $p^2 - p$ elements commuting with some involution. On the other hand, B^- consists of strongly real elements. Thus B^- contains the remaining elements of $\Omega_1(B_p)$, and is stabilized and hence centralized by σ.

Now we return to a consideration of $L = U_2(B^+)$. Let A_0 be a B-minimal subgroup of A. As the Prüfer p-rank of B is 2, the Prüfer p-rank of $C_B A_0$ is positive. The elements of order p in $C_B(A_0)$ lie in a conjugate under $\langle\sigma\rangle$ of B^+, by the analysis just above. As we may conjugate by any element of $\langle\sigma\rangle$, we may suppose that these elements lie in B^+. Fixing one such element x, we find as usual that $U_2(C(x)) = L$ again. This shows that $A_0 \leq L$. Also $w \in L$ so $A_0^w \leq L$. It follows that $A \cap L$ and $A^w \cap L$ are Sylow 2-subgroups of L. Therefore, there is a maximal torus T_1 of L which normalizes these two subgroups, and is therefore in $N(A) \cap N(A^w) = M \cap M^w$, that is $T_1 \leq T$. On the other hand, $B^- \leq L \cap N(A) \cap N(A^w) = T_1$, so $T_1 \leq B$ and finally $T_1 = B^-$.

Let $W = N_T(B)/B$. Recall that for $\alpha \in W^\times$, $C_B(\alpha)$ is finite (Lemma IV 2.3). Let $n = |W|$. For all sufficiently large primes ℓ, W acts freely on $\Omega_1(B_\ell) \simeq (\mathbb{Z}/\ell\mathbb{Z})^{d_\ell}$ with d_ℓ equal to 1 or 2. It follows that in any case n divides $\ell^2 - 1$ for all sufficiently large primes. Now n is odd. Take ℓ prime and congruent to 2 modulo n using Dirichlet's theorem. Then $\ell^2 - 1 \cong 3 \bmod n$ and thus $n|3$. So $|W| = 3$ and $p = 3$.

Now $\sigma^3 \in B$, and σ may be taken to have order a power of 3. Furthermore, since $C_B(\sigma)$ is finite, it follows from Lemma I 10.2 that the conjugates of σ are generic in $B\sigma$, and the same applies to any other element of this coset, so all elements of $B\sigma$ have the same order m; so m is a power of 3.

(4) σ has order 3.

Suppose the order of σ is 3^i with $i > 1$. We show first that $C_{B_3}(\sigma) \subseteq \langle\sigma\rangle$. Let $X = \langle C_{B_3}(\sigma), \sigma\rangle$, a finite abelian 3-group. By assumption every element of order 3 in $\langle\sigma\rangle B$ is contained in B, and the same applies to X, that is $\Omega_1(X) \leq B$. But $X \leq C(\langle\sigma\rangle)$, so $X \cap B_3$ is cyclic of order 3 by Lemma I 10.16. It follows that $\Omega_1(X)$ is cyclic, and hence X is cyclic, forcing $X = \langle\sigma\rangle$. So $C_{B_3}(\sigma) \subseteq \langle\sigma\rangle$.

Let $U = \Omega_1(B_3)$ and consider the action of σ on U. By our analysis so far, $[\sigma, U] = C_U(\sigma)$ and in particular taking $\tau = \sigma^{3^{i-1}}$, we have $[\sigma, u] = \tau$ for some $u \in U$. Then $\sigma^u \in \langle\sigma\rangle$ and u normalizes $C(\sigma)$, hence also $C_T{}^\circ(\sigma)$.

We consider the Borel subgroups of $C_T{}^\circ(\sigma)$. By the Frattini argument there is a Borel subgroup H_0 of $C_T{}^\circ(\sigma)$ normalized by an element ua with $a \in C_T{}^\circ(\sigma)$. Then H_0 is contained in a unique Borel subgroup H, of T, which is again normalized by σ and by ua. As $C_H(\sigma)$ is infinite, it follows from Lemma IV 2.3 that σ belongs to H. On the other hand, σ and u do not commute, so $u \notin H$.

We claim that the pair (H, ua) is conjugate in T to the pair (B, σ). We first conjugate H to B and ua goes to an element of σB or $\sigma^{-1}B$. If (H, uaH) is conjugate to $(B, \sigma^{-1}B)$ then we have the element w available. Clearly w does not commute with the action of σ on B, hence w carries σB to $\sigma^{-1}B$. So (H, uaH) is conjugate to $(B, \sigma B)$. After further conjugation by an element of B, ua will go to σ. So (H, ua) is conjugate to (B, σ).

Now in (H, ua) we have the element $\sigma \in H$ of order 3^i, with $[\sigma, ua] = \tau$. Here τ has order 3 and τ commutes with both u (in B) and h, so $\tau \in C(ua)$. Hence $\tau = (ua)^{\pm 3^{i-1}}$. So conjugating H to B and ua to σ, we find an element $\tau_1 \in B$ of order 3^i with $[\tau_1, \sigma] = \sigma^{\pm 3^{i-1}} = \tau^{\pm 1}$. In particular, $\tau_1^3 \in C(\sigma)$ and $C_{B_3}(\sigma) = \langle \tau_1^3 \rangle$.

Now commutation with σ defines a homomorphism which we restrict to the group $\langle \tau_1, u \rangle$, getting

$$\gamma_\sigma : \langle \tau_1, u \rangle \to \langle \tau \rangle .$$

The kernel of this map contains τ_1^3, and since $C_{B_3}(\sigma) = \langle \tau_1^3 \rangle$, this is the exact kernel. But the quotient $\langle \tau_1, u \rangle / \langle \tau_1^3 \rangle$ is isomorphic to $(\mathbb{Z}/3\mathbb{Z})^2$: as $u \in \Omega_1(B_3)$ and $[\sigma, u] \neq 1$, we have $u \notin \langle \tau_1 \rangle$. But the image of this group under the commutation map is cyclic, so we have a contradiction. This contradiction proves finally that σ is of order 3.

Now we observed earlier that w acts by inversion on $N_T(B)/B$, so in particular $C_{N_T(B)}(w) = B^+$. But by Lemma I 10.4 we have $N_T(B) = C_{N_T(B)}(w)N^-$ where $N^- = N_T(B)^-$ is the subset inverted by w. Thus $N_T(B) = B^+N^-$. It follows that the element σ may be taken to be inverted by w.

Let $X = \Omega_1(C_{B_3}(\sigma))$, a cyclic group of order 3. Then w inverts X, and hence acts by inversion on the group $Y = \langle X, \sigma \rangle \simeq (\mathbb{Z}/3\mathbb{Z})^2$. Take any irreducible Y-invariant subgroup A^* of A; then some nontrivial element $y \in Y$ centralizes A^*, and is inverted by w; this again is a contradiction. But this final contradiction goes all the way back to our initial assumption that $C_T{}^\circ(w)$ is nontrivial. Thus, under our present case assumption, the result is proved.

(Case II) $\qquad C_T(A)$ is infinite.

We let $K_1 = C_T{}^\circ(A)$ and $K_2 = C_T{}^\circ(A^w)$. Under our present assumption these groups are nontrivial. Now $[K_1, K_2] \leq K_1 \cap K_2 \leq C(A, A^w)$, and this is assumed finite. Since the group $[K_1, K_2]$ is connected it must therefore be trivial, and the central product $K_1 * K_2$ is a subgroup of T.

By Lemma 3.3 the Borel subgroups of K_1 are conjugate good tori; the same applies to K_2.

Let B_1 be a Borel subgroup of K_1. Then $B = B_1 B_1^w$ is an abelian group on which w acts, and this group contains no involutions, so we have the decomposition $B = B^+B^-$ with respect to the action of w.

Let A_w be the conjugate of w containing w, and let $V = C_{A_w}(B^+)$. We claim that V is an infinite proper subgroup of A_w.

Observe that V is nontrivial since $w \in V$. If $\langle V^{B^-} \rangle \leq A_w$, then it would follow that B^- normalizes A_w, but as w inverts B^-, this leads to the contradiction $[w, B^-] \leq A_w \cap B^- = 1$. So $\langle V^{B^-} \rangle \leq C^\circ(B^+)$ escapes from $N(A_w)$, and hence $L = U_2(C(B^+))$ is a group of type PSL_2, to which we will return below. On the other hand, by $(*)$ this group cannot contain a conjugate of A, so we find $V < A_w$.

Now suppose toward a contradiction that V is finite. Then conjugating A_w to A, we find that there is connected group R conjugate to B^+ such that $C_A(R)$ is finite and nontrivial; in particular, $R \le M$. Working in the quotient $\overline{M}^\circ = M^\circ/C_{M^\circ}(A)$, as in the previous case, we may suppose after conjugation that the action of R on A commutes with the action of B^-. Hence B^- normalizes the finite group $C_A(R)$ and centralizes it, producing a contradiction since the elements of B^- are strongly real. So V is infinite, and proper.

Now consider a Borel subgroup H of T containing B. Then H is a good torus by Lemma 3.3. We claim that H is almost self-normalizing in G. Let $H_1 = N^\circ(H) = C^\circ(H)$ (Lemma I 4.23). Then H_1 centralizes B_1 and hence normalizes $U_2(C(B_1))$. But B_1 centralizes A, and by hypothesis $(*)$ $C(B_1)$ cannot contain another conjugate of A, so $U_2(C(B_1)) = A$ and thus $H_1 \le N(A) = M$. Similarly, $H_1 \le M^w$ and thus $H_1 \le T$. As H is a Borel subgroup of T, we find $H_1 = H$, and H is almost self-normalizing in G.

Now H centralizes B^+ and hence normalizes L. Hence H acts so as to normalize two Sylow 2-subgroups of L, and also normalizes the two conjugates of A containing these two Sylow 2-subgroups. So take A_1, a conjugate of A, such that H normalizes A_1 and $A_1 \cap L$ is a Sylow 2-subgroup of L. Then $C_H(A_1)$ normalizes L and acts trivially on A_1, and contains no involutions, hence centralizes L; by condition $(*)$, this implies that $C_H(A_1)$ is finite.

To recapitulate, we have H, an almost self-normalizing good torus in G, and we have A and a conjugate A_1 of A such that $C_H(A)$ is infinite, $C_H(A_1)$ is finite, and H normalizes A and A_1. Conjugating A_1 to A, we have H and H_1, each an almost self-normalizing good torus in G, so that $C_H(A)$ is infinite, $C_{H_1}(A)$ is finite, and H and H_1 both lie in $M = N(A)$. It now follows from Lemma IV 1.5 that H and H_1 are conjugate in M; but as A is normal in M, this is a contradiction.

With this, all cases have been analyzed, and it follows, finally, that w inverts T, under hypothesis $(*)$. This will give us all we need to carry out a detailed analysis of the action of G on cosets of M. First, we record some direct consequences.

3.3. Consequences. We retain the same notation and hypotheses concerning G, M, A, w and T, and we now know, therefore, that w inverts T.

LEMMA 3.7. *Under the notation and hypotheses fixed above, we have the following.*

(1) *T is a good torus.*
(2) *T acts regularly on $A^\times$.*
(3) $\mathrm{rk}(T) = \mathrm{rk}(A)$.
(4) *For $i \in A$, $C_{M^\circ}(i) = C_M{}^\circ(A)$.*
(5) $M^\circ = C_M{}^\circ(A) \rtimes T$.
(6) *For any $x \notin M$, $(M \cap M^x)^\circ$ is abelian*

PROOF.

(1) T is a good torus:

As w inverts T, T is abelian, and the rest follows Lemma 3.3.

(2) T acts regularly on $A^{\times}$:

By Lemma 1.3 the action is transitive. On the other hand, as every element of $T^{\times}$ is strongly real, the action is free by Lemma 1.8.

(4) For $i \in A$, $C_{M^\circ}(i) = C_M{}^\circ(A)$.

Fix an involution $i \in A$. We have a decomposition

$$M^\circ = C_{M^\circ}(i) \rtimes T$$

by Lemma 1.3 and the regularity of the action.

This gives us a bijection between M°, which has Morley degree 1, and the definable set $C_{M^\circ}(i) \times T$, so the factor $C_{M^\circ}(i)$ has Morley degree 1 as well, and thus is connected, that is, $C_{M^\circ}(i) = C_M{}^\circ(i)$.

Now suppose $C_M{}^\circ(A) < C_M{}^\circ(i)$, and work in the quotient group $\overline{M}^\circ = M^\circ/C_{M^\circ}(A)$. The Borel subgroups of this quotient are conjugate by Lemma IV 1.6. One such Borel subgroup is formed by taking a Borel subgroup of $\overline{C_{M^\circ}(i)}$ and extending it. The action of this Borel subgroup on A is not regular. Therefore, to reach a contradiction it suffices to show that $\bar{T}$ is also a Borel subgroup of $\overline{M}^\circ$.

The Borel subgroups of $\overline{M}^\circ$ are abelian, again by Lemma IV 1.6. So it suffices to consider a connected abelian group $\bar{T}_1 \geq \bar{T}$. Now $\bar{T}_1 = \bar{T}_2 \times \bar{T}$ with $\bar{T}_2 = T_1 \cap \overline{C_M{}^\circ(i)}$. Since $\bar{T}_2$ commutes with $\bar{T}$, which acts transitively on A, it follows that $\bar{T}_2 \leq \bar{C}_{M^\circ}(A) = 1$. So $\bar{T}_1 = \bar{T}$ and $\bar{T}$ is a Borel subgroup of $\overline{M}^\circ$. Our claim follows; and the fifth claim is simply a reformulation of this one.

(6) For any $x \notin M$, $M^\circ \cap M^x$ is abelian:

By the preceding point, $M^{\circ\prime} \leq C(A)$; so $A, A^x \leq C([(M \cap M^x)^\circ]')$; by our hypothesis $(*)$ the commutator $((M \cap M^x)^\circ)'$ must then be finite, and being connected is trivial. □

LEMMA 3.8. *For any nontrivial subgroup H of T we have the following.*

(1) *w inverts $C^\circ(H)$.*

(2) *$N_M{}^\circ(H) = T$.*

PROOF. w inverts $C^\circ(H)$:

If not, then w centralizes a connected subgroup X of $C^\circ(H)$, by Lemma I 10.3. Let A_w be the conjugate of A containing w. Then X centralizes A_w by our fourth point above. As X is infinite, it follows from our condition $(*)$ that $U_2(C(X)) = A_w$. In particular, H normalizes A_w, giving the contradiction $[w, H] \leq H \cap A_w = 1$. So w inverts $C^\circ(H)$.

$N_M{}^\circ(H) = T$:

By Lemma I 4.23 we have $N_M{}^\circ(H) = C_M{}^\circ(H)$. Let $\hat{H} = C_M{}^\circ(H)$. Then $\hat{H} = C_{\hat{H}}(A) \rtimes T$ since $T \leq \hat{H} \leq M^\circ$, and as $\hat{H}$ is connected it follows that $C_{\hat{H}}(A)$ is also connected. Now $C_{\hat{H}}(A) = C_{M^\circ}(AH)$, and we need to prove that this group is finite, or in other words that $C^\circ(AH) = 1$.

But since w inverts $C^\circ(AH)$ and this group commutes with A we have $C^\circ(AH) = 1$ by Lemma 1.8. □

LEMMA 3.9. *For $x \in M^\circ \setminus N(T)$, the intersection $T \cap T^x$ is trivial.*

PROOF. If $H = T \cap T^x$ is nontrivial, then $N_M{}^\circ(H) = T$ and similarly $N_M{}^\circ(H) = T^x$, so $x \in N(T)$. □

LEMMA 3.10. *For any nontrivial subgroup H of T, $N_{M^\circ}(H) = T$.*

PROOF. Let $H_1 = N_{M^\circ}(H)$. Then $H_1{}^\circ = T$ by Lemma 3.8, and thus $H_1 \leq N(T)$. Now $H_1 = C_{H_1}(A) \rtimes T$ and $[C_{H_1}(A), T] \leq C(A) \cap T = 1$, so $C_{H_1}(A) \leq C_{M^\circ}(A, T)$.

By Lemma IV 2.3 applied to M° and T, for any $x \in N(T) \setminus T$, the centralizer $C_T(x)$ is finite. Thus $C_{M^\circ}(A, T) = 1$ and $H_1 = T$. □

LEMMA 3.11. *$M^\circ \cap (M^\circ)^w = T$.*

PROOF. Let $\hat{T} = M^\circ \cap (M^\circ)^w$. Then by definition $T = \hat{T}^\circ$. So $\hat{T} \leq N_{M^\circ}(T) = T$. □

4. Recognition: Strong embedding, II

In this section we will conclude the "recognition" phase of our classification theorem in the strong embedding case, under the the following favorable hypothesis, finally identifying the group.

As in the last section, G is a simple L^* group of finite Morley rank and even type, and M is a definable strongly embedded subgroup of G. $A = \Omega_1(O_2{}^\circ(M))$, and the standard hypothesis $(*)$ is assumed to hold.

$(*)$ Whenever A_1, A_2 are two distinct conjugates of $\Omega_1(O_2{}^\circ(M))$, the group $C_G(\langle A_1, A_2\rangle)$ is finite.

In the last section we began the study of intersections of the form $M \cap M^w$ with M strongly embedded and w a suitably chosen involution. We showed in fact that $T = M^\circ \cap (M^\circ)^w$ is a connected abelian group, and even a good torus, inverted by w, and we derived a number of structural consequences from this, notably $N_{M^\circ}(T) = T$, and that $M^\circ = C_{M^\circ}(A) \rtimes T$. This last point gives good control over M° (and eventually, M, once we prove that this group is connected), in spite of the necessary vagueness about the structure of $C(A)$ which persists right up to the end of the analysis.

We can now proceed fairly directly to a close analysis of the permutation representation of G acting on the cosets of M.

This proceeds by the following steps. First, one shows that a generic involution is "suitable" in the sense of the preceding section. Using this, one makes a fairly precise computation of the rank of the ambient group G: $\mathrm{rk}(G) = \mathrm{rk}(C(T)) + 2\,\mathrm{rk}(C(A))$. Here the term $\mathrm{rk}(C(T))$ has to be considered as unsatisfactory and a further argument is needed to show that $\mathrm{rk}(C(T)) = \mathrm{rk}(T)$, at which point the formula becomes really useful. One can then pass to less technical matters, arriving quickly at the conclusion that the action of G on $M\backslash G$ is doubly transitive, and then (finally) that M is connected. After that, given the structure of M already in hand, and the subsequent understanding of its embedding into G achieved at that point, one may verify directly that G is the sort of split Zassenhaus group identified by the DeBonis–Nesin classification of §III 2. In fact, the more technical hypotheses of that theorem have been proved already, apart from the connectedness of M: $M^\circ = U \rtimes T$ where $U = C_{M^\circ}(A)$ certainly contains a central involution. So what is mainly missing at this point is the basic permutation theoretic information that G acts doubly transitively, and that the stabilizer of three points is trivial. Note also that once we know that $M = M^\circ$, then this will also tell us that the 2-point stabilizer, which may be taken to be $M \cap M^w$, is in fact T.

We now take up the analysis.

4.1. The rank of G. In the previous section we made a detailed study of $M \cap M^w$ for w an involution in I_M^+. Here we must switch our attention briefly to the class I_M^+ itself. We made the following definitions, which depend not only on G but on a choice of strongly embedded subgroup M. For $w \in I(G)$ we defined $T(w)$ as

$$\{x \in M^\circ : x^w = w^{-1}\} .$$

We then defined I_M^+ as the set of involutions for which $\mathrm{rk}(T(w)) \geq \mathrm{rk}(I(M))$, which turns out to be also the set of involutions characterized by

$$\mathrm{rk}(T(w)) = \mathrm{rk}(A)$$

as follows from the analysis of the preceding section.

We will need also consider, very briefly, the complementary set I_M^- consisting of involutions w with $\mathrm{rk}(T(w)) < \mathrm{rk}(A)$. Our claim is that the set I_M^+ is generic.

All we have shown so far about I_M^+ is that it is nonempty; this is already useful, so for the present we will fix some $w \in I_M^+$ and work with

$$T = (M \cap M^w)^\circ = M^\circ \cap (M^\circ)^w$$

to elucidate the structure of M°, the main points being that $\mathrm{rk}(T) = \mathrm{rk}(A)$, and $M^\circ = C^\circ(A) \rtimes T = C^\circ(i) \rtimes T$ for any $i \in I(M)$. Note that in fact $T = T(w)$, as follows from the fact that w inverts T.

PROPOSITION 4.1. $\mathrm{rk}(I_M^+) = \mathrm{rk}(I(G))$.

PROOF. It suffices to show that $\mathrm{rk}(I_M^-) < \mathrm{rk}(I(G))$.

We consider the natural map $\pi : I_M^- \to G/M^\circ$. Our estimate for $\mathrm{rk}(I_M^-)$ will be a coarse one: $\mathrm{rk}(I_M^-) \leq \mathrm{rk}(G/M) + m$ where m is the maximal fiber rank; by definition of I_M^-, the number m is less than $\mathrm{rk}(A)$. So

$$\mathrm{rk}(I_M^-) < \mathrm{rk}(G) - \mathrm{rk}(M) + \mathrm{rk}(A) = \mathrm{rk}(G) - (\mathrm{rk}(M) - \mathrm{rk}(T)) = \mathrm{rk}(G) - \mathrm{rk}(C(i))$$

for any fixed $i \in I(M)$, and of course this last term is $\mathrm{rk}(I(G))$. □

Combined with the following, this result will give us a useful expression for the rank of G.

LEMMA 4.2. *Suppose w_1 and w_2 are involutions in I_M^+. Then $T_1 = T(w_1)$ and $T_2 = T(w_2)$ are conjugate under the action of $C^\circ(A)$.*

PROOF. T_1 and T_2 are almost self-normalizing good tori in M° by the results of the preceding section. By Lemma IV 1.5, they are conjugate in M°. On the other hand, $M^\circ = C^\circ(A) \rtimes T_1$ so the claim follows. □

PROPOSITION 4.3. $\mathrm{rk}(G) = \mathrm{rk}(C(T)) + 2\,\mathrm{rk}(C(A))$.

PROOF. Since $C^\circ(A) = C^\circ(i)$ for $i \in I(M)$, and $\mathrm{rk}(G) = \mathrm{rk}(I(G)) + \mathrm{rk}(C(i))$, while $\mathrm{rk}(I(G)) = \mathrm{rk}(I_M^+)$, what we actually need to show is

$$\mathrm{rk}(I_M^+) = \mathrm{rk}(C(T)) + \mathrm{rk}(C(A)) \ .$$

Furthermore, as $C^\circ(A) \cap C(T) = 1$ by Lemma 3.8, we have $\mathrm{rk}(C(T)) + \mathrm{rk}(C(A)) = \mathrm{rk}(C(T)C(A))$. So we need to compare the rank of the two sets I_M^+ and $C(T)C(A)$, which we do by introducing explicit maps.

We have fixed one involution $w \in I_M^+$, and set $T = T(w)$. We will now define a certain function

$$\Phi : I_M^+ \to w^{C(T)C^\circ(A)} \ .$$

For $i \in I_M^+$ we have $T(i) = T^f$ for some $f \in C^\circ(A)$. This element f is unique by Lemma 3.10. Now $i^{f^{-1}}$ inverts T and hence $i^{f^{-1}}w$ centralizes T. We set

$$\Phi(i) = w^{i^{f^{-1}}w \cdot f} \ .$$

Note that $T(\Phi(i)) = T(i)$.

We claim that the map Φ has finite fibers. Suppose therefore that $\Phi(i) = \Phi(j)$. Then $T(i) = T(j)$. Accordingly, there is $f \in C^\circ(A)$ for which we have

$$w^{i^{f^{-1}}wf} = w^{j^{f^{-1}}wf}$$

and in fact f is constant along this fiber. After conjugation we have $w^{i^{f^{-1}}} = w^{j^{f^{-1}}}$, which reduces to $w^{fi} = w^{fj}$ or $(ij) \in C(w^f)$; but $(ij) \in C(T^f)$ as well, so $(ij) \in C(T,w)^f$. Now by Lemma 3.8, w inverts $C^\circ(T)$, and thus $C(T,w)$ is finite. As the element f is fixed on this fiber, the fiber is finite.

So we conclude that $\mathrm{rk}(I_M^+) \le \mathrm{rk}(w^{C(T)C^\circ(A)} \le \mathrm{rk}(C(T)C^\circ(A))$, giving us the desired value for $\mathrm{rk}(G)$, but only as an upper bound.

For the matching lower bound it suffices to show that

$$\mathrm{rk}(w^{C(T)C(A)}) = \mathrm{rk}(C(T)C(A)) \ .$$

Let $X = C(T)C(A)$ and consider the natural map $X \to w^X$. It will suffice to check that the fibers of this map are also finite.

Note that for $c \in C(T)$, $f \in C(A)$, we have $T(w^{cf}) = T^f$; so if cf, $c'f'$ belong to the same fiber then $T^f = T^{f'}$ and again $f = f'$, $w^c = w^{c'}$, and $c'c^{-1} \in C(w,T)$, a finite group. □

To improve this estimate, we need to show $C^\circ(T) = T$.

4.2. $C^\circ(T) = T$**.** The idea of the proof that $C^\circ(T) = T$ is the following. If $C^\circ(T) > T$, then let $Y_0 = C^\circ(T) \setminus T$ and let $Y = C^\circ(A)Y_0C^\circ(A)$. We will show that this is a generic subset of G, and we will find another disjoint set which is already generic, to get a contradiction. So there are three things to do: find a "truly" generic set, show that it is disjoint from the "pretender" Y, and check that Y is either empty or generic. The truly generic set we will use is $I_M^+M^\circ$.

There is a certain amount of bootstrapping involved; we have a good estimate for $\mathrm{rk}(G)$, which yields structural information which gives us a better estimate.

Our first calculation is straightforward, and could have been carried out earlier.

LEMMA 4.4. $\mathrm{rk}(I_M^+M^\circ) = \mathrm{rk}(G)$.

PROOF. We know that any coset of M° meeting I_M^+ meets the latter set in a subset whose rank is $\mathrm{rk}(T(w)) = \mathrm{rk}(T)$. So

$$\mathrm{rk}(I_M^+M^\circ) = \mathrm{rk}(I_M^+) + \mathrm{rk}(M^\circ) - \mathrm{rk}(T) = \mathrm{rk}(I(G)) + \mathrm{rk}(C(A)) = \mathrm{rk}(G)\ .$$

□

LEMMA 4.5. *If $c \in C^\circ(T) \setminus M$ and $f \in C^\circ(A)$, then the coset fcM° contains no involutions.*

PROOF. We suppose toward a contradiction that fcb is an involution with $b \in M^\circ$ and since $C^\circ(T)$ and M° overlap in T we may take b to lie in $C^\circ(A)$. Then conjugating, the element bfc is also an involution, so replacing f by bf we may suppose $b = 1$ and $i = fc$ is an involution.

Take some nontrivial element $t \in T$ and consider the commutator $\gamma = [i, t]$. Then i inverts γ. On the other hand we compute $\gamma = (fc)(fc)^{-t} = (fc)(c^{-1}f^{-t}) = [f^{-1}, t] \in C(A)$. Now by Lemma 1.8, if γ is nontrivial, then γ is an involution, hence in A, and hence $i = fc \in M$, which is absurd since $f \in M$ and $c \notin M$. So $\gamma = 1$ and we conclude that $i \in C(T)$. But i is an involution, and T consists of strongly real elements, so we contradict Lemma 1.8. □

PROPOSITION 4.6. $C^\circ(T) = T$.

PROOF. If $C^\circ(T) \le M$, then by Lemma 3.8 we have the result. Suppose toward a contradiction that $C^\circ(T) \not\le M$ and let $Y_0 = C^\circ(T) \setminus M$. Then $\mathrm{rk}(Y_0) = \mathrm{rk}(C(T))$.

We showed in preceding lemmas that $I_M^+M^\circ$ is a generic subset of G and that it is disjoint from $Y = C^\circ(A)Y_0C^\circ(A)$. So it suffices to carry out a rank computation showing that Y is generic to reach a contradiction.

Since $\mathrm{rk}(G) = \mathrm{rk}(Y_0) + 2\,\mathrm{rk}(C(A_0))$ it suffices to show that the representation of the elements of Y is unique. So consider a relation $c = uc'v$ where $u, v \in C^\circ(A)$ and $c, c' \in Y_0$; we claim that $u = v = 1$.

Consider the group $X = [v, T]$. We will show that $X = 1$. We have $T^v \leq M$ and also $T^v = T^{u^{-1}c} \leq M^c$, and since also $T \leq M \cap M^c$ we have $X \leq M \cap M^c$. Now $c \notin M$ and so by Lemma 3.7, $\langle X, T\rangle \leq (M \cap M^c)^\circ$ is abelian. Hence by Lemma 3.8, w inverts X; as $X \leq C^\circ(A)$, by Lemma 1.8 $X \leq A$, and as $[X, T] = 1$ therefore $X = 1$. So $v \in C_{M^\circ}(A, T) = C_T(A) = 1$ in view of Lemma 3.10, and then as $c = uc'$ similarly $u = 1$, and so $c = c'$.

Thus $C^\circ(A)Y_0C^\circ(A)$ is a second generic subset disjoint from the first, and we have a contradiction. □

So the situation now is as follows.

LEMMA 4.7.

(1) $\mathrm{rk}(G) = \mathrm{rk}(T) + 2\,\mathrm{rk}(C(A))$.
(2) *For* $g \in G \setminus N_G(T)$ *we have* $T \cap T^g = 1$.
(3) *For any* $g \in G \setminus M$, $\mathrm{rk}(M \cap M^g) \geq \mathrm{rk}(T)$.

PROOF. As $\mathrm{rk}(C(T)) = \mathrm{rk}(T)$ we obtain the first point.

For the second, if $H = T \cap T^g$ is nontrivial, then w inverts $\langle T, T^g\rangle$ by Lemma 3.8 and thus $T^g \leq C^\circ(T) = T$.

For the last, we have $\mathrm{rk}(M\cap M^g) = 2\,\mathrm{rk}(M) - \mathrm{rk}(MM^g) \geq 2[\mathrm{rk}(C(A)) + \mathrm{rk}(T)] - \mathrm{rk}(G) = \mathrm{rk}(T)$ □

4.3. Double transitivity. We can now enter into the final stage of the argument which leads to a direct verification of the hypotheses of the classification theorem of DeBonis-Nesin, Theorem III 2.2.

The first issue is double transitivity of the permutation representation of G on $M\backslash G$. As M is the stabilizer of one point, double transitivity amounts to the "Bruhat decomposition"

$$G = M \cup MgM$$

for one, or any, element of $G \setminus M$. Since G is connected, an equivalent and more accessible statement is that MgM is generic in G for any $g \in G \setminus M$. In fact, we will show that $C^\circ(A)gM$ is generic for such g. We will need a little preparation first.

LEMMA 4.8. *For* $i \in I(G) \setminus M$, *i does not centralize* $(M \cap M^i)^\circ$.

PROOF. We suppose the contrary. Let $H = (M \cap C(A)^i)^\circ$. Then i centralizes H and H centralizes A^i and i, hence also A, and by hypothesis $(*)$ H is finite and hence trivial. That is, $M \cap C(A)^i$ is finite.

We consider the natural map $\theta : C^\circ(A)\times M \to C^\circ(A)iM$. As $M\cap C^\circ(A)^i$ is finite, one finds directly that this map has finite fibers and hence by a rank computation $C^\circ(A)iM$ is generic in G.

On the other hand, for $w \in I_M^+$, we also have $M \cap C^\circ(A)^w$ finite as the connected component of $(M\cap C^\circ(A))^w$ is inverted by w and commutes with A^w, so that Lemma 1.8 applies. So by the same token, $C^\circ(A)wM$ is generic in G and thus $MiM = MwM$.

We have $mi = wm'$ for some $m, m' \in M$ and hence $T^{mi} = T^{m'} \leq M \cap M^i$. So by hypothesis i centralizes T^{mi}, and as the elements of this group are strongly real, and are not involutions, we contradict Lemma 1.8. □

The following is essentially the statement of double transitivity.

LEMMA 4.9. *For any $g \in G \setminus M$, the group $C^\circ(A) \cap M^g$ is finite.*

PROOF. Suppose $H = (C^\circ(A) \cap M^g)^\circ \neq 1$ and let $i \in A^\times$. Then $H \leq M^g \cap M^{gi}$ and $A \leq C(H)$. By hypothesis $(*)$ we have $A = U_2(C(H))$.

Now $X = (M^g \cap M^{gi})^\circ$ is abelian by Lemma 3.7 so X centralizes H and normalizes A. On the other hand, i normalizes X, so $[i, X] \leq X \cap A = 1$ and this contradicts the preceding lemma, applied to M^g in place of M. □

LEMMA 4.10. *For $g \in G \setminus M$ we have $G = M \cup M^\circ g M^\circ$.*

PROOF. As $C^\circ(A) \cap (M^\circ)^g$ is finite, we find by direct computation that

$$\mathrm{rk}(C^\circ(A) g M^\circ) = \mathrm{rk}(C(A)) + \mathrm{rk}(M^\circ) = \mathrm{rk}(G)$$

and thus $M^\circ g M^\circ$ is generic. As G is connected, our claim follows. □

PROPOSITION 4.11.

(1) *The action of G on $M \backslash G$ is doubly transitive.*
(2) *For $g, h \in G \setminus M$, the groups $M \cap M^g$ and $M \cap M^h$ are conjugate in G.*

PROOF. Since M^g is the point stabilizer of Mg, these two statements are equivalent. On the other hand, as we have noted earlier, the direct translation of double transitivity into group theoretic language is the condition

$$G = M \cup MgM$$

for $g \in G \setminus M$, a special case of the preceding lemma. □

LEMMA 4.12. *M is connected.*

PROOF. Observe that $C^\circ(A) \cap M^w \leq N_{M^\circ}(T) = T$ by Lemma 3.10. So we have $C^\circ(A) \cap M^w \leq C_T(A) = 1$ and the map $C^\circ(A) \times M \to C^\circ(A) w M$ is injective. However, the image is generic in G and hence has Morley rank 1, so M has Morley rank 1. □

At this point, a great deal of our notation collapses; in particular, $T = M \cap M^w$ and $M = C(A) \rtimes T$. Since $M \cap M^g$ is a conjugate of T for any $g \in G \setminus M$, all intersections of this form are connected. Furthermore, $N_M(T) = N_{M^\circ}(T) = T$.

We now have ample information to conclude our analysis.

4.4. Recognition. Finally, we may prove a recognition theorem.

THEOREM 4.13. *Let G be a simple L^*-group of finite Morley rank and even type with a definable strongly embedded subgroup. Assume that G also satisfies hypothesis $(*)$. Then G is of the form $\mathrm{PSL}(2, K)$ for some algebraically closed field K of characteristic 2.*

PROOF. We consider the permutation representation of G on cosets $M\backslash G$. This is a doubly transitive action for which the point stabilizer M splits as $C(A) \rtimes T$, and as $T > 1$ it is not sharply 2-transitive. Thus to apply Theorem III 2.2 it will suffice to check that the stabilizer of any three points is trivial.

The stabilizer of two points may be taken to be $T = M \cap M^w$. Suppose that $t \in T^\times$ stabilizes the point Mg, that is, $t \in M^g$, and $g \notin M$. We claim then that $M^g = M^w$.

We have $t \in T$ and $t \in M \cap M^g$, which is a conjugate of T. So by Lemma 4.7, we have $g \in N(T)$. On the other hand, as $G \setminus M = MwC^\circ(A)$ we may also take $g \in wC^\circ(A)$. So taking $g = wf$ with $f \in C^\circ(A)$, we find $f \in N_M(T) = T$, hence $f = 1$ and $g = w$, as claimed. □

5. Recognition: Weak embedding, I

In this section we will begin the "recognition" phase of our classification theorem in the case of weak but not strong embedding.

At the outset we will *not* invoke the usual favorable hypothesis:

(∗) Whenever A_1, A_2 are two distinct conjugates of $\Omega_1(O_2{}^\circ(M))$, the group $C_G(\langle A_1, A_2\rangle)$ is finite.

Instead, we will first clarify the structure of the Sylow 2-subgroup of G, without making use of hypothesis (∗), and only introduce it afterward for the finer analysis.

So let G be a simple L^* group of finite Morley rank and even type, and M a definable weakly embedded subgroup of G which is not strongly embedded in G. Also, we fix an offending involution α, we set $A = U_2(C_M(\alpha))$.

Now by Proposition 1.1, $S = U_2(M)$ is a Sylow° 2-subgroup of M, and evidently unique. As we saw in §2, the Landrock-Solomon analysis applies here, to give a catalog of possible structures for S; namely, apart from two nonabelian possibilities, S is a homocyclic abelian group in which either $A = \Omega_1(S)$, or S is elementary abelian and A is a "diagonal" subgroup with respect to the action of α.

The first step of the analysis will be to eliminate all "unusual" possibilities for S, leaving the homocyclic abelian case with $A = \Omega_1(S)$. One expects S to be elementary abelian but this is neither proved, nor significant, during the main part of the analysis. In fact, the good configuration, after some analysis which appears to be headed toward recognition, lead to a situation in which it can be seen that the subgroup M is in fact strongly embedded, which in the present context is a contradiction, and thus ends the analysis abruptly.

This section deals only with the less plausible possibilities for S, and we will return to the main line in the following section.

5.1. The nonabelian case. In this section we operate under the following hypothesis:

(NA) S is nonabelian.

In this case, the Landrock-Solomon analysis gives two possible structures for S, explicitly.

We seek a contradiction, and we focus primarily on the nature of the conjugacy classes of involutions in M and in G. We aim to show in particular that the involutions in $S \setminus A$ are not conjugate in G to involutions of A, and that the involutions in $M \setminus S$ are all conjugate, at which point we will be able to get a contradiction, by comparing involutions outside M, but in the conjugacy class of A, with those inside M.

Recall that $S_1 = [\alpha, S]$ is abelian, and its involutions lie in A, while $S \setminus S_1$ also contains involutions. Also, $A = \Omega_1(S_1) = Z(S)$, and $S_1 \triangleleft S$. Furthermore, $C_S(\alpha) = Z(S)$ (Corollary III 4.6) and as this is not true for $i \in S$, the involutions of S are not offending. In particular, S is not a full Sylow 2-subgroup of M.

We begin with a few preparatory lemmas.

LEMMA 5.1. *All involutions in the coset αS are conjugate under the action of S.*

PROOF. This may be phrased as follows: if $s \in S$ and α inverts s, then $s \in [\alpha, S]$. Now we recall that in coordinates (from an associated field K) we parametrized S by triples (a, b, c), with the action of α given by

$$(a, b, c)^\alpha = (a, a + b, a + b + c + \sqrt{ab})$$

and the multiplication given by addition in the first two coordinates, so if $s = (a, b, c)$ is inverted by α, then $a = 0$; as we also checked that $[\alpha, S]$ consists of the triples $(0, b, c)$, our claim holds. □

LEMMA 5.2. *Let $\beta \in I(M)$ be an offending involution, that is, $C(\beta) \not\leq M$, and $L_\beta = U_2(C(\beta))$. If $L_\alpha \leq L_\beta$, then $L_\alpha = L_\beta$.*

PROOF. For any offending involution $\beta \in M$, the group L_β is a group of type PSL_2, and setting $A_\beta = U_2(L_\beta \cap M)$, we have $A_\beta = Z(S) = A$.

Now $\mathrm{rk}(L_\beta) = 3\,\mathrm{rk}(A_\beta) = 3\,\mathrm{rk}(A) = \mathrm{rk}(L_\alpha)$ and our claim follows. □

We will have to go beyond the Sylow° 2-subgroup S to a full Sylow 2-subgroup. So let $\hat{S}$ be a Sylow 2-subgroup of M containing $\langle S, \alpha \rangle$. Then $\hat{S}^\circ = S$.

LEMMA 5.3. *$\hat{S} = S \rtimes C_{\hat{S}}(L)$.*

PROOF. We first consider the subgroup $R \leq \hat{S}$ which is the preimage of $C_{\hat{S}/S}(\alpha)$. For $r \in R$, we have $\alpha^r \in \alpha S$ and hence by Lemma 5.1 we have $\alpha^r = \alpha^s$ for some $s \in S$, and $rs^{-1} \in C(\alpha)$. Then rs^{-1} acts on L by an inner automorphism, in view of CorollaryII 2.26, and being a 2-element it must act like an involution $i \in A$, that is, $rs^{-1}i \in C(L)$. So $r \in SC_R(L)$. On the other hand, $C_S(T) = 1$ by Proposition III 4.5. So we have $R = S \rtimes C_R(L)$. We will adjust α in the remainder of the argument to arrive at the same situation, but with $R = \hat{S}$.

Now consider the group R_1 defined as the preimage in $\hat{S}$ of $\Omega_1(Z(\hat{S}/S))$. Then $S < R_1 \leq R$. In particular, $R_1 = S \rtimes C_{R_1}(L)$. Take $\beta \in C_{R_1}(L)$, nontrivial. Then $\beta^2 \in C_S(L) = 1$, so β is an involution, and $L \leq L_\beta$, so by Lemma 5.2 we have $L = L_\beta$.

Now considering β in place of α, we repeat the first part of our argument with a new subgroup R_β replacing R, namely the preimage in $\hat{S}$ of $C_{\hat{S}/S}(\beta)$. But, by the choice of β, we now have $R_\beta = \hat{S}$, so we have the desired decomposition. □

LEMMA 5.4. *The only involution in $C_{\hat{S}}(L)$ is α.*

PROOF. We may suppose that α belongs to $Z(C_{\hat{S}}(L))$. We take $\beta \in I(C_{\hat{S}}(L))$. Now α and β commute with T and with each other. Furthermore, $S_1 = [\alpha, S]$ is the unique maximal proper $(\langle\alpha\rangle \times T)$-invariant subgroup of S, and hence is β-invariant. So if we let S_β be the maximal proper $(\langle\beta\rangle \times T)$-invariant subgroup of S, we have $S_1 \leq S_\beta$ and by the same argument $S_\beta \leq S_1$. But α inverts S_1 by Proposition III 4.9 and β inverts $S_\beta = S_1$, so $\alpha\beta$ centralizes S_1. Now if $\alpha\beta$ is an involution $L_{\alpha\beta} = L$ and we have a contradiction, so $\alpha\beta = 1$ and $\beta = \alpha$. □

LEMMA 5.5. *$I(M) \setminus I(S)$ consists of a single conjugacy class in M.*

PROOF. We noted earlier that the offending involutions lie outside S and hence the set in question is not empty.

It suffices to show that any offending involution β in $\hat{S}$ is conjugate to α. We have the decomposition $\beta = sb$ with $s \in S$ and $b \in C_{\hat{S}}(L)$. Then $b^2 \in C_S(L) = 1$, so b is an involution and hence $b = \alpha$, $\beta \in \alpha S$, and Lemma 5.1 applies. □

LEMMA 5.6. *The involutions of A are not conjugate to involutions in $S \setminus A$, under the action of G.*

PROOF. Suppose $i \in I(A)$, $j \in I(S)$, and $j^g = i$. Then $A^g \leq C(i) \leq M$ (Lemma 2.3), and thus $g \in M$ by weak embedding. Then as $A \lhd M$, we have $j \in A$. □

Now we have sufficient information to reach a contradiction in this case, eliminating the nonabelian case.

PROPOSITION 5.7. *Let G be a simple L^*-group of finite Morley rank and even type, with a definable weakly embedded subgroup that is not strongly embedded. Then the Sylow° 2-subgroups of G are abelian.*

Proof. Suppose the contrary. Consider two involutions $i, w \in I(G)$ chosen so that $i \in I(S)$, $i \notin A$, and $w \notin M$, but w is conjugate to an involution in A. Then as we have just shown, these two involutions are not conjugate in G. By the Basic Fusion Lemma I 2.20, there is a third involution j commuting with both.

Now as $i \in S$, the involution i is not offending, by Lemma 2.3, So $j \in M$. On the other hand, as $w \notin M$, j must be an offending involution in M, and hence $C_S(j) = A$; but $i \in C_S(j) \setminus A$, a contradiction. □

5.2. The diagonal case. There is a second unusual case to be eliminated before we can return to the main line, the diagonal case. Our assumption here is the following:

(Dg) $S = E \oplus E^\alpha$ with E and E^α definable T-invariant elementary abelian groups; $A = \{xx^\alpha : x \in E\}$. Moreover, T acts regularly on A, E, and E^α.

The analysis is different from the preceding case, and begins with some structural analysis in the vein of a recognition argument, terminating however with an early contradiction relating to the structure of the quotient $\overline{M}^\circ = M^\circ / C_{M^\circ}(S)$, which turns out to be extremely tight.

Lemma 5.8. *All elements of $S^\times$ are conjugate under the action of M°.*

Proof. For $i \in S^\times$ we have $\mathrm{rk}(i^{M^\circ}) = \mathrm{rk}(M) - \mathrm{rk}(C_M(i)) = \mathrm{rk}(i^M)$ so it suffices to show that $i^M = S^\times$, as then i^M has Morley degree one and can contain only one M°-conjugacy class of full rank. So we concern ourselves only with conjugacy in M.

So suppose $S^\times$ contains involutions not conjugate in M. If $i, j \in I(S)$ are conjugate in G, then they are conjugate in M: from $i^g = j$ we derive $S^g \le C(j) \le M$ (Lemma 2.3), and by weak embedding $g \in M$. Thus S meets at least two distinct conjugacy classes of involutions in G.

Now fix involutions i and j with $i \in A$ and with j conjugate to an involution in S which is not conjugate to i, but with $j \notin M$. Then there is a third involution w commuting with both i and j, by Lemma I 2.20. As w commutes with i, and i is not offending (Lemma 2.3), we have $w \in M$. On the other hand, $j \notin M$, so w is offending. Since w can play the role of α here, we will let $L = U_2(C(w))$, but only until the end of the present proof.

Now $j \notin L$ since j is not conjugate to i, and A is a Sylow 2-subgroup of L. On the other hand, j acts on L via an inner automorphism by Corollary II 2.26.

So there is an involution $j' \in L$ such that $jj' \in C(L)$. In particular, j and j' commute, and their product is an involution. Furthermore, $jj' \in M$.

Now j belongs to some conjugate S_1 of S, and so does j'. So each of these elements belongs to U_2 of its own centralizer, and neither can be offending, relative to any conjugate of M in which it sits. Let M_1 be the conjugate of M containing S_1. Then $j \in M_1$ and j commutes with j', so $j' \in M_1$. As

j' is not offending, there is a connected 2-subgroup of M_1 containing j' and thus $j' \in S_1$ as well, and finally $jj' \in S_1$, so jj' is not offending either; but jj' commutes with L and we have a contradiction. $\square$

NOTATION 5.9. *Fix an involution $w \in L$ which inverts T.*

LEMMA 5.10. *If A_1 is a 2-unipotent subgroup of S properly containing A, then $C_G(A_1, A^w) = 1$.*

PROOF. Let $x \in C_G(A_1, A^w)$ be nontrivial. As usual $L_x = U_2(C(x))$ is then a group of type PSL_2. This group contains $L = \langle A, A^w \rangle$ properly, since $A_1 \not\leq L$. Let T_1 be a maximal torus of L_x containing T. Then T_1 is the multiplicative group of a field, and T is the multiplicative group of a subfield, all of this interpretable within G. As the fields in question have finite Morley rank, they are equal by Lemma I 4.3. This forces $A = A_1$. $\square$

It is surprising that we can prove the following at this early stage.

LEMMA 5.11. *$C^\circ(L) = 1$; hence $C^\circ(\alpha) = L$.*

PROOF. We suppose $X = C^\circ(L) \neq 1$. Then X contains no involutions: any such involution would be offending in any conjugate of M containing it, forcing X to be of degenerate type in view of the structure of centralizers of offending involutions (Lemma 2.3). Thus by Theorem IV 4.1 the group X contains no involutions.

Now we wish to show that S can be viewed as a 2-dimensional vector space over some field whose multiplicative group can be identified with T. So we consider T as a subgroup of the group of units in $\mathrm{End}(S)$; as T acts regularly on various subgroups of S, this is at least a faithful representation.

Let R be the subring of $\mathrm{End}(S)$ generated additively by T. As the group E is T-invariant, it is also R-invariant. So consider the restriction map $\rho : R \to \mathrm{End}(E)$. On E, the group T acts like the multiplicative group of a field $K = \rho[R]$, so it suffices to check that ρ is injective.

Suppose $r \in R$ annihilates E. As α commutes with T, it commutes with R and thus r annihilates E^α. But as $S = EE^\alpha$ it follows that $r = 0$.

Thus writing K in place of R, S becomes a K-vector space and $K^\times = T$. Evidently S is 2-dimensional.

Now X commutes with T and hence acts linearly on S. We claim that this action is faithful. Indeed, $C_X(S) \leq C(S, A^w) = 1$ by the preceding lemma.

By Proposition II 4.5, X is a K-group. As a connected $2^\perp$-group X is therefore solvable. By Proposition I 11.7, X is a good torus.

We consider the abelian group XT and the subgroup $X_1 = C_{XT}(S/A)$. Now we have the structure of S as a T-module explicitly, and T acts regularly on the quotient S/A. As X is nontrivial, the group X_1 is nontrivial. Furthermore, $V = C_S(X_1)$ covers S/A by Proposition I 9.9.

Now in view of its definition, the group V is $(\langle \alpha \rangle \times T)$-invariant. In particular, $C_V(\alpha)$ is nontrivial and T-invariant; but $C_S(\alpha) = A$, so it follows that $C_V(\alpha) = A$ and $A \leq V$. Since V covers S/A, we have $V = S$, that is, X_1 centralizes S.

As X_1 commutes with X, it acts on $U_2(C(X)) = L$. Since X_1 contains no involutions and centralizes A, it acts trivially on L. As X_1 also centralizes S, the previous lemma shows that $X_1 = 1$. This contradiction shows that $C^\circ(L) = 1$.

As $C^\circ(\alpha) = L \times D$, we find $D = 1$, and $C^\circ(\alpha) = L$. □

Now we arrive at a conclusion parallel to one reached in the case of strong embedding, by another route.

LEMMA 5.12. *The involution w inverts $C^\circ(T)$.*

PROOF. Supposing the contrary, $X = C^\circ(w, T)$ is nontrivial by Lemma I 10.3. The connected group X is normalized by α and $X \cap L = 1$. Accordingly, $C_X(\alpha)$ is finite and α inverts X. In particular, X is abelian.

Let M_w be the conjugate of M containing w, and S_w the corresponding conjugate of S. We have $X \leq M_w$.

We will prove the following.

If $X_0 \leq X$ is nontrivial, definable, and connected, then $C_{S_w}(X_0)$ is finite.

Suppose the contrary. As α inverts X, it normalizes X_0, and as α commutes with w, it normalizes S_w, so α acts on $C_{S_w}(X_0)$ and hence centralizes a nontrivial connected subgroup of $C_{S_w}(X_0)$. Let A_w be the conjugate of A containing w; this is a Sylow 2-subgroup of L, and is $C_{S_w}{}^\circ(\alpha)$. Now X_0 centralizes $\langle T, C_{A_w}(X_0)\rangle$ and this group is clearly nonsolvable, as T does not normalize A_w, so by Fact II 4.6 this group is L. But $C^\circ(L) = 1$ by Lemma 5.11 and this would force $X_0 = 1$. This contradiction proves our claim.

It follows that $C_{S_w}(X)$ is a finite group, but nontrivial as it contains w. Now work in the group $\overline{M_w}^\circ = M_w{}^\circ / C_{M_w{}^\circ}(S_w)$, and specifically with the groups $\bar{X}$ and $\bar{T}_w$ acting on S_w, where T_w is a torus of L normalizing A_w. By Lemma IV 1.6 the Borel subgroups of $\overline{M_w}^\circ$ are abelian and conjugate, so there is a conjugate Y of X under the action of $M_w{}^\circ$ such that $\bar{Y}$ commutes with $\bar{T}_w$.

So $\bar{T}_w$ acts on $C_{S_w}(Y)$, and this group is nontrivial and finite. So T_w centralizes $C_{S_w}(Y)$. This contradicts the action of T on S. □

LEMMA 5.13. $N_M{}^\circ(T) = T$.

PROOF. By Proposition I 11.7, T is a good torus and thus $N_M{}^\circ(T) = C_M{}^\circ(T)$.

We let $H = C_M{}^\circ(T)$. By the previous lemma, w inverts H, and in particular H is abelian. It follows that $C_H(S) = C_H(S, S^w)$ and hence by Lemma 5.10, $C_H(S) = 1$. In view of the action of T on S, $H \cap S = 1$. So H is of degenerate type. Being connected, it has no involutions (Theorem IV 4.1—or more directly, as H is abelian). By Proposition I 11.7 again, H is a good torus.

Suppose $H > T$. As H is a good torus, it must contain torsion elements that are not in T. Moreover, T is a full algebraic torus in characteristic 2, and H has no involutions. So there is a prime p for which the Prüfer p-rank of H is at least 2.

We noted that $C_H(S) = 1$, so the action of H is faithful. Let V be an H-minimal subgroup of S. By Proposition I 4.11 H acts like a subgroup of the multiplicative group of a field on V and by our Prüfer rank condition it follows that $H_0 = C_H{}^\circ(V)$ is nontrivial.

Now w inverts H_0 and hence normalizes $C(H_0)$, so $\langle V, V^w\rangle \leq C^\circ(H_0)$. Then as usual $L_0 = U_2(C(H_0))$ is a group of type PSL_2 in characteristic 2, and w acts on this group, by an inner automorphism in view of Fact II 2.25. So w centralizes some Sylow 2-subgroup A_1 of L_0. Evidently A_1 is contained in the Sylow° 2-subgroup S_w containing w. So H_0 normalizes S_w. As w inverts H_0 we have $[w, H_0] \leq S_w \cap H_0 = 1$. As H_0 contains no involutions this is impossible. □

We are now ready to take up the consideration of $\overline{M}^\circ = M^\circ/C_{M^\circ}(S)$, and we intend to argue that for various reasons this group both does, and does not, contain involutions, which will provide a sufficient contradiction.

By Lemma IV 1.6 as usual, the Borel subgroups of M° are good tori, and are conjugate.

LEMMA 5.14. *$\bar{T}$ is a Borel subgroup of $\overline{M}^\circ$.*

PROOF. T is an almost self-normalizing good torus in M° and hence has the same properties in the subgroup $M_1 = TC_{M^\circ}(S)$. By Lemma IV 1.5 any almost self-normalizing good tori in M_1 are conjugate, so by a Frattini argument $N_M{}^\circ(M_1) \leq M_1 N_M{}^\circ(T) = M_1$. Hence $\bar{T}$ is almost self-normalizing in $\overline{M}^\circ$ and thus is again a Borel subgroup there. □

LEMMA 5.15. *$\overline{M}^\circ$ contains no involutions.*

PROOF. By the previous lemma $\overline{M}^\circ$ is of degenerate type, and Theorem IV 4.1 applies. □

Now we take matters from another direction and produce an involution in $\overline{M}^\circ$, by examining the action of this group on the cosets of $\bar{T}$.

LEMMA 5.16. *For any $i \in S^\times$, $C_{\overline{M}^\circ}(i)$ is finite.*

PROOF. If this group is infinite, extend a Borel subgroup of $C_{\overline{M}^\circ}(i)$ to a Borel subgroup of $\overline{M}^\circ$ and after conjugating one may assume this Borel subgroup is contained in $\bar{T}$, which however acts freely on S, giving a contradiction. □

LEMMA 5.17. *The intersection of two distinct Borel subgroups of $\overline{M}^\circ$ is finite.*

PROOF. Let $\bar{X}$ be a connected group contained in two distinct Borel subgroups of $\overline{M}^\circ$, and $\bar{H} = C_{\overline{M}^\circ}{}^\circ(\bar{X})$; then $\bar{H}$ contains these two Borel subgroups and is therefore nonsolvable. Let $V \leq S$ be $\bar{H}$-minimal. Recall that the Borel subgroups of $\overline{M}^\circ$ act freely on S, so $\bar{X}$ acts faithfully on V, and is central in $\bar{H}$. By Proposition I 4.11 we can view the action of $\bar{H}$ on V as linear. By Proposition II 4.5, since $\bar{H}$ is of degenerate type it follows that the faithful quotient $\bar{H}/C_{\bar{H}}(V)$ is solvable. Hence $C_{\bar{H}}{}^\circ(V)$ must be nontrivial.

Now extending a Borel subgroup of $C_{\bar{H}}{}^\circ(V)$ to a Borel subgroup of $\overline{M}^\circ$, which after conjugation we may take to be $\bar{T}$, we have a contradiction since $\bar{T}$ acts freely on S. □

LEMMA 5.18. $\mathrm{rk}(\overline{M}^\circ) = 2\,\mathrm{rk}(T)$

PROOF. $\overline{M}^\circ$ acts transitively on $S^\times$ by Lemma 5.8, and the centralizer of each element of S is finite. So $\mathrm{rk}(\overline{M}^\circ) = \mathrm{rk}(S) = 2\,\mathrm{rk}(T)$. □

LEMMA 5.19. $N_{\overline{M}^\circ}(\bar{T}) = \bar{T}$.

PROOF. As $\bar{T}$ is a Borel subgroup of $\overline{M}^\circ$, $W_0 = N_{\overline{M}^\circ}(\bar{T})/\bar{T}$ is finite. By Lemma IV 2.3, any element of $N_{\overline{M}^\circ}(\bar{T}) \setminus \bar{T}$ has finite centralizer in $\bar{T}$.

Now $\bar{T}$ is a full algebraic torus of dimension 1 in characteristic 2. By Lemma I 10.18 it has no automorphism of odd prime order with a finite centralizer. So W_0 is a finite 2-group. Again, $\overline{M}^\circ$ contains no involutions, so $W_0 = 1$. □

LEMMA 5.20. *$\overline{M}^\circ$ contains involutions.*

PROOF. We will show that the action of $\overline{M}^\circ$ on the cosets of $\bar{T}$ is doubly transitive; a doubly transitive permutation group of finite Morley rank must have involutions, in view of Lemma I 2.18.

By the preceding lemma, for any $\gamma \in \overline{M}^\circ \setminus \bar{T}$, the groups $\bar{T}$ and $\bar{T}^\gamma$ are distinct Borel subgroups, and hence have finite intersection by Lemma 5.17. Then a direct computation shows

$$\mathrm{rk}(\bar{T}\gamma\bar{T}) = 2\,\mathrm{rk}(\bar{T}) \ .$$

Since $2\,\mathrm{rk}(\bar{T}) = 2\,\mathrm{rk}(T) = \mathrm{rk}(\overline{M}^\circ)$, there is a unique double coset $\bar{T}\gamma\bar{T}$ apart from $\bar{T}$, and thus the action is doubly transitive. □

So as the diagonal case (Dg) has also been eliminated, we may state the following.

PROPOSITION 5.21. *Let G be a simple L^*-group of finite Morley rank and even type, with a definable weakly embedded subgroup that is not strongly embedded. Then the Sylow° 2-subgroups S of G are abelian homocyclic, and the subgroup A associated to an offending involution normalizing S is $\Omega_1(S)$.*

6. Recognition: Weak embedding, II

In this section we will complete the "recognition" phase of our classification theorem in the case of weak but not strong embedding, under the usual favorable hypothesis.

(∗) Whenever A_1, A_2 are two distinct conjugates of $\Omega_1(O_2{}^\circ(M))$, the group $C_G(\langle A_1, A_2\rangle)$ is finite.

This hypothesis was *not* invoked in the previous section, and is not immediately needed here either, so we *postpone* that assumption briefly, as we intend to make further use of results not depending on this assumption.

For the present, G is a simple L^* group of finite Morley rank and even type, and M a definable weakly embedded subgroup of G which is not strongly embedded in G. The previous section has whittled down the possibilities offered by the Landrock-Solomon analysis to the ordinary homocyclic abelian case, that is, where $A = \Omega_1(S)$ in our customary notation, in which S is a Sylow° 2-subgroup, α is an offending involution normalizing S, and $A = U_2(C(\alpha) \cap M)$.

We will carry along the rest of our usual notation: L for $U_2(C(\alpha))$, T for a torus in L normalizing A, and w for a fixed involution in L inverting T.

Since our target group $\mathrm{PSL}_2(K)$ contains no offending involutions this case will also arrive eventually at a contradiction, but in reality the bulk of the analysis simply drives us toward $\mathrm{PSL}_2(K)$, with a belated recognition of the incompatibility of the data with our initial assumption of the failure of strong embedding. In fact, it is clear that offending involutions in M lie outside M°, so that M cannot be connected, and if we follow our previous path we will eventually conclude that M is connected and reach a contradiction at this point. This is precisely what occurs.

6.1. Structure of M**.** As T acts transitively on $A^\times$, for any $i \in A^\times$ we have $M = C(i) \rtimes T$, and similarly for M°.

LEMMA 6.1. *For* $i \in A^\times$ *we have* $C_{M^\circ}(i) = C_M{}^\circ(A)$.

PROOF. As T acts transitively on $A^\times$ we have $M^\circ = C_{M^\circ}(i) \rtimes T$ and as M° has Morley degree 1 it follows that $C_{M^\circ}(i)$ is connected, that is,

$$C_{M^\circ}(i) = C_M{}^\circ(i) \; .$$

We claim that $C_M{}^\circ(i) = C_M{}^\circ(A)$.

If this fails, we work in $\overline{M}^\circ = M^\circ / C_{M^\circ}(A)$. This is a degenerate type group acting faithfully on A, so by Lemma IV 1.6 its Borel subgroups are conjugate good tori.

If $C_M{}^\circ(i) > C_M{}^\circ(A)$, then $\overline{C_M{}^\circ(i)}$ is nontrivial and thus some Borel subgroups of $\overline{M}^\circ$ do not act freely on A. To get a contradiction, it suffices to show that $\bar{T}$ is a Borel subgroup. Now $\bar{T}$ is contained in a Borel subgroup $\bar{B}$ which is again abelian, and fixing an involution $i \in A^\times$ we have $\bar{B} = C_{\bar{B}}(i) \times T$; since T acts transitively on A we find $C_{\bar{B}}(i) = C_{\bar{B}}(A) = 1$. So $\bar{B} = \bar{T}$ is a Borel subgroup and we have our contradiction, proving the claim. □

LEMMA 6.2. *For $i \in A^\times$ we have $C(i) = C(A)$. Hence $M = C(A) \rtimes T$.*

PROOF. As $M = C(i) \rtimes T$ we deal only with the first claim, and as $i \in U_2(C(i))$ it is not an offending involution, so $C(i) = C_M(i)$ and it suffices to consider the group $C_M(i)$.

Let $\bar{M} = M/C(A)$. By the preceding lemma we have $\bar{M}^\circ = \bar{T}$ acting on A like the multiplicative group of a field on a 1-dimensional vector space.

Put an additive structure on $\bar{T}$ by the condition $(s+t)a = sa + ta$ for $s, t \in \bar{T}$, $a \in A$ (clearly this is coherent). Then $\bar{M}/\bar{T}$ acts on $\bar{T}$ and respects addition since for $x \in C(i)$ we have $(s+t)^x i = (s+t)^x i^x = (si+ti)^x = (s^x + t^x)i$. So $\bar{M}/\bar{T}$ induces a group of field automorphisms, which is trivial by Lemma I 4.5. This means that $\bar{M}$ centralizes $\bar{T}$.

So now from $\bar{M} = C_{\bar{M}}(i) \times \bar{T}$ and the transitivity of $\bar{T}$ on $A^\times$, we find $\bar{M} = \bar{T}$. As $C_{\bar{T}}(i) = 1$ our claim follows. □

LEMMA 6.3. *Let $i, j \in I(G)$ be involutions conjugate to involutions in A, and let $g \in G$ be an element which commutes with i and is inverted by j. Then i and j commute, and $g^2 = 1$.*

PROOF. We may take $i \in A^\times$. Then $g \in C(i) = C(A)$. Now either $U_2(C(g)) = A$ or $U_2(C(g))$ will be a group of type PSL_2. In either case $U_2(C(g))$ is normalized by j and it follows that j centralizes some involution i' in a conjugate of A commuting with g, so we might as well assume that j centralizes i. Then again $j \in C(A)$ and if A_1 is the conjugate of A containing j, it follows symmetrically that A centralizes A_1. So $A_1 \leq \Omega_1(S) = A$ and j commutes with g; g is an involution. □

6.2. The hypothesis $(*)$. We now invoke our favorable hypothesis:

$(*)$ Whenever A_1, A_2 are two distinct conjugates of $\Omega_1(O_2{}^\circ(M))$, the group $C_G(\langle A_1, A_2 \rangle)$ is finite.

From this point to the end of the section, this will be assumed.

LEMMA 6.4. *For any $g \in G \setminus M$, the group $(M \cap M^g)^\circ$ is abelian.*

PROOF. We consider the commutator subgroup $H = [(M \cap M^g)^\circ]'$, which is connected. By Lemma 6.2, H is contained in $C(A)$ and similarly in $C(A^g)$, so by hypothesis $(*)$ H is finite, and being connected is trivial. □

In the next lemma, while one can certainly take $w_0 = w$ and $T_0 = T$, there will be other examples subsequently.

LEMMA 6.5. *Let w_0 be an involution lying in a conjugate of A, T_0 a connected subgroup of M inverted by w_0, and suppose that T_0 acts regularly on $A^\times$. Then for any nontrivial subgroup H of T_0 we have the following:*

(1) *w_0 inverts $C^\circ(H)$,*
(2) *$C_M{}^\circ(H) = T_0$.*

PROOF. Observe that our hypotheses imply $w_0 \notin M$.

We show w_0 inverts $C^\circ(H)$.

Otherwise, by Lemma I 10.3, the group $X = C^\circ(H, w_0)$ is nontrivial.

Let M_0 be the conjugate of M containing w_0, and A_0 the corresponding conjugate of A. Then $H \le C(A_0)$ by Lemma 6.2. Accordingly, the hypothesis $(*)$ implies $C(H) \le M_0$. Hence $H \le M_0$; but w_0 inverts H and thus $[w_0, H] \le A_0 \cap H = 1$, which forces H to consist of involutions. But then T_0 contains involutions, and being connected has an infinite Sylow 2-subgroup (either by Theorem IV 4.1 or more directly, since T_0 is abelian). As $T_0 \le M \cap M^{w_0}$ this contradicts weak embedding.

Next, we show that $C_M{}^\circ(H) = T_0$.

Otherwise, as T_0 is transitive on A and $C(i) = C(A)$ for $i \in A^\times$, we have $C_{C_M{}^\circ(H)}{}^\circ(A)$ nontrivial, and as w_0 inverts this group, by our first point, we get an infinite group commuting with A and A^{w_0}, violating our hypothesis $(*)$. □

LEMMA 6.6. *Any two distinct conjugates of T in M have trivial intersection.*

PROOF. If $H = T_1 \cap T_2$ is a nontrivial intersection of two conjugates of T in M, then as we just showed, $C_M{}^\circ(H) = T_1$ and $C_M{}^\circ(H) = T_2$. □

LEMMA 6.7. $N_{M^\circ}(T) = T$.

PROOF. Let $\hat{T} = N_{M^\circ}(T)$. Then $\hat{T} = C_{\hat{T}}(A) \rtimes T$. Let $H = C_{\hat{T}}(A) = C(A) \cap N(T)$. We claim $H = 1$.

Now $[H, T] \le C(A) \cap T = 1$. Once more we encounter the configuration of Lemma IV 2.3. Any nontrivial element of H would be required to have finite centralizer in T, so $H = 1$. □

LEMMA 6.8. *There are no offending involutions in M°.*

PROOF. It suffices to show that our (arbitrarily fixed) offending involution α is not in M°. Since it centralizes T, if α were in M°, then the last lemma would force α into T. □

Of course, at this point we also know that M is not connected, a fact to be held in reserve until it is eventually contradicted.

6.3. The rank of G. We have enough general structural information, and we can begin to focus more on involutions and a computation of $\mathrm{rk}(G)$.

LEMMA 6.9. *Let $g \in G \setminus M$. Then $M^\circ \cap M^g$ contains no involutions.*

PROOF. If i were such an involution, it would normalize both A and A^g and hence commute with nontrivial elements of each, forcing $i \in C(A, A^g)$. In particular, i is then an offending involution, hence outside M° by Lemma 6.8, a contradiction. □

LEMMA 6.10. *If $i \in I(G)$, $i \notin M$, and i is conjugate to an involution in A, then either i inverts $(M \cap M^i)^\circ$, or i centralizes $(M \cap M^i)^\circ$.*

PROOF. Let $H = (M \cap M^i)^\circ$. Then H is abelian by Lemma 6.4, and contains no involutions. So we have a factorization into connected subgroups $H = H^+ \times H^-$ where $H^\pm = \{h \in H : h^w = h^{\pm 1}\}$. We assume that the factor H^+ is nontrivial, and we show that the other factor is trivial.

Let A_i be the conjugate of A containing i. Then H^+ centralizes A_i by Lemma 6.2, and by the hypothesis $(*)$ we must have $C(H^+) \leq N(A_i)$, in particular $H^- \leq N(A_i)$ and thus $[i, H^-] \leq A_i \cap H^- = 1$, and $H^- = 1$. □

We now introduce the following sets of involutions.

NOTATION 6.11.

(1) *I_A is the set of involutions in G which are conjugate to elements of A.*
(2) *For $i \in I(G)$, $T(i) = \{x \in M^\circ : x^i = x^{-1}\}$.*
(3) *I_A^+ is the set of involutions i in I_A for which* $\mathrm{rk}(T(i)) \geq \mathrm{rk}(A)$.
(4) *$I_A^- = I_A \setminus I_A^+$.*

We will determine more precisely the action of involutions in I_A^+, show that this set of involutions is generic in I_A, and then compute the rank of G.

LEMMA 6.12. *For $i \in I_A^+$ we have the following:*

(1) *$T(i) = (M \cap M^i)^\circ$ and*
(2) *$T(i)$ acts regularly on A.*

PROOF. Consider the not necessarily connected group $H = M^\circ \cap (M^i)^\circ$. Here $H^\circ = (M \cap M^i)^\circ$. We show first that

$$H^\circ \leq T(i)\ .$$

We have shown in Lemma 6.9 that H contains no involutions, so we have a factorization $H = C_H(i)H^-$ where H^- is the *set* $\{h \in H : h^i = h^{-1}\}$, by Lemma I 10.4; we have a similar decomposition for H°. We have the corresponding rank computations: $\mathrm{rk}(H) = \mathrm{rk}(C_H(i)) + \mathrm{rk}(H^-)$, and similarly for H°. Of course $\mathrm{rk}(H) = \mathrm{rk}(H^\circ)$, and furthermore $\mathrm{rk}(C_H(i)) = \mathrm{rk}(C_{H^\circ}(i))$, so the remaining terms are equal as well:

$$\mathrm{rk}(H^-) = \mathrm{rk}((H^\circ)^-)\ .$$

Now our assumption on i is that $\mathrm{rk}(H^-) \geq \mathrm{rk}(A)$, hence we have $\mathrm{rk}((H^\circ)^-) \geq \mathrm{rk}(A)$. In particular, $(H^\circ)^-$ is nontrivial and hence by Lemma 6.10, i inverts H°. So $H^\circ \leq T(i)$.

Now $\mathrm{rk}(H^\circ) \geq \mathrm{rk}(A)$ and H° acts freely on A by Lemma 6.3. So as $A^\times$ has Morley rank 1, it must be a single orbit under H°, that is,

$$H^\circ \text{ acts regularly on } A^\times.$$

In particular, (A, H°) can be identified with a pair $(K_+, K^\times)$ with the action by multiplication. Now H acts on this situation and induces automorphisms of the field K, which by Lemma I 4.5 are trivial. That is, H° is central in H. If $g \in T(i)$ and $g \notin H^\circ$, then the group $\langle H^\circ, g\rangle$ contains some element that commutes with an involution in A, and as H° is centralized by g we find that the group $\langle H^\circ, g\rangle$ is inverted by i, contradicting Lemma 6.3. So $T(i) = H^\circ$ and all claims have been verified. □

LEMMA 6.13. $\mathrm{rk}(I_A) = \mathrm{rk}(I_A^+)$.

PROOF. As is customary in such cases, one aims at showing $\mathrm{rk}(I_A^-) < \mathrm{rk}(I_A)$.

We consider the natural map $I_A^- \to G/M^\circ$. The fibers of this map, by the definition of I_A^-, have ranks bounded strictly below $\mathrm{rk}(A)$, and so we have the estimate

$$\begin{aligned} \mathrm{rk}(I_A) &< \mathrm{rk}(A) + \mathrm{rk}(G) - \mathrm{rk}(M^\circ) \\ &= \mathrm{rk}(G) - [\mathrm{rk}(M^\circ) - \mathrm{rk}(T)] \\ &= \mathrm{rk}(G) - \mathrm{rk}(C(A))\ . \end{aligned}$$

So using $\mathrm{rk}(C(A)) = \mathrm{rk}(C(i))$ for $i \in A^\times$ (Lemma 6.2), we get $\mathrm{rk}(I_A) < \mathrm{rk}(i^G)$, and since $i^G = I_A$ this is as desired. □

What happens next was previously seen in §4.1.

LEMMA 6.14. *Suppose $w_1, w_2 \in I_A^+$. Then the following hold:*

(1) *If $T(w_1) \neq T(w_2)$ then $T(w_1) \cap T(w_2) = 1$,*
(2) *$T(w_1)$ and $T(w_2)$ are conjugate in M°, and hence under $C^\circ(A)$.*

PROOF. For the first point, suppose $H = T(w_1) \cap T(w_2) \neq 1$. Lemma 6.5 applies to both w_1 and w_2 with respect to T_1 and T_2, and we get $C_M{}^\circ(H)$ equal to both $T(w_1)$ and $T(w_2)$.

For the second point, the groups $T(w_1)$ and $T(w_2)$ are almost self-normalizing good tori in view of Lemma 6.5, and so Lemma IV 1.5 applies. This gives conjugacy in M°, and $M^\circ = C_M{}^\circ(A) \rtimes T(w_1)$. □

And now we have our rank computation.

LEMMA 6.15. $\mathrm{rk}(G) = \mathrm{rk}(C(T)) + 2\,\mathrm{rk}(C(A))$.

PROOF. As $\mathrm{rk}(G) = \mathrm{rk}(I_A) + \mathrm{rk}(C(A)) = \mathrm{rk}(I_A^+) + \mathrm{rk}(C(A))$, we need to show that

$$\mathrm{rk}(I_A^+) = \mathrm{rk}(C(T)C^\circ(A)) = \mathrm{rk}(C(T)) + \mathrm{rk}(C^\circ(A))\ .$$

As $C(T) \cap C^\circ(A) = 1$ by Lemma 6.7, the final equation is clear, so our claim is just

$$\mathrm{rk}(I_A^+) = \mathrm{rk}(C(T)C^\circ(A))$$

which is to be proved by making estimates in both directions. In the first direction, we define a map

$$\Phi : I_A^+ \to w^{C(T)C^\circ(A)}$$

with finite fibers as follows. For $i \in I_A^+$ we have seen that $T(i) = T^f$ for some $f \in C^\circ(A)$ and thus $i^{f^{-1}}w$ centralizes T, so we may define $\Phi(i) = w^{(i^{f^{-1}}w)f}$. Note that $T^{\Phi(i)} = T(i)$.

To see that Φ has finite fibers, if $\Phi(i) = \Phi(j)$, then $T(i) = T(j)$ and the value of f is constant on the fiber. So $\Phi(i) = \Phi(j)$ simplifies down to $w^{fi} = w^{fj}$ or $(ij)^{f^{-1}} \in C(T, w)$, and as w inverts $C^\circ(T)$, the latter is a finite group.

In the other direction, we let $X = C(T) \times C^\circ(A)$ and we consider the natural map $\Psi : X \to w^X$, whose fibers we claim are also finite.

If $w^{cf} = w^{c'f'}$ with $c, c' \in C(T)$ and $f, f' \in C^\circ(A)$, then $T^f = T^{f'}$ and again $f = f'$. Hence $c'c^{-1} \in C(T, w)$, and as this is finite we are done. □

6.4. $C^\circ(T)$**.** This subsection runs very much in parallel to §4.1. We show that $C^\circ(T) = T$ by showing that $C^\circ(T) \leq M$, arguing in the contrary case that we can find two disjoint generic sets. The first of these is $I_A^+ M^\circ$.

LEMMA 6.16. $\mathrm{rk}(I_A^+ M^\circ) = \mathrm{rk}(G)$

PROOF. As a result of Lemma 6.12, we know that the fibers of the natural map $I_A^+ \to G/M^\circ)$ have constant rank, equal to $\mathrm{rk}(A)$. Thus

$$\mathrm{rk}(I_A^+ M^\circ) = \mathrm{rk}(I_A) - \mathrm{rk}(A) + \mathrm{rk}(M^\circ) = \mathrm{rk}(I_A) + \mathrm{rk}(C(A)) = \mathrm{rk}(G)\ .$$

□

LEMMA 6.17. *If* $c \in C^\circ(T) \setminus M$, *then* $C^\circ(A)cM^\circ$ *contains no involutions in* I_A.

PROOF. We suppose toward a contradiction that fcb is an involution in I_A with $b \in M^\circ$, and since $C^\circ(T)$ and M° overlap in T we may take b to lie in $C^\circ(A)$. Then conjugating, the element bfc is also an involution, so replacing f by bf we may suppose $b = 1$ and $i = fc$ is in I_A.

Take some nontrivial element $t \in T$ and consider the commutator $\gamma = [i, t]$. Then i inverts γ. On the other hand, we may compute

$$\gamma = (fc)(fc)^{-t} = (fc)(c^{-1}f^{-t}) = [f^{-1}, t] \in C(A)\ .$$

Now by Lemma 6.3, if γ is nontrivial, then i is in M, which is false. □

PROPOSITION 6.18. $C^\circ(T) = T$.

PROOF. If $C^\circ(T) \leq M$, then by Lemma 6.7 we have the result. Suppose toward a contradiction that $C^\circ(T) \not\leq M$ and let $Y_0 = C^\circ(T) \setminus M$. Then $\mathrm{rk}(Y_0) = \mathrm{rk}(C(T))$.

We showed in preceding lemmas that $I_A^+ M^\circ$ is a generic subset of G and that it is disjoint from $Y = C^\circ(A)Y_0C^\circ(A)$. So it suffices to carry out a rank computation showing that Y is generic to reach a contradiction.

Since $\mathrm{rk}(G) = \mathrm{rk}(Y_0) + 2\,\mathrm{rk}(C(A_0))$ it suffices to show that the representation of the elements of Y is unique. So consider a relation $c = uc'v$ where $u, v \in C^\circ(A)$ and $c, c' \in Y_0$; we claim that $u = v = 1$.

Consider the group $X = [v, T]$. We will show that $X = 1$. We have $T^v \leq M$ and also $T^v = T^{u^{-1}c} \leq M^c$, and since also $T \leq M \cap M^c$ we have $X \leq M \cap M^c$. Now $c \notin M$ and so by Lemma 3.7, $\langle X, T\rangle \leq (M \cap M^x)^\circ$ is abelian. Hence by Lemma 6.5 w inverts X; as $X \leq C^\circ(A)$, by Lemma 6.3

$X \leq A$, and as $[X,T] = 1$, therefore $X = 1$. So $v \in C_{M^\circ}(A,T) = C_T(A) = 1$ in view of Lemma 6.7, and then as $c = uc'$, similarly $u = 1$, and so $c = c'$.

Thus $C^\circ(A)Y_0C^\circ(A)$ is a second generic subset disjoint from the first, and we have a contradiction. □

Our previous rank computation now takes on the following form.

LEMMA 6.19. $\mathrm{rk}(G) = \mathrm{rk}(T) + 2\,\mathrm{rk}(C(A))$.

6.5. M is connected. Now we reach a contradiction to the assumptions of this (and the last) section by showing that M is connected, after all.

We consider the natural map

$$\theta : C^\circ(A) \times M \to C^\circ(A)wM$$

and we claim that this is injective, which produces more than one subset of maximal rank if M is not connected.

For this, we have only to check that $C^\circ(A) \cap M^w = 1$. Suppose $x \in C^\circ(A) \cap M^w$. Then $x \in M^\circ$. Now x normalizes $T = (M \cap M^w)^\circ$, so $x \in N_{M^\circ}(T) = T$ by Lemma 6.7 and $x \in C_T(A) = 1$. This completes the argument.

One way to state what we have achieved in the last two sections is as follows.

PROPOSITION 6.20. *Let G be a simple L^*-group of finite Morley rank and even type, with a definable weakly embedded subgroup M. Assume the auxiliary hypothesis $(*)$ applies to G. Then M is strongly embedded.*

As shown earlier, G is then of type PSL_2.

7. ¬(∗), I: Toral blocks

In this section we will begin the proof of the hypothesis $(*)$ with which we have been working up to this point. So from this point on, we work under the opposite hypothesis, and aim at a contradiction.

$(\neg *)$ There are two distinct conjugates A_1, A_2 of $\Omega_1(O_2{}^\circ(M))$, such that the group $C_G(\langle A_1, A_2\rangle)$ is infinite.

This analysis will take up four sections, the remainder of this chapter. Our main tool will be the use of generic conjugacy theorems of the type we have exploited previously, in connection with good tori. This immediately yields a finiteness theorem for the number of conjugacy classes of a particular family of tori under the action of M°, after which it follows that exactly one of these conjugacy classes is generic in the whole family. In previous analyses it could be proved directly that the family reduced to a single conjugacy class; here we work similarly with the generic conjugacy class, and after carrying out a good deal of the structural analysis (and rank computations) on this basis, we arrive belatedly at the conclusion that the family does indeed consist of a single conjugacy class.

We will set this up in the present section and lay out the framework in which everything is proved. This material is related to both the rank computations we have seen in earlier sections, and the issue of control of $C^\circ(T)$ for suitable tori T, and as in the cases treated previously in the literature, once we have it in hand we can proceed to a structural analysis, which we will take up in the following two sections, going as far as double transitivity. What happens after that is a deviation into a contradiction via certain generators and relations, taking into account the elements of order 3. The purely computational arguments are all given in the final section.

In all of this analysis, we do not need to distinguish the cases of strong and weak embedding, in part because the condition $(\neg *)$ provides a copy of a "large" group L of type PSL_2 containing conjugates of $\Omega_1(O_2{}^\circ(M))$, which looks a good deal like an extreme form of the situation created by an offending involution, and in part because our prior results under either strong or weak embedding (that is, those independent of the hypothesis $(*)$) have led to parallel conclusions. But we will need to establish notation which can be used uniformly in both cases.

So G is just an L^*-group of finite Morley rank of even type with a definable weakly embedded subgroup M which may be strongly embedded, and we will assume $(\neg *)$ as well. Our goal then must be an ultimate contradiction.

7.1. Notation and basic facts. We write S for $U_2(M)$, which is $O_2(M)$ by §§1 and 2. In particular, S is a Sylow° 2-subgroup of M and M°/S is connected of degenerate type and hence contains no involutions, by Theorem IV 4.1. *A fortiori* $M^\circ/C_{M^\circ}(A)$ is of degenerate type.

We write A for $\Omega_1(S)$, and we remark that this is a definable, connected, elementary abelian subgroup central in S. This was shown in the strongly embedded case from Proposition 1.6, proved early on. In the case of weak but not strong embedding, this follows from the elimination of the nonabelian cases, which was carried out in §5, where as we noted the hypothesis $(*)$ was not in force. Before long we will show that $S = A$ and simplify the notation accordingly.

The following should also be retained.

LEMMA 7.1. *If $i \in I(S)$, then $C(i) \leq M$.*

This is part of a standard criterion in the case of strong embedding. In the case of weak but not strong embedding it is Lemma 6.2.

Now let us examine the consequences of $(\neg *)$. By assumption we have two conjugates A_1, A_2 of A, one of which we can assume to be A itself, such that $H = C^\circ(\langle A_1, A_2\rangle)$ is nontrivial. As usual $L = U_2(C^\circ(H))$ is a group of type PSL_2 in characteristic 2 and $L \cap M = A \rtimes T$ is a Borel subgroup where now $A = \Omega_1(S)$. Here we have T acting transitively on A.

Of course, $L = \langle A_1, A_2 \rangle$ in view of the structure of L. We also have the interesting group $H = C^\circ(L)$ which by assumption is nontrivial. We have $H \le C(A) \le M$ and $C^\circ(H) \not\le M$. The group H is of degenerate type since M is weakly embedded in G. This group contains no involutions, but it plays a role analogous to that of the offending involution in the weakly but not strongly embedded case. In particular, the group T plays a role very like that of the corresponding torus considered in the previous two sections, in many though not all respects.

The following structural points have been proved previously under other hypotheses; with T in hand the same proofs apply.

LEMMA 7.2.

(1) $C_{M^\circ}(i) = C_M{}^\circ(A)$.
(2) $M^\circ = C_M{}^\circ(A) \rtimes T$.

PROOF. By the transitivity of T on A we have $M^\circ = C_{M^\circ}(i) \rtimes T$, so the second point is a consequence of the first. And since M° is connected, this decomposition also shows that $C_{M^\circ}(i)$ is connected, that is, $C_{M^\circ}(T) = C_M{}^\circ(T)$. So our claim is simply the following:

$$C_M{}^\circ(i) = C_M{}^\circ(A) \ .$$

As usual, we work in $\overline{M}^\circ = M^\circ / C_{M^\circ}(A)$. If $C_M{}^\circ(i) > C_M{}^\circ(A)$, then we can construct a Borel subgroup of $\overline{M}^\circ$ in which $C(i)$ is infinite.

But the Borel subgroups of $\overline{M}^\circ$ are conjugate (Lemma IV 1.6), and thus no Borel subgroup of $\overline{M}^\circ$ acts freely on A. To conclude, it suffices to show that $\bar{T}$ is a Borel subgroup of $\overline{M}^\circ$.

Let $\bar{B} \ge \bar{T}$ be a Borel subgroup of $\overline{M}^\circ$. Then $\bar{B}$ is abelian and $\bar{T}$ acts regularly on A, while $\bar{B}$ acts faithfully, so $\bar{B} = \bar{T}$. □

LEMMA 7.3. *For $i \in A^\times$ we have $C(i) = C(A)$, and $M = C(A) \rtimes T$.*

PROOF. We may argue exactly as in the proof of Lemma 6.2. □

LEMMA 7.4. *If $i, j \in I(G)$ are conjugate to elements of $A^\times$ and $g \in G$ centralizes i while being inverted by j, then g is an involution, and i and j commute.*

This is Lemma 1.8 together with Proposition 1.6 in the case of strong embedding, and it is Lemma 6.3 in the case of weak but not strong embedding, and in fact it is the last result proved in that case prior to the introduction of hypothesis $(*)$ in §6.

DEFINITION 7.5. *I_1 is the set of involutions conjugate to elements of A.*

LEMMA 7.6.

(1) $I_1 \cap M = A$.
(2) *If L is a definable subgroup of G of type* SL_2, *then $I_1 \cap N(L) \subseteq L$.*

PROOF. For the first point, if $j \in I_1 \cap M$, then j acts on A and hence centralizes an involution $i \in A$. It then follows that j centralizes A and hence A centralizes the conjugate A_j of A containing j. Thus $A_j \leq C(A) \leq M$ and $A_j \leq U_2(M) = A$.

From this the second point follows. If $i \in I_1 \cap N(L)$ with L of type SL_2, then by Fact II 2.25, i acts as an inner automorphism and hence stabilizes a Sylow 2-subgroup of L, which we may take to be contained in A. It then follows that i normalizes A and lies in M, and the first point applies. □

7.2. SL_2-blocks and toral blocks. The fundamental tool in our analysis will be the study of groups LH of the type alluded to in the preceding section. We will now formalize the notions involved and note some of their agreeable properties.

DEFINITION 7.7.

(1) *An* SL_2-block *is a subgroup HL of G where $L = U_2(L)$ contains a conjugate of A and is of type* SL_2 *(=* PSL_2*, as we are in characteristic two), and $H = C^\circ(L)$ is nontrivial.*
(2) *If HL is an* SL_2*-block, then L is called the* SL_2-component *and H is called the* degenerate component.
(3) *A* toral block *is a group of the form HT which is contained in an* SL_2*-block HL with T a maximal torus in the* SL_2*-component $L = U_2(C(H))$.*

It is necessary to say something more about what these definitions actually mean. Strictly speaking, an SL_2-block or a toral block is more properly defined as a pair of groups (H, L) or (H, T) respectively. In the case of SL_2-blocks this is an inessential point, since we will see in a moment that the group HL determines the groups L and H individually. In the case of toral blocks it is really essential, and indeed situations in which a toral block HT also factors in a second way as a toral block H_1T_1 occur in practice and require analysis. To specify a toral block HT in a precise way one either specifies the pair (H, T), or the associated SL_2-block HL (that is, the pair (H, T) will be determined by the pair of groups HT and HL). So we now deal with these questions of uniqueness in a more precise way.

LEMMA 7.8. *Let HL be a toral block and let HT be an associated toral block, that is, let T be a maximal torus of L.*

(1) *$L = U_2(HL)$ and $H = C^\circ(L)$ are determined by HL. H has degenerate type.*
(2) *$U_2(C^\circ(X)) = L$ for any nontrivial subgroup X of H.*
(3) *The pair (H, T) determines the pair (HT, HL), and conversely.*

PROOF. For the first point, we may suppose that L contains A without loss of generality. Then $H \leq C(A) \leq M$ and L is not contained in M. If H contained a nontrivial unipotent 2-group, then $C(H)$ would be contained in

M by strong embedding, a contradiction. So H is of degenerate type and $L = U_2(HL)$.

For the second point, $U_2(C^\circ(X))$ has a weakly embedded subgroup, namely $M \cap U_2(C^\circ(X))$, and hence is of type PSL_2. So $\mathrm{rk}(U_2(C^\circ(X))) = 3\,\mathrm{rk}(A) = \mathrm{rk}(L)$ and hence $U_2(C^\circ(X)) = L$.

For the last point, the pair (H, T) determines HT, and H determines L, so one direction is clear. Conversely, the group HL determines L and H, and $T = L \cap HT$. □

So we will be fairly casual about our descriptions of SL_2-blocks, and more careful about our descriptions of toral blocks (often, by specifying the associated SL_2-block we have in mind). One can analyze the ambiguity a bit more closely: in a toral block HT, T is a good torus central in HT; if this does not determine T, then H contains a nontrivial central good torus. There is no reason why this should not be the case, and we will pursue this in its proper place.

In spite of this ambiguity, it should be possible in practice to speak of the *degenerate component* H and the *toral component* T of a toral block HT, as long as care is taken that this is determined by the context. For example, the following makes sense.

REMARK 7.9. *The toral component of a toral block is a good torus.*

Before coming to the main line of our analysis, we make one major clarification of the structure of M.

LEMMA 7.10. $S = A$.

PROOF. The action of T on $A = \Omega_1(S)$ shows that S is a free Suzuki 2-group, and hence abelian and homocyclic by Theorem III 3.2.

Suppose $S > A$. Let $S_1 = \Omega_2(S)$. Then squaring induces an isomorphism $S_1/A \simeq A$ which is compatible with the actions induced by M, that is these are isomorphic M-modules.

Fix an SL_2-block HL with $H \leq C(A)$. Then H centralizes S_1/A. Now let B be a Borel subgroup of H. As H is of degenerate type and B is solvable connected, B contains no involutions. So $V = C_{S_1}(B)$ covers S_1/A by Proposition I 9.9. Then $V \leq U_2(C(B)) = L$, a contradiction. □

The following is simple but useful.

LEMMA 7.11. *Let H be the degenerate component of an* SL_2*-block. Then distinct conjugates of H intersect trivially.*

PROOF. Suppose $X = H \cap H^g$ is nontrivial. Letting $L = U_2(C^\circ(H))$ we have $L = C^\circ(X) = L^g$ and thus $g \in N(L)$. As $H = C^\circ(L)$, we have $g \in N(H)$. □

LEMMA 7.12. *Any toral block is almost self-normalizing.*

PROOF. Apply Lemma I 10.20 to $H \lhd HT$, where of course H is associated with an SL_2-block HL containing HT. Then H contains a nontrivial subgroup Q which is continuously characteristic in HT. As such, $N^\circ(HT)$ normalizes Q and hence by the previous lemma, $N^\circ(HT)$ normalizes H. Hence $N^\circ(HT)$ acts on $L = U_2(C(H))$ and by Fact II 2.25 we have $N^\circ(HT) \leq LC^\circ(L) = HL$. So $N^\circ(HT) \leq N_{HL}(HT) = HT$ (Lemma 7.8). □

7.3. Covering by toral blocks.

LEMMA 7.13. *For any $g \in G \setminus N(H)$, the group $(HT) \cap (HT)^g$ is rigid abelian.*

PROOF. Let $A = (HT) \cap (HT)^g$. There are two projection maps $\pi_1 : HT \to T$ and $\pi_2 : (HT)^g \to T^g$, both of which can be restricted to A, and together they produce a map $\pi : A \to T \times T^g$. The image is a rigid abelian group, and we claim that this is an isomorphism.

Indeed, the kernel of the map is $H \cap H^g$ and if this is nontrivial, then $g \in N(H)$, a contradiction. □

In particular, M°/A contains no involutions, and in particular $I(M^\circ) = A^\times$.

PROPOSITION 7.14. *Let HT be a toral block, and let G_1 be a definable connected subgroup of G containing HT. Then HT contains a subgroup K which is almost self-normalizing and generically disjoint from its conjugates in G_1.*

PROOF. The group HT is almost self-normalizing and may possibly be generically disjoint from its conjugates in G_1, in which case there is nothing to prove.

Let us suppose therefore that the set

$$X = \bigcup_{g \in G_1 \setminus N_{G_1}(HT)} [(HT) \cap (HT)^g]$$

is generic in HT. For $g \in G$ let $X_g = (HT) \cap (HT)^g$.

Consider the following families of subgroups of HT:

$$\begin{aligned} \mathcal{F} &= \{X_g : g \in G_1 \setminus N_{G_1}(HT)\}, \\ \mathcal{F}_1 &= \{X_g : g \in N_{G_1}(H) \setminus N_{G_1}(HT)\}, \\ \mathcal{F}_2 &= \mathcal{F} \setminus \mathcal{F}_1 . \end{aligned}$$

By assumption, $\bigcup \mathcal{F}$ is generic in HT.

We show that $\bigcup \mathcal{F}_1$ is not generic in HT. For $g \in N_{G_1}(H) \setminus N_{G_1}(HT)$ we have $H \leq X_g < HT$. So X_g covers a proper subgroup of T. As this gives us a uniformly definable family of subgroups of T, this family is finite by Lemma I 4.23, so its union is nongeneric in T. Hence $\bigcup \mathcal{F}_1$ is not generic in HT.

Therefore $\bigcup \mathcal{F}_2$ is generic in HT. We now refine this generic covering further. Let $\hat{\mathcal{F}}_2$ be the family

$$\{X^\circ\langle a\rangle : X \in \mathcal{F}_2, a \in X\} \setminus \{T\} .$$

This is an inessential alteration in the sense that $\bigcup \hat{\mathcal{F}}_2 = \bigcup \mathcal{F}_2$ is still generic in HT; but the elements of $\hat{\mathcal{F}}_2$ are finite cyclic extensions of good tori.

One may take G to be sufficiently saturated by passing to an elementary extension. Now Theorem IV 3.1 applies and yield an element $K \in \hat{\mathcal{F}}_2$ with the property

$$K^\circ \text{ is a maximal good torus in } C_{HT}(K) .$$

Now by Lemma IV 1.20, all of the groups in $\hat{\mathcal{F}}_2$ are generically disjoint from their conjugates, so it suffices to check that K is almost self-normalizing in G_1. In any case, as K is rigid abelian, we have $N^\circ(K) = C^\circ(K)$. We must show that $C^\circ(K) = K^\circ$.

Fix $g \in G_1 \setminus (N_{G_1}(HT) \cup N_{G_1}(H))$ such that $[(HT) \cap (HT)^g]^\circ \leq K \leq (HT) \cap (HT)^g$. Now T is central in HT, so $T \leq C_{HT}(K)$. Hence $T \leq K^\circ$ by our maximality condition. By the choice of $\hat{\mathcal{F}}_2$, $K \neq T$. So $K = T \times (H \cap K)$ and $(H \cap K) \neq 1$.

Thus $C^\circ(K) \leq C^\circ(H \cap K) \leq N^\circ(H) \leq N^\circ(L) = HL$. Hence $C^\circ(K) \leq C_{HL}(T) = HT$. In particular, $T^g \leq HT$.

Now since $T^g \leq C_{HT}(K)$, our maximality condition on K implies that $T^g \leq K$. Hence arguing as we did for T, we find $C^\circ(K) \leq (HT)^g$. Thus $C^\circ(K) \leq [(HT) \cap (HT)^g]^\circ \leq K$, as claimed. □

This result implies the following.

LEMMA 7.15. *For any connected subgroup G_1 of G containing a toral block HT, the union of the conjugates of HT in G_1 is generic in G_1.*

PROOF. By the previous lemma there is a definable subgroup K of HT which is both almost self-normalizing and generically disjoint from its conjugates in G_1. By Lemma IV 1.2, the union of the conjugates of K in G_1 is generic in G_1. This then holds *a fortiori* for HT. □

7.4. Conjugacy. The following really begins the main line of our analysis of toral blocks.

LEMMA 7.16. *Any two toral blocks H_1T_1 and H_2T_2 such that $H_1, H_2 \leq C(A)$ and $T_1, T_2 \leq M$ are conjugate in M°.*

PROOF. We will make use of the associated groups $L_i = U_2(C(H_i))$ of type PSL_2 for $i = 1, 2$.

If the blocks in question are not conjugate in M°, then we will show first, by the argument used in the proof of Lemma 7.13, that any intersection of a conjugate of H_1T_1 with a conjugate of H_2T_2 under the action of M° is rigid abelian.

We may take this intersection without loss of generality to be $R = (H_1T_1) \cap (H_2T_2)$. We consider the map $\pi : R \to T_1 \times T_2$ induced by the projections from R to T_1 and T_2. The image is a rigid abelian group and we claim that the map is an isomorphism. The kernel is $K = H_1 \cap H_2$. If K is nontrivial, then $L_1 = U_2(C(K)) = L_2$ and accordingly $H_1 = H_2$. Then T_1 and T_2 are maximal tori in the Borel subgroup $L_1 \cap M = L_2 \cap M$ and thus are conjugate there. This contradiction proves that $K = 1$.

Now we apply Lemma IV 1.3 to the present situation: we have the two groups H_1T_1 and H_2T_2 contained in M°, each containing an almost self-normalizing subgroup generically disjoint from its conjugates, and with the union of its conjugates in M° generic in that group. The conclusion furnished is that the conjugates of H_2T_2 generically cover H_1T_1. In other words, setting $X_g = (H_1T_1) \cap (H_2T_2)^g$ for $g \in M^\circ$, the family $\mathcal{F} = \{X_g : g \in M^\circ\}$ is a generic covering of H_1T_1 by rigid abelian groups; the same holds if we delete the group T_1 from the family (if indeed it occurs).

So we may apply the covering principle Theorem IV 3.1, (after saturating) and find that one of the groups X_g is maximal in H_1T_1 in the sense that ${X_g}^\circ$ is a maximal torus in $C_{H_1T_1}(X_g)$. To lighten the notation we may take $g = 1$ here, and $X_g = X = (H_1T_1) \cap (H_2T_2)$.

Now since $T_1 \le Z(H_1T_1)$ we deduce that $T_1 \le X$ and thus $T_1 < X$, hence $X \cap H_1 > 1$. Now $X \cap H_1 \le C_{H_2T_2}(A) = H_2$, so $H_1 \cap H_2 > 1$, and hence $H_1 = H_2$. Accordingly, $L_1 = L_2$; so call this group "L". Then T_1, T_2 are conjugate within $N_L(A) \le M^\circ$ and we have shown after all that H_1T_1 is conjugate to H_2T_2 in M°. □

The foregoing leads to a number of additional useful facts of the same general type.

LEMMA 7.17.

(1) *Any two toral blocks in G are conjugate under the action of G.*
(2) *Any two* SL$_2$*-components in G are conjugate under the action of G.*
(3) *Any two toral components—that is, tori occurring as components of toral blocks—in G are conjugate under the action of G.*

PROOF. For the first point, we apply the preceding lemma. Consider $L = U_2(H)$. Then T is a maximal torus in L. Now we may conjugate a Sylow 2-subgroup of L to A, and then conjugate T into $M \cap L$. So after conjugation we have $H \le C(A) \le M$ and $T \le M$ and any two such blocks are conjugate in M.

The second point then follows as the toral blocks determine their associated SL$_2$-components, as we have stressed in Lemma 7.8).

For the last point, we view the toral blocks in question inside their corresponding SL$_2$-blocks, which after conjugation we take to be equal, and as noted their decompositions into degenerate and SL$_2$-type factors also coincide. So we may adopt the following notation: HL is an SL$_2$-block, and

T_1, T_2 are tori of L. Of course, T_1 and T_2 are then conjugate in L and our final point is proved. □

The last point in the preceding lemma is of particular importance, and we will incorporate it into an important piece of notation.

NOTATION 7.18.

(1) *$\mathcal{T}$ denotes the set of tori occurring as components of toral blocks in G.*

(2) *$\mathcal{T}_M$ denotes the set of tori in $\mathcal{T}$ which are contained in M.*

We have seen that $\mathcal{T}$ is a single conjugacy class of tori in G, and we will want to see the same for $\mathcal{T}_M$ under the action of M. As we have already seen $\mathcal{T}_M$ is nonempty. In addition the following finiteness theorem is a very useful first step.

THEOREM 7.19. *The set $\mathcal{T}_M$ breaks up into finitely many conjugacy classes under the action of M.*

PROOF. Corollary IV 1.16. □

7.5. Toral block types. Of very great importance for all that follows is the following classification of toral block types, for toral blocks whose toral components lie in M. Here we definitely view a toral block not as consisting of a single group of the form HT, but as a pair of specified groups H and T; in particular, a toral block determines its associated SL_2-block.

DEFINITION 7.20. *Let HT be a toral block with $T \leq M$, and let HL be the associated SL_2-block.*

(1) *HT is of type I if $C_H(A) \neq 1$.*

(2) *HT is of type II if $C_H(A) = 1$ and $(H \cap M)^\circ > 1$.*

(3) *HT is of type III if $H \cap M$ is finite.*

In this classification of toral block types, the three types are exhaustive and will be seen momentarily to be mutually exclusive. We transfer the notion of "type" to the toral components T as well: a torus $T \in \mathcal{T}_M$ is said to be of type I, II, or III, respectively, if and only if it belongs to some toral component of the corresponding type. One should not expect tori in $\mathcal{T}$ to determine their toral blocks, so the types of tori are not mutually exclusive—on the contrary. This is a useful point, in fact, we will show that tori of type III must also be of one of the other types, so that it suffices to analyze the first two types. Eventually, only type I will survive.

These notions are natural because the action of M by conjugation on $\mathcal{T}_M$ preserves type, and hence in approaching the question of the finiteness of the set of conjugacy classes in $\mathcal{T}_M$, we will work with the individual types.

We refine our view of the first two types as follows.

LEMMA 7.21.

(1) *If HT is a toral block of type I, then $H \leq C(A)$, and $HT \leq M$.*

(2) *If HT is a toral block of type II, then $(HT) \cap M$ is rigid abelian.*

PROOF. If HT is a toral block of type I, then $X = C_H(A) > 1$. Then $L = U_2(C(X)) = U_2(C(H))$. As $A \leq L \leq C(H)$, we have $H \leq C(A) \leq M$, and $HT \leq M$.

Now suppose HT is a toral block of type II. By Lemma 7.3, $M/C(A) \simeq T$ is a good torus. As $H \cap C(A) = 1$, $H \cap M$ is isomorphic with a subgroup of T and hence is rigid abelian. So $(HT) \cap M = (H \cap M)T$ is rigid abelian. □

In the case of tori of type II the picture may look odd. We have $M = C(A) \rtimes T$ with both T and H disjoint from $C(A)$, and $(H \cap M)^\circ$ nontrivial, so $(HT) \cap M = [(HT) \cap C(A)]T$, with $H \cap M$ disjoint from both factors $(HT) \cap C(A)$ and T. There is nothing immediately wrong with this, however.

We give further results on type I toral blocks.

LEMMA 7.22. *The type I tori in $\mathcal{T}_M$ form a single conjugacy class with respect to the action of M.*

PROOF. We consider H_1T_1 and H_2T_2 toral blocks of type I, with T_i the toral component of H_iT_i with respect to H_iL_i, where $L_i = U_2(C(H_i))$. By Lemma 7.16 these toral blocks are conjugate under the action of M, so we may suppose they are equal, $H_1T_1 = H_2T_2$.

Now by Lemma 7.21, we have $H_1 \leq C(A)$, and as T_1 is inverted by an involution we have $T_1 \cap C(A) = 1$ (Lemma 7.4). So $H_1T_1 \cap C(A) = H_1$. Similarly $H_2T_2 \cap C(A) = H_2$ and thus $H_1 = H_2$. Hence the SL_2-blocks H_1L_1 and H_2L_2 associated with H_1, T_1 and H_2, T_2 also coincide, and thus $AT_1 = L_1 \cap M = L_2 \cap M = AT_2$, and it follows that T_1 and T_2 are conjugate under the action of A. □

LEMMA 7.23. *Let HT be a toral block of type I. Then $C_M{}^\circ(T) = HT$.*

PROOF. Set $\hat{H} = C^\circ(A,T)$. Then $\hat{H} \geq H$, $C_M{}^\circ(T) = \hat{H} \cdot T$, and we claim $\hat{H} = H$.

Let HL be the SL_2-block associated with H, and $w \in L$ an involution inverting T. Let $\Gamma = \{[w,x] : x \in C^\circ(T)\}$. Our first claim is the following:

(1) $$\mathrm{rk}(C^\circ(T)) = \mathrm{rk}(\Gamma) + h\ .$$

There is a natural commutation map from $C^\circ(T)$ onto Γ, and we claim the fibers have rank h. For $x, y \in C^\circ(T)$ we have

$$[w,x] = [w,y] \iff w^{xy^{-1}} = w \iff xy^{-1} \in C(A_w)$$

where A_w is the conjugate of A containing w, in view of Lemma 7.3. On the other hand, $xy^{-1} \in C(T)$, and $\langle T, A_w \rangle = L$, so $xy^{-1} \in C(A_w)$ if and only if $xy^{-1} \in C(L)$; as $C^\circ(L) = H$ this gives us fiber rank equal to h and thus (1) holds.

Our next claim is as follows:

$$\mathrm{rk}(\hat{H} \cdot \Gamma) = \mathrm{rk}(\hat{H}) + \mathrm{rk}(\Gamma)\ .$$

For this we show that the natural map $\hat{H} \times \Gamma \to \hat{H}\Gamma$ is injective. So we consider the equation

$$\hat{h}\gamma_1 = \gamma_2$$

with $\hat{h} \in \hat{H}$ and $\gamma_1, \gamma_2 \in \Gamma$. We claim $\hat{h} = 1$.

Since w inverts every element of Γ, we find

$$\hat{h}^w \gamma_1^{-1} = (\hat{h}\gamma_1)^w = \gamma_1^{-1}\hat{h}^{-1}$$

and this may be rewritten as $\hat{h}^{w\gamma_1^{-1}} = \hat{h}^{-1}$.

Now $w\gamma_1^{-1}$ turns out to be conjugate to w: writing $\gamma_1 = [w, x]$ we have $w\gamma_1^{-1} = wx^{-1}wxw = w^{xw}$. In particular, $w\gamma_1^{-1}$ is conjugate to an element of A. Since $\hat{h}$ commutes with the involutions of A, it follows from Lemma 7.4 that $\hat{h}$ is an involution commuting with $w\gamma_1^{-1}$. Since we can replace the equation $\hat{h}\gamma_1 = \gamma_2$ by the equation $\hat{h}^{-1}\gamma_2 = \gamma_1$, we also have $\hat{h}$ commuting with $w\gamma_2^{-1}$ by the same token. It follows that $w\gamma_1^{-1}$ and $w\gamma_2^{-1}$ are commuting involutions in I_1. If these are equal, we are done. If they are unequal, then their product lies in I_1; but this is $w\gamma_2^{-1}w\gamma_1^{-1} = \gamma_2\gamma_1^{-1} = \hat{h}$, so $\hat{h} \in I_1 \cap M = A$. As $\hat{h} \in C(T)$, we find $\hat{h} = 1$ and our second claim is proved.

Now we combine the two claims. We have $\hat{H} \cdot \Gamma \leq C(T)$, so $\mathrm{rk}(\hat{H}) + \mathrm{rk}(\Gamma) \leq \mathrm{rk}(\Gamma) + h$ and $\mathrm{rk}(\hat{H}) \leq \mathrm{rk}(H)$; so $\hat{H} = H$. □

COROLLARY 7.24. *Let HT be a toral block of type I. Then any maximal good torus of HT is also a maximal good torus of M.*

We will take up the action of M on $\mathcal{T}_M$ in detail at the beginning of the next section. We show now that every torus in $\mathcal{T}_M$ is of type I or II (and we will not care greatly whether it is also of type III).

LEMMA 7.25. *Let $T \leq M$ be the toral component of a toral block HT. Then T is of type I or II. (We do not make the same claim about HT, however.)*

PROOF. Let H_1T_1 be a type I toral block, and T^* a maximal good torus of H_1T_1. Then T^* is a maximal good torus of M, and hence after conjugation we may suppose that $T \leq T^*$.

Thus we have a toral block HT with toral component $T \leq M$, and another toral block H_1T_1 of type I, with $T \leq H_1T_1$.

We have $H_1 \leq C(A)$ and T_1 is a complement to $C(A)$ in M. Now T, being a toral component of some toral block, is inverted by an involution. By Lemma 7.4, T is also disjoint from $C(A)$. By rank considerations, $H_1T = H_1T_1$.

We consider $R = C_{H_1T_1}{}^\circ(T)$. If $T_1 \leq T$ then $T = T_1$, and T is of type I. Suppose therefore that $T_1 \not\leq T$. We now claim

$$R = C_G{}^\circ(TT_1) \ .$$

We have $TT_1 > T$ and thus $TT_1 \cap H_1 > 1$, so $C^\circ(TT_1) \leq N^\circ(H_1) = H_1L_1$. So $C^\circ(TT_1) \leq C_{H_1L_1}(T_1) = H_1T_1$ and $C^\circ(TT_1) \leq C_{H_1T_1}(T) = R$.

Thus $R = C_G{}^\circ(TT_1)$. Since TT_1 is a good torus, it follows from Lemma I 4.24 that R is almost self-normalizing in G.

We now shift our point of view somewhat, and consider R as a subgroup of $C(T)$. We claim:

(2) R contains a connected definable subgroup K which is almost self-normalizing in $C(T)$ and is generically disjoint from it conjugates in $C(T)$.

Of course, R may itself be such a subgroup. If not, we will show that (after saturating) Lemma IV 3.2 applies to produce a good torus K in R which is almost self-normalizing in $C(T)$; then as K is a good torus it is also generically disjoint from its conjugates.

So suppose for the moment that $\bigcup_{g \in C(T) \setminus N(R)} (R \cap R^g)$ is generic in R. We would like to show that $R \cap R^g$ is rigid abelian for $g \in C(T) \setminus N(R)$, but this is not certain; we will show instead that there are at most finitely many exceptions to this claim, so that after discarding them the lemma applies.

So fix $g \in C(T) \setminus N(R)$. Lemma 7.13 applies, and this group is rigid abelian except possibly when $g \in N(H_1)$. On the other hand, as $N^\circ(H_1) = H_1L_1$, we have $N_{C(T)}{}^\circ(H_1) = C_{H_1L_1}{}^\circ(T)$, and as T normalizes both factors H_1 and L_1 we have $C_{H_1L_1}{}^\circ(T) = C_{H_1}{}^\circ(T) \times C_{L_1}{}^\circ(T) \leq C_{H_1}{}^\circ(T)T_1$ since $T \leq H_1T_1$; but $C_{H_1}{}^\circ(T)T_1 = R$, so we have $N_{C(T)}{}^\circ(H_1) \leq R$.

Thus there are only finitely many conjugates of R of the form R^g, for $g \in N_{C(T)}(H_1)$, and if we discard these then Lemma IV 3.2 applies to the remaining family. So our claim (2) is proved.

We now fix such a connected definable subgroup $K \leq R$, which is almost self-normalizing and generically disjoint from its conjugates when $C(T)$ is taken to be the ambient group. As K is almost self-normalizing, K contains T, and as T is not almost self-normalizing, we have $K > T$.

We now bring the toral block HT into the picture. Note that $HT \leq C(T)$. By Lemma 7.15, the conjugates of HT in $C^\circ(T)$ generically cover $C^\circ(T)$. We can apply Lemma IV 1.3 to this situation, taking $C^\circ(T)$ as the ambient group, and K, R, and HT as the relevant subgroups. The conclusion is that the conjugates of HT in $C^\circ(T)$ generically cover R.

We now consider the intersections $R_g = R \cap (HT)^g$ for $g \in C(T)$, and the family $\mathcal{F} = \{R_g : g \in C(T)\}$, which generically covers R. We must consider whether or not the groups R_g are rigid abelian.

Since $R \subseteq H_1T_1$, for each intersection R_g we have a natural map $\pi : R_g \to T_1 \times T^g$ induced by projections, and if the kernel $H_1 \cap H^g$ is nontrivial, one finds as usual that $H^g = H_1$, and thus $H_1T = (HT)^g$ is a toral block of type I in which T is the toral component, and T is of type I. So we may leave this case aside, and assume that each group R_g is rigid abelian, and $\mathcal{F}$ is a generic covering of R by rigid abelian groups.

So one such group R_g must be maximal in R in the sense that $R_g{}^\circ$ is a maximal good torus of $C_R(R_g)$; but $T_1 \leq Z(R)$ and hence $T_1 \leq R_g{}^\circ$. As $(HT)^g = H^gT$, we find $TT_1 \leq H^gT \cap M$. Thus $(H^gT \cap M)^\circ > T_1$, and H^gT

is therefore a toral block, with T as a toral component, which meets $C(A)$ in an infinite subgroup, and is therefore of type I or II, as claimed. $\square$

8. ¬(∗), II: Rank

We will now exploit our finiteness theorem to perform rank computations and structural analyses which are analogous to those which were carried out previously, by more direct methods, under the hypothesis (∗).

To begin with, we move to the study of involutions, specifically those conjugate to elements of A. By considering the involutions which invert a given torus in $\mathcal{T}$, we will show that there is a unique conjugacy class in $\mathcal{T}_M$ which is generic in $\mathcal{T}_M$. We aim eventually at showing that this generic conjugacy class consists exactly of the type I tori in $\mathcal{T}_M$, and then continuing this analysis we will see that all tori are in $\mathcal{T}_M$ are of type I, and in fact all toral blocks are of type I. But we do not proceed directly using the type classification, but indirectly by focussing initially on the generic conjugacy class and its position relative to the type classification.

This involves a considerable amount of rank computation of a familiar sort, and leads naturally into a conventional structural analysis.

8.1. Preliminaries.

LEMMA 8.1. *Let $T \leq M$ be a definable torus which is a complement to $C^\circ(A)$ in M°. Then $C_M(A,T)$ is infinite.*

PROOF. Let $K = C_M{}^\circ(T)$. Our claim is that $K > T$. As $T \simeq M^\circ/C(A)$, T is a good torus, and $N_M{}^\circ(T) = C^\circ(T)$. So if $K = T$, then T is an almost self-normalizing subgroup of M° which is generically disjoint from its conjugates in M°.

We know that there is also some toral block $H_1T_1 \leq M$ with $H_1 \leq C(A)$, and that the union of the conjugates of H_1T_1 in M° is also generic in M°, by Lemma 7.15. So by Lemma IV 1.3, with M° as the ambient group, we find that T is generically covered by its intersections with conjugates of H_1T_1, and as it is a good torus there are finitely many such intersections, one of which must be T. So we may suppose after conjugating that $T \leq H_1T_1$, which leads quickly to a contradiction.

Indeed, since $K = T$, we find $T_1 \leq T$ and by rank considerations $T_1 = T$, hence also $H_1 \leq K = T$, a contradiction. $\square$

This then leads to the following.

LEMMA 8.2. *Let $T \leq M$ be a definable torus which is a complement to $C^\circ(A)$ in M°, and suppose that T is inverted by some involution w conjugate to an involution in A. Then $C(T,w)$ is infinite.*

PROOF. We consider $\hat{T} = C^\circ(T)$, and the action of w on $\hat{T}$.

Suppose toward a contradiction that $C_{\hat{T}}(w)$ is finite. Then by Lemma I 10.3, w inverts $\hat{T}$. By Lemma 7.4, $\hat{T} \cap C(A) = 1$, and hence $\hat{T} \cap M^\circ = T$. This contradicts the preceding lemma. $\square$

8.2. Involutions. Recall that I_1 is the set of involutions conjugate to elements of A.

DEFINITION 8.3.

(1) *I_1^* is the set of involutions in I_1 which invert some torus in $\mathcal{T}_M$.*
(2) *For $T \in \mathcal{T}$, I_T is the set of involutions in I_1 which invert T.*

Our early goals include the following: (1) I_1^* is generic in I_1; (2) for $T_1, T_2 \in \mathcal{T}_M$ distinct, the corresponding sets of involutions I_{T_1} and I_{T_2} are disjoint.

Now we briefly consider the following set I_1^+ of involutions, which we will immediately prove coincides with I_1^*.

DEFINITION 8.4. *I_1^+ is the set of $w \in I_1$ for which the set*

$$\{g \in M : g^w = g^{-1}\}$$

has rank at least $\mathrm{rk}(A)$.

LEMMA 8.5. $I_1^+ = I_1^*$.

PROOF. We need to show that every involution w in I_1^+ inverts some torus in $\mathcal{T}_M$.

Let $X = \{g \in M : g^w = g^{-1}\}$ and let $K = d^\circ(\langle X \rangle)$. We will use two approaches, depending on whether K is abelian or not.

First suppose K is abelian. Then as the elements of X commute, K is inverted by w. By Lemma 7.4, K is disjoint from $C(A)$. It follows that K contains no involutions. By rank considerations, and the structure of M°, K is a complement to $C(A)$ in M°. So by Lemma 8.2, the group $Q = C^\circ(K, w)$ is nontrivial.

Let A_w be the conjugate of A containing w. By Lemma 7.3, $A_w \leq C^\circ(Q)$. Also $K \leq C^\circ(Q)$ and K does not normalize A_w, so it follows that $L = U_2(C(Q))$ must be a group of type PSL_2, and L is a component of an SL_2-block HL in which $Q \leq H$. Now K centralizes Q and hence lies in HL. As w inverts K and centralizes H, and $K \leq HL$ contains no involutions, we find $K \leq L$. As $\mathrm{rk}(K) \geq \mathrm{rk}(A)$ and K contains no involutions, it follows that K is a maximal torus of L. So $K \in \mathcal{T}_M$ is inverted by w. This disposes of the abelian case.

Now suppose that K is nonabelian. By Lemma 7.3, we have $K' \leq C(A)$, and similarly $K' \leq C(A^w)$. So $L = U_2(C(K'))$ is a group of type PSL_2 in characteristic 2, with A and A^w as Sylow 2-subgroups. Let T be a maximal torus in L which normalizes A and A^w. Now w normalizes K' and hence normalizes L, acting by an inner automorphism (Fact II 2.25) and interchanging the groups A and A^w. Accordingly, w acts by inversion on T, and $T \in \mathcal{T}_M$, as desired. □

LEMMA 8.6. *I_1^* is generic in I_1.*

PROOF. We work in fact with I_1^+, and we show that $I_1^- = I_1 \setminus I_1^+$ is nongeneric in I_1, by estimating the rank.

By definition, for $w \in I_1^-$, the set of elements of M inverted by w has rank strictly less than $\mathrm{rk}(A)$, and hence the set of involutions in wM has rank strictly less than $\mathrm{rk}(A)$, and therefore the rank of I_1^- is strictly less than

$$\mathrm{rk}(G/M) + \mathrm{rk}(A) = \mathrm{rk}(G) - [\mathrm{rk}(M) - \mathrm{rk}(A)] = \mathrm{rk}(G) - \mathrm{rk}(C(A)),$$

bearing in mind the relations $M = C(A) \rtimes T$ and $\mathrm{rk}(T) = \mathrm{rk}(A)$. But $C(A) = C(i)$ for $i \in A$ and thus $\mathrm{rk}(G) - \mathrm{rk}(C(A)) = \mathrm{rk}(i^G)$; since $i^G = I_1$, our claim follows. □

LEMMA 8.7. *If $T_1, T_2 \in \mathcal{T}_M$ are distinct, then $I_{T_1} \cap I_{T_2} = \emptyset$.*

PROOF. Suppose $w \in I_1$ inverts T_1 and T_2. Let $K = (M \cap M^w)^\circ$. Then $T_1, T_2 \leq K$, and T_1, T_2 act regularly on $A^\times$. If K is commutative, then T_1, T_2 commute and it follows directly via Lemma 7.4 that $T_1 = T_2$. Therefore we will suppose K is noncommutative.

Then $A, A^w \leq C(K')$, so $L = U_2(K')$ is a group of type PSL_2 in characteristic 2. Let $H = C^\circ(L)$. Then, as usual, $K \leq N^\circ(L) = HL$, and $K' \leq H$. Accordingly, the projection K_L of K into L is connected abelian, and also of degenerate type, hence K_L is contained in a maximal torus T of L, that is, $K \leq HT$, where T normalizes A and A^w. Now w normalizes L and hence lies in L by Lemma 7.6. So w inverts T and centralizes H. Now T_1 and T_2 are inverted by w so $T_1, T_2 \leq T$ and thus again $T_1 = T_2$. □

LEMMA 8.8. *There is a unique conjugacy class in $\mathcal{T}_M$, relative to the action of M, which is generic (i.e., of full rank) in $\mathcal{T}_M$.*

PROOF. We attach to any conjugacy class $\mathcal{C} \subseteq \mathcal{T}_M$, relative to the action of M, the set $I_1(\mathcal{C})$ of involutions in I_1 which invert some torus in $\mathcal{C}$.

As the tori in $\mathcal{T}$ are conjugate in G, the rank of I_T is independent of T; let this rank be r_0. Then $\mathrm{rk}(I_1(\mathcal{C})) = \mathrm{rk}(\mathcal{C}) + r_0$ since the various I_T are disjoint as T varies over $\mathcal{C}$.

On the other hand, as $\mathcal{C}$ varies over the conjugacy classes in $\mathcal{T}_M$, the sets $I_1(\mathcal{C})$ give a partition of I_1^*, which is generic in I_1 and hence has Morley degree 1. Hence there is exactly one conjugacy class $\mathcal{C}$ for which $I_1(\mathcal{C})$ has maximal rank, and hence exactly one conjugacy class $\mathcal{C}$ for which $\mathrm{rk}(\mathcal{C})$ is maximal. Again, as the classes are a finite partition of $\mathcal{T}_M$, our claim follows. □

NOTATION 8.9. *$\mathcal{T}_M^*$ is the generic conjugacy class in $\mathcal{T}_M$; a torus in $\mathcal{T}_M$ will be said to have* generic type.

8.3. Rank: Upper bounds.

LEMMA 8.10. *Let $\mathcal{C}$ be a conjugacy class in $\mathcal{T}_M$ with respect to the action of M, and $T \in \mathcal{C}$. Then $\mathrm{rk}(\mathcal{C}) = \mathrm{rk}(C(A)) - \mathrm{rk}(C(A,T))$.*

PROOF. We have

$$\mathrm{rk}(\mathcal{C}) = \mathrm{rk}(M) - \mathrm{rk}(N_M(T)) = \mathrm{rk}(M) - \mathrm{rk}(T) - \mathrm{rk}(N_{C(A)}(T)) .$$

Now $\text{rk}(M) - \text{rk}(T) = \text{rk}(C(A))$ since $M = C(A) \rtimes T$. Furthermore, $N_{C(A)}(T) = C(A,T)$ since $[T, N_{C(A)}(T)] \leq T \cap C(A) = 1$. So the formula holds. □

We note that the only variable term in the formula for $\text{rk}(\mathcal{C})$ is $\text{rk}(C(A,T))$, so the tori in $\mathcal{T}_M$ of generic type are those for which $\text{rk}(C(A,T))$ is *least*.

It is time to fix some numerical parameters for use in rank computations.

NOTATION 8.11.

(1) $g = \text{rk}(G)$; $c = \text{rk}(C(A))$; $a = \text{rk}(A) = \text{rk}(T)$.
(2) $h = \text{rk}(H)$ *where* H *is the degenerate component of an* SL_2*-block; as these groups are conjugate, the number* h *is well defined.*
(3) $c' = \text{rk}(C(T))$ *for* $T \in \mathcal{T}$*; as these tori are conjugate in* G*, this number is well defined.*

One may anticipate that c' equals $\text{rk}(T) + h$, which is certainly a lower bound. This will emerge eventually.

LEMMA 8.12. *Let* $T \in \mathcal{T}_M$ *be a torus of generic type. Then we have the following estimates:*

(1) $\text{rk}(C(A,T)) \leq h$.
(2) $\text{rk}(I_T) \leq c' - h$.
(3) $\text{rk}(G) \leq 2c - \text{rk}(C(A,T)) + c' - h$.

PROOF. For T a torus of type I we have $\text{rk}(C(A,T)) = h$, and for tori of generic type this rank is minimized. So the first point is clear.

For the second point, take an SL_2-block HL with T a maximal torus of L. For $w \in I_T$ we have $I_T = \{w' \in I_1 : w' \in wC(T)\}$. We will show that I_T meets each coset of H in $C(T)$ in *at most* one element.

So suppose $w, w' \in I_T$, and $w = w'h$, with $h \in H$ nontrivial. Then w inverts $h \in C(A)$. By Lemma 7.4, h is an involution, a contradiction. This proves the second point.

For the last, we work with the generic conjugacy class $\mathcal{T}_M^*$ in $\mathcal{T}_M$ and the associated set of involutions $I_1(\mathcal{T}_M^*)$, generic in I_1. Bearing in mind the equation $C(i) = C(A)$ for $i \in A^\times$ we have

$$\text{rk}(G) = \text{rk}(I_1) + c = \text{rk}(I_1(\mathcal{T}_M^*)) + c\ .$$

Now $\text{rk}(I_1(\mathcal{T}_M^*)) = \text{rk}(\mathcal{T}_M^*) + \text{rk}(I_T)$ for $T \in \mathcal{T}_M^*$ and $\text{rk}(\mathcal{T}_M^*) = c - \text{rk}(C(A,T))$ by Lemma 8.10. So

$$\text{rk}(G) = 2c - \text{rk}(C(A,T)) + \text{rk}(I_T)$$

and our claim follows from the second point. □

8.4. Rank: Lower bounds. In this subsection we consider the possibility that we have tori in $\mathcal{T}_M$ for which $C(T)$ is not contained in M. We compute the rank of the set $C^\circ(A)C(T)C^\circ(A)$ in this case and use it as a lower bound for the rank of G; we also show that this set is disjoint from a generic subset of G, making it a strict lower bound. Then in the case of a torus of generic type we arrive at a contradiction and conclude that $C(T) \leq M$ for T of generic type, which allows us to refine our previous rank computations. All of this is very much parallel to our earlier analysis under the hypothesis $(*)$.

We will begin with the "second" generic subset, which is, as usual, the set $I_1 M^\circ$.

LEMMA 8.13. *$I_1 M^\circ$ is a generic subset of G.*

PROOF. We consider the multiplication map $I_1^* \times M \to G$. The rank of the domain is $\mathrm{rk}(I_1) + \mathrm{rk}(M) = (g - c) + (c + a) = g + a$. We wish to show that the rank of each fiber of this map is a (actually an upper bound would suffice, but we can make this computation precisely).

The rank of the fiber containing $(w, x_0) \in I_1^* \times M$ is

$$\mathrm{rk}(\{x \in M : wx_0x^{-1} \in I_1^*\}) = \mathrm{rk}(\{x \in M : wx \in I_1^*\}) \ .$$

By the definition of I_1^*, w inverts a torus T in an SL_2-component of an SL_2-block and as $\mathrm{rk}(T) = \mathrm{rk}(A)$ the fiber rank is at least a. We turn to the upper bound.

Note that for $x \in M$, if w inverts x, then $x \in M \cap M^w$. We divide our analysis into two cases, depending on whether or not the latter group is abelian.

If $M \cap M^w$ is abelian, then the subset $X \subseteq M \cap M^w$ inverted by w is a subgroup and $X = (X \cap C(A)) \times T$. But by Lemma 7.4, $X \cap C(A) = 1$. So in this case $X = T$.

Now suppose $K = M \cap M^w$ is nonabelian. Then $L = U_2(C(K'))$ contains A and A^w and is a group of type PSL_2. Furthermore, K acts on L by inner automorphisms in view of Fact II 2.25, so $K \leq \hat{H}L$ with $\hat{H} = C(L)$. Furthermore, $w \in I_1 \cap N(L) \subseteq L$ by Lemma 7.6. So the subset $X \subseteq M$ inverted by w is $\{ht : h \in \hat{H}, h^2 = 1, t \in T\}$ with T the maximal torus of $L \cap M$ inverted by w (the one which normalizes A and A^w).

Now suppose w inverts $x \in M \cap M^w$ and $wx \in I_1$. Set $w_1 = wx$ and express x as ht with $h \in I(\hat{H})$ and $t \in T$. Note that $w_2 = wt \in I_1$ as well. Thus $h = ww_1t^{-1} = (w_1w_2)^w$, and as h commutes with w, we have $h = w_1w_2$; as this is an involution, w_1 and w_2 commute. So these involutions lie in the same conjugate of A, which therefore contains h as well, and this conjugate is contained in L since w_2 is. So $h \in L$ and as $h \in C(L)$, we find $h = 1$. Thus the fiber rank is a. □

We insert a small but useful lemma.

LEMMA 8.14. *Let HT be a type I toral block contained in M, $L = U_2(C(H))$, and $w \in L$ an involution inverting T. Then $(M \cap M^w)^\circ = HT$.*

PROOF. The group $M \cap M^w$ normalizes $\langle A, A^w \rangle = L$, hence lies in $LC(L)$, and the claim follows. □

Next we deal with the set $C^\circ(A)C(T)C^\circ(A)$, for $T \in \mathcal{T}_M$ for which $C(T) \not\leq M$.

LEMMA 8.15. *If* $T \in \mathcal{T}_M$ *and* $c \in C(T) \setminus M$, *then* $C^\circ(A)cC^\circ(A) \cap I_1 M^\circ = \emptyset$.

PROOF. Supposing the contrary, we have $ucv \in I_1$ for some $u \in C^\circ(A)$, $v \in M^\circ$, $c \in C(T) \setminus M$, and as $v \in TC^\circ(A)$ and $cT \subseteq C(T) \setminus M$, we may suppose $v \in C^\circ(A)$ as well. Then conjugating by u we find $ch \in I_1$ where $h = vu^{-1} \in C^\circ(A)$. Let $w = ch$.

Now $[w, T] = [h, T] \leq C^\circ(A)$. For $t \in T^\times$ we have $[w, t] \neq 1$ since $w \in I_1$, in view of Lemma 7.4. But $[w, t]$ is inverted by w and commutes with A, so by that same lemma we must have w commuting with A, hence $w \in M$, a contradiction. □

Now we concern ourselves with $\mathrm{rk}(C^\circ(A)C(T)C^\circ(A))$.

LEMMA 8.16. *Suppose* $T \in \mathcal{T}_M$, $h_1, h_2 \in C^\circ(A)$, *and* $c_1, c_2 \in C(T) \setminus M$, *with*

$$h_1 c_1 C^\circ(A) = h_2 c_2 C^\circ(A) \ .$$

Then $h_1 \in h_2[C^\circ(A) \cap C(T)]$ *and* $c_1 \in [C^\circ(A) \cap C(T)]c_2[C^\circ(A) \cap C(T)]$.

PROOF. Write $h_1 c_1 = h_2 c_2 v$ with $v \in C^\circ(A)$. Let $u = h_2^{-1} h_1 \in C^\circ(A)$. We have

$$v = c_2^{-1} u c_1, \quad u, v \in C^\circ(A), \quad c_1, c_2 \in C(T) \setminus M \ .$$

We aim to show that $v \in C(T)$. Given this, we then have $u \in C(T)$ and our claims follow. We consider the group $X = [T, v]$, generated by the corresponding set of commutators. We suppose $X \neq 1$, and we aim at a contradiction.

Now $X \leq C^\circ(A)$. Furthermore, $X = [T, c_1^{-1} u c_1] = [T, u c_1] = [T, u]^{c_1} \leq C(A^{c_1})$. So $L = U_2(C(X))$ contains $\langle A, A^{c_1} \rangle$ and is of type PSL_2 in characteristic 2.

Let $H = C^\circ(L)$ and let T_1 be the maximal torus of L contained in $M \cap M^{c_1}$. Let $w \in L$ invert T_1. Then $M^w = M^{c_1}$.

Now $T \leq M \cap M^{c_1}$ since $c_1 \in C(T)$, so $T \leq (M \cap M^w)^\circ = HT_1$.

As $A^{c_1 w} = A$, the element $x = c_1 w$ is in M. For $t \in T$ we have $t = t^{c_1} = t^{xw}$, $t^x = t^w$. Writing $t = ht_1$ with $h \in H$ and $t_1 \in T_1$, we find

$$h^x t^x = h t_1^{-1} \ ,$$

and reading this in $\bar{M} = M/C(A)$ we find $\bar{t}_1 = \bar{t}_1^{-1}$, which forces $t_1 = 1$. So $T \leq H \leq C(A)$, which is nonsense. This contradiction completes the proof. □

LEMMA 8.17. *Suppose* $T \in \mathcal{T}_M$, $h \in C^\circ(A)$, *and* $c \in C(T) \setminus M$. *Then the rank of the set of pairs* (h', c') *satisfying*

$$h' \in C^\circ(A), c' \in C(T) \setminus M, hcC^\circ(A) = h'c'C^\circ(A)$$

is $2\,\mathrm{rk}(C^\circ(A) \cap C(T))$.

PROOF. We apply the preceding lemma. A necessary condition for $hcC^\circ(A) = h'c'C^\circ(A)$ is

$$h' = hu_1; c' = u_2 c u_3$$

for suitable $u_1, u_2, u_3 \in C^\circ(A) \cap C(T)$. If $u_1 u_2 = 1$, then this condition is also sufficient. So all that remains to prove is that, indeed, $u_1 u_2 = 1$ here.

We may rewrite this as follows. Let $cC^\circ(A) = ucC^\circ(A)$ with $u \in C^\circ(A) \cap C(T)$ and $c \in C(T) \setminus M$. We claim $u = 1$.

As $cC^\circ(A) = ucC^\circ(A)$, we have $u^c \in C^\circ(A)$ and $u \in C(A, A^{c^{-1}})$. Suppose $u \neq 1$ and let $L = U_2(C(u))$, a group of type PSL_2.

Take an involution $w \in L$ with $A^w = A^{c^{-1}}$. Then $wc \in N(A) = M$. Let T_1 be the torus of L normalizing A and A^w. Then w inverts T_1.

Now $T \leq M \cap M^c = HT_1$. Also $[w, T] = [wc, T] \leq [M, T] \leq C(A)$. As $T \leq HT_1$ with w inverting T_1 and centralizing H, we have $[w, T] \leq T_1 \cap C(A) = 1$. Thus $T \leq H \leq C(A)$, a contradiction, since T is inverted by an involution. □

LEMMA 8.18. *Suppose that $T \in \mathcal{T}_M$ and $C(T)$ is not contained in M. Let $Y = C(T) \setminus M$. Then we have the following:*

(1) $\mathrm{rk}(C^\circ(A) Y C^\circ(A)) = c' + 2c - 2\,\mathrm{rk}(C(A, T))$.
(2) $g > c' + 2c - 2\,\mathrm{rk}(C(A, T))$.

PROOF. For the first point, we have $\mathrm{rk}(Y) = c'$ by assumption, and $\mathrm{rk}(C^\circ(A) \times Y \times C^\circ(A)) = 2c + c'$. By the two previous lemmas, the fiber ranks for the multiplication map $C^\circ(A) \times Y \times C^\circ(A) \to C^\circ(A) Y C^\circ(A)$ are all equal to $2\,\mathrm{rk}(C^\circ(A) \cap C(T))$, which agrees with $2\,\mathrm{rk}(C(A, T))$. So the first point is proved.

Now for the second point, we have a generic subset $I_1 M^\circ$ disjoint from the set $C^\circ(A) Y C^\circ(A)$. Hence the rank g of G is strictly greater than $\mathrm{rk}(C^\circ(A) Y C^\circ(A))$. □

8.5. Rank and $C(T)$. It is now an easy matter to combine the upper and lower bounds on rank from the preceding subsections, and derive structural consequences.

PROPOSITION 8.19.

(1) *For a torus $T \in \mathcal{T}_M$ of generic type, we have $C(T) \leq M$.*
(2) $g = 2c + a - h$.

PROOF. Let $T \in \mathcal{T}_M$ be a torus of generic type, and suppose that $C(T)$ is not contained in M. Then our two estimates are

$$c' + 2c - 2\,\mathrm{rk}(C(A, T)) < g \leq 2c - \mathrm{rk}(C(A, T)) + c' - h.$$

This simplifies to $h < \mathrm{rk}(C(A, T))$, which contradicts the estimate in Lemma 8.12. So this proves the first point.

For the second point, we first compute

$$c' = \mathrm{rk}(C(T)) = \mathrm{rk}(C_M(T)) = \mathrm{rk}(T) + \mathrm{rk}(C(A, T)) = a + \mathrm{rk}(C(A, T))\ .$$

and thus our upper bound becomes

$$g \leq 2c + a - h\ .$$

On the other hand, we can find a matching lower bound as follows.

Take $w \in I_1$ associated with a type I toral block $HT \leq M$, that is, $w \in L = U_2(C(H))$ inverts T. Then $g \geq \mathrm{rk}(MwM) = 2\,\mathrm{rk}(M) - \mathrm{rk}(M \cap M^w)$. As $M \cap M^w$ normalizes $\langle A, A^w \rangle = L$, we have $(M \cap M^w)^\circ \leq HL$ and $(M \cap M^w)^\circ = HT$, thus $g \geq 2(c+a) - (h+a) = 2c + a - h$. □

9. $\neg(*)$, III: Structure

We continue to explore the consequences of the hypothesis $(\neg *)$. We are beginning to have good control of the structure of G and proceed with a "recognition" phase in which G comes to look increasingly like PSL_2. However, there is one fundamental difference in structure from PSL_2, present throughout, namely the existence of the degenerate component in an SL_2-block, which affected the computation of $\mathrm{rk}(G)$ already. This will produce a contradiction by a very explicit computation in the next and last section of this chapter.

We recall that the various ranks we use in computations—mainly the parameters g, h, a, c—were defined in the previous section.

9.1. Double transitivity.

LEMMA 9.1. *Let $x \in G \setminus M$. Then* $\mathrm{rk}(M \cap M^x) = h + a$.

PROOF. We know $g = 2c + a - h$ and of course $g \geq \mathrm{rk}(MxM) = 2\,\mathrm{rk}(M) - \mathrm{rk}(M \cap M^x) = 2c + 2a - \mathrm{rk}(M \cap M^x)$, and this yields an inequality: $\mathrm{rk}(M \cap M^x) \geq h + a$. We need to prove the reverse inequality

$$\mathrm{rk}(M \cap M^x) \leq h + a \ .$$

Let $R = (M \cap M^x)^\circ$, and set $K_1 = C_R(A)$, $K_2 = C_R(A^x)$. Since $\mathrm{rk}(R) \geq h + a > a$, both K_1 and K_2 are infinite.

If $(K_1 \cap K_2)^\circ > 1$, then $L = U_2(C(K_1 \cap K_2))$ is of type PSL_2, and $H = C^\circ(L) \leq M$. Then $M^x = M^w$ for some $w \in L$, and $R \leq N^\circ(\langle A, A^w \rangle) = N^\circ(L) = HL$, so $R = H(R \cap L) \leq HT$ and $\mathrm{rk}(R) \leq h + a$ in this case. Accordingly, we will assume the contrary:

$$(K_1 \cap K_2)^\circ = 1 \ .$$

Now there is a natural map $R \to T \times T^x$ induced by the projections $M \to T$, $M^x \to T^x$, and the kernel is $K_1 \cap K_2$, which is finite and hence central in the connected group R. Since the image of R in $T \times T^x$ is a good torus, it follows that R' is finite and hence trivial, and R is a good torus as well.

We show now that R is almost self-normalizing. It suffices to show that $N^\circ(R) \leq M$, as a similar argument will then show $N^\circ(R) \leq M^x$, and thus $N^\circ(R) = R$.

We have $N^\circ(R) = C^\circ(R) \leq N(U_2(C(K_1)))$. If $U_2(C(K_1)) = A$, then this already shows $N^\circ(R) \leq M$. So suppose that $L_1 = U_2(C(K_1))$ is of type SL_2. Then $N^\circ(R)$ acts on L_1, with R acting like a maximal torus T_1 normalizing A , and hence $N^\circ(R)$ also acting like that maximal torus, and therefore normalizing A. So $N^\circ(R) \leq M$ as claimed, and it follows that R is almost self-normalizing.

Now Lemmas IV 1.3 and 7.15 apply to M° together with the subgroups R and HT, where HT is a type I toral block. The conclusion is that R is generically covered by the conjugates of HT in M°. As R is a good torus, we may therefore assume that $R \leq HT$, which yields the required estimate. □

PROPOSITION 9.2. *The action of G on the coset space $M\backslash G$ is doubly transitive.*

PROOF. This means that $G = M \cup MxM$ for any $x \in G \setminus M$, and it suffices to check that the double cosets MxM in question have full rank. Indeed, by the previous lemma we have

$$\mathrm{rk}(MxM) = 2\,\mathrm{rk}(M) - (h + a) = 2c + a - h = \mathrm{rk}(G) \ .$$

□

9.2. $\mathcal{T}_M$. We can now clarify the structure of $\mathcal{T}_M$: it is a single conjugacy class under the action of M, and there are no toral blocks of types II or III.

LEMMA 9.3. *Let HT be a toral block with $HT \leq M$. Then HT is of type I.*

PROOF. Let $L = U_2(C(H))$ and let $w \in L$ invert the torus T. By double transitivity, $L_1 = \langle A, A^w \rangle$ is also the SL_2-component of an SL_2-block, and is normalized by w. So $w \in I_1 \cap N(L_1) \subseteq L_1$ by Lemma 7.6. So w inverts the maximal torus T_1 of L_1 which normalizes A and A^w. By Lemma 8.7, since $w \in I_T \cap I_{T_1}$, we have $T = T_1$.

Again, $w \in L \cap L_1$ and hence the conjugate A_w of A containing w is contained in L and in L_1. So $L = \langle A_w, T \rangle = L_1$. Thus $A \leq L$ and $H \leq C(A)$; HT is of type I. □

LEMMA 9.4. *The tori of generic type in $\mathcal{T}_M$ are those of type I, and these tori are not of types II or III.*

PROOF. If $T \in \mathcal{T}_M$ is of generic type, then $C(T) \leq M$ by Proposition 8.19. So in an associated toral block HT we must have $H \leq M$, and the previous lemma shows that T is of type I. Furthermore, this argument shows that no associated toral block is of any other type.

Tori in $\mathcal{T}_M$ of type I are conjugate under the action of M by Lemma 7.22. Therefore, they are all of generic type, so the same conclusion applies. □

LEMMA 9.5. *$\mathcal{T}_M$ consists of a single conjugacy class under the action of M.*

PROOF. It suffices to show that all tori in $\mathcal{T}_M$ are of type I. As we have already shown that a torus in $\mathcal{T}_M$ of type III must also be of type I or II, it suffices to treat tori of type II. So fix $T \in \mathcal{T}_M$ of type II, and an associated SL_2-block HL. Fix $w \in L$ an involution inverting T.

By double transitivity and Lemma 8.14 we have $(M \cap M^w)^\circ = H_1T_1$ for some type I toral block H_1T_1.

If H_1^w meets H_1, then w normalizes H_1 and hence normalizes L_1. But $L_1 \cap HT_1 = T_1$ and thus w normalizes T_1. As $C_{T_1}(w) = 1$, w inverts T_1, and as w also inverts T it follows that $T = T_1$, and T is of type I.

Alternatively, suppose $H_1 \cap H_1^w = 1$. Then $H_1^w \leq (H_1T_1)$ projects injectively into T_1, and H_1 is a good torus. So H_1 is central in H_1T_1, and commutes with $H \cap H_1T_1$, hence normalizes H. As H is conjugate to H_1 in G, it is also a good torus and thus H_1 centralizes H. Hence H normalizes L_1.

Now w centralizes H and $H_1 \cap H_1^w = 1$, so $H \cap H_1 = 1$. Thus H acts faithfully on L_1. As $H \cap H_1T_1$ is nontrivial, H normalizes T_1 and hence centralizes T_1. But $T_1 \in \mathcal{T}_M$ is type I, and thus of generic type, and $C(T_1) \leq M$. So $H \leq M$ and now Lemma 9.3 shows that HT is of type I. □

LEMMA 9.6. *For $T \in \mathcal{T}_M$ we have $N_M(T) = C(T)$.*

PROOF. As T must be of generic type we have $C(T) \leq M$. Conversely, $N_M(T) = N_{C(A)}(T) \cdot T$ and $[T, N_{C(A)}(T)] \leq T \cap C(A) = 1$. □

PROPOSITION 9.7. *M is connected.*

PROOF. We take a toral block $HT \leq M$ and an involution $w \in U_2(C(H))$ inverting T. We claim

$$\mathrm{rk}(C^\circ(A)wM^\circ) = \mathrm{rk}(G) .$$

We have

$$\begin{aligned}\mathrm{rk}(C^\circ(A)wM^\circ) &= \mathrm{rk}(C^\circ(A)) + \mathrm{rk}(M) - \mathrm{rk}(C^\circ(A) \cap (M^\circ)^w) \\ &= 2c + a - \mathrm{rk}(C^\circ(A) \cap (M^\circ)^w) .\end{aligned}$$

Now $[C^\circ(A) \cap (M^\circ)^w]^\circ = [C^\circ(A) \cap (M \cap M^w)^\circ]^\circ = [C^\circ(A) \cap (HT)]^\circ = H$, so $\mathrm{rk}(C^\circ(A)wM^\circ) = 2c + a - h = g$.

Now we consider any $g \in G \setminus M$ and the corresponding double coset $C^\circ(A)gM^\circ$; the rank of this double coset is computed as above with $C^\circ(A) \cap (M^\circ)^g$ in place of $C^\circ(A) \cap (M^\circ)^w$. However, by double transitivity the pairs (M, M^w) and (M, M^g) are conjugate in G, and such conjugation carries $C^\circ(A)$ to $C^\circ(A)$, so the numbers involved are the same, and $C^\circ(A)gM^\circ$ is generic in G for all $g \in G \setminus M$.

Now suppose $x \in M$ is arbitrary. We conclude that $C^\circ(A)wM^\circ$ and $C^\circ(A)wxM^\circ$ are generic in G, and hence equal. So we arrive at an equation $cwM^\circ = wxM^\circ$ with $c \in C^\circ(A)$. Thus $c^w \in xM^\circ \subseteq M$, and $c \in C^\circ(A) \cap M^w \leq N(A) \cap N(A^w)$.

Consider $L = \langle A, A^w \rangle$. We have $c \in N(L)$ and $c \in C^\circ(A)$. So c acts on L like an element of A, and also normalizes A^w, forcing c to centralize L. In particular, c commutes with w. Hence our equation $cwM^\circ = wxM^\circ$ becomes $M^\circ = xM^\circ$ and thus $x \in M^\circ$.

So M is connected. □

LEMMA 9.8.

(1) $I(G) = I_1$.
(2) M *is strongly embedded in* G.

PROOF. We have $I(M) = I(M^\circ) = A^\times$ and from this both claims follow. □

LEMMA 9.9. $C(A) = C^\circ(A)$.

PROOF. $M = C(A) \rtimes T$ and as M has Morley degree 1, so does $C(A)$. □

LEMMA 9.10. *Let* $T \in \mathcal{T}_M$ *and let* HL *be a corresponding* SL_2*-block with* T *a maximal torus in* L, *and* $w \in L$ *an involution inverting* T. *Let* $\hat{H} = C(L)$. *Then we have the following:*

(1) $M \cap M^w = \hat{H}T$.
(2) $C(A) \cap M^w = \hat{H}$.

PROOF. For the first point, let $R = C_{M \cap M^w}(A)$. Then $M \cap M^w = RT$, and $\hat{H} \leq R$. We claim that R centralizes L.

Now as all toral blocks are of type I we have $(M \cap M^w)^\circ = HT$ and $R^\circ = H$. Thus R normalizes H, and L. Furthermore, R centralizes A, and acts on L like a subgroup of A, that is, $R \leq A\hat{H}$. Thus $R = \hat{H}(R \cap A)$. As $(R \cap A) \leq A \cap M^w = 1$, we have $R = \hat{H}$ as claimed.

Then for the second point we have $C(A) \cap M^w = C(A) \cap (\hat{H}T) = \hat{H}$. □

9.3. Final preparations. We have the structure of M and G firmly under control. We now put our structural remarks in final form, before embarking on the final computation of the next section.

LEMMA 9.11. *Let* $w \in I(G)$, $w \notin M$, *and set* $L = \langle A, A^w \rangle$. *Then we have the following:*

(1) L *is of type* PSL_2, *and* $N_L(A) \cap N_L(A^w)$ *is a torus inverted by* w *and contained in* $M \cap M^w$; $w \in L$.
(2) $G = M \sqcup C(A)wC(A)T$, *and the representation is unique, up to the following ambiguity:*
(3) *If* $c_1wc_2t_1 = c_1'wc_2't_1'$ *with* $c_1, c_2, c_1', c_2' \in C(A)$ *and* $t_1, t_1' \in T$, *then* $t_1 = t_1'$, *and for some* $x \in C(L)$ *we have* $c_1 = c_1'x$ *and* $c_2 = x^{-1}c_2'$.

PROOF. Most of the first point has been gone over in the past. L is of type PSL_2 by double transitivity of the action of G, and thus $N_L(A) \cap N_L(A^w)$ is a maximal torus, inverted by w since the action of w on L is inner, and w interchanges A and A^w. Finally, $w \in I_1 \cap N(L) \subseteq L$ by Lemma 7.6.

The second point is simply a repetition of double transitivity in the form $G = M \sqcup MwM$, taking into account $M = C(A)T$.

In the last point, the equation can be written as

$$x^w = c_2'tc_2^{-1}$$

where $x = c_1'^{-1}c_1$ and $t = t_1't_1^{-1}$. Thus $x \in C(A) \cap M^w = C(L)$ by Lemma 9.10. But $w \in L$, so the equation can be written as $x = c_2'tc_2^{-1}$ and hence

$t \in C(A)$. As $t \in T$ we have $t = 1$ and $t_1 = t_1'$, $x = c_2' c_2^{-1}$, and all claims have been verified. □

LEMMA 9.12. *For $T \in \mathcal{T}_M$ and $t \in T^\times$ we have $C(t) = C(T)$.*

PROOF. First we claim $C(t) \le M$. Let HL be an SL_2-block associated with T, and $w \in L$ an involution inverting T. As $G = M \cup C(A)wC(A)T$, it suffices to show that nothing in $C(A)wC(A)T$ commutes with t, and for this it suffices to deal with $C(A)wC(A)$.

So suppose that $c_1 w c_2 \in C(t)$ with $c_1, c_2 \in C(A)$. So $c_1 w c_2 = c_1^t w^t c_2^t = c_1^t w t^2 c_2^t$. This can be put in the form $xwt^2 = wy$ with $x, y \in C(A)$, or $x^w = yt^{-2}$; so $x^w \in C(A)^w \cap M = C(L)^w = C(L)$. As $y \in C(A)$ this implies $t^{-2} \in C(A)$ and hence $t^2 = 1$, $t = 1$. This contradiction shows that $C(t) \le M$.

Now w inverts t so also $C(t) \le M^w$. Furthermore, $C(t) = C(t, A) \cdot T$ and $C(t, A) \le C(A) \cap M^w = C(L) \le C(T)$, so $C(t) \le C(T)$. □

LEMMA 9.13. *Let $T \in \mathcal{T}$ and suppose T meets M nontrivially. Then $T \le M$.*

PROOF. Let $T_0 = T \cap M$. Take $w \in I(G)$ inverting T. As w inverts T_0, $w \notin M$. So $L = \langle A, A^w \rangle$ is a group of type PSL_2 and T_0 acts on L, normalizing A and A^w. So T_0 normalizes the maximal torus T_1 of L normalizing A and A^w, and as T_0 is a $2^\perp$-group acting by inner automorphisms, also T_0 centralizes T_1. So $T_1 \le C(T_0) \le C(T)$ by the preceding lemma, and $T \le C(T_1) \le M$ by Lemma 9.6. □

LEMMA 9.14. *Suppose $T_1, T_2 \in \mathcal{T}$, and $t_1 \in T_1$, $t_2 \in T_2$ are nontrivial elements, with t_1 and t_2 commuting. Then $T_1 = T_2$*

PROOF. As $C(t_1) = C(T_1)$ and $C(t_2) = C(T_2)$, the tori T_1 and T_2 commute. Let T_1 be associated with the toral block H_1T_1. We may suppose that $H_1T_1 \le M$.

Now $C(T_1) \le M$. Let $w \in L = U_2(H_1)$ invert T_1. Then similarly $C(T_1) \le M^w$. By Lemma 8.14, we have $(M \cap M^w)^\circ = H_1T_1$, so $C^\circ(T_1) = H_1T_1$. Hence $T_2 \le H_1T_1 \le M \cap M^w$ and then similarly $C^\circ(T_2) \le (M \cap M^w)^\circ = H_1T_1$, so $H_2T_2 \le H_1T_1$ and by symmetry $H_1T_1 = H_2T_2$.

As both H_1T_1 and H_2T_2 are type I toral blocks (there being no other possibilities), we have $H_1 = C_{H_1T_1}(A) = H_2$. Hence the associated SL_2-components L_1 and L_2 are equal, and the tori T_1, T_2 are commuting maximal tori in L_1, hence equal. □

LEMMA 9.15. $C(A) > AH$.

PROOF. Suppose $C(A) = A \times H$. Then as H is characteristic in AH, M normalizes H. Also $G = M \cup MwM$ with w centralizing H, so G normalizes H, and $H = 1$. But the hypothesis $(\neg *)$ contradicts this. □

10. ¬(*), IV: Contradiction

Operating under the hypothesis (¬*), with the structural information afforded by the last section, we will now reach a contradiction by an explicit calculation. In other words, we prove the following.

PROPOSITION 10.1. *Let G be a simple L^*-group of finite Morley rank and even type with a definable weakly embedded subgroup. Then the condition* (*) *holds.*

10.1. Elements of order 3. The following point is fundamental and will be invoked without explicit reference. We will fix an involution $w \in G \setminus M$; by what we have proved in the previous section, it will not matter which one we take. In particular, we have $G = M \sqcup MwM$. Bear in mind also that M and $C(A)$ are connected, which lightens the notation. We fix $T \in \mathcal{T}_M$ inverted by w.

LEMMA 10.2. *Let $x = wct$ with $c \in C(A)$ and $t \in T$. Then $x^3 = 1$ if and only if $c^w = c^{-t}wc^{-1}t$.*

PROOF. We expand $x^3 = 1$, bearing in mind $wt = t^{-1}w$. We find

$$(wct)^3 = c^w t^{-1} cwc^t = c^w (c^{-t}wc^{-1}t)^{-1} .$$

□

LEMMA 10.3. *Suppose that wct_1 and wct_2 are elements of order* 3 *with $c \in C(A)$ and $t_1, t_2 \in T$. Then $t_1 = t_2$.*

PROOF. The representation of c^w as c_1wc_2t is (sufficiently) unique (Lemma 9.11). □

We will make a close study of the set X_3 introduced in the following definition.

DEFINITION 10.4.

(1) *An element $t \in G$ is said to be* toral *if it belongs to some torus $T \in \mathcal{T}$.*
(2) $X_3 = \{c \in C(A)$: *For some $t \in T$, wct is toral and of order* $3\}$.

LEMMA 10.5. *X_3 is invariant under conjugation by HT.*

PROOF. Let wct be a toral element of order 3. For $h_1 \in H$ and $t_1 \in T$ we find

$$(wct)^{h_1t_1^{-1}} = w^{t_1^{-1}} c^{h_1t_1^{-1}} t = wt_1^{-2} c^{h_1t_1^{-1}} t = wc^{h_1t_1}(t_1^{-2}t)$$

and it follows that $c^{h_1t_1} \in X_3$. □

LEMMA 10.6. $X_3 \cap AC(L) = A^\times$

PROOF. Working inside SL_2 one can see that X_3 meets $A^\times$, and since $A^\times$ is a single conjugacy class under the action of T, it follows that $A^\times \subseteq X_3$.

Conversely, suppose $ax \in X_3$, where $a \in A$ and $x \in C(L)$. Then we have some $t \in T$ for which $waxt$ is a toral element of order 3, and thus

$$(ax)^w = (ax)^{-t}w(ax)^{-1}t = (a^{-1}wat)x^{-2} \ .$$

On the other hand, $(ax)^w = a^w x$, and comparison of the two yields $x^3 \in L$. But $x \in C(L)$, so $x^3 = 1$. Comparing the expressions again, we find

$$a^w = a^{-t}wat \ .$$

Thus wat is a toral element of order 3 (working in L, where any element of order 3 is toral).

As wat and $waxt$ are commuting toral elements, the tori involved are the same, and thus $x \in L$. As $x \in C(L)$ we find $x = 1$. Thus $ax = a \in X_3$, and then clearly $a \neq 1$. □

LEMMA 10.7. $\mathrm{rk}(X_3) = c - h$.

PROOF. Fix a toral element $t_0 \in T$ of order 3. Then the toral elements of order 3 belong to the conjugacy class t_0^G (recall that t_0 is inverted by an element of G). There is a bijection between $wM \cap t_0^G$ and X_3 defined by $wct \mapsto c$ for $c \in C(A)$, $t \in T$ with $wct \in t_0^G$. This is bijective because, as we have seen, c determines t.

We compute the ranks involved. First, $\mathrm{rk}(t_0^G) = \mathrm{rk}(G) - \mathrm{rk}(C_G(t_0)) = \mathrm{rk}(G) - \mathrm{rk}(C^\circ(T)) = g - (h + a) = 2(c - h)$.

Now $\mathcal{T}_M$ is a single conjugacy class in M, and any torus in $\mathcal{T}$ that contains a nontrivial element of M belongs to $\mathcal{T}_M$, so $t_0^G \cap M = \{t_0, t_0^{-1}\}^M$. Thus $\mathrm{rk}(t_0^G \cap M) = \mathrm{rk}(t_0^M) = \mathrm{rk}(M) - \mathrm{rk}(C_M(t_0)) = \mathrm{rk}(M) - \mathrm{rk}(C^\circ(T)) = (c + a) - (h + a) = c - h$.

From this it follows that t_0^G lies generically outside M. By double transitivity, this set is evenly distributed over the cosets of M in G. Thus writing r for $\mathrm{rk}(gM \cap t_0^G)$, where $g \in G \setminus M$, we find

$$2(c - h) = \mathrm{rk}(t_0^G) = \mathrm{rk}(G/M) + r = r + (2c + a - h) - (c + a) = r + c - h$$

and this yields $\mathrm{rk}(X_3) = r = c - h$, as claimed. □

NOTATION 10.8. $X_3' = X_3 \setminus A$.

LEMMA 10.9. *X_3' is generic in X_3.*

PROOF. By the last lemma, this means $a < c - h$, which is the content of Lemma 9.15. □

LEMMA 10.10. $\mathrm{rk}(X_3'C(L)) = \mathrm{rk}(C(A))$, *and the natural map $X_3' \times C(L) \to X_3'C(L)$ is a bijection.*

PROOF. As $\mathrm{rk}(X_3' \times C(L)) = (c - h) + h = c$, it suffices to check the second claim. Suppose therefore that we have

$$ch \in X_3'; c \in X_3', h \in C(L) \ .$$

We must show that $h = 1$.

We have $c^w = c^{-t_1}wc^{-1}t_1$ for some $t_1 \in T$, and similarly

$$(ch)^w = (ch)^{-t_2}w(ch)^{-1}t_2 = h^{-1}c^{-t_2}h^{-1}wc^{-1}t_2$$

for some $t_2 \in T$. Comparing this equation with

$$(ch)^w = c^w h = c^{-t_1}wc^{-1}ht_1$$

and using the uniqueness of the T-coordinates in such decompositions, we find $t_1 = t_2$ and hence

$$h^{-1}c^{-t_1}h^{-1}wc^{-1} = c^{-t_1}wc^{-1}h$$

which can be written as

$$(c^{t_1}c^{-t_1h}h^{-2})^w = c^{-1}hc \in C(A) \cap C(A^w) = C(L) .$$

So $h^c \in C(L)$. If $h \neq 1$, then easily $c \in N(L)$ and thus $c \in C(A) \cap N(L) = AC(L)$, and $c \in X_3' \cap AC(L) = \emptyset$ by Lemma 10.6. □

LEMMA 10.11. $C(L) = H$

PROOF. By the previous lemma the set $X_3C(L)$ is generic in $C(A)$, and since $C(A)$ has Morley degree 1, the same applies to $X_3C(L)$ and hence also to $X_3 \times C(L)$. So $C(L)$ has Morley degree 1, and $C(L)$ is connected. So $C(L) = C^\circ(L) = H$. □

10.2. A computation and a contradiction. We now have enough information in hand to make an explicit, and contradictory, computation. We compute the representation of a single element of $G = M \cup MwM$ in two ways.

As usual we fix the toral block $HT \leq M$ and an involution $w \in L = U_2(C(H))$ inverting T. We also choose a base point $a_0 \in A^\times$ satisfying $a_0^w = a_0wa_0$ (find a_0 in L so that a_0w has order 3).

Now $X_3'H$ is generic in $C(A)$ by Lemma 10.10, bearing in mind $C(L) = H$. We consider the intersections $cAH \cap X_3'H$ for $c \in C(A)$. It follows that for a generic set of $c \in C(A)$, the intersection of the coset cAH with $X_3'H$ is generic in cAH. Or in terms of the base point a_0, the set of $c \in C(A)$ satisfying

$$(\bullet) \qquad \{t \in T : ca_0^t \in X_3'H\} \text{ is generic in } T$$

is a generic subset of $C(A)$.

As $X_3'H$ is generic in $C(A)$, the set of $c \in X_3'H$ satisfying the same condition $(\bullet)$ is generic in $C(A)$. As this set is closed under multiplication by H on the right, and as the multiplication map $X_3' \times H \to X_3'H$ is bijective, the set X_3'' of $c \in X_3'$ satisfying the condition $(\bullet)$ is generic in X_3. Now X_3 is closed under inversion as can be seen by inverting both sides of the equation $c^w = c^{-t}wc^{-1}t$. So the following set is generic in X_3':

$$\{c \in X_3' : c, c^{-1} \in X_3''\} .$$

Accordingly, for a generic set of $c \in X_3'$, the following set $T_0(c)$ is generic in T:

$$T_0(c) = \{t \in T : ca_0^t \in X_3'H; c^{-1}a_0^{t^{-1}} \in X_3'H\} .$$

We now fix one such element $c \in X_3'$. We will write T_0 for $T_0(c)$.

As $c \in X_3'$ we have $c^w = c^{-t_1}wc^{-1}t_1$ for some $t_1 \in T$. Choose an element $t_0 \in T_0 \cap t_1^{-1}T_0$. Set $a = a_0^{t_0}$, and $t = t_0^{-2}$. We will carry out the calculation of the element $(ca)^w$ in two different ways.

First and most simply, by the choice of t_0 we have $ca \in X_3'H$, and $(cah_2) \in X_3'$ for some $h_2 \in H$, and hence for some $t_2 \in T$ we have

$$\begin{aligned}
(cah_2)^w &= (cah_2)^{-t_2}w(cah_2)^{-1}t_2 \\
&= h_2^{-1}a^{t_2}c^{-t_2}wh_2^{-1}ac^{-1}t_2 , \\
(ca)^w &= \left[a^{t_2}h_2^{-1}c^{-t_2}\right] w \left[ah_2^{-1}c^{-1}h_2^{-1}\right] t_2 .
\end{aligned}$$

For our second calculation we begin with $(ca)^w = c^w a^w$. We have $a_0^{t_0^{-1}} = a^t$, and hence

$$\begin{aligned}
a^w &= a_0^{wt_0^{-1}} = a^t wat , \\
c^w &= c^{-t_1}wc^{-1}t_1 , \\
(ca)^w &= c^w a^w = c^{-t_1}wc^{-1}t_1 a^t wat \\
&= c^{-t_1}\left[c^{-1}a^{tt_1^{-1}}\right]^w a^{t_1}t_1^{-1}t .
\end{aligned}$$

Since $c^{-1}a^{tt_1^{-1}} = c^{-1}a_0^{(t_0t_1)^{-1}} \in X_3'H$, we find $h_3 \in H$ and $t_3 \in T$ with

$$\begin{aligned}
(c^{-1}a^{tt_1^{-1}}h_3) &\in X_3' , \\
(c^{-1}a^{tt_1^{-1}}h_3)^w &= (c^{-1}a^{tt_1^{-1}}h_3)^{-t_3}w(c^{-1}a^{tt_1^{-1}}h_3)^{-1}t_3 \\
&= h_3^{-1}a^{tt_1^{-1}t_3}c^{t_3}wh_3^{-1}a^{tt_1^{-1}}ct_3 , \\
(c^{-1}a^{tt_1^{-1}})^w &= a^{tt_1^{-1}t_3}h_3^{-1}c^{t_3}wh_3^{-1}a^{tt_1^{-1}}ct_3h_3^{-1} .
\end{aligned}$$

So

$$\begin{aligned}
(ca)^w &= c^{-t_1}\left[c^{-1}a^{tt_1^{-1}}\right]^w a^{t_1}t_1^{-1}t \\
&= c^{-t_1}[a^{tt_1^{-1}t_3}h_3^{-1}c^{t_3}wh_3^{-1}a^{tt_1^{-1}}ct_3h_3^{-1}]a^{t_1}t_1^{-1}t \\
&= [a^{tt_1^{-1}t_3}c^{-t_1}h_3^{-1}c^{t_3}]w[a^{tt_1^{-1}+t_1t_3^{-1}}h_3^{-1}ch_3^{-1}]t_1^{-1}tt_3 .
\end{aligned}$$

Comparing this with $(ca)^w = \left[a^{t_2}h_2^{-1}c^{-t_2}\right] w \left[ah_2^{-1}c^{-1}h_2^{-1}\right] t_2$ we find

$$t_2 = t_1^{-1}tt_3 , \tag{1}$$

$$a^{tt_1^{-1}t_3}c^{-t_1}h_3^{-1}c^{t_3} = a^{t_2}h_2^{-1}c^{-t_2}h, \quad \text{some } h \in H , \tag{2}$$

$$a^{tt_1^{-1}+t_1t_3^{-1}}h_3^{-1}ch_3^{-1} = h^{-1}ah_2^{-1}c^{-1}h_2^{-1} . \tag{3}$$

Now let us put (3) in the form

$$\begin{aligned} a^{t^*} &= h^{-1}h_2^{-1}c^{-1}h_2^{-1}h_3c^{-1}h_3 \\ &= h'c^{-1}h''c^{-1} \end{aligned}$$

with $h', h'' \in H$ and $t^* = tt_1 + t_1t_3^{-1} + 1$ in $T \cup \{0\}$, where at the end we conjugate by h_3^{-1} before collecting terms. We can recast this further as

$$a^{t^*} = h^*(h''c^{-1})^2$$

with $h^* = h'h''^{-1}$.

Then the element $h''c^{-1}$ centralizes a^{t^*}, hence also centralizes h^*. If $h^* \neq 1$, then $h''c^{-1} \in N_{C(A)}(H) = AC(L)$ and hence $c \in AC(L)$, a contradiction to Lemma 10.6. We conclude that $h^* = 1$ and hence $a^{t^*} = (h''c^{-1})^2$, or after inversion:

$$a^{t^*} = (ch)^2$$

with $h = (h'')^{-1} \in H$. But M/A contains no involutions, so this implies $ch \in A$, a contradiction.

This contradiction proves that Hypothesis (∗) always holds in simple L^*-groups of finite Morley rank of even type with weakly embedded subgroups, and hence our recognition theorem holds generally in the following form.

THEOREM 10.12. *Let G be a simple L^*-group of finite Morley rank and even type with a definable weakly embedded subgroup. Then G is of the form* $\mathrm{SL}_2(K)$ *for some algebraically closed field K of characteristic* 2.

This has the following useful consequence.

PROPOSITION 10.13. *Let G be a simple L^*-group of finite Morley rank and even type, and suppose that the associated graph $\mathcal{U}(G)$ is disconnected. Then $G \simeq \mathrm{SL}_2(K)$ for some algebraically closed field K of characteristic* 2.

PROOF. This graph was introduced in the introduction to Chapter V. By Proposition V 1.2, if this graph is disconnected then there is a proper definable subgroup M of G containing the normalizer of each of its nontrivial unipotent subgroups, or in other words, a weakly embedded subgroup in the present context. So Theorem 10.12 applies. □

We find it useful to phrase this in yet another way.

DEFINITION 10.14. *Let G be a group of finite Morley rank of even type. Then we associate the following graph $\mathcal{U}^*(G)$ with G. The vertices are the Sylow° 2-subgroups of G, which in this context are the* maximal 2-*unipotent subgroups of G. We join two vertices by an edge if their intersection is infinite.*

LEMMA 10.15. *Let G be a group of finite Morley rank and of even type. Then the graph $\mathcal{U}(G)$ is connected if and only if the graph $\mathcal{U}^*(G)$ is connected.*

PROOF. If $\mathcal{U}(G)$ is connected, and S, T are vertices of $\mathcal{U}^*(G)$, then any path linking S to T in $\mathcal{U}(G)$ immediately gives rise to one in $\mathcal{U}^*(G)$, simply by extending each commuting pair of unipotent subgroups to a Sylow° 2-subgroup.

Conversely, suppose $\mathcal{U}^*(G)$ is connected. Every nontrivial unipotent subgroup U of G is connected to any Sylow° 2-subgroup S which contains U by the path $[U, Z^\circ(S), S]$ in $\mathcal{U}(G)$. So it suffices to show that any two Sylow° 2-subgroups S, T of G with infinite intersection are connected in $\mathcal{U}(G)$. But as just noticed, $(S \cap T)^\circ$ is connected to S and T in $\mathcal{U}(G)$. □

So the following is just a rephrasing of the previous proposition.

COROLLARY 10.16. *Let G be a simple L^*-group of finite Morley rank and even type, and suppose that the associated graph $\mathcal{U}^*(G)$ is disconnected. Then $G \simeq \mathrm{SL}_2(K)$ for some algebraically closed field K of characteristic 2.*

11. Notes

The material in this section has evolved considerably over more than a decade. In the tame (i.e., without bad fields) K^* case, it was dealt with in [**1, 4**]; then in the general K^* case (always, however, in even type) in [**125, 124**], which varied in interesting ways from the previous argument while keeping to the same overall strategy, and finally in a series of papers [**11, 12, 13, 14**] in the form given here, again keeping to the same overall strategy but with considerable adaptation to the new context—and noticeably greater length.

In the case of K^*-groups, the parallel to our §§1 and 2 is that the weakly embedded subgroup M has solvable connected component; here we prove that $U_2(M)$ is solvable, and since $U_2(M)$ is the part of M which is known to be a K-group in our context, this is a very natural generalization. The failure to control M more completely leads to considerable complications at later stages in the argument, and—particularly unfortunately—largely kills off the value of the theory of solvable groups to us in this context.

In the latter stages of the argument we replace the theory of Carter subgroups by generic conjugacy results involving good tori. Our use of Theorem IV 4.1 is inessential and in fact this material was worked out before that result was available. It does simplify some of the arguments.

The major case divisions used in the argument are similar to those occurring in [**87**].

§§1–2 give material worked out in [**11**].

§§3–6 give material worked out in [**12, 13**] first in the case of strong embedding and then in the weak but not strong context. The rank estimates in §3 are rather delicate but occur already in the early work of Nesin.

The analysis of the tori in the family $\mathcal{T}_M$ was a sticking point for the theory. In [**125, 124**] this was handled using the theory of solvable groups. Our analysis here uses properties of good tori via Theorem IV 1.15. In the first draft of [**14**] the argument was similar but more ad hoc. That work led to Proposition IV 1.15, which was proved in [**69**], and was used in the final version of [**14**], as here.

§4. The rank computation in the proof of Proposition 4.3 was introduced in [**81**], and exploited in [**1, 124**].

§§7–10 give material worked out in [**14**]. The structure of the argument follows [**124**], and our last section in particular reduces to a slight variant of the argument which concludes [**124**] in that case.

CHAPTER VII

Standard components of type SL_2

Introduction

In the present chapter we deal with groups with a *standard component* of type SL_2.

DEFINITION 1. *Let G be a group of finite Morley rank and even type, and L a quasisimple definable connected subgroup of G. Then L is said to be a* standard component *in—or "for", but not "of"—the group G if $C(L)$ contains an involution, and L is normal in $C^\circ(i)$ for all involutions $i \in C(L)$.*

THEOREM 1.1 (STANDARD SL_2). *Let G be a simple L^*-group of finite Morley rank and even type. Suppose that G has a standard component L of type* SL_2 *with $U_2(C(L)) > 1$. Let A be a Sylow 2-subgroup of L and U a Sylow$^\circ$ 2-subgroup of $C(L)$. Then AU is a Sylow$^\circ$ 2-subgroup of G.*

This is a purely technical result which merits a chapter of its own only because of the length of the analysis. It also has a tangled history, discussed in the notes to this chapter. There are aspects of finite group theory, which have parallels in groups of finite Morley rank, which explain the role of a theorem of this type, and in the process generalize it enormously. But as we do not need this material to carry through our classification project, we have reduced what is needed to a minimum, leaving this one technical result as the sole representative of a broader theory.

In the previous chapter we gave a characterization of SL_2, which will be followed in the next chapter, in rapid successions, by two more characterizations of the same group. Theorem VIII 1.1 characterizes SL_2 by the fact that its Sylow$^\circ$ 2-subgroups are abelian. Our final, most flexible characterization of SL_2 will be the "$C(G,T)$"-theorem, Theorem VIII 3.3, which says in essence that a simple L^*-group of even type is either generated by proper parabolic subgroups, or is of type SL_2. The present chapter prepares for the proof of the $C(G,T)$-theorem. The reader will very likely find it more natural to read the next chapter as far as that proof before looking into this one.

All of our characterizations of one very small group, SL_2, will eventually produce a very efficient recognition method for the general group of finite Morley rank and even type. The plan of attack will be seen at the end of the next chapter—rather late, but everything up to that point lays the basis for it.

We make use of a number of powerful tools in the present chapter, notably the theory of pseudoreflection groups and the Thompson rank formula. The basis of our strategy is to aim at a reduction to the classification of groups with weakly embedded subgroups. Just as in that classification, where an initially vague configuration is driven toward SL_2, we find that if we avoid the weak embedding configuration, a second configuration emerges, which eventually becomes extremely precise before succumbing to a final contradiction. The group looks locally a good deal like SL_3, but with some striking deviations. The route to a final contradiction passes through the Thompson rank formula; this gives a way to derive "global" contradictions from "local" discrepancies—that is, the presence of two distinct local structures which one would not expect to see in a single group.

Finite group theorists may notice that our use in §1 of "continuously characteristic" subgroups eliminates the detailed consideration of a number of configurations lying off the main line, which in finite group theory would very likely occupy the bulk of the analysis.

We give our analysis, broadly, in the next section, and then return in later sections to give each part in detail.

Overview

We will begin by outlining the proof of the following theorem, especially the computations that eventually provide a contradiction.

THEOREM 1.1. *Let G be a simple L^*-group of finite Morley rank and of even type, with a standard component L of type* SL_2 *over an algebraically closed field K of characteristic* 2. *Let A be a Sylow 2-subgroup of L, and let U be a Sylow° 2-subgroup of $C(L)$. Suppose that U is nontrivial.*

Then AU is a Sylow° 2-subgroup of G.

In the proof of this theorem we will first analyze the configuration by direct and elementary analysis, focussing on the structure of a Sylow 2-subgroup. We show that a counterexample mainly resembles SL_3, apart from the mysterious subgroup $U_2(C(L))$. Eventually we exploit the latter group in a fusion analysis, via the Thompson rank formula. Ultimately we compare the formulas $\mathrm{rk}(G) = \mathrm{rk}(C(i)) + \mathrm{rk}(i^G)$ for two involutions i in G, both of whose centralizers are fully controlled, and quite different, and conflicting inequalities result. One of these suggests that the rank of G should be something similar to $8f$ with f the rank of the base field for L (which is what one would expect if G resembles SL_3).

The proof. Our strategy is to use the criterion of Lemma V 1.3, which takes us from a slight generalization of weak embedding—namely, $N^\circ(U) \le M$ for $U \le M$ nontrivial 2-unipotent—to a weakly embedded subgroup.

We assume of course that

$$AU < S \text{ with } S \text{ a Sylow}^\circ \text{ 2-subgroup of } G.$$

In the first stage, one determines the structure of S. In such cases one works with what one has, namely AU at the moment, and the "next layer" of a Sylow° 2-subgroup S containing it, namely the preimage W in S of $\Omega_1{}^\circ(Z(S/AU))$. At this first stage, we identify a piece of S looking rather like a Sylow 2-subgroup of SL_3. It remains to show that $W = S$. For this, supposing the contrary, and looking at the next "slice" of S, we find the commutator structure is too tight to allow a further extension.

The basic facts are the following. Note that one proves this lemma first for W in place of S, after which applying this information at the next stage we show $W = S$; and afterward, we continue to make use of our detailed knowledge of what is now S.

LEMMA 1.5. *For $v \in AU \setminus A$ we have the following:*

(i) $C_S(v) = AU$.
(ii) $[S, v] = A$.
(iii) *S/A is elementary abelian.*
(iv) $Z(S) = A$.

Once we have this structure, we can deduce that $S \lhd C(A)$. With some further analysis we also arrive at $C(a) = C(A)$ for $a \in A^\times$, for which it suffices to have $C(a) \leq N(A) = C(A)T$, in view of the action of T on A.

This provides the raw ingredients for a fusion analysis leading to a contradiction.

Thompson maps. Our goal is to make two contradictory computations involving the Thompson rank formula. In our context, the Thompson rank formula arises whenever we have natural maps $I \times J \to K$ between sets of involutions for which we can calculate $\mathrm{rk}(I)$, $\mathrm{rk}(J)$, and $\mathrm{rk}(K)$ in terms of the rank of G and related parameters, and also estimate the fibers of the map. In this way we can exploit information about conjugacy classes of involutions and the structure of their centralizers. Our first result will have the form

$$\mathrm{rk}(G) = 6f + u + t_1 \tag{1}$$

where f is the rank of the base field (associated with a copy of SL_2), $u = \mathrm{rk}(U)$, and t_1 is the rank of another group introduced along the way. For the sake of intuition it is reasonable to think of u and t_1 as being about the same size as f, in which case this formula says $\mathrm{rk}(G) = 8f$, which is what would happen in the case of SL_3.

Our second calculation will give

$$\mathrm{rk}(G) = \mathrm{rk}(C(u)) + \mathrm{rk}(u^G) = 3f + u + \mathrm{rk}(u^G) \tag{2}$$

with u an involution in U, and combined with (1) this leads to

$$\mathrm{rk}(u^G) = 3f + t_1 .$$

Letting I_u be the union of the conjugacy classes in G of involutions in U, this will lead directly to the result

$$\mathrm{rk}(I_u) \leq 3f + u \ .$$

All of this is, in some sense, the "truth". But because of the presence of L in the centralizer of U, another similar computation yields

$$\mathrm{rk}(I_u) \geq 4f + u \ ,$$

and thus a contradiction.

A few details. We consider the following sets.

$$I_a = \{i \in I(G) : i^G \text{ meets } A\}; \ I_u = \{i \in I(G) : i^G \text{ meets } U\} \ .$$

Then I_a is a single conjugacy class, and the structure of I_u is unclear. However, we can easily see that these classes are disjoint, and that I_u covers $AU - A$.

With some more structural analysis one finds that $C^\circ(u) = C^\circ(U)$ for $u \in U^\times$, and hence the ranks of conjugacy classes of involutions in I_u are *constant*.

There is a minor bifurcation in the analysis at this point, depending on whether the subgroup A is strongly closed in S° or not. If A is not strongly closed in S°, one makes a little structural analysis of the resulting configuration, and one pins down some alternative information: $I(G) = I_u \cup I_a$; S° is a full Sylow° 2-subgroup of G. It turns out one gets much the same results along either branch of the analysis, once the details of each variant configuration have been clarified.

Now one considers the *Thompson map* $\theta : I_a \times I_u \to I(G)$ given by associating a pair $i \in I_a$, $j \in I_u$ to the unique involution in $d(\langle ij \rangle)$, which exists since i and j are not conjugate. First one must check the first order definability of this map, which is not obvious by the definition given. Next we show that the map takes its values generically in I_u. Here, we split into two cases, depending on whether A is strongly closed or not.

If A is strongly closed in S°, one argues directly that the entire range of θ is in I_u. If A is not strongly closed, then the values of θ lie in $I_a \cup I_u$, and it suffices to make a calculation showing that they do not lie, generically, in I_a. For this the following preliminary estimate is useful (and eventually, contradictory to our main calculation).

LEMMA 2.11. *Setting* $f = \mathrm{rk}(A)$, *and* $u = \mathrm{rk}(U)$, *we have*

$$\mathrm{rk}(I_u) \geq 4f + u \ .$$

This is not a direct estimate. What we do is to consider a second and less natural, partial, Thompson map

$$\tau : I_u \times I_u \to I_u$$

which is defined only for pairs $i, j \in I_u$ such that $d(\langle ij \rangle)$ meets I_u. By looking in the group UL we show that every fiber of this map has rank at least $u + 4f = \mathrm{rk}(U) + 2\,\mathrm{rk}(I(L))$. As the map is defined from a subset of $I_u \times I_u \to I_u$, some of these fibers must have rank at most I_u.

By this estimate, the rank of a generic fiber of the Thompson map θ should be at least $4f + u$, and one checks that over I_a the fiber ranks are at most $2f + u$.

So the map θ exhibits similar behavior in both cases. Next, we restrict it to be generically defined on $I_a \times I_u$, with image in I_u.

With this map in hand, and sufficiently well understood, we can use it for its intended purpose: the calculation of $\mathrm{rk}(G)$. This produces the following:

$$\mathrm{rk}(G) = \mathrm{rk}(C(A)) + 4f \ .$$

Namely, one computes the fiber ranks above I_u, finding that they are all of rank $4f$. As θ carries $I_a \times I_u$ generically onto I_u, we then find that $4f = \mathrm{rk}(I_a) = \mathrm{rk}(G) - \mathrm{rk}(C(a)) = \mathrm{rk}(G) - \mathrm{rk}(C(A))$ for $a \in A^\times$, whence our claim.

We are still not done. We need next to understand both $C(U)$ and $C(A)$. For $C(U)$, one shows quickly that $C^\circ(U) = LU$ and thus $\mathrm{rk}(C^\circ(U)) = 3f+u$.

For $C(A)$, one may show, again by close examination, that $C^\circ(A) = S^\circ \rtimes T_1$ with T_1 a torus acting freely on U. So we introduce the parameter

$$t_1 = \mathrm{rk}(T_1)$$

and we have $\mathrm{rk}(C(A)) = 2f + u + t_1$ as a result of our previous analysis, giving the not entirely perspicuous formula

$$\mathrm{rk}(G) = 6f + u + t_1 \ .$$

But since one might expect T_1 to be a 1-dimensional torus, and U a 1-dimensional unipotent group on which it acts, this formula suggests something like $\mathrm{rk}(G) = 8f$, a reasonable value.

Turning to $u \in U^\times$, we also have $\mathrm{rk}(G) = \mathrm{rk}(C(u)) + \mathrm{rk}(u^G) = 3f + u + \mathrm{rk}(u^G)$ and thus we have

$$\mathrm{rk}(u^G) = 3f + t_1$$

which is more interesting, since it is not so reasonable.

Consider the conjugacy classes contained in I_u as a definable set $\mathcal{C}_u$; then this set has a rank, which can be no greater than the rank of the set of conjugacy classes in U, under the action of T_1. The latter is $u - t_1$.

So $\mathrm{rk}(I_u) \le (u - t_1) + \mathrm{rk}(u^G) = u + 3f$. Recall, however, that $\mathrm{rk}(I_u) \ge 4f + u$, as shown above. So we have arrived at a contradiction.

1. Sylow structure

1.1. On standard components of type SL_2. Our goal in this and the next two sections is the following result, which is more than we need at this juncture, but will be useful subsequently as well.

THEOREM 1.1. *Let G be a simple L^*-group of finite Morley rank and of even type, with a standard component L of type* SL_2 *over an algebraically closed field K of characteristic* 2. *Let A be a Sylow* 2*-subgroup of L, and let U be a Sylow° 2-subgroup of $C(L)$. Suppose that U is nontrivial.*

Then AU is a Sylow° 2-subgroup of G.

What we will actually need, and what amounts to the same thing, is the same statement with the conclusion weakened as follows.

PROPOSITION 1.2. *Let G be a simple L^*-group of finite Morley rank and of even type, with a standard component L of type* SL_2 *over an algebraically closed field K of characteristic* 2. *Let A be a Sylow* 2*-subgroup of L, and let U be a Sylow° 2-subgroup of $C(L)$. Suppose that U is nontrivial.*

Then AU is a Sylow° 2-subgroup of $C_G(A)$.

The full theorem follows easily from the weaker proposition, as we will show immediately, but the latter result requires extensive analysis. Once we have Proposition 1.2, the classification of groups of even type with abelian normal subgroups (Theorem VIII 1.1) will also follow, as we shall see.

We now derive Theorem 1.1 from Proposition 1.2.

PROOF OF THEOREM 1.1. We assume Proposition 1.2 for the moment, that is, we assume that AU is a Sylow° 2-subgroup of $C(A)$. We wish to show that AU is also a Sylow° 2-subgroup of G, and for this it suffices to show that it is a Sylow° 2-subgroup of $N(AU)$, in view of the normalizer condition (Proposition I 5.3).

So consider a Sylow° 2-subgroup S of $N(AU)$. Note that $N(AU)$ contains a maximal torus T of L, commuting with U and normalizing A. Furthermore, by Proposition II 5.3, we have a Sylow° 2-subgroup S of $N(AU)$ invariant under the action of T; for this, we apply the proposition to the subgroup $U_2(N(AU))T$, which is a K-group. Now $AU \leq S$, and we claim $AU = S$. For this, it suffices to show that S lies in $C(A)$, in other words, that A is central in S.

Now $Z(S) \cap AU$ is infinite, by Lemma I 5.1, and is also T-invariant. If $Z(S)$ meets A nontrivially, then in view of the action of T it contains A, and we are done. The only alternative is

$$Z(S) \cap AU \leq U .$$

In this case, fix one element $u \in Z(S) \cap U^\times$. As L is a standard component, we find $L \lhd C^\circ(u)$. So S normalizes L, and hence also normalizes $L \cap (AU) = A$. So now $A \lhd S$ and hence $Z(S)$ meets A nontrivially, after all. □

1.2. C°(A): The 2-Sylow°. It remains only to prove Proposition 1.2. With the hypotheses and notation of the proposition in force, we will assume that *AU is not a Sylow° 2-subgroup of $C(A)$*, and we will analyze the structure of a Sylow° subgroup of $C(A)$ in considerable detail. We will find that this structure is well determined, and looks much like a Sylow subgroup of SL_3 in characteristic 2. We will also show that $U_2(C(A))$ is solvable, and that a Sylow° 2-subgroup of $C(A)$ is also a Sylow° 2-subgroup of the whole group G. This structural information will be exploited in the following section in a fusion analysis, via the Thompson rank formula, leading afterward to a contradiction.

We now fix a considerable body of notation all at once. We will recall the hypotheses and notation of Proposition 1.2, and add to it. Some of this will require further comment.

NOTATION 1.3.

(i) *G is a simple L^*-group of finite Morley rank and of even type.*
(ii) *L is a standard component in G, of type* SL_2 *over an algebraically closed field of characteristic* 2.
(iii) *A is a Sylow 2-subgroup of L, and T is a maximal torus of L which normalizes A.*
(iv) *U is a Sylow° 2-subgroup of $C(L)$, and is nontrivial.*
(v) *S is a Sylow 2-subgroup of $C(A)$, and S° is T-invariant.*
(vi) $S^\circ > AU$.
(vii) $\hat{U} = C_S(L)$ *and* $V = N_S(L)$.
(viii) *W is the preimage in S° of the group*

$$\bar{W} = \Omega_1{}^\circ(Z(N_{S^\circ}(AU)/AU)) .$$

The existence of all these objects calls for very little comment. The groups G, L, A, U, T are all part of our basic setup for Proposition 1.2. For the choice of S, one first selects S°, by applying Proposition II 5.3 to the K-group $U_2(C(A))T$, and one then extends this to a suitable group S. The assumption $S^\circ > AU$ is taken with a view toward an eventual contradiction, which will prove the proposition. Evidently $\hat{U}$, V, and W will all be of use in determining the structure of S, and indeed the main point will be that $S^\circ = W$, and also that the structure of W can be determined with considerable precision (with the former statement a consequence of the latter). Evidently W is intended to be the smallest interesting subgroup of S going beyond AU. There are a number of elementary relationships among the 2-groups which should be borne in mind, and treated as part of the basic setup; these are listed in the following lemma.

LEMMA 1.4. *With notation and hypotheses as formulated above, we have the following relationships:*

(i) $\hat{U}^\circ = U$.
(ii) $V = A \times \hat{U}$.
(iii) $V^\circ = AU$.
(iv) *V is elementary abelian.*
(v) $W > AU$.

PROOF.
(i) U is a Sylow° 2-subgroup of $C(L)$, and is certainly contained in $\hat{U}$.

(ii) S centralizes A, so V acts on L like a subgroup of A; this does not require connectedness as L has no graph automorphisms. As $A \le V$, we find $V = A \times C_V(L) = A \times \hat{U}$.

(iii) Since $V = A \times \hat{U}$, $V^\circ = AU$. (This brings V into a relationship with W as well.)

(iv) With $\phi(V)$ denoting the ordinary Frattini subgroup of V, the pull-back to V of $\{v^2 : v \in V/V'\}$, our claim can be expressed by the equation $\phi(V) = 1$. We have $\phi(V) = \phi(\hat{U})$. If this is nontrivial, then $N_S(V) \le N_S(\phi(V)) = N_S(\phi(\hat{U})) \le N_S(L)$, since L, being a standard component, is a quasisimple component of $C(\phi(\hat{U}))$, and S normalizes A. Thus $N_S(V) = V$ by definition of V. Then $S = V$, and $S^\circ = AU$, a contradiction to our current hypotheses.

(v) This is simply the normalizer condition; but it is also our point of departure. □

We can now analyze W satisfactorily.

Lemma 1.5. *With our standing hypotheses and notation, and with $v \in V \setminus A$, we have the following:*

(i) $C_S(v) = V$.
(ii) $[W, v] = A$.
(iii) *W/A is elementary abelian.*
(iv) $Z(W) = A$.

Proof. First, as far as v is concerned, we may write $v = au$ with $a \in A$ and $u \in \hat{U}$. Then $C_S(v) = C_S(u)$ and $[W, v] = [W, u]$, so we may suppose $v = u \in U^\times$. We now take up the points individually.

(i) As L is a standard component, it is a quasisimple component of $C^\circ(u)$. The group $C_S(u)$ permutes these components, and as S normalizes A it follows that $C_S(u)$ normalizes L, that is, $C_S(u) \le V$. On the other hand, as V is abelian, we have $V \le C(u)$.

(ii) First, $[W, u]$ is nontrivial since $C_S(u) = V$ and $V^\circ < W$. The group $[W, u]$ is also T-invariant, and in view of the action of T on A it suffices to show that $[W, u] \le A$. So consider $\gamma = [w, u]$ with $w \in W$. We may suppose $w \notin V$.

Suppose first that $\gamma \in V$ (this holds if $u \in U$, but is not yet clear for general $u \in \hat{U}$). By the definition of W and Lemma 1.4 (iii), we have $w^2 \in V^\circ$, hence $1 = [w^2, u] = \gamma^w \gamma$, so $w \in C(\gamma)$. So by the first point, if γ is not in A, then we find $w \in V$, a contradiction. So $\gamma \in A$ in this case.

Now we treat the general case. Take $u' \in U^\times$. Then $[w, u'] \in A$ by the case just treated. So we find $\gamma^{u'} = [w^{u'}, u] = [w, u] = \gamma$ since $w^{u'} \in wA$. So $\gamma \in C_S(u') = V$ and we conclude by the former case.

(iii) We claim that $w^2 \in A$ for $w \in W$. We have $w^2 \in AU$ and $w \in C(w^2)$, so by the first point, if $w^2 \notin A$, we find $w \in V$, and then $w^2 = 1$, a contradiction.

(iv) Take $v \in V \setminus A$. Then $Z(W) \leq C_S(v) = V$. If $Z(W) > A$, then we can take $v \in Z(W) \setminus A$ and conclude $W \leq C_S(v) = V$, hence $W = AU$, contradicting our current hypothesis. □

At this point we have adequate control of W. It remains to be seen that $W = S^\circ$, which involves a similar investigation of the next "layer" of S, leading this time to a contradiction. We set up the following notation.

NOTATION 1.6.
(i) $\hat{W}$ *is the inverse image in* S *of* $\Omega_1{}^\circ(Z(N_S(W)/W))$.
(ii) $W_1 = [Z(\hat{W} \bmod A) \cap W]^\circ$.
(iii) $V_1 = [Z(\hat{W} \bmod A) \cap V]$.

We introduced *continuously characteristic* subgroups (those invariant under *connected* groups of automorphisms) in Definition I 10.19, and we will again find the notion useful here.

If $S^\circ > W$, then $\hat{W} > W$ as well, and $\hat{W}$ is the "next layer" of S. While this notation is not particularly attractive, we will show shortly that $\hat{W} = W = S^\circ$. Similarly, our interest in V_1 and W_1 will be short-lived. Observe that in defining W_1 we take a connected component, and in defining V_1 we do not.

LEMMA 1.7. *With our accumulated hypotheses and notation, we have* $V_1 > A$.

PROOF. Suppose toward a contradiction that $V_1 = A$. We will aim to show by a structural analysis that V° is continuously characteristic in W, hence normal in $\hat{W}$, from which it follows by Lemma I 5.1, applied in $\hat{W}/A$, that $V_1 > A$ (even $V_1{}^\circ > A$ in this case), contradicting our initial assumption.

We first examine W and W_1. As W is normal in $\hat{W}$, applying Lemma I 5.1 in $\hat{W}/A$ we find that $W_1 > A$.

On the other hand, by assumption $V_1 = A$, so $W_1 \cap V = A$. Fixing $v \in V^\circ \setminus A$, it follows that $[W_1, v] > 1$. Now $[W_1, v] \leq W_1 \cap V^\circ = A$. As the group $[W_1, v]$ is both nontrivial and T-invariant, we find $[W_1, v] = A$. As $C_{W_1}(v) = V \cap W_1 = A$, we find $\mathrm{rk}(W_1/A) = \mathrm{rk}(A)$. Similarly, examining $[W, v]$ we find that $\mathrm{rk}(W/V^\circ) = \mathrm{rk}(A)$. It follows that $\mathrm{rk}(W) = \mathrm{rk}(V^\circ W_1)$ and by connectedness

$$W = V^\circ W_1 \ .$$

We claim also that W_1 is abelian. We have the sequence

$$1 \to A \to W_1 \to W_1/A \to 1$$

with T acting on each term. Now T acts on A as the multiplicative group of a field, and we claim that W_1/A is isomorphic to A as a T-module. For this, simply use any commutation map $W_1/A \to [W_1, u] = A$ with $u \in U^\times$. This puts us in a position to apply Proposition III 3.8, which settles the matter: W_1 is abelian.

We now have adequate structural information. We claim that V° is continuously characteristic in W, which then yields the contradiction alluded to at the outset.

First, if W_1 is elementary abelian, then since $W = V^\circ W_1$, all involutions in W are of the form vw with $v \in V^\circ$ and $w \in W_1$ commuting, which forces one of the two factors to lie in A. So the involutions of W all lie in V° or W_1, and it follows that V° and W_1 are the only maximal elementary abelian subgroups in W, so that each is continuously characteristic.

Suppose therefore that W_1 is not elementary abelian. Now as W has exponent at most 4, so does W_1. Since W_1 is abelian, with T acting transitively on the subgroup A. It follows easily that we have W_1 homocyclic of exponent four in this case, with

$$A = \Omega_1(W_1) \ .$$

There are again two possibilities, one of which is trivial: if all involutions of W lie in V, then V° is characteristic in W. So we will suppose that there is an involution $i = vw$ in W with $v \in V^\circ$ and $w \in W_1 \setminus A$. (Then $w \notin V$, since $\Omega_1(W_1) = A$.)

The element v may be supposed to lie in U. We will analyze the structure further, and show that V° is the unique maximal connected elementary abelian subgroup of W, and again it will be characteristic in W.

We have $1 = (wv)^2$ and thus $w^v = w^{-1}$. Suppose wv' is another involution in wV. Then similarly $w^{v'} = w^{-1}$, and $[w, vv'] = 1$. As $w \notin A$ we find $vv' \in A$. Thus the involutions in the coset wV are those of the form wvA. Now T acts transitively on $W/V^\circ \simeq W_1/A$, so every coset of V° in W contains a T-conjugate of wv, say $(wv)^t = w^t v$. So we see that involutions in W outside of V are those in $W_1 v$, where the element $v \in U^\times$ is fixed. Consider a commuting pair of such involutions wv and $w'v$ with $w, w' \in W_1$. Then $ww'^{-1} = (wv)(vw'^{-1})$ is an involution, hence in A under our current hypotheses. In short, wv and $w'v$ lie in the same coset of A. It follows that V° is the unique maximal connected elementary abelian subgroup of W in this case, and again is continuously characteristic.

We have considered all possibilities, under the hypothesis $V_1 = A$, and reached a contradiction in all cases. □

We can now settle the structure of S°.

LEMMA 1.8. *With our accumulated hypotheses and notations, we have* $S^\circ = W$. *In particular,* S°/A *is abelian.*

PROOF. We take $v_1 \in V_1 \setminus A$, and we make free use of Lemma 1.5. By definition, $[\hat{W}, v_1] \le A$, and $A = [W, v_1]$. Accordingly, we find $\hat{W} = W \cdot C_{\hat{W}}(v_1)$. As $C_{\hat{W}}(v_1) = V$, we find $\hat{W} = W$, and thus $S^\circ = W$.

We saw earlier that W/A is abelian (Lemma 1.5). □

1.3. Some structural consequences. It is beginning to seem, for the moment, that our ambient group G might be SL_3. We will now examine the structure of $C^\circ(A)$ further, showing that it has a normal (hence unique) Sylow° 2-subgroup S°; so $C^\circ(A)/O_2{}^\circ(C^\circ(A))$ is of degenerate type.

We retain our standing hypotheses and notation, though after the clarifying material of the last section some of the more esoteric features can be dropped. As always, the main assumption is that we have a standard component of type SL_2 in characteristic 2, commuting with a nontrivial 2-unipotent subgroup, and the objects of interest are G, A, U, and S°, primarily, along with L, T, $C^\circ(A)$, and $N^\circ(A)$.

LEMMA 1.9. *With notation and hypotheses as fixed in this section,* $S^\circ = O_2{}^\circ(C(A))$.

PROOF. We let $H = U_2(C^\circ(A))$, a K-group, and it suffices to show that this is solvable.

By Lemma 1.5, $\bar{H} = H/A$ has abelian Sylow° 2-subgroups, hence has the form $E(\bar{H}) \times \sigma(\bar{H})$ where $E(\bar{H})$ is a product of simple groups of type SL_2.

If $\bar{L}_1$ is a simple component of $\bar{H}$ of type SL_2, where as usual the base field is algebraically closed of characteristic 2, then $\bar{L}_1$ is covered by some definable perfect group L_1 (that is, $L_1' = L_1$), where $L_1 \cap A \leq Z(L_1)$. So by the theory of central extensions, L_1 is a quasisimple algebraic group covering SL_2, which in characteristic 2 must be SL_2 (Lemma II 2.23).

Accordingly, the decomposition of $\bar{H}$ lifts to a decomposition of H as $E(H) \times H_0$ with $E(H)$ a product of groups of type SL_2 in characteristic 2, and H_0 a 2-group, and we can decompose the Sylow° 2-subgroup S° in the same way, as a product of elementary abelian 2-groups with H_0.

However, this is incompatible with the known structure of S°. The center of S° is A, which is central in H, whereas the factor L_1 would contribute a noncentral subgroup of H to the center of S°. □

The next lemma is very general.

LEMMA 1.10. *Let G be a simple L^*-group of finite Morley rank and of even type, containing a definable subgroup of the form $A \rtimes T$ isomorphic to a Borel subgroup of* SL_2 *over an algebraically closed field of characteristic* 2, *split as usual with A a Sylow 2-subgroup and T a maximal torus.*

Then $N^\circ(A) = C^\circ(A) \cdot T$.

PROOF. We work with $H = N^\circ(A)$ and $\bar{H} = H/C(A)$, the latter acting faithfully on A.

The first point is that $\bar{H}$ is of degenerate type, which is seen as follows. We apply Proposition II 5.3 to $\bar{H}$, or rather to $U_2(\bar{H}) \cdot T$, and we conclude that there is a Sylow° 2-subgroup $\bar{S}$ of $\bar{H}$ which is invariant under the action of T. Then $C_A(\bar{S})$ is nontrivial (Lemma I 5.1), and T-invariant. In view of our assumptions, $C_A(\bar{S}) = A$, that is, $\bar{S}$ centralizes A, and is therefore trivial. So $\bar{H}$ is of degenerate type.

Now we can make an application of a very special case of Proposition IV 5.2. The image $\bar{T}$ of T in $\bar{H}$ acts as a group of pseudoreflections on A, although in this case $C_A(T) = 1$ and $A = [T, A]$. In this case, Proposition IV 5.2 implies that the normal closure $\hat{T}$ of $\bar{T}$ in $\bar{H}$ is abelian. Again, as $\bar{T}$ acts regularly on A it follows that it is self-centralizing, and thus $\bar{T}$ is normal in $\bar{H}$.

On the other hand, in view of its action on A, T is a good torus (Lemma I 11.7, and hence is central in $\bar{H}$. As $\bar{T}$ is self-centralizing, finally $\bar{T} = \bar{H}$, and this is our claim. □

LEMMA 1.11. *With notation and hypotheses as fixed in this section, S° is a Sylow° 2-subgroup of G.*

PROOF. As usual, it suffices to show that for any 2-unipotent subgroup S_1 containing S° as a normal subgroup, we have $S^\circ = S_1$.

As $S^\circ = W$, we have $A = Z(S^\circ)$ and hence $N^\circ(S^\circ) \le N^\circ(A)$. At the same time, by Lemma 1.10 we have $N^\circ(A) = C^\circ(A)T$ where T is our maximal torus of L normalizing A. Thus S° is a Sylow° 2-subgroup of $N^\circ(A)$. □

Next, we aim to show that $C(a) = C(A)$ for $a \in A^\times$, but we will break this down into several steps, beginning with an observation on the structure of S°.

LEMMA 1.12. *Let E be a definable elementary abelian subgroup of S°. Then* $\mathrm{rk}(E) \le 2\,\mathrm{rk}(A)$.

PROOF. We use the fact that $S^\circ = W$ and the detailed information about the structure of W afforded by Lemma 1.5. So in the present proof we write W rather than S° when using the relevant structural information. We may take E to be connected.

Fix $u \in U^\times$. Then $C_S{}^\circ(u) = AU$. We have a map $E \to A$ induced by commutation with u. If $\mathrm{rk}(C_E(u)) \le \mathrm{rk}(A)$, we have our estimate, and if $\mathrm{rk}(C_E(u)) > \mathrm{rk}(A)$, then $C_E{}^\circ(u)$ must meet U nontrivially.

So now take $u \in E \cap U^\times$. Then $E \le C_S{}^\circ(u) = AU$. Now fix $w \in W \setminus AU$. Again, we have a commutation map $E \to A$ induced by commutation with w, and since $E \le AU$ we have $C_E(w) \le A$, so again $\mathrm{rk}(C_E(w)) \le \mathrm{rk}(A)$ and our claim again follows. □

LEMMA 1.13. *With notation and hypotheses as fixed in this section, and with $a \in A^\times$, we have $U_2(C(a)) = U_2(C(A))$.*

PROOF. Let $H = U_2(C(a))$. If $A \le Z(H)$, then $H \le C(A)$ and our claim follows. So we suppose toward a contradiction that $A_0 = A \cap Z(H) < A$. Then we claim that

$$(*) \qquad A_0 \text{ is finite.}$$

Let $\check{T} = N_T(H)$. Evidently $\check{T}$ acts on A_0, and we claim that the action is transitive on $A_0^\times$. We have $a \in A_0$; if $b \in A_0^\times$, then we have $b = a^t$ with $t \in T$, and then $H \le C^\circ(b) = C^\circ(a^t) = H^t$, and by comparison of rank we have $H = H^t$, and $t \in \check{T}$. So the pair $(A_0, \check{T})$ represents the additive group of a field $\check{K}$ with the multiplicative group acting on it; the pair (A, T) represents a similar setup, with respect to a field K; and the embedding of the pair $(A_0, \check{T})$ into (A, T) gives an embedding of the field $\check{K}$ as a subfield of K, all definable. So $\check{K}$ is finite Lemma I 4.3, and hence A_0 is finite.

Let B be a Borel subgroup of H containing S°. Then B splits as $S^\circ \rtimes T_0$ with T_0 a definable connected $2^\perp$-group by Proposition I 9.8. As $A = Z(W) = Z(S^\circ)$, we have $T_0 \le N^\circ(A)$. Now $N^\circ(A) = C^\circ(A) \rtimes T$ (with $T \le L$ as usual) (Lemma 1.10). As T acts regularly on $A^\times$ it follows that $T_0 \le C(A)$. As $S^\circ \le C(A)$, we have $B \le C(A)$.

Now we will use the structure of the connected K-group H as given by Proposition II 4.8. So $\bar{H} = H/\sigma(H)$ is a product of simple algebraic groups, and T_0 covers a maximal torus in $\bar{H}$. As T_0 commutes with A, this forces A into $\sigma(H)$ (Fact II 1.8). But $\sigma^\circ(H) \le B \le C(A)$, so $A \le Z(\sigma^\circ(H))$.

Now we take an element $w \in H$ representing an involution of $H/\sigma(H)$ which carries the image of B to an "opposite" Borel subgroup. Then $H = \langle B, B^w \rangle$. Therefore $A \cap A^w \le Z(H)$, so by $(*)$ $A \cap A^w$ is finite, and $\mathrm{rk}(AA^w) = 2\,\mathrm{rk}(A)$. We observe then that AA^w is a maximal connected elementary abelian subgroup of S°, by the previous lemma.

So now $AA^w = \Omega_1{}^\circ(Z(\sigma^\circ(H))) \lhd H$. Now the maximal torus of the Borel subgroup B centralizes $AA^w \lhd H$, and therefore the conjugates of T_0 in H centralize this group; but taking into account the structure of H, the normal closure of T_0 in H covers H modulo $\sigma^\circ(H)$. Thus $AA^w \le Z(H)$ and we have a contradiction. □

Now we may "promote" this result to a fuller form.

LEMMA 1.14. *With notation and hypotheses as fixed in this section, and with $a \in A^\times$, we have $C(a) = C(A)$.*

PROOF. We claim first that

$$A = O_2{}^\circ(Z(U_2(C^\circ(a)))) \; .$$

The previous lemma implies that $A \le Z(U_2(C^\circ(a)))$. In the opposite direction, $O_2(Z(U_2(C^\circ(a)))) \le Z(S^\circ) = A$. So the claim holds, and it follows that $C(a) \le N(A)$.

Consider $H = N(A)$ and $\bar{H} = N(A)/C(A)$. Then $\bar{H}^\circ = \bar{T}$ by Lemma 1.10. Now $\bar{H}$ acts naturally on A and on $\bar{T}$, with both actions induced by conjugation in H and hence compatible with the action of T on A. It follows easily that $\bar{H}$ induces a finite group of field automorphisms on the field K associated with the pair (A, T). By Lemma I 4.5, these automorphisms are trivial, or in other words, $\bar{H}$ commutes with $\bar{T}$, and as T acts transitively on $A^\times$ we find that $\bar{H} = \bar{T}$, $N(A) = C(A) \rtimes T$.

So $C(a) \le N(A) = C(A) \rtimes T$, and as T acts regularly on $A^\times$ we find $C(a) \le C(A)$, as claimed. □

2. Fusion analysis

We enter the second, fusion-oriented, stage in the proof of Proposition 1.2, always under the contradictory hypothesis that AU is not a Sylow° 2-subgroup of $C(A)$.

2.1. The setup. In the previous section, assuming the failure of this proposition, we gathered a considerable amount of structural information, notably concerning the structure of a Sylow° 2-subgroup S° of $C_G(A)$. In the present section, we pursue this analysis, focussing on fusion (conjugacy of involutions) in G, and as a result we will reach a contradiction via the Thompson rank formula in the next section. This will prove Proposition 1.2, and will also prove Theorem VIII 1.1, which has previously been reduced to a configuration directly contradicting this proposition.

The relevant notation and facts, taken over from the previous section, are as follows.

NOTATION 2.1.

(i) *G is a simple L^*-group of finite Morley rank and of even type.*
(ii) *S° is a Sylow° 2-subgroup of G.*
(iii) *$A = Z(S^\circ)$ is an elementary abelian 2-group.*
(iv) *S°/A is elementary abelian.*
(v) *$U \le S$ is a unipotent, elementary abelian 2-group.*
(vi) *For $u \in U^\times$, we have $C_{S^\circ}(u) = AU$ and $[u, S^\circ] = A$.*
(vii) *L is a standard component of G of type SL_2 over an algebraically closed field of characteristic 2, and $[U, L] = 1$.*
(viii) *A is a Sylow 2-subgroup of L, and T is a maximal torus of L normalizing A (and acting regularly on $A^\times$).*
(ix) *$N^\circ(A) = C^\circ(A) \rtimes T$.*
(x) *$C(a) = C(A)$ for $a \in A^\times$*
(xi) *$S^\circ \lhd C^\circ(A)$, so $U_2(C^\circ(A)) = S^\circ$.*

As we are entering into a very detailed examination of this situation, terminating at a contradiction, we will keep these hypotheses in force to the end of the section, without further mention.

To this we add the following notation, with regard to involutions.

NOTATION 2.2.

(i) *I_a is the set of involutions in G conjugate to an involution in A.*
(ii) *I_u is the set of involutions in G conjugate to an involution in U.*

We observe that I_a is a single conjugacy class of involutions in G, and that the structure of I_u is considerably less clear.

LEMMA 2.3.

(i) *$I_a \cap I_u = \emptyset$.*
(ii) *I_u contains $AU \setminus A$.*

PROOF.

(i) For $a \in A^\times$, we have $U_2(C(a)) = S^\circ$, while for $u \in U^\times$ we have $L \leq U_2(C(u))$.

(ii) For $u \in U^\times$, we have $[u, S^\circ] = A$ and thus $uA \subseteq u^{S^\circ}$. □

This is our point of departure: we have some involutions in G known not to be conjugate.

2.2. $C(u)$**.** We carry out some further structural analysis, now aimed at understanding conjugacy classes of involutions. We begin with involutions in U.

LEMMA 2.4. *For $u \in U^\times$, we have $U_2(C^\circ(u)) = LU$, and $C^\circ(u) = C^\circ(U)$.*

PROOF. Set $H = C^\circ(u)$. As L is a standard component, we have $L \lhd H$, hence $H = L \times C_H(L)$. We have $C_H(L) \leq C^\circ(A)$ and hence $U_2(C_H(L)) \leq C_H(L) \cap U_2(C^\circ(A)) = C_H(L) \cap S^\circ \leq C_{S^\circ}(u) = AU$. So $U_2(C_H(L)) = U$ and $U_2(C^\circ(u)) = LU$.

Now any element $t \in C_H(L)$ centralizes A and hence normalizes $S^\circ = U_2(C(A))$. We examine the action of t on S°, recalling that t centralizes $u \in U^\times$.

For $w \in S^\circ$ we have $[u, w] \in A$ and $[u, w] = [u, w]^t = [u, w^t]$, thus $[w, t] \in C_{S^\circ}(u) = AU$. Hence for all $u' \in U$ we have $[u', w] = [u', w^t]$, and since $[u', w] = [u', w]^t = [u'^t, w^t] = [u'^t, w]$ we find $[u', t] \in C(w)$ as well; taking $w \in S^\circ \setminus AU$ we find $[U, t] \leq A$. It follows that t acts trivially on AU/A and on A.

Now $C_H(L)/U_2(C_H(L))$ has degenerate type and therefore contains no involutions by Theorem IV 4.1. Furthermore $U_2(C_H(L)) \leq C(AU)$ so the quotient $C_H(L)/U_2(C_H(L))$ acts on AU. By the above this group acts trivially on the factors of the chain $1 < A < AU$, so by Proposition I 10.7 it centralizes U. That is, $C_H(L) \leq C(U)$.

Thus $H = LC_H(L) \leq C(U)$. □

It follows from this that the rank of each conjugacy class of involutions contained in I_u is constant.

2.3. Failure of strong closure. Our analysis occasionally bifurcates, depending on whether A is strongly closed in S° or not. We have to prepare some extra information for use in the case in which strong closure fails, before returning to our general line of analysis. Our aim in this case is to show that all involutions fall into either I_a or I_u.

The next lemma tells us what happens when A is not strongly closed in S°: the hypothesis of the lemma holds if A is not strongly closed, and the converse is also valid, in view of point (i) of the lemma.

LEMMA 2.5. *Suppose that A_1 be a conjugate of A in G, with $A_1 \neq A$ and $A_1 \cap S^\circ > 1$. Set $B = \langle A^g : g \in G, [A, A^g] = 1 \rangle$.*

Then the following hold:

(i) $A_1 \leq S^\circ$,
(ii) $A_1 \cap A = 1$,
(iii) $S^\circ = B \rtimes U$,
(iv) $I(S^\circ) = (AU)^\times \cup B^\times$,
(v) $B = A \times A_1 = \langle A^{N(B)} \rangle$,
(vi) $N_G(A) \leq N(B)$,
(vii) AU *and* B *are normal in* $N(S^\circ)$.

PROOF.

(i) Fix $a_1 \in (A_1 \cap S^\circ)^\times$. Then $A \leq C^\circ(a_1) = C^\circ(A_1)$, so $A_1 \leq U_2(C^\circ(A)) = S^\circ$.

(ii) If $a \in (A \cap A_1)^\times$, then $C(A) = C(a) = C(A_1)$, and in particular, $A_1 \leq Z(S^\circ) = A$, and $A_1 = A$.

(iii) We will prove this with $B_0 = AA_1$ in place of B, leaving the verification that $B_0 = B$ until the proof of point (v). The group B_0 is elementary abelian of rank $2\,\mathrm{rk}(A)$, in view of (ii). If $B_0 \cap U > 1$, then taking $u \in (B_0 \cap U)^\times$ we have $B_0 \leq C_{S^\circ}(u) = AU$. In particular, $A_1^\times \subseteq (AU \setminus A) \subseteq I_u$, contradicting the fact that $I_u \cap I_a = \emptyset$. So $B_0 \cap U = 1$.

We remark that $B_0 \lhd S^\circ$ since $A \leq B_0$.

Now $\mathrm{rk}(S^\circ/AU) = \mathrm{rk}(A)$, as follows easily from our assumptions on S°. So by rank considerations, $S = UB_0$.

(iv) We consider an involution $i = ub$ with $u \in U$ and $b \in B_0$. As u and b are involutions, it follows that they commute, and then by our assumptions either u or b is in A.

(v) It follows easily from the preceding point that AU and B_0 are the unique maximal elementary abelian subgroups of S°. If $A_2 \neq A$ is a conjugate of A commuting with A, then $A_2 \leq U_2(C^\circ(A)) = S^\circ$, and A_2 can play the role of A_1; but then $AA_2 \neq AU$, so $AA_2 = B_0$. Thus $A_2 \leq B_0$, and $B = B_0$.

Evidently $\langle A^{N(B)} \rangle \leq B$. Now suppose $A_1 = A^g$. Then B^g is generated by A_1 and any conjugate of A_1 which commutes with A_1; one such conjugate is A, so $B^g = A_1 A = B$.

(vi) This point is clear from the definition of B.

(vii) We have seen that AU and B are the two maximal elementary abelian subgroups of S°, and that they are not conjugate in G. □

LEMMA 2.6. *Suppose that A is not strongly closed in S°. Set*

$$B = \langle A^g : g \in G, [A, A^g] = 1 \rangle \ .$$

Then the following hold:

(i) $N(B)$ *acts transitively on* $B^\times$.
(ii) $\mathrm{rk}(U) = \mathrm{rk}(A)$.

PROOF. We work with the group $H = N(B)$ and its quotient $\bar{H} = H/C_H(B)$ acting faithfully on B. We claim first that

(1) B is irreducible under the action of $\bar{H}$.

We consider $B_1 \leq B$ H-irreducible, and in particular normal in S°. Thus B_1 meets A nontrivially. Furthermore, $T \leq H$, so in view of the action of T on A, we have $A \leq B_1$. But then $B = \langle A^H \rangle \leq B_1$, so $B_1 = B$. This proves (1).

Now by the irreducibility and the faithfulness of the action, we find $O_2(\bar{H}) = 1$. In particular, by Propositions II 5.10 and II 5.13, we find $U_2(\bar{H}) = E(U_2(\bar{H}))$ is a product of algebraic groups of type SL_2 in characteristic 2. Let $\bar{L}_H = U_2(\bar{H})$; then $\bar{H} = \bar{L}_H C_{\bar{H}}(\bar{L}_H)$ in view of Fact II 2.25.

Let $\bar{H}_1 = C_{\bar{H}}(\bar{L}_H)$. We have $\bar{U} \leq \bar{L}_H$ and $\bar{T}$ acts on $\bar{L}_H$ centralizing $\bar{U}$, which is a Sylow° 2-subgroup, hence $\bar{T}$ belongs to $\bar{H}_1$. Now $\bar{H}_1$ acts on $C_B(\bar{U}) = A$, with $\bar{T}$ acting transitively on $A^\times$. Applying Lemma 1.10 we find that $N^\circ(A) = C^\circ(A) \cdot T$ and hence $\bar{H}_1 = C_{\bar{H}_1}(A) \cdot T$. Now $C_{\bar{H}_1}(A)$ is normalized by $C_{\bar{H}}(\bar{L}_H)$ and by $\bar{L}_H$, hence by $\bar{H}$, and thus $C_B(C_{\bar{H}_1}(A))$ is H-invariant; since this group contains A it is nontrivial, hence equal to B, or in other words, by faithfulness, $C_{\bar{H}_1}(A) = 1$. Thus $\bar{H}_1 = \bar{T}$ and $\bar{H} = \bar{L}_H \times \bar{T}$.

We can derive a field K from the action of $\bar{T}$ on B with respect to which the action of $\bar{H}$ is linear, by Proposition I 4.11. Furthermore, in view of the action of $\bar{T}$ on A, T is the full multiplicative group of the field, bearing in mind Lemma I 4.3, and thus A is 1-dimensional. On the other hand, $\mathrm{rk}(B) = 2\,\mathrm{rk}(A)$, so B is 2-dimensional.

Let L_0 be a quasisimple component of $E(\bar{H})$ (we drop the bars now, as we are not much concerned with the relation to G). As B is 2-dimensional, L_0 is a copy of $\mathrm{SL}_2(K_1)$ contained in $\mathrm{GL}_2(K_2)$ where K_1, K_2 are algebraically closed fields of characteristic 2. So $L_0 \leq \mathrm{SL}_2(K_2)$ (the commutator subgroup) and by rank considerations $L_0 = \mathrm{SL}_2(K_2)$. That is, we have now identified the action of L_0 on B. So we see that $\bar{H}$ acts transitively on B. Furthermore, U is a Sylow 2-subgroup of L_0 and thus has the same rank as the field K, which is $\mathrm{rk}(A)$. All claims are verified. □

LEMMA 2.7. *If A is not strongly closed in S°, then we have the following:*

(i) $I(G) = I_u \cup I_a$.
(ii) S° *is a Sylow 2-subgroup of G.*

PROOF.
(i) We will reduce this point to the next.
By Lemma 2.5, we have the group

$$B = \langle A^g : g \in G, [A, A^g] = 1 \rangle$$

to work with, and $I(S^\circ) = I(AU) \cup I(B)$. By the previous lemma $I(B) \subseteq I_a$, and we have already seen that $I(AU) \setminus I(A) \subseteq I_u$ (Lemma 2.3). So $I(S^\circ) \subseteq$

$I_u \cup I_a$. Accordingly, once we know that S° is a Sylow 2-subgroup of G we will be done.

(ii) Our point of departure is an element $s \in N_G(S^\circ)$ satisfying $s^2 \in S^\circ$. We must show that $s \in S^\circ$. Again, we make use of the subgroup

$$B = \langle A^g : g \in G, [A, A^g] = 1 \rangle \ .$$

Since $C_A(s) > 1$, we have $s \in C(C_A(s)) = C(A)$. Now s normalizes AU. In particular, one can find $u \in U^\times$ with $[s, u] \in A$. But $[B, u] = A$ and thus we may replace s by some element sb with $b \in B$ so as to obtain $[s, u] = 1$. The standard component L is a simple component of $C^\circ(u)$, and these are permuted by s, while s centralizes A. Hence s normalizes L and acts on it like an element of A. We may therefore make a second adjustment of s by an element of A, and now we have $[s, L] = 1$. So $s^2 \in C_{S^\circ}(L) = U$.

As s also acts on B, there is $b \in B \setminus A$ such that $[s, b] \in A$. Then $1 = [s, b]^2 = [s^2, b]$, and as $s^2 \in U$ and $b \notin A$, we find $s^2 = 1$.

Now $\mathrm{rk}(U) = \mathrm{rk}(A)$, and it follows easily that $[U, b] = A$. Thus there is $u \in U$ with $[u, b] = [s, b]$, and adjusting s by u we find $[s, b] = 1$. Now if we have $s = 1$ after all these adjustments, then the original element was in S°. So assume that $s \neq 1$.

Since $L \leq C^\circ(s)$ and s is an involution, while L is a standard component, we have $L \lhd C^\circ(s)$. Furthermore, b permutes the components of $C^\circ(s)$. As b centralizes A, it normalizes L, and hence also normalizes $C^\circ(L) \cap S^\circ = U$. But this is not the case, and we have a contradiction. □

2.4. The Thompson map. We are going to define a partial function $I_a \times I_u \to I_u$, and show that it is generically defined (that is, defined for a generic element of the domain). Then by comparing the ranks of the domain, range, and fibers of this map, we will get useful information, first numerical and then structural, and finally derive a contradiction from the present configuration.

The starting point is Lemma I 2.20: for any pair of involutions i, j in G which are not conjugate, there is an involution $k \in d(\langle ij \rangle)$ commuting with both. This applies in particular when $i \in I_a$ and $j \in I_u$, and our first point will be that this gives rise to a definable function from $I_a \times I_u$ into $I(G)$; our second point will be that this function has an image which lies, generically, in I_u. With that in hand, we will be ready to undertake our actual computations; these computations aim at getting some sharp (and ultimately, unreasonable) estimate for the rank of G.

LEMMA 2.8. *Let G be a group of finite Morley rank of even type. Then the following hold:*

(i) *For any pair i, j of involutions in G, there is at most one involution in $d(\langle ij \rangle)$.*
(ii) *The function with domain all pairs of involutions (i, j) in G, for which $d(\langle ij \rangle)$ contains an involution k, given by taking (i, j) to that involution k, is a definable function.*

PROOF.

(i) Let $a = ij$. The group $d(a)$ has the form $D \oplus C$ with D divisible and C finite cyclic (Lemma I 2.16). In particular, D is a $2^{\perp}$-group and our claim follows.

(ii) Definability issues are handled by referring to the definable function $\hat{d}$ introduced in Lemma IV 4.2, which for these purposes has the same properties as the function d. □

NOTATION 2.9. *We use the preceding lemma to define a function θ, called the* Thompson map*:*

$$\theta : I_a \times I_u \to I(G) .$$

Here we use Lemma I 2.20 to guarantee the existence of the desired involution.

Now we claim that for generic elements of $I_a \times I_u$ we will have $\theta(i,j) \in I_u$. At this point we divide into cases, depending on whether or not A is strongly closed in S°.

LEMMA 2.10. *If A is strongly closed in S°, then the image of the Thompson map θ is contained in I_u.*

PROOF. This means that we consider an involution i (in the image) commuting with involutions a and u in I_a and I_u respectively, where we may suppose that u is actually in U. We then claim that i is in I_u.

By Lemma 2.4 we have $U_2(C(u)) = LU$ and it follows that i, which acts on $C(u)$, normalizes L. Now i acts as an inner automorphism on L and hence centralizes a conjugate of A; we may suppose this conjugate to be A itself. At the same time we have a lying in a conjugate A_1 of A, and $i \in C(a) = C(A_1)$.

Suppose first that A and A_1 generate a 2-subgroup Q of $C^{\circ}(i)$. Then by strong closure in Q (Lemma II 5.15) these two groups coincide. But then ua is an involution and $i \in d(\langle ua \rangle) = \langle ua \rangle$, $i = ua$. As $u \neq 1$, this belongs to I_u (Lemma 2.3).

So we suppose now that A and A_1 do not generate a 2-subgroup of $C^{\circ}(i)$. In particular, neither of these groups is normal in $C^{\circ}(i)$ and hence by Lemma II 6.6 we get subgroups L_1 and L_2 normal in $C^{\circ}(i)$, of type SL_2, associated with A_1 and A respectively. That is, L_1 is normalized by A_1, and $A_1 \cap L_1$ is a Sylow 2-subgroup of L_1, while L_2 and A are similarly related. As A normalizes L_2, and centralizes the Sylow 2-subgroup $A \cap L_2$, the group A acts on L_2 like $A \cap L_2$ and therefore

$$A = L_2 \times C_A(L_2) .$$

On the other hand, $C(a) = C(A)$ for $a \in A^{\times}$, so if $C_A(L_2) > 1$ we take $a \in C_A(L_2)^{\times}$ and conclude $(A \cap L_2) \leq Z(L_2)$, which is nonsense. It follows that $C_A(L_2) = 1$ and thus $A \leq L_2$. Similarly, $A_1 \leq L_1$. As A and A_1 do not commute, we find $L_1 = L_2 = \langle A, A_1 \rangle$.

Now $u \in C(i)$ and u centralizes A, hence normalizes L_1. Also $a \in L_1$ and $ua \in \langle u \rangle L_1$, so $i = \theta(a, u) \in \langle u \rangle L_1$. Since L_1 centralizes i, we cannot have $i \in L_1$; so i is in the coset uL_1.

Now as u centralizes A and acts on L_1, it acts like an element a' of A. Thus $iua' \in L_1 \cap C(L_1) = 1$, and $i \in uA$. So $i \in AU \setminus A \subseteq I_u$. □

We prepared some additional information in the preceding subsection to deal with the case in which A is not strongly closed in S°. We will however need one estimate before proceeding to this case.

LEMMA 2.11. *With* $f = \mathrm{rk}(A)$ *and* $u = \mathrm{rk}(U)$, *we have* $\mathrm{rk}(I_u) \geq 4f + u$.

PROOF. We consider another partial Thompson map,

$$\tau : I_u \times I_u \to I_u \ ,$$

defined for pairs of involutions $u_1, u_2 \in I_u$ for which the group $d(\langle u_1 u_2 \rangle)$ contains an element of I_u. This is a definable map, and the element in question is unique when it exists.

Let $u_1, u_2 \in U^\times$ be arbitrary and consider the set

$$X = \{(a, b) \in I(L) \times I(L) : d(\langle ab \rangle) \text{ is a } 2^\perp\text{-subgroup of } L\} \ .$$

The elements of L are either semisimple $2^\perp$-elements lying in tori, or involutions. The set $I(L) \times I(L)$ has rank $4f$, and a generic pair $(a, b) \in I(L) \times I(L)$ belongs to the set X. Accordingly, with $u \in U^\times$ fixed, the set of pairs $(u_1 a, u_2 b)$ subject to

$$u_1, u_2 \in U^\times; u_1 u_2 = u; (a, b) \in X$$

has rank $u + 4f$. We claim that all such pairs belong to the fiber $\tau^{-1}(u)$. Certainly $ua, ub \in I_u$, as a, b are conjugate to elements of A in L and $AU \setminus A \subseteq I_u$. Furthermore, $u_1 a \cdot u_2 b = u(ab)$ and thus $d(\langle (u_1 a)(u_2 b) \rangle) \leq \langle u \rangle \times d(\langle (ab) \rangle)$. As $(uab)^2 = (ab)^2 \in d(\langle (ab) \rangle)$ and $(uab) \notin d(\langle (ab) \rangle)$, there is an involution in $d(\langle (uab) \rangle)$; but $d(\langle (uab) \rangle) \leq \langle u \rangle \times d(\langle (ab) \rangle)$ and this involution can only be u. So $\tau(u_1 a, u_2 b) = u$. In other words, above every point of $U^\times$, and hence above every point of I_u, the fiber of τ has rank at least $4f + u$. At the same time, since τ carries a subset of $I_u \times I_u$ to I_u, in view of Lemma I 2.1 (Fubini property) we must have some fibers of rank at most $\mathrm{rk}(I_u)$ (which bounds the difference between the rank of the domain and the rank of the image), giving the stated inequality $\mathrm{rk}(I_u) \geq 4f + u$. □

LEMMA 2.12. *Suppose that* A *is not strongly closed in* S°. *Then the Thompson map*

$$\theta : I_a \times I_u \to I(G)$$

maps generically into I_u; *that is, the image of a generic pair in* $I_a \times I_u$ *is in* I_u.

PROOF. The main point is the exhaustive analysis of involutions in G in this case: $I(G) = I_a \cup I_u$. So it suffices to show that the set of pairs $(a, u) \in I_a \times I_u$ with $\theta(a, u) \in I_a$ is *not* generic in $I_a \times I_u$.

Now I_a is a single conjugacy class in G, so for $i \in I_a$, the rank of the fiber $\theta^{-1}(i) \subseteq I_a \times I_u$ is independent of the choice of i. So if $\theta^{-1}(I_a)$ is generic in $I_a \times I_u$, then the rank of the fiber is the difference of the rank of domain and image, namely $\mathrm{rk}(I_u)$.

But we can in fact compute this fiber rank. We fix $a \in I_a$, and we may suppose $a \in A^\times$. We consider a pair (a_1, u) in $I_a \times I_u$ with $\theta(a_1, u) = a$. In particular, a commutes with both a_1 and u.

Now we have the subgroup $B = \langle A^g : g \in G, [A, A^g] = 1\rangle$ which we have analyzed. The element $a_1 \in C(a) = C(A)$, and a_1 belongs to a conjugate A_1 of A, with $A \leq C(a_1) = C(A_1)$; so $a_1 \in B$. Similarly, $u \in C(a) = C(A)$. Now $S^\circ = U_2(C(A))$, and by Lemma 2.7, S° is a full Sylow 2-subgroup of G. Since $S^\circ \lhd C(A)$, it follows that $u \in S^\circ$. Furthermore, we have determined that $I(S^\circ) = I(B) \cup I(AU)$ and $I(B) \subseteq I_a$, hence $u \in AU \setminus A$. Then $(a_1u)^2 = [a_1, u] \in A$ is an involution, and thus $a = [a_1, u]$; conversely, given such elements a_1 and u we will have $\theta(a_1, u) = a$, and thus the fiber rank is determined: $\mathrm{rk}(\theta^{-1}(a)) = \mathrm{rk}(B) + \mathrm{rk}(AU) - \mathrm{rk}(A) = 2f + u$ (or $3f$, since in fact $u = f$ in this case).

Now the fiber rank for a map from a generic subset of $I_a \times I_u$ to I_a (with constant fiber rank) should be $\mathrm{rk}(I_u) \geq 4f + u$, so we have a contradiction. □

NOTATION 2.13. *The Thompson map $\theta_0 : I_a \times I_u \to I_u$ is the restriction of $\theta : I_a \times I_u \to I(G)$ to the generic subset $\theta^{-1}(I_u) \subseteq I_a \times I_u$.*

2.5. The rank of G. With the Thompson map θ_0 in hand, we can carry out our computation of the rank of G.

LEMMA 2.14. $\mathrm{rk}(G) = \mathrm{rk}(C(A)) + 4f$.

PROOF. We claim that the rank of the fibers of the Thompson map θ_0 is constant, and equal to $4f$. As $\mathrm{rk}(I_a \times I_u) - \mathrm{rk}(I_u) = \mathrm{rk}(I_a)$, we then find $4f = \mathrm{rk}(I_a) = \mathrm{rk}(G) - \mathrm{rk}(C(A))$.

We compute the ranks of these fibers much as we have just done above points of I_a. We fix $u \in U^\times$ and consider $r = \mathrm{rk}(\theta^{-1}(u))$.

On one hand, for $a, b \in I(L)$, we will have, generically, that $d(\langle(ab)\rangle)$ is a $2^\perp$-group, as in our previous analysis, and in such cases we will have $\theta(a, ub) = u$, much as before. So this gives us a lower bound for the size of the fiber rank: $r \geq 2\,\mathrm{rk}(I(L)) = 4f$.

In the reverse direction, consider any pair $(a, v) \in I_a \times I_u$ with $\theta(a, v) = u$. Here $a \in A_1$ for some conjugate of A and $u \in C(a) = C(A_1)$. Hence $a \in U_2(C^\circ(u)) = LU$ by Lemma 2.4. Since $a \in I_a$ it follows that $a \in L$. On the other hand, as $v \in C(u)$, it follows that v normalizes $U_2(C^\circ(u))$ and in particular v normalizes L. Hence $d(\langle av\rangle) \leq L\langle v\rangle$ and as $\theta(a, v) = u$ we have $u \in L\langle v\rangle \setminus L = Lv$, $v \in Lu$ and thus $v = ub$ with $b \in I(L)$ (since u commutes with L). So $r \leq 4f$, and finally $r = 4f$ as claimed. □

We will see later that $\mathrm{rk}(C(A)) \leq 4f$, which gives $8f$ as an upper bound for $\mathrm{rk}(G)$, a possibility which is again suggestive of SL_3.

3. Final analysis

3.1. On $C^\circ(A)$. With the rank of G in hand we may return to our structural analysis; a contradiction is not far off.

LEMMA 3.1.
(1) $\langle S^\circ, L\rangle = G$,
(2) $C(S^\circ) = A$,
(3) $C^\circ(AU) = AU$,
(4) $C^\circ(U) = LU$.

PROOF.
Ad (1) We set $H = U_2(\langle S^\circ, L\rangle)$, and we suppose $H < G$. Then H is a K-group, and by Proposition II 5.23 we have $E(C_H(U)) \lhd E(H)$. But $E(C_H(U)) = L$ by Lemma 2.4, so $L \lhd E(H)$. As H is connected, it follows that L is normal in H, and hence S° normalizes L, which is false.

Ad (2) We have $C(S^\circ) \le C(U) = L \times C(LU)$. So $C(S^\circ) \le C_{L\times C(LU)}(A) = A \times C(LU)$, and hence $C(S^\circ) = A \times C_{S^\circ}(LU)$. But the intersection $C(S^\circ) \cap C(LU)$ centralizes $\langle S^\circ, L\rangle = G$, so $C(S^\circ) = A$.

Ad (3) We have $C(AU) \le N(A) \le N(S^\circ)$ (Lemma 1.9). For $g \in C(AU)$ and $w \in S^\circ$, $u \in U^\times$ we have $[u, w] = [u, w]^g = [u, w^g]$ and hence $[w, g] \in C_{S^\circ}(u) = AU$. Thus g acts trivially on S°/AU as well as on AU. On the other hand, AU is a Sylow° 2-subgroup of $C(AU)$ and thus for a Borel subgroup B of $C(AU)/AU$, B is a $2^\perp$ group and hence acts trivially on S° by Proposition I 10.7. But $C(S^\circ) = A$ so B is trivial and $C(AU)/AU$ is finite.

Ad (4) $C^\circ(U) = LC_{C(U)}{}^\circ(L)$ and $C_{C(U)}{}^\circ(L) \le C^\circ(AU) = AU$, whence the claim. □

LEMMA 3.2. *$C^\circ(A) = S^\circ \rtimes T_1$ for some torus T_1, where T_1 acts freely on U.*

PROOF. We consider the groups $H = N^\circ(A)$ and $\bar{H} = N^\circ(A)/S^\circ$. Since S° acts trivially on both S°/A and on A, and $N^\circ(A)$ normalizes S°, the group $\bar{H}$ has a natural action on each of these groups. Our claim is mainly concerned with the structure of $C_{\bar{H}}(A)$.

Note that $\bar{H}$ is of degenerate type.

Let $\widetilde{W} = S^\circ/A$. We claim that $C_{\bar{H}}(\widetilde{W}) = 1$. For this, we will compute in H. Suppose $h \in N^\circ(A)$ acts trivially on $\widetilde{W}$. Then for $w \in S^\circ$ we have $(w^2)^h = w^2$ and hence h acts trivially on A as well. Now $C_{\bar{H}}(\widetilde{W})$ is a $2^\perp$-group, hence 2-divisible, and acts trivially on A and on $\widetilde{W}$, and thus by Proposition I 10.7 this group acts trivially on S° itself. But $C(S^\circ) = A$ and thus $C_{\bar{H}}(\widetilde{W}) = 1$.

We consider a torus T in L normalizing A. We claim that $\bar{T}$ acts on $\widetilde{W}$ as a pseudoreflection subgroup.

The $\bar{T}$-module $\widetilde{W}/\widetilde{U}$ is isomorphic to A via any commutation map with an element of $U^\times$. In particular, $\widetilde{W} = \widetilde{U} + [\bar{t}, \widetilde{W}]$ for any $t \in T$, and it follows from this that $[\bar{T}, \widetilde{W}] = [\bar{t}, \widetilde{W}]$ as well. Now by rank considerations $U \cap [\bar{t}, \widetilde{W}]$ must be finite. We must, however, consider the possibility that $U_0 = U \cap [\bar{T}, \widetilde{W}]$ is a finite group centralized by T. But $[\bar{T}, \widetilde{W}]$ is connected, and by Theorem III 4.1 the sequence

$$1 \to U_0 \to [\bar{T}, \widetilde{W}] \to A \to 1$$

(or one may prefer to write "0" for "1" here) splits definably. Thus $U_0 = 1$. So $\widetilde{W} = \widetilde{U} \oplus [T, \widetilde{W}]$. As $U \cap A = 1$ and $A \leq [T, S^\circ]$ we see that $S^\circ = [T, S^\circ] \rtimes U$.

Returning to $\widetilde{W}$, we have seen that $\bar{T}$ acts as a pseudoreflection subgroup. We will show that $\bar{T}$ is central in $\bar{H}$. We may consider the normal closure $\hat{T}$ of $\bar{T}$ in $\bar{H}$. As this is a degenerate type group, it follows by Proposition IV 5.2 that this normal closure is abelian, after factoring out the annihilator of $\widetilde{W}$, which is a finite and hence central subgroup. It follows that $\hat{T}$ is itself abelian, being connected. Furthermore, in view of its almost faithful action on $\widetilde{W}$, $\hat{T}$ is a good torus; being normal in H, it is central in H. In particular, $\bar{T}$ is central in $\bar{H}$ (so, in fact, $\hat{T} = \bar{T}$).

Now consider the group $B = [T, S^\circ]$. Then $\bar{T}$ acts on B/A as it acts on A, and $\bar{H}$ acts on B/A commuting with $\bar{T}$, so $\bar{H} = C_{\bar{H}}(B/A) \times T$.

There are two possibilities for the structure of B. If $B \setminus A$ contains an involution b, then the coset bA consists of involutions, and by transitivity of the action of T it follows that B is an elementary abelian 2-group. Otherwise, the elements of $B \setminus A$ are all of order 4, and the pair (B, T) constitutes a free Suzuki 2-group, hence is abelian homocyclic by Theorem III 3.2.

In either case, taking $b \in B \setminus A$, representing $\bar{b} \in B/A$, we have $\bar{H} = C_{\bar{H}}(\bar{b}) \times \bar{T}$, and by the action of T we have $C_{\bar{H}}(\bar{b}) = C_{\bar{H}}(B/A)$; also $C_{\bar{H}}(\bar{b})$ is connected since H is. Now in the homocyclic abelian case, $C_H(\bar{b}) \leq C_{\bar{H}}(b^2) = C_{\bar{H}}(A)$ and $C_{\bar{H}}(\bar{b})$ centralizes B/A and A, hence $C_{\bar{H}}(\bar{b})$ centralizes B (apply Proposition I 10.7 to a Borel subgroup).

One shows easily that $C^\circ(B) = B$ using the commutator map

$$B/A \times U \to A$$

(compare the proof that $C^\circ(AU) = AU$). So in the homocyclic abelian case, exponent 4, we find $\bar{H} = \bar{T}$ and $C^\circ(A) = S^\circ$, a very strong form of our claim.

Suppose therefore that $B = [T, S^\circ]$ is elementary abelian. Again, with $b \in B \setminus A$, consider the decomposition $\bar{H} = C_{\bar{H}}(b) \times \bar{T}$. Let $\bar{T}_0 = C_{\bar{H}}(b)$. In view of the action of $\bar{T}$, $C_B(\bar{T}_0)$ covers B/A; if $C_B(\bar{R}_0)$ meets A nontrivially, then we fall back into the same situation as in the exponent 4 case and our claims hold in a strong form. Consequently, we may suppose that $A_1 = C_B(\bar{T}_0)$ is a complement to A in B.

We claim that T_0 acts freely on U. Suppose $x \in T_0$ and $u \in U^\times$ with $u^x = u$. Then taking $b \in B^\times$ we have $[u, b]^x = [u, b] \in A^\times$ and thus B meets $A^\times$, a contradiction.

Now for $x \in T_0$ we can relate the action of x to the action of T as follows. Fix $u \in U^\times$, $b \in B^\times$. We have $[u^x, b] = [u, b]^x = [u, b]^t = [u, b^t]$ for some unique $t \in T$. Furthermore, for $b_1 = b^{t'} \in B^\times$, $t \in T$, we have

$$[u^x, b_1] = [u^x, b^{t'}] = [u^x, b]^{t'} = [u, b^t]^{t'} = [u, (b^{t'})^t] \ .$$

and thus the choice of t depends only on x and u.

So if $[u^{x_1}, b] = [u, b^{t_1}]$ and $[u^{x_2}, b] = [u, b^{t_2}]$ with $x_1, x_2 \in T_0$ and $t_1, t_2 \in T$, we find:

$$[u^{x_1 x_2}, b] = [u^{x_1}, b]^{x_2} = [u, b^{t_1}]^{x_2} = [u, b^{t_1 t_2}]$$

Thus T_0 is commutative and acts freely on U. By Proposition I 11.7, T_0 is a good torus, and $\mathrm{rk}(T_0) \leq u$ by the freeness of the action.

Now $\bar{H} = T_0 T$ and $C_H(A)$ is disjoint from $\bar{T}$. So $C^\circ(A)/S^\circ$ is a good torus of rank at most u, and by Proposition I 9.6 $C^\circ(A)$ splits over S° as stated, with the complement definable by Proposition I 9.8. Since it is disjoint from T it acts freely on U. □

3.2. A final contradiction. We come to the final step in the proof of Proposition 1.2: we can now kill off our configuration, using the rank information coming from the Thompson map. Let $t_1 = \mathrm{rk}(T_1)$ in the notation of the previous lemma.

We have

$$\mathrm{rk}(G) = \mathrm{rk}(C(A)) + 4f \ .$$

Now $\mathrm{rk}(S^\circ) = 2f + u$ (recall the setup in Notation 2.1, point (vi)). So by Lemma 3.2 we have $\mathrm{rk}(C(A)) = 2f + u + t_1$ and $\mathrm{rk}(G) = 6f + t_1 + u$.

Compare this with the corresponding formula for $i \in I_u$. Then $\mathrm{rk}(G) = \mathrm{rk}(i^G) + \mathrm{rk}(C(i)) = \mathrm{rk}(i^G) + \mathrm{rk}(C(U))$, so

$$6f + u + t_1 = 3f + u + \mathrm{rk}(i^G)$$

and $3f + t_1 = \mathrm{rk}(i^G)$. Now each such conjugacy class i^G has rank $\mathrm{rk}(G) - \mathrm{rk}(C(i)) = \mathrm{rk}(G) - \mathrm{rk}(C(U))$ and thus these ranks are constant. On the other hand the rank of the set of conjugacy classes in I_u is at most the rank of the set of conjugacy classes in U with respect to the action of T_1, which is $u - t_1$, and hence $\mathrm{rk}(I_u) \leq \mathrm{rk}(i^G) + (u - t_1)$. By Lemma 2.11 we have the inequality $\mathrm{rk}(I_u) \geq 4f + u$, so this yields

$$\mathrm{rk}(i^G) \geq 4f + t_1 \ .$$

As $\mathrm{rk}(i^G) = 3f + t_1$, we have a contradiction.

With this contradiction, the proof of Proposition 1.2 is complete.

As we see throughout, the group G "wants" to be SL_3, but the group U interferes with this. Much of the time U simply plays the part of a subgroup of S° which should be of rank f, but the assumption that this group commutes with L leads finally to a contradictory result.

4. Notes

General remarks

We deal here with groups with a standard component of type SL_2. Our approach to the classification of groups of even type avoids extensive work with standard components, except at two extremes: the very small (SL_2, treated in this chapter to the limited degree actually needed later), and the very large (components of parabolic subgroups).

Our definition of *standard component*, given in the introduction to the present chapter, is not a faithful analog of the notion in the finite case, where one would also assume a further "largeness" condition: L does not commute with any of its conjugates. This additional condition can in fact be derived from our definition, but only by using the machinery we develop here. By the time we are in a position to recover the customary definition we have already completed our analysis. So we merely emphasize that our terminology is best viewed as a significant deviation from that customary in the finite case, though eventually reconcilable.

It should be mentioned that the possibility of using the theory of pseudoreflection groups in our context has no natural analog in finite group theory. This was one of the original incentives for taking up our project, as it was clear that here, at least, some dramatic simplification occurs. Other drastic simplifications encountered later (notably, on the way to the theory of amalgams), were not anticipated at the outset.

The Thompson rank computations of §2 are a direct analog of the so-called Thompson order formula in finite group theory. For such purposes, one should think of $\mathrm{rk}(X)$ as giving the order of magnitude of $\ln(|X|)$, where however the base of the logarithm is in some sense infinite. This can be made more precise in the case of Chevalley groups over finite fields, but in general is only intended to be suggestive.

Strategic shifts

The material in this chapter (and the beginning of the next) has had a complicated history. In addition to the substantial modifications involved in passing from a K^* context to an L^* context, it has been greatly affected by shifts in our overall strategy.

At this point in the analysis, finite simple group theory offers an embarrassment of riches, and we were initially seduced by it into a number of developments that turned out not to be essential in our particular setting.

In finite group theory, standard components are an important point of departure for the recognition of simple groups. We have developed a standard component theory for groups of finite Morley rank (unpublished) but soon thereafter we found a shorter way to reach our goal, and we did not pursue this—once one succeeds in classifying the simple groups of even type, by any method, this theory becomes vacuous. But we require various fragments of that theory. In particular, the proof of Theorem VIII 5.2 developed out of a treatment of the *existence of standard components* parallel to [**20**], though this is no longer visible. The material on standard components also exists in a more general form (cf. §VIII 7) in the K^*-case, and can easily be adapted to the L^*-case.

But in the present chapter we deal not with the existence of standard components, but with a vestige of the analysis of groups having particular groups as standard components, but only in the case of a component of type SL_2. And we deal with only the first part of this analysis, namely the size of the Sylow 2-subgroup in the centralizer of the standard component, Theorem 1.1. (cf. [**22**]).

Theorem 1.1 has its own peculiar history, and for some time our proof was entangled with the proof of a more general form of Theorem VIII 1.1 from the next chapter, which treats the case of abelian Sylow 2-subgroups. In finite group theory one should treat the more general case of Sylow 2-subgroups with strongly closed abelian 2-subgroups, and we initially followed that path (with Sylow° 2-subgroups in place of Sylow 2-subgroups), but as we use the amalgam method rather than fusion analysis in our identification of quasithin groups it turns out that this falls away.

In our present treatment Theorem 1.1 will only be needed once, in the proof of the $C(G,T)$ theorem in the next chapter; namely, within the proof of Theorem VIII 3.1.

While we have eliminated most of the standard component analysis (apart from the case of SL_2 as a standard component) by invoking the amalgam method at a relatively early stage, this approach has not yet been made to work in finite group theory, though something similar is under active investigation. There are various reasons that our approach could be unworkable in the case of finite simple groups, notably the obstructions associated with groups defined over fields of order 2.

CHAPTER VIII

The $C(G,T)$ Theorem and a Plan of Attack

Introduction

We arrive, in this chapter, at the third and last of our characterizations of SL_2: the $C(G,T)$ theorem.

DEFINITION 1. *Let G be a group of finite Morley rank and S a definable subgroup of G. Then $C(G,S)$ is the subgroup of G generated by all subgroups of the form $U_2(N(X))$ as X varies over definable connected subgroups of S which are invariant under the action of $N_G{}^\circ(S)$.*

Here the "C" stands for "characteristic", but we have drifted away from the definition used in finite groups, and the notion we use looks more like a notion from algebraic group theory; the condition of invariance under $N_G{}^\circ(S)$ can be read on the one hand as saying that the normalizer of X is parabolic, or on the other hand as an approximation to "characteristic". One could call this the "parabolic-generated core" but in practice the notation $C(G,S)$ is sufficient.

One case interests us here: S is a Sylow° 2-subgroup, and in an L^*-group of even type, it will follow eventually that $N_G{}^\circ(S)$ is a Borel subgroup, so we consider the normalizers of unipotent subgroups of S which contain this Borel subgroup, or in more suggestive language: $C(G,S)$ is generated by the *parabolic subgroups* containing $B = N^\circ(S)$.

THEOREM 3.3. *Let G be a simple L^*-group of finite Morley rank and of even type, and S a Sylow° 2-subgroup. If $C(G,S) < G$, then $G \simeq \mathrm{SL}_2(K)$ for some algebraically closed field K of characteristic 2.*

The "T" in "$C(G,T)$" comes from the usage in finite group theory, where T would be a (full) Sylow 2-subgroup. It seemed best to leave the name of the theorem intact while adapting the content.

This theorem relies on two prior results. One is the main result of the preceding chapter, Theorem VII 1.1. The other is the classification of groups with abelian Sylow° 2-subgroups, given in §1.

Another important topic we take up, after dealing with the $C(G,T)$ theorem, concerns the structure of the parabolic subgroups.

DEFINITION 2. *Let G be a group of finite Morley rank of even type.*

(1) *A Borel subgroup B of G is* standard *if it contains a Sylow° 2-subgroup.*

(2) *A definable subgroup H of G is* parabolic *if it contains a standard Borel subgroup.*

The structural result is as follows.

THEOREM 5.2. *Let G be a connected simple L^*-group of finite Morley rank, and of even type, and P a proper parabolic subgroup. Then $F^*(P) = O_2(P)$.*

Among other things, this prepares us for the use of the amalgam method in the next chapter. More generally, the structure of parabolic subgroups together with the $C(G,T)$ theorem give us the key to the identification of an L^*-group G as an algebraic group in all cases, with the amalgam method used to handle the smallest cases.

We lay this all out at the end of the chapter, where we present the plan of attack for the identification phase of the proof of the classification theorem (§6).

The first three sections of the chapter lead to the proof of the $C(G,T)$ theorem, beginning with the classification of groups with abelian Sylow° 2-subgroups, followed by an analog of a theorem of Baumann. The next two sections analyze the structure of proper parabolic subgroups, after which we give our plan of attack for the recognition phase in the classification of simple groups of even type.

After our lengthy preparations, the recognition phase goes very quickly for the most part, reducing to the "quasithin" cases treated by the amalgam method in the next chapter, which require an extensive analysis. The amalgam method is only loosely tied to finite group theory in the first place, and goes over quite directly to our context, modulo the usual additional attention to be paid to connected components, and some technical issues of interpretability.

We should note that the amalgam method does not actually produce an identification of the desired groups, but provides all the data needed for an identification via the theory of Moufang polygons—which may be considered, from our point of view, as simply the logical continuation of the amalgam method in favorable cases.

1. Abelian Sylow° 2-subgroups

In this section we provide the last ingredient needed for the proof of the $C(G,T)$ theorem, namely the classification of groups with abelian Sylow° 2-subgroups.

THEOREM 1.1 (Abelian Sylow°). *Let G be a simple L^*-group of finite Morley rank and even type. Suppose that G has an abelian Sylow° 2-subgroup S. Then $G \simeq \mathrm{SL}_2(K)$ for some algebraically closed field of characteristic 2.*

Well-placed SL_2-subgroups. Our first lemma shows that under the hypothesis that our group is not a copy of SL_2, we arrive at once at a configuration involving subgroups of type SL_2 which are "well-placed" with respect to certain subgroups of S, namely $A \le S$ nontrivial, definable, and $N(S)$-invariant. What will interest us in practice is the case in which the group A is *minimal* among infinite definable $N(S)$-invariant subgroups.

LEMMA 1.2 (SL_2-components). *Let G be a simple L^*-group of finite Morley rank and even type. Let S be a Sylow° 2-subgroup of G, assumed abelian, and $A \le S$ definable, connected, and $N(S)$-invariant. Then either G has a weakly embedded subgroup, or G contains a proper definable subgroup L of type* SL_2 *with the following properties:*

(i) *$A \cap L$ is a Sylow 2-subgroup of L.*
(ii) *A normalizes L.*
(iii) $U_2(C(L)) > 1$.

PROOF. The group S is a Sylow° 2-subgroup of $N(A)$. If for every nontrivial unipotent 2-subgroup U of S, we have $N^\circ(U) \le N(A)$, then Lemma V 1.3 applies and G has a weakly embedded subgroup.

So we fix $U \le S$ nontrivial and unipotent for which $N^\circ(U) \not\le N(A)$. Notice that $S \le N(U)$. By Lemma II 5.16 the group A is strongly closed in S, and by Lemma II 6.6 as A is not normal in $N^\circ(U)$ there is a component L of type SL_2, normal in $N^\circ(U)$, for which $A \cap L$ is a Sylow 2-subgroup of L.

This already gives conditions (i) and (ii). Furthermore, as $U, L \lhd N^\circ(U)$, we have $[L, U] = 1$ and condition (iii) is satisfied. □

1.1. $\mathbf{A \le L}$. We aim to show that when the group A occurring in Lemma 1.2 is minimized, subject to the condition of $N(S)$-invariance, then it is forced inside the associated group L of SL_2 type.

We first insert a general lemma which will be used repeatedly.

LEMMA 1.3. *Let G be a simple L^*-group of finite Morley rank and even type with an infinite definable strongly closed abelian 2-subgroup A, which normalizes a subgroup L of type* SL_2, *and contains a Sylow subgroup of L. Suppose $AL \le H < G$ with H definable. Then $L \lhd H^\circ$.*

PROOF. Let $H_1 = U_2(H)$, a K-group. Then $A \le H_1$ and H_1 factors as $E(H_1) \times C_{H_1}(E(H_1))$, using Lemma II 5.18. So easily $L \le E(H_1)$.

As A is strongly closed, A breaks up as a product of Sylow 2-subgroups of components of $E(H_1)$, and a factor in the centralizer of $E(H_1)$. Hence $L = [A, L]$ lies in the corresponding product of components of $E(H_i)$. So L is normalized by a Sylow 2-subgroup in that product (namely, the corresponding subgroup of A). It follows that L is a single component of $E(H_1)$, by Proposition II 6.8, and is therefore normal in H°. □

COROLLARY 1.4. *Let G be a simple L^*-group of finite Morley rank and even type with an infinite definable strongly closed abelian 2-subgroup A, which is also the Sylow 2-subgroup of a definable subgroup L of type* SL_2. *Suppose $L \le H < G$ with H definable. Then $L \lhd H^\circ$.*

We now come to the minimization of the subgroup A. This next step requires a very substantial argument, and here we use the theory of pseudoreflection groups.

PROPOSITION 1.5. *Let G be a simple L^*-group of finite Morley rank and of even type with an abelian Sylow° 2-subgroup S, and let $A \leq S$ be minimal among infinite definable $N(S)$-invariant abelian 2-subgroups of S. Let L be a definable subgroup of G of type* SL_2 *over an algebraically closed field of characteristic 2, normalized by A, such that its intersection with A is a Sylow 2-subgroup of L, and such that $C_G(L)$ contains a nontrivial 2-unipotent subgroup U. Then $A \leq L$.*

PROOF. This proof is quite long. We assume

$$A \cap L < A \ .$$

From this we extract a definable weakly embedded subgroup in some lines, or a direct contradiction in others.

Once we have a definable weakly embedded subgroup then by Theorem VI 10.12 we have $G \simeq \mathrm{SL}_2(K)$ with K algebraically closed in characteristic 2. Then easily $L = G$, $U = 1$, and we have a contradiction.

By Lemma II 5.16 we have

$$A \text{ is strongly closed in } S.$$

Our hypotheses relating L and A yield the following.

$$A = (A \cap L) \times C_A(L) \ .$$

We consider a maximal torus T in L normalizing $A \cap L$. Then in view of the structure of L (of type SL_2), the torus T acts transitively on $(A \cap L)^\times$, and $A \cap L = [T, A]$. That is, T acts as a *pseudoreflection group* on A.

We now pass to the group $H = N(A)/C(A)$, thought of as a group of automorphisms of A. Note that $N(S) \leq N(A)$. By our choice of A, any proper definable H-invariant subgroup of A is finite, hence contained in $A_0 = C_A(H^\circ)$. In view of the presence of T, $A_0 \neq A$ and thus A_0 is finite. Thus

$$\bar{A} = A/A_0 \text{ is an irreducible } H\text{-module}$$

Let $A_1 \leq A$ be minimal among definable H°-invariant infinite subgroups of A. Then the pseudoreflection subgroup T of H° either acts trivially on A_1 or as a pseudoreflection group. As A is the sum of the conjugates of A_1 under H, T will act as a pseudoreflection group on one of these conjugates, which we may suppose to be A_1. Then applying Theorem IV 5.3 to A_1 with the action induced by H°, $A_1/C_{A_1}(H^\circ)$ carries a vector space structure definably, relative to an interpretable algebraically closed field K, and H° acts linearly (in fact, as $\mathrm{GL}(A_1)$). By Lemma III 4.2, A_1 splits definably as $C_{A_1}(H^\circ) \oplus \tilde{A}_1$ as a K-module. Here the first factor is finite and fixed by K. It is easy to see that the second factor is H°-invariant, and hence by the choice of A_1, coincides with A_1. So A_1 is H°-irreducible. Thus A is completely reducible as an H°-module. It follows that $A_0 = 1$.

Now let us write $A = \oplus_i A_i$, with the A_i all H°-irreducible, and H-conjugate, and definable. All maximal tori in L act on one and the same factor A_i as pseudoreflection subgroups, and trivially on any others (since they do not commute), and hence L centralizes all the remaining factors. It follows that $L \cap A \leq A_i$ for some i.

We split the analysis up into two cases, depending on whether there is more than one factor A_i, or, alternatively, H° acts irreducibly on A.

(Case I:) The number of factors A_i is at least two.

We may suppose $L \cap A \leq A_1$. We will show in this case that $L \cap A = A_1$, and more generally that there is exactly one factor L_i conjugate to L and satisfying $L_i \cap A = A_i$, at which point one may check that the factors L_i commute and that the normalizer of their product (equivalently, the normalizer of any one of them) satisfies the criterion of Lemma V 1.3, which produces a weakly embedded subgroup.

We have noticed already that L centralizes A_2. So $L \leq C^\circ(A_2)$, and thus L is a quasisimple component of $C(A_2)$ (Lemma 1.3). On the other hand, H normalizes $C(A_2)$ and permutes its quasisimple components, so

(I.1) H° normalizes L.

So H° normalizes $A \cap L$ and thus $L \cap A = A_1$. Furthermore, there can be only one such component containing A_1.

Now as the factors A_i are conjugate under the action of H, we can similarly attach a unique H-conjugate L_i of L to A_i, with $A \cap L_i = A_i$ for each i. Here L_i acts trivially on A_j for $j \neq i$.

We claim next that the various factors L_i commute. First consider a Sylow° 2-subgroup S containing A. Then S centralizes A and acts on each of the groups L_i, centralizing A_i. So it follows that $S = A \cdot C_S(\langle L_i : \text{all } i\rangle)$.

Now if $S > A$, then the group $K = \langle L_i : \text{all } i\rangle$ is a proper subgroup of G and thus each L_i is a component of K, and they commute with one another in this case.

Suppose alternatively that $S = A$. Then every proper connected definable subgroup of G has the form given by Proposition II 5.13. To lighten the notation, we will consider the groups L_1 and L_2.

The group L_1 is generated by the subgroups A_1, a torus T_1 normalizing A_1, and $N_{L_1}(T_1)$. We already know A_1 centralizes L_2 and in particular $L_2 \leq N(A_1)$. Since L_2 is normalized by A, it follows that L_2 is a component of $N^\circ(A_1)$ and thus T_1 normalizes L_2; then since T_1 centralizes A_2 it follows that T_1 centralizes L_2, and similarly T_2 centralizes L_1.

Now we pass to $C^\circ(T_1)$, which contains L_2. Let S_1 be a Sylow° 2-subgroup of $C^\circ(T_1)$ containing $C_A(T_1)$. Then $S_1 \leq C(A_2)$ and hence S_1 normalizes L_1. As S_1 centralizes T_1, it follows that S_1 centralizes L_1 and hence A_1. So AS_1 is a connected 2-subgroup of G and hence by our current assumption $S_1 \leq A$, that is, $S_1 = C_A(T_1)$ is a Sylow° 2-subgroup of $C^\circ(T_1)$.

This group normalizes L_2 and it follows that L_2 is a component of $C^\circ(T_1)$, hence is normalized by $N_{L_1}(T_1)$. But $N_{L_1}(T_1)$ acts trivially on both A_2 and the corresponding torus T_2, hence centralizes L_2. So, finally, L_1 centralizes L_2.

So in either case, the group K generated by the groups L_i is their product. Observe that for each i we have $A \leq N(L_i)$ and in view of Lemma II 6.6, we find that $L_j \lhd N^\circ(L_i)$ for all i, j, in particular $N^\circ(K) = N^\circ(L_i)$ for all i.

In order to get a weakly embedded subgroup in G, we use Lemma V 1.3, relative to the subgroup $N(K)$. We must check that for $U \leq N(K)$ nontrivial and 2-unipotent, we have $N^\circ(U) \leq N(K)$. Here we may restrict U to any fixed Sylow° 2-subgroup of $N(K)$.

Extend A to a Sylow° 2-subgroup S of $N(K)$; then $S = AC_S(K) \leq C(A)$. We may suppose $U \leq S$. Then $A \leq N^\circ(U)$. Accordingly, $U_2(N^\circ(U))$ has the structure afforded by Proposition II 5.13. Let $K_1 = U_2(N^\circ(U))$. If $E(K_1) > 1$, then any component of $E(K_1)$ is one of the components L_i, and thus $N^\circ(U) \leq N(L_i) = N^\circ(K)$.

Suppose finally $E(K_1) = 1$, so $N^\circ(U)$ has a normal Sylow° 2-subgroup. It follows that $A \lhd N^\circ(U)$. Hence as we have seen before $N^\circ(U)$ normalizes the components L_i, and again normalizes K.

Hence, if the number of factors A_i is at least two, we arrive at a weakly embedded subgroup via $N^\circ(K)$.

We may pass to the second case.

(Case II:) A is H°-irreducible.

As H° contains a pseudoreflection subgroup, H° must act on A as $\mathrm{GL}(A)$ with respect to some definable vector space structure on A making A finite dimensional. But the Sylow° 2-subgroups of G are abelian, so the Sylow 2-subgroup of $\mathrm{GL}(A)$ must be trivial, and A is one-dimensional. This can only occur if A is one-dimensional relative to the relevant field structure, that is, H° is acting like the multiplicative group of a field. But we supposed that our pseudoreflection group T in L fixes points of A, and we arrive at a contradiction.

So this case is eliminated quickly, and with this Proposition 1.5 is fully proved. □

1.2. Standard components. Now we return to the main argument, and we take a new tack. We continue to work with the strong "standard component" condition furnished by Corollary 1.4. We first review the configuration reached in our analysis up to this point.

Suppose then that G is our simple L^*-group of finite Morley rank and even type, with an abelian Sylow° 2-subgroup S, and not of the form $\mathrm{SL}_2(K)$. We now take a subgroup A of S which is minimal among its nontrivial definable $N(S)$-invariant subgroups. As we have seen in Lemma II 5.16, the

$N(S)$-invariance is equivalent to strong closure, and we use this fact freely for the remainder of the argument.

As G is assumed to be not of the form SL_2, it follows that G has no weakly embedded subgroup, by Theorem VI 10.12. Hence Lemma 1.2 applies and produces a subgroup L of type SL_2 normalized by A, with $A \cap L$ a Sylow 2-subgroup of L, and with $U_2(C(L))$ nontrivial. Then by Lemma 1.5, the group A is a Sylow 2-subgroup of L.

Let U be a Sylow° 2-subgroup of $C(L)$. Our main tactic now will be to compare the group AU to a Sylow 2-subgroup of G. We show first that AU is in fact a Sylow 2-subgroup of G; and then we show, at greater length, that it is not. This contradiction then proves Theorem 1.1.

LEMMA 1.6. *Let G be a simple L^*-group of finite Morley rank and even type with an abelian Sylow° 2-subgroup S, and suppose $L \le G$ is a group of the form $\mathrm{SL}_2(K)$ over an algebraically closed field K of characteristic 2, with a Sylow 2-subgroup A normalized by $N(S)$. Let U be a Sylow° 2-subgroup of $C(L)$, and suppose $U > 1$. Then AU is a Sylow° 2-subgroup of G.*

PROOF. We may suppose that $AU \le S$. We have A strongly closed in S and thus $L \lhd C^\circ(U)$ by Lemma 1.3. As S is abelian we have $S \le C^\circ(U)$ as well. Thus S normalizes L and commutes with A, so S acts on L like A. Since S also contains A, we find $S = A \times C_S{}^\circ(L)$. Now $U \le C_S{}^\circ(L)$ so $U = C_S{}^\circ(L)$ and $S = AU$. □

Now we aim to prove the precise opposite of the foregoing result, always under the assumption that our group G contains a group L of the specified type.

1.3. The contradiction. As the analysis now becomes longer, we fix the notation.

NOTATION 1.7. *Let G be a simple L^*-group of even type of finite Morley rank, with an abelian Sylow° 2-subgroup S. Assume that G is not of type SL_2, hence contains no definable weakly embedded subgroup. Fix a minimal nontrivial $N(S)$-invariant definable subgroup A of S. Applying Lemma 1.2 and Proposition 1.5, we obtain a definable subgroup L of G with the following properties:*

1. *L is of type SL_2 over an algebraically closed field of characteristic 2.*
2. *A is a Sylow 2-subgroup of L.*
3. *$U_2(C(L)) > 1$.*

Let U be a Sylow° 2-subgroup of $C(L)$.

By Corollary 1.4, whenever $L \le H < G$ with H definable and connected, we have L normal in H. Furthermore, by Lemma 1.6, AU is a Sylow° 2-subgroup of G.

Now we will show, in two stages, that the group AU is not a Sylow° 2-subgroup of $C^\circ(A)$, thus reaching a contradiction.

LEMMA 1.8. *Let G be a simple L^*-group of even type with an abelian Sylow° 2-subgroup S and with an infinite definable subgroup A strongly closed in S. Suppose that L is a definable subgroup of G of type* SL_2 *over an algebraically closed field of characteristic 2, with A a Sylow 2-subgroup of L, and that a Sylow° 2-subgroup U of $C(L)$ is nontrivial. Suppose further that AU is a Sylow° 2-subgroup of $C^\circ(A)$.*

Then $N^\circ(A) \leq N(L)$.

PROOF. Since G has an abelian Sylow° 2-subgroup, we have $U_2(C^\circ(A)) = F^*(C^\circ(A))$; in other words, a product of copies of groups of type SL_2 with an abelian 2-group (Proposition II 5.13).

Fix T a maximal torus of the group L, normalizing A. Consider the group $K = E(U_2(C(A)))$.

If K is nontrivial, then $Q = AU \cap K$ is a Sylow 2-subgroup of K, and is T-invariant. Since, furthermore, $Q \leq AU$ and $Q \cap A = 1$, we find $Q \leq U$. Hence $N_K{}^\circ(Q) \leq N(L)$. Let R be a maximal torus of K normalizing Q. We claim $N(R) \leq N(L)$.

Now R normalizes Q and $L \lhd C^\circ(Q)$. Since R is connected, R normalizes L. As R centralizes A and is a $2^\perp$-group, R acts trivially on L. In particular, $L \leq N^\circ(R)$ and hence $L \lhd N^\circ(R)$, that is, $N^\circ(R) \leq N(L)$.

Now if $g \in N(R)$ and $L^g \neq L$, then L^g is another component of $N^\circ(R)$. In particular, $A \cap A^g = 1$ and AA^g is contained in a Sylow° 2-subgroup of G. But A is strongly closed in any Sylow 2-subgroup that contains it, by Lemma II 5.15, so we get a contradiction. Thus $N(R) \leq N(L)$.

Now we have enough to conclude $K \leq N(L)$. As K centralizes A we find $K \leq AC(L)$, and $N^\circ(A) \leq N^\circ(K) \leq N^\circ(L)$ (Lemma 1.3).

There remains the possibility that $K = E(U_2(C(A)))$ is trivial, and $U_2(C(A)) = AU$. So $N(A) \leq N(AU)$. By Lemma VII 1.10, we have $N^\circ(A) = C^\circ(A)T$.

Let H_0 be $C_{N(A)}{}^\circ(AU/A)$. Then $T \leq H_0 \lhd N(A)$. So $H_0 = (C_{H_0}(A) \cdot T)$. Now $C_{H_0}(A)$ contains U and acts trivially on the factors of the chain

$$1 < A < AU \ .$$

Since AU is abelian, we can also consider the action of the quotient group $C_{H_0}(A)/AU$ on this chain, and as the quotient has degenerate type it contains no involutions. So by Proposition I 10.7 this group acts trivially on AU. In other words, $C_{H_0}(A) \leq C(AU) \leq C(U) \leq N(L)$. So $H_0 = C_{H_0}(A)T \leq N^\circ(L)$.

On the other hand, by a Frattini argument, letting T_0 be a maximal decent torus of H_0 containing T, we find

$$N^\circ(A) \leq H_0 N_{N^\circ(A)}(T_0) = H_0 C^\circ{}_{N^\circ(A)}(T_0) \leq H_0 N^\circ{}_{N^\circ(A)}(T) \ .$$

Now $U_2(C^\circ(AT)) = U$ and thus $N^\circ(AT) \leq N^\circ(U) \leq N(L)$, and we conclude $N^\circ(A) \leq H_0 N^\circ(AT) \leq N(L)$. □

LEMMA 1.9. *Let G be a simple L^*-group of even type with an infinite abelian Sylow° 2-subgroup S, and A a strongly closed abelian subgroup of S. Suppose that there is a definable subgroup L of G of type* SL_2 *over an algebraically closed field of characteristic* 2, *with A a Sylow 2-subgroup of L, and that a Sylow° 2-subgroup U of $C(L)$ is nontrivial. Suppose further that $N^\circ(A) \le N^\circ(L)$. Then G has a weakly embedded subgroup.*

PROOF. Under our hypotheses, the group AU is a Sylow° 2-subgroup of G, by Lemma 1.6.

We examine the group $H = N^\circ(L)$ and we aim at the following:

$$(*) \qquad N^\circ(V) \le H \text{ for } V \le AU \ .$$

Indeed, AU is a Sylow° 2-subgroup of H, so the condition $(*)$ is exactly what is needed to get a weakly embedded subgroup using Lemma V 1.3.

So we take up the proof of condition $(*)$. In the special case $V = A$, this condition is one of our hypotheses.

We have $V \le AU$ and we have $A \le N^\circ(V)$. If A is normal in $N^\circ(V)$, then $N^\circ(V) \le N^\circ(A) \le H$ and we are done.

So we assume that A is not normal in the group $N^\circ(V)$. We then have some component $L_1 \lhd N^\circ(V)$ of type SL_2 over some algebraically closed field of characteristic 2, where A normalizes L_1 and $A \cap L_1$ is a Sylow 2-subgroup of L_1 (Lemma II 6.6). As V and L_1 are both normal in $N^\circ(V)$, they commute.

We take a maximal torus T_1 in L_1 normalizing A. So now $V, T_1 \le N^\circ(A) \le N^\circ(L)$. We consider the action of V and T_1 on L. The torus T_1 acts faithfully on A, more particularly, regularly on $A \cap L_1$, and hence acts on L like a subgroup of T, that is, freely on A. Hence $A \le L_1$, and T_1 acts regularly on A.

The group V centralizes A and hence acts on L_1 like a subgroup of A. At the same time V commutes with T_1. So the action of V on L must be trivial, and $L \le N^\circ(V)$, so $N^\circ(V) \le N^\circ(L)$. This proves $(*)$ in this last case, and completes the argument. □

COROLLARY 1.10. *Let G be a simple L^*-group of even type with an abelian Sylow° 2-subgroup S. Suppose that L is a definable subgroup of G of type* SL_2 *over an algebraically closed field of characteristic* 2, *with A a Sylow 2-subgroup of L, and that a Sylow° 2-subgroup U of $C(L)$ is nontrivial, and A is strongly closed in S. Then AU is not a Sylow° 2-subgroup of $C^\circ(A)$.*

PROOF. Suppose AU is a Sylow° 2-subgroup of $C^\circ(A)$. By Lemma 1.8, we have

$$N^\circ(A) \le N(L) \ .$$

Then by Lemma 1.9, G has a weakly embedded subgroup. By Theorem VI 10.12, G must be PSL_2. But then this contradicts our hypotheses. □

PROOF OF THEOREM 1.1. Under the hypotheses of Theorem 1.1, and in terms of Notation 1.7, Corollary 1.10 contradicts Lemma 1.6. □

2. Baumann's Pushing Up Theorem

DEFINITION 2.1. *Let M be a connected group of finite Morley rank and of even type. We will say that M is of* minimal parabolic type *if $F^*(M) = O_2(M)$ and $M/O_2(M) \simeq \mathrm{SL}_2(K)$ for some algebraically closed field of characteristic 2.*

The main result of this section is the following.

PROPOSITION 2.2. *Let G be a group of finite Morley rank and of even type, and M a definable connected subgroup of minimal parabolic type. Let S be a Sylow 2-subgroup of M. Then there is a nontrivial definable connected subgroup P of S such that $P \lhd M$ and P is $N_G{}^\circ(S)$-invariant.*

The proof we give is not very direct. In Theorem III 5.3 we gave a set of conditions that hold when a slightly stronger condition (involving $N_G(S)$ in place of $N_G{}^\circ(S)$) is *not* met. Analyzing this situation further yields the foregoing conclusion.

There is a closely related result, with a similar proof, which is also useful, In this case, we drop the requirement that P be connected.

PROPOSITION 2.3. *Let G be a group of finite Morley rank and of even type, M a definable connected subgroup of minimal parabolic type, and S a Sylow 2-subgroup of M. Suppose that there is no nontrivial definable subgroup of S normalized by both M and $N_G(S)$. Then there is a definable automorphism α of S such that $S = O_2(M) \cdot Z(O_2(M))^\alpha$.*

Let us now fix the notation for the remainder of this section: G has finite Morley rank and even type, M is a definable subgroup of minimal parabolic type, and S is a Sylow 2-subgroup of M. Under the hypotheses of either Proposition 2.2 or 2.3, we have the assumption of Theorem III 5.3, as follows:

$(*)$ No nontrivial connected definable subgroup of S is normalized by both M and $N_G(S)$.

So we will assume at least $(*)$ throughout this section. We adopt the notation introduced in the statement of Theorem III 5.3: $Q = O_2(M)$, $L_0 = O^2(M)$, $V = [Q, L_0]$, and $D = C_Q{}^\circ(L_0)$.

We then have the following facts, by Theorem III 5.3.

III 5.3.1. V is an elementary abelian 2-group which is central in Q.
III 5.3.2. $V/V \cap Z(M)$ is a natural (2-dimensional) module for $M/O_2(M)$.
III 5.3.3. $Q = D * V$, a central product.
III 5.3.4. $S/\Omega_1{}^\circ(Z(S))$ is an elementary abelian 2-group.
III 5.3.5. $Z^\circ(Q)$ is an elementary abelian 2-group.

NOTATION 2.4.

(1) *f is the rank of the field K over which $M/O_2(M)$ is defined, that is,* $M/O_2(M) \simeq \mathrm{SL}_2(K)$.
(2) $W = VZ^\circ(S)$.

LEMMA 2.5. *With the hypotheses and notation above, W has the following properties:*

(1) *W is an elementary abelian 2-group contained in $Z^\circ(Q)$.*
(2) $\mathrm{rk}(W/Z^\circ(S)) = f$.
(3) *For $i \in W \setminus Z(S)$, we have $C_S(i) = Q$.*

PROOF. As $Q = F^*(M)$ and W centralizes Q, by Proposition I 7.3 we have $W \leq Z(Q)$. Thus $W \leq Z^\circ(Q)$, which is an elementary abelian 2-group. The first point follows.

For the rank $\mathrm{rk}(W/Z^\circ(S))$, it is necessary to evaluate $\mathrm{rk}(V/(V \cap Z^\circ(S)))$. By III 5.3.4, we have

$$[S, V] \leq Z(S) \ .$$

Now by III 5.3.2, $V/(V \cap Z(M))$ is a natural $M/O_2(M)$-module. Hence $C_{V/Z(M)}(S)$ is the image in $V/Z(M)$ of $[S,V]$. In particular, $V \cap Z(S) \leq [S,V]Z(M)$ and thus $V \cap Z(S) = [S,V](V \cap Z(M))$. Again, as we know the module $V/Z(M)$, we find that $\mathrm{rk}(V/(V \cap Z(S))) = f$. Thus $\mathrm{rk}(V/(V \cap Z^\circ(S))) = f$ and $\mathrm{rk}(W/Z^\circ(S)) = f$ as well.

Now let $i \in W \setminus Z(S)$. We have $Q \leq C_S(i) \leq S$, and we know the action of S/Q on $V/Z(M)$. If i is centralized by S modulo $Z(M)$, then it lies in $[S,V] \leq Z(S)$, a contradiction. So i is not centralized by S modulo $Z(M)$, and hence, in view of the module structure, is not centralized by any element of S/Q. Thus $C_S(i) = Q$. □

In a more technical vein, the following lemma will serve as a complement to the third point above.

DEFINITION 2.6. *Let $X \subseteq Y$ be definable subsets of a group of finite Morley rank. Then the* co-rank $\text{co-rk}_X(Y)$ *is defined as* $\mathrm{rk}(Y) - \mathrm{rk}(X)$; *if X and Y are groups, then this is* $\mathrm{rk}(Y/X)$ *(the latter being a coset space).*

LEMMA 2.7. *With the notation and hypotheses as above, let $i \in I(S)$, and suppose that* $\text{co-rk}_S(C_S(i)) = f$. *Then $i \in Q \cup (S \cap L_0)Z(M)$.*

PROOF. We will suppose $i \notin Q$. We have $S = (S \cap L_0)Q = (S \cap L_0) * D$, the first equality by definition of L_0, and the second by definition of D, considering the action of S on L_0. So we write $i = i_0 i_1$ with $i_0 \in (S \cap L_0)$ and $i_1 \in D$. Then $i_0 \notin Q$. Now i_0 acts on the natural module $V/(V \cap Z(M))$ for $L_0/O_2(L_0)$, nontrivially since $i_0 \notin Q$, so the co-rank of i_0 on V is at least f. But $C_V(i) = C_V(i_0)$, so the same applies to i.

Now we have $Q = VD$ and we claim that we have a corresponding decomposition

$$C_Q(i) = C_V(i)C_D(i) \ . \tag{1}$$

Suppose therefore that $[vd,i]=1$ with $v\in V$ and $d\in D$. Then $[v,i]=[d,i]\in D\le C(L_0)$. Now as $V/(V\cap Z(M))$ is a natural module for $M/O_2(M)$, it follows that $[v,i]\in Z(M)$ and that $v\in[S,V](V\cap Z(M))$ (as $i\in S$). But $[S,V]\le Z(S)$ by III 5.3.4 and thus $v\in C(i)$, and hence also $d\in C(i)$. So (1) holds.

Furthermore, $V\cap D\le Z(S)\le C(i)$. So we can work modulo $V\cap D$; working in $\bar{Q}=Q/(V\cap D)$ we find that $\text{co-rk}_{\bar{Q}}(C_{\bar{Q}}(i))=\text{co-rk}_{\bar{V}}(C_{\bar{V}}(i))+\text{co-rk}_{\bar{D}}(C_{\bar{D}}(i))\ge f+\text{co-rk}_{\bar{D}}(C_{\bar{D}}(i))$. Since we assume $\text{co-rk}_S(C_S(i))=f$, we find $D\le C(i)$; hence also $D\le C(i_1)$. So i_1 commutes with $L_0DV=L_0Q=M$. Our claim follows. □

LEMMA 2.8. *With the above hypotheses and notation, suppose that α is a definable automorphism of S, and $W^\alpha\not\le Q$. Then $S=W^\alpha Q$.*

PROOF. As S/Q is a Sylow 2-subgroup of M/Q, we have

(1) $$\text{rk}(S/Q)=f\ .$$

Our aim is to show that $W^\alpha/(W^\alpha\cap Q)$ also has rank f.

By Lemma 2.5, for $i\in W\setminus Z(S)$ we have $C_S(i)=Q$. Thus the co-rank in S of $C_S(i)$ is f. It follows that for $i\in W^\alpha\setminus Z(S)$ we also have $\text{co-rk}_S(C_S(i))=f$. By Lemma 2.7 it follows that $W^\alpha\setminus Z(S)\subseteq(S\cap L_0)Z(M)$, and as $W^\alpha\setminus Z(S)$ generates W^α, we find $W^\alpha\le(S\cap L_0)Z(M)$ and also $W^\alpha\le(S\cap L_0)Z^\circ(M)$.

Now we have $W^\alpha\cap Q\le Q\cap[(S\cap L_0)Z^\circ(M)]=(Q\cap L_0)Z^\circ(M)$. Now since $Q-DV$, $V\subseteq L_0$, and $[D,L_0]=1$ we have $Q\cap L_0=V$, and thus $W^\alpha\cap Q\le VZ^\circ(M)$. Recalling, however, that W^α meets $S\setminus Q$, any element of $W^\alpha\cap V$ commutes with a nontrivial element of S/Q, and hence lies in $V\cap Z(S)$; so we see that $W^\alpha\cap Q\le Z(S)$. Hence $\text{rk}(W^\alpha/(W^\alpha\cap Q))\ge f$, and our claim follows. □

Now both versions of Baumann's Pushing-Up Theorem, Propositions 2.2 and 2.3, follow easily.

PROOF OF PROPOSITION 2.2. We will show eventually that the normal closure of V under $N_G{}^\circ(S)$ is normal in M. We begin with the further study of W:

(1) If α,β are automorphisms of S for which $W^\alpha,W^\beta\not\le Q$, then $W^\alpha=W^\beta$.

We have $S=W^\alpha Q=W^\beta Q$. Take $i\in W^\alpha\setminus Q$, and take $j\in W^\beta$ such that $iQ=jQ$. Then $ij\in Q=VD$. Write $ij=vd$ with $v\in V$ and $d\in D$. Then $(vd)^i=(vd)^{-1}=vd^{-1}$, so $v^iv\in V\cap D\le Z(M)$, and i acts trivially on the element $\bar{v}$ represented by v in $\bar{V}=V/(V\cap Z(M))$. This is a natural module, and $[i,\bar{V}]$ is covered by $Z(S)$, so $v\in Z(S)$ and thus $ij=vd\in C(i)$.

But $ij \in Q \cap (L_0 Z(S)) = VZ(S)$, and $C_{VZ(S)}(i) = Z(S)$, so $ij \in Z(S)$, that is, $i \in W^\beta Z(S)$. Thus $W^\alpha \setminus Q \leq W^\beta Z(S)$ and thus $W^\alpha \leq (W^\beta Z(S))^\circ = W^\beta$. Our claim (1) follows.

Now we will show the following:

(2) If X is a connected definable group of automorphisms of S, then $\bigcup W^X \subseteq Q$.

Suppose on the contrary $W^\alpha \not\leq Q$ for some $\alpha \in X$. We consider the orbit $\mathcal{O}_W$ of W under X: $\mathcal{O}_W = \{W^\beta : \beta \in X\}$. This contains a unique element which does not commute with W, namely W^α. Hence, more generally, for any $W_1 \in \mathcal{O}_W$, there is a unique $W_1^* \in \mathcal{O}_W$ not commuting with W_1. Furthermore, since the orbit $\mathcal{O}_W$ is nontrivial and X is connected, this orbit is infinite. This leads to a contradiction as follows.

There is a finite $X_0 \subseteq X$ such that $C(\bigcup W^X) = C(\bigcup W^{X_0})$. There is a noncommuting pair of elements $W_1, W_1^* \in \mathcal{O}_W$ such that neither lies in W^{X_0}. Hence $W_1^* \leq C(\bigcup W^{X_0}) = C(\bigcup W^X)$ and thus W_1^* should commute with W_1, a contradiction. So (2) follows.

Now we conclude as follows. Taking $X = N_G{}^\circ(S)$, the normal closure of W under $N_G{}^\circ(S)$ is contained in Q. Hence the normal closure $\hat{V} = \langle V^{N_G{}^\circ(S)} \rangle$ of V under $N_G{}^\circ(S)$ is also contained in Q. We claim that $\hat{V}$ is normalized by M.

Certainly $\hat{V}$ is normalized by S. Furthermore, $[\hat{V}, L_0] \leq [Q, L_0] = V$, so $\hat{V}$ is normalized by L_0 and hence by M. As V is connected, the group $\hat{V}$ is definable. This completes the proof. □

Proof of Proposition 2.3. We claim that $S = W^\alpha Q$ for some definable automorphism α of S; Lemma 2.8 then applies.

Assuming the contrary, the normal closure of W under $N_G(S)$ is contained in Q, and hence the same applies to the normal closure $\hat{V} = \langle V^{N_G(S)} \rangle$ of V. Then as above (in the proof of Proposition 2.2), M normalizes $\hat{V}$ and $\hat{V}$ is definable. □

3. The $C(G,T)$ Theorem

3.1. Pushing up 2-local subgroups. The following "pushing-up" theorem will be useful for the proof of the $C(G,T)$ theorem.

Theorem 3.1. *Let G be a simple L^*-group of finite Morley rank and of even type, Q a definable 2-subgroup of G, and $H = N^\circ(Q)$. Suppose that $Q = O_2(H)$ and that $U_2(H/Q) \simeq \mathrm{SL}_2(K)$ for some algebraically closed field K of characteristic 2. Then H contains a Sylow° 2-subgroup of G.*

Proof. By Lemma IV 4.19, Q is connected. We may suppose that Q is nontrivial.

Set $M = U_2(N^\circ(Q))$. Let S be a Sylow° 2-subgroup of M, and extend S to a Sylow° 2-subgroup T of G. It suffices to prove that

$$N_T{}^\circ(S) = S\ .$$

We have $Q = O_2(M)$ and $M/Q \simeq \mathrm{SL}_2(K)$. We consider two cases, according to whether the conclusion of Proposition 2.2 applies to M or not.

Suppose first that *the conclusion of Proposition* 2.2 *applies to* M, so we have some nontrivial definable connected subgroup X of S which is normalized both by M and by $N_G{}^\circ(S)$. Let $\hat{H} = N_G{}^\circ(X)$. Then $\hat{H}$ is an L-group and $M \leq \hat{H}$, with $M = N_{\hat{H}}{}^\circ(O_2(M))$. So by Lemma II 6.7, M contains a Sylow° 2-subgroup of $\hat{H}$. Thus S is a Sylow° 2-subgroup of $N_G{}^\circ(S)$ and hence also of G.

Now suppose that *no such subgroup X exists.* Then M is not of minimal parabolic type, so in view of our hypotheses we have $M = F^*(M) = L \times Q$ with $L \simeq \mathrm{SL}_2(K)$. We can refine the structure of M further. The group Q is abelian, as otherwise we could set $X = Q'$, and is even elementary abelian, as otherwise we could take $X = \Omega_1(Q)$.

Now we claim that

$$Q \text{ is a Sylow}^\circ \text{ 2-subgroup of } C_G(L)\ .$$

Extend Q to a Sylow° 2-subgroup U of $C(L)$. Now $N_U{}^\circ(Q) \leq M = LQ$ and $N_U{}^\circ(Q) \cap L = 1$, so $N_U{}^\circ(Q) = Q$ and thus Q is a Sylow° 2-subgroup of $C(L)$.

Now we claim that

$$L \text{ is a standard component of } G\ .$$

In the first place, as Q is nontrivial, $C(L)$ contains involutions. We fix an involution $i \in C(L)$, and we claim that L is a component of $C^\circ(i)$.

Suppose first that $i \in Q$. Then $M \leq C^\circ(i)$, and S is a Sylow° 2-subgroup of $C^\circ(i)$ by Lemma II 6.7. So L is normalized by a Sylow° 2-subgroup of $C^\circ(i)$, and hence by $C^\circ(i)$ in view of Proposition II 6.8.

Now let i be any involution in $C(L)$. As Q is a Sylow° 2-subgroup of $C(L)$, we may suppose that i normalizes Q. Let $V = C_Q{}^\circ(i)$. Then L is a component of $C^\circ(V)$. Apply Proposition II 5.23 to the K-group $H_0 = U_2(C^\circ(i))$ and the 2-subgroup V of H_0. Then $E^\circ(C_{H_0}(V)) \lhd E^\circ(C^\circ(i))$. Thus L is a component of $C^\circ(i)$.

Thus L is a standard component of G. Now by Theorem VII 1.1, S is a Sylow 2-subgroup of G, and our claim again follows. □

3.2. $C(G,T)$**.** Our intent is to derive a weakly embedded subgroup from the failure of the $C(G,T)$ theorem, using the graph $\mathcal{U}^*(G)$ introduced at the end of §VI 10. So the essential point is the following.

LEMMA 3.2. *Let G be a simple L^*-group of finite Morley rank and of even type, and S,T two Sylow° 2-subgroups, with $S \cap T$ infinite. Then $C(G,S) = C(G,T)$.*

PROOF. We suppose the contrary, and we fix a pair S,T with $C(G,S) \neq C(G,T)$, and with the rank of $S \cap T$ maximal. Let $Q = (S \cap T)^\circ$, and

$H = N^\circ(Q)$. Our first claim is that we may choose the pair S, T so that each of $S \cap H$ and $T \cap H$ contains a Sylow° 2-subgroup of H.

Notice that H contains $N_S{}^\circ(Q)$ and $N_T{}^\circ(Q)$, so $S \cap H$ and $T \cap H$ properly contain Q. Let $S_0 = (S \cap H)^\circ$, and let S_1 be a Sylow° 2-subgroup of G containing S_0. Then $S_1 \cap S \geq S \cap H$, so $\mathrm{rk}(S \cap S_1) > \mathrm{rk}(S \cap T)$. It follows by our choice of S and T that $C(G,S) = C(G,S_1)$. Furthermore, since $S_1 \cap T \geq S \cap T$, we may replace S by S_1. Similarly, we may replace T by a Sylow° 2-subgroup of G which contains a Sylow° 2-subgroup of H, so we may assume that our initial choice of S, T was made appropriately.

Now as $O_2(H)$ is contained in any Sylow 2-subgroup of H, our choice of S and T ensures that $O_2{}^\circ(H) \leq (S \cap T)^\circ = Q$; so $O_2{}^\circ(H) = Q$. By Lemma II 6.4, $O_2{}^\circ(M) = O_2(M)$ and so we have shown that $H = N^\circ(Q)$ with $Q = O_2(H)$. This is the first hypothesis of Theorem 3.1, and we now turn to the second, which concerns the structure of $U_2(H/Q)$.

So let $\bar{H} = H/Q$. Consider the graph $\mathcal{U}^*(\bar{H})$ with vertices the Sylow° 2-subgroups. Setting $S_0 = (S \cap H)^\circ$ and $T_0 = (T \cap H)^\circ$, the groups $\bar{S}_0$ and $\bar{T}_0$ represent two vertices of $\mathcal{U}^*(\bar{H})$, and we observe that they lie in distinct components. Indeed, if there were a chain $\bar{P}_i$ of Sylow° 2-subgroups of $\bar{H}$ linking $\bar{S}_0$ and $\bar{T}_0$ in $\mathcal{U}^*(\bar{H})$, these could be lifted back to connected 2-subgroups P_i of H, which extend to Sylow° 2-subgroups $\hat{P}_i$ of G, and then we find for each successive pair that $\hat{P}_i \cap \hat{P}_{i+1} \geq P_i \cap P_{i+1} > Q$, and hence $C(G, P_i)$ is independent of i, by the choice of Q, finally forcing $C(G,S) = C(G,T)$, a contradiction.

So $\mathcal{U}^*(\bar{H})$ is disconnected, and hence by Lemma V 1.4 we conclude that $U_2(H) \simeq \mathrm{SL}_2(K)$ for some algebraically closed field K of characteristic 2. With this, the hypotheses of Theorem 3.1 are verified, and we conclude that H contains a Sylow° 2-subgroup of G. On the other hand, $S \cap H$ and $T \cap H$ are Sylow° 2-subgroups of H, so we now have $S, T \leq H$.

As in the proof of Theorem 3.1, we must now consider whether or not the conclusion of Baumann's Theorem 2.2 applies to the group $U_2(H)$. If it does, then S has a nontrivial definable connected subgroup X which is normalized by $N_G{}^\circ(S)$ as well as by $U_2(H)$, and the latter includes T. In this case, by definition of $C(G,S)$, we find $T \leq C(G,S)$. But then the Sylow° 2-subgroups S and T are conjugate by an element g of $C(G,S)$, and this yields a contradiction to the choice of S and T:

$$C(G,T) = C(G,S^g) = C(G,S)^g = C(G,S) \ .$$

So we may suppose that the conclusion of Theorem 2.2 is not applicable to $U_2(H)$, and thus $U_2(H) = L \times Q$ with $L \simeq \mathrm{SL}_2(K)$, K is algebraically closed of characteristic 2, and, furthermore, Q elementary abelian, as noted in the proof of Theorem 3.1 at the corresponding point. So we see that the Sylow° 2-subgroups of H are elementary abelian.

But H contains Sylow° 2-subgroups of G, so G has elementary abelian Sylow° 2-subgroups. So Theorem 1.1 applies, and $G \simeq \mathrm{SL}_2(K)$ for some algebraically closed field of characteristic 2. But in this case, the groups S and T must coincide, a contradiction. □

We now come to the $C(G,T)$-theorem.

THEOREM 3.3 (C(G,T)). *Let G be a simple L^*-group of finite Morley rank and of even type, and S a Sylow° 2-subgroup. If $C(G,S) < G$, then $G \simeq \mathrm{SL}_2(K)$ for some algebraically closed field K of characteristic* 2.

PROOF. Consider the graph $\mathcal{U}^*(G)$ whose vertices are the Sylow° 2-subgroups of G, with two of them joined by an edge if their intersection is infinite. This graph is disconnected, as otherwise the previous lemma implies that the group $C(G,S)$ is independent of the choice of S, and hence normal in G, contradicting the simplicity of G.

Now Corollary VI 10.16 says that $G \simeq \mathrm{SL}_2(K)$ for some algebraically closed field K of characteristic 2, as claimed. □

4. 2-local subgroups

In this section we begin the study of 2-local° subgroups, most especially those which are "large".

DEFINITION 4.1. *Let G be a group of finite Morley rank. A* 2-local° subgroup *of G is a group of the form $N_G{}^\circ(U)$ with U 2-unipotent and non-trivial.*

Our goal in the present section is to show that $\hat{\mathrm{O}}(H)$ is trivial for H 2-local.

LEMMA 4.2. *Let H be a group of finite Morley rank and $U \leq H$ a definable 2-subgroup. Then $\hat{\mathrm{O}}(N_H(U)) = \hat{\mathrm{O}}(C_H(U))$.*

PROOF. It suffices to show that $\hat{\mathrm{O}}(N(U)) \leq C(U)$. We have

$$[\hat{\mathrm{O}}(N(U)), U] \leq \hat{\mathrm{O}}(N(U)) \cap U = 1,$$

since this is an intersection of a $2^{\perp}$-group and a 2-group. □

We begin with the L-group version of our result.

PROPOSITION 4.3. *Let H be an L-group of even type and $P \leq H$ a unipotent 2-group. Then $\hat{\mathrm{O}}(C_H(P)) \leq \hat{\mathrm{O}}(H)$.*

PROOF. We will proceed by induction on $\mathrm{rk}(H)$. Set $X = \hat{\mathrm{O}}(C_H(P))$.

Suppose first that $\hat{\mathrm{O}}(H) > 1$. Let $K = C_H{}^\circ(P \bmod \hat{\mathrm{O}}(H))$. Then K normalizes $\hat{\mathrm{O}}(H)P$, and this is a central product by Proposition I 10.13. So K normalizes $O_2(\hat{\mathrm{O}}(H)P) = P$, and $[K,P] \leq P \cap \hat{\mathrm{O}}(H)$ hence K centralizes P and $K = C_H{}^\circ(P)$.

Applying induction to the quotient $\bar{H} = H/\hat{\mathrm{O}}(H)$, we have $\bar{X} \leq \hat{\mathrm{O}}(\bar{K}) = 1$, that is, $X \leq \hat{\mathrm{O}}(H)$ as claimed. So we may assume the contrary:

(1) $$\hat{\mathrm{O}}(H) = 1\ .$$

Next we will show the following:

$$(2) \qquad [X, O_2(H)] = 1 \ .$$

We consider the action of $X \times P$ on $Q = O_2(H)$. We have $[X, C_Q(P)] \leq X \cap Q = 1$, so by Lemma I 12.4 we have $[X, Q] = 1$, as claimed.

Now if $C_H(O_2(H)) < H$, then by induction we find $X \leq \hat{\mathrm{O}}(C_H(O_2(H))) \leq \hat{\mathrm{O}}(H)$, as claimed. So we may suppose that

$$(3) \qquad O_2(H) \leq Z(H) \ .$$

Now $H_0 = U_2(H)$ is a K-group with $O(H_0) = 1$ and $O_2(H_0)$ central. It follows from Proposition II 5.10, together with Proposition I 9.6 and the theory of central extensions that $U_2(H) = E(U_2(H)) * O_2(H)$.

Let L be a quasisimple component of $E(U_2(H))$. Then P acts on L like a unipotent subgroup $\hat{P}$ of L, and $C_L(P) = C_L(\hat{P})$. Similarly, X acts on L like a subgroup $\hat{X}$ of L, and $\hat{X} \leq O(N_L(\hat{P}))$. By Proposition II 1.25, $O(N_L(\hat{P})) = 1$. This means that X centralizes L, and hence

$$(4) \qquad X \text{ centralizes } E(U_2(H)) \ .$$

Thus X centralizes $U_2(H)$. So $X \leq C_H(U_2(H)) \leq C_H(P)$ and hence $X \leq \hat{\mathrm{O}}(C_H(U_2(H)) \leq \hat{\mathrm{O}}(H)$. □

Now we treat the simple L^*-case.

PROPOSITION 4.4. *Let G be a simple L^*-group of finite Morley rank and of even type, and let $P \leq G$ be a nontrivial 2-unipotent subgroup. Then $\hat{\mathrm{O}}(N(P)) = 1$.*

PROOF. Supposing the contrary, let X be a maximal subgroup of G of the form $\hat{\mathrm{O}}(N_G(Q))$, with Q 2-unipotent. Let $M = N_G(X)$. We claim that M is weakly embedded in G.

By Proposition I 10.13 we have $[X, Q] = 1$. So $Q \leq M$ and thus M contains a nontrivial Sylow° 2-subgroup. Let $U \leq M$ be any nontrivial unipotent 2-group; our claim is that $N(U) \leq M$.

Note that U centralizes X, again by Proposition I 10.13. Conjugating by an element of M, we may suppose that QU is a unipotent 2-group. Then $X \leq C(QU) \leq C(Q)$ and therefore $X \leq \hat{\mathrm{O}}(C(QU))$. On the other hand, by Proposition 4.3, $\hat{\mathrm{O}}(C(QU)) \leq \hat{\mathrm{O}}(C(U))$. So $X \leq \hat{\mathrm{O}}(C(U))$ and by maximality of X we have $X = \hat{\mathrm{O}}(C(U)) = \hat{\mathrm{O}}(N(U))$ and thus $N(U) \leq M$.

So M is weakly embedded in G, and G must be of the form $\mathrm{SL}_2(K)$ for some algebraically closed field K of characteristic 2. In this case the desired conclusion follows by inspection. □

5. Parabolic subgroups

DEFINITION 5.1. *Let G be a simple group of finite Morley rank and of even type. A* parabolic *subgroup of G is a definable connected subgroup containing the normalizer of a Sylow° 2-subgroup of G.*

Note that by Corollary I 8.4, the normalizer of a Sylow° 2-subgroup contains a Borel subgroup of G, so that parabolic subgroups will contain Borel subgroups according to our definition.

We aim here at the following property of parabolic subgroups.

THEOREM 5.2. *Let G be a simple L^*-group of finite Morley rank and of even type, and P a proper parabolic subgroup of G. Then $F^*(P) = O_2(P)$.*

5.1. Components.

PROPOSITION 5.3. *Let G be a simple L^*-group of finite Morley rank and of even type, and P a proper parabolic subgroup of G. Then $E(U_2(P)) = 1$.*

PROOF. Let $P_0 = U_2(P)$. Suppose toward a contradiction that A is a quasisimple component of P_0. Let S be a Sylow° 2-subgroup such that $N^\circ(S) \leq P$. Our goal will be to prove that $C(G,S) \leq N(A)$.

As P_0 is a K-group, A is algebraic, of even type. Let $S_A = S \cap A$, a Sylow 2-subgroup of A. Let $B_A = N_A(S_A)$, a Borel subgroup of A, and factor B_A as usual as $S_A \rtimes T_A$, with T_A a maximal torus in A. By Fact II 2.25 it follows easily that $S = S_A \times C_S(A)$. Let $S_2 = C_S(A)$.

We now turn to $C(G,S)$. Let $X \leq S$ be a nontrivial definable connected subgroup normalized by $N_G(S)$, and let $K = U_2(N_G(X))$. Our claim is that $K \leq N(A)$.

We consider the structure of K. Set $Q = O_2{}^\circ(K)$ and $\bar{K} = K/Q$. By Proposition II 5.10, bearing in mind $K = U_2(K)$, we have $\bar{K} = E(\bar{K})$, a central product of quasisimple algebraic groups in characteristic 2.

Now we analyze the structure of K more closely, in its relationship to A and, in particular, to T_A. Let $Q_A = Q \cap A$, $Q_2 = C_Q(A)$. We claim that Q decomposes much as S does:

$$Q = Q_A \times Q_2 \ . \tag{1}$$

For this, it is enough that $Q \leq S$ is T_A-invariant, as we shall see.

By Corollary I 9.10, $Q = [Q, T_A]C_Q(T_A)$. Now $[Q, T_A] \leq Q_A$ and $Q_A = [Q_A, T_A]$ as T_A is a maximal torus of A (Fact II 1.11). Thus $[Q, T_A] = Q_A$. On the other hand $C_Q(T_A)$ acts on A as a 2-group commuting with a maximal torus, so $C_Q(T_A) = C_Q(A)$ (Fact II 1.8). So $Q = Q_A Q_2 = Q_A \times Q_2$. This proves (1).

Now we work in the quotient group $\bar{K}$. In view of the decomposition of S and of Q we have

$$\bar{S} = \bar{Q}_A \times \bar{S}_2 \ , \tag{2}$$

and $\bar{S}$ is a Sylow 2-subgroup of $\bar{K}$. Accordingly, by Lemma II 2.11 we have a corresponding decomposition of $\bar{K}$:

$$\begin{aligned} &\bar{K} = \bar{K}_1 * \bar{K}_2 \ , \\ &S_A \text{ and } S_2 \text{ are Sylow° 2-subgroups of } \bar{K}_1 \text{ and } \bar{K}_2, \text{ respectively,} \\ &\bar{K}_1 = E(C_{\bar{K}}(\bar{S}_2));\ \bar{K}_2 = E(C_{\bar{K}}(\bar{S}_A)) \ . \end{aligned} \tag{3}$$

We let K_1, K_2 be the full preimages of $\bar{K}_1$, $\bar{K}_2$ in K; in particular $K_1, K_2 \lhd K$.

After these preliminaries, we may show that $K \leq N(A)$. We make a case division, according to whether $\bar{K}_2$ is, or is not, trivial, and we begin with the latter case.

(Case I) $\bar{K}_2 \neq 1$.

Let $K_2^* = C_{K_2}(T_A)$. We show first that

(I.1) $$K_2 = Q_A \times K_2^*.$$

As T_A centralizes $\bar{S}_2$, which is a Sylow 2-subgroup of $\bar{K}_2$, and T_A is a $2^\perp$-group, it follows that T_A centralizes $\bar{K}_2$. Then by Corollary I 9.10 we have $K_2 = QK_2^*$; since $Q_2 \leq K_2^*$, this yields

$$K_2 = Q_A K_2^* \ .$$

On the other hand, $Q_A \cap K_2^* = 1$ so we have a semidirect product $Q_A \rtimes K_2^*$, and we claim that K_2^* acts trivially on Q_A.

Now $S \cap K_2$ is a Sylow 2-subgroup of K_2, and contains Q, so $S \cap K_2^*$ is a Sylow 2-subgroup of K_2^*. But $S \cap K_2^* = C_S(T_A) = C_S(A)$, in view of the structure of S, so $S \cap K_2^*$ acts trivially on A. As a Sylow 2-subgroup of K_2^* acts trivially on A, and $K_2^*/Q_2 \simeq \bar{K}_2$ is a central product of quasisimple groups, it follows that K_2^* centralizes A, and hence Q_A. Thus (I.1) holds.

We claim that

(I.2) S_2 is a Sylow° 2-subgroup of $N_G(T_A)$

Recall that $N_G{}^\circ(T_A) = C_G{}^\circ(T_A)$ as T_A is a good torus.

Now $S_2 \neq 1$ and $S \leq N^\circ(S_2)$. Working in the L-group $N^\circ(S_2)$, we have A a normal quasisimple subgroup normalized by the Sylow° 2-subgroup S, and thus $A \lhd N^\circ(S_2)$, that is, $N^\circ(S_2) \leq N^\circ(A)$.

Thus $N_{C(T_A)}{}^\circ(S_2) \leq N(A) \cap C(T_A) = T_A C(A)$. Now S_2 is a Sylow° 2-subgroup of $C(A)$, and hence of $T_A C(A)$, so S_2 is a Sylow° 2-subgroup of its own normalizer in $C(T_A)$, and hence is a Sylow° 2-subgroup of $C(T_A)$, and hence of $N(T_A)$, as claimed.

Now let $H = N_{C(T_A)}(Q_2)$. Then S_2 is a Sylow° 2-subgroup of H, and $K_2^* \leq H$. Let $\bar{H} = H/Q_2$. Using the bar notation now in this sense, we have $\bar{K}_2^*$ a quasisimple subgroup of $\bar{H}$ and containing a Sylow° 2-subgroup of $\bar{H}$, so by Proposition II 6.8, $\bar{K}_2* \leq E(\bar{H}^\circ)$. But since $\bar{S}_2$ is a Sylow° 2-subgroup of $\bar{H}$, we conclude that $\bar{K}_2^* = E(\bar{H}^\circ)$.

Now $N_A(T_A)$ centralizes Q_2 and normalizes H, hence acts on the quotient $\bar{H}$, and in particular normalizes $\bar{K}_2^*$. So $N_A(T_A)$ normalizes K_2^*, and hence also the iterated commutator subgroup $(K_2^*)^{(\infty)}$; and the latter group is $K_2^{(\infty)}$ by (I.1).

Now by assumption $S \leq K$, so S normalizes K_2 and $K_2^{(\infty)}$. In particular, we now have $K_2^{(\infty)}$ normalized by both S_A and $N_A(T_A)$; as these two groups generate A, it follows that

(I.3) $$A \text{ normalizes } K_2^{(\infty)}.$$

So now we work in $N^\circ(K_2^{(\infty)})$. Here we have the quasisimple subgroup A normalized by the Sylow° 2-subgroup S, and hence by Proposition II 6.8 we have $A \lhd N^\circ(K_2^{(\infty)})$. But the latter group contains K, so $K \leq N^\circ(A)$, as claimed. This completes the analysis of Case I.

(Case II) $$\bar{K}_2 = 1.$$

So we have $\bar{K} = \bar{K}_1$. As $\bar{S}_2$ acts trivially on $\bar{K}_1 = \bar{K}$, it follows that $S_2 \leq Q$, and $\bar{S}_A$ is a Sylow 2-subgroup of $\bar{K}$.

Now $\bar{K}$ is a central product of quasisimple algebraic groups. Let $B_1 = N_K(S)$, so that $\bar{B}_1$ is a Borel subgroup of $\bar{K}$. Using Proposition I 9.6, split $N_K(S)$ as $S \rtimes T_1$ with $\bar{T}_1$ a maximal torus of $\bar{B}_1$ (so here we do not take T_1 to be the full preimage of $\bar{T}_1$).

The group $\bar{K}$ is generated by the Levi factors (SL_2) of minimal parabolic subgroups containing $\bar{B}_1$. So it will suffice to prove the following: If $P_1 \leq K$ is the full preimage of a minimal parabolic subgroup $\bar{P}_1$ of $\bar{K}$ containing $\bar{B}_1$, then $P_1^{(\infty)} \leq N(A)$. So fix such a subgroup P_1 of K. Before dealing directly with P_1, we need to compare root systems in A and K.

Now T_1 acts on A like a subgroup of $N_A(S_A)$, and as T_1 is a $2^\perp$-group it acts like a subgroup of T_A; in particular, T_1 and T_A commute. Let $T = T_A T_1$. Thus the image of T in $\mathrm{Aut}(S_A)$ can be identified with T_A.

On the other hand, T also acts on K and induces an action on $\bar{K}$ and $\bar{S}_A$. Here this action will agree with the one induced by $T_1 \simeq \bar{T}_1$ on $\bar{S}_A$. So the image of T in $\mathrm{Aut}(\bar{S}_A)$ can be identified with T_1. (In particular, we have a homomorphism from T_A to T_1, but it is not useful to think of this as an identification, as two different points of view are involved.)

Now we have root systems associated to S_A and $N_A(S_A)$ on the one hand, and to $\bar{S}_A$ and $N_{\bar{K}}(\bar{S}_A)$ on the other, and we wish to compare them. The root subgroups of S_A are the minimal T_A-invariant subgroups, or equivalently the minimal nontrivial T-invariant subgroups. These may be indexed as S_α with α varying over the set Φ of positive roots for a suitable root system attached to A. On the other hand, the root subgroups of $\bar{S}$, relative to $\bar{T}_1$, are the minimal nontrivial T_1-invariant subgroups of $\bar{S}$, or equivalently the minimal nontrivial T-invariant subgroups of $\bar{S}_A = S_A/Q_A$, or the minimal T_A-invariant subgroups of S_A properly containing Q_A. As T_A-invariant subgroups of S_A are directly spanned by the root groups they contain, Q_A corresponds to a certain subset Φ_0 of Φ, and the minimal T_A-invariant subgroups of S_A properly containing Q_A correspond to sets of the form $\Phi_0 \cup \{\alpha\}$, with $\alpha \in \Phi \setminus \Phi_0$. Thus if Φ_1 is the set of positive roots in $\bar{K}$, corresponding to $\bar{S}$, then we have a natural bijection $\iota : \Phi \setminus \Phi_0 \leftrightarrow \Phi_1$, where $(\bar{S})_{\iota(\alpha)} = \overline{S_{\{\alpha\}\cup\Phi_0}} = \overline{S_\alpha Q}$.

Now this bijection does not necessarily preserve any interesting structure on the roots, such as the length, Dynkin diagram, or linear relations. But we claim that under this map, the simple roots in Φ_1 correspond to the simple roots of Φ lying outside Φ_0. Indeed, by Fact II 2.3, the simple roots β in Φ_1 are those for which $\bar{S}_\beta$ is not contained in $[\bar{S}, \bar{S}] = \overline{[S, S]Q}$, and these correspond to the roots α for which S_α is contained neither in $[S, S]$ nor in Q; in other words, the simple roots outside Φ_0.

Now we return to the group P_1. Let $R \leq P_1$ be the preimage of $O_2(\bar{P}_1)$. As $R \leq S = S_A Q$, we have $R = R_A Q$ with $R_A = R \cap A$. Corresponding to $\bar{P}_1$, there is a simple root $\iota(\delta)$, with $\delta \in \Phi \setminus \Phi_0$ simple, such that $\bar{R}$ is spanned by the root groups $\bar{S}_\alpha$ with $\alpha \in \Phi_1$, $\alpha \neq \iota(\delta)$. Hence R_A is spanned by the root groups S_α in S_A with $\alpha \neq \delta$, and $P_A = N_A(R_A)$ is a minimal parabolic subgroup of A with $O_2(P_A) = R_A$. We consider the relationship of $P_1^{(\infty)}$ and P_A, working in $N(R)$.

Now $(P_A Q/R)^{(\infty)} \simeq (P_A/R_A)^{(\infty)}$ is a simple subgroup of $N^\circ(R)/R$, and contains the Sylow° 2-subgroup S/R of $N^\circ(R)/R$, so by Proposition II 5.20 we have $(P_A Q/R)^{(\infty)} \lhd N^\circ(R)/R$. On the other hand, $(P_1/R)^{(\infty)}$ is a simple subgroup of $N^\circ(R)/R$, meeting the normal subgroup $(P_A Q/R)^{(\infty)}$ nontrivially (in S/R), and thus $P_1^{(\infty)} \leq P_A R$. But $P_A R \leq N(A)$, so $P_1^{(\infty)} \leq N(A)$, as claimed. □

5.2. 2-locality.

PROPOSITION 5.4. *Let G be a simple L^*-group of finite Morley rank and of even type, and P a proper parabolic subgroup of G. Then P is a 2-local° 2-subgroup of G.*

PROOF. By Fact IV 4.19, $Q = O_2(P)$ is connected. If $Q = 1$ then by Proposition II 5.10 we have $U_2(P) = E(U_2(P))$. But $E(U_2(P)) = 1$ by Proposition 5.3. So $U_2(P) = 1$, a contradiction.

So Q is nontrivial and P is contained in the 2-local° subgroup $H = N^\circ(Q)$. We claim $P = H$.

Let $P_0 = U_2(P)$ and $H_0 = U_2(H)$. Now P_0 contains the normalizer of a Sylow° 2-subgroup S of G, so $O_2(H) \leq S \leq P$ and it follows that $O_2(H) = Q$. Now in $\bar{H} = H/Q$ we have $\bar{P}_0 = E(\bar{P}_0)$ and $\bar{H}_0 = E(\bar{H}_0)$. As $S \leq P_0$ it follows from Proposition II 5.20 that $\bar{P}_0 \lhd \bar{H}_0$ and as $\bar{P}_0$ contains a Sylow° 2-subgroup of $\bar{H}_0$, we deduce $\bar{P}_0 = E(\bar{H}_0) = H_0$. Thus $H_0 \leq P$. Now $H = H_0 N^\circ(S)$ by the Frattini argument, and $N^\circ(S) \leq P$, so $P = H$. □

5.3. $O_{2\perp}$.

PROPOSITION 5.5. *Let G be a simple L^*-group of finite Morley rank and of even type, and P a proper parabolic subgroup of G. Then $F^*(U_2(P)) = O_2(P)$.*

PROOF. Let $P_0 = U_2(P)$. By Proposition 5.3, $E(P) = 1$. Hence $F^*(P) = F(P)$. Now $O(P) = 1$ by Proposition 4.4, in view of Proposition 5.4. Hence $F^\circ(P) = O_2(P_0)$. Thus $F(P_0) = O_2(P_0)O_{2^\perp}(P_0)$, with the factor $A = O_{2^\perp}(P_0)$ finite, hence central in P.

Let S be a Sylow° 2-subgroup of G with $N^\circ(S) \le P$. We claim that $C(G,S) \le C(A)$. Let $X \le S$ be $N^\circ(S)$-invariant, definable, and connected, and set $H = N^\circ(X)$. We claim $H \le C(A)$. Let $H_0 = U_2(H)$. Then by a Frattini argument $H = H_0 N_H{}^\circ(S)$ and $N_H{}^\circ(S) \le P \le C(A)$, so it suffices to check that $H_0 \le C(A)$.

We show first that $A \le H$. Certainly $A \le N_{P_0}(S)$, and we claim this is connected. In $\bar{P} = P/O_2(P)$, we have $\bar{P}_0 = E(\bar{P}_0)$ (with algebraic components) by Proposition II 5.10. Thus $N_{\bar{P}_0}(\bar{S})$ is a Borel subgroup of $\bar{P}_0$ and $N_{P_0}(S)$ is connected. Thus $A \le H$.

Let $\bar{H}_0 = H_0/O_2(H)$. Then $\bar{H}_0 = E(\bar{H}_0)$ and $N_{\bar{H}_0}(S)$ is a Borel subgroup of $\bar{H}_0$. As $N_{H_0}(S) \le P$, A centralizes a Borel subgroup of $\bar{H}_0$ and hence acts trivially on $\bar{H}_0$, that is,

$$[H_0, A] \le O_2(H) \; .$$

On the other hand, $[O_2(H), A] = 1$. By Proposition I 10.7, we have $[H_0, A] = 1$, as claimed.

Thus $C(G,S) \le C(A)$. So if $A > 1$, then $C(G,S) < G$ and by Theorem 3.3 we have $G \simeq \mathrm{SL}_2(K)$ with K algebraically closed of characteristic 2. Then the only proper parabolic subgroups are the Borel subgroups, and the theorem holds by inspection. □

5.4. $N^\circ(S)$.

DEFINITION 5.6. *Let H be a group of finite Morley rank. Then $H^{\mathcal{K}}$ denotes the smallest connected normal subgroup K of H for which H/K is a K-group.*

LEMMA 5.7. *Let G be a simple L^*-group and S a Sylow° 2-subgroup. Then $N_G{}^\circ(S)$ is solvable.*

PROOF. Let $K = [N_G{}^\circ(S)]^{\mathcal{K}}$. It suffices to show that $K = 1$.

Let P be a proper definable connected subgroup of G containing $N^\circ(S)$, and $L = U_2(P)$. Then P acts on $\bar{L} = L/O_2(L)$, via inner automorphisms. Thus if K_0 is the kernel of this action, we have $P^{\mathcal{K}} \le K_0$, or in other words, $[P^{\mathcal{K}}, L] \le O_2(L)$. In particular, $P^{\mathcal{K}} \le N^\circ(S)$. Hence $P^{\mathcal{K}} = K$, and K is P-invariant.

Since $C(G,S)$ is generated by groups P of this type, also $C(G,S)$ normalizes K. So if $C(G,S) = G$, then $K = 1$ as required, while if $C(G,S) < G$ then by Theorem 3.3 we have $G \simeq \mathrm{SL}_2(K)$ with K algebraically closed of characteristic 2, in which case our claim is clear. □

LEMMA 5.8. *Let G be a simple L^*-group and S a Sylow° 2-subgroup of G. Let P be a proper definable connected subgroup of G containing $N^\circ(S)$. Then*

(1) *P is a K-group.*
(2) $E(P) = 1$.
(3) $F^*(P) = O_2(P)$.

PROOF. By the Frattini argument $P = U_2(P)N^\circ(S)$, so P is a K-group. Hence $E(P) \leq E(U_2(P)) = 1$.

We turn to the third point. We have $F^*(P) = F(P) = O_2(P)O_{2^\perp}(P)$. Since $O(P) = 1$, it suffices to show that P has no finite central $2^\perp$-subgroup A. Let $L = U_2(P)$, and $\bar{L} = L/O_2(L)$. Let $P_0 = C_P(\bar{L})$. Then $A \leq P_0$. Furthermore, $P/O_2(L) = \bar{L} \times \bar{P}_0$, so P_0 is connected. But $P_0 \leq N(S)$, so $A \leq N^\circ(S)$. Now if P_1 is any proper definable subgroup of G containing $N^\circ(S)$, then $A \leq P_1$ and A commutes with S. As A is a $2^\perp$-group, it follows easily that A commutes with $U_2(P_1)$. As groups of the form $U_2(P_1)$ generate $C(G, S)$, we have $C(G, S)$ centralizing A. So if A is nontrivial, then $C(G, S) < G$, and $G \simeq \mathrm{SL}_2(K)$ for some algebraically closed field K of characteristic 2, in which case our claim holds by inspection. □

6. The classification: Plan of attack

We may now present a detailed plan of attack for the proof of our main theorem. As the mixed type case has been dealt with, it remains to prove the following.

THEOREM 6.1. *A simple group of finite Morley rank of even type is isomorphic to a Chevalley group over a field of characteristic* 2.

Of course, it suffices to prove this in the L^* context.

PROPOSITION 6.2. *A simple L^*-group of finite Morley rank of even type is isomorphic to a Chevalley group over a field of characteristic* 2.

6.1. The case division.

DEFINITION 6.3. *Let G be a group of finite Morley rank of even type, and S a Sylow° 2-subgroup.*

(1) *We define $\mathcal{M}(S)$ as the set of proper definable connected subgroups P of G containing $N_G{}^\circ(S)$ such that $U_2(P)/O_2(P) \simeq \mathrm{SL}_2(K)$ for some algebraically closed field K of characteristic* 2.
(2) *G is said to be*
   (a) **thin** *if $\mathcal{M}(S) = \emptyset$,*
   (b) **quasithin** *if $G = \langle U_2(P_1), U_2(P_2)\rangle$ for some $P_1, P_2 \in \mathcal{M}(S)$,*
   (c) **generic** *if it is neither thin nor quasithin.*

Our plan of attack becomes three plans of attack. In the thin case, we aim to show that $G \simeq \mathrm{SL}_2(K)$ for some algebraically closed field K of characteristic 2, using the $C(G, T)$ theorem. In the quasithin case, we aim at identifying G as a group of Lie type in Lie rank 2, using the amalgam method. In the generic case, we identify G as a group of Lie type in Lie rank at least 3, using the version of Niles' theorem adapted to the finite

Morley rank context. The thin and generic cases can be dealt with here; the amalgam method requires a long analysis, given in the following chapter.

The groups $U_2(P)$ for $P \in \mathcal{M}(S)$ have a clear structure. Hence the following considerations may be suggestive.

LEMMA 6.4. *Let G be a group of finite Morley rank of even type, and S a Sylow° 2-subgroup. Let $P_1, \ldots, P_r \in \mathcal{M}(S)$, and $L_i = U_2(P_i)$ for $i = 1, \ldots, r$. Set $H = \langle P_1, \ldots, P_r \rangle$ and $G_0 = \langle L_1, \ldots, L_r \rangle$. Then:*

(1) $P_i = L_i \cdot N^\circ(S)$.
(2) $G_0 = U_2(H)$.

In particular, if G is simple and $H = G$, then $G_0 = G$.

PROOF. The first point follows from the Frattini argument as $S \leq L_i \lhd P_i$. Furthermore, this has the following consequence:

$$P_i \leq N(G_0)\ . \tag{1}$$

Indeed, $L_i \leq G_0$, while $N^\circ(S) \leq P_i \leq N(L_i)$ for each i, and hence $N^\circ(S) \leq N(G_0)$. So (1) holds, and, in particular,

$$H \leq N(G_0)\ . \tag{2}$$

On the other hand, G_0 contains S, a Sylow° 2-subgroup of H. As G_0 is normal in H, G_0 contains all Sylow° 2-subgroups of H, and hence $G_0 = U_2(H)$. □

6.2. The thin case.

LEMMA 6.5. *Let G be a simple L^*-group of finite Morley rank and of even type, and S a Sylow° 2-subgroup of G. If $\mathcal{M}(S) = \emptyset$, then $G \simeq \mathrm{SL}_2(K)$ for some algebraically closed field of characteristic 2.*

PROOF. It suffices to show that $C(G, S) < G$; we show in fact that $C(G, S) = S$.

Suppose that X is a $N_G{}^\circ(S)$-invariant subgroup of S and $H_0 = U_2(N(X))$. We claim that $H_0 = S$. Let $\bar{H}_0 = H_0/O_2(H_0)$. Then $\bar{H}_0 = E(\bar{H}_0)$ is a central product of quasisimple algebraic groups in characteristic 2, over algebraically closed fields. If $\bar{H}_0 = 1$, then $H_0 = O_2(H_0)$, and as $S \leq H_0$ it follows that $H_0 = S$.

Suppose toward a contradiction that $\bar{H}_0$ is nontrivial. Let $\bar{L}$ be a quasisimple component of $\bar{H}_0$, let P_0 be the preimage in H_0 of a minimal parabolic subgroup $\bar{P}_0$ in $\bar{L}$ (possibly equal to L) and set $Q = O_2(P_0)$. It suffices to show that $N^\circ(Q) \in \mathcal{M}(S)$.

Now $N^\circ(S)$ acts on L via inner automorphisms. These will normalize P_0 as P_0 is parabolic. So $N^\circ(S) \leq N^\circ(Q)$.

Let $P = N^\circ(Q)$. Then $O_2(P) \leq S \leq P_0$, so $O_2(P) \leq Q$ and thus $O_2(P) = Q$. Working in $\bar{P} = P/Q$, it follows from Proposition II 5.20 that $U_2(\bar{P}) = \overline{U}_2(P_0)$, and thus $U_2(P) = U_2(P_0)$. So $P \in \mathcal{M}(S)$, a contradiction. □

6.3. The generic case.

LEMMA 6.6. *Let G be a simple L^*-group of finite Morley rank and of even type, and not thin. Then G is generated by the groups $U_2(P)$ for $P \in \mathcal{M}(S)$.*

PROOF. Let $G_0 = \langle U_2(P) : P \in \mathcal{M}(S)\rangle$. We claim that G_0 contains $C(G, S)$.

Let $X \leq S$ be $N^\circ(S)$-invariant, and set $H_0 = U_2(N(X))$. Let $Q = O_2(H_0)$, and $\bar{H}_0 = H_0/Q$. Then $\bar{H}_0 = E(\bar{H}_0)$ is a central product of quasisimple algebraic groups, and is generated by the Levi factors of its minimal parabolic subgroups. So it suffices to show that if $\bar{P}_0$ is $U_2(\bar{M})$ for some minimal parabolic subgroup of $\bar{H}_0$, and P_0 is the full preimage in H_0, then $P_0 \leq G_0$.

As in the proof of Lemma 6.5, we let $P = N^\circ(O_2(P_0))$ and we find that P is parabolic, $O_2(P) = O_2(P_0)$, and finally $U_2(P) = P_0$. Hence $P \in \mathcal{M}(S)$ and $P_0 \leq G_0$. □

LEMMA 6.7. *Let G be a simple L^*-group of finite Morley rank and of even type, and of generic type. Then G is isomorphic to a Chevalley group over an algebraically closed field of characteristic 2, in Lie rank at least 3.*

PROOF. We check the hypotheses of Theorem III 8.1 (a form of Niles' Theorem).

We have $G = \langle U_2(P) : P \in \mathcal{M}(S)\rangle$ by Lemma 6.6, and finitely many subgroups $L_i = U_2(P_i)$ will suffice for this. Let them be indexed by $i = 1, \ldots, r$. Then we have the following:

(1) $G = \langle P_i : i = 1, \ldots, r\rangle$.
(2) $N_{P_i}{}^\circ(S) = N_G{}^\circ(S)$ is solvable.
(3) $L_i = U_2(P_i)$ and $L_i/O_2(L_i) \simeq \mathrm{SL}_2(K_i)$ with K_i an algebraically closed field of characteristic 2.
(4) Let $G_{ij} = \langle L_i, L_j\rangle$. Then $G_{ij}/O_2(G_{ij})$ is either a Chevalley group of Lie rank 2 over an algebraically closed field of characteristic 2 (hence of type SL_3, PSL_3, Sp_4, or G_2) or a product of two Chevalley groups of Lie rank 1, that is, $\mathrm{SL}_2(K_1) \times \mathrm{SL}_2(K_2)$, where K_1 and K_2 are algebraically closed fields of characteristic 2.

Only the last point requires verification. As G is of generic type, we have $G_{ij} < G$, and as $G_{ij} = U_2(G_{ij})$, the group G_{ij} is a K-group. In the quotient $\overline{G}_{ij} = G_{ij}/O_2(G_{ij})$, we have $\overline{G}_{ij} = E(\overline{G}_{ij})$ and $\overline{G}_{ij}$ is generated by groups contained in two minimal parabolic subgroups. Hence either $\overline{G}_{ij}$ has Lie rank 2, or it is a central product of two groups of Lie rank 1.

Thus we have the hypotheses of Theorem III 8.1, and under these conditions Corollary III 8.2 gives the result, mediated by Fact III 7.11. □

So the thin and generic cases are handled using the $C(G,T)$ theorem, and a combination of Niles' theorem and the theory of buildings, respectively, and only the quasithin case remains.

6.4. The generic case, again. At the end of our analysis we invoked the classification of BN pairs of finite Morley rank of spherical type and Tits rank at least 3. One does not need the full force of this classification; since there are very few Moufang polygons of finite Morley rank, an inductive approach gives the classification of Moufang buildings of spherical type for Tits rank at least 2 (cf. [**130**]). But for this approach one does need to know that the Moufang property holds automatically above Tits rank 3, a substantial result. In fact, at this point in the analysis one can also conclude with just a little more argument, using the generic identification theorem of §III 10, Theorem III 10.2.

Recall that when we applied Niles' theorem, the group B was a standard Borel subgroup (normalizer of a Sylow° 2-subgroup, which factors as $S \rtimes T$ with T a torus), and it is on this torus that the associated Weyl group W acts. As in §I 13, the reflections of W act as reflections on the subgroup $T_p = T[p]$ defined by $t^p = 1$, and faithfully, at least for p odd and not the characteristic. So the Prüfer p-rank of this torus is at least 3 for every odd p other than the characteristic.

The next lemma is hypothesis (G) of Theorem III 10.2.

LEMMA 6.8. *Let T be a maximal torus of B. Then*

$$G = \langle U_2(C^\circ(x)) : x \in T_p \rangle \ .$$

PROOF. Let $G_0 = \langle C^\circ(x) : x \in T_p \rangle$.

Recall that $C(G,S)$ is generated by the subgroups $U_2(N(X))$ with $X \leq S$ definable, connected, and B-invariant. In particular, T_p acts on each $U_2(N(X))$. By Lemma II 5.26, these groups $U_2(N(X))$ are contained in G_0. So $C(G,T) \leq G_0$.

So if $G_0 < G$, then $G = \mathrm{SL}_2(K)$ by Theorem 3.3, which is not the case. □

So to complete the identification using Theorem III 10.2, we must deal with hypotheses (R.1) and (R.2), on the structure of $U_2(C^\circ(x))$ for $x \in T_p^\times$. By Proposition III 9.4, the group $U_2(C^\circ(x))$ is reductive, By Proposition II 5.10 it follows that $U_2(C^\circ(x)) = F^*(C^\circ(x))$.

Finally, hypothesis (R.1) says that the groups $U_2(C^\circ(x))$ contain no unipotent p-subgroups (p is odd). Since $H = U_2(C^\circ(x))$ is a K-group and $H/O_2(H)$ is a central product of Chevalley groups in characteristic 2, this is immediate.

7. Notes

This is a key transitional chapter. Our series of characterizations of SL_2 terminates in §2 with the most flexible version of all, after which it turns out we are in a position to move quickly in the direction of an identification theorem, by a strategy which is outlined, finally, in §5.

§1. Abelian Sylow° 2-subgroups

This material was worked out in the K^*-context in [**5**], making essential use of the theory of groups generated by pseudoreflections, which has no obvious analog in the finite case. In addition, the classification of groups with weakly embedded subgroups is very powerful here, much more powerful than the strong embedding classification would be. We dealt in [**5**] with groups with strongly closed abelian subgroups, the analog of Goldschmidt's theorem (cf. [**98**]). This turned out to be more than is actually needed for our classification theorem.

The methods of this chapter would suffice to generalize all of [**5**] to the L^*-context. It may be of some independent interest that the treatment of strongly closed abelian subgroups can be treated by a mixture of finite group theoretic methods and more geometrical arguments, along the same lines as the case of abelian Sylow 2-subgroups. But once we have the full classification of simple groups of even type this result follows anyway as a special case.

The classification of groups with strongly closed abelian subgroups was given in the finite case in [**98**]. The essential point in passing to L^*-groups is to develop the theory of groups generated by pseudoreflection groups in a context which allows for the presence of nonsolvable connected degenerate type groups.

The treatment of groups of finite Morley rank whose Sylow 2-subgroups contain *finite* strongly closed abelian subgroups would be different, and more direct. This is given in [**4**] for the K^*-case, modulo applications of tameness that were removed by Jaligot's work. The L^*-case goes beyond this, and has not been published. Again, we will arrive at the full classification in even type without relying on this particular result.

§2. Baumann's Pushing Up Theorem

The versions of Baumann's Pushing Up Theorem given in §2 are taken from [**6**].

§3. The $C(G,T)$ Theorem

The $C(G,T)$ theorem was presented in [**6**], with a slightly different definition of $C(G,T)$, more suitable for the K^*-case. With the definition given here the result is slightly sharper, but the proof is the same. One can also give a quick a priori proof via a Frattini argument that with either definition the two versions of the theorem are equivalent, but in any case the natural line of argument gives the stronger version.

§4. 2-local subgroups

The elimination of $\hat{\mathrm{O}}(N_G(P))$ generalizes the elimination of the core $O(N_G(P))$ as given in [**4**]. It can be carried out as soon as one has the weak embedding classification (and could perhaps be used more heavily at an earlier point).

§5. Parabolic subgroups

This material corresponds to [**8**] in the K^*-context. That paper evolved over several versions and at first included a general result on the existence of standard components. However, that turned out not to be needed when working with the amalgam method and was suppressed, as we were in a very utilitarian frame of mind at the time. Of course the suppressed material follows from our main classification theorem, but this is a little beside the point.

The solvability of $N^\circ(S)$ in the L^*-context comes from [**37**]. Though quick, the argument is delicate. We actually gave Theorem III 8.1 in a version which would allow us to bypass this particular result, but we will not expand on this.

§6. The classification: Plan of attack

While this material has not appeared previously in print, it is the basis for the approach we took in the K^* context. The suggestion to take an approach which heads as rapidly as possible toward the amalgam method was made by Stellmacher and Stroth.

The generic identification theorem of §III 10 (or [**37**]) was devised as a more direct approach to identification than the corpus of material associated with Tits' theory of buildings. But as we have previously noted, even if we may avoid the Tits classification in higher ranks, we will continue to rely on the classification of Moufang polygons, which is invoked in the next chapter.

CHAPTER IX

Quasithin groups

Als Hegel auf dem Totbette lag, sagte er: "nur Einer hat mich verstanden", aber gleich darauf fügte er verdrießlich hinzu: "und der hat mich auch nicht verstanden".

— Heine, 1834

Introduction

In the present chapter we prove the following.

THEOREM **QT**. *Let G be a simple L^*-group of finite Morley rank and of even type, and suppose that G is quasithin. Then G is isomorphic to a Chevalley group of Lie rank* 2*, that is, to* $\mathrm{PSL}_3(K)$*,* $\mathrm{Sp}_4(K)$*, or* $\mathrm{G}_2(K)$ *with K some algebraically closed field of characteristic* 2.

The hypothesis that G is quasithin means that if S is a Sylow° 2-subgroup of G, then there are two subgroups P_1, P_2 of minimal parabolic type in G containing $N^\circ(S)$, such that G is generated by the subgroups L_1, L_2, where $L_i = U_2(P_i)$.

The basic tool used throughout most of the chapter, is the *amalgam method*, in which a graph of groups Γ is associated with the configuration (G, P_1, P_2).

This graph of groups Γ has a universal cover $\hat{\Gamma}$, a tree of groups, which corresponds to the free amalgam $P_1 *_B P_2$ with $B = P_1 \cap P_2$. In the amalgam method, one works in this graph, which provides a geometrical notation for working with $\hat{G} = P_1 *_B P_2$. This last group need not be of finite Morley rank, but we work mainly with definable subgroups of one of the conjugates of P_1 and P_2, which can be considered as subgroups of G.

In §1 we introduce this method, and the associated notation, including a critical parameter s. After a lengthy analysis we find that s is 4, 5, or 7, values which turn out eventually to correspond to the groups of type PSL_3, Sp_4, and G_2, respectively.

In §2 we have some preparatory material. We then deal separately with the two possibilities s even and s odd in separate sections (beginning with the observation that $s \geq 4$ in any case). We aim at a determination of the relevant values of s and a sufficiently clear picture of what s means in terms of the action of P_i on $O_2(P_i)$ for $i = 1, 2$. In §3 we introduce some fundamental concepts, in particular a parameter b_i ($i = 0, 1$) whose analysis plays a crucial role. This analysis is carried out in §§4–8.

With this information in hand, in §9 we construct a generalized n-gon Γ^* as a natural quotient of $\hat{\Gamma}$, where $n = s - 1$: the values of n are therefore 3, 4, and 6, and thus we have the generalized triangle (projective plane) associated with PSL_3, the generalized quadrangle associated with Sp_4, and the generalized hexagon associated with G_2; or rather, that is what we must show. For this, we verify the *Moufang property* and then apply the classification theorem of [**130**].

We can then show that the group G^* induced on this generalized polygon by $\hat{G}$ is interpretable in the original group G, and thus has finite Morley rank. This allows us to determine G^*. On the other hand, G^* is only a quotient of a free amalgam associated with G, and it is not immediately clear how G may be determined. However, the pattern of groups (P_1, P_2, B) is visible both in G and in $G*$, as it corresponds to the labeling of a single edge in each of the graphs Γ, $\hat{\Gamma}$, and Γ^*. Thus the determination of G^* yields a determination of P_1, P_2, B *inside* G. Now to conclude we use a theorem of Tits, Fact II 2.28. After a little more computation to verify the hypotheses of this Fact, in §10, we finally arrive at the identification of G.

1. The amalgam method

We will use the *amalgam method* at considerable length to prove the following result.

PROPOSITION 1.1. *Let G be a an L^*-group of finite Morley rank and of even type, and S a Sylow° 2-subgroup of G. Suppose that G is generated by two definable subgroups L_1, L_2 satisfying the following conditions for $i = 1, 2$.*

(A) $L_i = U_2(L_i)$.
(B) $\bar{L}_i = L_i/O_2(L_i) \simeq SL_2(K_i)$ *for some algebraically closed field K_i of characteristic 2.*
(C) $F^*(L_i) = O_2(L_i)$.
(D) $N^\circ(S)$ *normalizes L_i.*

Then there is a simple Chevalley group of Lie rank 2 G^*, and there are definable subgroups G_i of G containing L_i for $i = 1, 2$, such that the triple $(G_1, G_2, G_1 \cap G_2)$ is isomorphic to a triple (G_1^*, G_2^*, B^*) in G^*, where G_1^* and G_2^* are minimal parabolic subgroups of G^* containing the Borel subgroup B^*.

We express the conclusion of the result as follows: G and G^* are *parabolic isomorphic.* We will say at once what the groups G_1 and G_2 are, and rephrase the result accordingly. However, the hypotheses of Proposition 1.1 are those which are actually given to us if we assume G_0 is quasithin. As we have discussed above, the conclusion of this proposition does not immediately provide the identification of G_0, but brings us very close to the criterion given in Fact II 2.28, and it will be easy to close that gap when we come to it.

NOTATION 1.2. *Under the hypotheses of Proposition* 1.1, *let* $B_i = N_{G_i}(S)$ *for* $i = 1, 2$, *let* $B = \langle B_1, B_2 \rangle$, *and set* $G_i = L_i B$ *for* $i = 1, 2$.

LEMMA 1.3. *Under the hypotheses of Proposition* 1.1 *we have the following:*

(1) $G = \langle G_1, G_2 \rangle$.
(2) $S \triangleleft B$ *and* B/S *is abelian.*
(3) $U_2(G_i) = L_i$.
(4) $G_1 \cap G_2 = B$.
(5) $F^*(G_i) = O_2(G_i)$.

PROOF. We have $G = \langle G_1, G_2 \rangle$ and $S \triangleleft B$ in view of our hypotheses and our definitions, but these points are nonetheless worth recording.

Now B_2 acts on $\bar{L}_1$ by inner automorphisms normalizing $\bar{S}$, hence B_2 commutes with $\bar{B}_1$ modulo $\bar{S}$, or, in other words, $[B_1, B_2] \leq S$, and the second point is established. In particular, B/S is a divisible abelian $2^\perp$-group normalizing L_1 and L_2, and our third point follows.

Now $G_1 \cap G_2 = L_1 B \cap L_2 B = (L_1 \cap L_2)B$ and as $L_1 \cap L_2$ is a proper subgroup of L_1 containing S, the intersection must be contained in B_1 and hence in B, so the fourth point follows.

Finally, taking for definiteness $i = 1$, we claim $F^*(G_1) = O_2(G_1)$. Let $P_1 = N^\circ(L_1)$. Evidently $E(G_1) \leq E(L_1) \leq E(P_1) = 1$ as P_1 is parabolic, using Proposition VIII 5.3.

So we are left with $F(G_1) = O_2(L_1)O_{2'}(G_1)$. Now $O_{2'}(G_1)$ commutes with $L_1 = U_2(G_1)$, and in particular with S. On the other hand, $O_{2'}(G_1)$ acts on $\bar{L}_2$, centralizing $\bar{S}$, and hence acts trivially. So $[O_{2'}(L_1), L_2] \leq O_2(L_2)$, and $[O_{2'}(L_1), O_2(L_2)] = 1$, so for any Sylow° 2-subgroup Q of L_2 we have $[O_{2'}, Q] = 1$, and thus $[O_{2'}(L_1), L_2] = 1$. So $O_{2'}$ centralizes $\langle L_1, L_2 \rangle = G$, a contradiction. □

The main result may be reformulated as follows.

PROPOSITION 1.4. *Let* G_0 *be a group of finite Morley rank and of even type, and* S *a Sylow*° 2*-subgroup of* G_0. *Suppose that* G_0 *is generated by two definable connected L-subgroups* G_1, G_2 *satisfying the following conditions for* $i = 1, 2$, *where we write* L_i *for* $U_2(G_i)$, *and* B *for* $G_1 \cap G_2$:

(B′) $\bar{L}_i = L_i/O_2(L_i) \simeq \mathrm{SL}_2(K_i)$ *for some algebraically closed field* K_i *of characteristic* 2.
(C′) $F^*(G_i) = O_2(L_i)$.
(D) $N_{G_i}(S) = G_i \cap B$ *for* $i = 1, 2$.
(E) $O_2(G_0) = 1$.
(F) $B = (L_1 \cap B)(L_2 \cap B)$

Then there is a simple Chevalley group of Lie rank 2 G^*, *and there are groups* G_i *containing* L_i *for* $i = 1, 2$, *such that the triple* $(G_1, G_2, G_1 \cap G_2)$ *is isomorphic to a triple* (G_1^*, G_2^*, B^*) *in* G^*, *where* G_1^* *and* G_2^* *are minimal parabolic subgroups containing the Borel subgroup* B^*.

LEMMA 1.5. *Under the hypotheses above, no nontrivial subgroup of* $G_1 \cap G_2$ *is normal in* G_0.

PROOF. Let $K \leq G_1 \cap G_2$ be normal in G_0. Then $O_2(K) = 1$ by hypothesis (E). Hence $K \leq O(G_1) = 1$, by hypothesis (C′). □

An equivalent statement is that the action of G_0 on the graph Γ defined below is faithful.

1.1. The graph Γ. In the context of Proposition 1.4, we have $G_0 = \langle G_1, G_2 \rangle$ and we wish to identify the triple (G_1, G_2, B) where $B = G_1 \cap G_2$. For the most part we work in the group

$$G = G_1 *_B G_2,$$

the free product of G_1 and G_2 over B. This has the defect that it is not itself a group of finite Morley rank, a fact which we will simply have to live with.

We let Γ be the associated *coset graph*, whose vertices are the cosets of G_1 and G_2 in G, with an edge whenever two cosets meet. In particular, G_1 and G_2 are vertices of Γ, connected by an edge. The graph Γ is a tree, and is the universal cover of the coset graph Γ_0 for G_0. The virtues of the graph Γ will provide ample compensation for the problems caused by the loss of finite Morley rank in passing from G_0 to G.

Vertices of Γ are denoted by small Greek letters (reserving the letter γ, however, for paths). There is a natural action of G on Γ, and the stabilizer of the vertex δ is denoted G_δ. These stabilizers are conjugates of G_1 and G_2. More generally, if V is a set of vertices, G_V will denote the pointwise stabilizer of V. In particular, we write $G_{\delta\delta'}$ for $G_{\{\delta,\delta'\}}$.

We let α and β denote G_1 and G_2, respectively, when considered as vertices of Γ.

Write $B = S \rtimes K$ with K a torus. More precisely, apply property (F) and Schur-Zassenhaus to split B as $S \rtimes K$, and then observe that K is the product of two commuting tori, from L_1 and L_2.

Let $T = T_K$ be the fixed point set of K in $V(\Gamma)$; we often think of T also as the graph induced on this set of vertices by Γ.

For δ a vertex of Γ, we write L_δ for $U_2(G_\delta)$, and $\bar{G}_\delta$ for $G_\delta/O_2(L_\delta)$. Let $K_\delta = K \cap L_\delta$ for $\delta \in V(\Gamma)$. Then $K = K_\alpha K_\beta$ by our initial assumptions.

Recall that $\bar{L}_\delta \simeq \mathrm{SL}_2(F)$ for some field F of characteristic 2.

We write $\Delta(u)$ for the neighbors of u in Γ.

LEMMA 1.6. *For δ a vertex of Γ, L_δ acts transitively on the neighbors of δ, and the action factors through the action of $\bar{L}_\delta$ on the projective line. In particular, the action of G on Γ is edge-transitive.*

PROOF. Consider the action of G_1 on the neighbors of α. The neighbors of G_1 are, by definition, the cosets $G_2 g$ with $g \in G_1$. Thus the action of G_1 is transitive on $\Delta(\alpha)$, and as the action of G on Γ has two orbits on the vertices, and every edge meets both orbits, it follows that the action is transitive on the edges.

The stabilizer of β in G_1 is $G_{\alpha\beta} = B$, and $G_1 = L_1 B$. So L_1 also acts transitively on $\Delta(\alpha)$, and the stabilizer of β is a Borel subgroup. As this is also the stabilizer of a point in the natural action of L_1 on the projective line (factoring through $\bar{L}_1$), our claims follow. □

LEMMA 1.7. *Let λ, μ be adjacent vertices of Γ, and $Q \leq G_{\lambda\mu}$ a subgroup whose normalizers in G_λ and in G_μ act transitively on the neighbors of the respective vertices. Then $Q = 1$.*

PROOF. Let $H_\delta = N_{G_\delta}(Q)$ for any vertex δ of Γ for which $Q \leq G_\delta$.

Let Γ_0 be the graph induced on the union of the orbits of $H = \langle H_\lambda, H_\mu \rangle$ on λ and μ. Then H acts transitively on the neighbors of any vertex of Γ_0 and it follows that $\Gamma_0 = \Gamma$. As H normalizes Q, it follows that Q fixes all the points of Γ. But the kernel of the action of G on Γ is trivial, by Lemma 1.5, so Q is trivial. □

LEMMA 1.8. *T is a 2-way infinite path in Γ.*

PROOF. In the first place, T is connected, since the path connecting any two vertices of T must be fixed pointwise by K. Also T contains α and β. So it suffices to show that the degree of any vertex of T, relative to Γ_K, is 2.

Let δ be a vertex of T. Then δ has at least one neighbor δ' in T. Then $G_{\delta\delta'}$ is a conjugate of B, and contains K. Hence K must be a complement to $O_2(G_{\delta\delta'})$ (a conjugate of S), and covers a maximal torus of $\bar{L}_\delta$. So K has two fixed points in the action of $\bar{L}_\delta$ on the projective line, hence two fixed points among the neighbors of δ, as claimed. □

LEMMA 1.9. *Let $\delta \in T$. Then $C_{G_\delta}(K) \leq G_T$.*

PROOF. $G_\delta = L_\delta K$, so $C_{G_\delta}(K) \leq K Q_\delta$ with

$$Q_\delta = O_2(G_\delta)\ .$$

As the latter group fixes both neighbors of δ in T, we can continue inductively along all vertices of T. □

LEMMA 1.10. *There are elements w_α, w_β in L_α, L_β respectively, representing nontrivial Weyl group elements in $\bar{L}_\alpha$, $\bar{L}_\beta$ acting on $\bar{K}$ (in the corresponding quotient), such that w_α and w_β normalize K.*

PROOF. In L_α, for example, we have $N(KO_2(L_\alpha)) = O_2(L_\alpha)N(K)$ by the Frattini argument, so there is an appropriate w_α normalizing K. □

Observe that in the foregoing lemma the element w_α acts on T, fixing α and switching its two neighbors in T, and that w_β acts similarly with center β. So the composite $w_\alpha w_\beta$ acts like a shift on T. We will identify the vertex set of T with $\mathbb{Z}$, and the vertices α, β with $0, 1$ (rather than $1, 2$, as earlier notation might suggest). The shift map is then $i \mapsto i + 2$.

1.2. Definability. As we will be making use of Morley rank, and the structures we work with (Γ and the various stabilizers associated with it, or equivalently $G_\alpha *_B G_\beta$) do not fit neatly into a ranked universe, we make some comments on general issues of definability. The main point is that the point stabilizers G_δ that we work with are all definable in a group of finite Morley rank, and as we work locally with finitely many at a time, for the most part, we can use the notion of Morley rank very freely.

DEFINITION 1.11. *Let H be a group acting on a graph A.*

(1) *For $\delta \in V(A)$ and $k \geq 0$, let $\Delta_k(\delta)$ be the set of vertices at distance at most k from δ, let $F_k(\delta) = \bigcup_{\delta' \in \Delta_k(\delta)} H_{\delta'}$ and let $H_k(\delta) = F_k(\delta)^k$, the set of products of at most k elements, each of which lies in $F_k(\delta)$.*

(2) *H is* locally of finite Morley rank *on A if for each $\delta \in V(A)$ and each $k \geq 0$, the structure $(H_k(\delta), \Delta_k(\delta))$ has finite Morley rank. Here $H_k(\delta)$ is viewed as a partial group with a partial action on the graph induced on $\Delta_k(\delta)$. In other words, viewing (H, A) as a relational structure, $(H_k(\delta), \Delta_k(\delta))$ is the corresponding substructure.*

(3) *H is* locally of uniformly finite Morley rank *on A if for each $\delta \in V(A)$ and each $l \geq k \geq 0$, the structure induced on $(H_k(\delta), \Delta_k(\delta))$ by all definable relations on $(H_l(\delta), \Delta_l(\delta))$ is of finite Morley rank, depending on k and δ but not on l.*

LEMMA 1.12. *Let H be a group of finite Morley rank operating on a coset graph A with respect to two definable subgroups H_1, H_2 with intersection H_0, let $\hat{A}$ be the universal cover, and $\hat{H} = H_1 *_{H_0} H_2$ acting naturally on $\hat{A}$. Then $\hat{H}$ is locally of uniformly finite Morley rank on $\hat{A}$.*

PROOF. The claim is that the structures $(\hat{H}_k(\delta), \Delta_k(\delta))$ are interpretable in (H, A), and that the restriction from $(\hat{H}_l(\delta), \Delta_l(\delta))$ to $(\hat{H}_k(\delta), \Delta_k(\delta))$ of a definable relation is definable over (H, A). It suffices to prove the first point as the second point amounts to the trivial point that under the appropriate interpretations the embedding of one such structure into another is also interpretable.

Let $X = P_1 \cup P_2$. Everything comes down to the definability in H of the k-place relation $R_k(x_1, \ldots, x_k)$ defined by "$x_1 \cdot \cdots \cdot x_k = 1$ in $\hat{H}$". This is proved by induction on k beginning with $k = 1$. For $k = 1$ it suffices to note that the natural maps from H_0 and H_1 to $\hat{H}$ are embeddings. For $k > 1$ we use the following property of free products with amalgamation: if $x_1, \ldots, x_k$ are alternately in $H_1 \setminus H_0$ and $H_2 \setminus H_0$, then the product is nontrivial. In the remaining cases, since $k > 1$ the product can be shortened and induction applies. $\square$

2. Preparation

2.1. The natural module. We give some variations on the results in §III 5.

LEMMA 2.1. *Let V be a $\mathbb{Z}[L]$-module where $L \simeq \mathrm{SL}_2(K)$ with K a field. Suppose the following:*

(1) $C_V(L) = 0$ *and* $[V, L] = V$.
(2) $[V, S, S] = 0$, *where S is a maximal unipotent subgroup of L.*

Then V is a direct sum of natural modules for L.

PROOF. Let $W \leq V$ the sum of the irreducible submodules of V which can be viewed as natural modules for G with an appropriate K-vector space structure. By Fact II 5.28, $C_V(S)$ is contained in W. Hence $[V, S] \leq W$. This applies to any Sylow° 2-subgroup of L, and as W is L-invariant it follows that $[V, L] \leq W$. Thus $W = V$. □

LEMMA 2.2. *Let V be an elementary abelian 2-group, $G \simeq \mathrm{SL}_2(K)$ with K an algebraically closed field of characteristic 2, G acting on V, and let S be a Sylow 2-subgroup of G. Suppose the following conditions hold:*

(1) $C_V(G) = 0$.
(2) $V = \langle C_V(S)^G \rangle$.
(3) $[V, S, S] = 0$.

Then $V = [V, G]$ is a sum of natural modules.

PROOF. It suffices to show that $V = [V, G]$.

Let $W = [V, G]$. Then $W + C_V(S)$ is a G-submodule of V containing $C_V(S)$, so by the second hypothesis $V = WC_V(S)$.

Now $2\,\mathrm{rk}(C_W(S)) = \mathrm{rk}(W)$, so if $V > W$, then we have $2\,\mathrm{rk}(C_V(S)) > \mathrm{rk}(V)$. Therefore if S_1 is another Sylow 2-subgroup of G, the intersection $C_V(S) \cap C_V(S_1)$ is infinite; but this is $C_V(G) = 0$, a contradiction. □

LEMMA 2.3. *Let V be an elementary abelian 2-group, $G \simeq \mathrm{SL}_2(K)$ with K an algebraically closed field of characteristic 2, G acting on V, and let S be a Sylow 2-subgroup of G. Suppose the following conditions hold:*

(1) $C_V(G) = 0$.
(2) $V = \langle C_V(S)^G \rangle$.
(3) $\mathrm{rk}(V) = 2\,\mathrm{rk}(C_V(S))$

Then $V = [V, G]$ is a sum of natural modules.

PROOF. Let $S_1 \neq S$ be another conjugate of S. For $x \in S$, we have $C_V(x) \cap C_V(S_1) = 1$ and hence $\mathrm{rk}(C_V(s)) \leq \mathrm{rk}(V/C_V(S)) = \mathrm{rk}(C_V(S))$. Hence $C_V{}^\circ(x) = C_V{}^\circ(S)$. On the other hand, $[x, V] \leq C_V{}^\circ(x)$, and it follows that the action of S on V is quadratic, so the previous lemma applies. □

LEMMA 2.4. *Let H be a group of finite Morley rank with $\bar{H} = H/O_2(H) \simeq \mathrm{SL}_2(K)$ for some algebraically closed field K of characteristic 2. Suppose that $\sigma \in H$ has order 3, and acts without fixed points on $O_2(H)$. Then $O_2(H)$ is elementary abelian, and is a direct sum of natural modules.*

PROOF. The analogous statement for finite groups is proved in [**111**, 8.2], taking K finite and of order greater than 2.

Working with finite subgroups of our group H, it follows that $O_2(H)$ is elementary abelian and that the action of a Sylow° 2-subgroup on $O_2(H)$ is quadratic. Our assumption implies that $C_{O_2(H)}(\bar{H}) = 1$.

Again, Higman's Lemma implies that $O_2(H) = [O_2(H), S]$. So Lemma 2.1 applies. □

LEMMA 2.5. *Let H be a group such that $H/O_2(H) \simeq \mathrm{SL}_2(K)$ for some field K of characteristic 2, where $|K| > 2$, with $Z(H)$ elementary abelian, and $V = O_2(H)/Z(H)$ elementary abelian. Suppose that V affords a natural module for $H/O_2(H)$. Then $O_2(H)$ is elementary abelian.*

PROOF. If $O_2(H)$ is not elementary abelian then the ordinary Frattini subgroup $\phi(O_2(H))$ is nontrivial, and contained in $Z(H)$. We can pass to a quotient group by factoring out a subgroup of $Z(H)$, and arrange to have $|Z(H)| = 2$ and $\phi(H)$ still nontrivial. So we consider this case.

Then there is an involution $z \in O_2(G) \setminus Z(G)$, and all elements of the coset $zZ(G)$ are involutions. But G acts transitively on $(O_2(G)/Z(G))^\times$, and it follows that $O_2(G)$ has exponent 2. □

2.2. Involutions acting on L. We give a couple of minor technical lemmas that will be useful later.

LEMMA 2.6. *Let G be a group of finite Morley rank of the form $AK\langle z\rangle$ with A and K definable in G, A a normal abelian 2-subgroup, and K the multiplicative group of an algebraically closed field of characteristic 2, interpretable in G. Suppose that AK is normal in G, and z is an involution which inverts the image of K in AK/A. Then A has a complement in G.*

PROOF. Let $K_0 \leq K$ be a finite subgroup. Note that AK_0 is z-invariant. By Fact I 9.5, taking $H = A\langle z\rangle$, the group $AK_0\langle z\rangle$ splits over A. Without loss of generality, the complement contains K_0, so there is an involution i in zA which inverts K_0. By compactness, we may suppose there is an involution $i \in Az$ which inverts all of the torsion in K. Thus the subgroup of K consisting of elements inverted by i contains the multiplicative group of an infinite field, and hence is all of K by Fact I 4.16. In particular i normalizes K and $K\langle i\rangle$ is the desired complement. □

LEMMA 2.7. *Let $H = QK\langle z\rangle$ be a group of finite Morley rank with $Q = O_2(H)$, K the multiplicative group of a field interpreted in H, and z an involution which inverts QK/Q. Then z inverts a conjugate of K.*

PROOF. We work inductively, relative to $\mathrm{rk}(Q)$. If $Q_0 < Q$ is infinite and $K\langle z\rangle$-invariant, we apply induction to H/Q_0 and reduce to the case $Q = Q_0$.

If Q is finite, then $[z, K] \leq (QK)^\circ = K$, and our claim holds. So we may suppose that Q is minimal infinite $K\langle z\rangle$-invariant, and in particular abelian.

If K commutes with Q, then $QK = Q \times K$ and z normalizes K already. If $C_Q(K)$ is nontrivial and finite, we may factor it out. So we may suppose that Q is irreducible as a $K\langle z\rangle$-module.

If Q is irreducible as a K-module, then the additive subgroup of $\mathrm{End}(A)$ generated by the image of K is a definable field Φ, and z induces an action on Φ, a field automorphism of order at most 2. The fixed field is definable and infinite, hence equal to Φ by Lemma I 4.3. But z inverts K, a contradiction.

So $Q = V \oplus V^z$ with Q an irreducible K-module. Hence $C_Q(z) = \{[q, z] : q \in Q\}$. It follows that all involutions in the coset Qz are conjugate. On the other hand, by Lemma 2.6 at least one such involution normalizes K, so every such involution normalizes a conjugate of K. □

2.3. Regular paths. A path is a sequence of distinct adjacent vertices in Γ; it is ordered, so that its reversal is another path.

DEFINITION 2.8.

(1) *A path $\gamma = (\delta_0, \ldots, \delta_n)$ in Γ is* regular *if G_γ acts transitively on the two sets $\Delta(\delta_0) \setminus \{\delta_1\}$ and $\Delta(\delta_n) \setminus \{\delta_{n-1}\}$; in other words, on the neighbors of each endpoint, omitting the neighbors in γ.*

(2) *A path is* right *or* left *regular if the regularity condition holds at the right or left endpoint, respectively.*

(3) *A path γ is* singular *if it is not regular; similarly,* right *or* left *singular if the regularity condition fails on the right or left side.*

LEMMA 2.9. *Every vertex is an endpoint of a singular finite path.*

PROOF. If we fix the vertex δ, then the stabilizers G_γ for paths beginning at δ are all contained in G_δ. Therefore by the descending chain condition on definable subgroups of G_δ, there is a path γ with initial point δ for which G_γ is minimal. Then G_γ fixes all neighbors of the other endpoint, hence γ is singular. □

NOTATION 2.10. *Let s be the length of the shortest singular path.*

LEMMA 2.11. *Let γ, γ' be two paths of length m, where $m \leq s$, suppose that their initial vertices are in the same orbit under G. Then γ and γ' lie in the same orbit under the action of G.*

PROOF. Proceeding by induction, we may suppose γ and γ' agree on an initial segment γ_0 of length $m - 1$. Then γ_0 is regular, so we may conjugate the remaining endpoint of γ to γ' while fixing γ_0. □

In particular, G has two orbits on the paths of length s, depending only on the type of the first vertex. There is a distinction to be observed between the cases in which s is odd and s is even. In the former case, reversal of the path changes its orbit, so all paths of length s are singular.

NOTATION 2.12. *Let O be an orbit of G in $V(\Gamma)$, and $m \geq 1$ an integer. A path γ is of type (O, m) if the length is m and the initial vertex is in O, and of type (m, O) if the length is m and the terminal vertex is in O.*

LEMMA 2.13. *Let O_0 and O_1 be the two orbits of G on $V(\Gamma)$.*

(1) *If paths of type (O_0, s) are left regular, then paths of type (O_1, s) are right regular.*
(2) *If s is even, then any path of length at least s is singular on both sides.*
(3) *If s is odd, then any path of odd length at least s is singular.*

PROOF.
(1) We consider a path γ of type (O_1, s) and two paths γ', γ'' extending γ by an additional vertex. We claim that γ' and γ'' are conjugate. We consider the paths $\hat{\gamma}', \hat{\gamma}''$ obtained from γ' and γ'' by deleting the first vertex in each case. Then $\hat{\gamma}'$ and $\hat{\gamma}''$ are paths of the same type and length s, so $(\hat{\gamma}')^g = \hat{\gamma}''$ for some $g \in G$.

So $(\gamma')^g$ and γ'' are extensions to the left of a path of type (O_0, s), and hence are conjugate under the action of G.

(2) Let s be even. We may suppose that paths of type (O_0, s) are singular. Since the reversal of such a path γ is of the same type, γ must be singular on both sides. Then by (1) also paths of type (O_1, s) are singular on both sides.

(3) Let s be odd, and let γ be a path of length s which is singular. Reversing the path if necessary, we may suppose that it is left singular. Let its type be (O_0, s). Let γ' be any path of odd length m greater than s. Reversing the path if necessary, let its type be (O_0, m). Then this path is also left singular, since an initial segment is conjugate to γ. Hence the original path was singular. □

2.4. $Q_\alpha \cap Q_\beta$. It is important to eliminate the following special configuration.

LEMMA 2.14. *Let λ, μ be adjacent vertices in Γ. Then $Q_\lambda \cap Q_\mu$ is not normal in G_λ.*

PROOF. We may take the edge in question to be (α, β). So we will assume $G_\alpha \cap Q_\beta \triangleleft G_\alpha$.

We cannot have $Q_\beta \leq Q_\alpha$, as our hypothesis would become $Q_\beta \triangleleft G_\alpha$, while also $Q_\beta \triangleleft G_\beta$.

Let L be the normal closure of Q_β in L_α. As Q_β is not contained in Q_α, but is K-invariant, we see that Q_β covers a Sylow 2-subgroup of $\bar{L}_\alpha$, and hence L covers $\bar{L}_\alpha$. That is, $L_\alpha = Q_\alpha L$.

We claim that Q_β is a Sylow 2-subgroup of L. Now $O_2(L) \leq Q_\alpha$, so $[O_2(L), Q_\beta] \leq Q_\alpha \cap Q_\beta$. Conjugating this relation within L_α, we find $[O_2(L), L] \leq Q_\alpha \cap Q_\beta$. Working in the quotient $\bar{L} = L/(Q_\alpha \cap Q_\beta)$, we have

$$[O_2(\bar{L}), \bar{L}] = 1$$

and by the theory of central extensions, $\bar{L}^{(\infty)} \simeq \mathrm{SL}_2$ and thus $\bar{L} = E(\bar{L}) \times O_2(\bar{L})$, since $L = U_2(L)$. Now $\bar{L}$ is generated by conjugates of $\bar{Q}_\beta$ under L_α, so in view of its structure we find $O_2(\bar{L}) \leq \bar{Q}_\beta$ and $\bar{Q}_\beta$ is a Sylow subgroup of $\bar{L}$, so Q_β is a Sylow 2-subgroup of L.

Now $F^*(L) = O_2(L)$. By Proposition VIII 2.2 we have a nontrivial definable connected subgroup P of Q_β normalized by both L and $N_{G_\beta}{}^\circ(Q_\beta) = G_\beta$. There is a question here as to what the "ambient group" for the application of Proposition VIII 2.2 should be. Recall that the configuration (G_α, G_β, B) is assumed to occur in a group of finite Morley rank G_0. We can apply the Baumann theorem directly in that group, and the conclusion has the same meaning whether we work in G_0 or in $G_\alpha *_B G_\beta$.

At this point, P is normalized by the groups L and G_β, stabilizing adjacent vertices α, β and acting transitively on the neighbors of those vertices, respectively. By Lemma 1.7, P is trivial, a contradiction. □

LEMMA 2.15. *Let $\gamma = (\delta_0, \ldots, \delta_s)$ be a right singular path of length s. Then $O_2(G_\gamma) \leq Q_{\delta_s}$.*

PROOF. As the length is s, γ is conjugate to a path in T, in other words, one which is K-invariant. We will assume γ is a path in T. Observe that by definition $s \geq 2$ and thus $G_\gamma = O_2(G_\gamma)K$.

Now K has two orbits on $\Delta(\delta_s) \setminus \{\delta_{s-1}\}$, one of length 1; we call that vertex δ_{s+1}. Let $Q = O_2(G_\gamma)$. If Q moves δ_{s+1} then γ is right regular, a contradiction.

So $Q \leq G_{\delta_{s+1}}$, and since $Q \leq G_{\delta_{s-1}}$ as well, and these two groups intersect G_{δ_s} in subgroups which map to opposite Borel subgroups in $\bar{G}_{\delta_s}$, the 2-group Q must belong to $O_2(G_{\delta_s})$, as claimed. □

LEMMA 2.16. *The invariant s is at least 4.*

PROOF. Fix a singular path $\gamma = (\delta_0, \ldots, \delta_s)$ of length s, reversing it if needed to get right singularity. Let $Q = Q_{\delta_1} \cap \ldots Q_{\delta_{s-1}}$. Then $Q \leq G_\gamma$, so by Lemma 2.15 we have

$$(*) \qquad Q \leq Q_{\delta_s} \ .$$

Now if $s = 2$ the relation $(*)$ becomes $Q_{\delta_1} \leq Q_{\delta_2}$ and in particular $Q_{\delta_1} \cap Q_{\delta_2} \lhd G_{\delta_1}$, contradicting Lemma 2.14.

If $s = 3$ then the relation $(*)$ becomes $Q_{\delta_1} \cap Q_{\delta_2} \leq Q_{\delta_3}$, which can be written

$$Q_{\delta_2} \cap Q_{\delta_1} = Q_{\delta_2} \cap Q_{\delta_3} \ .$$

Furthermore, in this case all paths of length 3 with δ_2 as middle vertex are conjugate, so the intersection $Q_{\delta_2} \cap Q_\delta$ is independent of the choice of $\delta \in \Delta(\delta_2)$, and hence is normal in G_{δ_2}, that is, $Q_{\delta_1} \cap Q_{\delta_2} \lhd G_{\delta_2}$, again contradicting Lemma 2.14. □

3. Z_δ

3.1. The group Z_δ.

DEFINITION 3.1. *For δ a vertex of Γ we define*

$$Z_\delta = \langle Z^\circ(S) : S \text{ a Sylow 2-subgroup of } G_\delta \rangle \ .$$

LEMMA 3.2. *Z_δ is a commutative subgroup of Q_δ.*

PROOF. Observe that Z_δ centralizes Q_δ, hence $Z_\delta \leq Q_\delta$. Hence Z_δ is commutative. □

In particular, Z_δ is contained in $G_{\delta'}$ for all neighbors δ' of δ. We note also that as Z_α is K-invariant, either $Z_\alpha \leq Q_\beta$ or Z_α covers a Sylow 2-subgroup of $\bar{L}_\beta$, and of course the same applies to any pair of adjacent vertices of Γ. In the first case, if $Z_\alpha \leq Q_\beta$, then Z_α is contained in all G'_β for β' adjacent to β, and in particular for that choice of β' lying on $T = T_K$; this style of argument can be iterated, and eventually leads to Z_α covering some $\bar{L}_\delta$ with $\delta \in T$. We will return to this more systematically later.

At present we return to the notation in which α, β are called $0, 1$, in terms of the labeling of $T = T_K$.

The next lemma gives some information about Z_δ in general.

LEMMA 3.3. *One of the following holds:*
(i) $Z_0 = Z^\circ(L_0)$.
(ii) $Z_1 = Z^\circ(L_1)$.
(iii) *For $i = 0, 1$ we have:*
  (a) $Z_i = Z^\circ(Q_i)$.
  (b) *$Z^\circ(Q_i)/Z^\circ(L_i)$ is a natural module for $\bar{L}_i$.*
  (c) *If $Z_0 \leq Q_1$ then $Z^\circ(L_0) \neq 1$.*

PROOF. This argument will be a long one.
We assume

(1) $$Z_0 > Z^\circ(L_0);\ Z_1 > Z^\circ(L_1).$$

We will prove (iii).

Let $\mathcal{A}(S)$ be the set of connected definable abelian subgroups of S of maximal rank, and define $\mathcal{A}(Q_i)$ similarly. Let $J(S) = \langle A : A \in \mathcal{A}(S) \rangle$.

If $J(S) \leq Q_0$, then $J(S) = J(Q_0) \lhd L_0$. Thus if $J(S) \leq Q_0, Q_1$, then $J(S)$ is normal in both L_0 and L_1, a contradiction. We will therefore suppose:

(2) $$J(S) \not\leq Q_0 \ .$$

Now we prove:

(3) $Z^\circ(Q_0)/[Z^\circ(Q_0) \cap Z^\circ(L_0)]$ is elementary abelian, and affords a natural module for $\bar{L}_0 = L_0/Q_0$.

Let $A \in \mathcal{A}(S) \backslash \mathcal{A}(Q_0)$, using (2). Taking $g \in L_0$ so that $\langle A, A^g \rangle Q_0 = L_0$, using Corollary II 5.29, we have $(A \cap A^g \cap Z^\circ(Q_0))^\circ = Z^\circ(L_0)$. Thus

$$\begin{aligned} \mathrm{rk}([A^g \cap Z^\circ(Q_0)]/Z^\circ(L_0)) &= \mathrm{rk}(A^g \cap Z^\circ(Q_0)/[A^g \cap A \cap Z^\circ(Q_0)]) \\ &\leq \mathrm{rk}(Z^\circ(Q_0)/A \cap Z^\circ(Q_0)) \\ &= \mathrm{rk}((A \cap Q_0)Z^\circ(Q_0)/A \cap Q_0) \\ &\leq \mathrm{rk}(A/A \cap Q_0). \end{aligned}$$

For the final inequality note that $(A \cap Q_0)Z^\circ(Q_0)$ is abelian, and hence $\mathrm{rk}((A \cap Q_0)Z^\circ(Q_0)) \leq \mathrm{rk}(A)$. In summary, writing f_0 for the rank of the base field of $\bar{L}_0$:

$$\mathrm{rk}([A^g \cap Z^\circ(Q_0)]/Z^\circ(L_0)) \leq \mathrm{rk}(Z^\circ(Q_0)/A \cap Z^\circ(Q_0)) \leq \mathrm{rk}(A/A \cap Q_0) \leq f_0 \,.$$

Hence

$$\mathrm{rk}(Z^\circ(Q_0)/Z^\circ(L_0)) \leq \mathrm{rk}(Z^\circ(Q_0)/A \cap Z^\circ(Q_0)) + \mathrm{rk}(A/A \cap Q_0) \leq 2f_0 \,.$$

Now $\bar{L}_0$ acts faithfully on $Z^\circ(Q_0)/Z^\circ(L_0)$, and hence $Z^\circ(Q_0)/Z^\circ(L_0)$ is a natural module for $\bar{L}_0$ by Lemma II 5.31. This proves (3).

With (3) in hand we will deduce more from the foregoing calculation:

(4)
(a) $Z_0 = Z^\circ(Q_0)$;
(b) $\mathrm{rk}(Z(S)/Z(L_0)) = f_0$;

and for $A \in \mathcal{A}(S) \setminus \mathcal{A}(Q_0)$,

(c) $AQ_0 = S$;
(d) $(A \cap Q_0)^\circ Z_0 \in \mathcal{A}(Q_0) \subseteq \mathcal{A}(S)$.

As we assume $Z_0 > Z(L_0)$, and as $Z^\circ(Q_0)/Z^\circ(L_0)$ is irreducible, this implies (4a): $Z_0 = Z^\circ(Q_0)$.

For $A \in \mathcal{A}(S)$, the proof of (3), together with the fact that $\mathrm{rk}(Z(Q_0)/Z(L_0)) = 2f_0$, yields the following:

$$\mathrm{rk}(Z_0/A \cap Z_0) = \mathrm{rk}(A/A \cap Q_0) = f_0$$

from which it follows that $AQ_0 = S$, and that $\mathrm{rk}((A \cap Q_0)Z_0) = \mathrm{rk}(A \cap Q_0) + f_0 = \mathrm{rk}(A)$. Hence $(A \cap Q_0)^\circ Z_0 \in \mathcal{A}(S)$ and as the latter group is contained in Q_0 it follows that $\mathcal{A}(Q_0) \subseteq \mathcal{A}(S)$. Thus we have (4c,d). Finally, $(A \cap Z_0)^\circ = Z^\circ(S)$, and thus $\mathrm{rk}(Z(S)/Z(L_0)) = f_0$. This proves (4).

Next, by a lengthy analysis we will show that

(5) $$Z^\circ(J(S)) = Z^\circ(S) \,.$$

Set $R = Z^\circ(J(S))$. Fix $T \neq S$ another Sylow 2-subgroup of L_0, and take $A \in \mathcal{A}(T) \setminus \mathcal{A}(Q_0)$.

As $(A\cap Q_0)^\circ Z_0 \in \mathcal{A}(Q_0)$, we have $Z^\circ(J(Q_0)) \le (A\cap Q_0)^\circ Z_0$, and hence we have the following, using (4b) in the middle.

$$\begin{aligned}
\mathrm{rk}(Z^\circ(J(Q_0))/A\cap Z^\circ(J(Q_0))) &\le \mathrm{rk}((A\cap Q_0)^\circ Z_0/(A\cap Q_0)^\circ) \\
&= \mathrm{rk}(Z_0/A\cap Z_0) \\
&= f_0 \\
&= \mathrm{rk}(Z(S)/Z(L_0)) \\
&= \mathrm{rk}(Z(S)/A\cap Z(S)) \\
&\le \mathrm{rk}(R/A\cap R) \\
&\le \mathrm{rk}(Z^\circ(J(Q_0))/A\cap Z(J(Q_0))) \,.
\end{aligned}$$

Hence all inequalities here are equalities, and

$$\mathrm{rk}(R/A\cap R) = \mathrm{rk}(Z(S)/A\cap Z(S)) = f_0 \,.$$

In particular, $R = (R\cap A)^\circ Z^\circ(S)$.

Set $X = (R\cap A)^\circ$. Then $\langle J(S), A\rangle \le C(X)$, so $L_0 = Q_0C(X)$. Let $B = C_S{}^\circ(R)$, and $\hat{L}_0 = \langle B^{L_0}\rangle$. Then $O^2(L_0) \le \hat{L}_0$. Now Q_0 normalizes B, so

$$\hat{L}_0 = \langle B^{Q_0C(X)}\rangle = \langle B^{C(X)}\rangle \le C(X) \,.$$

Hence $X \le C_R(O^2(L_0))$. Since $Z(S)\cap C_R(O^2(L_0)) = Z(L_0)$, and $Z^\circ(L_0) \le X$, we get $X = C_R{}^\circ(O^2(L_0))$. Hence Q_0 normalizes X, and $X \lhd L_0$. Thus we have

$$Z^\circ(J(S)) = XZ^\circ(S) \text{ with } X \lhd L_0 \,.$$

Thus $B = C_S{}^\circ(R) = C_S{}^\circ(X)$. Now $(S\cap \hat{L}_0)^\circ \le C_S{}^\circ(X) = B$ so $B = (S\cap \hat{L}_0)^\circ$ is a Sylow° 2-subgroup of $\hat{L}_0$.

Now $[S,X]$ is normalized by S and A, hence by L_0. However, $[S,X] = [S, Z^\circ(J(S))]$. If we also have $J(S) \not\le Q_1$, then similarly $[S, Z^\circ(J(S))] \lhd L_1$, hence $[S, Z^\circ(J(S))] = 1$, so $Z^\circ(J(S)) = Z^\circ(S)$ as claimed.

Suppose finally that $J(S) \le Q_1$. Then $J(S) = J(Q_1)$. If L_1 centralizes $Z^\circ(J(S))$, then $Z^\circ(S) = Z^\circ(L_1)$, a contradiction. So $C_{L_1}(Z^\circ(J(S))) \le Q_1 \le S$, and we have

$$C_{L_1}{}^\circ(Z^\circ(J(S))) = C_S{}^\circ(Z^\circ(J(S))) = B,$$

and $B \lhd L_1$. Now B is a Sylow 2-subgroup of $\hat{L}_0$. As $F^*(\hat{L}_0) \le F^*(L_0)$, the Baumann theorem, Fact VIII 2.2, applies and yields a normal subgroup of $\hat{L}_0$ which is normalized by L_1 as well, providing a contradiction. Thus (5) is proved.

Now we can prove, outright that

$$(2') \,. \qquad J(S) \not\le Q_1$$

Otherwise, $Z^\circ(S) = Z^\circ(J(S)) = Z^\circ(J(Q_1)) \lhd L_1$, and $Z_1 = Z^\circ(S) = Z^\circ(L_1)$, contradicting our hypothesis.

Now from (2′) the analogs (3′), (4′) of (3), (4) follow. In particular (iii.a), (iii.b) have now been verified. Accordingly for the remainder of the argument we may assume

$$Z_1 \le Q_0 \,.$$

(6) Z_1 is elementary abelian.

We know that Z_1 is abelian, and $Z_1/Z^\circ(L_1)$ is a natural module. Let $U = \phi(Z^\circ(S))$. Then $U \le Z(L_1)$. Now Z_1 is the union of the conjugates of $Z^\circ(S)$ in L_1, so $U = \phi(Z_1)$. Similarly, $U = \phi(Z_0)$. Hence U is normal in both G_0 and G_1, and therefore $U = 1$.

(7) For $z \in Z_i \setminus Z(S)$, $i = 0$ or 1, we have $C_S(z) = C_S(Z_i) = Q_i$.

As L_i acts transitively on $Z_i/Z^\circ(L_i)$, we have $z \in Z(S^g)$ for some g. If $C_S(z) \not\le Q_i$, and $x \in C_S(z) \setminus Q_i$, then $\langle x, S^g \rangle = L_i$, in consequence of Fact II 4.6, so $z \in Z(S)$, a contradiction. Thus $C_S(z) = Q_i$ and our claim follows.

(8) For $x \in S \setminus Q_i$, $Z^\circ(S) = [x, Z_i]Z^\circ(L_i)$ and $[x, Z_i] \cap Z^\circ(L_i)$ is finite.

We have $\mathrm{rk}([x, Z_i]) = \mathrm{rk}(Z_i/C_{Z_i}(x)) = \mathrm{rk}(Z_i/Z^\circ(L_i)) = f_i$. Hence $f_i = \mathrm{rk}([x, Z_i]) \ge \mathrm{rk}([x, Z_i/Z^\circ(L_i)]) = f_i$. It follows that $[x, Z_i] \cap Z^\circ(L_i)$ is finite, and (8) follows.

Now define

$$V_i = [L_i, Z_i] \ .$$

Then $V_i/V_i \cap Z(L_i)$ is a natural module covering $Z_i/Z^\circ(L_i)$,

(9) $$[S, Z_i] = (V_i \cap Z(S))^\circ \ .$$

Take $x \in L_i \setminus N(S)$ a 2-element. Let $R = [S, Z_i]$ and set $U = RR^x$, a group normalized by x.

Then

$$\begin{array}{ccccccc} f_i & = & \mathrm{rk}(R/R \cap Z(L_i)) & \le & \mathrm{rk}(R/R \cap R^x) & \le & \mathrm{rk}(R/C_R(x)) \\ & \le & \mathrm{rk}(U/C_U(x)) & = & \mathrm{rk}([U, x]) & \le & \mathrm{rk}([Z_i, x]) \\ & = & f_i & & & & \end{array}$$

by (8), so $[Z_i, x] = [U, x] \le U$. As $L_i = \langle x, S \rangle$, we have $V_i = [L_i, Z_i] \le U$. Thus $V_i \cap Z(S) = U \cap Z(S) = R(R^x \cap Z(S)) \le RZ(L_i)^x$ and $(V_i \cap Z(S))^\circ = R$.

(10) $$(Z_0 \cap Z_1) \le Z(S) \ .$$

If $z \in Z_0 \cap Z_1 \setminus Z(S)$, then $Q_0 = C_S(z) = Q_1$ is normal in both L_0 and L_1, a contradiction.

(11) $$Z_0 Z_1 \ntriangleleft L_0 \ .$$

If $Z_0 Z_1 \triangleleft L_0$, then $Q_0 \cap Q_1 = C_{L_0}(Z_0 Z_1) \triangleleft L_0$, contradicting Lemma 2.14.

(12) $$(Z_0 \cap V_1)^\circ \le Z(L_0) \ .$$

Take $A \in \mathcal{A}(S) \setminus \mathcal{A}(Q_0)$, and $g \in L_0 \setminus N(S)$, and set $B = (A \cap Q_0)^\circ Z_0$. Then $B \in \mathcal{A}(Q_0)$, and $[A, B] \le Z_0$.

If $[A, Z_1^g] \le Z_0$, then $[A^{g^{-1}}, Z_0 Z_1] \le Z_0$ and hence $L_0 = \langle A^{g^{-1}}, S\rangle \le N(Z_0 Z_1)$, a contradiction. So $[A, Z_1^g] \not\le Z_0$, and hence $[A, V_1^g] \not\le Z_0$, and so $V_1^g \not\le B$. Thus $[B, V_1^g] \ne 1$.

Now $B^{g^{-1}} \le Q_0 \le S$ so $B^{g^{-1}} \in \mathcal{A}(S)$. But $[B^{g^{-1}}, V_1] \ne 1$ so $B^{g^{-1}}$ is not contained in Q_1, and hence $B^{g^{-1}} Q_1 = S$, $BQ_1^g = S^g$, and

$$[B, V_1^g] = [S^g, V_1^g] = (V_1^g \cap Z(S^g))^\circ \ .$$

Suppose for some choice of g as above we have

(I) $$C_{V_1}(V_1^g) \le Z(S) \ .$$

Then $V_1 Q_1^g = S^g$ as $\mathrm{rk}(V_1/V_1 \cap Q_1^g) \ge \mathrm{rk}(V_1/V_1 \cap Z(S)) = f_1$, so

$$(V_1^g \cap Z(S^g))^\circ = [S^g, V_1^g] = [V_1, V_1^g],$$

and similarly

$$[V_1, V_1^g] = (V_1 \cap Z(S))^\circ \ .$$

Thus $(V_1 \cap Z(S))^\circ = (V_1^g \cap Z(S^g))^\circ \le Z(S) \cap Z(S^g) = Z(L_0)$. But $(Z_0 \cap V_1)^\circ \le (Z_0 \cap Z_1)^\circ = Z^\circ(S)$, so this yields $(Z_0 \cap V_1)^\circ \le Z(L_0)$, as claimed.

Now suppose, alternatively, that for all g as above we have

(II) $$C_{V_1}(V_1^g) \not\le Z(S);$$

taking $x \in C_{V_1}(V^g) \setminus Z(S)$, this yields

$$V_1^g \le C(x) = Q_1, \text{ and } [V_1^g, V_1] = 1 \ .$$

Let $C = C_B(V_1^g)V_1^g$. We have seen that $\mathrm{rk}(V_1^g/B \cap V_1^g) = f_1$. Thus

$$\mathrm{rk}(C) = \mathrm{rk}(B \cap Q_1^g) + f_1 \ge \mathrm{rk}(B)$$

and $C \in \mathcal{A}(S)$.

Now $C \le C_A{}^\circ(V_1^g)Z_1$, which is abelian, so $C = C_A{}^\circ(V_1^g)Z_1$, and A normalizes C. Thus for $x \in A \setminus Q_0$ we have

$$[B, V_1^g] = [B, C] = [B, C^x] = [B, V_1^{gx}]$$

and $[B, V_1^g] \le Z(S^g) \cap Z(S^{gx}) = Z(L_0)$.

Now $(V_1^g \cap Z(S^g))^\circ = [B, V_1^g] \le Z(L_0)$, so $(V_1 \cap Z(S))^\circ \le Z(L_0)$, and as above $(Z_0 \cap V_1)^\circ \le Z(L_0)$.

This proves (12).

Now if $Z^\circ(L_0) = 1$, we have $(Z_0 \cap V_1)^\circ = 1$, and in particular $(Z(S) \cap V_1)^\circ = 1$, while $\mathrm{rk}(V_1/Z(S) \cap V_1) = f_1$, a contradiction.

This completes the proof of part (iii.c). □

3.2. The parameters b_i and r.

DEFINITION 3.4. *Let $i \in T$ (which we identify with $\mathbb{Z}$). Then b_i is*

$$\max(|j - i| : j \in T, Z_i \leq G_j) \ .$$

As T falls into two orbits under the stabilizer of the set T in G, in view of the shift automorphism mentioned at the end of §1.1, there are at most two values for b_i, namely b_0 and b_1.

As $Z_0 \leq Q_0 \leq G_1$, we have $b_0 \geq 1$ and similarly $b_1 \geq 1$. As $Z_0 \leq G_1$, we have $Z_0 \leq G_2$ if and only if $Z_0 \leq Q_1$. Thus $b_0 > 1$ if and only if $Z_0 \leq Q_1$.

These parameters are important for our case division. There is another parameter whose relation to s is a critical question. We first make some remarks on stabilizers of paths in T.

LEMMA 3.5. *Let γ be a path of length at least 2 in T. Then*
(1) *$G_\gamma = O_2(G_\gamma)K$.*
(2) *$G_\gamma > G_T$ if and only if $O_2(G_\gamma) > O_2(G_T)$.*

PROOF. If γ has length at least 2, then we may suppose that it contains the edge $(0, 1)$ and thus $G_\gamma \leq B$ and $G_\gamma = O_2(G_\gamma)K$.

The second point then follows at once. □

LEMMA 3.6. *There is a maximal positive integer r with the property that there is some regular path γ of length r contained in T with $G_\gamma > G_T$.*

PROOF. It suffices to consider paths beginning with either 0 or 1, and once the initial vertex is fixed, the associated stabilizers G_γ form a decreasing sequence, which must therefore be constant from some point on. □

DEFINITION 3.7. *The parameter r is defined in accordance with the preceding lemma as the maximal integer so that there is some regular path γ of length r contained in T with $G_\gamma > G_T$.*

We note that there may be longer regular paths in Γ (though this does not actually occur in reality). In the long run, we aim at the following:

$$r = s - 1 \ .$$

Initially, since all paths of length $s - 1$ are regular, we have $r \geq s - 1$.

It follows from Lemma 2.13 that for s even, we have $r = s-1$, as desired. However, s is expected to be 4, 5, or 7, thus frequently odd. When s is odd, it follows that r is even since we either have $r = s - 1$ or the last clause of Lemma 2.13 applies.

Our main case division will be according to the parity of s. If s is even the analysis will end quickly, while for s odd it passes through a number of phases. In addition to the three cases corresponding to simple algebraic groups of Lie rank 2, a variety of other potential configurations are examined and eliminated in the course of the analysis, sometimes quickly and sometimes after considerable elaboration. The three cases that survive are described by the following results.

Let f_δ be the rank of the field F_δ underlying $\bar{L}_\delta$.

PROPOSITION 4.3. *Suppose s is even. Then $s = 4$ and $r = 3$, and the stabilizers of paths of length s are $2^\perp$-groups.*

LEMMA 4.4. *Suppose $s = 4$. Then we have the following:*

(1) $f_0 = f_1$.
(2) *For any vertex δ, Q_δ is elementary abelian, and is a natural module for $\bar{L}_\delta$.*

LEMMA 8.2. *Suppose that $[Z_1, Z_{b_1}] \neq 1$, and b_1 is odd. Then $b_0 = b_1 = 1$, $s = 5$, $r = 4$, and in addition we have the following:*

(A) $f_0 = f_1 =: f$.
(B) *Q_0 and Q_1 are elementary abelian, of rank $3f$.*
(C) *$Z(L_1)$ and $Z(L_2)$ both have rank f.*
(D) *$Q_i/Z(L_i)$ is a natural module for $\bar{L}_i$.*

In particular, $r = s - 1$ in this case.

LEMMA 8.3. *Suppose that $[Z_1, Z_{b_1}] \neq 1$, and b_1 is even. Then $b_0 = 3$, $b_1 = 2$, $s = 7$, $r = 6$, and in addition we have the following:*

(A) *$Z_0 = Z(L_0)$ has rank f_1, and Q_0 has rank $2f_0 + 3f_1$.*
(B) *Z_1 is a natural module for L_1, and $Q_1/Z(L_1/Z_1)$ is a direct sum of natural modules for L_1.*
(C) *$\phi(Q_0) = Z_0$, and Q_0/Z_0 is an irreducible $\bar{G}_0$-module.*
(D) $f_1 \leq f_0$.

In particular, $r = s - 1$ and $Z_0 \leq Z_1$.

With regard to the last two results, note that the hypothesis $[Z_1, Z_{b_1}] \neq 1$ is proved eventually (Lemma 8.8).

4. Even s

In the present section, we show mainly that $s = 4$ and $r = 3$ when s is even. We first determine explicitly a small number of possible values for s.

4.1. A theorem of Weiss.

PROPOSITION 4.1. *Let Γ be a tree, and G a group acting on Γ. Suppose that G is locally of finite Morley rank in its action on Γ. Suppose the following hold for some parameter $s > 0$:*

(A) *For each vertex α of Γ, the stabilizer G_α is connected, and acts transitively on the neighbors of α.*
(B) *For any path $\gamma = (\delta_0, \ldots, \delta_m)$ of length $m < s$, with $m \geq 1$, if S is a Sylow 2-subgroup of G_{δ_0,δ_1}, then S_γ acts transitively on $\Delta(\delta_m) \setminus \{\delta_{m-1}\}$ (the right-hand neighbors of γ).*
(C) *With the notation of part (B), we have $Z(S) \leq O_2(G_{\delta_0})$.*
(D) *The connected component of the pointwise stabilizer of any path of length s is a $2^\perp$-group.*

Then we have

$$s \in \{1, 2, 3, 4, 5, 7, 9, 13\}\ .$$

PROOF. We will use natural numbers as names for vertices in Γ. Fix $0, 1$ adjacent vertices of Γ and let $H = G_{01}$. Let S be a Sylow 2-subgroup of H. Let e be the edge $(0, 1)$. For a vertex δ, the distance $d(\delta, e)$ is $\min(d(\delta, 0), d(\delta, 1))$.

We may suppose that $s > 1$ and hence S is nontrivial. Let $b \in Z(S)$. Let $m = \lfloor s/2 \rfloor - 1$. We claim:

(1) b fixes every vertex δ lying at distance at most m from e.

Supposing the contrary, we can find a path $\gamma = (0, 1, \ldots, n)$ with $n \leq m+1$, so that b moves n and fixes the other vertices of γ. Extend γ to a path $\tilde{\gamma} = (0, 1, \ldots, s-1)$ of length $s - 1$ and let $a \in S_{\tilde{\gamma}}$, $a \neq 1$. As a commutes with b, a also fixes $\tilde{\gamma}^b$ pointwise; but $\gamma \cup \gamma^b$ contains a path of length $2(s-n)$ passing through the vertex $n - 1$. As this path is fixed pointwise by $S_{\tilde{\gamma}}$, it follows by condition (D) that $2(s - n) < s$ and hence $2(m + 1) > s$, a contradiction. This proves (1).

We make a case division. Suppose first:

(2a) For every vertex δ with $d(\delta, e) = m + 1$, $b(\delta) \neq \delta$.

Consider any path γ of the form $(-m \ldots, 0, \ldots, s)$ passing through $0, 1$. Consider the edge $e' = (m + 1, m + 2)$ contained in γ. Clearly G operates edge-transitively on Γ and hence there is an element b' conjugate to b, $b' \in O_2(G_{m+1})$, which fixes all vertices at distance at most m from e', and none at distance exactly $m+1$. Thus b' fixes $(1, \ldots, m+1, \ldots, 2m+2)$ and moves 0. We examine the action of the commutator $[b', b]$ on these vertices. Note that $[b', b] \in O_2(G_{m+1})$.

As b' fixes 1, $d((-m + 1)^{b'^{-1}}, 1) \leq m$ and thus b fixes $(-m + 1)^{b'^{-1}}$. Therefore, $[b', b]$ fixes the vertex $-m+1$. Similarly, as b fixes $m+1$, $[b', b]$ fixes $2m+1$. However, $(2m+2)^{[b',b]} = (2m+2)^{b^{-1}b'b}$ and $d((2m+2)^{b^{-1}}, e') = m+1$ since b moves $m + 2$ and fixes $m + 1$. Hence b' moves $(2m + 2)^{b^{-1}}$ and $[b', b]$ moves $2m + 2$. This shows in particular that $[b', b] \neq 1$.

As the 2-element $[b', b]$ fixes the path $(-m+1, \ldots, 2m+1)$ of length $3m$, we have $3m < s$. For s even this yields $s \leq 4$ and for s odd this yields $s \leq 7$; thus the lemma holds in these cases.

Accordingly, we may suppose, without loss of generality:

(2b) For some vertex δ with $d(\delta, e) = d(\delta, 0) = m + 1$, $b(\delta) = \delta$.

We may suppose that $\gamma = (-m - 1, \ldots, 0)$ is a path of length $m + 1$ fixed pointwise by b, and not containing 1. Take $a \in S$ moving -1. Then b fixes $\gamma \cup \gamma^a$, a path of length $2(m + 1)$. Hence $2(m + 1) < s$ and it follows that s is odd.

We claim now:

(3) If $d(0, \delta) \leq m + 1$, then $\delta^b = \delta$; if $d(0, \delta) = m + 2$, then $\delta^b \neq \delta$.

The second point is proved as above, by observing that in the contrary case b fixes a path of length at least s pointwise.

For the first point, suppose $d(0,\delta) \le m+1$. If $d(e,\delta) \le m$ then b fixes δ, so we need only consider vertices for which: $d(0,\delta) = m+1$, $d(1,\delta) = m+2$. As $m+2 < s$ it follows easily from our hypotheses that S acts transitively on the set of all such vertices, and as b is central in S and fixes one such vertex, it fixes them all. Thus (3) holds.

Suppose now that $s \equiv 3 \mod 4$. Consider any path passing through $(0,1)$ of the form $(-(s-5)/2, \ldots, 0, \ldots, s)$ The vertices 0 and $(s+1)/2$ are conjugate. Let b' be an element of $O_2(G_{(s+1)/2})$ which fixes all vertices at distance at most $m+1$ from $(s+1)/2$ and moves all vertices at distance $m+2$. Again we consider the commutator $[b', b]$ and in this case it fixes the path $(-(s-5)/2, \ldots, s-2)$ and moves vertex $s-1$. This gives the estimate $(3s-9)/2 < s$, so $s \le 7$ as s is odd. This is again in accordance with the lemma.

Thus we may suppose that

$$s \equiv 1 \mod 4\ . \tag{4}$$

In this case we work with a path of the form $(-(s-9)/2, \ldots, 0, \ldots, s-2)$ passing through $(0,1)$. As $(s+3)/2$ is even there is an element $b' \in O_2(G_{(s+3)/2})$ fixing all vertices of distance at most $(s-1)/2$ from $(s+3)/2$, and none at distance $(s+1)/2$. The commutator $[b', b]$ will move $s-2$ and fix the remaining points on this path, yielding the estimate $(3s-15)/2 < s$, $s < 15$. As $s \equiv 1 \mod 4$ this is the desired conclusion. □

4.2. Application. We now revert to the notation used in conjunction with the amalgam method, so that s is the length of the shortest singular path in our graph Γ. Of course, Proposition 4.1 is intended to apply here.

LEMMA 4.2. *With the notation of previous sections, if the pointwise stabilizers of paths of length s are $2^{\perp}$-groups, then*

$$s \in \{4, 5, 7, 9, 13\}\ .$$

PROOF. We only have to verify the hypotheses of Proposition 4.1, bearing in mind Lemma 2.16.

Clauses (A) and (C) are part of our initial setup, and for the moment we are assuming clause (D).

For clause (B), we note that paths of length at most s are conjugate to paths in T, and for those paths the group S_γ is K-invariant. If the path is of length less than s, hence regular, then $\bar{S}_\gamma$ in $\bar{G}_{\delta_m}$ cannot be trivial as $G_\gamma = S_\gamma K$, hence S_γ covers a Sylow 2-subgroup of G_{δ_m}, namely the one contained in $G_{\delta_{m-1}} \cap G_{\delta_m}$, which acts transitively on the remaining neighbors of δ_m. □

4.3. The even case.

PROPOSITION 4.3. *Suppose s is even. Then $s = 4$ and $r = 3$, and the stabilizers of paths of length s are $2^{\perp}$-groups.*

PROOF. By Lemma 2.13, $r = s - 1$.

We wish to apply Lemma 4.2, and for this it is sufficient to check that the pointwise stabilizers of paths of length s are $2^{\perp}$-groups. So we may confine ourselves to this point.

Let γ be a path of length s, which we may suppose, without loss of generality, to consist of $(0, 1, \ldots, s)$ inside T. Let $Q = O_2(G_\gamma)$. Then $G_\gamma = Q \rtimes K$. We claim that $Q = 1$.

By Lemma 2.13, any path of length at least s is singular on both sides. Hence by Lemma 2.15 and induction, Q fixes every vertex of Γ. As the action is faithful, $Q = 1$.

So now Lemma 4.2 together with the assumption that s is even yields $s = 4$. □

LEMMA 4.4. *Suppose $s = 4$, and let f_δ be the rank of the field F_δ underlying $\bar{L}_\delta$. Then we have the following:*

(1) $f_0 = f_1$.
(2) *For any vertex δ, Q_δ is elementary abelian, and is a natural module for $\bar{L}_\delta$.*

PROOF. Let $S = O_2(L_0 \cap L_1)$, a Sylow 2-subgroup of both groups. As G acts transitively on the paths of fixed length $m \leq 4$ beginning with the path $(0, 1)$, and also on those beginning with $(1, 0)$, we have two ways to compute the rank of S, working along a path of length 4 starting from 0 or from 1. We have $O_2(G_\gamma) = 1$ once γ has length 4, while $O_2(G_\gamma)$ covers a Sylow 2-subgroup of L_δ for δ an endpoint of γ, when γ has length less than 4. This yields the following information:

$$\begin{aligned} \operatorname{rk}(S) = & \operatorname{rk}(O_2(G_{(0,1)})) = & f_0 + f_1 + f_0 \\ = & \operatorname{rk}(O_2(G_{(1,0)})) = & f_1 + f_0 + f_1 \ , \end{aligned}$$

or $f_0 = f_1$, as claimed. So writing $f = f_0 = f_1$, we have

$$\operatorname{rk}(S) = 3f \ .$$

In particular, $\operatorname{rk}(Q_0) = 2f$.

There is a normal series for Q_0 such that $\bar{L}_0$ acts naturally on each induced quotient, with each factor an elementary abelian 2-group, and this can be refined to a composition series for the action of $\bar{L}_0$. If all of these $\bar{L}_0$-modules are trivial, then any maximal torus of $\bar{L}_0$ acts trivially on Q_0, and hence gives rise to a maximal torus of L_0 commuting with Q_0, contradicting $C_{L_0}(Q_0) \leq Q_0$.

So let V be a nontrivial composition factor for the action of $\bar{L}_0$ on a section of Q_0. Then $\operatorname{rk}(V) \geq 2f$ by Lemma II 5.30. Thus $V = Q_0$ since Q_0 is connected, and $\operatorname{rk}(V) = 2f$, so Q_0 is a natural module by Lemma II 5.31.

The same applies equally to Q_1. □

5. Odd s, $\mathcal{S}^*_{\gamma,K}$

In this and the following three sections we take s to be odd. Our initial goal, which occupies us in the first two of these sections, is to show that $O_2{}^\circ(G_T)$ is trivial. The present section is preparatory. We establish some technical properties of $O_2{}^\circ(G_T)$, and more particularly, of its normalizer. In the notation established below, this can all be expressed by the condition

$$O_2{}^\circ(G_T) \in \mathcal{S}^*_{\gamma,K}$$

with γ an appropriate path of length 2 in T.

5.1. $S_{\gamma,K}$.

DEFINITION 5.1. *Let G be a group of automorphisms of a graph Γ, $K \leq G$, γ a path in Γ of length 2, and δ_1 the midpoint of γ. Then $\mathcal{S}_{\gamma,K}$ is the set of nontrivial subgroups X of G such that the group $M = N_G(X)$ satisfies the following conditions:*

(1) *$X \leq G_\gamma$.*
(2) *$K \leq M$.*
(3) *For either endpoint δ of γ, M_δ acts transitively on all of $\Delta(\delta)$.*
(4) *M_{δ_1} leaves γ setwise invariant, but some element of M_{δ_1} reverses γ.*

The point will be that $O_2{}^\circ(G_T)$ belongs to this class, and this will lead us to consider other groups in the class which are in some sense maximal.

DEFINITION 5.2. *Let G be a group of automorphisms of a tree Γ, K a subgroup of G, γ a path of length 2 in Γ, $X \in \mathcal{S}_{\gamma,K}$, and $M = N_G(X)$. Then we define Γ' as the graph whose vertices are those in the orbit of an endpoint of γ under the action of M, with edges between those vertices whose distance in Γ is 2.*

LEMMA 5.3. *With Γ, K, γ, X, and M as above, and δ a vertex of Γ', the actions of M_δ on $\Delta^\Gamma(\delta)$ and on $\Delta^{\Gamma'}(\delta)$ are equivalent; in fact, every edge of Γ with endpoint δ is the initial edge of a unique path of length 2 leading to a neighbor of δ in Γ'.*

PROOF. We label the vertices of γ $(0, 1, 2)$. We may suppose that $\delta = 0$. As M_0 is assumed to act transitively on the neighbors of 0, we may suppose that the edge in question is $(0, 1)$. Our claim is then the following: if $\delta' \in 0^M$, and δ' is a neighbor of 1, then δ' is on γ.

As 0 is conjugate to δ' under M there is a corresponding conjugate γ' of γ beginning with δ'. Conjugating further by $M_{\delta'}$, we may suppose that $\gamma' = \gamma^g$ is of the form $(\delta', 1, \delta^*)$; so here $g \in M_1$. Now by assumption M_1 leaves invariant the set $\{0, 2\}$, so δ' is in this set, as claimed. □

5.2. $O_2{}^\circ(G)$**.** We now return to the framework of the amalgam method, with s odd, and in particular $s \geq 5$. Let $\tilde{\gamma} = (0, \ldots, r)$. This is a regular path of maximal length in Γ, subject to $G_\gamma > G_T$. By Lemma 2.13, r is even. We remark that with this choice of notation, we are now assuming, in particular, that the vertex called "0" is of suitable type, so that *the symmetry between the two orbits on vertices which has existed up to this point may now be broken.*

We assume for the remainder of the section that

$$O_2{}^\circ(G_T) \neq 1,$$

aiming at a contradiction.

NOTATION 5.4.

(1) *We set* $Q = O_2{}^\circ(G_T)$.
(2) *We continue to label* T *as* $\mathbb{Z}$*, and we set* $\gamma = (0, 1, 2) \subseteq T$.

LEMMA 5.5. *The group* Q *belongs to the class* $\mathcal{S}_{\gamma,K}$.

PROOF. Let $H = N_G(Q)$. We show first that $H_0{}^\circ$ acts transitively on $\Delta(0)$.

As $\tilde{\gamma}$ is regular it follows that $O_2(G_{\tilde{\gamma}})$ is not contained in Q_0, and $O_2{}^\circ(G_{\tilde{\gamma}}) > Q$.

We claim that

$$(*) \qquad O_2(G_{\tilde{\gamma}}) \cap Q_0 = Q\ .$$

Suppose the contrary and let $Q^* = O_2(G_{\tilde{\gamma}}) \cap Q_0$. Let T_{Q^*} be the fixed point set of Q^* in T. This is a path $\hat{\gamma}$ containing $\tilde{\gamma}$, and of finite length. Now $KQ^* \leq G_{\hat{\gamma}}$ acts transitively on the neighbors of the endpoints of $\hat{\gamma}$ lying off $\hat{\gamma}$, and so $\hat{\gamma}$ is regular. By maximality of $\tilde{\gamma}$, we have $\hat{\gamma} = \tilde{\gamma}$, but as $Q^* \leq Q_0$ this is a contradiction.

By $(*)$, Q is normal in $G_{\tilde{\gamma}}$, and so $H_{\tilde{\gamma}} = G_{\tilde{\gamma}}$.

Since r is even, there is a shift map τ induced by an element g of G, carrying r to 0. Then g leaves Q and H invariant. Now the group

$$\langle H_{\tilde{\gamma}}{}^\circ, (H_{\tilde{\gamma}}{}^\circ)^g \rangle$$

acts transitively on $\Delta(0)$. So $H_0{}^\circ$ acts transitively on $\Delta(0)$. Hence $H_0{}^\circ$ covers $\bar{G}_0$.

Now we will find a reflection on γ lying in H. We can take an element x of H_0 switching the vertices 1 and -1. If τ is an element acting as a shift of order 2 on T, then $\tau \in H$ and $\tau^{-1} \cdot x$ swaps 0 and 2, as required.

Finally, we claim that H_1 leaves the set $\{0, 2\}$ invariant. Now H_1 contains the reflection just constructed, as well as K, so if it does not leave this set invariant, then it acts transitively on the neighbors of 1 in Γ. In this case, Q is normalized by the groups H_0, H_1, each acting transitively on the corresponding set of neighbors, and hence $Q = 1$ by Lemma 1.7. This contradiction proves our claim.

Thus $Q \in \mathcal{S}_{\gamma,K}$. □

5.3. $\mathcal{S}^*_{\gamma,K}$**.** Now we introduce the narrower class $\mathcal{S}^*_{\gamma,K}$.

NOTATION 5.6. *Let $\mathcal{S}^*_{\gamma,K}$ be the set of groups $X \in \mathcal{S}_{\gamma,K}$ for which the Sylow° 2-subgroup of the stabilizer in $N(X)$ of the vertex 0 is maximal. (By our conventions, 0 is the left endpoint of γ.)*

We will show below that Q is also in $\mathcal{S}^*_{\gamma,K}$.

LEMMA 5.7. *Let X be a group in $\mathcal{S}_{\gamma,K}$, and set $M = N_G(X)$, $S = M \cap Q_1$. Then:*

(1) *S is a Sylow 2-subgroup of M_0, and of M_2.*
(2) *If $X \in \mathcal{S}^*_{\gamma,K}$, then no nontrivial definable subgroup of S is normal in both M_0 and $N_G(S)$.*

PROOF. Let R be a Sylow 2-subgroup of M_0. Then R fixes some neighbor of 0 in Γ, which after conjugation in M_0 we may suppose to be the vertex 1, that is, $R \le M_1$. In this case R leaves the set $\{0, 2\}$ invariant, and fixes the vertex 0, so in G_1 R is a 2-subgroup fixing two points, and hence lies in Q_1. Thus $R \le M \cap Q_1$. This proves the first point, for M_0, and the same holds for M_2 by symmetry.

Now we suppose that $X \in \mathcal{S}^*_{\gamma,K}$, and we suppose toward a contradiction that $Y \le S$ is definable, nontrivial, and normal in both M_0 and in $N_G(S)$.

We claim

$$Y \in \mathcal{S}_{\gamma,K} .$$

As $M_1 \le N(S)$, Y is normal in M_1, and hence after applying a reflection in M_1 switching 0 and 2, we find $Y \lhd M_2$ as well. Thus writing H for $N(Y)$, we have $M_i \le H_i$ for $i = 0, 1, 2$.

Thus to see that $Y \in \mathcal{S}_{\gamma,K}$, it suffices to check that H_1 leaves the set $\{0, 2\}$ invariant. If this fails, then H_1 acts transitively on $\Delta(1)$, and then by Lemma 1.7 we have a contradiction.

Now we may reach a contradiction. We have $M_0 \le H_0$, and as by hypothesis $X \in \mathcal{S}^*_{\gamma,K}$, it follows that S is a Sylow 2-subgroup of H_0. Let U be the Sylow 2-subgroup shared by G_0 and G_1 (in other words, $O_2(G_{01})$). We have $S \le N_U{}^\circ(Y) \le H_0$, so $N_U{}^\circ(Y) \le M_0$, and thus $N_U{}^\circ(Y) \le M_1$. So $N_U{}^\circ(Y)$ is a connected group acting on the set $\{0, 2\}$, and hence this set is fixed pointwise. As this is also a 2-group, we have $N_U{}^\circ(Y) \le Q_1$. Thus $N_U{}^\circ(Y) \le M \cap Q_1 = S$. However, $N_U{}^\circ(S) \le N^\circ(Y)$, so we find $U = S \le Q_1$, contradicting the structure of G_1. □

We summarize the state of knowledge relative to $\mathcal{S}^*_{\gamma,K}$ as follows.

PROPOSITION 5.8. *Let $X \in \mathcal{S}^*_{\gamma,K}$ and $M = N_G(X)$. Let Γ' be the associated graph on 0^M, with edges between vertices whose distance in Γ is 2. Let $\tilde{\Gamma}$ be the connected component of Γ' containing 0. For $\delta \in \tilde{\Gamma}$, let $\tilde{L}_\delta = U_2(M_\delta)$, and $\tilde{Q}_\delta = O_2(M_\delta)$. Let δ, δ' be adjacent in $\tilde{\Gamma}$.*

Then the following hold:

(1) *M_δ and $M_{\delta'}$ are conjugate in M.*

(2) $\tilde{L}_\delta/\tilde{Q}_\delta \simeq \mathrm{SL}_2(F)$ *for some field* F, *independent of* δ.
(3) $O_2{}^\circ(M_\delta \cap M_{\delta'})$ *is a Sylow* 2*-subgroup of* M_δ *and* $M_{\delta'}$.
(4) *There is no nontrivial subgroup normal in both* L_δ *and* $L_{\delta'}$.

PROOF. The first point is part of the setup.

For the second, we know that L_δ acts transitively on the neighbors of δ in Γ, so the first assertion follows; as the groups in question are conjugate, the field F is independent of δ.

For the remaining points, we may take δ, δ' to be the points $0, 2$.

For the third point, we know $M \cap Q_1$ is a Sylow 2-subgroup common to both M_0 and M_2. Furthermore, $M_0 \cap M_2 \leq G_{01}$, whose Sylow 2-subgroup is normal.

For the last point, if a subgroup is normalized by L_0 and L_2, then is also normalized by a subgroup of L_1 acting transitively on the neighbors $\Delta(1)$ in Γ, namely the subgroup generated by the Sylow 2-subgroups of L_1 contained respectively in L_0 and L_2. This forces the subgroup to be trivial by Lemma 1.7. □

5.4. About M. The following preparatory lemmas deal with situations that arise when we pass from our original amalgam to $M = N^\circ(X)$ with $X \in \mathcal{S}^*_{\gamma,K}$.

LEMMA 5.9. *Let* M *be a group of finite Morley rank, and let* $M_1, M_2 \leq M$ *be definable connected subgroups. Let* $L_i = U_2(M_i)$ *and* $Q_i = O_2(M_i)$ *for* $i = 1, 2$. *Suppose*

(A) $L_i/Q_i \simeq \mathrm{SL}_2(F_i)$ *with* F_i *algebraically closed of characteristic* 2, *for* $i = 1, 2$.
(B) $S = O_2(M_1 \cap M_2)$ *is a Sylow* 2*-subgroup of* M_1 *and of* M_2, *and* $N_{M_1}(S) = N_{M_2}(S) = M_1 \cap M_2$.
(C) *There is no nontrivial definable subgroup normal in both* L_1 *and* L_2.
(D) $C_{L_1}(Q_1) \not\leq Q_1$.

Then S *is elementary abelian,* L_i *splits as* $Q_i \rtimes L_i^*$ *with* $L_i^* \simeq \mathrm{SL}_2(F_i)$, *and* $\mathrm{rk}(S)$ *is either* f_1 *or* $f_1 + f_2$.

PROOF. If Q_1 or Q_2 is trivial, then S is elementary abelian. In this case Q_i is central in M_i for $i = 1, 2$ and by the theory of central extensions, these extensions split. If Q_1 is trivial, then the rank of S is f_1. If Q_1 is nontrivial, but Q_2 is trivial, then Q_1 is a proper subgroup of S which is invariant under $M_1 \cap M_2$, and hence under $N_{M_2}(S)$. However, looking in M_2, this is impossible.

Now suppose Q_1 and Q_2 are both nontrivial. Then by (C), these two groups are distinct, and as they are invariant under the action of $M_1 \cap M_2$, we find $S = Q_1 Q_2$.

Now $C_{L_1}(Q_1)$ covers L_1/Q_1 and is a central extension of $\mathrm{SL}_2(F_1)$ by $Z(Q_1)$, hence splits. Hence $L_1 = Q_1 \times L_1^*$ with $L_1^* \simeq \mathrm{SL}_2(F_1)$. Let $\phi(Q_2)$ be the ordinary Frattini subgroup, as in §I 5.5. Then $\phi(S) = \phi(Q_1)$ since the Sylow subgroups of L_1^* are elementary abelian. Thus $\phi(Q_2) \le Q_1$ is normalized by both L_2 and L_1^*. Furthermore, as $Q_1 \le L_2$, Q_1 normalizes $\phi(Q_2)$ and thus $\phi(Q_2)$ is normalized by $L_1^*Q_1 = L_1$. By our hypothesis (D), we have $\phi(Q_2) = 1$ and Q_2 is elementary abelian.

Let K_1 be a complement to S in $N_{L_1}(S)$. As L_1/L_1^* is a 2-group, we have $K_1 \le L_1^*$. Hence $[K_1, Q_1] = 1$. So we find $[K_1, S] = [K_1, Q_2] \le Q_2$. Hence

$$(*) \qquad [K_1, L_2] \le Q_2 \ .$$

Let $V_2 = Z(Q_2)$. We claim that

$$C_{V_2}(K_1) = 1 \ .$$

For this, we argue that $C_{V_2}(K_1)$ is normal in L_1 and L_2.

For $g \in L_2$, we have $K_1^g \le K_1Q_2$ by $(*)$ It follows that $C_{V_2}(K_1)$ is normal in L_2.

Now $C_{V_2}(K_1) \le C_S(K_1) = Q_1$, and the latter is centralized by L_1^*. Thus $C_{V_2}(K_1)$ is normalized by L_1^*, and also by $Q_1 \le L_2$, so by L_1. Thus $C_{V_2}(K_1) = 1$.

In particular, $V_2 \cap Q_1 = 1$. Thus $(Z(S) \cap Q_1) \cap Q_2 = 1$ and $Z(S)$ is not contained in Q_2. So $C_{L_2}(Q_2)$ covers L_2/Q_2, and so as above we find $L_2 = Q_2 \times L_2^*$ with $L_2^* \simeq \mathrm{SL}_2(F_2)$.

In particular, $Q_1 \cap Q_2$ is normal in L_1 and in L_2, forcing $Q_1 \cap Q_2 = 1$. It follows easily that $\mathrm{rk}(S) = f_1 + f_2$ and everything is proved. □

LEMMA 5.10. *Let G be a group of finite Morley rank, and $M_1, M_2 \le G$ two definable connected subgroups. Assume that M_1 and M_2 are conjugate in G. Set $L_i = U_2(M_i)$, $Q_i = O_2(M_i)$. Assume the following:*

(A) *$L_i/Q_i \simeq \mathrm{SL}_2(F)$ with F algebraically closed of characteristic 2. Set $f = \mathrm{rk}(F)$.*
(B) *$S = O_2(M_1 \cap M_2)$ is a Sylow 2-subgroup of M_1 and of M_2, and $N_{M_1}(S) = N_{M_2}(S) = M_1 \cap M_2$.*
(C) *There is no nontrivial definable subgroup of G normal in both L_1 and L_2.*
(D) *$N(S)$ leaves the pair $\{M_1, M_2\}$ invariant.*

Then one of the following holds for $i = 1, 2$:

(1) *$L_i = Q_i \times L_i^*$ with $L_i^* \simeq \mathrm{SL}_2(F)$, and S is elementary abelian of rank f or $2f$; or*
(2) *Q_i is elementary abelian of rank $2f$ or $3f$, and the quotient $Q_i/Z(Q_i)$ is a natural module for $\bar{L}_i = L_i/Q_i$.*

PROOF. Take $w \in G$ so that $M_1^g = M_2$. Adjusting by an element of M_2, we may suppose that $w \in N(S)$. Then w interchanges M_1 and M_2.

If $C_{L_i}(Q_i) > Z(Q_i)$ for some i, then the previous lemma produces the first case. So we will assume

$$C_{L_i}(Q_i) = Z(Q_i) \ .$$

Now we claim the following:

No nontrivial definable subgroup X of S is normalized by both L_1 and $N_G(S)$.

Indeed, such a subgroup X would be invariant under the element w, and hence normalized by both L_1 and L_2, a contradiction.

Now Theorem III 5.3 applies, and yields, in particular, the following.

(a) $[L_i, Q_i] \leq Z(Q_i)$
(b) $Z^\circ(Q_i)/Z(L_i)$ is a natural module for L_i/Q_i.

Let $V_i = Z(Q_i)$. We claim $V_1 \not\leq Q_2$ and $V_2 \not\leq Q_1$. Suppose for example $V_1 \leq Q_2$. Then by (a) we have $[L_2, V_1V_2] \leq V_2$, and after conjugating by w we find $[L_1, V_1V_2] \leq V_1$, and hence V_1V_2 is normalized by both L_1 and L_2, a contradiction. So our claim follows. In view of the action of $N_{L_i}(S)$ on the groups involved, we find $S = V_1Q_2 = V_2Q_1$. Hence $Q_1 = V_1(Q_1 \cap Q_2)$ and $Q_2 = V_2(Q_1 \cap Q_2)$, and $V_1 \cap Q_2 = V_2 \cap Q_1 = Z(S)$.

Let $x \in L_1$ be a semisimple element that does not normalize S. Then $L_1 = \langle V_2, V_2^x \rangle Q_1$. Set

$$Q_0 = Q_2 \cap Q_1 \cap Q_2^x \ .$$

Then $Q_0 \cap V_1$ commutes with the generators of L_1 and hence $Q_0 \cap V_1 = Z(L_1)$. As $\mathrm{rk}(Q_1/Q_0) \leq 2f$, this must be a natural module for $\bar{L}_1 = L_1/Q_1$, and in particular irreducible. So

$$\mathrm{rk}(Q_1/Q_0) = 2f; \quad Q_1 = Q_0V_1 \ .$$

Now we claim that $Q_0 \cap Q_0^w = 1$. The groups V_2, V_2^x centralize Q_0, and $Q_1 = Q_1^x$ normalizes Q_0, so S normalizes Q_0. Applying w, S normalizes $Q_0 \cap Q_0^w$. Now V_2 and V_2^x centralize $Q_0 \cap Q_0^w$, so $L_1 = \langle V_2, V_2^x \rangle S$ normalizes $Q_0 \cap Q_0^w$, and applying w, also L_2 normalizes $Q_0 \cap Q_0^w$. So this intersection is trivial.

Now $Q_0, Q_0^w \leq Q_1 \cap Q_2$, and $\mathrm{rk}(Q_1 \cap Q_2/Q_0) = f$ (a 1-dimensional space in a natural module). If Q_0 is nontrivial, then Q_0^w covers this quotient and hence $Q_1 \cap Q_2 = Q_0Q_0^w$, with $\mathrm{rk}(Q_1 \cap Q_2) = 2f$ and $\mathrm{rk}(Q_0) = f$. If Q_0 is trivial, then $\mathrm{rk}(Q_1 \cap Q_2) = f$.

Then correspondingly the rank of Q_1 is $2f$ or $3f$. The conditions of case (b) have been recovered. □

5.5. $\tilde{\Gamma}$**.** In this subsection, we deal with groups $X \in S^*_{\gamma,K}$ and the associated graph $\tilde{\Gamma}$ in the sense of Proposition 5.8.

LEMMA 5.11. *Suppose that $X \in S^*_{\gamma,K}$, and set $M = N_G(X)$. Let $\tilde{\Gamma}$ be the associated graph as in Proposition* 5.8, *and let $\Delta = V(\tilde{\Gamma})$, which is the connected component of* 0 *in the graph Γ' on 0^M. Let M^* be the setwise stabilizer of Δ in M, and let M_Δ be the pointwise stabilizer of Δ in M.*

Then the following hold.

(1) *$M_\Delta \leq KQ_0$.*
(2) *For $\delta \in \Delta$, M^*_δ/M_Δ acts on the neighbors of δ in $\tilde{\Gamma}$ as G_δ acts on the neighbors of δ in Γ.*

PROOF.

(1) M_Δ fixes $0, 2, -2$ and hence also $0, 1, -1$ in Γ, and $G_{0,1,-1} = KQ_0$ by Lemma 3.5.

(2) For $\delta \in \Delta$, we have $M_\delta^* = M_\delta$. So this is contained in Lemma 5.3. □

LEMMA 5.12. *Retain the hypotheses and notations of the previous lemma, and consider $M^\Delta = M^*/M_\Delta$ acting on $\tilde{\Gamma}$. Then the maximal regular paths in Δ, with respect to the action of M^Δ, have length at most 4. Furthermore, if k is the maximal length of such paths (so $1 \leq k \leq 4$), then the following hold:*

(1) *The rank of a Sylow subgroup of M_0 is $kf_0 + \mathrm{rk}(M_\Delta)$.*
(2) *If $k \leq 2$, then the Sylow 2-subgroups of M_0 are elementary abelian.*
(3) *If $k > 2$, then $O_2(M_0)/Z(U_2(M_0))$ is a natural module under the action of $U_2(M_0)/O_2(M_0)$.*
(4) *$O_2(M_0)$ is elementary abelian.*
(5) *If $L = O^2(U_2(M_0))$, then $[L, O_2(M_\Delta)] = 1$.*

PROOF. Let $S = Q_1 \cap M$ and let S^Δ be its image in M^Δ, a Sylow 2-subgroup of M_0^Δ and M_2^Δ, by Lemma 5.7.

The hypotheses of Lemma 5.10 are satisfied, and this leads to two cases.

(a) S^Δ is elementary abelian of rank f_0 or $2f_0$.
(b) $O_2(M_i^\Delta)$ is elementary abelian of rank $2f_0$ or $3f_0$, and the quotient $O_2(M_i^\Delta)/Z(U_2(M_i^\Delta))$ is a natural module for $U_2(M_i^\Delta)/O_2(M_i^\Delta)$.

In particular, the rank of S is of the form $k'f_0$ where $1 \leq k' \leq 4$.

We claim that it follows from the structure of $O_2(M_i^\Delta)$ that $k = k'$. For this it is necessary to work out the stabilizers of paths γ of length $r \leq k'$, showing that each acts transitively on the neighbors of the endpoints of γ. This forces the rank to decrease by f_0 at each stage; it also implies that all paths up to length r are regular, and hence can be conjugated into the set $T = \Gamma_K$, simplifying the picture.

Initially $r = 1$, we may take $\gamma = (0, 2)$, and then $M_\gamma = S$. For $r = 2$ (and $k' \geq 2$) we may take $\gamma = (-2, 0, 2)$ and $M_\gamma = O_2(M_0)$. If $O_2(M_0) = O_2(M_1)$ then $O_2(M_\delta) = O_2(M_{\delta'})$ for all adjacent pairs, and hence $O_2(M_0) = M_\Delta$, contradicting $k' \geq 2$. So the claim follows for $r = 2$.

For $r = 3$ or 4 only case (b) is relevant. In this case, consider the path $(-2, 0, 2, 4)$ with stabilizer $O_2(M_0) \cap O_2(M_2)$. This covers a 1-dimensional subspace of $O_2(M_0^\Delta)/Z(U_2(M_0^\Delta))$, namely the centralizer in this module of the Sylow 2-subgroup S in $M_0 \cap M_2$. On the other side, $O_2(M_0) \cap O_2(M_{-2})$ covers a different 1-dimensional subspace, so M_γ does not stabilize the neighbors of -2 off γ, and in view of the action of K, paths of length 3 are regular. In particular, the stabilizer of a path of length 4 will be the preimage of $Z(U_2(M_0^\Delta))$. Again, if this covers $Z(U_2(M_2^\Delta))$, then it lies in M_Δ, and $k' = 3$. So for $k' = 4$ we again have regularity.

It follows that $k' = k$.

For the rest, it suffices to prove the last two claims, as the remaining structural assertions can then be deduced from the structure of S^Δ.

We have $M_0 = LKS$. If $[L, O_2(M_0)] = 1$, that is, $L \simeq \mathrm{SL}_2(F_0)$, then $\phi(S)$ is both characteristic in S and normal in M_0, hence trivial by Lemma 5.7. Thus all our claims follow in this case. We therefore assume the following:

$$(1) \qquad [L, O_2(M_0)] \neq 1 \ .$$

Set $V = [L, O_2(M_0)]$.

Returning to Γ, we have defined

$$Z_1 = \langle Z(S)\colon\ S \text{ a Sylow 2-subgroup of } G_1\rangle$$

Note that $[L, Z_1] \neq 1$, as otherwise Z_1 is normalized by both L and G_1, contradicting Lemma 2.14.

Now in view of Lemma 5.7, and the assumption $V \neq 1$, Theorem III 5.3 applies to M_0. The structural information given there implies that $[L, Z_1] = V \leq Z(O_2(M_0))$. Furthermore, if $[L, O_2(M_\Delta)] \neq 1$, we will also have $V = [L, O_2(M_\Delta)]$.

Now $S \leq Q_1$. If $S = Q_1$, then $G_1 \leq N(S)$ and then Proposition VIII 2.2 gives a nontrivial subgroup of Q_1 which is normal both in G_1 and in M_0, contradicting Lemma 2.14. So we will assume

$$S < Q_1 \ .$$

Take $t \in N_{M_1}(K)$ interchanging 0 and 2. We may suppose $t^2 \in S$. Let S_1 be the preimage in Q_1 of the center of $C_{Q_1/S}(t)$. Let $B_1 = \langle S_1, K\rangle$. We claim that B_1 normalizes $O_2(M_0)$. Both S_1 and K are contained in G_0, and $O_2(M_0) = Q_0 \cap S$, while S_1 and K normalize S.

We will now prove

$$[L, O_2(M_\Delta)] = 1 \ .$$

Assuming the contrary, we have $[L, O_2(M_\Delta)] = V$. In particular, $V \leq O_2(M_\Delta)$.

Consider $R = \langle (VV^t)^{B_1}\rangle$. Then $R \leq O_2(M_\Delta)$. We claim that

$$R \in S_{\gamma,K} \ .$$

Now $R \leq G_\gamma$, $K \leq N(R)$, and L normalizes R since $[L, R] \leq V \leq R$, so $N(R)_0$ acts transitively on the neighbors of 0 in $\tilde{\Gamma}$ other than 2. Finally, $t \in N(R)$ since t normalizes VV^t and B_1. So $R \in S_{\gamma,K}$.

On the other hand, $S_1 \leq N(R)_0$, and S_1/S is infinite. This violates the assumption that $X \in S^*_{\gamma,K}$. This contradiction shows that $[L, O_2(M_\Delta)] = 1$.

Finally, we claim that $O_2(M_0)$ is elementary abelian. Let $H = \phi(O_2(M_0))$, the ordinary Frattini subgroup. Assume toward a contradiction that $H \neq 1$. We know $H \leq M_\Delta$, so $[L, H] = 1$.

The group S_1 introduced above normalizes $O_2(M_0)$ and hence normalizes H. Thus the group HH^t is normalized by t, S_1, M_0, and K, and therefore belongs to $S_{\gamma,K}$. Again, as S_1/S is infinite, this produces a contradiction. □

5.6. Conclusion. We return to the consideration of $Q = O_2{}^\circ(G_T)$.

PROPOSITION 5.13. $Q \in S^*_{\gamma,K}$.

PROOF. We take $X \in S^*_{\gamma,K}$ so that the Sylow 2-subgroup of $N(Q)$ is contained in $N(X)$. Let $H = N(X)$. By Lemma 5.12, $O_2(H_0)$ is elementary abelian. Now $Q \leq O_2(H_0)$, in view of its action. So $O_2(H_0) \leq N(Q)$ and hence $H_0 \leq N(Q)$. Thus $Q \in S^*_{\gamma,K}$. □

6. Odd s, $O_2{}^\circ(G_T)$

As in the previous section we suppose s is odd. Our goal now, building on the work of the preceding section, is to show that $O_2{}^\circ(G_T)$ is trivial.

6.1. The parameter $\tilde{s}$. From the previous section, we retain mainly the notation $Q = O_2{}^\circ(G_T)$ and the fact (Proposition 5.13) expressed briefly by

$$Q \in S^*_{\gamma,K} \tag{$*$}$$

with $\gamma = (0, 1, 2)$. Now that we have fixed the group of interest, namely Q, we set correspondingly

$$M = N(Q)$$

and we retain this notation for the remainder of the present section.

We may reformulate, and sharpen, the analysis carried out in the previous section as follows.

LEMMA 6.1. *For some $\tilde{s}$, which is either 4 or 5, we have the following:*

(1) *A Sylow 2-subgroup of M_0 has rank $(\tilde{s}-1)f_0 + \mathrm{rk}(Q)$.*
(2) *Maximal regular subpaths of T have length $2\tilde{s}-2$.*
(3) $s \leq 2\tilde{s}-3$.
(4) *$O_2(M_0)/Z(U_2(M_0))$ is a natural module for $U_2(M_0)/O_2(M_0)$.*
(5) *$O_2(M_0)$ is elementary abelian.*

PROOF. The last point was established in Lemma 5.12, in view of $(*)$.

The first point was also established in Lemma 5.12, taking $\tilde{s} = k+1$ in the notation of that lemma, but with a wider range of possible values at that point: $2 \leq \tilde{s} \leq 5$.

We take up the second point. Notice that $2\tilde{s}-2$ is $2k$ in the notation of Lemma 5.12, and that paths of length up to $2k$ beginning with 0 (or one of its conjugates) were shown there to be regular.

Now consider a maximal regular path $\tilde{\gamma}$ in T, of length r. Then by Lemma 2.13, r is even. If $r = s-1$, then all paths of this length are regular, and we may suppose $\tilde{\gamma}$ begins with 0. If $r > s$, then by Lemma 2.13, the endpoints of $\tilde{\gamma}$ are necessarily conjugate to 0. So in either case we may suppose that $\tilde{\gamma}$ begins with 0, and hence corresponds to a path $\hat{\gamma}$ of length

$r/2$ in the associated graph $\tilde{\Gamma}$. As $\tilde{\gamma}$ is maximal regular, $Q_0 \cap G_{\tilde{\gamma}} = Q$, and hence Q is normal in $G_{\tilde{\gamma}}$. Hence this path is again regular relative to the group M^Δ acting on $\tilde{\Gamma}$. By the analysis of Lemma 5.12 this forces $r/2 \le k$. So point (2) is established.

Now $s \le r+1 = 2\tilde{s}-1$, and s is odd. So for the third point, we must eliminate the possibility $r = s-1$.

If $r = s-1$, then the path $\tilde{\gamma}_1 = (1, \ldots, 2\tilde{s}-1)$ is a maximal regular path in T. Again, Q is normal in $G_{\tilde{\gamma}_1}$, so $G_{\tilde{\gamma}_1} \le M_1$. But $G_{\tilde{\gamma}_1}$ does not normalize the set $\{0, 2\}$, a contradiction. Thus we have our inequality:

$$s \le 2\tilde{s} - 3 \ .$$

Now $s \ge 5$, so $\tilde{s} \ge 4$, and we have the possibilities $\tilde{s} = 4$ or 5. So in the notation of Lemma 5.12, we have $k > 2$ and thus the fourth point follows as well. □

LEMMA 6.2. *For $i \in T$ we we have $Q \cap Z_i = 1$.*

PROOF. We may suppose $i = 0$ or $i = 1$.

Now $Q \le Z(U_2(M_0))$, by the preceding lemma, and $Z_0 \le Z(Q_0)$, so $Q \cap Z_0 \le Z(U_2(G_0)) \le Z_1$. That is, $Q \cap Z_0 \le Q \cap Z_1$, so it suffices to treat the latter. Suppose toward a contradiction that

$$R = Q \cap Z_1 \ne 1 \ .$$

As $R \le Z(U_2(M_0))$, and R is invariant under an element $t \in G_1$ interchanging G_0 and G_2 and normalizing K, it follows that $N(R)$ belongs to $S_{\gamma,K}$. But $Q_1 \le N(R)_0$, so Q_1 is a Sylow 2-subgroup of $N(R)_0$. Then Proposition VIII 2.2 provides a nontrivial subgroup of Q_1 which is normal both in $N_G(R)_0$ and in $N_G{}^\circ(Q_1)$, and in particular in G_1. This then contradicts Lemma 2.14. □

6.2. The parameter b_0. Recall that b_i is $\max(|j-i| : j \in T, Z_i \le G_j)$. In the present subsection we show $b_0 \ge 3$.

LEMMA 6.3. *Suppose that $i \in T$, and $Q_{i-1} \cap Q_{i+1}$ is normal in G_i. Then the following hold:*

1. *$Q_i/(Q_{i-1} \cap Q_{i+1})$ is elementary abelian, of rank $2f_{i-1}$.*
2. *$Q_i = [Q_i, Q_{i-1}][Q_i, Q_{i+1}](Q_{i-1} \cap Q_{i+1})$.*

PROOF. To lighten the notation, let us write Q_i^+, Q_i^- for the intersections $Q_i \cap Q_{i\pm 1}$. Write f for $f_{i-1} = f_{i+1}$, and let $Q_* = Q_i^- \cap Q_i^+$.

Considering Q_i within $G_{i\pm 1}$, we see that the Frattini subgroup $\phi(Q_i)$ is contained in Q_*, and thus Q_i/Q_* is elementary abelian.

We claim

$$Q_i^\pm > Q_* \ .$$

If, for example $Q_i^- = Q_*$, consider the path $\gamma_0 = (i-2, i-1, i, i+1)$. Then we have $G_{\gamma_0} = K(Q_i^-) = KQ_*$, so this is not right regular, contradicting $s \ge 4$. So $Q_i^- > Q_*$, and similarly for Q_i^+.

In view of the action of K on Q_i^-/Q_*, we find that Q_i^- covers a Sylow 2-subgroup of G_{i+1}/Q_{i+1}. Thus Q_i^-/Q_* has rank f, and $Q_i \leq Q_i^- Q_i^+$. Thus

$$Q_i/Q_* = Q_i^-/Q_* \times Q_i^+/Q_* \ .$$

In particular, the rank of Q_i/Q_* is $2f$, as claimed. This disposes of the first point.

Now we claim

$$R = [Q_{i+1}, Q_i^-] \not\leq Q_* \ .$$

The group $Q_i^- = Q_i \cap Q_{i-1}$ is normalized by Q_i and Q_{i-1}. If it is normalized by Q_{i+1} as well, then we have $L_0 = \langle Q_{i-1}, Q_{i+1}\rangle Q_i$ in the normalizer of Q_i^-. There is an element in L_0 switching G_{i-1} and G_{i+1}, so we find $Q_i^- Q_i^+ = Q_*$, a contradiction. Our claim follows.

Now $R \leq Q_i^+$ and $R \not\leq Q_{i-1}$. In view of the action of K on $[Q_{i+1}, Q_i^-]$, we have $Q_i \leq Q_{i-1}R$, in particular, $Q_i^+ \leq Q_i^- R$ and

$$Q_i^+ = Q_* R = Q_*[Q_{i+1}, Q_i^-]$$

Similarly, $Q_i^- = Q_*[Q_{i-1}, Q_i^+]$. Since $Q_i = Q_i^- Q_i^+$, our second claim follows. □

LEMMA 6.4. *If $b_0 = 2$, then $Z_1 = Z(L_1)$, $b_1 = 3$, and Z_0 is a natural module for $\bar{L}_0 = U_2(G_0)/O_2(G_0)$.*

PROOF. We claim that

(1) $$Z_0 Z_1 \text{ is normal in } G_0 \ .$$

If Z_0 is contained in $Z(U_2(M_0))$, then Z_0 centralizes $M_0 \cap Q_1 = M_2 \cap Q_1$, a Sylow subgroup of M_2, and in view of Lemma 6.1, applied in M_2, we find that Z_0 lies in $O_2(M_2)$. But $O_2(M_2) \leq Q_2$, so this contradicts our hypothesis $b_0 = 2$. So Z_0 is not contained in $Z(U_2(M_0))$.

In view of the structure of $O_2(M_0)$, given by Lemma 5.9, we find that $Z_0/(Z_0 \cap Z(U_2(M_0)))$ is a natural module for $U_2(M_0)/O_2(M_0)$, and

$$[U_2(M_0), O_2(M_0)] \leq Z_0 \ .$$

Now Z_1 centralizes Z_0 and from the structure of Z_0 (as a module) it follows that $Z_1 \leq O_2(M_0)$. Thus $[U_2(M_0), Z_1] \leq Z_0$ and $U_2(M_0)$ normalizes $Z_0 Z_1$. As Q_0 and K also normalize $Z_0 Z_1$, our claim (1) follows.

We will show that

(2) $$Z_1 = Z(L_1) \text{ and } b_1 = 3 \ .$$

Suppose $Z_1 > Z(L_1)$. Then $C_{L_1}(Z_1) = Q_1$. Hence $C_{Q_0}(Z_0 Z_1) = Q_0 \cap Q_1$, and $Q_0 \cap Q_1$ is normal in G_0. Conjugating in G_0, we find $Q_0 \cap Q_1 = Q_0 \cap Q_{-1}$ and hence the path $(-1, 0, 1, 2)$ is left singular, which contradicts the fact that $s \geq 4$. So $Z_1 = Z(L_1)$. Thus $Z_1 \leq Z_0 \leq Z_{-2}$, and $b_1 \geq 3$.

On the other hand, if $b_1 > 3$, then as Z_0 is generated by Z_δ for δ adjacent to 0, and $s \geq 4$, we would have $Z_\delta \leq Q_3$ for all such δ, and thus $Z_0 \leq Q_3$, a contradiction. So $b_1 = 3$ and claim (2) is proved.

Now $Z(L_0) \le Z_1$ centralizes L_0 and L_1, and is trivial by Lemma 2.14. Thus $Z_0 \cap Z(U_2(M_0)) \le Z(L_0) = 1$, and so the structure of Z_0 simplifies:

(3) $$Z_0 \text{ is a natural module for } \bar{L}_0 \ .$$

This proves all our claims. □

LEMMA 6.5. $b_0 > 2$.

PROOF. We have $Z_0 \le O_2(M_0)$, and $O_2(M_0) \le Q_1$ by Lemma 5.9. So $Z_0 \le Q_1$ and this means $b_0 \ge 2$. So we may suppose, toward a contradiction that

(1) $$b_0 = 2 \ .$$

Then we have the structure afforded by the previous lemma. We let S be the Sylow 2-subgroup common to G_0 and G_1.

We have $Z_1 \le Z_0$ and $[S, Z_1] = 1$. Furthermore, these groups are K-invariant, so we find $[S, Z_0] = Z_1$. Similarly, $[Q_1, Z_0] = Z_1$. Let $V = \langle Z_0^{G_1} \rangle$. Then $[Q_1, V] = Z_1$.

We claim that

(2) $$(Q_{-1} \cap Q_1) \lhd G_0 \ .$$

As $b_0 = 2$, the group V covers a Sylow 2-subgroup of L_0. Let $\tau \in G$ operate as a shift degree -2 on T, and let $Q_* = Q_{-1} \cap Q_1$. As $[V, Q_1] \le Z_1$, we have $[V^\tau, Q_{-1}] \le Z_{-1}$, and thus

$$[\langle V, V^\tau \rangle, Q_*] \le Z_0 \le Q_* \ .$$

Now $\langle V, V^\tau \rangle$ covers $\bar{L}_0$, and Q_* is normal in Q_0, so (2) follows.

Now we can apply Lemma 6.3. Thus Q_0/Q_* is elementary abelian, of rank $2f_1$, and

$$Q_0 = [Q_0, Q_{-1}][Q_0, Q_1]Q_*$$

or $Q_0/Q_* = [Q_0, Q_{-1}]/Q_* \times [Q_0, Q_1]/Q_*$.

Now V contains Z_2, and $Z_2 \le G_0$, but $Z_2 \not\le Q_0$, so Z_2, and hence V, covers the Sylow subgroup of G_{01} modulo Q_0. Similarly, V^τ covers the Sylow subgroup of $G_{0,-1}$ modulo Q_0. Hence we can also write Q_0 as

$$[Q_0, V^\tau][Q_0, V]Q_*$$

or $(Q_0 \cap V^\tau)(Q_0 \cap V)Q_*$.

Now $V \le Z(Q_1)$ is elementary abelian, so the action of V on Q_0 is quadratic.

Now let Q_0^* be the centralizer in Q_0/Q_* of $\bar{L}_0$. Looking at Q_0^* as a subgroup of G_1, consider the action of K. As this is trivial, we find that $Q_0^* \le Q_1$. Similarly $Q_0^* \le Q_{-1}$, and $Q_0^* = Q_*$.

Similarly, considering the action of K on $R = Q_0 \cap Q_{-1}$ in G_1, we find that $R \le [K, R]Q_1$ and thus $R = [K, R]Q_*$. So the conditions of Lemma 2.1 are satisfied, and

$$Q_0/Q_* \text{ is a direct sum of natural modules} \ .$$

Recall the condition $B = (L_0 \cap B)(L_1 \cap B)$ imposed at the outset. Let K_i denote $K \cap L_i$.

Now take $t \in N_{U_2(M_1)}(K)$ interchanging the vertices $0, 2$, with $t^2 \in Q_1$. Then K_1 centralizes Z_1. On the other hand, $Z_1 \leq Z_0$, a natural module for $\bar{L}_0$, so K_0 acts on Z_1 without fixed points. Hence $K_0 \cap K_1 = 1$, and $K = K_0 \times K_1$.

Let $\tilde{K} = K_0 K_0^t \cap K_1$. We claim that

(3) $$\tilde{K} \neq 1, \text{ and } \tilde{K} \text{ centralizes } Q \,.$$

If $\tilde{K} = 1$, then $K_0^t = K_0$, and $[K_0, t] \leq K_0 \cap L_1 = 1$. Thus K_0 acts trivially on an involution of $\bar{L}_1$, and hence trivially on $\bar{L}_1$, that is, $[K_0, L_1] \leq Q_1$. So $[K_0, Q_0] \leq Q_0 \cap Q_1$. Conjugating in G_0, we find $[K_0, Q_0] \leq Q_*$, contradicting the structure of Q_0. So $\tilde{K} \neq 1$.

Now Q is a 2-subgroup of G_0 fixing two neighbors of 0, so Q fixes $\Delta(0)$. As $O_2(M_0)/Z(U_2(M_0))$ is a natural module by Lemma 5.10, we find that $Q \leq Z(U_2(M_0))$. Thus K_0 centralizes Q, hence so does R.

Now we may reach a contradiction to the assumption $b_0 = 2$.

Let $W = V \cap Q_0$. Then $[Q, W] \leq [Q_1, V] = Z_1$. Thus $[Q, W, \tilde{K}] \leq [Z_1, L_1] = 1$. Also $[Q, \tilde{K}, W] = 1$ by (2) and thus, by the three subgroups lemma, we have

$$[\tilde{K}, W, Q] = 1 \,.$$

So $[\tilde{K}, W] \leq C_{Q_0}(Q) \leq O_2(M_0) \leq Q_{-1}$. Now WQ_{-1} is a Sylow 2-subgroup of L_{-1}, and $\tilde{K} \leq K_1$, so $[\tilde{K}, L_{-1}] \leq Q_{-1}$. On the other hand, as $b_1 = 3$, $Z_1 Q_{-2}$ is a Sylow 2-subgroup of L_{-2}, and as $[\tilde{K}, Z_1] = 1$ we have $[\tilde{K}, L_{-2}] \leq Q_{-2}$. Hence $C(\tilde{K})$ covers both $\bar{L}_{-1}$ and $\bar{L}_{-2}$, contradicting Lemma 2.14. □

6.3. $Q = 1$. We complete the proof that $Q = O_2{}^{\circ}(G_T) = 1$ in this subsection. We first make a detailed analysis of the situation.

LEMMA 6.6. *Suppose* $Q > 1$. *Then:*

(1) $Z_0 = Z(L_0)$.
(2) $\mathrm{rk}(Z_0) = f_0$.
(3) $\tilde{s} = 5$.
(4) $b_0 = 4$.
(5) $b_1 = 3$.

PROOF. We have proved previously that $b_0 \geq 3$ and $\tilde{s} = 4$ or 5.

We show $Z_0 = Z(L_0)$. If $Z_0 \leq Z(U_2(M_0))$, then as Z_0 is central in Q_0, our claim follows. So suppose that Z_0 is not contained in $Z(U_2(M_0))$. Then Lemma 6.1 implies that $Z_0 = O_2(M_0)$. Now $b_0 \geq 3$, so $Z_0 \leq Q_2 \cap M = O_2(M_2)$, and thus $Z_0 = Z_2$. So Z_0 is normal in both G_0 and G_2, hence also in G_1, contradicting Lemma 2.14. So $Z_0 = Z(L_0)$.

Now $Q \cap Z_0 = 1$ by Lemma 6.2. So by Lemma 6.1 we find

$$\mathrm{rk}(Z_0) \leq [(\tilde{s} - 1)f_0 + \mathrm{rk}(Q)] - [3f_0 + \mathrm{rk}(Q)] = (\tilde{s} - 4)f_0 \,.$$

Hence

$$\tilde{s} = 5, \mathrm{rk}(Z_0) \le f_0 \ .$$

As Z_0 covers a nontrivial K-invariant subspace of the natural module

$$O_2(M_2)/Z(U_2(O_2(M_2)))$$

we find

$$\mathrm{rk}(Z_0) = f_0 \ .$$

Furthermore, as $\tilde{s} = 5$, maximal regular subpaths of T have length 8, and $s = 5$ or 7.

We show $b_0 = 4$. We have $O_2(M_0) \le Q_1$ by Lemma 5.7. Similarly, $O_2(M_2) \le Q_3$. As $b_0 \ge 3$, we have

$$Z_0 \le O_2(M_2) \le Q_3 \ .$$

So $b_0 \ge 4$.

Now Z_0 covers a 1-dimensional subspace of $O_2(M_2)/Z_2$, and centralizes the Sylow 2-subgroup $Q_1 \cap M$ of M_{02}. If Z_0 lies in Q_4, then $Z_0 \le Q_4 \cap M = O_2(M_4)$ and Z_0 centralizes Q_4, which covers the Sylow 2-subgroup of $M_{2,4}$. It follows that Z_0 centralizes $U_2(M_2)$, and hence lies in $Z(L_2)$, which is not the case. Thus $b_0 = 4$, as claimed.

Now Z_1 is generated by Z_δ for δ adjacent to 1, and $s \ge 5$, so all such Z_δ are contained in G_4 as $b_0 = 4$, and since Z_0 is not contained in G_5, we have $b_1 = 3$. All claims have been verified. □

LEMMA 6.7. *Suppose $Q > 1$. Then we have the following:*

(1) $f_0 = f_1$.
(2) $Q = C_{Q_0}(K)$.
(3) *Z_1 is a natural module for $\bar{L}_1$.*

PROOF. The setting is provided by Lemma 6.6.

Let S be the Sylow 2-subgroup common to G_0 and G_1. Then $S \cap M = Q_1 \cap M$ by Lemma 5.7. So $Z_1 \le Z(S \cap M_0)$. Thus $\mathrm{rk}(Z_1/Z_1 \cap Z(U_2(M_0))) = f_0$. On the other hand, a Sylow 2-subgroup of M_0 has rank $4f_0 + \mathrm{rk}(Q)$, so $O_2(M_0)$ has rank $3f_0 + \mathrm{rk}(Q)$, and $Z(U_2(M_0))$ has rank $f_0 + \mathrm{rk}(Q)$. Furthermore, $Z_1 \cap Q = 1$, so $Z_1 \cap Z(U_2(M_0)) = Z_0$, and $\mathrm{rk}(Z_1) = 2f_0$. So we have $Z_1 = Z_0 \times Z_2$. By the same rank calculation, $Z(U_2(M_0)) = Z_0 Q$.

In view of Lemma 6.6 (2), (4), Z_0 is elementary abelian. Now by Lemma 2.2, it follows easily that Z_1 is a direct sum of natural modules for $\bar{L}_1$. In particular, we have

$$f_1 \le f_0 \ .$$

There is an element $t \in N_{M_1}(K)$ switching the vertices $0, 2$. Now K_0 centralizes $Z(U_2(M_0))$, and K_0^t centralizes $Z(U_2(M_2))$. The intersection $Z(U_2(M_0)) \cap Z(U_2(M_2))$ stabilizes a subpath of T of length 10: $(-4, \ldots, 6)$. Thus this intersection is Q. Thus K_0 acts fixed point freely on $Z(U_2(M_0))/Q$, and $K_0 \cap K_0^t = 1$.

On the other hand, K_1 acts fixed point freely on Z_0, so $K_0 \cap K_1 = 1$ and $K = K_0K_1$. Thus we have $\mathrm{rk}(K) = f_0 + f_1 \geq 2f_0$, and it follows that $f_1 \geq f_0$, so finally

$$f_0 = f_1 \ .$$

Let us call the common value f. Now $K = K_0 \times K_0^t$. Hence K centralizes Q, and $Q = C_{Q_0}(K)$. □

PROPOSITION 6.8. $O_2{}^{\circ}(G_T) = 1$.

PROOF. The setting is provided by Lemmas 6.6 and 6.7.

We consider $V = [K_4, O_2(M_0)]$. We have $V = [K_4, Z_1Z_{-1}] \leq Z_1Z_{-1}$. As $b_0 = 4$, the group $Z_4O_2(M_0)$ is a Sylow 2-subgroup of M_0, and as K_4 centralizes Z_4 we find that $[K_4, M_0] = V$.

For $x \in M_0$, we have $V^x = [K_4^x, O_2(M_0)] = V$ since $[K_4, x] \leq O_2(M_0)$ and $O_2(M_0)$ is abelian. Thus V is normal in M_0.

Suppose $V \leq Z_0$. Then $V \leq Q_2$. Hence $[K_4, M_{02}] \leq Q_2$, and as this contains a Sylow 2-subgroup of M_2 we conclude $[K_4, M_2] \leq Q_2$. Hence $C(K_4)$ covers $\bar{M}_0$ and $\bar{M}_2$, and acts transitively on neighbors of 0 or 2 in $\tilde{\Gamma}$. It then follows that K_4 fixes all the vertices of $\tilde{\Gamma}$. The same applies to K_0 and to K_0^t, hence to K. But K has only two fixed points adjacent to a given vertex. So $V \not\leq Z_0$.

Hence V covers $Z_{-1}Z_1/Z_0$, and in particular Z_1V is normal in M_0.

Now consider $W = \langle Z_1^{L_0} \rangle$. As $L_0 = Q_0M_0$, we have $W = \langle Z_1^{M_0} \rangle$, and we have $Z_1^{N)0} \leq Z_1V \leq Z_1Z_{-1}$. Thus $W = Z_1Z_{-1}$. Let τ be a shift on T of degree 2. Then $W^\tau = Z_1Z_3$.

As $b_0 = 4$, we have $L_0 = \langle Z_{-4}, Z_4 \rangle Q_0$, and as $b_1 = 3$ we have $Q_1 \leq Z_{-1}Q_2$. Let $Q_* = Q_{-1} \cap Q_1$. Then we have $Q_* \leq Z_{-1}(Q_{-1} \cap Q_2)$, and hence

$$\begin{aligned} [Q_*, Z_4] &\leq [Z_{-1}(Q_{-1} \cap Q_2), Z_4] &\leq [Z_{-1}, Z_3][Q_2, W^\tau] \\ &= [Z_{-1}, Z_3][Q_2, Z_1Z_3] &\leq [Z_{-1}, Z_3]Z_2 \ . \end{aligned}$$

So $[Q_*, Z_4] \leq V \leq Q_*$. We conclude that

$$(1) \qquad Q_* \text{ is normal in } L_0 \ .$$

By Lemma 6.3, the quotient Q_0/Q_* is elementary abelian of rank $2f$. On the other hand, $Q_0 \cap Q_1$ is not normal in G_0, by Lemma 2.14, and hence is not normal in L_0. So this quotient is a nontrivial module, and is natural by Lemma II 5.31.

Let $D = C_{Q_0}(K_0)$. We claim

$$(2) \qquad Q_* = DV \ .$$

We have $[\langle Z_{-4}, Z_4 \rangle, Q_*] \leq V$ as shown above. Hence $[K_0, Q_*] \leq V$. Thus $Q_* \leq C_{Q_0}(K_0)V = DV$. As K_0 operates fixed point freely on Q_0/Q_*, we find that $DV = Q_*$.

We analyze D. We will show that $D =\leq L_4$ and $D \cap Q_4 = Q$.

We claim first that

$$D \leq Q_2 \ .$$

If this fails, then $[K_0, L_2] \leq Q_2$, and $L_2 = C_{L_2}(K_0)Q_2$. Then as $[K_0, Z_1] = [K_0, Z_2] \leq Z_2$, we find $[K_0, W^\tau] = [K_0, \langle Z_1^{C_{L_2}(K_0)}\rangle] \leq [K_0, Z_1]^{L_2} \leq Z_2$. Hence $[K_0, Z_4] \leq Z_2 \leq Q_0$, and similarly $[K_0, Z_{-4}] \leq Q_0$, so $[K_0, L_0] \leq Q_0$, which is false. So $D \leq Q_2$. In particular, $D \leq L_3$.

We claim that

$$D \leq Q_3 \ .$$

Supposing the contrary, we have $[K_0, L_3] \leq Q_3$, and $L_3 = C_{L_3}(K_0)Q_3$. As $b_0 = 4$, the group Z_0Q_4 is a Sylow 2-subgroup of L_4. Thus $[K_0, L_4] \leq Q_4$ as well, and $C(K_0)$ is transitive on the neighbors of 3 and 4 in Γ, contradicting Lemma 2.14. So $D \leq Q_3$.

Thus $D \leq L_4$, and $D \leq Z_0Q_4$. But $Z_0 \leq D$, so we have

$$D = Z_0(D \cap Q_4) \ .$$

We claim that $D \cap Q_4 = Q$.

We have $Q \leq D \cap Q_4$. Now $N_{D\cap Q_4}(Q) \leq O_2(M_0)$, and $D \cap O_2(M_0) \leq QZ_0$. Thus $Q \leq N_{D\cap Q_4}(Q) \leq QZ_0$, and $Z_0 \cap Q_4 = 1$, so $N_{D\cap Q_4}(Q) = Q$, and thus $D \cap Q_4 = Q$.

Putting everything together, we have $D = Z_0Q$, and hence $Q_* = DV = QV$. We have proved that

$$Q_* = QV \ . \tag{3}$$

We consider the $\bar{L}_0$-group Q_0/V, with the submodule QV/V. By Lemma 6.1, QV/V is a trivial module, while $Q_0/QV = Q_0/Q_*$ is a natural module. By Lemma 2.5, Q_0/V is elementary abelian. In particular, we have

$$[Q, Q_0] \leq W \ . \tag{4}$$

Now we claim that

$$[Q, Q_1] \leq Z_1 \ . \tag{5}$$

We have $Q_1 = Z_4(Q_0 \cap Q_1)$ and $(Q_0 \cap Q_1) \leq Z_{-1}(Q_0 \cap Q_2)$. As $Z_4 = Z(L_4)$ and $Z_{-1} - Z(L_0)Z(L_2)$, we have $[Q, Q_1] = [Q, Q_0 \cap Q_1] \leq [Q, Q_0 \cap Q_2]$. Thus $[Q, Q_1] \leq [Q, Q_0] \cap [Q, Q_2] \leq W \cap W^\tau$.

Now $W \cap W^\tau = Z_{-1}Z_1 \cap Z_1Z_3 = Z_1$ (examine this in L_0). So $[Q, Q_1] \leq Z_1$.

We now consider the M_2-module $[Q, Q_2]$. We claim that

$$[Q, Q_2]Z_2 = W^\tau \ . \tag{6}$$

We have $[Q, Q_2] \leq W^\tau$. If claim (6) fails, then by Lemma 6.1 we have $[Q, Q_2] \leq Z_2$. Then $[Q, Q_2, K_2] = 1$ and $[Q, K_2, Q_2] = 1$ since in fact $[Q, K] = 1$. So by the three subgroups lemma we have $[Q_2, K_2, Q] = 1$ and hence also $[Q_0, K_0, Q] = 1$. But $Q_0 = [Q_0, K_0]Q_* = [Q_0, K_0]QV$, and we deduce $[Q, Q_0] = [QV, Q] = 1$ by Lemma 6.1. But Q centralizes $U_2(M_0)$, hence Q is central in a Sylow 2-subgroup of L_0, and hence $Q \leq Z_0$, contradicting Lemma 6.2. This proves (6).

Now we may reach a contradiction. Consider the action of K on L_1. Let K^1 be the centralizer of $\bar{L}_1 = L_1/Q_1$ in K. Then $\mathrm{rk}(K^1) = f$. Now

$[L_1, K^1, Q] \le [Q_1, Q] \le Z_1$, and $[K^1, Q, L_1] = 1$, so by the three subgroups lemma we have $[Q, L_1, K^1] \le Z_1$. In particular, $[Q, Q_2, K^1] \le Z_1$. By (6) we find

$$[W^\tau, K^1] \le Z_1 \,.$$

In particular, $[Z_4, K^1] \le Q_0$. Hence $[L_0, K^1] \le Q_0$, and $C(K^1)$ acts transitively on the neighbors of 0 and of 1 in Γ, contradicting Lemma 2.14. □

7. Odd s: Initial analysis

In this section and the next, s is odd, and we know in consequence that $O_2{}^\circ(G_T) = 1$. Our aim is to show that s is 5 or 7, and that $r = s - 1$. We will also analyze the composition factors of Q_i with respect to the action of L_i for $i = 0, 1$. In the present section we make some further preparations, reserving the detailed analysis for the following section.

NOTATION 7.1. *Fix a path γ contained in T of maximal length with $O_2{}^\circ(G_\gamma) \neq 1$. Set $Q_\gamma = O_2{}^\circ(G_\gamma)$.*

Note that this notation varies from that adopted in the previous section. The path γ can also be described as a maximal regular path in T, so its length is r.

We will assume that the orbits of G on the vertices of Γ are labeled so that γ begins with 0. Note that if $r = s - 1$, then we can interchange 0 and 1 without loss of generality, as all paths of length r will be regular in this case.

7.1. Estimates for b_0 and b_1.

LEMMA 7.2.

(1) *r is even.*
(2) *Either $r = s - 1$, or every maximal regular path contained in T is of type (O_0, r).*
(3) *$Q_\gamma \cap Q_0 = 1$.*
(4) *$\mathrm{rk}(Q_\gamma) = f_0$, and K acts transitively on $Q_\gamma^\times$.*

PROOF. The first two points are covered by Lemma 2.13. It follows from Lemma 2.13 (3) that r is even, since if $r < s$, then $r = s - 1$. Given that r is even, our point (2) follows from Lemma 2.13 (1).

For the last two points examine Q_γ in G_0. K operates on Q_γ. By the choice of γ, we have $Q_\gamma \cap Q_0$ finite, hence centralized by K. As Q_γ is not contained in Q_0, it covers a Sylow 2-subgroup of $\bar{L}_0$ and K_0 acts regularly on $[Q_\gamma/(Q_\gamma \cap Q_0)]^\times$. Looking at the orbits of K_0 on $Q_\gamma^\times$, as Q_γ is connected it follows that there is just one orbit and $Q_\gamma \cap Q_0 = 1$. From this, both of the last points follow. □

LEMMA 7.3.

(1) *$r/2 - 1 \le b_1 \le r/2$.*
(2) *$r/2 - 2 \le b_0 \le r/2$.*

PROOF. The set of fixed points of Z_1 in T is $(1 - b_1, \ldots, 1 + b_1)$, and the length of this path is $2b_1$. As $Z_1 > 1$, the definition of r gives $r \geq 2b_1$. Similarly, $r \geq 2b_0$.

Next we show that $r/2 - 1 \leq b_1$. Let $j = 1 + b_1$. We claim $r \leq 2j$. Now $j + b_1$ is an odd number, so the set of fixed points of Z_{j+b_1} in T is the path $(j, \ldots, j + 2b_1)$, centered at $j + b_1$. Suppose $r \geq 2j$. We have $Q_\gamma \leq Q_1$ and $Q_\gamma \leq Q_{2j-1}$, so Q_γ centralizes Z_1 and $Z_{2j-1} = Z_{j+b_1}$. Hence the group

$$H = \langle Z_1, Z_{j+b_1}, K \rangle$$

normalizes Q_γ. Note that H is a subgroup of G_j acting transitively on the neighbors of j.

Take $z \in Z_1 \setminus Q_j$. By Lemma 2.7, the involution z normalizes some conjugate K^u of K with $u \in N(Q_\gamma)$. Then T^u is the corresponding K^u-track, and z reflects T^u about j. Now $Q_\gamma^u = Q_\gamma$ fixes $(j, \ldots, r)^u$ and hence also fixes the reflection of this path under z, so Q_γ fixes a path of length $2(r - j)$ centered at j, contained in T^u. By the maximality of r, we have $2(r - j) \leq r$, or $r \leq 2j$, as desired.

Finally, we claim that $r/2 - 2 \leq b_0$. As $Q_\gamma \leq Q_2$, we can apply the foregoing argument beginning at Z_2 in place of Z_1, which yields the slightly weaker inequality indicated. □

We elaborate on the earlier Lemma 6.3, repeating its conclusions for ease of reference.

LEMMA 7.4. *Suppose that $Q_{-1} \cap Q_1$ is normal in G_0. Set $Q_* = Q_{-1} \cap Q_1$. Then the following hold:*

(1) *Q_0/Q_* is elementary abelian, of rank $2f_1$.*
(2) *$Q_0 = [Q_0, Q_{-1}][Q_0, Q_1]Q_*$.*
(3) *$C_{(Q_0/Q_*)}(L_0) = 1$.*
(4) *If Z_0 is a natural module for $\bar{L}_0$, and $[Q_*, L_0] \leq Z_0$, then Q_* is elementary abelian.*

PROOF. The first two points were given in Lemma 6.3. We will make use of the analysis given in the proof of that lemma as well.

We take up the third point. The groups $Q_0 \cap Q_\delta$, with δ varying over the neighbors of 0, are pairwise disjoint modulo Q_*. Fix a Sylow 2-subgroup S of $\bar{L}_0$, fixing the vertex 1. The conjugates of $(Q_{-1} \cap Q_0)/Q_*$ then cover a generic subset of Q_0/Q_*, so the fixed point set of S is a K_1-invariant subgroup of smaller rank. It follows that S fixes only the points of $(Q_0 \cap Q_1)/Q_*$, and similarly a conjugate of S fixes only the points of $(Q_0 \cap Q_{-1})/Q_*$, and our claim follows.

Finally, suppose that Z_0 is a natural module, and $[L_0, Q_*] \leq Z_0$. Then $[L_0, Q_*] = Z_0$, and by Proposition I 9.9, we have $Q_* = C_{Q_*}(K_0)Z_0$. Furthermore, we have $Z_0 \leq Z(Q_*)$, so $Q_* = C_{Q_*}(K_0) \times Z_0$.

Thus $\phi(Q_*) = \phi(C_{Q_*}(K_0))$. So $[L_0, \phi(Q_*)] \leq \phi(Q_*) \cap Z_0 = 1$, and $\phi(Q_*) = 1$. So Q_* is elementary abelian. □

7.2. A commutator condition. The next result again involves a long analysis.

LEMMA 7.5. *If* $[Z_1, Z_{1+b_1}] \not\leq Z_1 \cap Z_{1+b_1}$, *then we have the following:*

(1) $r = s - 1$.

(2) $1 < [Z_0, Z_{b_0}] \leq Z_0 \cap Z_{b_0}$.

In particular, in this case one may interchange 0 *and* 1, *and after this change of notation we have* $[Z_1, Z_{1+b_1}] \leq Z_1 \cap Z_{b_1}$.

PROOF. The final remark will follow at once from $(1, 2)$. The relation $r = s-1$ allows us to replace the path $\gamma = (0, \ldots, r)$ by the path $(1, \ldots, r+1)$, and hence to switch the two orbits O_0 and O_1. So we turn to the verification of our main claims.

Let $j = 1 + b_1$. We have $Z_1 \leq G_j$, hence $[Z_1, Z_j] \leq Z_j$. If $b_j \geq b_1$, then similarly $[Z_1, Z_j] \leq Z_1$, contradicting our assumption. Thus we have $b_j < b_1$, so j is even, and we have

(1) $$b_0 < b_1;\ b_1 \text{ is odd}\ .$$

Now since $b_0 < b_1$, our initial argument shows

$$[Z_0, Z_{b_0}] \leq Z_0 \cap Z_{b_0}\ .$$

We invoke the estimates of Lemma 7.3:

(2) $$r/2 - 2 \leq b_0 < b_1 \leq r/2\ .$$

We will need to distinguish two cases: b_0 is odd, or even.

(**Case I**) b_0 is odd.

In this case we reach a contradiction.

Now as b_1 is also odd, our estimates reduce to the following:

$$b_0 = r/2 - 2; b_1 = r/2\ .$$

Now Z_1 stabilizes a path of length r centered at 1, so $Z_1 \leq Q_\gamma$, and as Q_γ has rank f_0, we find that K acts transitively on Z_1, and hence

$$Z_1 = Z(L_1) \text{ has rank } f_0, \text{ and } [Z_0, Z_{b_0}] = 1\ .$$

Indeed, for each Sylow 2-subgroup S of L_1 we find that $Z(S)$ has rank f_0 and thus $Z(S) = Z_1 = Z(L_1)$. As b_0 is odd, we have $Z_{b_0} - Z(L_{b_0})$ and thus $[Z_0, Z_{b_0}] = 1$.

Now if $r \leq s$, then $r = s - 1$ and we may interchange the orbits represented by 0 and by 1. But Lemma 7.3 again applies, and in terms of our original notation it now says: $r/2 - 1 \leq b_0$, which is false. Accordingly, we are in the following case:

$$r > s\ .$$

The group Z_0 is generated by the groups $Z_\delta = Z(L_\delta)$ for δ adjacent to 0. For any such vertex δ, any arc beginning with δ and of length at most r is right regular, by Lemma 2.13, and can be conjugated into the path

$(-1,\ldots,m)$ for some m. In particular, the path from $\delta \neq 1$ to $b_1 - 1 \in T$ is conjugate to $(-1,\ldots,b_1-1)$ and thus $Z_\delta \leq L_{b_1-1}$ for neighbors of 0, and $Z_0 \leq L_{b_1-1}$, thus $b_0 \geq b_1 - 1$, a contradiction.

This disposes of the case of odd b_0.

(Case II) b_0 is even.

In this case we have the estimates

$$r/2 - 2 \leq b_0 = b_1 - 1 < r/2.$$

In particular, $Z_0 \neq Z(L_0)$, and $C_{L_0}(Z_0) = Q_0$. Now $Z_{b_0} \not\leq Q_0$, so

$$[Z_0, Z_{b_0}] \neq 1 \ .$$

If $r \leq s$, then $r = s-1$, and the roles of 0 and 1 can be interchanged, and as $[Z_0, Z_{b_0}]$, this suffices. So we assume that

$$r > s \ .$$

We claim that

(II.1) $Z_0/Z(L_0)$ is a natural module for $\bar{L}_0$.

As $[Z_0, Z_{b_0}] \leq Z_{b_0}$, the group Z_0 affords a quadratic module with respect to the group Z_{b_0}, which covers a Sylow 2-subgroup of $\bar{L}_0$.

Let $V = Z_0/Z(L_0)$. Let $W = C_V(O^2(L_0))$. As $[O^2(L_0), W] \leq Z(L_0)$ and $O^2(L_0)$ is generated by $2^\perp$-groups, we find that $[O^2(L_0), W] = 1$ and as $[V, Q_0] = 1$, $[L_0, W] = 1$ and $W = (0)$ (in V).

By Lemma 2.1, it follows that $[O^2(L_0), V]$ is a direct sum of natural modules, and as we have seen, is also nontrivial.. We are interested in $\mathrm{rk}(C_{Z_0}(Z_{b_0}))$, from the point of view of the module Z_0. We can, however, look at this also from the point of view of the action of Z_0 on Z_{b_0}, and from this point of view the rank is seen to be f_0. It follows easily that $[O^2(L_0), V]$ is a single natural module for $\bar{L}_0$.

Now by Lemma 2.2 it follows that V is itself a natural module for L_0. We claim that

(II.2) $Z(L_0) = 1$, and thus Z_0 is a natural module for $\bar{L}_0$.

Suppose on the contrary that $Z(L_0)$ is nontrivial; then so is $Z(L_j)$ ($j = 1 + b_1$). Now $Z(L_j) \leq Z_{j-1} \cap Z_{j+1}$, so $Z(L_j)$ stabilizes the path $(0, 2j)$ in T. Furthermore, $2j \geq r$, and hence $Z(L_j) \leq Q_\gamma$, and $r = 2j$. As K_0 acts on Q_γ, we find $Z(L_j) = Q_\gamma$, and $\mathrm{rk}(Z(L_j)) = f_0$, hence also $\mathrm{rk}(Z(L_0)) = f_0$.

Let S be a Sylow 2-subgroup of G_{01}. Then $\mathrm{rk}(Z(S) \cap Z_0 = 2f_0$. But $Z(S) \cap Z(L_2) \leq Z(L_1) \cap Z(L_2) = 1$, in view of Lemma 2.14. But $Z(L_2) \leq Z_1$. Hence $\mathrm{rk}(Q_0) \geq 3f_0$.

Consider Z_1 as a subgroup of $G_{1\pm b_1}$. We find $\mathrm{rk}(Z_1 \cap Q_j \cap Q_{2-j}) \geq f_0$. In particular, this intersection is nontrivial. However, it stabilizes the path $(1-j,\ldots,j+1)$ of length $2j = r$, and furthermore the endpoints of this path have odd parity, which contradicts Lemma 2.13. Thus $Z(L_0) = 1$, as claimed.

As b_0 is even, we have $Z_0 \le Q_1$ and by Lemma 3.3, we have

(II.3) $$Z_1 = Z^\circ(L_1) \ .$$

Looking at Z_1 inside the module Z_0, we get

(II.4) $$\mathrm{rk}(Z_1) = f_0 \ .$$

As Z_0 is not contained in Q_{b_0}, the commutator $[Z_0, Z_{b_0}]$ is nontrivial, and looking inside the natural module Z_{b_0}, it must be Z_{b_0-1}. Similarly, $[Z_0, Z_{b_0}] = Z_1$. In particular $Z_1 = Z_{b_0-1}$.

Now $[Z_0, Z_{b_0}] \le Z_0 \cap Z_{b_0}$ stabilizes the arc $(-b_0, 2b_0)$ of length $3b_0$. Thus $3b_0 \le r$. By the estimate above, we find $r/2 - 2 \le r/3$, $r \le 12$, and $b_0 \le 4$. If $b_0 = 4$, we have $Z_1 = Z_3$ and similarly $Z_1 = Z_i$ for i odd, which is impossible. So $b_0 = 2$, $r \ge 6$, and by our estimates above

(II.5) $$b_0 = 2, b_1 = 3, r = 6 \text{ or } 8 \ .$$

These two cases must be eliminated.

We consider the groups $A = Q_{-1} \cap Q_1$, and $V = \langle Z_0^{G_1} \rangle$. As $b_0 = 2$, we have $Z_0 \le A$, and $V \le Q_1$. We aim to show that A is normal in L_0.

By the structure of the module Z_0, we have $[Z_0, Q_1] = Z_1$, and hence after conjugating by G_1, $[V, Q_1] = Z_1 \le A$. Letting τ be the shift by $+2$ along the track T, we have $[V^{\tau^{-1}}, Q_{-1}] \le Z_{-1} \le A$. Thus $\langle V, V^{\tau^{-1}} \rangle$ normalizes A. On the other hand, Z_0 covers a Sylow 2-subgroup of L_2 modulo Q_2 and under the action of G_1 it follows that V covers a subgroup of L_0 modulo Q_0, specifically the one in $G_{0,1}$; similarly $V^{\tau^{-1}}$ covers a Sylow 2-subgroup of $G_{0,-1}$. Thus $L_0 = \langle V, V^{\tau^{-1}} \rangle Q_0$, and hence L_0 normalizes A.

Lemma 7.4 applies. In particular, $Q_0 = [Q_0, Q_{-1}][Q_0, Q_1]A$ and thus $Q_0 \cap Q_1 = [Q_0, Q_1]A$. But $Q_1 \le VQ_0$ and thus

$$[Q_0, Q_1] \le [Q_0, VQ_0] \le VQ_0' \le VA \ .$$

Hence

(II.6) $$Q_0 \cap Q_1 \le VA \ .$$

Also Q_0/A is elementary abelian, of rank $2f_1$. We will show that V acts quadratically on Q_0/A. We have $[Q_0, V] \le Q_0 \cap V \le Q_0 \cap Q_1$ and $[Q_0 \cap Q_1, V] \le [VA, V] \le V'A$. Furthermore $V' \le [V, Q_1] = Z_1 \le A$. So

(II.7) $$[Q_0, V, V] \le A$$

and V acts quadratically on Q_0/A.

We claim furthermore that $C_{Q_0/A}(L_0) = 1$. Writing Q^+, Q^- for $Q_0 \cap Q_{0\pm1}$, we have $Q_0/A = Q^+/A \times Q^-/A$. In particular, if L_0 centralizes an element $q \in Q_0$ modulo A, then writing $q/A = q^+/Aq^-/A$ with $q^\pm \in Q^\pm$, and applying an element g from the Sylow 2-subgroup S of $G_{0,1}$, we find $(q^+)^g(q^-)^g \in A$, and since $q^- \in Q_{(i-1)^g}$ while $q^+ \in Q_1$, it follows that $(q^\pm)^g \in A$. However, the groups $Q_\delta \cap Q_0/A$ are disjoint, for δ adjacent to

0, and S acts transitively on $\Delta(0) \setminus \{1\}$. So this means $q^- \in A$. Similarly, $q^+ \in A$ and $q \in A$, as claimed.

So Lemma 2.2 applies, and we conclude that

(II.8) $$Q_0/A \text{ is a direct sum of natural } \bar{L}_0\text{-modules.}$$

We claim that

(II.9) $$A = Z_0 \times C_{Q_0}(X) \text{ for } 1 < X \leq K_0 \ .$$

This has two cases of interest: $X = K_0$ and $X = \langle d \rangle$ cyclic. Of course it also will imply that the centralizer is independent of the choice of X.

We have $C_{Q_0/A}(X) = 1$ and in particular $C_{Q_0}(X) \leq A$. Furthermore, $C_{Z_0}(X) = 1$ and thus $Z_0 \times C_{Q_0}(X) \leq A$.

On the other hand, we have $[V, A] \leq [V, Q_1] = Z_1 \leq Z_0$. Hence $[V^{G_0}, A] \leq Z_0$ and thus $[O^2(L_0), A] \leq Z_0$. In particular, $[K_0, A] \leq Z_0$. So by Proposition I 9.9, the centralizer $C_A(K_0)$ covers A/Z_0, and (II.9) follows.

Now if $r = 6$, then by tracing the action of the Sylow 2-subgroup S of $G_{0,1}$ along the path $(0, \ldots, 6)$, computing the successive stabilizers of its initial segments, and using their invariance under K, we find $\mathrm{rk}(S) = 3f_0 + 2f_1$. We will now see that this is too small.

In the first place we find that $A = Q_1 \cap Q_{-1}$ has rank $2f_0$ since Q_1 has rank $2f_0 + 2f_1$ and we lose $f_0 + f_1$ by intersecting first with Q_0 and then with Q_{-1}. It follows that $A = Z_0$ and in view of (II.9), we have $C_{Q_0}(d) = 1$ for any nontrivial element of K_0; what interests us here is an element of order 3. It follows from Lemma 2.4 that Q_0 is elementary abelian and is a direct sum of natural modules.

But in view of the definition of Z_0 and the structure of a natural module, it then follows that $Q_0 = Z_0$, which contradicts both our formula for the rank of S, and the condition $b_0 > 1$. So the case $r = 6$ is excluded and by default we have

$$r = 8 \ .$$

Now take $w \in N_{O^2(L_0)}(K)$ switching ± 1, and take w_+ similarly in $N_{L_1}(K)$, inverting K_1. We claim that

(II.10) $$[w, C_{Q_0}(K_0)] = 1 \ .$$

Let $D = C_{Q_0}(K_0)$. The element w normalizes D and thus $[w, D] \leq D \cap [O^2(L_0), D]$. As noted above, $[O^2(L_0), A] \leq Z_0$ and hence $[w, D] \leq D \cap Z_0 = 1$. This proves (II.10).

Now w_+ normalizes $D \cap D^{w_+}$, and w centralizes it. So the $+2$-shift $\tau = ww_+$ along T normalizes this intersection, which therefore fixes T pointwise. Thus

(II.11) $$(D \cap D^{w_+})^\circ \leq O_2{}^\circ(G_T) = 1 \ .$$

Recall the group $Q_\gamma = O_2{}^\circ(G_{0,\ldots,r}) = (Q_1 \cap \cdots \cap Q_{r-1})^\circ$, of rank f_0. Now $Q_\gamma^{\tau^{-1}}$ and $Q_\gamma^{\tau^{-2}}$ are both contained in A, as $r \geq 6$. These two groups are

K_0-invariant, and have rank f_0. In view of the structure of A, such groups are contained either in Z_0 or in $C_{Q_0}(K_0)$.

Suppose that $Q_\gamma^{\tau^{-1}}$ or $Q_\gamma^{\tau^{-2}}$ is contained in Z_0. Then $Q_\gamma \leq Z_2$ or Z_4 is a K_0-invariant subgroup of rank f_0. It follows that $Q_\gamma = Z_\delta$ for some δ adjacent to 2 or 4. Then $Z_\delta = Q_\gamma \leq G_0, G_8$, which contradicts the fact that $b_1 = 3$. So we have a contradiction. We conclude that

$$Q_\gamma^{\tau^{-1}}, Q_\gamma^{\tau^{-2}} \leq C_{Q_0}(K_0) \ .$$

So $Q_\gamma^{\tau^{-1}} = Q_\gamma^{\tau^{-2}ww_+} = Q_\gamma^{\tau^{-2}w_+} \leq D \cap D^{w_+}$, which is a final contradiction. Thus the treatment of Case II is complete. □

8. Odd s: Detailed analysis

We now take up the detailed analysis of concrete cases under the hypotheses that s is odd. We prove more than is strictly necessary, as it is of some interest to see what can be done by this kind of direct argument before invoking the very powerful classification of Moufang generalized polygons.

We choose our notation as permitted by Lemma 7.5:

$$[Z_1, Z_{b_1}] \leq Z_1 \cap Z_{b_1} \ .$$

This may break the symmetry between 0 and 1, even if $r = s - 1$.

In the remaining analysis, we deal separately with the cases in which $[Z_1, Z_{b_1}]$ is, or is not, trivial.

We will also rely on our estimates from Lemma 7.3: $r/2 - 1 \leq b_1 \leq r/2$ and $r/2 - 2 \leq b_0 \leq r/2$.

8.1. First case: $[Z_1, Z_{1+b_1}]$ nontrivial.

LEMMA 8.1. *Suppose that $[Z_1, Z_{b_1}] \neq 1$. Let j be an index of opposite parity to b_1. Then*

$$(*) \qquad 2b_1 + b_{1+b_1} \leq r \ .$$

PROOF. We may take $j = 1 + b_1$. Setting $R = [Z_1, Z_j] \leq Z_1 \cap Z_{b_1}$, we see that R stabilizes the path $(1 - b_1, \ldots, j + b_j)$ or length $j - 1 + b_1 + b_j = 2b_1 + b_j$, whence the inequality. □

LEMMA 8.2. *Suppose that $[Z_1, Z_{b_1}] \neq 1$, and b_1 is odd. Then $b_0 = b_1 = 1$, $s = 5$, $r = 4$, and in addition we have the following:*

(A) *$f_0 = f_1 =: f$.*
(B) *Q_0 and Q_1 are elementary abelian, of rank $3f$.*
(C) *$Z(L_1)$ and $Z(L_2)$ both have rank f.*
(D) *$Q_i/Z(L_i)$ is a natural module for $\bar{L}_i$.*

In particular, $r = s - 1$ in this case.

PROOF. The inequality $(*)$ takes the form

$$(*_1) \qquad 2b_1 + b_0 \leq r \ .$$

We let $j = 1 + b_1$ and $R = [Z_1, Z_j]$.

In particular, $2(r/2-1)+(r/2-2)\le r$, and thus $r\le 8$. With $r=6$ and b_1 odd, we find $b_1=3$ and thus $(*_1)$ is violated. So $r=4$ or $r=8$, and we begin by disposing of the latter case.

Suppose then that $r=8$, and correspondingly $b_0=2$, $b_1=3$, $R=[Z_1,Z_4]$.

Then R fixes the path $\tilde{\gamma}=(-2,\ldots,6)$ of length 8, and under a shift τ of $+2$ along the track T we have $R^\tau=Q_\gamma$, in view of the action of K on both groups.

Let t be an involution in G_2 which reflects the track T about the vertex 2. Then t preserves the arc $\tilde{\gamma}$ and hence normalizes R.

We claim that $R=[Z_2,Z_4]$. We have Z_2 contained in G_4 but not in Q_4, hence Z_2 covers the same Sylow 2-subgroup of G_4 as Z_1 does. Therefore, $[Z_2,Z_4]=[Z_1,Z_4]=R$.

Therefore, $R\le Z_2\cap Z_4$ and R commutes with Q_2 and Z_4, while Z_4Q_2 is a Sylow 2-subgroup of G_2. Now applying the flip t, R centralizes a second Sylow 2-subgroup of G_2 and hence lies in $Z(L_2)$.

Now $[\langle Z_4,Z_0\rangle Q_2,Z_2]\le R\le Z(L_2)$, so $[O^2(L_2),Z_2]=1$. Also $[Q_2,Z_2]=1$, so we find $Z_2=Z(L_2)$. But $Z_4\le L_2$ and hence $R=[Z_2,Z_4]=1$, a contradiction. So the case $r=8$ is eliminated and we are left with

$$r=4, b_1=1, b_0\le 2, \tag{1}$$

and since $s\ge 5$ we have $s=5$.

In this case, we have $R=[Z_1,Z_2]$.

Suppose $b_0=2$. Then Z_2 fixes the path $(0,\ldots,4)$, and in view of the action of K on Z_2, we find that $Z_2=Q_\gamma$ has rank f_0, and $Z_2\cap Q_4=1$. In particular, $[Z_2,Z_4]=1$. Thus Z_2 centralizes the Sylow 2-subgroup Z_4Q_2 of L_2, and similarly Z_0Q_2. So $Z_2\le Z(L_2)$, so $R=1$, a contradiction. We have shown that

$$b_0=1\ . \tag{2}$$

Now we consider the module structure. We have $R=[Z_1,Z_2]\le Z_1\cap Z_2$, and hence each of Z_1 and Z_2 is a quadratic module with respect to the other. So in the first place for $i=1,2$ (or $0,1$) we have $Z_i/Z(L_i)$ is a sum of natural modules for L_i, and by considering both actions, and in particular the subgroup R in both cases, we see that $f_0=f_1$ and each quotient module is natural. So we set $f:=f_0=f_1$.

Now $s=r+1=5$ and so we can compute the rank of the Sylow 2-subgroup S by working with the path $(0,\ldots,5)$, and we find $\mathrm{rk}(S)=2f_0+2f_1=4f$. Thus $\mathrm{rk}(S/Z_i)\le 2f$ and $\mathrm{rk}(Q_i/Z_i)\le f$ for any i. Hence Q_i/Z_i has a trivial composition series and in particular $[O^2(L_i),Q_i]\le Z_i$. Thus $Q_i=C_{Q_i}(K_i)Z_i$. Now $C_{Q_i}(K_i)\cap Z_i=Z(L_i)$, so $\mathrm{rk}(C_{Q_i}(K_i))=\mathrm{rk}(Q_i/Z_i)+\mathrm{rk}(Z(L_i))=\mathrm{rk}(Q_i)-\mathrm{rk}(Z_i/Z(L_i))=f$.

Now suppose toward a contradiction that for some i we have $Z(L_i)=1$, and let i' be a vertex adjacent to i. Then the intersection $Q_{i'}\cap Q_i$ is a K-invariant group of rank $2f$, and $Q_{i'}\cap Z_i$ has rank f, by our hypothesis. It then follows from the structure of Q_i that $C_{Q_i}(K_i) \leq Q_{i'}$, that is, $C_{Q_i}(K_i) \leq C_{Q_{i'}}(K_{i'})Z_{i'}$.

Now K_i acts fixed point freely on $Z_i/Z(L_i)$, while $K_{i'}$ has fixed points there, so $K_i \cap K_{i'} = 1$ and $K = K_iK_{i'}$. Hence $C_{Q_i}(K_i) \cap C_{Q_{i'}}(K_{i'}) = 1$, as otherwise this group would fix the track T pointwise.

Since $C_{Q_i}(K_i) \leq C_{Q_{i'}}(K_{i'})Z_{i'}$ is K-invariant and disjoint from $C_{Q_{i'}}(K_{i'})$, it follows that $C_{Q_i}(K_i) \leq Z_{i'}$. Hence $\mathrm{rk}(Z_{i'} \cap Q_i) \geq 3f$, that is, $Z_{i'} = Q_i$. In particular, $\mathrm{rk}(Z(L_{i'})) = f$. Then looking at a Sylow 2-subgroup S of $G_{i,i'}$, from the point of view of $L_{i'}$, we see that $\mathrm{rk}(Z(S)) = 2f$. But by our hypothesis $\mathrm{rk}(Z_i) = 2f$, so now $Z_i = Z(S)$. But then applying conjugation in L_i we find $Z_i \leq Z(L_i)$, a contradiction. Thus we may conclude that

(3) $$Z(L_i) \text{ is nontrivial, all } i\ .$$

Now with i, i' adjacent, we have $Z(L_i) \cap Z(L_{i'}) = 1$ and $Z(L_i) \leq Z_{i'}$. So the action of $K_{i'}$ on $Z(L_i)$ is fixed point free and $\mathrm{rk}(Z(L_i)) = f$, and $Z_i = Q_i$.

Now by Lemma 2.5, Q_i is elementary abelian for all i. With this, all of our claims are proved. □

The next case involves a more elaborate analysis.

LEMMA 8.3. *Suppose that* $[Z_1, Z_{b_1}] \neq 1$, *and* b_1 *is even. Then* $b_0 = 3$, $b_1 = 2$, $s = 7$, $r = 6$, *and in addition we have the following:*

(A) $Z_0 = Z(L_0)$ *has rank* f_1, *and* Q_0 *has rank* $2f_0 + 3f_1$.

(B) Z_1 *is a natural module for* $\bar{L}_1$, *and* $(Q_1/Z_1)/Z(L_1/Z_1)$ *is a direct sum of natural modules for* L_1.

(C) $\phi(Q_0) = Z_0$, *and* Q_0/Z_0 *is an irreducible* $\bar{G}_0$*-module.*

(D) $f_1 \leq f_0$.

(E) $Q_0 \cap Q_2 = C_{Q_1}(K_1) \times Z_1$.

In particular, $r = s - 1$ *and* $Z_0 \leq Z_1$.

PROOF. Now $j = 1 + b_1$ is odd, so $b_1 = b_j$ and the inequality $(*)$ from Lemma 8.1 becomes

$(*_2)$ $$3b_1 \leq r\ .$$

As $r/2 - 1 \leq b_1$, we find $r \leq 6$. On the other hand $b_1 \geq 2$, as b_1 is even, hence by $(*_2)$ also $r \geq 6$. Thus

(1) $$r = 6, \quad b_1 = 2\ .$$

So $R = [Z_1, Z_j] = [Z_1, Z_3]$ stabilizes the path $\tilde{\gamma} = (-1, \ldots, 5)$ of length 6, which is therefore regular. But $(0, \ldots, 6)$ is also regular, by our hypotheses and choice of notation, and by Lemma 2.13, it follows that $r < s$, that is, $r = s - 1$ and $s = 7$.

Now we can compute the rank of the Sylow 2-subgroup S of $G_{0,1}$ by working along the track T, and we find

(2) $$\mathrm{rk}(S) = 3f_0 + 3f_1\ .$$

Now we consider $R = [Z_1, Z_3] \leq Z_1 \cap Z_3$. We have $[R, Z_1] = 1$, and Z_1 covers a Sylow 2-subgroup of $G_{2,3}$ modulo Q_3. Since $[R, Q_3] = 1$, the group R centralizes the Sylow 2-subgroup $O_2(G_{2,3})$.

Take $t \in N_{G_3}(K)$ inverting K. Then t acts on the path $\tilde{\gamma}$ and hence normalizes R. Thus R is central in two Sylow 2-subgroups of G_2 and hence $R \leq Z(L_2)$. Thus $R \cap Z(L_1) = 1$.

We will show

$$(4) \qquad Z_2 = Z(L_2) \ .$$

For the moment, suppose the contrary. Then Lemma 3.3 applies, and we have

$$(4^{\perp}) \qquad Z_i/Z(L_i) \text{ is natural for } i = 2 \text{ or } 3 \ .$$

Also, since $b_1 > 1$ here, the same lemma shows that $Z^{\circ}(L_0) > 1$.

Now since $Z(L_2) \cap Z(L_3) = 1$, looking at $Z(L_2)$ inside the module Z_3, or at $Z(L_3)$ inside Z_2, we find

$$(5) \qquad \mathrm{rk}(Z(L_2)) = f_1; \mathrm{rk}(Z(L_3)) = f_0 \ .$$

Thus $\mathrm{rk}(Z_2) = 2f_0 + f_1$ and $\mathrm{rk}(Z_3) = f_0 + 2f_1$, and therefore $\mathrm{rk}(Q_2/Z_2) = 2f_1$, $\mathrm{rk}(Q_3/Z_3) = 2f_0$.

By considering the ranks, we find that $Q_i = Z_{i-1}Z_iZ_{i+1}$. With $i = 2$ we find $[Z_1, Q_2] = [Z_1, Z_3] = Z(L_2)$.

Suppose toward a contradiction that $[O^2(L_2), Q_2] \leq Z_2$. In particular, we have $[O^2(L_2), Z_1] \leq Z_2$, and as $[Q_2, Z_1] \leq Z_2$ by the preceding, we find $[L_2, Z_1] \leq Z_2$. Similarly, $[L_2, Z_3] \leq Z_2$ and hence $[L_2, Q_2] \leq Z_2$. Then $Q_2 \cap Q_3$ is normal in L_2, contradicting Lemma 2.14.

This shows that there is some nontrivial composition factor in Q_2/Z_2, with respect to the action of $\bar{L}_2$. A similar argument applies to Q_3/Z_3. In view of the ranks involved in the two cases, it follows that $f_0 = f_1$, and that both quotients Q_i/Z_i are natural modules for the respective $\bar{L}_i$ (Lemma II 5.31). All of this is under the hypothesis $(4^{\perp})$, and we are approaching a contradiction.

Now as above we have $[Z_2, Q_1] \leq Z(L_1)$ and $[Z_2, Q_3] \leq Z(L_3)$, hence $[Z_2, L_2] \leq Z(L_1)Z(L_3)$. In particular, $Z(L_1)Z(L_3) \lhd L_2$.

By rank considerations we have $Z_2 = Z(L_1)Z(L_2)Z(L_3)$, and hence $Z_2/Z(L_1)Z(L_3)$ is central in $L_2/Z(L_1)Z(L_3)$. As Q_2/Z_2 is a natural module, it follow from Lemma 2.5 that $Q_2/Z(L_1)Z(L_3)$ is (elementary) abelian. Hence $Z(L_2) = [Z_1, Z_3] \leq Z(L_1)Z(L_3)$, a contradiction.

This contradiction refutes $(4^{\perp})$, and proves (4). Then as $Z(L_3) \leq Z_2 = Z(L_2)$ we conclude that

$$(5) \qquad Z(L_3) = 1 \ .$$

Now we claim that

$$(6) \qquad Z_3 \text{ is a natural module for } L_3 \ .$$

Now Z_1 acts quadratically on Z_3, so Z_3 is a direct sum of natural modules. Z_2 fixes the path $(-1, \ldots, 5)$ pointwise and thus $\mathrm{rk}(Z_2) = f_1$. Thus $Z_3 = Z_2Z_4$ is a natural module, as claimed.

Since $Z_1 = Z_0Z_2$ acts nontrivially on Z_3, it is also clear that

(7) $$b_0 = 3 .$$

We still need to work out the structure of Q_1/Z_1 and Q_0.

We claim that $Q_0 \cap Q_2$ is normal in L_1. Let

$$V = \langle Z_1^{G_0} \rangle .$$

As Z_1 is a natural module, we find $[Z_1, Q_0] = Z_0$, and thus $V/Z_0 \leq Z(Q_0/Z_0)$.

As $b_0 = 3$, Z_4Q_1 is a Sylow 2-subgroup of L_1, and $\langle Z_{-2}, Z_4\rangle Q_1 = L_1$. Let τ be a shift by $+2$ along the track T. Then $[Z_4, Q_0 \cap Q_2] \leq [Z_4, Q_1 \cap Q_2] \leq [V^\tau, Q_1 \cap Q_2] \leq [V, Q_0]^\tau = Z_2$. Similarly, $[Z_{-2}, Q_0 \cap Q_2] \leq Z_0$. As Q_1 normalizes $Q_0 \cap Q_2$, it now follows that L_1 normalizes $Q_0 \cap Q_2$.

So Lemma 7.4 applies. Thus $\bar{Q}_1 = Q_1/(Q_0 \cap Q_2)$ is elementary abelian of rank $2f_0$. Furthermore, $C_{\bar{Q}_1}(L_1) = 1$. Our calculations above show that $[O^2(L_1), Q_0 \cap Q_2] \leq Z_1$.

Now we consider the action of Z_4 on $Q_1/(Q_0 \cap Q_2)$. We claim that this is a quadratic module. Indeed, $[Q_1, Z_4] \leq Q_1 \cap Q_2$, and $[Q_1 \cap Q_2, Z_4] \leq Z_2 \leq Q_0 \cap Q_2$. In view of Lemma 2.2, we have:

(8) The module $Q_1/(Q_0 \cap Q_2)$ is a direct sum of natural $\bar{L}_1$-modules .

In view of Lemma 7.4, the rank of this quotient is $2f_0$, and in particular

$$f_1 \leq f_0 .$$

Now $[O^2(L_1), Q_0 \cap Q_2] \leq Z_1$, and Z_1 is a natural module. It follows that

$$Q_0 \cap Q_2 = C_{Q_1}(K_1) \times Z_1 ,$$

which is point (E).

We set

$$D = C_{Q_1}(K_1) .$$

We claim that

(9) $$Q_0 = DV .$$

Let $\widetilde{V} = \langle Z_1^{G_2} \rangle$ (in other words, V "on the opposite side") and set $W = (V \cap Q_1)(\widetilde{V} \cap Q_1)$. We have $[Q_1, Z_3] \leq [Q_1, \widetilde{V}] \leq Q_1 \cap \widetilde{V} \leq W$. Similarly, $[Q_1, Z_{-1}] \leq W$. It follows that $[Q_1, O^2(L_1)] \leq W$.

If $W \leq Q_0 \cap Q_2$, then we find $Q_0 \cap Q_1 \lhd L_1$, contradicting Lemma 2.14. Now the two factors $V \cap Q_1$ and $\widetilde{V} \cap Q_1$ are conjugate under the action of L_1, so it follows that

$$V \cap Q_1 \not\leq Q_0 \cap Q_2 .$$

In view of the action of K, it follows that $Q_0 \cap Q_1 = (V \cap Q_1)(Q_0 \cap Q_2) \leq VDZ_1 = VD$. But $Q_0 = Z_{-1}(Q_0 \cap Q_1)$, so $Q_0 \leq Z_{-1}VD = VD$, and thus (9) follows.

Next we show that D is elementary abelian. We have $\phi(D) = \phi(DZ_1) = \phi(Q_0 \cap Q_2) \lhd L_1$. If $\phi(D) > 1$, then $\phi(D)$ must meet $Z(S)$, with S a Sylow 2-subgroup of L_1, and hence D meets Z_1, a contradiction. So $\phi(D) = 1$ and D is elementary abelian.

So $Q_0' = [Q_0, V] \le Z_0$. As D is elementary abelian and V is generated by elementary abelian subgroups, Q_0/Q_0' is elementary abelian. So $\phi(Q_0) \le Z_0$. But Q_0 is nonabelian, since it acts nontrivially on Z_1. So $\phi(Q_0) > 1$ and in view of the action of K we have $\mathrm{rk}(\phi(Q_0)) \ge f_1$, $\phi(Q_0) = Z_0$.

Now consider the group $Q_* = Q_0 \cap Q_2$. This has rank $f_0 + 2f_1$, and it fixes the path $(-1, \ldots, 3)$. The maximal regular path $(-2, \ldots, 4)$ is fixed by a K-invariant subgroup of Q_* of rank f_0. This cannot meet Z_1, so it must be contained in D, and hence must be D itself.

Now $Q_0 \cap Q_2 \lhd Q_1$ so $Q_0 \cap Q_2/Z_1 = DZ_1/Z_1$ meets $Z(Q_1/Z_1)$, and in view of the action of K we find $DZ_1/Z_1 \le Z(Q_1/Z_1)$. As $[O^2(L_1), Q_0 \cap Q_2] \le Z_1$ we find $DZ_1/Z_1 = Z(L_1/Z_1)$. Thus

(10) $(Q_1/Z_1)/(Z(L_1/Z_1))$ is a direct sum of natural modules for $\bar{L}_1$.

We claim that

$$L_0 = O^2(L_0) .$$

Let $P = [L_0, Q_0]$. It suffices to show that $P = Q_0$. We have $P \lhd L_0$ and $D = [D, K_0] \le P$. Hence $Z_1 = [D, Q_1] \le P$ and $V = Z_1^{G_0} \le P$. So $Q_0 = DV \le P$, and $P = Q_0$. Our claim follows.

We come to the last point:

(11) Q_0/Z_0 is an irreducible $\bar{G}_0$-module .

We fix $N \le Q_0$ properly containing Z_0 with $\bar{N} = N/Z_0$ minimal normal in L_0/Z_0.

If $\mathrm{rk}(\bar{N}) < 2f_0$, then $[N, L_0] = [N, O^2(L_0)] = 1$ and $N \le Z_0$, a contradiction. Similarly, if $\mathrm{rk}(Q_0/N) < 2f_0$, then $Q_0 = [Q_0, L_0] \le N$, and we have our claim.

Supposing this fails, then, we have $3f_1 + 2f_0 = \mathrm{rk}(Q_0) = \mathrm{rk}(Q_0/N) + \mathrm{rk}(N/Z_0) + \mathrm{rk}(Z_0) \ge 4f_0 + f_1$, that is, $f_0 \le f_1$ and thus $f_0 = f_1$. We will set $f := f_0 = f_1$. Note that now $\mathrm{rk}(N/Z_0) = \mathrm{rk}(Q/N) = 2f$ and N is a natural module.

If $N \ge Z_1$, then $N \ge V$ and $\mathrm{rk}(Q_0/N) \le f$, a contradiction. If $N \cap Z_1 > Z_0$, then $N \cap Z_1/Z_0$ is K_1-invariant and hence has rank at least f, forcing $N \ge Z_1$, a contradiction. We conclude that $N \cap Z_1 = Z_0$.

Now $\mathrm{rk}(N) = 3f$ and hence $\mathrm{rk}(N \cap Q_1) \ge 2f$. If $\mathrm{rk}(N \cap Q_1) > 2f$, then $\mathrm{rk}(N \cap Q_0 \cap Q_2) > f$. In view of the action of K_1, we then have $N \cap D > 1$. On the other hand, D covers L_{-2} modulo Q_{-2}, and has rank f, so $[D \cap N, K] = D$ and hence $D \le N$.

Now $D \le L_{-2}$, so $[D, Z_{-2}] = 1$. Here Z_{-2} covers a Sylow 2-subgroup of L_1. If $[D, Q_1] = 1$, it follows that $D \le Z_1$, whereas $D \cap Z_1 = 1$. Hence $[D, Q_1]$ is a nontrivial subgroup of L_1; and it is also normal. It follows that $[D, Q_1] = Z_1$. But then since $D \leq N$ we have $Z_1 = [D, Q_1] \leq N$, a contradiction.

So from all of this we conclude that

$$\text{rk}(N \cap Q_1) = 2f \ . \tag{12}$$

In particular, N is not contained in Q_1, and $N \cap Q_0 \cap Q_2 = Z_0$.

Now take $t \in L_1$ switching the vertices 0 and 2, and set $\widetilde{N} = N^t$, and $A = (N \cap Q_1)(\widetilde{N} \cap Q_1)$. Then $N \cap \widetilde{N} \cap Q_1 = N \cap \widetilde{N} \cap Q_0 \cap Q_2 = Z_0 \cap Z_2 = 1$. Hence $A = (N \cap Q_1) \times (\widetilde{N} \cap Q_2)$.

Now $L_1 = \langle N, \widetilde{N} \rangle Q_1$ and hence A is L_1-invariant. Furthermore, $A/Z_1 \simeq Q_1/Q_0 \cap Q_2$ as $\bar{L}_1$-module, a natural module. So an element of order 3 in K_1 will act on A without fixed points, and it follows from Lemma 2.4 that A is abelian

On the other hand, $\text{rk}(A \cap Q_{-1}) \geq 2f$, and as A acts transitively on $\Delta(0) \setminus \{1\}$, and is abelian, $A \cap Q_{-1}$ is contained in Q_δ for all δ adjacent to 0, that is, $A \cap Q_{-1} = A \cap G_0^{(2)}$, where $G_0^{(2)}$ denotes the pointwise stabilizer of the set of vertices at distance at most 2 from 0.

So $\text{rk}(W \cap G_0^{(2)}) \geq 2f$ and in particular $G_0^{(2)} > Z_0$. So $Z_0 < G_0^{(2)} < Q_0$, and therefore in view of our previous analysis $G_0^{(2)}/Z_0$ is also a minimal normal subgroup of G_0/Z_0, that is we could have chosen $N = G_0^{(2)}$ from the beginning. But we already showed that N is not contained in Q_0, so this is a contradiction. □

8.2. 2nd case: $[Z_1, Z_{1+b_1}]$ trivial.

LEMMA 8.4. *Suppose that* $[Z_1, Z_{1+b_1}] = 1$. *Then, without loss of generality, we may suppose that* b_1 *is odd.*

PROOF. Suppose that b_1 is in fact even. Then we claim that b_0 is odd, and that after interchanging 0 and 1, we retain all of our hypotheses. In particular we will need $r = s - 1$ to justify this change of notation, so all in all we make the following three claims:

$$r = s - 1;\ [Z_0, Z_{b_0}] \leq Z_0 \cap Z_{b_0};\ b_0 \text{ is odd} \ . \tag{$*$}$$

The second condition may seem too weak, as our strategy seems to require $Z_0 \cap Z_{b_0} = 1$, but in the case in which $[Z_0, Z_{b_0}] \neq 1$ and at the same time $[Z_0, Z_{b_0}] \leq Z_0 \cap Z_{b_0}$, with $r = s - 1$ holding as well, we are free to interchange 0 and 1 and find ourselves in the setting of the previous subsection. This in itself is sufficient—but according to our previous results, in this case b_1 (which corresponds to b_0 in the notation of the previous subsection) would then be 1 or 3, so that in this case b_1 already is odd.

We begin the analysis.

Let $j = 1 + b_1$, which we have assumed is odd. As $[Z_1, Z_j] = 1$ and Z_1 is not contained in Q_j, it follows that

$$Z_j = Z(L_j) \ . \tag{1}$$

In particular, $Z_1 \leq Z_2$ and hence

$$b_1 \geq b_0 + 1 \ .$$

Suppose first that

(I) $$b_1 = r/2 \ .$$

Then Z_j fixes the path $(1, \ldots, j + b_1)$ of length $2b_1 = r$, and has odd endpoints. So Lemma 7.2 implies that $r = s - 1$, giving our first point.

Now suppose that $[Z_0, Z_{b_0}] \not\leq Z_0 \cap Z_{b_0}$. Since $r = s - 1$, Lemma 7.5 may be applied, with the indices 0 and 1 interchanged, and this yields, among other things,

$$1 < [Z_1, Z_{1+b_1}] \ ,$$

which contradicts our case hypothesis in this subsection. So $[Z_0, Z_{b_0}] \leq Z_0 \cap Z_{b_0}$, and as we have explained, this allows us to conclude that $Z_0 \cap Z_{b_0} = 1$.

Finally, we claim that b_0 is odd. If this fails, then we may in any case *interchange* 0 *and* 1, and find ourselves once more in the setting of the present lemma. Then our claim (1) applies with $j' = 0 + b_0$ and yields

$$Z_0 = Z(L_0) \ .$$

Then Z_0 meets Z_1 and this is normal in both L_0 and L_1, a contradiction. So b_0 is odd, and after interchanging 0 and 1 we have the desired state of affairs.

So we now consider the alternative:

(II) $$b_1 = r/2 - 1 \ .$$

As $b_1 > b_0$, we have $b_0 = r/2 - 2$. In particular b_0 is odd.

Let $L = \langle Z_2, Z_{r-2} \rangle$. Observe that $Q = O_2(G_{0,\ldots,r})$ centralizes L. Also Z_j belongs to Q_2 and Q_{r-2} and hence centralizes L as well.

Now L covers $\bar{L}_j$ and is normalized by K, so $K_j \leq L$. So K_j centralizes Q_γ and Z_j.

Q_γ covers a Sylow 2-subgroup of L_0, while Z_j covers a Sylow 2-subgroup of L_1. Since K_j centralizes both, it acts trivially on the quotients $\bar{L}_0$, $\bar{L}_1$, that is, $[K_j, L_i] \leq Q_i$ for $i = 0, 1$. Thus $C(K_j)$ covers $\bar{L}_i$ for $i = 0, 1$, and K_j is normalized, and even centralized, by transitive subgroups of L_0 and L_1 contradicting Lemma 2.14.

This disposes of the case $b_1 = r/2 - 1$, and completes the proof. □

LEMMA 8.5. *Suppose that* $[Z_1, Z_{1+b_1}] = 1$ *with* b_1 *odd. Then we have the following:*

(A) $b_0 = r/2$ *and* $b_1 = r/2 - 1$.
(B) $Z_0 = Z(L_0)$, $\mathrm{rk}(Z_0) = f_0$.

PROOF. We set $j = 1 + b_1$. As $[Z_1, Z_j] = 1$ and Z_1 is not contained in Q_j, we have $Z_j = Z(L_j)$. Since j is even, it follows that $Z_0 = Z(L_0)$. Thus

$Z_0 \le Z_1$ and hence $b_0 \ge b_1 + 1$. In view of our estimates on b_0 and b_1 we find

$$b_0 = r/2; \quad b_1 = r/2 - 1 .$$

As j is even, Z_j fixes $(0, \dots, r)$ and it follows easily that $Z_j = Q_\gamma$ (i.e., $O_2(G_\gamma)$) and $\mathrm{rk}(Z_j) = f_0$, hence also $\mathrm{rk}(Z_0) = f_0$. This completes the proof. □

LEMMA 8.6. *Suppose that* $[Z_1, Z_{1+b_1}] = 1$ *with* b_1 *odd. Let* $j = 1 + b_1$. *Then* $Z_1 \cap Q_j \le Q_{j+1}$.

PROOF. We assume the contrary:

(1) $$Z_1 \cap Q_j \not\le Q_{j+1} .$$

Set $H = Z_1 \cap Q_j$.

Now $Z(L_{j+1}) = 1$ since j is even and hence $[H, Z_{j+1}] \ne 1$. Fix $[h, z]$ a nontrivial element with $h \in H$, $z \in Z_{j+1}$.

Suppose $b_1 \ge 4$. As z fixes the vertex 2, the vertex $\delta = -1^{z^{-1}}$ has distance at most 4 from the vertex 1. As $s \ge 5$ and $b_1 \ge 4$, we find that Z_1 fixes the vertex δ. So h fixes δ as well as -1, and we have

$$-1^{[h,z]} = \delta^{hz} = \delta^z = -1$$

and thus $[h, z]$ fixes the path $(-1, \dots, j{+}1{+}b_1)$ of length $j{+}2{+}b_1 = r{+}1 > r$, a contradiction. So we conclude that

(2) $$b_1 \le 3 ,$$

or, in other words, $b_1 = 1$ or 3.

We first treat the case $b_1 = 3$. Then $b_0 = 4$, $r = 8$, and $H = Z_1 \cap Q_4$. Now as above $[H, Z_{j+1}]$ is nontrivial, and $[H, Z_{j+1}]$ fixes the path $(0, \dots, 8)$ of length r pointwise, so in view of the action of K we find $[H, Z_{j+1}] = Q_\gamma = Z_4$.

Now $[H, Z_4] = 1$, so H acts quadratically on Z_{j+1}, with $\mathrm{rk}([Z_{j+1}, H]) = f_0$. By Lemma 2.1, $Z_5 = Z_{j+1}$ is a sum of natural modules, and thus has rank $2f_0$. In particular, $Z_5 = Z_4 Z_6$, and similarly $Z_1 = Z_0 Z_2$.

Now Z_0 covers a Sylow 2-subgroup of $\bar{L}_4$ and has rank f_0, so $Z_0 \cap Q_4 = 1$ and $H = Z_2$. But then $H \le Q_5 = Q_{j+1}$, a contradiction to our assumption.

This contradiction eliminates the case $b_1 = 3$ and leaves the case

(3) $$b_1 = 1$$

with which we will be concerned for the remainder of the proof. So we have $b_0 = 2$ and $r = 4$. As $s \ge 5$ we have $s = 5$ and in particular $r = s - 1$. So we can compute the rank of a Sylow 2-subgroup S of $G_{0,1}$ by working with the path $(0, \dots, 5)$. We find

(4) $$\mathrm{rk}(S) = 2f_0 + 2f_1 .$$

We claim that

(5) Q_1 is elementary abelian.

The group Z_2Q_0 is a Sylow 2-subgroup of L_0, and $Z_2 \cap Q_0 = 1$ since $r = 4$. As $b_1 = 1$, we have $Q_1 \leq Z_1Q_2$ and Q_1 normalizes Z_2. So $Q_1 = Z_2 \times (Q_0 \cap Q_1)$. Similarly, $Q_0 \cap Q_2 \leq Z_0Q_2$, so $Q_0 \cap Q_1 = Z_0 \times (Q_0 \cap Q_2)$. So

$$Q_1 = Z_0 \times Z_2 \times (Q_0 \cap Q_2) \ .$$

Now $\phi(Q_0 \cap Q_2) \leq Q_{-1} \cap Q_3$ stabilizes $(-2, \ldots, 4)$, which has length $6 > r$, so $\phi(Q_0 \cap Q_2) = 1$ and $Q_0 \cap Q_2$ is elementary abelian, from which (5) follows.

Now we claim that $\mathrm{rk}(Q_1) \geq 2f_1$. Otherwise, in a composition series for Q_1 relative to the action of $\bar{L}_1$, all factors are trivial, and in particular the tori of L_1 centralize Q_1, contradicting our hypothesis. So $\mathrm{rk}(Q_1) \geq 2f_1$. But we know the rank of Q_1, namely $2f_0 + f_1$, and hence $2f_0 + f_1 \geq 2f_1$, and

$$(6) \qquad f_1 \leq 2f_0 \ .$$

Now $Q_{-1} \cap Q_1$ has rank f_0 and hence $Q_{-1} \cap Q_1 = Z_0$ which is normal in G_0. So Lemma 7.4 applies, and Q_0/Z_0 is elementary abelian of rank $2f_1$, with $C_{Q_0/Z_0}(\bar{L}_0) = 1$. As Q_1 is abelian, its action on Q_0 is quadratic, and it follows by Lemma 2.2 that Q_0/Z_0 is a direct sum of natural modules for $\bar{L}_0$. As $f_1 \leq 2f_0$ this sum has at most two factors, and $f_1 = f_0$ or $2f_0$.

If Q_0/Z_0 is a natural module, then by Lemma 2.5 it follows that Q_0 is elementary abelian. Then $Q_0 \cap Q_1$ is centralized by Q_0Q_1, and this is a Sylow 2-subgroup of $G_{0,1}$. So $Q_0 \cap Q_1 \leq Z_0 = Z(L_0)$. This contradicts Lemma 2.14. So we conclude that

(7) $\qquad f_1 = 2f_0$ and Q_0/Z_0 is the sum of two natural modules.

In particular, $\mathrm{rk}(Q_1) = 2f_1$ and it follows that Q_1 is a natural module for $\bar{L}_1$, in view of Lemma II 5.31.

Thus $[Q_1, Q_0, Q_0] = 1$. As $[Q_1, Z_2] = 1$ we find that Z_2Q_0 centralizes $[Q_1, Q_0]$, and as Z_2Q_0 is a Sylow 2-subgroup of L_0, this implies that $[Q_0, Q_1] \leq Z_0$. Then Q_1 acts trivially on Q_0/Z_0, contradicting the structure of this $\bar{L}_0$-module.

We have reached a contradiction in all cases, and our claim is proved. □

LEMMA 8.7. *Suppose that* $[Z_1, Z_{1+b_1}] = 1$ *with* b_1 *odd. Then we have the following:*

(A) $f_0 = f_1$.
(B) $Z_1 \cap Q_{1+b_1} = Z_2$.
(C) $Z_1 = Z_0Z_2$ *is a natural module for* $\bar{L}_1$.
(D) $b_1 \geq 3$.

PROOF. Let $j = 1 + b_1$. The group $H = Z_1 \cap Q_j \leq Q_{j+1}$ fixes the path $\tilde{\gamma} = (1 - b_1, \ldots, j + 2)$ of length $j + b_1 + 1 = 2b_1 + 2 = r$. Thus we have $Z_2 \leq H \leq Q_\gamma$ with $Q_\gamma = O_2(\tilde{\gamma})$ as usual, and by rank considerations and the action of K we find $Z_2 = Q_\gamma$. In particular, point (B) follows.

Next we show $Z_1 = Z_0Z_2$. We have $Z_1 \leq Z_0Q_j$ and therefore $Z_1 = Z_0(Z_1 \cap Q_j) = Z_0Z_2$. It follows also that Z_1 is elementary abelian.

Now $Z(L_1) = 1$ and $\mathrm{rk}(Z_1)$ is $2 \cdot \mathrm{rk}(C_{Z_1}(S))$, with S a Sylow 2-subgroup of $L_{1,2}$. By Lemma 2.3 it follows that Z_1 is a sum of natural $\bar{L}_1$-modules. As $\mathrm{rk}(Z_1) = 2f_0$, it also follows that

$$f_1 \leq f_0 \ .$$

Our next point is

$$f_0 = f_1, \text{ and } Z_1 \text{ is a natural module.}$$

We have $K = K_0 \times K_1$ since K_0 centralizes $Z_0 = Z(L_0)$, and the action of K_1 on Z_0 is fixed point free. Now $Z_j Q_0$ is a Sylow 2-subgroup of $\bar{L}_0$, and $[K_j, Z_j] = 1$, so $[K_j, L_0] \leq Q_0$. Hence $K_j \cap K_0 = 1$. So $f_0 = \mathrm{rk}(K_j) \leq \mathrm{rk}(K_1) = f_1$. So $f_0 = f_1$ and Z_1 is natural.

Finally we deal with b_1. If $b_1 < 3$, then as b_1 is odd we have $b_1 = 1$.

Now $\mathrm{rk}(Z_1 \cap Q_0) = f$, and hence $Z_1 \cap Q_0 = Z_0$. Hence with $b_1 = 1$, we have $[Z_1, Q_0] = Z_0$. As $Z_1 Q_0$ covers a Sylow 2-subgroup of $\bar{L}_0$, we find $[O^2(L_0), Q_0] = 1$, contradicting our hypothesis that $C_{L_0}(Q_0) \leq Q_0$. So $b_1 \geq 3$, and all our claims are proved. □

PROPOSITION 8.8. *The case* $[Z_1, Z_{1+b_1}] = 1$ *is impossible.*

PROOF. The remaining analysis is quite substantial.

We set $f := f_0 = f_1$. Let $V_0 = \langle Z_1^{L_0} \rangle$. As $b_1 \geq 3$, V_0 is commutative and $V_0 \leq Q_0 \cap Q_1$. As Z_1 is a natural $\bar{L}_1$-module, we find $[Z_1, Q_0] = Z_0$. Hence also $[V_0, Q_0] = Z_0$.

With $j = 1 + b_1 = b_0$, let $R = [Z_{-1}, Z_j]$. As $Z_j \leq G_0$ acts transitively on $\Delta(0) \setminus \{1\}$, we have $V_0 = \langle Z_{-1}^{Z_j} Z_0 \rangle = Z_{-1} Z_0 R$. We claim, in fact, that

$$V_0 = Z_{-1} Z_1$$

or, in other words, $R \leq Z_{-1} Z_1$; indeed, we claim $R \leq Z_1$.

With $V_{j-2} = \langle Z_{j-1}^{L_{j-2}} \rangle$, we have $Z_j \leq Z_{j-1} \leq V_{j-2}$ and $Z_{-1} \leq Z_{j-2}$, hence $R \leq V_0 \cap V_{j-2}$. Then $[R, V_j] = 1$ since V_{j-2} is abelian. Hence $[V_0, Z_j, Z_j] \leq [Z_{-1}, Z_j, Z_j] = 1$ and V_0 is a quadratic module with respect to the action of Z_j. Hence $V_0 / C_{V_0}(O^2(L_0))$ is a direct sum of natural modules.

As $[R, Z_j] = 1$, it follows that R is contained in $Z_1 C_{V_0}(O^2(L_0))$. As R is K_0-invariant, we find $R = (R \cap Z_1) \cdot C_R(O^2(L_0))$. Let $R_0 = C_R(O^2(L_0))$.

Take an involution $t \in O^2(L_0)$ acting as a reflection on T. Now $R_0 \leq V_{j-2} \leq L_j$ and t centralizes R_0, so $R_0 Z_0$ stabilizes the path $(-j, \ldots, j)$ and hence $\mathrm{rk}(R_0 Z_0) = \mathrm{rk}(Z_0)$ and $R_0 \leq Z_0$. Thus $R \leq Z_1$ and $V_0 = Z_{-1} Z_1$. In particular, V_0 has rank $3f$ and V_0 / Z_0 has rank $2f$, and it follows that

$$(8) \qquad V_0/Z_0 \text{ is a natural module.}$$

Now $R \neq 1$, and $R \leq V_0 \cap V_{j-2}$ fixes the path $(1 - b_1, \ldots, j - 3 + b_1)$, of length $j - 4 + 2b_1 = 3r/2 - 6$. So $3r/2 - 6 \leq r$ and hence

$$(9) \qquad r \leq 12 \ .$$

We deal first with the case

$$r = 12;$$

then

$$b_0 = 6 \text{ and } b_1 = 5.$$

Now R stabilizes the aforementioned maximal regular path $(-4, \ldots, 8)$, as does Z_2, and thus $R = Z_2$ has rank f. That is, $[Z_{-1}, Z_6] = Z_2$. Now $Z_5 = Z_4 Z_6$ so we find also that $[Z_{-1}, Z_5] = Z_2$. So in general for i odd we have $[Z_i, Z_{i+6}] = Z_{i+3}$. This refers to paths contained in the track T, and we wish to prove the same statement for paths not necessarily contained in this track, that is, for vertices at distance 6 in the orbit of the vertex 1.

We claim that any path of length 6 and endpoints conjugate to 1 can be conjugated into T. Let λ be such a path and denote its vertices as follows: $(\delta_{-3}, \ldots, \delta_3)$. As $s \geq 5$ we may suppose that $(\delta_{-2}, \ldots, \delta_3)$ already coincides with the path $(0, \ldots, 5)$; but this path is regular on the left, so we may move λ into T. So under the assumption $r = 12$, we find $[Z_{\delta'}, Z_{\delta''}] = Z_\delta$ whenever δ', δ'' are vertices in the orbit of 1, at distance 6, with midpoint δ.

Now consider $z \in Z_0 \setminus G_7$ and $z' \in Z_{10} \setminus G_3$. Then z fixes the vertex 6, and the path $(10^z, \ldots, 6, \ldots, 10)$ has length 8. Now 9^z and 9 are vertices at distance 6, so we have $[Z_{9^z}, Z_9] = Z_6$. Similarly, $[Z_{1^{z'}}, Z_1] = Z_4$. Thus $[z, z^{z'}] \in Z_4$, and $[z', z'^z] \in Z_6$.

Now z and z' are involutions, so $[z, z^{z'}] = (zz')^4 = (z'z)^{-4} = [z', z'^z] \in Z_4 \cap Z_6 = 1$, that is, $[z, z'^z] = 1$.

Now look at the action of $[z, z^{z'}]$ on the element $-3 \in T$. Let $t = z^{z'}$. We have $z \in Z_1$. The distance between the vertices 1 and $2^{z'}$ is 5. As $b_1 = 5$ and $s \geq 5$, it follows that z fixes the vertex $2^{z'}$. So t fixes the vertex 2.

Now z fixes the path $(-6, \ldots, 4, 3^{z'}, 2^{z'})$ of length r (i.e., 12), and hence cannot fix the immediate neighbor $1^{z'}$. So t moves the vertex 1, and fixes the vertex 2.

Therefore the distance between the vertices 0 and $(-3)^t$ is 7. Now z fixes the path $(-6, \ldots, 0)$ and hence cannot fix the vertex $(-3)^t$. Now $(-3)^{[z,t]} = [(-3)^{tz}]^t \neq [(-3)^t]^t = -3$. So $[z, t] \neq 1$, a contradiction.

We conclude that

$$r < 12 .$$

Recalling that $b_1 \geq 3$ is odd and $b_1 = r/2 - 1$, we arrive at

$$(10) \qquad r = 8; \quad b_0 = 4; \quad b_1 = 3 .$$

This is the key configuration to eliminate.

We will show first

$$(11) \qquad Q_* = Q_{-1} \cap Q_1 \text{ is normal in } G_0 .$$

We work with V_{-2}, V_0, V_2 where $V_i = \langle Z_{i+1}^{L_i} \rangle$ for i even. Recall that $V_i = Z_{i-1} Z_{i+1}$. So $V_2 \cap Q_0 = Z_1 \leq V_0$, and we find $[V_2, Q_0 \cap Q_1] \leq V_2 \cap Q_0 \leq V_0$. In particular $[V_2, Q_*] \leq V_0$ and similarly $[V_{-2}, Q_*] \leq V_0$. Also $V_0 \leq Q_*$ and thus Q_* is normalized by $\langle V_{-2}, V_2\rangle$; since Q_* is also normalized by Q_0 and K, it is normalized by G_0, as claimed. We have also seen that

$$(12) \qquad [O^2(L_0), Q_*] \le V_0 \ .$$

In particular, Lemma 7.4 applies, and the rank of the quotient Q_0/Q_* is $2f$. Hence

$$(13) \qquad Q_0/Q_* \text{ is a natural } \bar{L}_0\text{-module.}$$

Now it follows from the last two points and Proposition I 9.9 that

$$(14) \qquad Q_* = C_{Q_0}(K_0)V_0 \ .$$

We will show that

$$(15) \qquad C_{Q_0}(K_0) = Z_0 \ .$$

Set $D = C_{Q_0}(K_0)$. Fix $t \in N_{O^2(L_0)}(K_0)$ operating as a reflection about 0 on the track T. Then t normalizes every subgroup of D containing Z_0, since $[t, D] \le V_0 \cap D = Z_0$.

Now $D \cap L_4$ fixes the arc $(0, \ldots, 4)$ and contains Z_0. Hence this group is invariant under the reflection t and fixes $(-4, \ldots, 4)$ of length $r = 8$. As Z_0 already has rank f it follows that $D \cap L_4 = Z_0$, and in particular $D \cap Q_3 \le Z_0$.

Whenever D meets some L_i in a subgroup not contained in Q_i, then K_0 acts trivially on $\bar{L}_i$, and thus $C(K_0) \cap L_i$ acts transitively on the neighbors of i. This cannot happen for two consecutive values of i, by Lemma 2.14. However, $Z_0 \le D$ and Z_0 covers a Sylow 2-subgroup of L_4. Hence $D \cap L_3 \le Q_3$, and hence $D \cap L_3 = Z_0$. So if $D \le L_3$, then $D = Z_0$, as claimed. Assume therefore that D is not contained in L_3.

We claim $N_{L_2}(Z_0) \le G_0$. Otherwise, fixing $x \in N_{L_2}(Z_0) \setminus G_0$, we have $Z_0 = Z_{0^x}$ where the vertex 0^x has distance d at most 4 from 0. We may conjugate the path $(0, \ldots, 0^x)$ into the path $(0, \ldots, d)$, by an element of G_0, and find $Z_0 = Z_d$. Then Z_0 stabilizes the path $(-b_0, \ldots, d + b_0)$ or length $r + d$, a contradiction. So $N_{L_2}(Z_0) \le G_0$.

Let us see that $C_{Q_2}(K_0) = Z_0$. If $C_{Q_2}(K_0)$ is not contained in Q_1, then $C_{L_1}(K_0)$ covers $\bar{L}_1$ and acts transitively on $\Delta(1)$. On the other hand since D is not contained in L_3 it covers a Sylow 2-subgroup of $\bar{L}_2$ and hence $C_{L_2}(K_0)$ acts transitively on $\Delta(2)$, and we have a contradiction. So $C_{Q_2}(K_0) \le C_{Q_1}(K_0) \le C_{Q_0}(K_0) = D$ and $C_{Q_2}(K_0) = D \cap Q_2 = Z_0$.

It follows that $C_{L_2}(K_0) \le N_{L_2}(Z_0) \le G_0$. As $C_{L_2}(K_0)$ covers $\bar{L}_2$ we find that $L_2 \le G_0Q_2$. This is incompatible with the action of L_2 on the tree Γ. So we have a contradiction, and it follows that $D = Z_0$, as claimed.

In particular, we now have the following:

$$(16) \qquad Q_* = V_0; \quad \mathrm{rk}(Q_0) = 5f \ .$$

As $Q_0' \le Z_0 \le Q_2$, it follows that Q_0 normalizes $Q_0 \cap Q_2$. Similarly, Q_2 normalizes the same intersection, and then G_1 normalizes $Q_0 \cap Q_2$. So Lemma 7.4 applies. Hence $Q_1/(Q_0 \cap Q_2)$ has rank $2f$ and is a natural module. Recall that $(Q_0 \cap Q_2)/Z_1$ is the center of L_1/Z_1. So by Lemma 2.5 we find that Q_1/Z_1 is abelian. On the other hand, V_0/Z_0 is a natural $\bar{L}_0$-module and Q_1 represents a Sylow 2-subgroup of $\bar{L}_0$, so $\mathrm{rk}(Q_1') \geq f$. Thus $Q_1' = Z_1$.

Now Q_0/V_0 is natural, and V_0/Z_0 is natural. Via Lemma 2.4 it follows that

Q_0/Z_0 is elementary abelian, and is a direct sum of natural modules.

So we may write

$$Q_0/Z_0 = V_0/Z_0 \oplus W_1/Z_0$$

with W_1/Z_0 a natural module for $\bar{L}_0$.

Then W_1 covers a Sylow 2-subgroup of $\bar{L}_1$. Take $g \in L_1 \setminus G_0$. Then $L_1 = \langle W_1, W_1^g \rangle Q_1$. Let $X = (W_1 \cap Q_1)(W_1^g \cap Q_1)/Z_1$. Notice that $W_1 \cap Q_1$ is normal in Q_1, and hence the same applies to $W_1^g \cap Q_1$. In particular these two factors normalize one another and as $Z_1 = Z_0 Z_0^g$ is contained in their product, the quotient at least makes sense.

On the other hand, $[W_1, W_1^g \cap Q_1] \leq W_1 \cap Q_1$ and thus W_1 normalizes X; similarly, W_1^g normalizes X and hence so does L_1.

Now $W_1 \cap Q_1/(W_1 \cap Q_1 \cap Z_1) = W_1 \cap Q_1/Z_0$ has rank f, so X has rank at most $2f$, and g acts nontrivially, so X is a natural $\bar{L}_1$-module.

Now L_1 centralizes the quotient $Q_1/(W_1 \cap Q_1)(W_1^g \cap Q_1)$ since $L_1 = \langle W_1, W_1^g \rangle Q_1$ and in particular K_1 centralizes this quotient.

Finally, we consider $[V_0, K_1]$. As Z_1 is a natural module we have $[Z_1, K_1] = Z_1$. Now Z_{-1} covers a Sylow 2-subgroup of $\bar{L}_2$, and $Z_{-1} \cap Q_2 = Z_0$, while $[Z_{-1}, K_1]$ is a K-invariant subgroup. So either $[Z_{-1}, K_1] = Z_{-1}$ or $C_{L_2}(K_1)$ covers $\bar{L}_2$. In the latter case $[K_1, Q_1] \leq Q_2$ and hence $[K_1, Q_1] \leq Q_0 \cap Q_2$. But $Q_1/(Q_0 \cap Q_2)$ is a natural module, so this is a contradiction. So $[Z_{-1}, K_1] = Z_{-1}$ and $[V_0, K_1] = V_0$.

Since K_1 centralizes $Q_1/(W_1 \cap Q_1)(W_1^g \cap Q_1)$, it follows that $V_0 \leq (W_1 \cap Q_1)(W_1^g \cap Q_1)$. Intersecting with Q_0, we have $V_0 \leq (W_1 \cap Q_1) \cdot (W_1^g \cap Q_0 \cap Q_1)$.

Now $[Z_{-1}, W_1] \leq Q_0' \leq Z_0$, and W_1 covers a Sylow 2-subgroup of $\bar{L}_1$. As X is a natural module and Z_{-1} commutes with the action of W_1 on it, we find that $Z_{-1} \leq W_1 \cap Q_1$ in view of the structure of X. So $V_0 = Z_{-1} Z_1 \leq (W_1 \cap Q_1) Z_1$. But $\mathrm{rk}(V_0) = \mathrm{rk}((W_1 \cap Q_1) Z_1) = 3f$, so we find $V_0 = (Q_1 \cap Q_1) Z_1$ and hence $V_0 \cap W_1$ has corank at most f in V_0, whereas $V_0 \cap W_1 = Z_0$, a contradiction. □

9. A generalized polygon

After all this preparation, we have a limited number of configurations which survive, and in all of them $r = s - 1$. One would like to prove at this point that the original graph Γ_0 associated to the group G was the incidence graph for a Moufang generalized quadrangle. It is characteristic of the amalgam method that the actual line of argument is less direct. We argue that one can construct a Moufang generalized quadrangle as a quotient of the universal cover Γ, and that this is associated with another group of finite

Morley rank with the same parabolic subgroups. In the following section we will show that one can identify this second group, read off the structure of its parabolic subgroups, and then use this information to identify the original group.

9.1. The surviving cases. There are three cases which survived analysis so far: these were encountered in Lemmas 4.3, 8.2, and 8.3, and are collected at the end of §3. These correspond in fact to the three types of Chevalley group in Lie rank 2, as we will show in this section and the next.

First we adjust our notation. Until now we have largely followed the notation of [**174**], but we are going to shift our notation in order to bring it more closely into line with the notation of [**84**]. There are three points to take note of here, only one of which actually requires an adjustment.

First, in the three cases which survived analysis so far, we found that Z_δ is an elementary abelian 2-group for all δ; this is given in Lemmas 4.4, 8.2, and 8.3. In [**84**], the definition of Z_δ is different from that used here (it is generated by $\Omega_1(Z(S))$ as S varies over Sylow 2-subgroups of G_δ), but when our Z_δ is elementary abelian the two coincide. So this does not require any alteration of notation.

Secondly, the definition of the b_i in [**84**] varies substantially from the definition in [**174**]. Most importantly, in [**84**] the definition does not privilege the K-track T. However, since in all of our cases $b_i \leq s$, all relevant paths can be conjugated into T, and again the two definitions yield the same result, and our notation remains satisfactory.

The final point refers again to the parameters b_i and the way the symmetry between 0 and 1 is (or is not) broken. The convention adopted in [**84**] labels the vertices so that

$$(*) \qquad b_0 \leq b_1$$

(and in the case of equality the symmetry can be broken in other ways). As it happens, this actually conflicts with the conventions adopted here. From this point on, *we will switch our labeling of vertices if necessary* to conform to the usage in [**84**]. So condition $(*)$ will hold. This affects the notation in the last of our three cases.

With this proviso, our three cases may be listed according to the following chart:

Case	r	b_0	b_1	Lemma	Example
I	3	1	1	4.3	PSL_3
II	4	1	1	8.2	Sp_4
III	6	2	3	8.3	G_2

In all cases we found $r = s - 1$, and the connected component of the stabilizers of a path of length s is a $2^\perp$-groups.

We also found $f_0 \leq f_1$ in all cases, bearing in mind the reversal of labels in the third case. In fact, we found $f_0 = f_1$ in the first two cases. In Case

(III), we must still leave open a formal possibility corresponding to ${}^3D_4$, which will disappear later as a result of the classification of the relevant (B, N)-pairs.

Concerning the modules involved, we found the following.

(I) ($r = 3$), Lemma 4.4: For any i, Q_i is elementary abelian, and affords a natural module for $\bar{L}_i$; $f_0 = f_1$.

(II) ($r = 4$), Lemma 8.2: For any i, Q_i is elementary abelian and $Q_i/Z(L_i)$ is a natural module for $\bar{L}_i$; $f_0 = f_1 =: f$; $\mathrm{rk}(Z(L_i)) = f$.

(III) ($r = 6$), Lemma 8.3: After switching the labels $0, 1$ to conform to our current conventions:

$f_0 \leq f_1$;
Z_0 is a natural module for $\bar{L}_0$;
$(Q_0/Z_0)/Z(L_0/Z_0)$ is a direct sum of natural modules for $\bar{L}_0$;
$Z_1 = Z(L_1)$;
$\mathrm{rk}(Z_1) = f_0$;
$\phi(Q_1) = Z_1$;
$\mathrm{rk}(Q_1) = 3f_0 + 2f_1$;
Q_1/Z_1 is an irreducible $\bar{G}_1$-module.

9.2. The Uniqueness Condition. The idea in what follows is that our graph Γ should be the universal cover of the incidence graph of a generalized polygon associated with our simple group G, once G is properly identified, and, in particular, the conjugates of the track T should be the universal covers of the apartments of that generalized polygon. Since we have considerable control over the track T and its associated point stabilizers, we begin at this point, essentially rewriting the axioms for generalized n-gons in terms of the universal cover. Of course, we exploit heavily the fact that we already have a group acting on the geometry. We will introduce two properties, the *uniqueness* and *exchange* properties for apartments, verify that they apply in the cases under consideration, and carry out the remainder of the analysis simply using these two conditions.

DEFINITION 9.1.

(1) *An* apartment *in* Γ *is a* G*-conjugate of the* K*-track* T.

(2) T *satisfies the* uniqueness condition *if every path of length* $r+1$ *in* Γ *is contained in a unique* G*-conjugate of* T.

In the next lemma, we need two facts: $r = s - 1$, and the connected component of the stabilizer of a path of length $r + 1$ is a $2^\perp$-group. The further structural information is not yet relevant.

LEMMA 9.2. T *satisfies the uniqueness condition.*

PROOF. Since $s = r + 1$ in all cases, any path of length at most s is conjugate to one contained in T, by Lemma 2.11.

Now suppose that γ is a path of length s contained both in T and in a conjugate T^g. As a matter of notation, we may suppose that $0, 1 \in$

γ. Adjusting g, we may also suppose that g fixes γ pointwise (this may involve a shift along T^g, and possibly a reflection). That is, $g \in G_\gamma = KO_2(G_\gamma)$, by Lemma 3.5. Now $O_2(G_\gamma)$ is a finite group normalized by K and hence centralized by K, so g commutes with K and therefore leaves T invariant. □

9.3. Root groups. The exchange condition will be formulated in terms of the following groups.

NOTATION 9.3.
(1) *We label the vertices of T by integers, as usual.*
(2) $\gamma_i^{-r} = (i-r,\ldots,i)$; $\gamma_i^{+r} = (i,\ldots,i+r)$.
(3) $R_i^\pm = O_2{}^\circ(G_{\gamma_i^{\pm r}})$.
(4) $R_i = \langle R_i^-, R_i^+\rangle$.

A general remark is in order concerning stabilizers of paths. For paths γ of length at most s, the rank of the group $O_2(G_\gamma)$ (and a good deal of its structure as well) can be computed by working along the path from one end; one loses a "1-dimensional" piece at each stage (in particular, the rank drops by f_0 or f_1 as the case may be). Applying this to paths of length s, where $O_2(G_\gamma) = 1$, one gets the rank of $S = O_2(G_{0,1})$, a technique we have used repeatedly in our earlier analysis. Applying this to paths of length $r = s-1$, one sees that the $R_i^\pm$ cover a 2-Sylow subgroup of $G_{i\pm r}$, as well as a Sylow 2-subgroup of G_i, exactly, and these groups are in some sense "1-dimensional". These groups are expected to be "root groups" once the group G is properly identified as a Chevalley group, and in particular "opposite root groups" in G_i. We record the essential points as follows.

LEMMA 9.4.
(a) $R_i^- \cap G_{i+1} = 1$, $R_i^+ \cap G_{i-1} = 1$.
(b) $R_i^\pm$ *is elementary abelian.*
(c) $R_iQ_i = L_i$.
(d) $K_i \le R_i$.
(e) $O_2(R_i) = R_i \cap Q_i$.

PROOF. For (a), argue as in the proof of Lemma 7.2.
So R_i^- is elementary abelian; similarly for R_i^+. So (b) holds.
R_i^-, R_i^+ cover Sylow 2-subgroups of L_i/Q_i so (c) holds.
K normalizes R_i^- and R_i^+. Working in R_iK, (d) follows.
(e) is clear. □

LEMMA 9.5. *There is an element $t_i \in R_i$ normalizing K and acting as a reflection on T, with $t_i^2 \in O_2(G_T)$.*

PROOF. There is a 2-element $t \in KR_i$ such that, taken modulo $O_2(R_i)$, t is an involution normalizing K and inverting K_i. Adjusting by an element of $O_2(R_i)$, we may suppose in addition that t normalizes K outright, and in particular inverts K_i.

Then $t^2 \in C_{G_i}(K) \le G_T$ by Lemma 1.9, and $G_T = KO_2(G_T)$ by Lemma 3.5. Also, $t^2 \in Q_i$ so $t^2 \in O_2(G_T)$.

Since t normalizes K and T is the fixed point set of K in Γ, t leaves T invariant. Since t fixes i and interchanges $i \pm 1$, t induces the reflection on T with center i. $\square$

9.4. The exchange condition. Now we may introduce our key condition, the exchange condition.

DEFINITION 9.6.

(1) *T satisfies the* exchange condition *at i if for all nontrivial $x \in R_i^+$ there is $y \in R_i^-$ so that $(i-r)^{xy} = i+r$.*

(2) *T satisfies the* exchange condition *if it satisfies the exchange condition at all i in T (or in other words, at 0 and 1).*

The next three lemmas give criteria for the exchange condition to hold. The third of these is an important special case of the second.

LEMMA 9.7. *Suppose that $O_2(R_i) = 1$. Then T satisfies the exchange condition at i.*

PROOF. Let $x \in R_i^+$, $x \neq 1$. Then as $R_i^+ \cap G_{i-1} = 1$, x moves $i-1$. Now R_i^- acts transitively on $\Delta(i)\backslash\{i-1\}$, so there is $y \in R_i^-$ with $(i-1)^{xy} = i+1$. Thus $(R_i^-)^{xy} \le R_i \cap G_{i+1}$. Since $O_2(R_i) = 1$, we have $R_i \cap G_{i+1} = R_i^+ K$, so $(R_i^-)^{xy} = R_i^+$.

Now with t_i as in Lemma 9.5, we have $(R_i^-)^{xyt_i} = R_i^-$ and in view of the structure of R_i we have $xyt_i \in R_i^- K$, so $(i-r)^{xyt_i} = i-r$, and $(i-r)^{xy} = (i-r)^{t_i} = i+r$. $\square$

LEMMA 9.8. *Suppose that for any $x \in R_i^+$, $x \neq 1$, there is $y \in R_i^-$ so that*

$$\langle x, y\rangle \simeq \mathrm{SL}_2(2) \text{ and } C_{O_2(R_i)}(\langle x, y\rangle) = 1 \ .$$

Then T satisfies the exchange condition at i.

PROOF. Set $L = \langle x, y\rangle$. Let $t = y^x$. Then $x^t = y$, $y^t = x$. Working in $\bar{R}_i = R_i/O_2(R_i)$, we find that as t switches x and y, $\bar{t}$ must interchange $\overline{R_i^{\pm}}$. With t_i chosen as in Lemma 9.5 it follows that $\overline{tt_i}$ normalizes both $\bar{R}_i^-$ and $\bar{R}_i^+$, so $\overline{tt_i} \in \bar{K}_i$ and

$$tt_i \in K_i O_2(R_i);$$

choose $k \in K_i$ so that $tt_i k \in O_2(R_i)$.

Let $u = tt_i k$. Then $x^u = y^{t_i k} \in R_i^+$ since t_i reflects R_i^- to R_i^+. Thus $[x, u] \in R_i^+ \cap Q_i = 1$, and similarly $[y, u] = 1$. So by hypothesis $u = 1$, that is,

$$t \in t_i K O_2(G_T) \ .$$

Then $(i-r)^{xy} = (i-r)^{tx} = i+r$, and T satisfies the exchange condition at i. $\square$

LEMMA 9.9. *If $O_2(R_i)$ is a natural module for $R_i/O_2(R_i)$, then T satisfies the exchange condition at i.*

PROOF. For $x \in R_i^+$ we take $y \in R_i^-$ so that in $\bar{R}_i = R_i/O_2(R_i)$, we have $\overline{\langle x,y\rangle} \simeq \mathrm{SL}_2(2)$. As $\langle x,y\rangle$ is a dihedral group and the elements of order 3 act fixed point freely on $O_2(R_i)$, we have $\langle x,y\rangle \cap O_2(R_i) = 1$, $\langle x,y\rangle \simeq \mathrm{SL}_2(2)$, and $C_{O_2(R_i)}(\langle x,y\rangle) = 1$. Thus Lemma 9.8 applies in this case. □

LEMMA 9.10. *If $r = 3$, then T satisfies the exchange condition.*

PROOF. In this case Q_i is a natural module for $\bar{L}_i$ (Lemma 4.4) and hence $O_2(R_i)$ is either trivial or a natural module for $\bar{R}_i$, for each i. □

We will show in the following subsections that Lemmas 9.7 and 9.8 cover the other two cases, $r = 4$ and $r = 6$, as well.

9.5. The case $r = 4$. We introduce some general notation that will be of use in both of the remaining cases.

NOTATION 9.11. *$G_i^{(j)}$ is the pointwise stabilizer in G of $\{\delta : d(i,\delta) \le j\}$. In particular, $G_i^{(1)}$ is the pointwise stabilizer of $\Delta(i)$, namely $Q_iK_{\Delta(i)}$.*

LEMMA 9.12.
(a) $C_K(R_i^-) = K \cap G_i^{(1)}$.
(b) $[R_i, K \cap G_i^{(1)}] = 1$.
(c) $O_2(R_i) \le G_i^{(2)}$.

PROOF. Let $\bar{G}_i = G_i/Q_i$. As $\bar{R}_i^-$ is a Sylow 2-subgroup of $\bar{G}_i$, we have $C_K(R_i^-) \le C_K(\bar{R}_i^-) = C_K(\bar{G}_i) \le G_i^{(1)}$. Conversely, $[R_i^-, K \cap G_i^{(1)}] \le R_i^- \cap G_i^{(1)} = 1$, so (a) follows.

From (a) and the analog for R_i^+, (b) follows.

For (c), suppose toward a contradiction that $O_2(R_i) \not\le G_i^{(2)}$. As R_i is transitive on $\Delta(i)$, we have $O_2(R_i) \not\le Q_{i+1}$, so $O_2(R_i)$ covers a Sylow 2-subgroup of $\bar{L}_{i+1}$. As $K \cap G_i^{(1)}$ centralizes $O_2(R_i)$, it follows that $K \cap G_i^{(1)}$ centralizes $\bar{L}_{i+1}$, and $C_{L_{i+1}}(K \cap G_i^{(1)})$ covers $\bar{L}_{i+1}$. So in view of (b), $N(K \cap G_i^{(1)})$ acts transitively on both $\Delta(i)$ and $\Delta(i+1)$, contradicting Lemma 1.7. □

Now we turn to the case $r = 4$.

LEMMA 9.13. *If $r = 4$ and $x \in (R_i^-)^\times$, then $x^{L_i} \cap xZ(L_i) = \{x\}$.*

PROOF. In other words, we claim that $C_{L_i}(x \bmod Z(L_i)) = C_{L_i}(x)$. As R_i^- is elementary abelian, it centralizes x. Furthermore, $C_{L_i}(x \bmod Q_i) = R_i^-Q_i$, so it suffices to show that $C_{Q_i}(x \bmod Z(L_i))$ centralizes x.

Recall that in the case $r = 4$, we have $Q_i = Z_i$ elementary abelian, and $Q_i/Z(L_i)$ is a standard module (p. 489, II). Accordingly, $C_{Q_i}(x \bmod Z(L_i)$ is $[x, Q_i]Z(L_i)$ and it suffices to check that x commutes with $[x, Q_i]$. As $x^2 = 1$, the element x inverts $[x, Q_i]$, and as Q_i is elementary abelian, our claim follows. □

LEMMA 9.14. *If $r = 4$, then $O_2(R_i) = 1$ for all i, and in particular T satisfies the exchange condition.*

PROOF. By Lemma 9.12, $O_2(R_i) \leq O_2(G_i^{(2)})$, which in the case at hand is $Z(L_i)$. Thus R_i is a central extension, hence splits as $R_i \simeq \mathrm{SL}_2(F_i) \times Z(R_i)$ (we are in characteristic two here, though the argument is easily adapted to arbitrary characteristic, taking $Z^\circ(R_i)$ in place of $Z(R_i)$).

However, $[K_i, R_i^\pm] = R_i^\pm$ and hence $[K_i, R_i] = R_i$, forcing $Z(R_i) = 1$. So $O_2(R_i)$ is trivial. □

9.6. The case $r = 6$.

LEMMA 9.15. *Suppose that $r = 6$ and $j = i - 3$. Then $[K_j, R_i] = 1$.*

PROOF. We show first that

$$(*) \qquad [K_j, R_i^-] = 1\ .$$

Here j is to be thought of as the middle vertex of a path of length 6.

If j is odd, then easily R_i^- is $Z_j = Z^\circ(L_j)$, by a rank computation, and in this case our claim is obvious.

If j is even, say $j = 0$, we go back into the analysis as carried out in Lemma 8.3; however as we have switched the labels 0 and 1 in the meantime, we will quote this material using our current conventions; cf. §9.1. Thus we find

$$Q_{-1} \cap Q_1 = C_{Q_0}(K_0) \times Z_0$$

with Z_0 a natural module.

Now R_3^- is K_0-invariant and disjoint from Z_0, so it follows that $R_3^- \leq C_{Q_0}(K_0)$, as claimed.

So $(*)$ is proved, and similarly $[K_j, R_i^+] = 1$, so finally $[K_j, R_i] = 1$. □

LEMMA 9.16. *Suppose that $r = 6$. Then for each i, one of the following holds:*

(a) $O_2(R_i) = 1$.
(b) $O_2(R_i)$ *is a natural module.*

In particular, T satisfies the exchange condition.

PROOF. We assume $O_2(R_i) \neq 1$.

From Lemma 9.12 we get $[R_i, K \cap G_i^{(1)}] = 1$ and $O_2(R_i) \leq G_i^{(2)}$.

Let $j = i - 3$. As $[K_j, R_i] = 1$ while $R_j^+ Q_j$ is a Sylow 2-subgroup of L_j, R_j^+ is not contained in $O_2(R_i)$.

If i is odd and j even, by inspection and a rank calculation we have $R_j^+ = Z_i$; from the structure of Q_i (p. 489, III), if $Z_i \not\leq O_2(R_i)$, then $O_2(R_i) = 1$. So we suppose that

$$i \text{ is even.}$$

From the structure in this case, $G_i^{(2)} = R_j^+ Z_i$, so $O_2(R_i) \leq R_j^+ Z_i$. By Lemma 9.15, switching the roles of the indices, $[K_i, R_j^+] = 1$. As $O_2(R_i)$

is K-invariant and does not contain R_j^+, $O_2(R_i)$ is disjoint from R_j^+. So if $O_2(R_i)$ is nontrivial, then it meets Z_i. As $O_2(R_i)$ is R_i-invariant, it then contains Z_i, so $O_2(R_i) = Z_i$, which is a natural module. □

LEMMA 9.17. *T satisfies the exchange condition.*

PROOF. Lemmas 9.10, 9.14, 9.16. □

9.7. The Moufang property. Now that we have verified the uniqueness and exchange conditions in Γ, the rest of the analysis leading to a generalized n-gon (with $n = r$) is purely algebraic. This was described in Proposition III 6.10. There is some slight variation in notation here; we followed the notation of [**84**] closely in §III 6. In particular, our torus K is the "Cartan" subgroup referred to there, where a broad definition was given which also works well in extreme finite cases. We also phrased the uniqueness condition more concretely above; the two versions are equivalent, and in any case the form we have actually verified is the stronger of the two, formally speaking. So the terminology is consistent at this point and we are free to apply Proposition III 6.10.

Now the pattern of parabolic subgroups (G_0, G_1, B) with which we began is associated with both the tree Γ and its quotient $\tilde{\Gamma}$, as according to Proposition III 6.10 the induced homomorphism $G \to G/G_{\tilde{\Gamma}}$ induces an isomorphism between the associated triples.

So at this stage G is known to be *parabolic isomorphic* to some group with a (B, N)-pair of rank 2 with our B and N, where $B = KS$ is split. This, however, is not enough; we will need the Moufang property. In the case of generalized n-gons, that is buildings of Tits rank 2, this may be defined as follows.

DEFINITION 9.18. *The generalized n-gon Γ is* Moufang *if for each path $\tilde{\gamma} = (0, 1, \ldots, n-1)$ of length $n - 1$, the set of automorphisms fixing $\Delta(i)$ for all vertices i of $\tilde{\gamma} \setminus \{0\}$ acts transitively on $\Delta(0) \setminus \{1\}$.*

LEMMA 9.19. *The generalized r-gons constructed from the trees Γ considered above are Moufang.*

PROOF. It suffices to consider a path $(0, 1, \ldots, r-1)$ of length $r - 1$ in the tree Γ, and the group $Q_1 \cap \ldots \cap Q_{r-1}$. Extend this path to a path $\tilde{\gamma} = (0, 1, \ldots, r)$ of length r. Then $O_2(G_{\tilde{\gamma}})$ acts transitively on the neighbors at both ends, and is contained in $Q_1 \cap \ldots Q_{r-1}$. □

With this, we are in range of the final identification of the group G_0 with which we began, in the statement of Proposition 1.4.

10. Identification

We have all the ingredients for a proof of Proposition 1.4, and it is now time to assemble them.

Proof of Proposition 1.4. We began with a centerless group G_0 generated by certain subgroups G_1, G_2 with normal subgroups L_1, L_2 and intersection B satisfying various conditions appropriate for minimal parabolic subgroups in a group of Lie rank 2.

From these data, we constructed a graph Γ_0 whose universal cover Γ was the subject of an extended analysis. In §III 6 we saw that Γ has a quotient $\tilde{\Gamma}$ on which a group G^* of finite Morley rank acts, with $\tilde{\Gamma}$ a Moufang generalized r-gon, for some r (in fact, as expected, $r = 3$, 4, or 6). Furthermore, the pattern of (so-called) parabolic subgroups (G_1, G_2, B) with common "Borel" (where the term "Borel" is used loosely) is known to be isomorphic to the corresponding data in G^*, involving the vertex stabilizers in Γ^*. We summarize this by the expression: G_0 and G^* are *parabolic isomorphic*.

Now by Proposition III 6.3, the group G^* is a Chevalley group, $\mathrm{SL}_3(F)$, $\mathrm{PSp}_4(F)$, or $\mathrm{G}_2(F)$, with F an algebraically closed field (of characteristic 2, in our context), corresponding respectively to the cases $r = 3$, $r = 4$, and $r = 6$. So G_0 has the same parabolic structure as one of these groups. □

At the end of the last proof, note that we apply the classification of Moufang n-gons theorem in one of three specific cases, corresponding to $n = 3, 4, 6$ respectively, and that in the difficult case $n = 4$ we have a good deal more information that could be used to cut down the analysis considerably, notably $f_0 = f_1$, which radically restricts the possibilities. In fact, the rest of the analysis amounts to locating the various root groups within a single Sylow 2-subgroup, and working out the (Chevalley) commutator formula there.

As shown at the outset, Proposition 1.1 reduces to Proposition 1.4.

Now we return to the main result, Theorem **QT**.

Proposition 1.1 applies to the quasithin group G_0 and shows that it is parabolic isomorphic with a Chevalley group G^*.

We wish to derive an isomorphism of G_0 with G^*. By Fact II 2.28, we need to examine the triple (G_1, G_2, N) with $N = N(K)$ (viewed as an amalgam of three groups, with specified intersections), showing that the embedding of (G_1, G_2) into G^* extends coherently to N. Our strategy is to show that $C_{G_0}(K) = K$, and that this is sufficient. These are not abstract results: they involve the parabolic isomorphism that we already have in hand, which we apply freely.

Lemma 10.1. *If G_0 is an L^*-group of finite Morley rank and even type which is parabolic isomorphic with G^*, then $C_{G_0}(K) = KZ(G_0)$.*

Proof. Each parabolic subgroup G_i of G_0 has a Levi decomposition of the form $Q_i \rtimes (\hat{L}_i \times \hat{K}_i)$, with $\hat{L}_i \simeq \mathrm{SL}_2(F)$, where F is the base field of G^*. More exactly, $\hat{L}_i \simeq \mathrm{SL}_2(F_i)$ with F_1 and F_2 definably isomorphic, but this amounts to the same thing.

Take $a \in K$ of order greater than 3. By Corollary II 1.16, $C_{G^*}(a)$ is either a torus, or the product of a torus with $\mathrm{SL}_2(F)$.

In particular, the rank of $C_S(a)$ is at most $f = \mathrm{rk}(F)$ for any such element a. Accordingly, the same applies to $C_Q(a)$ for any Sylow° 2-subgroup Q of G_0, and any a normalizing Q of order greater than 3. Let U be a Sylow° 2-subgroup of $C_{G_0}(K_i)$ ($i = 1$ or 2). It follows that $\mathrm{rk}(U) \leq f$. As $\mathrm{rk}(S \cap \hat{L}_i) = f$, we conclude that $S \cap \hat{L}_i$ is a Sylow° 2-subgroup of $C_{G_0}{}^\circ(K_i)$.

Let $U_i = S \cap \hat{L}_i$. Then we have

$$(*) \qquad U_i \leq \hat{L}_i \leq C_{G_0}{}^\circ(K_i)$$

and $C_{G_0}{}^\circ(K_i)$ is a connected L-group, with U_i as a Sylow 2-subgroup.

By $(*)$ we have $O_2(C_{G_0}{}^\circ(K_i)) = 1$, and by Lemma II 6.10 it follows that $C_{G_0}{}^\circ(K_i) = E * O$ with $E = U_2(C_{G_0}{}^\circ(K_i)) = E(U_2(C_{G_0}{}^\circ(K_i)))$ and $O = \hat{O}(C_{G_0}{}^\circ(K_i))$ of degenerate type. Here E is a central product of quasisimple algebraic groups, U_i is a Sylow 2-subgroup of E, and $U_i \leq \hat{L}_i \leq E$. It is then easy to see that $\hat{L}_i = E$. As a result, $\hat{L}_i$ is normalized by $C_{G_0}(K_i)$ for $i = 1, 2$ and hence:

Both L_1 and L_2 are normalized by $C_{G_0}(K)$.

The groups $\hat{L}_i \simeq \mathrm{SL}_2(F)$, $i = 1, 2$, do not allow definable groups of outer automorphisms by Corollary II 2.26. Hence $C_{G_0}(K)$ must act on $\hat{L}_i$ via inner automorphisms commuting with $K \cap \hat{L}_i$ and hence $C_{G_0}(K) = (K \cap \hat{L}_i) \times C_{G_0}(K\hat{L}_i)$. Let $H_i = C_{G_0}(K\hat{L}_i)$. Since $(K \cap L_1)(K \cap L_2) \leq K$, it follows that $C_{G_0}(K) = K(H_1 \cap H_2)$.

Now $H = H_1 \cap H_2$ centralizes $\langle U_1, U_2 \rangle = S$ and H centralizes each $\hat{L}_i$, hence also each P_i, hence G_0. So $H \leq Z(G_0)$. □

In our final argument we return to the notation of Theorem **QT**, so the group G_0 of finite Morley rank with which we have been working throughout this chapter now once more becomes a simple group, and is denoted G.

Proof of Theorem **QT**. The group G is assumed to be simple and quasithin of finite Morley rank, and by the analysis so far is parabolic isomorphic to a rank 2 Chevalley group G^*, with $C(K) = K$ since G is centerless. By Fact II 2.28, it suffices to check now that $N(K)$ is the same in both groups (identifying K with its image under the parabolic isomorphism).

We have $G_i = O_2(G_i) \rtimes (L_i \times K_i)$ with $L_i \simeq \mathrm{SL}_2(F)$ and $K = (K \cap L_i)K_i$. Let $w_i \in L_i$ be an involution inverting $K \cap L_i$ and let $W = \langle w_1, w_2 \rangle$, $a = w_1 w_2$. Evidently the structure of G_1 and G_2 determines the map $W \to \mathrm{Aut}(K)$, so as G and G^* are parabolic isomorphic, W acts on K like the dihedral group D_r of order $2r$. In particular, $a^r \in C_G(K) = K$, and a is inverted by both w_1 and w_2. It follows that $a^r = 1$. Thus $KW \simeq N_{G^*}(K)$.

Now G^* is the universal amalgam of $(G_1^*, G_2^*, N_{G^*}(K))$ relative to their intersections. Hence the subgroup of G generated by (G_1, G_2, KW) is isomorphic to G^*. But this subgroup is G, since it is already generated by G_1 and G_2. □

11. Notes

The amalgam method is laid out in detail in [**100**]. In the form we need, the paradigms are [**174**] and [**84**], particularly the former, which we follow closely through much of the argument, though for the construction of the generalized n-gon, and some earlier points, we follow [**84**].

For the final recognition phase, the theorem of Tits used is taken from [**32**], where an elegant proof is given.

This chapter follows the finite case closely. The amalgam method is largely a method of abstract group theory, though issues of finiteness or finite Morley rank intervene eventually in the recognition phase, and affect the analysis along the way in various minor ways.

The paradigm of [**84**] is very attractive, and we explored it first, but we encountered some technical difficulties in carrying that approach through, in the case of one particular configuration. We would still like to see a full treatment along those lines, as it would probably be a little more transparent than the one we give. We do come back from [**174**] to [**84**] at the end, for the construction of the associated Moufang polygon.

§2. **Borel subgroups**

Lemma 2.7 is 3.1 in [**174**].

Lemma 2.13 is a fundamental lemma of Goldschmidt.

Lemma 2.14 corresponds to Lemma 6.5 of [**100**].

§3. **Borel subgroups**

Lemma 3.3 corresponds to Lemma 1.11 of [**174**]. The proof follows the method of [**27**].

§§4-8. **Borel subgroups**

Proposition 4.1 was given in [**174**] as a variation on a result given in [**191**].

Throughout §§4-8 we are following [**174**] closely, with some excursions into [**84**] in §8. The line of argument in [**84**] is more transparent, but we encountered some difficulties in the adaptation of one argument there. The main difference in the two approaches relates to the class of paths considered in the definition of the parameters r and s. Following [**174**] we consider paths in T; in [**84**] one considers arbitrary paths. In consequence, we need the arguments of §§5 and 6, which reproduce a line of argument in one section of [**174**].

In §9 we follow the method of Delgado and Stellmacher [**84**], especially §14. Once one verifies their uniqueness and exchange conditions, the rest of the argument is formal, though we also need to retain control over issues of definability, not for the construction itself, but in order to be able to exploit the result subsequently.

CHAPTER X

Conclusion

We also know there are known unknowns; that is to say we know there are some things we do not know.
— *D. Rumsfeld, 2002*

Introduction

At this point we have proved the following.

MAIN THEOREM. *Let G be a simple group of finite Morley rank. Then G satisfies one of the following two conditions.*

(1) *G is an algebraic group over an algebraically closed field of characteristic 2.*
(2) *G has finite 2-rank.*

A minimal counterexample to this result would be a simple L^*-group of finite Morley rank, of even or mixed type, and not algebraic. In Chapter V we showed that such a group could not be of mixed type. In succeeding chapters we discussed the possibility of even type, and after considerable preparation we showed in §§VIII 6.5–VIII 6.7 that such a group is neither *thin* nor *generic* in the sense of §VIII 6, and hence is *quasithin*. The identification theorem in the generic case used a version of Niles' theorem, followed by an appeal to the full classification theorem of buildings of spherical type and Tits rank at least 3, or the alternate method of §VIII 6.4. In Chapter IX we discussed the quasithin case, using the amalgam method and following closely the methods of Delgado and Stellmacher, followed by an application of the classification of Moufang buildings in Tits rank 2 and an identification theorem of Tits (Fact II 2.28).

Much of this long proof consists of various characterizations of SL_2, culminating in the $C(G,T)$-theorem, Theorem VIII 3.3. Each of these characterizations reduces to an earlier one, going back ultimately to the classification of split Zassenhaus groups with a suitably placed involution, Theorem III 2.2, from which generators and relations for SL_2 are recovered.

As we have remarked, the two classification theorems used from the theory of buildings are massive pieces of work in their own right. In a sense, these theorems simply pick up the problem from the point to which we have reduced it, which from some points of view could be considered the half-way mark. Unfortunately, in passing to these classification results we lose all the information we have accumulated along the way, and this information is reconstructed *ab initio* in the classification of the relevant buildings. So there is a certain inefficiency in proceeding this way. On the other hand, this approach shows that our model theoretic hypotheses become largely irrelevant at this stage, and the remainder of the analysis is purely algebraic.

We noted at the end of Chapter VIII that the more onerous classification theorem, in Tits rank at least 3, can be sidestepped with just a little more work. In the case of Tits rank 2, the classification of Moufang polygons is just the logical continuation of the amalgam analysis, and is very natural. The underlying idea is that we have enough information to determine the analog of the Chevalley commutator formula in all cases, and that this (with just a little more information) determines the ambient group uniquely (if it exists, which is a separate question!). This classification can be shortened considerably by invoking our model theoretic hypotheses, as only a fragment of that classification is actually relevant to our context. But even this fragment remains a very substantial body of work.

The study of 2-local structure in groups of finite Morley rank divides naturally into four cases: mixed, even, odd, and degenerate type. We have concentrated here on mixed and even types for two related reasons: (1) the results are complete, and (2) the state of knowledge appears to be relatively stable. In degenerate type, Theorem IV 4.1 disposes of the question of 2-local structure. So it is tempting to aim at a complete proof of the Algebraicity Conjecture for groups containing involutions, something which did not seem reasonable when this project began. In fact, this project still does not seem reasonable, in view of the difficulties arising in the analysis of groups of odd type, particularly in the presence of nonsolvable degenerate sections, but even in some configurations of low 2-rank without nonsolvable degenerate sections.

In spite of these difficulties there are now very substantial results on groups of odd type. We will discuss some of these matters in §1; much of this part of the story was not known when we began work on this text, and no doubt some details of our account will rapidly become out of date.

In our account we have also given some rudiments of a general theory of groups of finite Morley rank, notably Proposition IV 1.15. We will show in §2 that our explicit structure theory in even and mixed type can be combined with this more rudimentary general structure theory to solve a general problem in the theory of permutation groups of finite Morley rank, an area which is now ripe for further exploration.

In §5 we will discuss a variety of open problems, some related to unfinished work in odd type and others relating to unexplored directions suggested by a comparison of our work with other themes in the theory of finite simple groups.

We had hoped at the outset that the theory presented here, and the companion theory in odd type, would cast some light, at least by analogy, on the structure of the classification theorem for finite simple groups. Our conclusions at this point are mixed. We have arrived at a proof whose strategy is certainly different from the one we envisioned initially, and one which we feel is less "canonical" than we anticipated. We will consider this point in §3. In particular, our theory has more in common with the so-called *third generation* approach, as well as ideas of Timmesfeld, than we had expected. As a result a number of lines which had been developed for use in this project turned out to be dispensable—notably the Alperin-Goldschmidt fusion theorem, most of the theory of standard components, and the full classification of groups with strongly closed abelian 2-subgroups. In our initial project this would have been followed by very extensive developments in the same direction, parallel to the case of finite simple groups, but this line was abandoned as the availability of the amalgam method became clear.

On reflection, we have the impression that the proof we give for our classification theorem is parallel to a proof we have never actually seen, corresponding perhaps to some self-contained theory contained within the theory of finite simple groups, one that was bypassed on the way to results of greater generality. One possible form for such a theory will be indicated in §4.2; if that theory can in fact be developed at the final level in a self-contained way, then it would be a faithful model of the theory we give here (or conversely, the theory we give here could be read as a faithful model of it).

We observe also that the two existing proofs (one still in progress) of the classification of the finite simple groups are resolutely "semisimple" in their approach, in the sense that one seeks to identify the groups ultimately via centralizers of elements of prime order distinct from the characteristic, while the "third generation" or "amalgam" method is resolutely unipotent in its approach, as is the classification of the Moufang polygons. But in studying simple groups of finite Morley rank, we have worked systematically with involutions, letting the characteristic determine the methods used, from the start: in even type, as we have seen, we use unipotent methods. In odd type there is another body of work which uses semisimple methods. Note, however, that at the end of Chapter VIII, our detour via the generic identification theorem of §III 10 represents a very brief excursion into semisimple methods as the concluding phase of the analysis, and follows the usual identification scheme of finite simple group theory exactly. But here this approach is reduced to the status of a "punch line".

Consequences. There are a number of other ways of viewing our main result. While we focus our attention on simple groups, our results have consequences for the general structure theory of groups of finite Morley rank, beginning with the following reformulation.

THEOREM 1. *Every group of finite Morley rank is an L-group.*

This says no more or less than the Main Theorem. But it does make some of its consequences more transparent, since we have included a section on *L*-groups in Chapter II. Now everything proved in that chapter becomes a fact about groups of finite Morley rank. Some of these facts are worth quoting again at this point.

The first of these is Lemma II 6.3, which takes on the following form.

PROPOSITION 1. *Let G be a group of finite Morley rank. Then $U_2(G)$ is a K-group of even type, and $G/U_2(G)$ has odd or degenerate type.*

The next is Lemma II 6.10.

PROPOSITION 2. *Let G be a connected group of finite Morley rank containing no nontrivial 2-torus. Then $O_2(G) = O_2{}^\circ(G)$ and it is a definable unipotent subgroup of G. The quotient $G/O_2{}^\circ(G)$ has the form $E * D$ with E a central product of quasisimple algebraic groups over algebraically closed fields of characteristic 2, and D a connected group without involutions.*

PROOF. Evidently $O_2{}^\circ(G)$ is unipotent and we may factor it out, assuming therefore that G is reductive. By hypothesis G is either of even or of degenerate type, and in the degenerate case G contains no involutions (Theorem IV 4.1). So we suppose G has even type. So by Lemma II 6.10 we have $G = E(U_2(G)) * \hat{O}(G)$ and this is of the desired form.

Note finally that $O_2(G) = 1$ at this point; this holds for $E(U_2(G))$ since the characteristic in each factor is 2, and in $\hat{O}(G)$ since there are no involutions present. □

Of course, this includes Theorem IV 4.11, but for this we have already given a direct proof. For groups of odd and degenerate type, the structure theory is not so well developed, but as we saw in Chapter IV one can sometimes make good use of the conjugacy theorem for maximal decent tori, Proposition IV 1.15.

We conclude with one more variation on the same theme, at a greater level of generality.

PROPOSITION 3. *Let G be a connected group of finite Morley rank and set $\bar{G} = G/U$ where U is the largest normal unipotent 2-subgroup. Then*

$$\bar{G} = E(\bar{G}) * \bar{H}$$

a central product with finite intersection, where $E(\bar{G})$ is a central product of quasisimple algebraic groups over algebraically closed fields in characteristic 2, and $\bar{H}$ is a connected group of finite Morley rank containing no nontrivial 2-unipotent subgroup.

PROOF. We may suppose $U = 1$ and omit the bars from our notation.

Then $U_2(G)$ is a K-group of even type with trivial $O_2{}^\circ$, hence $U_2(G) = E(G)$ is a central product as described. Then Corollary II 2.26 yields

$$G = E(G) * H$$

with $H = C_G(E(G))$. It follows that $H \cap E(G)$ is finite, and as $E(G) = U_2(G)$ it follows that H contains no nontrivial 2-unipotent subgroup. □

In the notation of Proposition 3, it is less clear what the preimage H of $\bar{H}$ looks like, as this may involve a central extension of a group of odd or degenerate type by a unipotent 2-group.

1. Odd type

We have shown that nonalgebraic simple groups of finite Morley rank have finite 2-rank. In other words, they are of *odd* or *degenerate* type: odd type if Sylow 2-subgroup is a finite extension of a nontrivial 2-torus, and degenerate type if a Sylow 2-subgroup is finite, in which case it is trivial. If the Algebraicity Conjecture holds, then in odd type we have Chevalley groups over fields of characteristic other than 2 (possibly 0), and in degenerate type there should be none; this last point remains decidedly obscure.

A large body of work gives considerable information about nonalgebraic simple K^*-groups of odd type, and in particular bounds the Prüfer rank, as we shall see below. One would like to bound the 2-rank absolutely, at least, and indeed go considerably beyond that. As we have had occasion to note already, the tools used in even and mixed type are very different from those used in the treatment of odd and degenerate type groups (while we have given the treatment of degenerate type groups, we have not given the more general theory which motivated some of the arguments we used). In addition, the odd type analysis is restricted to K^*-groups at present, and is not complete as yet even under that restrictive hypothesis. And since it would take another volume to present that material in detail in any case, we confine ourselves to an indication of the present state of affairs.

1.1. Odd type.

THEOREM 1.1 ([**60, 46, 48, 63, 66**]). *Let G be a simple K^*-group of finite Morley rank and odd type. Then either G is algebraic, or G has Prüfer rank at most 2.*

The proof goes via the generic identification theorem III 10.2, or rather a close analog. As we are working with K^*-groups, and in odd characteristic, we can work with the subgroups $C^\circ(a)$ where a varies over a suitable class of involutions. The main tool is signalizer functor theory, which tends to work well in Prüfer 2-rank (or normal 2-rank) at least 3, and the main problem is getting suitable signalizer functors. The difficulty is the following. The natural goal of signalizer functor theory would be to prove $O(C(i)) = 1$ for all involutions i, a strong form of reductivity of centralizers. However, this is completely unreasonable: in the presence of bad fields, a torus T can easily have $O(T)$ nontrivial, and since tori occur as components of centralizers in the algebraic case, the condition $O(C(i)) = 1$ can fail *even in algebraic groups*, if the underlying field carries additional structure. There are, however, two ways of getting around this problem. On the one hand, we only want to prove that $C^\circ(i) = F^*(C(i))$, and nontrivial $O(C(i))$ is not necessarily an obstacle to this; working with modified signalizer functors one can kill off enough of $O(C(i))$ to prove the desired result. Alternatively, one can get by with less: for the proof of the generic identification theorem it is enough to have $E(C^\circ(i))$ reasonably large (large enough that the associated set Σ of "root SL_2" subgroups generates G, in particular). For this one can exploit a fairly trivial observation: the rank of a proper subgroup of the multiplicative group of a field is less than the rank of the additive group. This turns out to be quite useful when one has fields of characteristic zero involved (and fields of positive characteristic can be handled more simply). The signalizer functor theory of [**61**] exploits this observation at length, and one can use this idea to kill off "large" parts of $C(i)$, and to show that "small" parts are reasonably harmless.

What the signalizer functor theory delivers (under the hypothesis that the Prüfer rank is sufficiently large) is a *proper* 2*-generated core*. This is a "uniqueness" condition which lies between strong and weak embedding, and can in fact be strengthened to strong embedding; this is done in [**48**], and along the way it is shown that the groups in question are minimal connected simple.

In odd type groups the Sylow 2-subgroups have fewer involutions than in even or mixed type groups, and as a result the force of the strong embedding condition is weaker. Still, in the minimal connected simple context, in odd type, strong embedding leads to the desired bound on the Prüfer rank by a style of argument specific to minimal connected simple groups, making extensive use of the "characteristic zero" unipotence theory introduced in [**61, 60**], one of the few methods that also seems to lend itself to further exploitation in the degenerate case.

This brief sketch is quite schematic, but we hope it can be seen here that the analysis is relatively direct, and follows general principles; furthermore, its structure is more straightforward than the analysis in even type, with its elaborate sequence of preliminary "uniqueness type" characterizations of SL_2.

At the present writing (Summer 2007), this analysis has gone about as far as general methods allow, and what seems to be needed is the close, and possibly lengthy, analysis of the various configurations that arise in low Prüfer rank. An explicit program for carrying this out has been given in [**60**, Chapter 11], in the form of a dozen well-defined, and independent problems, some of them notoriously difficult, some of them undoubtedly well within reach of known methods, and all meriting further investigation. Among recent progress on these problems, we cite [**85, 86**], dealing with the minimal simple case, treated first under a tameness hypothesis in [**70**]; the elimination of this hypothesis involves some very recent technical developments, which take us increasingly away from the usual techniques of finite simple group theory, and into more geometrical, or model theoretic, territory.

1.2. Degenerate type. According to Theorem IV 4.1 there are no involutions in this case. While this result is predicted by the Algebraicity Conjecture, it was not generally considered to be within reach, and indeed was found only after this text was largely written. But it will not end our interest in degenerate type groups.

Apart from this result, the most striking results in degenerate type are among the earliest in the study of simple groups of finite Morley rank, and concern so-called *bad groups*.

DEFINITION 1.2. *A* bad group *is a simple group of finite Morley rank in which every proper definable connected subgroup is nilpotent.*

We observe that an equivalent definition is the following: a bad group is a minimal connected simple group of finite Morley rank in which Borel subgroups coincide with Carter subgroups. We mention this because we are not at all sure we understand properly what the notion of "bad group" really should be, and we believe that this is a topic meriting further consideration.

Bad groups are not assumed *a priori* to be of degenerate type, but this can be proved, and in fact the following striking result goes back to the beginning of the subject.

FACT 1.3 ([**52**]). *Let G be a bad group of finite Morley rank. Then the following hold:*

(1) *The Borel subgroups are conjugate, their union is G, and their pairwise intersections are trivial.*
(2) *A bad group has no nontrivial involutive automorphisms.*

These two points go in two different directions, though their proofs are intertwined. One begins with the first point. One then shows that bad groups contain no involutions, at which point one can show that the Borel subgroups are self-normalizing. After that, one can eliminate involutive outer automorphisms. The elimination of involutions from within a bad group is the most subtle point, going via the construction of an associated projective space through the geometry of involutions.

In fact, minimal connected simple degenerate type groups which satisfy the first point also satisfy the second [**121**], and apart from this seem to lend themselves to little more in the way of group theoretic analysis.

We remark that in the study of minimal connected simple groups of odd type, the study of intersections of Borel subgroups (which may be considered the *Bender method* in this context) has proved to be very powerful; the foregoing result relates to the case when this technique is unavailable.

There are a couple of fundamental issues still not resolved in the theory of degenerate type groups, which we will return to in §5, notably the Genericity Conjectures and the problem of finding the broadest useful notion of "bad group" (ideally, as part of a meaningful dichotomy).

2. Permutation groups

In this section we consider permutation groups (G, X) of finite Morley rank: here the group G is given together with its action on a set X, and the whole structure is supposed to have finite Morley rank. Usually these actions are taken to be faithful as well, but we will include this hypothesis explicitly.

In finite group theory, the classification of the finite simple groups produced a revolution in permutation group theory, and seems essential for the solution of many problems of a general character. Here, even though we have no complete classification, we can proceed in an entirely parallel fashion to get some results of considerable generality, notably the following.

THEOREM 2.1. *If (G, X) is a faithful and definably primitive permutation group of finite Morley rank, then the rank of G is bounded as a function of the rank of X. That is, there is a function*

$$\rho : \mathbb{N} \to \mathbb{N}$$

such that $\mathrm{rk}(G) \leq \rho(\mathrm{rk}(X))$ *for all such (G, X).*

Here a permutation group (G, X) is called *definably primitive* if there is no nontrivial definable G-invariant equivalence relation on X. Examples show that the hypothesis of definable primitivity is needed [**107, 49**].

A noteworthy feature of our proof is that it is very "soft," and the bounds obtained are very loose, leaving a great deal more to be done in this area.

A full account of this result is in [**49**]. Here we aim mainly to see how our structure theory may be brought to bear on the problem. But this theory only becomes relevant after a number of reductions, which we will indicate briefly.

2.1. Generic n-transitivity.

DEFINITION 2.2. *Let (G, X) be a permutation group and $t \geq 1$. Then the action of G on X is* generically t-transitive *if there is a generic orbit O for the induced action of G on the Cartesian power X^t; that is,* $\mathrm{rk}(O) = \mathrm{rk}(X^t)$.

As an example, consider the natural action of $\mathrm{GL}(n)$. This is generically n-transitive since a generic n-tuple consists of linearly independent vectors. Similarly, the natural action of $\mathrm{PGL}(n)$ is $(n+1)$-transitive.

We note that if the action of G is t-transitive then it is generically t-transitive, as the set of t-tuples with distinct entries is generic in X^t. But generic t-transitivity is a much looser notion.

NOTATION 2.3. *If (G, X) is a permutation group of finite Morley rank, let*

$$\tau(G, X) = \max(t : \textit{The action of } G \textit{ is generically } t\textit{-transitive.})$$

In practice we write $\tau(G)$ for $\tau(G, X)$.

The following result turns out to be the key to the proof of Theorem 2.1.

THEOREM 2.4. *Let (G, X) be a faithful transitive permutation group of finite Morley rank with G simple. Then there is a bound on $\tau(G)$ in terms of* $\mathrm{rk}(X)$. *That is, there is a function $\rho : \mathbb{N} \to \mathbb{N}$ so that*

$$\tau(G, X) \leq \rho(\mathrm{rk}(X))$$

for such pairs (G, X).

We claim that *Theorem* 2.4 *is a special case of Theorem* 2.1. Indeed, specializing Theorem 2.1 to the case of simple groups G, it is easy to show that the bound which applies to definably primitive actions also applies to faithful transitive actions, as definable imprimitivity would allow us to make a simple inductive argument when the group acting is simple. On the other hand, a generically t-transitive permutation group (G, X) satisfies

$$(*) \qquad \mathrm{rk}(G) \geq \tau(G)\,\mathrm{rk}(X)\ .$$

So a bound on $\mathrm{rk}(G)$ gives a bound on $\tau(G)$.

But more to the point, *Theorem* 2.1 *can be reduced to Theorem* 2.4. This takes considerable argument. First, one can show that X has Morley degree 1. This point will not be very visible in our sketch but as the reader will appreciate it does simplify matters whenever we work with generic subsets of X.

Second, one needs to find an analog of $(*)$ in the opposite direction.

FACT 2.5. *Let (G, X) be a definably primitive faithful permutation group with* $\mathrm{rk}(X) = r$. *Then*

$$(*') \qquad \mathrm{rk}(G) \leq r\tau(G) + \binom{r}{2}\ .$$

This is one of the critical results, and the one where definable primitivity is exploited.

The essential point here is that if one takes a generic point p of $X^{\tau(G)+r}$, the point stabilizer of the sequence p will be finite, and this allows a bound like $(*')$ to be obtained. For a more precise version of this statement, let G_k be the connected component of the pointwise stabilizer of $(p_1, \ldots, p_k)$ for $0 \leq k \leq r$, and let o_k be the "generic rank" of an orbit of G_k on X. That is, let o_k be chosen so that the union of all orbits in X with respect to G_k which have rank o_k forms a generic subset of X; since there are only finitely many such distinct ranks possible, there must be such a generic value o_k. Now our claim becomes the following:

$$(\dagger) \qquad \text{If } r > o_k > 0, \text{ then } o_k > o_{k+1}\ .$$

Now by unwinding the definitions one can see that $k = \tau(G)$ is the least value for which $r > o_k$—the stabilizer of $\tau(G)$ independent points no longer acts generically transitively on X—and then $(\dagger)$ pushes down o_k to 0 in less than r further steps.

We will not enlarge much on the proof of $(\dagger)$. The whole thrust of the proof is to construct a non-trivial invariant equivalence relation on X if condition (†) fails, one whose classes have rank o_k; the conditions $r > o_k > 0$ then guarantee that this is a nontrivial equivalence relation.

At this point, we can bound $\mathrm{rk}(G)$ in Theorem 2.1 in terms of $\tau(G)$. That is, we have reduced Theorem 2.1 to the variant of Theorem 2.4 in which G is not necessarily simple.

At this point, the reduction of our problem to the case of simple groups follows reasonably well-travelled paths of finite group theory, but with some interesting detours. The main line is provided in finite group theory by the O'Nan-Scott-Aschbacher theorem, giving incisive information about both the socle of a primitive permutation group, and intersection of that socle with a point stabilizer. An exact analog of this theorem in the finite Morley rank context is given by Macpherson and Pillay in [**139**].

In applying this theory there are a number of interesting points that come up, touching on many of the topics we have considered here, including such points as Wagner's theorem on fields of finite Morley rank, Proposition I 4.20, which is needed in the so-called "affine" case in which the socle A of G is abelian, and more specifically when A is an elementary abelian p-group. But we will not go into this further here.

2.2. Generically t-transitive actions of simple groups. We have sketched the reduction of Theorem 2.1 to Theorem 2.4 and we now take up the proof of the latter. It suffices at this point to bound either $\mathrm{rk}(G)$ or $\tau(G)$, and we will take whichever is most convenient at a given point.

As we deal with simple groups the relevance of our classification theory is not in doubt, though as we lack a complete classification the way forward is still unclear. The following result casts some light on the situation.

FACT 2.6 ([**49**]). *Let (G, X) be a definably primitive permutation group of finite Morley rank, T a definable divisible abelian subgroup of G, T_0 its torsion subgroup, and $O(T)$ the largest definable torsion free subgroup of T. Then* $\mathrm{rk}(T/O(T)) \leq \mathrm{rk}(X)$.

PROOF. Take a point $\alpha \in X$ generic over the torsion subgroup T_0 of T. Suppose that some torsion element $t \in T_0$ fixes α. Then t fixes a generic subset of X pointwise. Now using the definable primitivity—or just the transitivity of the action together with the fact that X has Morley degree 1—we can show $t = 1$ (a general lemma—we omit the details).

In other words, the point stabilizer T_α is torsion free and thus is contained in $O(T)$. Hence

$$\mathrm{rk}(T/O(T)) \leq \mathrm{rk}(T/T_\alpha) = \mathrm{rk}(\alpha^T) \leq \mathrm{rk}(X) .$$

□

That is, the rank of X limits the structure of T; if the torsion subgroup T_0 of T were definable, we would be tempted to say that $\mathrm{rk}(T_0) \leq \mathrm{rk}(X)$, but as this is meaningless we will have to stick with the version we have actually proved. Still, for working out the rest of the proof one would do well to suppose $\text{rk}(T) \leq \text{rk}(X)$, as the role of $O(T)$ turns out to be marginal.

In particular, if G is an algebraic group in characteristic 2, possibly with additional structure, we can look at a maximal torus T of G, and in this case $O(T) = 1$, since T is a good torus in this case. So Fact 2.6 delivers a bound on the Lie rank of G and hence on the rank of G, and in particular on the rank of G, and we conclude in this case.

This leaves us in the less well charted waters of simple groups of odd or degenerate type. We may dispose of the degenerate case at once.

LEMMA 2.7. *Let (G, X) be a generically 2-transitive permutation group. Then G contains an involution.*

PROOF. If (α, β) is a generic point of X^2, then there is $g \in G$ interchanging α and β, and it follows easily that $d(g)$ contains an involution. □

So what concerns us now is the case of simple groups G of *odd type*. If T_2 denotes a maximal 2-torus in G and $T = d(T_2)$ is its definable hull, what would be ideal at this point would be a bound on $\text{rk}(G)$ in terms of $\text{rk}(T/O(T))$, as the latter is already bounded by Fact 2.6. But this we do not have. So we will have to work harder, and assuming that $\tau(G)$ is very large we will have to force $\text{rk}(T/O(T))$ to be large as well, by a more direct argument.

The general thrust of this argument will be as follows. Just as in the proof of Lemma 2.7, if G is a generically t-transitive group for some large value of t, and $\alpha = (\alpha_1, \ldots, \alpha_t)$ is a generic point of X^t, then G induces the full action of the symmetric group Sym_t on α. Then a Frattini argument will deliver a similar action on a maximal 2-torus T_2^* of G_α, and hence also on $T^* = d(T_2^*)$. At this point we need only two small miracles to reach our target: it would be very nice if $T^* = T$, that is, T_2^* is still a maximal 2-torus of the original group. And it would also be very nice if we knew that our action of Sym_t on T^* were nontrivial!

Given these two implausible conditions, the rest of the argument is straightforward. The action of Sym_t on T^* can be used to pump up

$$\text{rk}(T^*/O(T^*))$$

above r, if t is large enough. And of course if we also have $T^* = T$ then this contradicts our prior estimate.

It remains then to be seen how to extract sufficiently good approximations to our two implausible hypotheses. In the language of Lewis Carroll, we must believe two impossible things before breakfast.

2.3. Breakfast. We recapitulate. G is a simple group of finite Morley rank acting on a set X of rank r, and T is the definable hull of a maximal 2-torus. The point $\alpha \in G^t$ is generic, and T^* is the definable hull of T_2^*, a maximal 2-torus of G_α. We would like $T = T^*$, which is obviously unreasonable.

The solution is to replace G by a connected definable subgroup H for which something of the desired sort does occur when H and H_α are compared, but with α now generic over a name for H, and with H generically t'-transitive, with t' still large. It suffices for this to consider the groups $H_k = G_{\alpha_1,\dots,\alpha_k}$ for $k \le rt'$, with $t' = \lfloor t/r \rfloor$. One may consider 2-tori $T_{2,k}^*$ and their definable hulls T_k for $k \le r$ and look at the ranks $\mathrm{rk}(T_k/O(T_k))$, all bounded by r. There will be stretches of length t' over which these ranks are constant, say from $k = k_0$ to $k = k_1$. So if we let H be $G_{k_0}{}^\circ$ and $\alpha' = (\alpha_{k_0+1}, \dots, \alpha_{k_1})$, we can replace G and α by H and α', and have a situation close to the one we began with, but having lost simplicity, and with a slightly lower value of t. However, we are now closing in on the configuration necessary for a contradiction, since our maximal 2-torus T_2 in $H_{\alpha'}$ is still maximal in H, and simplicity will play no further role.

Let us turn to the other half of our problem. The point of the sequence α', of length t', is to witness the generic t'-transitivity of H, which is now playing the role previously held by G. This will give us something like an action of $\mathrm{Sym}_{t'}$ on our torus T; or anyway, it will give us an of action H'_α on T, which among other things induces Sym_t on α'. The main technical point now is to take a 2-element in H'_α representing an involution of $\mathrm{Sym}_{t'}$, and to show that the induced action on T_2 is nontrivial. Here we use the fact that T_2 is maximal not only in $H_{\alpha'}$, but in the ambient connected group H. We can then bring into play the following result, which refines our results on groups of degenerate type.

FACT 2.8 ([**49**]). *Let G be a connected group of finite Morley rank and odd type, and let T be a maximal 2-torus of G. Then T contains all the involutions in $C(T)$.*

While this is stated for odd type only, it is really a result about connected groups of finite Morley rank with no nontrivial 2-unipotent subgroups, and it generalizes the nonexistence of involutions in groups of degenerate type. This again makes use of the "generic covering" arguments we met in Chapter IV, and it suffices to look at the corresponding part of the analysis in degenerate type to see the flavor of it. The reference for the general theorem is [**64**].

This brings our sketch to an end. The analysis makes use of our classification theorem. Indeed, the presence of nontrivial unipotent subgroups is actually an obstruction to the more abstract line of argument with which we concluded, which focusses on the behavior of 2-tori in the absence of 2-unipotent subgroups. So our classification theorem disposes of a case which would otherwise cause serious problems. With that out of the way, we can exploit general properties of decent tori. Of course, the initial reduction of our problem (Theorem 2.1) to a problem involving simple groups (Theorem 2.4) involves many more applications of the tools we have seen throughout this work, used in their own right, outside any particular classification result, and this includes such results as Wagner's Proposition I 4.20, whose usefulness is not restricted to the context of classification problems. While this result is very distant from anything encountered in finite group theory, it is part of our toolbox of general methods on much the same footing as something like Carter subgroup theory, at this point.

3. Lessons learned

Leaving aside model theoretic issues, one motivation for the present work (and companion work in odd type groups) was to extract from the methods used in the classification of finite simple groups a "skeleton" theory, relevant to the finite Morley rank case, which would allow a reading of the original theory as an elaboration dealing with a wider range of issues and technical complications, not all connected directly with sporadic groups.

3.1. Finite groups vs. groups of finite Morley rank. The original proof of the classification of the finite simple groups, extending roughly from the 1960s to the 1980s and to some extent beyond [**18**], is estimated as taking up 15,000 journal pages and about 100 papers. The "second generation" proof of Gorenstein, Lyons, and Solomon, is to appear in a series of 12 volumes (AMS, in course of publication since 1994). A "third generation" approach has emerged, in work of Meierfrankenfeld, Stellmacher, Stroth, and many others, which takes a very different line; this is not yet part of a complete approach to the problem.

The "canonical" proof at present is the second generation one, which aims both at systematizing the approach taken and profiting as far as possible from the inductive nature of the proof (knowing at the outset that the proof will in fact terminate at some point, and hence nothing is to be gained by working outside an inductive framework). This proof involves a mixture of "unipotent" and "semisimple" methods, with semisimple methods dominating the identification phase. Generally speaking we associate unipotent methods with groups of characteristic 2, and semisimple methods with groups of odd characteristic. One reason for the mixture of methods is the behavior of some small and exceptional cases, for example, the exceptional isomorphism $\mathrm{PSL}_2(\mathbb{F}_5) \simeq \mathrm{PSL}_2(\mathbb{F}_4)$ ($\simeq \mathrm{Alt}_5$).

The third generation idea is to work in a unipotent manner with elements of order p, where p is (or "will be") the characteristic.

In our context, groups of finite Morley rank, there is a clear separation of semisimple and unipotent methods. Odd type groups are handled by semisimple methods throughout and aim at the Curtis-Tits Theorem by a direct route. This requires the signalizer functor method and encounters difficulties in low 2-ranks, which will have to be approached in other ways (in some cases, by direct analysis of specific configurations). Even type is approached in a unipotent manner and can terminate at the classification of buildings, combining [**183**] with [**130**]. We can, however, jump from "unipotent" to semisimple methods at the end in order to eliminate the classification of buildings from the analysis. This makes use of a global $C(G,T)$ theorem, signalizer functors, and a p-uniqueness theorem.

However, by the law of conservation of difficulty, if a theorem has a substantial list in its conclusion, something in the proof should be hard. If so, how could the classification of buildings be swept away? In the end, we arrive at the Curtis-Tits theorem, which goes back to a description of (most) algebraic groups as amalgams of copies of SL_2. We do make use of the classification of Moufang polygons, and it seems that the general classification of spherical buildings of higher rank reduces efficiently to this case, as shown in [**185**]. The classification of Moufang polygons in turn reduces to working directly with the commutation relations among related root groups (including opposite pairs of root groups) and in a sense is the logical continuation of the amalgam method, though as we have noted, a great deal of the information given by the amalgam method is thrown away at this point, and then rederived from the relations retained, and not very easily.

From this perspective there are two key arguments: the initial characterization of SL_2 as a Zassenhaus group, providing the basis of an inductive analysis, and the use of the amalgam method to handle the Lie rank 2 case—where, incidentally, we deviate least from the original model.

3.2. The skeleton. Our reading of the situation at this point , viewing the theory of finite simple groups through the distorting lens of the theory of groups of finite Morley rank, is as follows:

There is no canonical skeleton for CFSG.

Specifically, every possible combination of pieces in the CFSG jigsaw puzzle can be mapped faithfully into the setting of groups of finite Morley rank.

The present text does not, and is not intended to, support this conclusion very well—which reflects both our desire to present a coherent (and finite) story, and our initial expectation that some well-defined skeleton would emerge naturally.

Let us illustrate this with a concrete example. At the outset we intended to construct "standard components" and identify our groups at considerable length in terms of these standard components, as in the finite case. Early drafts of [**8**] consisted of a proof of the existence of standard components (which had made a brief appearance in the context of components of type SL_2 already in [**5**]). About this time, our attention was drawn to third generation methods, and this paper eventually was converted into a very different, almost unrecognizable, treatment of parabolic subgroups, as needed for the amalgam method. In retrospect it seems like a mistake to suppress this line of development. No doubt it is also a legitimate part of the theory, even though all such results trivialize once one has any proof of the classification of the simple groups of even type. The point is that there are several proofs of the classification, making use of several different theories, some of which belong to classical group theory, and others which were only developed when they were needed as tools for the classification. Some of these theories are beginning to crop up in contexts lying entirely outside group theory; for example, some "sporadic" local subgroup configurations which provably do not lead to sporadic finite groups are known to occur in topology [**57**]. We allowed ourselves in fact only one variation in the proof, alluded to above, where one may circle around the theory of buildings.

On the other hand, a comparison of Chapter VI with the papers [**1**], [**4**], [**124**] gives another instance of such variation, particularly toward the end of the argument; here one can use the theory of solvable groups in a very classical vein in the K^*-context, but we were forced to replace all of this by a purely model theoretic result of Wagner in the L^*-context; if we already had the degenerate case under control, then the distinction would be moot and the solvable theory would provide the more direct route. The mechanics of this example are reasonably clear: we need some decent conjugacy theorems, and it does not seem to matter too much which ones we have. In other words, the problem is highly overdetermined.

Another case in which the classification theory for finite simple groups can be "miniaturized" in a self-contained way is seen in work of Paul Flavell [**88**]. In this work results of Timmesfeld play a key role, and as well shall see in §4.2 the natural finitization of our own work leads back to the same results, or close analogs.

3.3. Things not done. Or rather, things that have been done, but have not been done here; the existence of standard components was alluded to above. As noted in Chapter VII, we have suppressed the theory of simple groups with strongly closed abelian 2-subgroups in favor of the simpler theory of groups with abelian Sylow° 2-subgroups, and a rudimentary analysis of groups with a standard component of type SL_2. The case of *finite* strongly closed abelian 2-subgroups actually follows from the elimination of cores in 2-local subgroups, and in particular the proof of the Z^*-theorem requires no special apparatus, a remarkable circumstance if one thinks of the proof in the finite case. This material was treated in [**4**] and is omitted here as irrelevant to the proof of the classification theorem.

We have also passed over the Alperin-Goldschmidt theorem [**79**], though it seems that the amalgam method incorporates some of the force of the underlying argument, though not the theorem itself.

We have also been selective in our coverage of the theory of permutation groups of finite Morley rank. In particular, the fundamental O'Nan-Scott theorem has a useful analog in this context [**139**], which would merit incorporation into Chapter I alongside such matters as Schur-Zassenhaus theorems—but from that point of view, one might prefer to know the simple groups first. So in discussing the theory of permutation groups of finite Morley rank, one is discussing, for the most part, open questions, and we will come to this in §5.

3.4. Bad fields. Generally speaking, "bad fields" tend to be viewed as a model theoretic aberration. In some sense, however, all finite fields are "bad", and the finite theory has learned to cope with them.

We have also learned gradually to cope with bad fields in our context of finite Morley rank, first in even type and then in odd type. In even type, we rely on Wagner's theorem exclusively and one cannot say that one sees much connection with the finite approach (perhaps they overlap only in the Sylow theorems).

In odd type the parallels become much clearer. Signalizer functor theory is essential in both the finite and finite Morley rank cases, and is enormously complicated by bad fields. Signalizer functor aims at controlling $O(C(i))$ with i an involution. The crucial example from our point of view is the following.

Let $G = \mathrm{PSL}_3(K)$ with K an algebraically closed field of odd characteristic, and t an involution. Then

$$C_G(t) = K^\times \times \mathrm{SL}_2(K) .$$

The question is:

$$\text{Is } O(C_G(t)) = 1?$$

And the answer is: *it depends.* We are simply asking whether $O(K^\times) = 1$. If not, then $(K, O(C_G(t)))$ is a bad field.

In the classification of the finite simple groups, Gorenstein, Walter, et al. wrote hundreds of pages which *overcame bad fields.* In the context of finite Morley rank, **[60]** introduced a "rebalancing" which goes quickly around bad fields. This depends on having a robust notion of unipotence, something which tends to be missing in the finite context, in odd characteristic.

3.5. Complex reflection groups. In the semisimple approach to groups of finite Morley rank (essential in odd characteristic, and available in even characteristic), the classification of the *finite complex reflection* groups enters in. This is one point where an "external" list enters in as a key ingredient toward the final list of algebraic groups. These are involved essentially because of the possible presence of a generalized kind of bad field considered by Poizat, a nonalgebraic definable subgroup of a split torus. Otherwise, we could show directly that the Weyl groups we construct are crystallographic Coxeter groups (the relevant $\mathbb{Z}$-lattice would be encoded in the set of definable subtori of a maximal torus).

3.6. Black box groups. Black box groups are large finite simple groups presented probabilistically, and black box algorithms are randomized algorithms that work with such groups, and attempt to determine their structure. Such algorithms form an integral part of standard group theoretic software today (GAP, MAGMA). It is assumed that one can generate elements of these (large) groups randomly and independently, and in favorable cases that one can also multiply and invert these elements, or failing that at least determine some of their properties (such as cycle structure, if they are permutations). We will deal here with the case in which one can in fact multiply and invert the elements, as this runs parallel to the finite Morley rank theory.

Here it is the underlying measure that provides a parallel to the rank notion, particularly in those very common cases where we deal with "generic" elements of the group, which are practically speaking the only ones the black box algorithms see initially (nongeneric elements may certainly arise via group theoretic operations on generic ones).

The notion of "definable subset" has an analog in black box theory: these are the subsets for which one can construct elements systematically in such a way as to generate a sequence of uniformly distributed and independent elements. In particular, a definable subgroup, in this sense, is again a black box subgroup. Let us call these subgroups "constructible".

Some notions of black box group theory turn out to be directly applicable to the finite Morley rank context, as we have seen in §IV 4. This material combines very smoothly with the geometrical lines of argument of Chapter IV, and seems indeed to form a natural component of that theory.

In the finite Morley rank context, black box methods show under suitable internal hypotheses that the connectivity of the ambient group passes to centralizers of involutions; in the black box context under analogous hypotheses it shows that the uniformly distributed measure on the ambient group passes to centralizers of involutions.

Of course, centralizers of involutions are more useful in odd and degenerate type (as long as there are involutions!) than in even and mixed type.

3.7. Back to finite groups. We have already alluded to our feeling that there is nothing "canonical" in the approach taken here, though we adopted it because it seemed to be the most efficient one available. What is noticeable in retrospect is that while we take a very linear approach to our subject, and all the individual ingredients have parallels in finite group theory, the line actually taken does not bring to mind any particular line of analysis in finite group theory.

We have tried to mine the repertoire of ideas in the classification of the finite simple groups systematically, both from the first generation of papers and from later revisionism, especially, as has been seen, the Third Generation, with the proviso that wherever possible, we work with *connected subgroups*. For this reason we have adopted a relatively unobtrusive notation. Indeed, when we write "Sylow$^\circ$" or "N°", this could be interpreted as "Sylow" or "N" in a category of connected groups; we would like to write $X \cap^\circ Y$ for the $(X \cap Y)^\circ$ as well, but this is pushing the limits of notation. If one traces through the arguments, one finds that a great deal of the relative simplicity of our analysis is traceable directly to this point.

Bearing all of this in mind, we ask whether there is a fragment of the classification theory of finite simple groups which actually does correspond reasonably closely to what we do here. It is entirely possible that there is such a self-contained theory which is not necessarily part of any proof of the full classification. We will make one concrete proposal in this direction in §4.2. One will observe some connection with the line followed by Timmesfeld.

3.8. Major differences. The theory of simple groups of finite Morley rank is generally a couple of orders of magnitude simpler than the theory of finite simple groups, at least in the presence of a healthy supply of involutions. The main reasons for this are the following.

- The fields involved are algebraically closed; in particular, they have no quadratic extensions.
- There is a notion of connected component, and in particular of Sylow° 2-subgroup.

Complicating factors, on the other hand, are the following.

- Rank provides a rough measure of size, but the order of a set cannot be assigned a "parity".
- There is no "transfer" map.
- There is no useful group algebra and no way to bring linear algebra to bear.
- Representation theory, even for simple algebraic groups, is extremely rudimentary in this category.

Among the simplifying factors, we have not listed the absence of sporadic groups, which in any case is merely conjectural. In particular, our work on mixed and even type allows unknown sporadic groups to be carried along almost indefinitely—for practical purposes they disappear in Lemma VIII 5.8. In the case of odd type groups we do not know how to work around them, and hence we work only with K^*-groups. So this remains an unresolved issue of fundamental importance.

The fact that the fields involved are algebraically closed has a number of consequences, though the possible existence of bad fields continues to raise issues of the type associated with finite fields. Still, this results in very striking simplifications in a wide variety of contexts; in particular, it yields the elements of order three which play such a strong role at the end of Chapter VI, and quadratic closure is essential to the theory of Suzuki groups in §III 3. Furthermore, quadratic closure simplifies the representation theory, and while this is not expressed by a rigorous result it is somewhat in view in the deeper reaches of the amalgam method analysis in Chapter IX.

The notion of connectivity is spectacularly effective, and has led us to give virtually every notion of group theory, and many of the theorems, in a "connected" form. Among the more subtle manifestations of this is the notion of "continuously characteristic" subgroups which allows us to obtain sufficiently characteristic subgroups in contexts which would have to be treated as distinct exceptions in the finite case. By working in the connected Sylow subgroup rather than the full Sylow subgroup we achieve many simplifications, and in particular it will be noticed that configurations reminiscent of wreath products, to which vast numbers of pages are devoted in the finite case, do not even arise for us as distinct cases. This absence of wreath products is probably a greater simplification than our avoidance of cases associated with sporadic groups.

As far as the complicating factors are concerned, it would be very pleasant to be able to distinguish "even" and "odd" orders. We do measure this in the case of groups by the presence or absence of involutions, but this begs a number of questions. The clearest evidence of parity, or more generally residue modulo n, is given by a definable equivalence relation with all but finitely many classes of order n. There is no reason why a group of finite Morley rank should not exhibit a variety of parities, and indeed it follows from results of Hrushovski on strongly minimal sets that an algebraic group can be enriched, without changing its rank, to one whose order is both even and odd in this sense. There are other finiteness principles that would be useful, notably the surjectivity of injective maps. The Algebraicity Conjecture implies that the theory of a simple group of finite Morley rank should satisfy this principle with respect to sets and functions definable in the language of the group, a point which is unlikely to have an independent proof. Using finiteness in a more precise way, via actual calculation, one can eliminate bad groups instantly, or more generally any group covered disjointly by a conjugacy class of proper subgroups. The use of *transfer* in finite group theory can be viewed as another application of elementary notions of arithmetic, and its absence in our context is particularly deplorable.

The most powerful technique for which we have no analog is character theory, or in other words the linear algebra of the group ring. It is noteworthy that we can achieve results that are normally achieved by character theory in the finite case, using either the presence of toral elements of order three, or considerations connected with connectivity (notably, weak embedding as opposed to strong embedding). Where we really feel the lack of this technique is in the treatment of groups of degenerate type, where after the elimination of involutions we become comparatively helpless.

Finally, representation theory is an essential ingredient in the amalgam method, and increasingly so in its more recent manifestations. Fortunately, the fragment of the theory with which we deal relies only on characterizations of the natural module for SL_2, and these we can squeeze out, though in some cases with effort.

The recent development of the theory has tended in practice to move away from the methods of finite group theory, and to head in a more geometric direction, notably along lines now represented by Chapter IV, and the ongoing work in odd type.

4. New directions

We propose two lines of research which are suggested by our experience here: a proposed independent fragment of the classification of the finite simple groups, and the study of fusion systems of finite Morley rank. At the end we take note of a third line which is more model theoretic, and was suggested by Hrushovski.

4.1. Fusion systems. In this book one recurrent theme has been a systematic transfer of ideas from finite group theory to the theory of groups of finite Morley rank. The general feeling was that the theory of groups of finite Morley rank would pick out a "generic" or "regular" component from the classification of finite simple groups, leaving behind "irregular", "sporadic" bits of the theory—among them, the sporadic simple groups.

In this connection it is interesting to look at a dramatic new departure, the theory of *fusion systems* or *p-local groups* in which finite groups are replaced by structures which are no longer groups but which capture essential features of the local structure of a finite group. Given the prominence the local analysis has in our book, it might be interesting to look at such fusion systems also in the finite Morley rank context.

Let p be a prime number. A *fusion system* on a finite p-group S is a category $\mathcal{F}$ whose objects are the subgroups of S, and whose morphisms are injective group homomorphisms, subject to certain axioms. The notion of a *saturated* fusion system is designed to axiomatize the p-local structure of a finite group G which contains S as a Sylow p-subgroup. Every such group G gives rise to a fusion system $\mathcal{F}_S(G)$ on S, and we say that G *realizes* $\mathcal{F}$ if $\mathcal{F}_S(G) = \mathcal{F}$.

It is known that there are saturated fusion systems $\mathcal{F}$ which are not realized by any finite group G, although showing that this is the case is very delicate. In the case when $p = 2$, the only known examples are certain systems discovered by Ron Solomon in connection with his theorem on the characterization of Conway's sporadic simple group Co_3 [**33, 134, 169**]. Without a doubt, nonrealizable fusion systems–and their intimate relations with algebraic topology—is the most intriguing aspect of the theory.

There is a school of thought which suggests that fusion systems provide the proper and right setting for an understanding of the sporadic simple groups and that, in effect, sporadic simple groups are "sporadic" saturated fusion systems which happened to be groups, almost incidentally.

Now we will give a formal definition of a (finite) fusion system and a saturated fusion system. We introduce only the minimal formalism needed to give the flavor of the subject.

A *fusion system* over a finite p-group S is a category $\mathcal{F}$, where $\mathrm{Ob}(\mathcal{F})$ is the set of all subgroups of S, and which satisfies the following three properties for all $P, Q \leq S$:

- All morphisms in $\mathrm{Hom}_{\mathcal{F}}(P, Q)$ are injective homomorphism of groups;

- $\mathrm{Hom}_{\mathcal{F}}(P, Q)$ includes all homomorphisms from P to Q induced by conjugation in S;
- Each $\alpha \in \mathrm{Hom}_{\mathcal{F}}(P, Q)$ is the composite of an isomorphism in $\mathcal{F}$ followed by an inclusion.

Given a Sylow p-subgroup S in a finite group G, all subgroups in S together with all homomorphisms induced by conjugation in G form a fusion system, denoted $\mathcal{F}_S(G)$.

We need some further notation. In a fusion system $\mathcal{F}$:

- $\mathrm{Iso}_{\mathcal{F}}(P, Q) = \mathrm{Hom}_{\mathcal{F}}(P, Q)$ if $|P| = |Q|$;
- $\mathrm{Aut}_{\mathcal{F}}(P) = \mathrm{Iso}_{\mathcal{F}}(P, P)$; and
- $\mathrm{Out}_{\mathcal{F}}(P) = \mathrm{Aut}_{\mathcal{F}}(P)/\mathrm{Inn}(P)$ where $\mathrm{Inn}(P)$ is the group of inner automorphisms of P.

If $\mathcal{F}$ is a fusion system over a finite p-subgroup S, then two subgroups $P, Q \leq S$ are said to be *$\mathcal{F}$-conjugate* if they are isomorphic as objects of the category $\mathcal{F}$.

The next group of definitions continues to mimic the behavior and properties of subgroups of a Sylow p-group in a finite group.

Let $\mathcal{F}$ be a fusion system over a finite p-subgroup S.

- A subgroup $P \leq S$ is *$\mathcal{F}$-centric* if $C_S(P') = Z(P')$ for all $P' \leq S$ which are $\mathcal{F}$-conjugate to P.
- A subgroup $P \leq S$ is *$\mathcal{F}$-radical* if $\mathrm{Out}_{\mathcal{F}}(P)$ is p-reduced; i.e., if $O_p(\mathrm{Out}_{\mathcal{F}}(P)) = 1$.

Saturated fusion systems. Let $\mathcal{F}$ be a fusion system over a p-group S.

- A subgroup $P \leq S$ is *fully centralized* in $\mathcal{F}$ if $|C_S(P)| \geq |C_S(P')|$ for all $P' \leq S$ which is $\mathcal{F}$-conjugate to P.
- A subgroup $P \leq S$ is *fully normalized* in $\mathcal{F}$ if $|N_S(P)| \geq |N_S(P')|$ for all $P' \leq S$ which is $\mathcal{F}$-conjugate to P.
- $\mathcal{F}$ is a *saturated* fusion system if the following two conditions hold:
  (I) For all $P \leq S$ which is fully normalized in $\mathcal{F}$, P is fully centralized in $\mathcal{F}$ and $\mathrm{Aut}_S(P) \in Syl_p(\mathrm{Aut}_{\mathcal{F}}(P))$.
  (II) If $P \leq S$ and $\alpha \in \mathrm{Hom}_{\mathcal{F}}(P, S)$ are such that αP is fully centralized, and if we set

  $$N_\alpha = \{g \in N_S(P) \mid \alpha c_g \alpha^{-1} \in \mathrm{Aut}_S(\alpha P)\}$$

  (where c_g denotes conjugation by g, $x \mapsto gxg^{-1}$), then there exists $\bar{\alpha} \in \mathrm{Hom}_{\mathcal{F}}(N_\alpha, S)$ such that $\bar{\alpha}|_P = \alpha$.

If G is a finite group and $S \in Syl_p(G)$, then the category $\mathcal{F}_S(G)$ is a saturated fusion system [**56**, Proposition 1.3]. Regarding the realizability of saturated fusion systems, the best result so far is a theorem by Leary and Stancu [**133**] which states that any saturated fusion system over S can be realized in a (possibly infinite) group G which contains S, and has the property that every p-subgroup in G is conjugate to a subgroup of S.

Moving to the finite Morley rank domain, the generalization of saturated fusion systems appears to be immediate and obvious: we take for S a p-unipotent group of finite Morley rank (or possibly a 0-unipotent subgroup in the sense of Burdges), and simply demand that, in the definitions above, all groups are definable and all $\mathrm{Hom}(P, Q)$ are uniformly definable families of definable homomorphisms. In the case $p = 2$, where we have a good Sylow theory, the fact that a group G of finite Morley rank and even type together with a Sylow 2-subgroup S give rise to a saturated fusion system appears to be not much different from the Alperin Fusion Theorem which we already have due to Corredor [**79**], while realizability issues are surprisingly close to gluing a group of finite Morley rank from amalgam data.

Restricting ourselves to the case $p = 2$, may we conjecture that every saturated fusion system of finite Morley rank on a connected 2-unipotent group S comes from a connected group of finite Morley rank and even type? This would provide an even stronger version of killing of sporadic configurations and sporadic simple groups than we have in our Even Type Theorem.

What is perhaps more important, a study of fusion systems may shed some useful light on the structure of our theory, and, in particular, on the interaction between the direct study of fusion and the amalgam method. If the shortcut we took via the amalgam method in Chapter IX is no longer available, there are other well-established methods in finite group theory that could be developed here.

4.2. Atomic 2-groups: Definitions. The other question that we wish to explore is whether there may be a reasonable self-contained theory of *finite* simple groups of *even type* which is more *closely parallel* to the theory presented here. In other words, have we perhaps been imitating a certain line of development that never actually emerged distinctly in the classical theory? A loose analogy exists with work of Timmesfeld, but we are looking for something sharper.

In the finite Morley rank context, even type means that the Sylow 2-subgroups are infinite, definable, and of bounded exponent. In particular, their connected components are *unipotent* (definable connected 2-groups).

Let G be a simple group of finite Morley rank and of even type, and let A be a *minimal* unipotent 2-subgroup of G. Then the following hold:

(1) For all $g \in G$, either $A \cap A^g$ is finite, or g normalizes A.
(2) [Frécon] If g normalizes A, then either $C_A(g) = 1$, or g centralizes A.
(3) If $Q > 1$ is a definable connected 2-group, then $Z(Q)$ contains a minimal unipotent 2-subgroup.

Let us consider, by analogy, the following conditions in the finite context.

DEFINITION 4.1. *For G a finite group, and $\mathcal{A}$ a G-invariant set of elementary abelian 2-subgroups of G, we say that $\mathcal{A}$ is a family of* atomic *2-subgroups of G if the following axioms are satisfied:*

(A1) *For any* $g \in G$, $A \cap A^g$ *is* A *or* 1. *(One says* A *is a "TI-subgroup", where "TI" abbreviates "trivial intersection".)*
(A2) $C_G(A) = C_G(a)$ *for* $a \in A^\times$.
(A3) *If* $A, B \in \mathcal{A}$ *and* $[A, B] \neq 1$, *then* $[A, B]$ *contains some element of* $\mathcal{A}$.
(1) $G = \langle \mathcal{A} \rangle$.

This is a good time to make a comparison with Timmesfeld's axioms for *abstract root subgroups* [**181**].

DEFINITION 4.2. Σ *is a* G*-invariant set of abelian subgroups of* G, *generating* G, *such that for any pair* $A, B \in \Sigma$ *one of the following three possibilities holds:*
(1) $[A, B] = 1$.
(2) $\langle A, B \rangle$ *is a* rank 1 *group (defined below).*
(3) $[A, B] = [a, B] = [A, b] \in \Sigma$ *for* $a \in A^\times$, $b \in B^\times$, *and* $[A, B] \leq Z(\langle A, B \rangle)$.

A rank 1 group is a group with a split BN-pair of rank 1. This can be written equivalently as follows.

DEFINITION 4.3. *A* rank 1 *group* X *is a group of the form* $\langle A, B \rangle$ *with* A *and* B *distinct and nilpotent, such that for all* $a \in A^\times$ *there is* $b \in B^\times$ *satisfying* $A^b = B^a$.

4.3. Atomic 2-groups: A proposal. The proposal is to classify the finite simple groups G which are generated by a system of atomic 2-groups, and more specifically to do so along very much the same lines that apply in the finite Morley rank context. We find it convenient to introduce a very coarse notion of connectedness.

DEFINITION 4.4. *Let* G *be a finite group generated by a family of atomic* 2*-subgroups, and* H *a subgroup. Then* H° *denotes the subgroup of* H *generated by all atomic* 2*-subgroups of* H *contained in* H, *together with all elements of odd order. The group* H *is said to be* connected *if* $H = H^\circ$, *and we transfer the conventions of the finite Morley rank context to this context* ($N_G{}^\circ(X)$, *Sylow*$^\circ$ 2*-subgroups, and so forth).*

However, some of our target groups have disconnected Sylow 2-subgroups ($\mathrm{SU}_3(2^{2n})$, $\mathrm{Sz}(2^{2n+1})$) .

ATOMIC GROUP CONJECTURE. *Let* G *be a finite simple group generated by a system of atomic* 2*-subgroups. Then* G *is a group of Lie type in characteristic* 2.

We will now state ten problems in this area. Each of these is the direct translation of one step in our analysis in finite Morley rank. Remarkably, each such problem also makes good sense as a problem in finite group theory, under our present conventions.

We would be happy to have solutions even with a K^*-hypothesis (as was long the case in the finite Morley rank context). However, with this hypothesis removed (or reduced to some suitable L^*-hypothesis) we would then have a self-contained chapter of finite group theory exactly parallel to what we do here.

Of course, as we are dealing with finite group theory, it might turn out to be necessary to introduce some explicit exceptions at some stage. One could, and one should, use the existing classification to check the accuracy of our formulations; this is not a triviality, and it has not been carried through. The axioms for atomic 2-groups are loose enough that we do not expect an explicit classification of all such systems, but only of the groups in which such systems occur.

4.4. Atomic 2-groups: The problems. In the present section, G is a finite simple group with a generating family of atomic 2-subgroups. We begin with the uniqueness theorems associated with strong and weak embedding, strongly closed abelian subgroups, weakly closed abelian subgroups, and the global $C(G,T)$ theorem. In our presentation in this book, we suppressed the strongly closed abelian case in favor of the simpler case of abelian Sylow subgroups, and one could also ask whether that route could be followed here.

We must first define weak embedding.

DEFINITION 4.5. *Let S be a Sylow° 2-subgroup of G. A proper subgroup M of G is* weakly embedded *if*

$$M = \langle N_G{}^\circ(U) : 1 < U = U^\circ \leq S \rangle \ .$$

We remark that G has a weakly embedded embedded proper subgroup if and only if the graph on $\mathcal{A}$, in which A, B are connected by an edge in $\mathcal{U}(G)$ if they commute, is disconnected.

PROBLEM 1. *If G contains a weakly embedded subgroup, show that G is one of the groups* $\mathrm{PSL}_2(2^n)$, $\mathrm{SU}_3(2^{2n})$, *or* $\mathrm{Sz}(2^{2n+1})$.

In an extreme case, the graph $\mathcal{A}$ has no edges! Then we are dealing with a TI-subgroup weakly closed in its centralizer, a case handled by Timmesfeld [**179**]. This generalizes the Z^*-theorem.

FACT 4.6. *Let A be an elementary abelian TI 2-subgroup of a finite group G and assume that $[A, A^g] = 1$ implies $A = A^g$. Set $G^* = \langle A^G \rangle$. Then either G^* is solvable, or G contains a normal 2-subgroup N such that G^*/N is a covering group of* $\mathrm{PSL}_n(2^m)$, $\mathrm{Sz}(2^{2m+1})$, $\mathrm{SU}_3(2^{2m})$, *an alternating group* $\mathrm{Alt}_6, \mathrm{Alt}_7, \mathrm{Alt}_8, \mathrm{Alt}_9$, *or a Mathieu group M_{22}, M_{23}, or M_{24}.*

PROBLEM 2. *Assuming an affirmative solution to Problem 1, show that $O(N_G{}^\circ(U)) = 1$ for every connected 2-subgroup U.*

Here O denotes the maximal normal subgroup of odd order (under our definitions, it is automatically connected).

A helpful point here is the following: If a connected 2-group U normalizes a group R of odd order, then U centralizes R.

PROBLEM 3. *Suppose the 2-Sylow° subgroup S contains a nontrivial strongly closed and connected abelian subgroup A. Prove that $N_G{}^\circ(A)$ is weakly embedded in G.*

This would simplify the analysis relative to the more general result of Goldschmidt.

DEFINITION 4.7. *For S a Sylow° 2-subgroup of G we write $\mathcal{A} \cap S$ for $\{a \in \mathcal{A} : A \leq S\}$.*

PROBLEM 4. *Let $B = \langle \mathcal{A} \cap S \rangle$. Suppose B is abelian. Show that $N_G{}^\circ(B)$ is weakly embedded in G.*

At this point, we expect *pseudoreflection subgroups* of $N_G{}^\circ(B)/C_G{}^\circ(B)$ to come into play.

Next we come to $C(G, S)$.

DEFINITION 4.8.

(1) *Let $Q \leq P$ be connected 2-subgroups of G. Then Q is* continuously characteristic *in P if Q is invariant under $N_G{}^\circ(P)$.*
(2) *Let $C(G, S) = \langle N_G{}^\circ(Q) : Q$ is continuously characteristic in $S \rangle$.*

This last definition was replaced by a more subtle one in our text (but this is not really necessary, as our previous definition can be linked to this one by a Frattini argument).

PROBLEM 5. *If $C(G, S) < G$, show that G contains a weakly embedded subgroup.*

After the uniqueness theorems, we come to pushing up and an analysis of parabolic subgroups.

DEFINITION 4.9. *Let H be a subgroup of G.*

(1) *The* unipotent radical *$R_u(H)$ of H is its maximal connected 2-subgroup.*
(2) *$U_2(H)$ is the subgroup generated by $\langle \mathcal{A} \cap H \rangle$.*

PROBLEM 6. *Let Q be a connected 2-subgroup of G satisfying:*

(1) $Q = R_u(N_G{}^\circ(Q))$.
(2) *$U_2(N_G{}^\circ(Q))/Q$ is isomorphic to* $\mathrm{SL}_2(2^n)$, $\mathrm{SU}_3(2^{2n})$, *or* $\mathrm{Sz}(2^{2n+1})$.

Show that $N_G{}^\circ(Q)$ contains a Sylow° 2-subgroup of G.

DEFINITION 4.10. *A subgroup P of G is* parabolic *(relative to S) if it is connected and contains S.*

PROBLEM 7. *For P parabolic show the following:*

(1) $R_u(P) \neq 1$.
(2) $C_G{}^\circ(R_u(P)) \leq R_u(P)$.

(The first condition is a special case of the second, but worth noting separately.)

It is at this point that we would expect to bring in the strongly closed abelian case and some rudimentary component analysis, along with the global $C(G,T)$ theorem. After this, we aim at identification.

Let $\mathcal{M}$ be the set of minimal parabolic subgroups properly containing $N_G{}^\circ(S)$.

PROBLEM 8. *Prove that if $|\mathcal{M}| \leq 1$, then G has a proper weakly embedded subgroup.*

PROBLEM 9. *Prove that if G is generated by two elements of $\mathcal{M}$, then G is a rank 2 Lie group over a field of characteristic 2.*

This would be an analog of Delgado and Stellmacher.

PROBLEM 10. *If $|\mathcal{M}| \geq 3$ and G is not generated by any two elements of $\mathcal{M}$, show that G has a BN-pair of rank at least three, and thus is a group of Lie type over a field of characteristic 2.*

This last is an analog of Niles' theorem.
We must add to this list one broad question.

PROBLEM 11. *Can one simplify any of the above by invoking the existing theory of groups generated by abstract root subgroups?*

Finally, we alluded to the theory of pseudoreflection groups above. In the finite Morley rank theory, this was tied up with the strongly closed abelian analysis, and proved to be very convenient. This subject has not been developed in the finite context.

PROBLEM 12. *Let X be a finite group acting irreducibly on an elementary abelian group of order 2^n. Assume that X is generated by a conjugacy class of* pseudoreflection groups*; that is, groups K of odd order acting irreducibly on $W = [K, V]$.*

(1) *Can these groups be classified without assuming the classification of finite simple groups for their factors?*
(2) *... or indeed, assuming this classification?*

In our discussion, we have said nothing about the Feit-Thompson Theorem. In the finite case it is perfectly reasonable to take this as given, without, however, invoking the entire classification of finite simple groups. In the finite Morley rank context we do not have this luxury, and hence in our text we have worked around the absence of the Feit-Thompson theorem (it follows for definable sections of groups of even type from their classification, but only by using Wagner's theorem yet again). Judging by this analogy it may not be absolutely necessary to invoke Feit-Thompson; that is, it might be possible to keep the entire treatment *strictly* self-contained. But at the moment the methods we have used for this purpose in the model theoretic context have no known analogs in the finite case, and have every appearance, at present, of being tied up with properties of infinite fields.

4.5. Generic automorphisms. Our last line is an old suggestion of Hrushovski, not much explored. The idea is to consider the group of fixed points of a "generic automorphism" of a simple group of finite Morley rank (e.g., induced by a generic automorphism of a strongly minimal subset), which should carry the structure of a measurable group, and to develop measurable group theory in this particular context. One test of the theory, but perhaps not a fair one, is whether it can tell us more about groups of degenerate type.

In any case it would be very welcome to have another model theoretic tool which can contribute anything to the analysis of any of the concrete, and very resistant, configurations which have emerged to date in the close study of simple groups of odd or degenerate type.

5. Other open problems

We take this opportunity to present a number of open problems which are either relevant to the development of the subject, or similar in spirit. We begin with some problems which are in some sense classical, or at least of long standing. We then point out that there is a good deal we do not know about the theory of simple groups of finite Morley rank and even type, even though their classification is complete. In odd type, there is already a detailed program in place for further analysis, given in [**60**], to which we must refer for a detailed discussion. In degenerate type, on the other hand, there is no such program, and we indicate some possible lines of inquiry.

Other lines of investigation within the context of groups of finite Morley rank involve their representation theory, cohomological questions, and the further consideration of relations with finite group theory.

Finally, we take note of analogs and other directions: o-minimal groups, and the various other stability classes. Much of the qualitative theory of groups of finite Morley rank does go over to considerably broader classes, and this is a line which has played a considerable role in model theory. This applies to some degree also to theories which seem to have a more "concrete" flavor (such as Hall theory), but much of the theory exists at present only in the finite Morley rank context. (That part of the theory which depends directly on [**188**] seems firmly wedded to the finite Morley rank context.)

5.1. Classical problems. We must begin with the following.

PROBLEM 13. *Are there any* bad groups*?*

Bad groups are minimal connected simple groups all of whose Borel subgroups are nilpotent. It is known that they contain no involutions, so analysis of 2-local structure is certainly not going to accomplish anything. These groups have the property that G is the union of the conjugates of a Borel subgroup, and this property already implies that the group is not finite by a direct counting argument:

$$|G| \le |\bigcup_{g \in G} B^g| \le 1 + (|B| - 1) \cdot |G/B| \le |G| - (|G|/|B| - 1) .$$

There is one standard method for producing exotic structures of finite Morley rank, Hrushovski's amalgamation method, and so far it lacks the capacity to produce structures of this kind. In technical terms, it produces only CM-trivial structures, and they cannot be simple groups as was proven by Pillay [**153**]. But there is no reason as yet to believe that this limitation on the method is intrinsic to the method, which is not rigidly constrained to any particular framework.

Among the possibilities that must be considered is that of a group whose Borel subgroups are elementary abelian p-groups for some p, and in particular the group has exponent p. Here we rejoin combinatorial group theory and the Burnside problem, and this is very reasonable. The problem appears to be intrinsically combinatorial. One could hope that the progress made in the theory of hyperbolic groups could some day be brought to bear on this problem. Indeed, there is one result going in this direction: free groups are stable.

It would of course be useful if one could formulate rigorously what properties a category of groups would have to possess in order to serve as the context for a useful Hrushovski amalgamation, leaving to the combinatorial group theorists the problem of deciding whether such a category exists. The recent works [**152**] and [**126**] are relevant in this context.

Next in line, of almost as venerable vintage, we have.

PROBLEM 14. *Are there any bad fields?*

In fact a solution has been announced; these exist in characteristic zero [**26**]. In particular, it seems the complexities of the degenerate type will not be eliminated by a model theoretic *deus ex machina.*

At an earlier point a consensus had emerged that there should be bad fields of characteristic zero, and none of positive characteristic. At this point only the negative half has been proved. Elimination of positive characteristic, or at least characteristic 2, would simplify our own work somewhat, though it may be noticed that large parts of our analysis would be unaffected. What we need in this direction is already furnished by Wagner's results.

In positive characteristic, the results are due to Wagner [**189**]. If a bad field exists in characteristic p, then one exists for which the field is the algebraic closure of $\mathbb{F}_p$, and then it follows that there are only finitely many p-Mersenne primes (of the form $(p^\ell - 1)/(p-1)$); and a good deal more in the same vein follows. Since Wagner's results show that in positive characteristic a bad field is an elementary extension of a locally finite bad field, this tends to reinforce the connection with finite, or at least locally finite, group theory.

This indeed was the case in [**159**], a work much relevant to the story told in this book.

On the characteristic zero side, the first phase of the Hrushovski amalgamation process is given in [**158**]. This is the "amalgamation" phase. Already in this phase some algebraic geometry comes into play. This produces examples of infinite Morley rank. The second phase involves "collapsing", and according to [**26**] it can be carried out using the same information from algebraic geometry.

The next problem has a different character. It would be largely trivialized by a successful classification, and stands in the meantime as a symbol of our ignorance.

PROBLEM 15. *Suppose that a group of finite Morley rank generically satisfies an identity of the form*

$$x^n = 1 \ .$$

Show that it satisfies this identity.

Since first making this list, we have managed to treat the case of exponent a power of 2, and this is given in Chapter IV. So the symbol of our ignorance is perhaps less striking. Still, one may take $n = 5$ to reduce us to helplessness.

This problem has been solved for solvable by finite stable groups by Jaber [**120**], who extends the result to arbitrary identities in nilpotent by finite stable groups. It follows for groups of finite Morley rank of even or mixed type by the classification. It is worth mentioning that Frécon's recent theorem on the conjugacy of Carter subgroups in connected minimal simple groups of finite Morley rank [**90**] has as one of its consequences the solution to the generic equations problem in the context of minimal simple groups.

We remark that the first draft of [**14**] ran straight into this problem (with an unknown value of n), and handled it not by eliminating it, but by treating the configuration that arose much as other configurations are treated. Subsequently, the general result of [**69**] was used to bypass this analysis, and in particular there is no trace of it in the present text.

PROBLEM 16. *Let G be a simple group of finite Morley rank. Show that G contains a proper subgroup M such that the union of its conjugates is generic in G. In particular, treat the case in which G is minimal connected simple, with M a suitably chosen Borel subgroup.*

It follows from Lemma IV 1.14 that this holds whenever G contains a good torus (and by a slight generalization given in [**69**]), the same applies whenever there is a nontrivial p-torus for some p.

This problem is of broader significance for the degenerate case.

In the special case of minimal connected simple groups, the conjugacy theorem of Frécon mentioned above and the methods used in its proof are relevant.

A problem with some similarity to Problem 15 is the following.

PROBLEM 17. *Let G be a finitely generated subgroup of a group of finite Morley rank. Show that G is residually finite.*

Note that if G is linear, that is, has a finite dimensional representation over some commutative ring, then it is indeed residually finite. It would be satisfying to approach this problem via linearity, but this seems a bit too strong in the nilpotent case in view of an example due to Baudisch [**25**].

The simple case is unlikely to succumb in the absence of a classification, but the solvable case also remains completely open.

The following problem sits in a similar line of inquiry, though from a different context. A group G is called *pseudofinite* if it is a model of the theory of finite groups. Evidently the classification of the finite simple groups has far-reaching consequences for this class. Sabbagh has posed the following problem.

PROBLEM 18. *Let G be a finitely generated pseudofinite group. Then G is finite.*

This is false for finitely generated subgroups of pseudofinite groups (even in the abelian case), so there is some essential difference from the previous problem. In fact bringing the two hypotheses to bear in itself poses something of a problem. Sabbagh has proved this in the solvable case, and Khelif has introduced a technique which seems likely to prove it in the case of a group which is a model of the theory of finite *simple* groups. We sketch his argument for the case of a group which is a model of the theory of the alternating groups (so elementarily equivalent to a nonstandard alternating group).

First, the natural representation of Alt_n as a permutation group is interpretable in the theory of the group; so our finitely generated group G acts in a similar way on some definable set Ω. Now take the generators of G to define a Cayley graph on Ω, and consider the definably connected sets of minimal size containing two arbitrary points. These are easily seen to be finite, and there is no largest one, contradicting the theory of finite sets.

However, in this argument we are using subsets of Ω, rather than elements of Ω, in an essential way. In particular, to compare the sizes of two sets, one conjugates by G so that one is contained in the other. So one must argue also that the action of G on the power set of Ω is interpreted in G. (More precisely, this holds in Alt_n with Ω finite of order n, and hence in G there is a "definable power set" with the same encoding, and similar properties.) It turns out that this interpretation is classical, due to Hodges.

This is a powerful argument, but seems to afford no hold on groups which are models of the theory of finite nilpotent groups (note that such groups need not be solvable). So in the present instance the pseudonilpotent case seems to be critical.

5.2. Even type. As we have remarked, the classification theorem does not really complete the theory of groups of even type (though we think the mixed type theory has been adequately treated at this point).

PROBLEM 19. *Develop an approach to groups of even type via standard components.*

The existence of standard components is known, and will perhaps be documented at some point. Everything after that (including the theorem of Aschbacher and Seitz [**22**]) has been left aside. One can read everything off from the classification but it would be instructive to see this material worked out in the finite Morley rank context, given that it is the workhorse of the analysis in the finite case.

PROBLEM 20. *Work out the rank* 2 *amalgam analysis in the style of Delgado and Stellmacher.*

It turns out, perhaps surprisingly, that there is at least one obstacle to doing this. On the whole that analysis resembles the analysis given by Stellmacher, operating more cleanly at a greater level of generality. We fell back to Stellmacher's approach after running into some difficulty with the other. It would be preferable to reconcile this material with the other approach.

Another topic which could have been useful in the even type analysis is representation theory, which is in a pitiful state.

PROBLEM 21. *Let G be a simple algebraic group over a field of characteristic p (= 2, if one prefers). Let V be an elementary abelian p-group on which G acts irreducibly and definably (with V also of finite Morley rank). Then V can be equipped with a vector space structure over a field in such a way that the action is rational.*

This can be phrased in purely algebraic terms: V satisfies at least the descending chain condition on centralizers, and one can state the same problem with this weaker hypothesis.

In the amalgam method, in its classical form, one needs a little representation theory for SL_2: some characterizations of the natural module suffice. These seem not so easy to obtain, though they are available, so any real progress toward the solution of Problem 21 would be a major change in the picture. More recent versions of the amalgam method rely heavily on more substantial representation theory, and this material also has a direct bearing on local group theoretic analysis.

5.3. Odd type. We recall that a simple nonalgebraic group of finite Morley rank of odd type has Prüfer rank at most 2. There are some known difficult configurations in Prüfer rank 1 and 2. At present these consist of the first three of the four configurations described in [**70**], present already in the minimal connected simple tame case, and some configurations associated with an interesting attempt to characterize SL_3 in characteristic not 2 [**16**]. There may very well be others lying in wait.

A full account of what remains to be done in odd type was given in [**60**, Chap. 11]. He gives eleven conjectures into which the problem can be neatly divided, with the following properties: they can be considered independently of one another, and do not interact; their complete solution would complete the problem; there is some line of attack visible for each, except possibly the worst of those identified in [**70**]. We will give these briefly here; the discussion in [**60**] is detailed.

The first has been treated, and was incorporated into our statement of known results.

PROBLEM 22.1 [SOLVED]. *Show that a minimal connected simple group of finite Morley rank and odd type has Prüfer rank at most* 2.

The next three concern Prüfer rank 2.

PROBLEM 22.2. *Let G be a simple K^*-group of finite Morley rank and odd type. Suppose the Prüfer rank is* 2 *and the* 2*-rank $m(G)$ is at least* 3. *If G has a proper* 2*-generated core, reach a contradiction.*

Here it should be recalled that in higher Prüfer rank we get a proper 2-generated core in the nonalgebraic case, and also the hypothesis $m(G) \geq 3$ points in this direction. This is not actually known in the present case, but a 2-generated core tends to be the "default" configuration in pathological cases, and needs to be treated separately.

PROBLEM 22.3. *Let G be a simple K^*-group of finite Morley rank and odd type. Suppose the Prüfer rank is* 2 *and the* 2*-rank $m(G)$ is at least* 3, *and G does not have a proper* 2*-generated core. Show that G is either* PSp_4 *or* G_2, *over an algebraically closed field of characteristic not* 2.

This case is based on a close consideration of components of centralizers of involutions, more specifically centralizers of toral involutions (i.e., involutions in a torus). There are altogether five configurations to be considered, three to be eliminated and two to be characterized; see [**60**, p. 196].

PROBLEM 22.4. *Let G be a simple K^*-group of finite Morley rank and odd type. Suppose the Prüfer rank is* 2 *and the* 2*-rank $m(G)$ is also* 2. *Show that $G \simeq PSL_3(K)$ over an algebraically closed field of characteristic not* 2.

This is one of the difficult cases. One would like to show that $C^\circ(i)$ is nonsolvable for *some* involution i (one alternative is that G is minimal connected simple). One would then like gradually to rejoin the Altseimer analysis, which as we have noted leads to some potentially troublesome configurations.

Now we come to Prüfer rank 1. Again, the value of the 2-rank $m(G)$ seems significant. (In particular, for low 2-rank, the notion of 2-generated core becomes largely meaningless.)

PROBLEM 22.5. *Let G be a simple K^*-group of finite Morley rank and odd type, and Prüfer rank* 1. *Suppose $m(G) \geq 3$. If the weak 2-generated core is proper, then the 2-generated core is proper.*

Here the 2-generated core is associated with elementary abelian 2-subgroups of a fixed Sylow 2-subgroup of 2-rank at least 2, and the weak 2-generated core is associated with elementary abelian 2-subgroups of a fixed Sylow 2-subgroup which not only have 2-rank at least 2, but are contained in elementary abelian subgroups of S of 2-rank at least 3. In low Prüfer rank this distinction becomes increasingly significant. (The configurations to be considered at this point become increasingly explicit.)

PROBLEM 22.6. *Let G be a simple K^*-group of finite Morley rank and odd type, and Prüfer rank* 1. *Suppose $m(G) \geq 3$. Show that the 2-generated core is not proper.*

Problems 22.5 and 22.6 form a unit; if one is unable to treat them as independent problems, then in any case a contradiction must be reached from a proper weak 2-generated core.

Finally, since we are aiming at a contradiction, in this subcase, we have.

PROBLEM 22.7. *Let G be a simple K^*-group of finite Morley rank and odd type, and Prüfer rank* 1. *Suppose $m(G) \geq 3$. Show that G has a proper weak 2-generated core.*

Here we still have some of the force of the signalizer functor theory, since $m(G) \geq 3$, though one also has to bring in the sort of component analysis mentioned in connection with Problem 22.4.

Finally we have the cases associated with Prüfer rank 1 and 2-rank at most 2, where we approach the thinnest configurations. We expect G to be minimal connected simple, but this must be proved.

PROBLEM 22.8. *Let G be a simple K^*-group of finite Morley rank and odd type, and Prüfer rank* 1. *Suppose $m(G) \leq 2$. Show that G is minimal connected simple.*

Evidently the contrary assumption leads to some fairly definite configurations involving SL_2 or PSL_2 (two distinct cases).

PROBLEM 22.9. *Let G be a minimal connected simple K^*-group of finite Morley rank and odd type, and Prüfer rank* 1. *Suppose $m(G) \leq 2$. Let S be a Sylow* 2*-subgroup of G. Then one of the following holds:*

(1) *S is connected.*
(2) $S \simeq \mathbb{Z}(2^\infty) \rtimes (\mathbb{Z}/2\mathbb{Z})$, *as in* PSL_2.

This result is proved in [**70**] in the tame context, and tools have emerged since for working in the nontame case.

These last two configurations are difficult to come to grips with, as there is little internal group theoretic structure remaining. They would be treated in the finite case by radically different methods.

5.4. Degenerate type. This is the Wild West of our subject. There are two distinct goals that one might wish to achieve: an understanding of the degenerate case for its own sake, or results which limit the impact of the degenerate case on the analysis of odd type groups, which to date functions only under inductive hypotheses, in large part because of the possibility of uncontrolled sections of degenerate type.

In the tame case, the minimal connected simple degenerate groups are bad groups, and have been analyzed fairly thoroughly from that point of view. However, there is a large range of unexplored possibilities. We note that the very substantial group theoretic portion of the analysis in the proof of the Feit-Thompson theorem [**30**] explores the rich pattern of structure found in the maximal subgroups and their various intersections (we would take the maximal connected subgroups, which in the present context are referred to as Borel subgroups).

Problem 16 is certainly of general interest here. The fundamental problem was the following, solved while the present text was in preparation (Theorem IV 4.1).

PROBLEM 23 [SOLVED]. *Show that a connected degenerate type group contains no involutions.*

At one point this seemed too much to ask; now that we have it, we still need more. One should look for stronger properties which neutralize the impact of degenerate sections on the odd type theory. Either of the following could be useful; we would much prefer the first.

One such is the following, which follows from Theorem IV 4.1 [**47**].

FACT 5.1. *If an elementary abelian 2-group A of order 4 acts definably on G, then $G = \langle C_G(a) : a \in A^\times \rangle$.*

It would be better to eliminate involutory automorphisms in connected groups of degenerate type, or at least to show the following.

PROBLEM 24. *Let G be a simple group of finite Morley rank and of degenerate type. If a 2-torus acts definably on G, then it centralizes G.*

Without waiting for this, it would also be of considerable interest to take up an L^*-theory for groups of odd type in which a positive solution to Problem 24 is simply assumed for connected degenerate simple sections of the group in question.

In another direction, the conjugacy of maximal good tori (or more generally, definable tori with definably dense torsion) gives some control over the torsion, but we have no control in the torsion free case. The natural analog of "standard torus" here would be a nontrivial connected nilpotent subgroup with minimal possible reduced rank in the sense of [**61**].

Is the natural analog also the right one? Recently, Frécon introduced the notion of *pseudotorus* [**91**] which generalizes almost all known notions of tori in the context of groups of finite Morley rank. He has obtained strong conjugacy results for the maximal ones. In the light of this, one can formulate the following problem:

PROBLEM 25. *Prove the conjugacy of maximal tori which are standard in the above very weak sense. Investigate the relationships between standard and pseudo-tori.*

Let us close the discussion of tori with the following remark. The very notion of torus is a "semisimple" one. This semisimplicity has frequently been measured using abstract notions of unipotence ([**61**], [**95**]). A now old exception to this was the notion of ∞-*element* introduced by Frécon in [**89**]. Frécon developed a substantial body of results around this notion mostly about groups with solvable aspects. [**93**] contains the only published introduction to this approach forced to remain largely unpublished as well as a Hall theory developed thereof.

Lastly, we consider lines suggested by the Feit-Thompson analysis. The difficulty here is that that analysis is not only long and complex, but leads to a picture which apparently requires character theory to resolve. Perhaps more troublesome is the use in that analysis of the focal subgroup theorem, for which we have no analog.

In any case, a "dry run" for that analysis was furnished by the treatment of the special case in which the centralizer of every nontrivial element is nilpotent. One should at least explore this in degenerate groups, and we can ask for a little more.

PROBLEM 26. *Analyze intersections of Borel subgroups in degenerate type groups in which $C^\circ(a)$ is nilpotent for every nontrivial a.*

In the tame case we easily get the following, in general.

PROBLEM 27. *Let G be a minimal connected simple group, and let B_1, B_2 be Borel subgroups. Then $B_1 \cap B_2$ is abelian.*

This was explored in [**60**, Lemma 9.2], and a weak form was achieved which was adequate for the purposes [**66**]. It would seem that that lemma cannot be much improved in general, that is, in degenerate type groups, without involutions. What is more likely is that a result of this general character can be achieved if the 2-rank is at least 2; such a result could become an ingredient in a solution to Problem 23.

In the Feit-Thompson context, every element turns out to belong to a maximal proper subgroup. The analog in our context would be that G is a union of Borel subgroups; but then G would necessarily contain no involutions. So all of these issues are intertwined.

PROBLEM 28. *Let G be a simple group of finite Morley rank without involutions. Prove that the Borel subgroups are maximal subgroups, and that G is their union.*

Some relevant conclusions in this direction can be found in [**9**].

Finally, we should say something about general torsion.

PROBLEM 29. *Let G be a minimal connected simple group which is not algebraic. Show that the Borel subgroups of G are torsion free.*

Another way of putting this is that every solvable torsion subgroup of G is finite.

5.5. Cohomology. There are a few cohomological issues which are essential to our developments and have been treated successfully at this point: the Schur-Zassenhaus theory, the theory of central extensions of algebraic groups in the larger category of groups of finite Morley rank, and Lemma III 4.2, which uses a genericity argument in a cohomological context. Beyond that, the theory is little developed.

PROBLEM 30. *Develop definable cohomology theory using definable actions and definable cocycles.*

There are other cohomological theorems related to the Schur-Zassenhaus theorem (and a theorem of Gaschütz). This has been explored by Derek Robinson [**162**] in the context of locally nilpotent groups by methods which are roughly parallel to the style of proof used to derive the Schur-Zassenhaus theorem in the finite Morley rank context.

PROBLEM 31. *Prove vanishing theorems for cohomology parallel to those in* [**162**] *in a finite Morley rank context.*

5.6. Other theories. The dominant "other theory" here is the theory of finite simple groups, but we have discussed this sufficiently. Here the question that requires clarification is the relationship of the case of finite Morley rank to Timmesfeld's theory (possibly along the lines of §4.2).

There are a number of model theoretic theories to consider as well. In the theory of *o*-minimality, which relates to the real field in somewhat the same manner that finite Morley rank relates to algebraically closed fields, the Algebraicity Conjecture is elegantly (and efficiently) proved in [**154**], by a mixture of substantial model theory and algebra. One of the advantages of this setting is that a notion of tangent space is available. There is considerably more to this theory than the simple case, including features very different from the finite or algebraic cases. For example, in definably compact groups G in this category, there is a smallest type definable subgroup of bounded index, denoted $G^{\circ\circ}$, and the quotient $G/G^{\circ\circ}$ carries a natural "logic topology" which makes it a compact Lie group. (For example, a nonstandard "circle group" S has the "infinitesimals" as $S^{\circ\circ}$ and the quotient is a true circle.) The main open problem in this area is to show that this quotient is the "right" Lie group, in the sense that the *o*-minimal dimension of the original group should match the real dimension of the quotient group.

There is also a model-theoretic framework for dealing with p-adic fields and generalizations, but the question as to whether there is a theory of pseudo-p-adic groups parallel to the *o*-minimal theory is entirely open. (This would have to cover at least groups definable in the p-adic fields themselves.)

Closer in spirit to our subject is the stability theory hierarchy above the finite Morley rank context: in ascending order $\aleph_0$-stable, superstable, stable, or simple (with supersimple as well as some classes with finite ranks forking off along the way). It has been established that with some model theoretic sophistication, large chunks of the general theory go over to stable groups (and beyond); cf. [**156, 186**] for the core theory. There is considerably more grist for this mill here.

PROBLEM 32. *Analyze simple stable groups of mixed and even type.*

Note that we do not have any control of bad fields in this context. In fact, a pair $(K, L^\times)$ with $K \geq L$ algebraically closed fields is a bad field of Morley rank ω. This is not a completely daunting example (the tori involved are still good), but suggests there may be serious problems. One can of course fall back to the K^*-case to handle them. In any case, some component theories (particularly around solvable groups) have been worked out in the stable case, and even beyond, and tend to take on a sharper form when worked out fully (for example, the generalized theory of Carter subgroups in [**62**] works with the methods Wagner used in [**186**] to push the classical Carter theory to a general setting). It is clear in any case that from a model theoretic point of view, we are working much of the time in an unnecessarily restrictive context, though it may in the end be necessary to return to this context to put all the pieces together.

One particular corner of this enterprise seems particularly attractive.

PROBLEM 33. *Classify stable Moufang polygons.*

As yet, we cannot even classify stable fields (another major problem), but we mean here that one should either classify stable Moufang polygons modulo the classification of stable fields, or construct some interesting new examples (and some interesting simple stable groups). Since there is an explicit classification of all Moufang polygons, one begins with that, and the most interesting case involves quadrangles of mixed type [**185**]. These are associated with groups which are in some sense algebraic over a pair of intertwined fields. The question is simply whether the associated algebraic structures can be stable without being trivial.

In Spring 2005, Zoé Chatzidakis answered this question to our satisfaction by finding a number of striking examples of stable quadrangles of mixed type in which the coordinatizing "indifferent sets" are quite distant from the associated fields, which are separably closed and of characteristic 2. So it may now be said that this class is about as rich as it could possibly be.

One can even give model theoretically interesting *rank* 1 groups in Timmesfeld's sense by this construction, which are indeed the rank 1 analogs of SL_2 from which this rank 2 group is built.

5.7. Permutation groups. We discussed a "soft" proof of the following two results in §2.

THEOREM 2.1. *Let G be a definably primitive permutation group on a set X of rank r. Then the rank of G is bounded by a function of r.*

THEOREM 2.4. *Let G be a simple group of finite Morley rank acting generically t-transitively on the set X. Then t is bounded as a function of* $\mathrm{rk}(X)$.

Our methods give large bounds on the parameter $t = \tau(G, X)$ and correspondingly large bounds for the rank of G. The passage from τ to $\mathrm{rk}(G)$ is not too sloppy; but our control of the parameter τ is extremely loose. One may pose a variety of problems here aiming at sharp bounds on τ, expecting PSL_n in its natural projective action to play the role of an extreme case.

Even when G is a simple algebraic group, the classification of multiply transitive actions is far from complete. Bounds on t in the case of *rational actions* in *characteristic* 0 have been given by Popov [**161**]. But even in the case of algebraic groups of characteristic 0, once one allows an enrichment of the language the question of bounds is open again.

PROBLEM 34. *Determine, for each simple algebraic group G, the maximal t such that G has a generically t-transitive action of finite Morley rank.*

PROBLEM 35. *Determine, for a fixed t (preferably $t = 2$) all actions of simple algebraic groups which are generically t-transitive.*

This problem has been solved for $t = 3$ in the case of characteristic 0 and algebraic actions. Two attractive variants which lend themselves to an inductive analysis are the following.

PROBLEM 36. *Let G be a connected group of finite Morley rank acting faithfully, definably, transitively and generically $(n + 2)$-transitively on a set Ω of Morley rank n. Then the pair (G, Ω) is equivalent to the projective linear group $\mathrm{PGL}_{n+1}(F)$ acting on the projective space $\mathbb{P}^n(F)$ for some algebraically closed field F.*

PROBLEM 37. *Let G be a connected group of finite Morley rank acting faithfully, definably, and generically n-transitively on a connected abelian group V of Morley rank n. Show that V has the structure of an n-dimensional vector space over an algebraically closed field F of Morley rank 1, and G is $\mathrm{GL}_n(F)$ in its natural action on F^n.*

For more problems in a related vein we refer to [**49**].

6. Notes

§1. Odd type

A full account of the classification of simple K^*-groups of finite Morley rank of odd type, as it stood in Spring 2004, along with a good deal of the history, can be found in Jeff Burdges' doctoral dissertation [**60**]. At present it is known that a nonalgebraic simple K^*-group of finite Morley rank must have Prüfer rank at most 2. More recent developments in low Prüfer rank can be found in the work of Adrien Deloro [**85, 86**], based on his thesis (2007). We will not attempt to bring the story fully up to date here. For the benefit of the reader who has gotten acclimatized to the setting of the present book, we should emphasize that this is indeed a K^*-group theory rather than some kind of L^*-theory, and there is at present very little machinery which would support similar results in the presence of degenerate sections, and no clear remedy for that situation.

The work in odd type began with Borovik's [**42**], in the tame case, using signalizer functor theory. Later Berkman and Borovik gave the generic identification theorem for the K^* case [**36**], which streamlines the analysis considerably; this was further adapted to the L^* context in [**37**]. The theory developed in a number of other directions, staying mainly in the tame case, until Burdges' thesis [**60, 61**] gave a route to the signalizer functor theory in the non-tame case, via a formal notion of "characteristic zero unipotence" which complements the more straightforward notion of p-unipotence for p prime. This was incorporated into [**46**], which both improves on and simplifies the treatment in [**42**], and was followed up in [**48**] to give the results modulo the vexing issue of "$O(C(i)) = 1$", which was still bound up with tameness; while the signalizer functor theory had been worked out satisfactorily, it was still not clear where the signalizer functors would come from in the tame case. One approach to the construction of useful signalizer functors is given in [**60**, Chap. 10], cf. [**63**], and there is an alternative following more closely the line used in finite group theory (Borovik, unpublished). The analysis of the minimal connected case to which this reduces was given first in [**70**] with a very strong use of the hypothesis of tameness, and then adapted to the general case in [**66**] using, among other things, the new unipotence theory.

For degenerate type, the results on bad groups were given originally in [**143**] (in the rank 3 case) and [**78, 54**] in general.

In minimal connected simple groups one works with the pattern of intersection of Borel subgroups (this is equally true of the group theoretic analysis prior to the application of character theory in the proof of the Feit-Thompson theorem, as in [**30**]). This line of analysis has been considerably developed ([**60**, Chap. 9], [**62**]), and was exploited heavily in [**66**], but only in the very tight configuration produced by the odd type analysis, in Prüfer rank at least 2. The method can also be used in Prüfer rank 1 [**85**].

Fact IV 1.15 was extracted from the analysis in the first version of [**14**], first for the case of good tori. In the published version, this result is quoted and applied directly at a fairly early stage of the argument. The extension from good tori to decent tori, sketched in [**69**], was suggested by Borovik. This whole line of thought was sparked by [**188**], which implies that the multiplicative group of a field of finite Morley rank and positive characteristic is a good torus. To date this has had considerably more impact in even type than in odd or degenerate type, for obvious reasons.

§2. Permutation groups

This follows [**49**]. Fact 2.8 makes its first appearance there, but turns out to be better viewed as part of a more general study of "semisimple torsion" pursued in [**64**], with further ramifications for the structure of the "Weyl group."

§3. Lessons learned

We would stress the parallels between our own experience and the line of argument taken in the work of Paul Flavell mentioned [**88**]. In this work results of Timmesfeld play a key role, and the self-contained fragment of finite simple group theory he develops flows in similar channels to our own.

Throughout this section, references are given in the text.

§4. New directions

Fusion systems have not been looked at to date in a model theoretic context. For a general discussion see [**57**]. There is a body of material, some of it published, which could presumably be rephrased within this setting. By passing rapidly to the amalgam method, our presentation has obscured this line of thought.

The material in §4.2 has benefited from comments by various finite group theorists, including Lyons and Aschbacher. Of course it also owes a good deal to the line of research pursued by Timmesfeld. It seems that what we have been doing here goes in a similar direction to Timmesfeld, and that this impression can perhaps be made precise. In particular, we would stress the problem posed (Problem 11) of applying his theory directly to our context.

See also [**45**] for a discussion of this theme.

As we remarked in §4.5, the line of thought put forward there was suggested to us by Hrushovski many years ago, and has not yet been much followed up, but is currently attracting some interest.

§5. Other open problems

The work of Chatzidakis mentioned was carried out at the Newton Institute, Cambridge, during the month devoted to groups of finite Morley rank within the larger semester program on model theory and its applications, in Spring 2005.

Bibliography

[1] T. Altınel, Groups of finite Morley rank with strongly embedded subgroups. *J. Algebra* 180:778-807, 1996.

[2] T. Altınel, Groupes de rang de Morley fini de type pair avec des sous-groupes faiblement inclus. Habilitation, *Institut Desargues, Université Claude Bernard (Lyon I)*. June 2001.

[3] T. Altınel, A. Borovik, and G. Cherlin, Groups of mixed type. *J. Algebra* 192:524-571, 1997.

[4] T. Altınel, A. Borovik, and G. Cherlin, On groups of finite Morley rank with weakly embedded subgroups. *J. Algebra* 211:409-456, 1999.

[5] T. Altınel, A. Borovik, and G. Cherlin, Groups of finite Morley rank of even type with strongly closed abelian subgroups. *J. Algebra* 232:420–461, 2000.

[6] T. Altınel, A. Borovik, and G. Cherlin, Pushing Up and $C(G,T)$ in groups of finite Morley rank of even type. *J. Algebra* 247:541-576, 2002.

[7] T. Altınel, A. Borovik, and G. Cherlin, Classification of simple K^*-groups of finite Morley rank and even type: geometric aspects. 1–12, in *Groups, combinatorics & geometry (Durham, 2001)* World Sci. Publishing, River Edge, NJ, 2003.

[8] T. Altınel, A. Borovik, G. Cherlin, and L.-J. Corredor, Parabolic subgroups in groups of finite Morley rank of even type. *J. Algebra* 269:250–262, 2003.

[9] T. Altınel and J. Burdges, On analogies between algebraic groups and groups of finite Morley rank. To appear ca. 2008.

[10] T. Altınel and G. Cherlin, On central extensions of algebraic groups. *J. Symbolic Logic* 64:68-74, 1999.

[11] T. Altınel and G. Cherlin, On groups of finite Morley rank of even type. *J. Algebra* 264:155-185, 2003.

[12] T. Altınel and G. Cherlin, Simple L^*-groups of even type with strongly embedded subgroups *J. Algebra* 272:95–127, 2004.

[13] T. Altınel and G. Cherlin, Simple L^*-groups of even type with weakly embedded subgroups *J. Algebra* 286:76-106, 2005.

[14] T. Altınel and G. Cherlin, Limoncello *J. Algebra* 291:371–413, 2005.

[15] T. Altınel, G. Cherlin, L.-J. Corredor, and A. Nesin, A Hall theorem for ω-stable groups. *J. London Math. Society* 57:385-397, 1998

[16] C. Altseimer, *Groups of finite Morley rank with small Prüfer* 2*-rank*, Ph.D. thesis, UMIST, 1998.

[17] C. Altseimer and A. Borovik, Probabilistic recognition of orthogonal and symplectic groups, in *Groups and Computation. III*, pp. 1–20, W. Kantor and A. Seress, eds., de Gruyter, Berlin, 2001; Corrections (and GAP code): `math.GR/0110234`

[18] M. Aschbacher, The status of the classification of the finite simple groups. *AMS Notices* 51:736-740 (2004).

[19] M. Aschbacher, *Finite Group Theory*, Cambridge Studies in Advanced Mathematics, 10. Cambridge University Press, Cambridge, 1986. x+274 pp. ISBN: 0-521-30341-9; Second edition, 2000: xii+304 pp. ISBN: 0-521-78145-0; 0-521-78675-4 20-01.

[20] M. Aschbacher, On finite groups of component type. *Illinois J. Math.* 19:87–115, 1975.

[21] M. Aschbacher, G. Seitz, Involutions in Chevalley groups over fields of even order. *Nagoya J. Mathematics* 63:1-91 (1976); errata, 72:135-136 (1978).

[22] M. Aschbacher, G. Seitz, On groups with a standard component of known type. *Osaka J. Math.* 13:439–482, 1976.

[23] J. Baldwin and J. Saxl, Logical stability in group theory. *J. Austral. Math. Soc. Ser. A* 21:267–276, 1976.

[24] H. Bass and J. Tate, The Milnor ring of a global field. In *Algebraic K-Theory. II*, Lecture Notes in Mathematics, no. 342, pages 349–446. Springer Verlag, 1973.

[25] A. Baudisch, A new uncountably categorical group. *Trans. Amer. Math. Soc.* 348:3889–3940, 1996.

[26] A. Baudisch, M. Hils, A. Martin-Pizarro, and F. O. Wagner, Die böse Farbe. To appear ca. 2008.

[27] B. Baumann, Über endliche Gruppen mit einer zu $L_2(2^n)$ isomorphen Faktorruppe. *Proc. Amer. Math. Soc.* 74:215-22, 1979.

[28] O. Belegradek, On groups of finite Morley rank. In *Abstracts of the Eight International Congress of Logic, Methodology and Philosophy of Science LMPS'87,* pages 100–102, Moscow, 1987.

[29] H. Bender, Transitive Gruppen gerader Ordnung, in dene jede Involution genau einen Punkt festlägt. *J. Algebra* 17:527-554, 1971.

[30] H. Bender and G. Glauberman, Local analysis for the odd order theorem. With the assistance of Walter Carlip. London Mathematical Society Lecture Note Series, 188. Cambridge University Press, Cambridge, 1994. xii+174 pp. ISBN 0-521-45716-5.

[31] C. D. Bennett and S. Shpectorov, A remark on a theorem of J. Tits, *Proc. Amer. Math. Soc.* 129:2571–2579, 2001.

[32] C. D. Bennett, R. Gramlich, C. Hoffman and S. Shpectorov, Curtis-Phan-Tits theory. 13–29 in *Groups, combinatorics & geometry (Durham, 2001)*, World Sci. Publishing, River Edge, NJ, 2003.

[33] D. J. Benson, Conway's group Co3 and the Dickson invariants. *Manuscripta Math.* 85:177–193, 1994.

[34] A. Berkman. The classical involution theorem for groups of finite Morley rank. *J. Algebra* 243(2):361–384, 2001.

[35] A. Berkman and A. Borovik, An identification theorem for groups of finite Morley rank and even type. *J. Algebra* 266:375–381, 2003.

[36] A. Berkman and A. Borovik, A generic identification theorem for groups of finite Morley rank. *J. London Math. Soc.* 69:14–26, 2004.

[37] A. Berkman, A. Borovik, J. Burdges, and G. Cherlin, A generic identification theorem for L^*-groups of finite Morley rank. *J. Algebra* To appear.

[38] G. Birkhoff, *Lattice Theory.* American Mathematical Society, Providence, Rhode Island, third edition, 1967.

[39] A. Borel, Groupes linéaires algébriques. *Ann. of Math.* 64:20–82, 1956.

[40] A. Borel, Essays in the history of Lie groups and algebraic groups. History of Mathematics, 21. American Mathematical Society, Providence, RI; London Mathematical Society, Cambridge, 2001. xiv+184 pp. ISBN 0-8218-0288-7.

[41] A. Borel and J. Tits, Eléments unipotents et sous-groupes paraboliques de groupes réductifs. *Inventiones Math.* 12:95-104, 1971.

[42] A. V. Borovik, Simple locally finite groups of finite Morley rank and odd type. In *Finite and locally finite groups (Istanbul, 1994)* 247–284, NATO Adv. Sci. Inst. Ser. C Math. Phys. Sci., 471, Kluwer Acad. Publ., Dordrecht, 1995.

[43] A. V. Borovik, Tame groups of odd and even type. pp. 341–366, in *Algebraic Groups and their Representations*, R. W. Carter and J. Saxl, eds. (NATO ASI Series C Mathematical and Physical Sciences, vol. 517), Kluwer Academic Publishers, Dordrecht, 1998.

[44] A. V. Borovik, Centralisers of involutions in black box groups. *Contemporary Mathematics*, 298:7–20, 2002.

[45] A. V. Borovik, Variations on a theme of Timmesfeld: A finite group theoretic analogue of the classification of groups of finite Morley rank and even type. Preprint of the Manchester Insitute of Mathematical Sciences, Manchester, UK, 2006, #2006-300. `http://eprints.ma.man.ac.uk/111/`.

[46] A. Borovik and J. Burdges, A new trichotomy theorem for groups of finite Morley rank. *J. London Math. Soc.,* To appear ca. 2007.

[47] A. Borovik, J. Burdges, and G. Cherlin, Involutions in groups of finite Morley rank of degenerate type. *Selecta Mathematica,* 13:1–22, 2007.

[48] A. Borovik, J. Burdges, and A. Nesin, Uniqueness cases in odd type groups of finite Morley rank. *J. London Math. Society,* To appear.

[49] A. Borovik and G. Cherlin. Permutation groups of finite Morley rank. In *Model Theory with Applications to Algebra and Analysis*, Z. Chatzidakis, D. Macpherson, A. Pillay, A. Wilkie, eds. Cambridge University Press, Cambridge, 2008.

[50] A. Borovik and A. Nesin, On the Schur-Zasssenhaus theorem for groups of finite Morley rank. *J. Symbolic Logic*, 57:1469–1477, 1992.

[51] A. Borovik and A. Nesin, Schur-Zassenhaus theorem revisited. *J. Symbolic Logic*, 59:283–291, 1994.

[52] A. V. Borovik and A. Nesin, *Groups of Finite Morley Rank.* Oxford University Press, 1994.

[53] A. V. Borovik and B. Poizat, Tores et p-groupes. *J. Symbolic Logic*, 55:565-583, 1990.

[54] A. V. Borovik and B. Poizat, Simple groups of finite Morley rank without non-nilpotent connected subgroups (in Russian). VINITI, Moscow, N2062-B, 1990, mimeographed.

[55] J. Bray, An improved method for generating the centralizer of an involution. *Arch. Math. (Basel)*, 74:241-245 (2000).

[56] C. Broto, R. Levi, and B. Oliver, The homotopy theory of fusion systems. *J. Amer. Math. Soc.*, 16 779–856, 2003.

[57] C. Broto, R. Levi, and R. Oliver, The theory of p-local groups: A survey. 51–84, in *Homotopy theory: Relations with algebraic geometry, group cohomology, and algebraic K-theory.* Contemp. Math., 346, Amer. Math. Soc., Providence, RI, 2004.

[58] R. H. Bruck and E. Kleinfeld, The structure of alternative division rings. *Proc. Amer. Math. Soc.*, 2:878-890, 1951.

[59] R. Bryant and B. Hartley, Periodic locally soluble groups with the minimal condition on centralizers. *J. Algebra*, 61:328-334, 1979.

[60] J. Burdges, Odd and degenerate type groups of finite Morley rank. Doctoral thesis, Rutgers, June 2004.

[61] J. Burdges, A signalizer functor theorem for groups of finite Morley rank. *J. Algebra*, 274:216–229, 2004.

[62] J. Burdges, The Bender method in groups of finite Morley rank, *J. Algebra*, 312:33–55, 2007.

[63] Jeffrey Burdges, Signalizers and balance in groups of finite Morley rank. Submitted, 2008.

[64] J. Burdges and G. Cherlin, Semisimple torsion in groups of finite Morley rank. Submitted, 2008.

[65] J. Burdges and G. Cherlin, A generation theorem for groups of finite Morley rank Submitted, 2008.

[66] J. Burdges, G. Cherlin, E. Jaligot, Minimal connected simple groups of finite Morley rank and odd type with strongly embedded subgroups. *J. Algebra*, 314:581–612, 2007.

[67] R. Carter, *Simple groups of Lie type.* Pure and Applied Mathematics Vol. XXVIII, John Wiley and Sons, London, 1972,

[68] G. Cherlin, Groups of small Morley rank. *Ann. Math. Logic*, 17:1–28, 1979.

[69] G. Cherlin, Good tori in groups of finite Morley rank. *J. Group Theory,* 8:613-621, 2005.

[70] G. Cherlin and E. Jaligot, Tame minimal simple groups of finite Morley rank. *J. Algebra*, 276:13–79, 2004.

[71] S. N. Chernikov, Complete groups with ascending central series (in Russian). *Mat. Sbornik*, 18:307-422, 1946.

[72] S. N. Chernikov, On the theory of complete groups. *Mat. Sbornik*, 22:319-348, 455-456, 1948.

[73] C. Chevalley Sur certain groupes simples. *Tohoku Math. J.*, 7:14–66, 1955.

[74] C. Chevalley, Théorie des groupes de Lie, Tome II: Groupes algébriques. *Actualités Sci. Indust., No. 1152,* Hermann Paris, 1951.

[75] C. Chevalley, Une démonstration d'un théorème sur les groupes algébriques, *J. Mathématiques Pures et Appliquées* 30:307–317, 1960.

[76] A. Cohen, Finite complex reflection groups. *Ann. Sci. Ecole Norm. Sup.*, 9:379–436, 1976.

[77] B. Conrad, A modern proof of Chevalley's theorem on algebraic groups. *J. Ramanujan Math. Soc.*, 17:1–18, 2002.

[78] L.-J. Corredor, Bad groups of finite Morley rank. *J. Symbolic Logic*, 54:768–773, 1989.

[79] L.-J. Corredor, Fusion of 2-elements in groups of finite Morley rank. *J. Symbolic Logic*, 66:722–730, 2001.

[80] M. Davis and A. Nesin, On Suzuki 2-groups of finite Morley rank. *J. Group Theory*, 6:229–243, 2003.

[81] M. DeBonis and A. Nesin, On CN-groups of finite Morley rank. *J. London Mathematical Society*, 50:532-546, 1994.

[82] M. DeBonis and A. Nesin, On split Zassenhaus groups of mixed characteristic and of finite Morley rank. *J. London Math. Society*, 50:430–439, 1994.

[83] F. Delahan and A. Nesin, On Zassenhaus groups of finite Morley rank, *Comm. Algebra*,23:455–466, 1995.

[84] A. Delgado and B. Stellmacher, Weak (B, N)-pairs of rank 2. Part II in *Groups and Graphs: New Results and Methods*, eds. A. Delgado, D. Goldschmidt, and B. Stellmacher, pp. 1-57. Birkhäuser Verlag, Basel, 1985.

[85] A. Deloro, Groupes simples minimaux de 2-rang de Prüfer 1 *J. Algebra*, 317:877–923, 2007.

[86] A. Deloro, Groupes simples connexes minimaux non-algébriques de type impair. *J. Algebra,* To appear.

[87] W. Feit, A characterization of the simple groups SL(2, 2^a). *Amer. J. Math.*, 82:281–300, 1960.

[88] P. Flavell, A Hall-Higman-Shult type theorem for arbitrary finite groups. *Invent. Math.*, 164:361-397, 2006.

[89] O. Frécon, Etude des groupes résolubles de rang de Morley fini. Thesis, Lyon, IGD, November 2000.

[90] O. Frécon, Conjugacy of Carter subgroups in groups of finite Morley rank. Submitted, 2008.

[91] O. Frécon, Pseudo-tori and subtame groups of finite Morley rank. To appear ca. 2008.

[92] O. Frécon, Sous-groupes anormaux dans les groupes de rang de Morley fini résolubles *J. Algebra*, 229:118–152, 2000.

[93] O. Frécon, Sous-groupes de Hall généralisés dans les groupes résolubles de rang de Morley fini. *J. Algebra*, 233:253–286, 2000.

[94] O. Frécon and E. Jaligot, The existence of Carter subgroups in groups of finite Morley rank. *J. Group Theory*, 8:623–633, 2005.
[95] O. Frécon and E. Jaligot, *Conjugacy in groups of finite Morley rank*. In *Model Theory with Applications to Algebra and Analysis*, Z. Chatzidakis, D. Macpherson, A. Pillay, A. Wilkie, eds. Cambridge University Press, Cambridge, 2008.
[96] L. Fuchs, *Infinite abelian groups. Vol. I. Pure and Applied Mathematics,* Vol. 36, Academic Press, New York-London 1970 xi+290 pp.
[97] A. Gardiner, B. Hartley, and M. Tomkinson. Saturated formations and sylow structure in locally finite groups *J. Algebra*, 17:177-211, 1971.
[98] D. M. Goldschmidt, 2-fusion in finite groups. *Ann. Math.*, 99:70–117, 1974.
[99] D. Goldschmidt, Automorphisms of trivalent graphs. *Ann. Math.*, 111:377-404, 1980.
[100] D. Goldschmidt, Graphs and Groups. Part I in *Groups and Graphs: New Results and Methods* eds. A. Delgado, D. Goldschmidt, and B. Stellmacher, pp. 1-57. Birkhäuser Verlag, Basel, 1985.
[101] D. Gorenstein, *Finite Groups*. Chelsea Publishing Company, New York, 1980.
[102] D. Gorenstein and R. Lyons, The local structure of finite groups of characteristic 2 type. *Memoirs American Math. Society* **276** 1983.
[103] D. Gorenstein, R. Lyons, and R. Solomon, *The classification of the finite simple groups*. Vol. 40, Number 1, American Math. Society, Mathematical Surveys and Monographs, 1994.
[104] D. Gorenstein, R. Lyons, and R. Solomon, *The classification of the finite simple groups*. Vol. 40, Number 2, American Math. Society, Mathematical Surveys and Monographs, 1996.
[105] D. Gorenstein, R. Lyons, and R. Solomon, *The classification of the finite simple groups*. Vol. 40, Number 3, American Math. Society, Mathematical Surveys and Monographs, 1999.
[106] R. Gramlich, Phan Theory. Habilitation, Darmstadt, October 2004.
[107] Ursula Gropp, There is no sharp transitivity on q^6 when q is a type of Morley rank 2. *J. Symbolic Logic 57:1198–1212*, 1992.
[108] P. Hall, A contribution to the theory of groups of prime-power order. *Proc. London Mathematical Society*, 36:29–62, 1933.
[109] B. Hartley, Sylow subgroups of locally finite groups. *Proc. London Math. Society*, 23:159-192, 1971.
[110] B. Hartley, Sylow theory in locally finite groups. *Compositio Math.*, 25:263-280, 1972.
[111] G. Higman, Odd characterizations of finite simple groups. University of Michigan, 1968, cited in [**84**, 5.17].
[112] G. Higman, Suzuki 2-groups. *Illinois J. Math.*, 7:79–96, 1963.
[113] E. Hrushovski and B. Zilber, Zariski geometries. *J. Amer. Math. Soc.*, 9:1–56, 1996.
[114] D. R. Hughes and F. C. Piper, *Projective Planes*. Springer-Verlag, Berlin, 1973.
[115] J. E. Humphreys, *Linear Algebraic Groups*. Springer-Verlag, Berlin-New York, second edition, 1981.
[116] J. E. Humphreys, *Reflection Groups and Coxeter Groups*. Cambridge Studies in Advanced Mathematics, 29. Cambridge University Press, Cambridge, 1990. xii+204 pp. ISBN 0-521-37510-X.
[117] J. E. Humphreys, *Linear Algebras and Representation Theory*. Graduate Texts Math. 9, Springer, New York, 1972.
[118] B. Huppert, *Endliche Gruppen. I*. Springer, Berlin, 1967.
[119] B. Huppert and N. Blackburn, *Finite groups. II*. Grundlehren der Mathematischen Wissenschaften 242. AMD, 44. Springer-Verlag, Berlin-New York, 1982. xiii+531 pp. ISBN 3-540-10632-4.
[120] Kh. Jaber, Equations génériques dans un groupe stable nilpotent. *J. Symbolic Logic*, 64:761–768, 1999.

[121] E. Jaligot, Full Frobenius groups of finite Morley rank and the Feit-Thompson theorem. *Bull. Symbolic Logic*, 7:315–328, 2001

[122] E. Jaligot, Generix never gives up. *J. Symbolic Logic*, 71:599–610, 2006.

[123] E. Jaligot, Groupes de type mixte. *J. Algebra*, 212:753-768, 1999.

[124] E. Jaligot, Groupes de type pair avec un sous-groupe faiblement inclus. *J. Algebra*, 240:413–444, 2001.

[125] E. Jaligot, *Contributions à la classification des groupes simples de rang de Morley fini.* Ph.D. thesis, Université Claude Bernard-Lyon I, 1999.

[126] E. Jaligot, A. Muranov, and A. Neman, Independence property and hyperbolic groups. *Bull. Symbolic Logic*, 14:88–98, 2008.

[127] W. M. Kantor and A. Seress, Black box classical groups. *Memoirs American Math. Society*, 149 no. 708, AMS, Providence, RI, 2000.

[128] O. Kegel and B. Wehrfritz, *Locally Finite Groups* North-Holland, Amsterdam, 1973.

[129] E. Kleinfeld, Alternative division rings of characteristic 2. *Proc. Nat. Acad. Sci.*, 37:818-820, 1951.

[130] L. Kramer, K. Tent, and H. van Maldeghem, Simple groups of finite Morley rank and Tits buildings. *Israel J. Mathematics*, 109:189–224, 1999.

[131] T. Y. Lam, *A first course in noncommutative ring theory*, Graduate Texts in Mathematics 131, Springer-Verlag, New York. 1st ed. 1991, 2nd ed. 2001.

[132] P. Landrock and R. Solomon, A characterization of the Sylow 2-subgroups of $PSU(3, 2^n)$ and $PSL(3, 2^n)$. Technical Report 13, Aarhus Universitet, Matematisk Institut, January 1975.

[133] I. J. Leary and R. Stancu, Realising fusion systems. arXiv:math.GR/0701892v1, 30 Jan 2007.

[134] R. Levi and R. Oliver, Construction of 2-local finite groups of a type studied by Solomon and Benson. *Geometry and Topology*, 6:917–990, 2002.

[135] D. H. McLain, On locally nilpotent groups. *Proc. Cambridge Phil. Soc.*, 52:5-11, 1956.

[136] J. McLaughlin, Some groups generated by transvections, *Arch. Math.*, 18:364–368, 1967.

[137] A. Macintyre, On ω_1-categorical theories of fields. *Fundamenta Mathematicae*, 71:1–25, 1971.

[138] A. Macintyre, On ω_1-categorical theories of abelian groups. *Fundamenta Mathematicae*, 70:253–270, 1971.

[139] D. Macpherson and A. Pillay Primitive permutation groups of finite Morley rank. *Proc. London Math. Society*, 30:481-504, 1995.

[140] H. van Maldeghem, *Generalized Polygons.* Monographs in Mathematics 93, Birkhäuser.

[141] R. Moufang, Die Schnittpunktsätze des projektiven speziellen Funfecksnetzes in ihrer Abhängigkeit voneinander (Das *A*-Netz). *Math. Annalen*, 106:755-795, 1932.

[142] Mumford, David, *Abelian varieties.* Tata Institute of Fundamental Research Studies in Mathematics, No. 5., Published for the Tata Institute of Fundamental Research, Bombay; Oxford University Press, London 1970; viii+242 pp.

[143] A. Nesin, On bad groups, bad fields, and pseudoplanes. *J. Symbolic Logic*, 56:915–931, 1991.

[144] A. Nesin, On sharply n-transitive superstable groups. *J. Pure Appl. Algebra*, 69:73–88, 1990.

[145] A. Nesin, On solvable groups of finite Morley rank. *Trans. Amer. Math. Soc.*, 321:659–690, 1990.

[146] A. Nesin, Generalized Fitting subgroup of a group of finite Morley rank. *J. Symbolic Logic*, 56:1391–1399, 1991.

[147] A. Nesin, Poly-separated and ω-stable nilpotent groups. *J. Symbolic Logic*, 56:694–699, 1991.

[148] L. Newelski, On type definable subgroups of a stable group, *Notre Dame J. of Formal Logic*, 32:173–187, 1991.

[149] R. Niles, BN-pairs and finite groups with parabolic-type subgroups. *J. Algebra*, 75:484–494, 1982.

[150] G. Pickert, *Projektive Ebenen* Springer-Verlag, Berlin, 1955.

[151] A. Pillay, *An Introduction to Stability Theory.* Number 8 in Oxford Logic Guides. Oxford University Press, 1983.

[152] A. Pillay, Forking in the free group. Preprint, 2007

[153] A. Pillay, The geometry of forking and groups of finite Morley rank. *J. Symbolic Logic*, 60:1251–1259, 1995.

[154] Y. Peterzil, A. Pillay, and S. Starchenko, Definably simple groups in o-minimal structures. *Trans. Amer. Math. Soc.*, 352:4397–4419, 2000.

[155] A. Pillay and W. Y. Pong, On Lascar rank and Morley rank of definable groups in differentially closed fields. *J. Symbolic Logic*, 67:1189–1196, 2002.

[156] B. Poizat, *Groupes Stables.* Nur Al-Mantiq Wal-Ma'rifah, Villeurbanne, France, 1987; English translation in [**157**].

[157] B. Poizat, *Stable Groups.* Translated from the 1987 French original by Moses Gabriel Klein. Mathematical Surveys and Monographs, 87, American Mathematical Society, Providence, RI, 2001. xiv+129 pp., ISBN 0-8218-2685-9.

[158] B. Poizat, L'égalité au cube *J. Symbolic Logic*, 66:1647–1676, 2001.

[159] B. Poizat, Quelques modestes remarques à propos d'une conséquence inattendue d'un résultat de Monsieur Frank OLaf Wagner. *J. Symbolic Logic*, 66:1637–1646, 2001.

[160] B. Poizat and F. Wagner, Lift the Sylows! *J. Symbolic Logic*, 53:703–704, 2000.

[161] V. Popov, Generically multiply transitive algebraic group actions. arXiv:math/0409024v2, 5 Mar 2005.

[162] D. Robinson, Vanishing theorems for cohomology of locally nilpotent groups. 120–129, in *Group theory (Bressanone, 1986)* Lecture Notes in Math., 1281.

[163] B. Rose, Model theory of alternative rings. *Notre Dame J. Formal Logic*, 19:215–243, 1978.

[164] J. Rosenberg, *Algebraic K-Theory and its Application.* Number 147 in Graduate Texts in Mathematics. Springer-Verlag, 1994.

[165] M. Rosenlicht, Some basic theorems on algebraic groups. *American J. Mathematics*, 88:492–517, 1968.

[166] G. Seitz, The root subgroups for maximal tori in finite groups of Lie type. *Pacific J. Math.*,106:153–244, 1983.

[167] G. Seitz, Algebraic groups. In *Finite and locally finite groups (Istanbul, 1994)* vol. 471 of *NATO Adv. Sci. Inst. Ser. C Math. Phys. Sci.*, 45–70. Kluwer Acad. Publ., Dordrecht, 1995.

[168] G. C. Shephard and J. A. Todd, Finite unitary reflection groups. *Canadian J. Math.*,6:274–304, 1954.

[169] R. Solomon, Finite groups with Sylow 2-subgroups of type 3. *J. Algebra*, 28:182–198, 1974.

[170] R. Solomon, A brief history of the classification of the finite simple groups. *Bulletin American Math. Society*, 38:315–352, 2001.

[171] T. A. Springer, *Linear algebraic groups,* Second edition. Progress in Mathematics, 9. Birkhäuser Boston, Inc., Boston, MA, 1998. xiv+334 pp. ISBN 0-8176-4021-5.

[172] T. A. Springer and R. Steinberg, Conjugacy classes. in *Seminar on Algebraic Groups and related finite groups* (A. Borel et al., eds.), Lecture Notes Math. 131, Springer-Verlag, New York, 1970, 167–266.

[173] R. Steinberg, *Lectures on Chevalley Groups.* Yale University, 1967.

[174] B. Stellmacher, On graphs with edge-transitive automorphism groups. *Illinois J. Mathematics*, 28:211-266, 1984.

[175] B. Stellmacher, Pushing up. *Arch. Math.*, 46:8-17, 1986.

[176] B. Stellmacher, A pushing up result. *J. Algebra*,83:484-489, 1983.

[177] M. Suzuki, *Group Theory II.* Number 248 in *Grundlehren der mathematischen Wissenchaften*, Springer-Verlag, 1986.

[178] S. Thomas, An identification theorem for the locally finite nontwisted Chevalley groups, *Arch. Math. (Basel)*, 40:21–31, 1983.

[179] F. G. Timmesfeld, Groups with weakly closed TI-subgroups, *Math. Z.*, 143:243–278, 1975.

[180] F. G. Timmesfeld, Groups generated by k-transvections. *Invent. Math.*, 100:173-200, 1990.

[181] F. G. Timmesfeld, *Abstract Root Subgroups and Simple Groups of Lie-Type* Monographs in Mathematics 95, Birkhäuser, Basel, 2001, xiii+389 pp.

[182] F. G. Timmesfeld, The Curtis–Tits–Presentation, *Advances in Mathematics*, 189:38–67, 2004.

[183] J. Tits, *Buildings of Spherical Type and finite BN-pairs.* Lecture Notes in Mathematics 386, Springer-Verlag, Berlin, 1974.

[184] J. Tits, Ensembles ordonnés, immeubles et sommes amalgameés. *Bulletin Societé Math. Belg.*, A38:367–387, 1986.

[185] J. Tits and R. Weiss, *Moufang Polygons* Springer, Berlin, Monographs in Mathematics, 2002.

[186] F. Wagner, *Stable groups.* London Mathematical Society Lecture Note Series **240**, Cambridge University Press, 1997, *ix*+309 pp.

[187] F. Wagner, Subgroups of stable groups. *J. Symbolic Logic*, 55:151-156, 1990.

[188] F. Wagner, Fields of finite Morley rank. *J. Symbolic Logic*, 66:703–706, 2001.

[189] F. Wagner, Bad fields in positive characteristic. *Bull. London Math. Soc.*, 35:499–502, 2003.

[190] J. H. Maclagan-Wedderburn, A theorem on finite algebras. *Trans. Amer. Math. Soc.*, 6:349-352, 1905.

[191] R. Weiss, Über lokal s-reguläre Graphen. *J. Combinatorial Theory Ser. B*, 20:124-127, 1976.

[192] R. Weiss, The nonexistence of certain Moufang polygons. *Inventiones Mathematicae*, 51:261-266, 1979.

[193] R. Weiss, The structure of spherical buildings. Princeton University Press, Princeton, NJ, 2003. xiv+135 pp. ISBN: 0-691-11733-0.

[194] B. I. Zilber, pp. 63–68, in *Abstracts of papers presented at the Fourth All-Union Symposium on Group Theory (Novosibirsk, 1973)* (Russian), *Mat. Inst. Sibirsk. Otdel. Akad. Nauk SSSR,* Novosibirsk, 1973.

[195] B. I. Zilber. Groups and rings whose theory is categorical. *Fundamenta Mathematicae*, 55:173–188, 1977.

[196] B. I. Zilber The structure of models of categorical theories and the problem of finite axiomatizability. Kemerov. Univ., Kemerovo, 1977; manuscript deposited at VINITI on July 12, 1977, Deposition No. 2800–77.

[197] B. I. Zilber. Some model theory of simple algebraic groups over algebraically closed fields. *Colloq. Math.*, 48:173–180, 1984.

Index of Notation

Index of Terminology

Index of Names